GUIDELINES

ANALYSES AND TOOLS

Methods, Standards, and Work Design

TENTH EDITION

Benjamin W. Niebel

Professor Emeritus of Industrial Engineering
The Pennsylvania State University
Niebel & Associates

Andris Freivalds

Professor of Industrial Engineering
The Pennsylvania State University

Boston Burr Ridge, IL Dubuque, IA Madison, WI New York San Francisco
St. Louis Bangkok Bogotá Caracas Lisbon London Madrid Mexico City
Milan New Delh Seoul Singapore Sydney Taipei Toronto

WCB/McGraw-Hill 

A Division of The ***McGraw-Hill*** *Companies*

This book was previously published under the title, "Motion and Time Study."

METHODS, STANDARDS, AND WORK DESIGN

This book is printed on acid-free paper.

1 2 3 4 5 6 7 8 9 0 QPF/QPF 9 3 2 1 0 9 8

ISBN 0-256-19507-2

Vice president and editorial director: *Kevin T. Kane*
Publisher: *Thomas Casson*
Executive editor: *Eric M. Munson*
Marketing manager: *John T. Wannemacher*
Senior project manager: *Jean Lou Hess*
Production supervisor: *Michael R. McCormick*
Freelance design coordinator: *JoAnne Schopler*
Photo research coordinator: *Sharon Miller*
Supplement coordinator: *Rose M. Range*
Compositor: *GAC Shepard Poorman Communications*
Typeface: *10.5/12 Times Roman*
Printer: *Quebecor Printing Book Group/Fairfield*

Library of Congress Cataloging-in-Publication Data

Niebel, Benjamin W.
Methods, standards, and work design / Benjamin W. Niebel, Andris Freivalds.—10th ed.
p. cm.
Includes bibliographical references and index.
Previous eds. published under title: Motion and time study.
ISBN 0-256-19507-2
1. Motion study. 2. Time study. I. Title. II. Freivalds, Andris III. Niebel, Benjamin W. Motion and time study
T60.7N54 1999
658.5'42—dc21 98-36671

http://www.mhhe.com

McGraw-Hill Series in Industrial Engineering and Management Science

CONSULTING EDITORS

Kenneth E. Case, *Department of Industrial Engineering and Management, Oklahoma State University*

Philip M. Wolfe, *Department of Industrial and Management Systems Engineering, Arizona State University*

Barnes
Statistical Analysis for Engineers and Scientists: A Computer-Based Approach

Bedworth, Henderson, and Wolfe
Computer-Integrated Design and Manufacturing

Black
The Design of the Factory with a Future

Blank
Engineering Economy

Blank
Statistical Procedures for Engineering, Management, and Science

Bridger
Introduction to Ergonomics

Denton
Safety Management: Improving Performance

Ebeling
Reliability and Maintainability Engineering

Grant and Leavenworth
Statistical Quality Control

Hillier and Lieberman
Introduction to Mathematical Programming

Hillier and Lieberman
Introduction to Operations Research

Juran and Gryna
Quality Planning and Analysis: From Product Development through Use

Kelton, Sadowski, and Sadowski
Simulation with ARENA

Khoshnevis
Discrete Systems Simulation

Kolarik
Creating Quality: Concepts, Systems, Strategies, and Tools

Law and Kelton
Simulation Modeling and Analysis

Marshall and Oliver
Decision Making and Forecasting

Moen, Nolan, and Provost
Improving Quality through Planned Experimentation

Nash and Sofer
Linear and Nonlinear Programming

Nelson
Stochastic Modeling: Analysis and Simulation

Pegden
Introduction to Simulation Using SIMAN

Riggs, Bedworth, and Randhawa
Engineering Economics

Sipper and Bulfin
Production: Planning, Control, and Integration

Steiner
Engineering Economic Principles

Taguchi, Elsayed, and Hsiang
Quality Engineering in Production Systems

Wu and Coppins
Linear Programming and Extensions

PREFACE

BACKGROUND

As we approach the 21st century with a widely expanded market and top manufacturing capability in Europe, the South Pacific Rim, and China, both the opportunities and the need for technical competence are growing dramatically. American companies are feeling the pressure of global competition and the impact of the information explosion. These changes have resulted in a dramatic growth in computerization and delayering of American industry and business. Ten years ago, foreign competition was centered in only a few industries—electronics and automotive in particular. But today this competition is both industrywide and worldwide. Almost every industry, business, and service organization is restructuring itself to operate more effectively in an increasingly competitive world. Each segment of these organizations is increasing the intensity of its cost reduction and quality improvement efforts, while at the same time working with a reduced labor force. Downsizing is becoming a trend. Therefore, cost-effectiveness and product reliability without excess capacity are the keys to successful activity in all areas of business, industry, and government. And cost-effectiveness with improved quality under restricted plant capacity is the end result of methods engineering, equitable time standards, and improved employee motivation through the introduction of modern management reward systems.

These tools are the keys to productivity improvement in any business, industry, or service organization, whether in a bank, a hospital, a department store, a railroad, or the postal system. Furthermore, success in a given product line or service leads to new products and innovations. It is this accumulation of successes that drives hiring and the growth of an economy. These are the tools that the Japanese have used so effectively in connection with their lean production concepts. This concept emphasizes methods engineering and the employment of teams of multiskilled workers at all levels of the organization and uses highly flexible, increasingly automated facilities to produce volumes of product in large variety. The tremendous success of the Japanese auto, camera, electronics, and appliance industries has been attributed to their lean production concept.

The reader should be careful not to be swayed or intimidated by some of the relatively new jargon offered as a cure-all for the lack of competitiveness of an enterprise. Often these fads destroy sound engineering and management procedures that, when properly utilized, represent the key to continued success. Thus, today we hear a good deal about re-engineering, use of cross-functional terms, etc., as business leaders reduce cost, inventory, cycle time, and nonvalue activities. However, experience in the past few years has proven that cutting people from the payroll just for the sake of automating their jobs is not always the wise procedure. The authors, with many years of experience in more than 100 industries, strongly recommend the application of sound methods engineering, realistic standards, and equitable wage payment as the keys to success in both manufacturing and business.

WHY THIS BOOK WAS WRITTEN

The tenth edition has been written for several reasons. Foremost, it emphasizes the importance of ergonomics and work design as parts of methods engineering. Far too often industrial engineers have focused solely on increasing productivity through methods changes and job simplification, resulting in overly repetitive jobs for the operators and increased incidence rates of musculoskeletal injuries. Any cost reductions obtained are more than offset by the increased medical and Workers Compensation costs, especially considering today's ever-escalating health care costs.

More importantly, this tenth edition updates existing material and examples that may have become obsolete because of technological change. A survey was taken of 100 recent graduates of the industrial engineering programs at Penn State University and Kansas State University and the 41 members of the Industry Professional Advisory Committees at both of these institutions. The responders identified the top 10 items they utilized on their jobs (with percent responses) as: flow process charts (88%), flow diagrams (83%), teamwork (79%), costing (76%), facilities layout (76%), training (72%), safety principles (71%), OSHA regulations (64%), workplace layout (64%), and job evaluations (64%). Surprisingly, some traditional work measurement topics (time study, standard data, work sampling) were no longer at the top of the list (although methods tools were). On the other hand, several nontraditional work organizational items (teamwork, job evaluations, and training) jumped to the top 10.

This tenth edition features a continued reliance on work sampling, time study, facilities layout, and various flow process charts in the industrial engineering profession. A greater emphasis is placed on standard data, costing, and the use of checklists. Work design and ergonomics topics have been expanded because of greater concern for occupational health and safety issues. A completely new chapter on training and other management practices, and new sections on problem identification and problem-solving techniques, have been added. Finally, traditional topics, such as simo charting, micromotion and memomotion study, and certain predetermined time systems that have apparently fallen from favor, have been completely eliminated.

However, the objectives of this edition have remained the same as for the first nine—to provide a practical, up-to-date college text describing engineering methods including work design, time study, and wage payment, and to give practicing labor and management analysts an authentic source of reference material.

ORGANIZATION OF THE TEXT AND COURSE MATERIAL

The tenth edition is laid out to provide roughly one chapter of material per week of a semester-long introductory course. Although there are a total of 18 chapters, Chapter 1 is short and introductory, Chapter 12 on formula construction may typically be covered in a statistics course, and Chapter 15 on standards for indirect and expense work may not need to be covered in an introductory course, leaving only 15 chapters to be covered in the semester.

A typical semester plan, chapter by chapter, might be as follows:

Chapter	Lectures	Coverage
1	1	Quick introduction on the importance of productivity and work design, with a bit of historical perspective
2	3	A few tools from each area (Pareto analysis, job analysis/worksite guide, flow process charts, worker–machine charts) with some quantitative analysis on worker–machine interactions. Line balancing may be covered in other courses.
3	3	Eight of the nine operation analysis approaches with some examples for each.
4	4–5	Full, but can gloss over basic muscle physiology and energy expenditure.
5	4–5	Full.
6	4–5	Basics on illumination, noise, temperature, and, perhaps, two other topics as desired. Safety and OSHA may be covered in another course.
7	3	Three tools: value engineering, cost-benefit analysis, and cross-over charts; job analysis and evaluation, and interaction with workers. Other tools may be covered in other classes.
8	2–3	Basics of time study.
9	1	One form of rating.
10	2	First half of the allowances that are well established.
11	1–2	Vary coverage of standard data depending on instructor's interest.

12	0	Formula construction may be covered in other engineering classes.
13	3	Only one predetermined time system in depth.
14	2	Work sampling.
15	0	Vary coverage of indirect and expense labor standards depending on instructor's interest.
16	2	Overview and costing.
17	3	Daywork and standard hour plan.
18	3	Learning curves, motivation, and people skills.

The recommended plan covers 40–45 lectures. Some instructors may wish to spend more time on any given chapter, for which additional material is supplied, e.g., work design (Chapters 4–6), and less time on traditional work measurement (Chapters 8–16) or vice versa. The text allows for this flexibility.

SUPPLEMENTARY MATERIAL

This edition also assists the educator by providing an updated instructor's manual with tear-out sheets of necessary forms, additional practice problems, case studies, suggested laboratory exercises, and ready-to-use software for work sampling, standard data, costing, etc., packaged on a CD. It continues to focus on the ubiquitous use of personal computers as well as the Internet to establish standards, conceptualize possibilities, evaluate costs, and disseminate information.

ON-LINE HELP

A Web site (*http://indy.ie.psu.edu/classes/ie327/index.htm*) is available for on-line background material, electronic versions of the forms available in the instructor's manual, and up-to-date information on any errors found or corrections needed in this new edition. Suggestions received from individuals at the universities, colleges, technical institutes, industries, and labor organizations that regularly use this text have helped materially in the preparation of this tenth edition. Further suggestions are still welcome, especially if any errors are noticed. Please simply respond to the *OOPS!* button on the Web site or by e-mail: *axf@psu.edu*. As with any Web site, this one will also continually evolve.

HOW THIS BOOK DIFFERS FROM OTHERS

Most textbooks on the market deal strictly either with the traditional elements of motion and time study or with human factors and ergonomics. Few textbooks integrate both topics into one book, or for that matter, one course. In this day and age, the industrial engineer needs to consider both productivity issues and their effects on the health and safety of the worker simultaneously. Few of the books on the market are formatted for use in the classroom setting. This text includes additional questions, problems, and sample laboratory exercises to assist the educator. Finally, no text has gone on-line to the Internet to provide electronic forms, current information, and changes as this edition does.

ACKNOWLEDGMENTS

The authors wish to acknowledge Professor Kenneth Knott for stimulating discussions and guidance in work measurement and Professor Stephan Konz for initiation of the survey leading to some of the major changes in this edition. Thanks also to the following reviewers for their invaluable input: Dr. Bala Subramaniam, CalPoly; Brian Kettler, General Motors Corp.; Dr. Patrick Patterson, Iowa State University; Dr. Terrence Stobbe, West Virginia University and Dr. Carter Kerk, South Dakota School of Mines & Technology. Finally, the authors wish to express their considerable gratitude to Dace Freivalds in the production of the final manuscript, without whom this edition may not have been completed.

Benjamin W. Niebel
Andris Freivalds

BRIEF CONTENTS

TABLE OF CONTENTS

CHAPTER 1

Methods, Standards, and Work Design: Introduction

KEY POINTS:

- Increasing productivity drives U.S. industry.
- Worker health and safety are just as important as productivity.
- Methods engineering simplifies work.
- Work design fits work to the operator.
- Time study measures work and sets standards.

PRODUCTIVITY IMPORTANCE

Certain changes continually taking place in the industrial and business environment must be considered both economically and practically. These include the globalization of both the market and the producer, the delayering of corporations in an effort to become more competitive without deteriorating quality, the growth of computerization in all facets of an enterprise, and the ever expanding applications of the information highway. The only way a business or enterprise can grow and increase its profitability is by increasing its productivity. Productivity improvement refers to the increase in output per work-hour or time expended. The United States has long enjoyed the world's highest productivity. Over the last 100 years, productivity in the United States has increased approximately 4 percent per year. However, in the last decade, the U.S. rate of productivity improvement has been exceeded by that of Japan, Korea, and Germany, and it has been challenged by Italy, France, and China.

The fundamental tools that result in increased productivity are: methods, time study standards (frequently referred to as work measurement), and work design. Of the total cost of the typical metal products manufacturing enterprise, 12 percent is direct labor, 45 percent is direct material, and 43 percent is overhead. All aspects of a business or industry—sales, finance, production, engineering, cost, maintenance, and management—provide fertile areas for the application of methods, standards, and work design. Too often, people consider only the production function when applying these tools. Important as the production function is, other aspects of the enterprise also contribute substantially to the cost of operation and are equally valid areas for the application of cost improvement techniques. In sales, for example, modern information retrieval methods usually result in more reliable information, leading to greater sales at less cost. Product quotas for specific territories provide a base or standard that individual salespeople endeavor to exceed, and payment for results always produces above-standard performance.

Today, most U.S. businesses and industries are, by necessity, restructuring themselves by downsizing, in order to operate more effectively in an increasingly competitive world. With more intensity than ever before, they are addressing cost reduction and quality improvement through productivity improvement. They are also critically examining all business components that do not contribute to their profitability.

Since the field of production within manufacturing industries utilizes the greatest number of young men and women in methods, standards, and work design efforts, this text will treat that field in more detail than any other. However, examples from other areas of the manufacturing industry, such as maintenance, transportation, sales, and management, as well as the service industry, will be provided.

The production areas of opportunity for students enrolled in engineering, industrial management, business administration, industrial psychology, and labor–management relations are: (1) work measurement, (2) work methods and design, (3) production engineering, (4) manufacturing analysis and control, (5) facilities planning, (6) wage administration, (7) ergonomics and safety, (8) production and inventory control, and (9) quality control. Other position areas, such as personnel or industrial relations, cost, and budgeting, are closely related to, and dependent on, the production group. These areas of opportunity are not confined to manufacturing industries. They exist, and are equally important, in such enterprises as department stores, hotels, educational institutions, hospitals, banks, airlines, insurance offices, military service centers, government agencies, and retirement complexes. Today, in the United States, only about 20 percent of the total labor force is employed in manufacturing industries. The remaining 80 percent is engaged in service industries or staff-related positions. As the United States becomes more service-industry oriented, the philosophies and techniques of methods, standards, and work design must be utilized in the service sector. Wherever people, materials, and facilities interact to obtain some objective, productivity can be improved through the intelligent application of methods, standards, and work design.

The production section of an industry may well be called its heart; if the activity of this section is interrupted, the whole industry ceases to be productive. The production department includes methods engineering, time study standards, and work design activity, offering the young technical graduate one of the most satisfying fields of endeavor.

In the production department, material to produce is requisitioned and controlled; the sequence of operations, inspections, and methods is determined; tools are ordered; time values are assigned; work is scheduled, dispatched, and followed up; and customers are kept satisfied with quality products delivered on time. Training in this field demonstrates how production is accomplished, where it is done, when it is performed, and how long it takes to do. A background that includes such training will prove invaluable, whether one's ultimate objective is sales, production, or cost.

If the production department is considered the heart of an industrial enterprise, the methods, standards, and work design activity is the heart of the production group. Here more than in any other place, people determine whether a product is going to be produced on a competitive basis. Here is where they use initiative and ingenuity to develop efficient tooling, worker and machine relationships, and workstations on new jobs in advance of production, thus assuring that the product will stand the test of stiff competition. Here is where they are creative in improving existing methods and products to help the company attain leadership in its product line. In this activity, good labor relations may be maintained through establishing fair labor standards, or may be impeded by setting one inequitable rate.

Methods, standards, and work design offer real challenges. Industries with competent engineers, business administrators, industrial relations personnel, specially trained supervisors, and psychologists all using methods, standards, and work design techniques are inevitably better able to meet competition and better equipped to operate profitably.

The objective of the manufacturing manager is to produce a quality product, on schedule, at the lowest possible cost, with a minimum of capital investment and a maximum of employee satisfaction. The focus of the reliability and quality control manager is to maintain engineering specifications and satisfy customers with the product's quality level and reliability over its expected life. The production control manager is principally interested in establishing and maintaining production schedules with due regard for both customer needs and the favorable economics obtainable with careful scheduling. The manager of methods, standards, and work design is mostly concerned with combining the lowest possible production cost and with the maximum employee satisfaction. The maintenance manager is primarily concerned with minimizing facility downtime due to unscheduled breakdowns and repairs. Figure 1–1 illustrates the relationship of the manager of the methods, standards, and work design department to the staff and line departments under the general manager.

METHODS AND STANDARDS SCOPE

Methods engineering includes designing, creating, and selecting the best manufacturing methods, processes, tools, equipment, and skills to manufacture a product based on the working drawings that have been developed by the product engineering section. When the best method interfaces with the best skills

FIGURE 1–1
Typical organization chart showing the influence of methods, standards and work design on the operation of the enterprise.

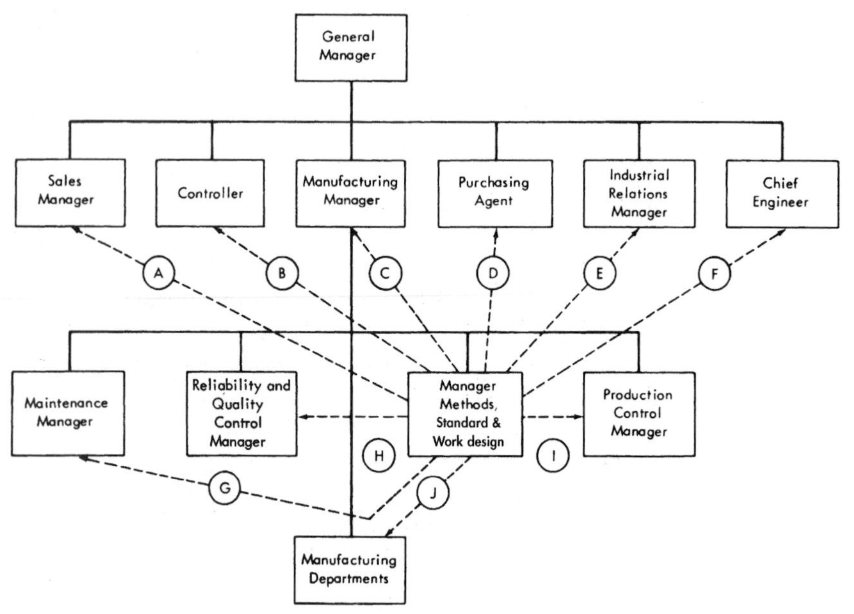

A—Cost is largely determined by manufacturing methods.
B—Time standards are the bases of standard costs.
C—Standards (direct and indirect) provide the bases for measuring the performance of production departments.
D—Time is a common denominator for comparing competitive equipment and supplies.
E—Good labor relations are maintained with equitable standards and a safe work environment.
F—Methods work design and processes strongly influence product designs.
G—Standards provide the bases for preventive maintenance.
H—Standards enforce quality.
I—Scheduling is based on time standards.
J—Methods, standards and work design provide how the work is to be done and how long it will take.

available, an efficient worker–machine relationship exists. Once the complete method has been established, the responsibility for determining the standard time required to produce the product falls within the scope of this work. Also included is the responsibility for following through to see that: (a) predetermined standards are met; (b) workers are adequately compensated for their output, skills, responsibilities, and experience; and (c) workers have a feeling of satisfaction from the work that they do.

The overall procedure includes: defining the problem; breaking the job down into operations; analyzing each operation to determine the most economical manufacturing procedures for the quantity involved, with due regard for operator safety and job interest; applying proper time values; and then following through to assure that the prescribed method is put into operation. Figure 1–2 illustrates the opportunities for reducing manufacturing time through the application of methods engineering and time study.

FIGURE 1–2
Opportunities for savings through the applications of methods engineering and time study.

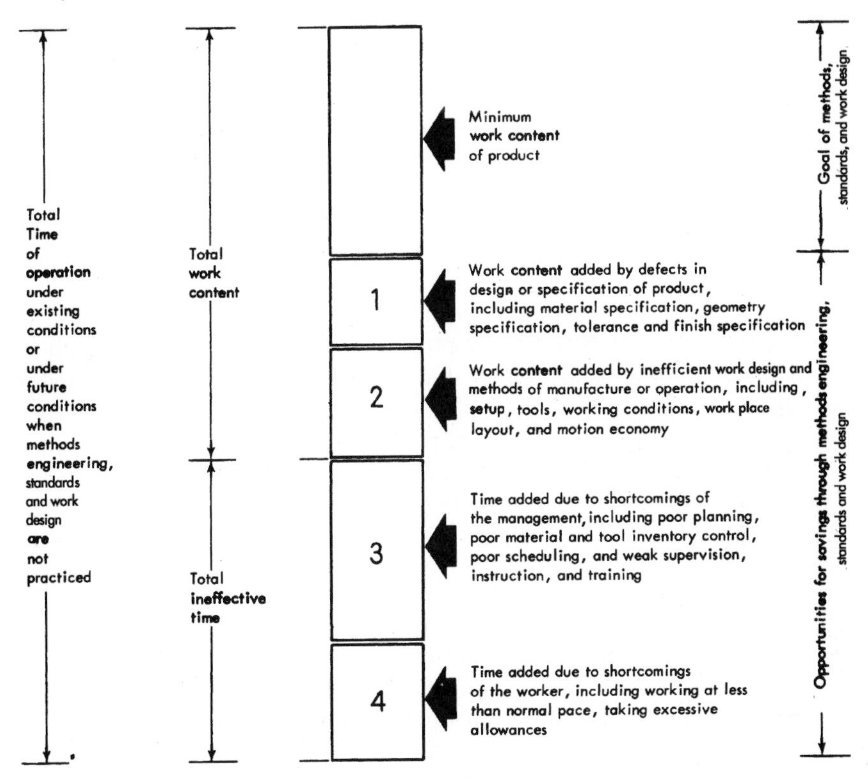

Methods Engineering

The terms *operation analysis*, *work design and simplification*, and *methods engineering and corporate re-engineering* are frequently used synonymously. In most cases, the person is referring to a technique for increasing the production per unit of time or decreasing the cost per unit output—in other words, *productivity improvement*. However, methods engineering, as defined in this text, entails analyses at two different times during the history of a product. The methods engineer is first responsible for designing and developing the various work centers where the product will be produced. Second, that engineer must continually restudy the work centers to find a better way to produce the product and/or improve its quality.

In recent years, this second analysis has been called corporate re-engineering. In this regard, we recognize that a business must introduce changes if it is to continue profitable operation. Thus, it may be desirable to introduce changes outside of the manufacturing area. Often, profit margins may be enhanced through positive changes in such areas as accounting, inventory management, materials requirements planning, logistics, and human resource management. Information automation can provide dramatic rewards in all of these areas. The more thorough the methods study during the planning stages, the less the necessity for additional methods studies during the life of the product.

Methods engineering implies the utilization of technological capability. Primarily because of methods engineering, improvements in productivity are never-ending. The productivity differential resulting from technological innovation can be of such magnitude that developed countries will always be able to maintain competitiveness with low-wage developing countries. Research and development (R&D) leading to new technology is therefore essential to methods engineering. The 10 countries with the highest R&D expenditures per worker, as reported by the United Nations Industrial Development Organization (1985), are: United States, Switzerland, Sweden, Netherlands, Germany, Norway, France, Israel, Belgium, and Japan. These countries are among the leaders in productivity. As long as they continue to emphasize research and development, methods engineering through technological innovation will be instrumental in their ability to provide high-level goods and services.

Methods engineers use a systematic procedure to develop a work center, produce a product, or provide a service. This procedure is outlined here, and it summarizes the flow of the text. Each step is detailed in a later chapter. Note that Steps 6 and 7 are not strictly part of a methods study, but are necessary in a fully functioning work center.

1. *Select the project.* Typically, the projects selected represent either new products or existing products that have a high cost of manufacture and a low profit. Also, products that are currently experiencing difficulties in maintaining quality and are having problems meeting competition are logical projects for method engineering. (See Chapter 2 for more details.)
2. *Get and present the data.* Assemble all the important facts relating to the product or service. These include drawings and specifications, quantity requirements, delivery requirements, and projections of the anticipated life of the product or service. Once all important information has been acquired, record it in an orderly form for study and analysis. The development of process charts at this point is very helpful. (See Chapter 2 for more details.)
3. *Analyze the data.* Utilize the primary approaches to operations analysis to decide which alternative will result in the best product or service. These primary approaches include: purpose of operation, design of part, tolerances and specifications, materials, process of manufacture, setup and tools, working conditions, material handling, plant layout, and principles of motion economy. (See Chapter 3 for more details.)
4. *Develop the ideal method.* Select the best procedure for each operation, inspection, and transportation by considering the various constraints associated with each alternative, including productivity, ergonomics, and health and safety implications. (See Chapters 3–6 for more details.)
5. *Present and install the method.* Explain the proposed method in detail to those responsible for its operation and maintenance. Consider all details of the work center, to insure that the proposed method will provide the results anticipated. (See Chapter 7 for more details.)

6. *Develop a job analysis*. Conduct a job analysis of the installed method to insure that the operators are adequately selected, trained, and rewarded. (See Chapter 7 for more details.)
7. *Establish time standards*. Establish a fair and equitable standard for the installed method. (See Chapters 8–15 for more details.)
8. *Follow up the method*. At regular intervals, audit the installed method to determine if the anticipated productivity and quality are being realized, whether costs were correctly projected, and whether further improvements can be made. (See Chapter 16 for more details.)

In summary, methods engineering is the systematic close scrutiny of all direct and indirect operations to find improvements that make work easier to perform and allow work to be done in less time with less investment per unit. In other words, the real objective of methods engineering is profit improvement.

Work Design

As part of developing or maintaining the new method, the principles of work design must be used to fit the task and workstation ergonomically to the human operator. Unfortunately, work design is typically forgotten in the quest for increased productivity. Far too often, overly simplified procedures result in machine-like repetitive jobs for the operators, leading to increased rates of work-related musculoskeletal disorders. Any productivity increases and reduced costs are more than offset by the increased medical and Workers' Compensation costs, especially considering today's ever-escalating health-care trends. Thus, it is necessary for the methods engineer to incorporate the principles of work design into any new method, so that it will not only be more productive but will also be safe and injury-free for the operator. (Refer to Chapters 4–6.)

Standards

Standards are the end result of time study or work measurement. This technique establishes a time standard allowed to perform a given task, based on measurements of the work content of the prescribed method, with due consideration for fatigue and for personal and unavoidable delays. Time study analysts use several techniques to establish a standard: a stopwatch time study, computerized data collection, standard data, fundamental motion data, work sampling, and estimates based on historical data. Each technique is applicable to certain conditions. Time study analysts must know when to use a given technique, and must then use that technique judiciously and correctly.

The functions of time study analysts and methods engineers are closely allied. Although the objectives of the two differ, good time study analysts are good methods engineers, since their techniques include methods engineering as a basic component. In small industries, these two activities are often handled by the same individual. Establishing time values is a step in the systematic development of new work centers and the improvements in methods used in existing work centers. To compete in a worldclass market, the methods engineer must also consider the demands of just-in-time quality control and time-compressed management.

The resulting standards are used to implement a wage payment scheme. In many companies, particularly in smaller enterprises, the wage payment activity is performed by the same group responsible for the methods and standards work. Also, the wage payment activity is performed in concert with those responsible for conducting job analyses and job evaluations, so that these closely related activities function smoothly.

Production control, plant layout, purchasing, cost accounting and control, and process and product design are additional areas closely related to both the methods and standards functions. To operate effectively, all of these areas depend on time and cost data, facts, and operational procedures from the methods and standards department. These relationships are briefly discussed in Chapter 16.

Objectives of Methods, Standards, and Work Design

The principal objectives of methods, standards, and work design are: (a) to increase productivity and product reliability safely; and (b) to lower unit cost, thus allowing more quality goods and services to be produced for more people. The ability to produce more for less will result in more jobs for more people for a greater number of hours per year. Only through the intelligent application of the principles of methods, standards, and work design can producers of goods and services increase, while, at the same time, the purchasing potential of all consumers grows. Through these principles, unemployment and relief rolls can be minimized, thus reducing the spiraling cost of economic support to nonproducers.

Corollaries to the principal objectives are:

1. Minimize the time required to perform tasks.
2. Continually improve the quality and reliability of products and services.
3. Conserve resources and minimize cost by specifying the most appropriate direct and indirect materials for the production of goods and services.
4. Take the availability of power into careful consideration.
5. Maximize the safety, health, and well-being of all employees.
6. Produce with an increasing concern for protecting the environment.
7. Follow a humane program of management that results in job interest and satisfaction for each employee.

HISTORICAL DEVELOPMENTS

The Work of Taylor

Frederick W. Taylor is generally conceded to be the founder of modern time study in this country. However, time studies were conducted in Europe many years before Taylor's time. In 1760, Jean Rodolphe Perronet, a French engineer, made extensive time studies on the manufacture of No. 6 common pins, while 60 years later, an English economist, Charles W. Babbage, conducted time studies on the manufacture of No. 11 common pins.

Taylor began his time study work in 1881 while associated with the Midvale Steel Company in Philadelphia. Although born in a wealthy family, he disdained his upbringing and started out serving as an apprentice. After 12 years' work, he evolved a system based on the "task." Taylor proposed that the work of each employee be planned out by the management at least one day in advance. Workers were to receive complete written instructions describing their tasks in detail and noting the means to accomplish them. Each job was to have a standard time, determined by time studies made by experts. In the timing process, Taylor advocated breaking up the work assignment into small divisions of effort known as "elements." Experts were to time these individually and use their collective values to determine the allowed time for the task.

Taylor's early presentations of his findings were received without enthusiasm, because many of the engineers interpreted his findings to be a new piece-rate system rather than a technique for analyzing work and improving methods. Both management and employees were skeptical of piece rates, because many standards were either typically based on the supervisor's guess or inflated by bosses to protect the performance of their departments.

In June 1903, at the Saratoga meeting of the American Society of Mechanical Engineers (ASME), Taylor presented his famous paper, "Shop Management," which included the elements of scientific management: time study, standardization of all tools and tasks, use of a planning department, use of slide rules and similar timesaving implements, instruction cards for workers, bonuses for successful performance, differential rates, mnemonic systems for classifying products, routing systems, and modern cost systems. Taylor's techniques were well received by many factory managers, and by 1917, of 113 plants that had installed "scientific management," 59 considered their installations completely successful, 20 partly successful, and 34 failures (Thompson, 1917).

In 1898, while at the Bethlehem Steel Company (he had resigned his post at Midvale), Taylor carried out the pig-iron experiment that came to be one of the most celebrated demonstrations of his principles. He established the correct method, along with financial incentives, and workers carrying 92-pound pigs of iron up a ramp onto a freight car were able to increase their productivity from an average of 12.5 tons per day to between 47 and 48 tons per day. This work was performed with an increase in the daily rate of $1.15 to $1.85. Taylor claimed that workmen performed at the higher rate "without bringing on a strike among the men, without any quarrel with the men and were happier and better contented."

Another of Taylor's Bethlehem Steel studies that gained fame was the shoveling experiment. Workers who shoveled at Bethlehem owned their own shovels and would use the same one for any job—lifting heavy iron ore to lifting light rice coal. After considerable study, Taylor designed shovels to fit the different loads: short-handled shovels for iron ore, long-handled scoops for light rice coal.

Another of Taylor's well-known contributions was the discovery of the Taylor-White process of heat treatment for tool steel. Studying self-hardening steels, he developed a means of hardening a chrome–tungsten steel alloy without rendering it brittle, by heating it close to its melting point. The resulting "high-speed steel" more than doubled machine cutting productivity and remains in use today all over the world. Later, he developed the Taylor equation for cutting metal.

Not as well known as his engineering contributions is the fact that in 1881, he was a U.S. tennis doubles champion. Here he used an odd-looking racket he had designed with a spoon curved handle. Taylor died of pneumonia in 1915, at the age of 59. For more information on this multitalented individual, the authors recommend Kanigel's biography (1997).

In the early 1900s, the country was going through an unprecedented inflationary period. The word efficiency became passé, and most businesses and industries were looking for new ideas that would improve their performance. The railroad industry also felt the need to increase shipping rates substantially to cover general cost increases. Louis Brandeis, who at that time represented the eastern business associations, contended that the railroads did not deserve, or in fact need, the increase because they had been remiss in not introducing the new "science of management" into their industry. Brandeis claimed that the railroad companies could save $1 million a day by introducing the techniques advocated by Taylor. Thus, Brandeis and the Eastern Rate Case (as the hearing came to be known) first introduced Taylor's concepts as "scientific management."

At this time, many people without the qualifications of Taylor, Barth, Merrick, and other early pioneers, were eager to make names for themselves in this new field. They established themselves as "efficiency experts" and endeavored to install scientific management programs in industry. They soon encountered a natural resistance to change from employees, and since they were not equipped to handle problems of human relations, they met with great difficulty. Anxious to make a good showing and equipped with only a pseudoscientific knowledge, they generally established rates that were too difficult to meet. Situations became so acute that some managers were obliged to discontinue the whole program in order to continue operation.

In other instances, factory managers would allow the establishment of time standards by the supervisors, but this was seldom satisfactory. Once standards were established, many factory managers of that time, interested primarily in the reduction of labor costs, would unscrupulously cut rates if some employee made what the employer felt was too much money. The result was harder work at the same, and sometimes less, take-home pay. Naturally, violent worker reaction resulted.

These developments spread in spite of the many favorable installations started by Taylor. At the Watertown Arsenal, labor objected to such an extent to the new time study system that in 1910, the Interstate Commerce Commission (ICC) started an investigation of time study. Several derogatory reports on the subject influenced Congress to add a rider to the government appropriations bill in 1913, stipulating that no part of the appropriation should be made available for the pay of any person engaged in time study work. This restriction applied to the government-operated plants where government funds were used to pay the employees.

It wasn't until 1947 that the House of Representatives passed a bill that rescinded the prohibition against using stopwatches and the use of time study. It is of interest that even today the use of the stopwatch is still prohibited by unions in some railroad repair facilities. It is also interesting to note that Taylorism is very much alive today, in contemporary assembly lines, in lawyer's bills that are calculated in fractional hours, and in the recommended arrangement of kitchen appliances.

Motion Study and the Work of the Gilbreths

Frank and Lilian Gilbreth were the founders of the modern motion study technique, which may be defined as the study of the body motions used in performing an operation, to improve the operation by eliminating unnecessary motions, simplifying necessary motions, and then establishing the most favorable motion sequence for maximum efficiency. Frank Gilbreth originally introduced his ideas and philosophies into the bricklayer's trade in which he was employed. After introducing methods improvements through motion study, including an adjustable scaffold that he had invented, as well as operator training, he was able to increase the average number of bricks laid to 350 per worker per hour. Prior to Gilbreth's studies, 120 bricks per hour was considered a satisfactory rate of performance for a bricklayer.

More than anyone else, the Gilbreths were responsible for industry's recognition of the importance of a detailed study of body motions to increase production, reduce fatigue, and instruct operators in the best method of performing an operation. They developed the technique of filming motions to study them, in a technique known as micromotion study. The study of movements through the aid of the slow-motion moving picture is by no means confined to industrial applications.

In addition, the Gilbreths developed the cyclegraphic and chronocyclegraphic analysis techniques for studying the motion paths made by an operator. The cyclegraphic method involves attaching a small electric light bulb to the finger or hand or part of the body being studied and then photographing the motion while the operator is performing the operation. The resulting picture gives a permanent record of the motion pattern employed and can be analyzed for possible improvement. The chronocyclegraph is similar to the cyclegraph, but its electric circuit is interrupted regularly, causing the light to flash. Thus, instead of showing solid lines of the motion patterns, the resulting photograph shows short dashes of light spaced in proportion to the speed of the body motion being photographed. Consequently, with the chronocyclegraph it is possible to compute velocity, acceleration, and deceleration, as well as to study body motions. The world of sports has found this analysis tool, updated to video, invaluable as a training tool to show the development of form and skill.

As an interesting side note, the reader may wish to read about the extreme lengths to which Frank Gilbreth went to achieve maximum efficiency even in his personal life. His eldest son and daughter recount vignettes of their father shaving with razors simultaneously in both hands or using various communication signals to assemble all of the children, of which there were twelve. Hence the title of their book, *Cheaper by the Dozen* (Gilbreth and Gilbreth, 1948)!

Early Contemporaries

Carl G. Barth, an associate of Frederick W. Taylor, developed a production slide rule for determining the most efficient combinations of speeds and feeds for cutting metals of various hardnesses, considering the depth of cut, size of tool, and life of the tool. He is also noted for his work in determining allowances. He investigated the number of foot-pounds of work a worker could do in a day. He then developed a rule that equated a certain push or pull on a worker's arms with the amount of weight that worker could handle for a certain percentage of the day.

Harrington Emerson applied scientific methods to work on the Santa Fe Railroad and wrote a book, *Twelve Principles of Efficiency*, in which he made an effort to inform management of procedures for efficient operation. He reorganized the company, integrated its shop procedures, installed standard costs and a bonus plan, and transferred its accounting work to Hollerith tabulating machines. This effort resulted in annual savings in excess of $1.5 million and the recognition of his approach, termed *efficiency engineering*.

In 1917, Henry Laurence Gantt developed simple graphs that would measure performance while visually showing projected schedules. This production control tool was enthusiastically adopted by the shipbuilding industry during World War I. For the first time, this tool made it possible to compare actual performance against the original plan, and to adjust daily schedules in accordance with capacity, backlog, and customer requirements. Gantt is also known for his invention of a wage payment system that rewarded workers for above-standard performance, eliminated any penalty for failure, and offered the boss a bonus for every worker who performed above standard. Gantt emphasized human relations and promoted scientific management as more than an inhuman "speedup" of labor.

Motion and time study received added stimulus during World War II when Franklin D. Roosevelt, through the U.S. Department of Labor, advocated establishing standards for increasing production. The stated policy advocated: greater pay for greater output but without an increase in unit labor costs, incentive schemes to be collectively bargained between labor and management, and the use of time study or past records to set production standards.

Emergence of Work Design

Work design is a relatively new science that deals with designing the task, workstation, and working environment to fit the human operator better. In the United States, it is more typically known as *human factors,* while internationally it is better known as *ergonomics*, which is derived from the Greek words for work (erg) and laws (nomos).

In the United States, after the initial work of Taylor and the Gilbreths, the selection and training of military personnel during World War I and the industrial psychology experiments of the Harvard Graduate School at Western Electric (see the Hawthorne studies in Chapter 7) were important contributions to the work design area. In Europe, during and after World War I, the British Industrial Fatigue Board performed numerous studies on human performance under various conditions. These were later extended to heat stress and other conditions by the British Admiralty and Medical Research Council. However, during and after World War II, the complexity of military equipment and aircraft led to the development of the U.S. military engineering psychology laboratories and a real growth of the profession. The start of the race to space with the launch of Sputnik in 1957 only accelerated the growth of human factors, especially in the aerospace and military sectors. From the 1970s on, the growth has shifted to the industrial sector and, more recently, into computer equipment, user-friendly software, and the office environment. Other driving forces for the growth in human factors are the rise in product liability and personal injury litigation cases and also, unfortunately, tragic, large-scale technological disasters, such as the nuclear incident at Three-Mile Island and the gas leak at the Union Carbide Plant in Bhopal, India. Obviously, the growth of computers and technology will keep human factors specialists and ergonomists busy designing better workplaces and products and improving the quality of life and work for many years to come.

Organizations

Since 1911, there has been an organized effort to keep industry abreast of the latest developments in the techniques inaugurated by Taylor and Gilbreth. Technical organizations have contributed much toward bringing the science of time study, motion study, work simplification, and methods engineering up to present-day standards. In 1915, the Taylor Society was founded to promote the science of management, while in 1917 the Society of Industrial Engineers was organized by those interested in production methods. The American Management Association (AMA) traces its origins back to 1913, when a group of training managers formed the National Association of Corporate Schools. Its various divisions sponsor courses and publications on productivity improvement, work measurement, wage incentives, work simplification, and clerical standards. Together with the American Society of Mechanical Engineers (ASME), AMA annually presents the Gantt Memorial Medal for the most distinguished contribution to industrial management as a service to the community.

The Society for the Advancement of Management (SAM) was formed in 1936 by the merger of the Society of Industrial Engineers and the Taylor Society. This organization emphasized the importance of time study and methods and wage payment. Industry has used SAM's time study rating films over a long period of years. SAM annually offers the Taylor key for the outstanding contribution to the advancement of the science of management and the Gilbreth medal for noteworthy achievement in the field of motion, skill, and fatigue study. In 1972, SAM combined forces with AMA.

The Institute of Industrial Engineers (IIE) was founded in 1948 with the purposes of: maintaining the practice of industrial engineering on a professional level; fostering a high degree of integrity among the members of the industrial engineering profession; encouraging and assisting education and research in areas of interest to industrial engineers; promoting the interchange of ideas and information among members of the industrial engineering profession (e.g., publishing the journal *IIE Transactions*); serving the public interest by identifying persons qualified to practice as industrial engineers; and promoting the professional registration of industrial engineers. IIE's Society of Work Science (the result of merging the Work Measurement and Ergonomics Divisions in 1994) keeps the membership up to date on all facets of this area of work. This Society annually gives the Phil Carroll Award and M. M. Ayoub Award for achievement in work measurement and ergonomics, respectively.

In the area of work design, the first professional organization, the Ergonomics Research Society, was founded in the United Kingdom in 1949. It started the first professional journal, *Ergonomics*, in 1957. The U.S. professional organization, The Human Factors and Ergonomics Society, was founded in 1957. In the 1960s, there was rapid growth in The Society, with membership increasing from 500 to 3,000. Currently, there are well over 5,000 members organized in 20 different technical groups. Their primary goals are to: (1) define and support human factors/ergonomics as a scientific discipline and in practice, with the exchange of technical information among members; (2) educate and inform business, industry and government about human factors/ergonomics; and (3) promote human factors/ergonomics as a means for bettering the quality of life. The Society also publishes an archival journal, *Human Factors*, and holds annual conferences where members can meet and exchange ideas.

With the proliferation of national professional societies, an umbrella organization, the International Ergonomics Association, was founded in 1959 to coordinate ergonomic activities at an international level. At present, there are 33 individual societies encompassing over 16,000 members worldwide.

Present Trends

Practitioners of methods, standards, and work design have come to realize that such factors as sex, age, health and well-being, physical size and strength, aptitude, training attitudes, job satisfaction, and motivation response have a direct bearing on productivity. Furthermore, present-day analysts recognize that workers object, and rightfully so, to being treated as machines. Workers dislike and fear a purely scientific approach and inherently dislike any change from their present way of operation. Even management frequently rejects worthwhile methods innovations because of a reluctance to change.

Workers tend to fear methods and time study, for they see that the results are an increase in productivity. To them, this means less work and consequently less pay. They must be sold on the fact that they, as consumers, benefit from lower costs, and that broader markets result from lower costs, meaning more work for more people for more weeks of the year.

Some fears of time study today are due to unpleasant experiences with efficiency experts. To many workers, motion and time study is synonymous with the speedup or the stretch-out. These terms denote using incentives to spur employees to higher levels of output, followed by establishing new levels as normal production, thus forcing the workers to still greater exertions to maintain even their previous earning power. In the past, shortsighted and unscrupulous managers did resort to this practice.

Even today, some unions oppose the establishment of standards by measurement, the development of hourly base rates by job evaluation, and the application of incentive wage payment. These unions believe that the time allowed to perform a task and the amount that an employee should be paid represent issues that should be resolved by collective bargaining arrangements.

Today's practitioners must use the "humane" approach. They must be well versed in the study of human behavior and accomplished in the art of communication. They must also be good listeners, respecting the ideas and thinking of others, particularly the worker at the bench. They must give credit where credit is due. In fact, they should habitually give the other person credit, even if there is some question of that person deserving it. Also, practitioners of motion and time study should always remember to use the questioning attitude emphasized by the Gilbreths, Taylor, and the other pioneers in the field. The idea that there is "always a better way" needs to be continually pursued in the development of new methods that improve productivity, quality, delivery, worker safety, and worker well being.

Today, there is a greater intrusion by the government in the regulation of methods, standards, and work design. For example, military equipment contractors and subcontractors are under increased pressure to document direct labor standards as a result of MIL-STD 1567A (released 1975; revised 1983 and 1987). Any firm awarded a contract exceeding $1 million is subject to MIL-STD 1567A, which requires a work measurement plan and procedures, a plan to establish and maintain engineered standards of known accuracy and traceability, a plan for methods improvement in conjunction with standards, a plan for the use of the standards as an input to budgeting, estimating, planning, and performance evaluation, and detailed documentation for all of these plans.

Similarly, in the area of work design, Congress passed the OSHAct establishing the National Institute for Occupational Safety and Health (NIOSH), a research agency for developing guidelines and standards for worker health and safety, and the Occupational Safety and Health Administration (OSHA), an enforcement agency to maintain these standards. With the sudden increase in repetitive motion injuries in the food processing industry, OSHA established the Ergonomics Program Management Guidelines for Meatpacking Plants in 1990. Similar guidelines for general industry and an eventual standard were to follow, but the plans were derailed with the shift in the political balance in Congress in 1994.

With increasing numbers of individuals with different abilities, Congress passed the Americans with Disabilities Act (ADA) in 1990. This regulation has a major impact on all employers with 15 or more employees, affecting such employment practices as recruiting, hiring, promotions, training, laying off, firing, allowing leaves, and assigning jobs.

While work measurement once concentrated on direct labor, methods and standard development have increasingly been used for indirect labor. This trend will continue as the number of traditional manufacturing jobs decreases and the number of service jobs increases in the United States. The use of computerized techniques will also continue to grow. Several of the predetermined time systems are fully computerized today. Notable among these are MOST, and WOCOM. Many companies have also developed time study and work sampling software, using electronic data collectors for compiling the information required.

Table 1–1 illustrates the progress made in methods, standards, and work design.

TABLE 1–1
Progress Made in Connection with Methods, Standards and Work Design

Year	Event
1760	Perronet makes time studies on No. 6 common pins.
1820	Charles W. Babbage makes time studies on No. 11 common pins.
1832	Charles W. Babbage publishes *On the Economy of Machinery and Manufactures.*
1881	Frederick W. Taylor begins his time study work.
1901	Henry L. Gantt develops the task and bonus wage system.
1903	Taylor presents paper on Shop Management to ASME.
1906	Taylor publishes paper *On the Art of Cutting Metals.*
1910	Interstate Commerce Commission starts an investigation of time study.
	Gilbreth publishes *Motion Study.*
	Gantt publishes *Work, Wages, and Profits.*
1911	Taylor publishes text on *The Principles of Scientific Management.*
1912	Society to Promote the Science of Management is organized.
	Emerson estimates $1 million per day can be saved if eastern railroads apply scientific management.
1913	Emerson publishes *The Twelve Principles of Efficiency.*
	Congress adds rider to government appropriation bill stipulating that no part of this appropriation should be made available for the pay of any person engaged in time study work.
	Henry Ford unveils the first moving assembly line in Detroit.
1915	Taylor Society formed to replace the Society to Promote the Science of Management.
1917	Frank B. and Lillian M. Gilbreth publish *Applied Motion Study.*
1923	American Management Association formed.
1927	Elton Mayo begins Hawthorne study at Western Electric Company's plant in Hawthorne, Illinois.
1933	Ralph M. Barnes receives the first Ph.D. granted in the United States in the field of Industrial Engineering from Cornell University. His thesis leads to the publication of "Motion and Time Study."
1936	Society for the Advancement of Management organized.
1945	Department of Labor advocates establishing standards to improve productivity of supplies for the war effort.
1947	Bill passed allowing the War Department to use time study.

TABLE 1–1
(concluded)

Year	Event
1948	The Institute of Industrial Engineers is founded in Columbus, Ohio.
	Eiji Toyoda and Taichi Ohno at Toyota Motor Company pioneer the concept of lean production.
1949	Prohibition against using stopwatches dropped from appropriation language.
	The Ergonomics Research Society founded in the U.K.
1957	The Human Factors and Ergonomics Society founded in the U.S.
	E.J. McCormick publishes *Human Factors Engineering.*
1959	International Ergonomics Association founded to coordinate ergonomics activities worldwide.
1970	Congress passes the OSHAct, establishing the Occupational Safety and Health Administration.
1972	Society for the Advancement of Management combines with the American Management Associations.
1975	MIL-STD 1567 (USAF), Work Measurement, released.
1981	NIOSH lifting guidelines first introduced.
1983	MIL-STD 1567A, Work Measurement, released.
1986	MIL-STD 1567A, Work Measurement Guidance Appendix, finalized.
1988	ANSI/HFS Standard 100-1988 for Human Factors Engineering of Visual Display Terminal Workstations released.
1990	Americans with Disabilities Act (ADA) passed by Congress.
	Ergonomics Program Management Guidelines for Meatpacking Plants established by OSHA.
1993	NIOSH lifting guidelines revised.
1995	Draft ANSI Z-365 Standard for Control of Work-Related Cumulative Trauma Disorders released.

SUMMARY

Industry, business, and government are in agreement that the untapped potential for increasing productivity is the best hope for dealing with inflation and competition. The principal key to increased productivity is a continuing application of the principles of methods, standards, and work design. Only in this way can more output from people and machines be realized. American labor expects, and has the bargaining strength, to get a continuing increase in wage levels. The American government has pledged itself to an increasingly paternalistic philosophy of providing for the disadvantaged—housing for the poor, medical care for the aged, jobs for minorities, and so on. To accommodate the spiraling costs of labor and government taxes and still stay in business, we must get more from our productive elements—people and machines.

QUESTIONS

1. What is another name for time study?
2. What is the principal objective of methods engineering?
3. List the eight steps in applying methods engineering.
4. Where were time studies originally made and who conducted them?
5. Explain Frederick W. Taylor's principles of scientific management.
6. What is meant by motion study, and who are the founders of the motion study technique?
7. Was the skepticism of management and labor toward rates established by "efficiency experts" understandable? Why?
8. Which organizations are concerned with advancing the ideas of Taylor and the Gilbreths?
9. What psychological reaction is characteristic of workers when methods changes are suggested?
10. Explain the importance of the humanistic approach in methods and time study work.
11. How are time study and methods engineering related?
12. Why is work design an important element of methods study?
13. What important events have contributed to the need for ergonomics?

REFERENCES

Barnes, Ralph M. *Motion and Time Study: Design and Measurement of Work.* 7th ed. New York: John Wiley & Sons, 1980.

Eastman Kodak Co., Human Factors Section. *Ergonomic Design for People at Work.* New York: Van Nostrand Reinhold, 1983.

Gilbreth, F., and L. Gilbreth. *Cheaper by the Dozen.* New York: T.W. Crowell, 1948.

Kanigel, R. *One Best Way.* New York: Viking, 1997.

Konz, Stephan. *Work Design.* 4th ed. Columbus, OH: Grid, 1995.

Mundell, Marvin E. *Motion and Time Study: Improving Productivity*. 5th ed. Englewood Cliffs, NJ: Prentice-Hall, 1978.

Nadler, Gerald. "The Role and Scope of Industrial Engineering." In *Handbook of Industrial Engineering*, 2nd ed. Ed. Gavriel Salvendy. New York: John Wiley & Sons, 1992.

Niebel, Benjamin W. *A History of Industrial Engineering at Penn State*. University Park, PA: University Press, 1992.

Salvendy, G., ed. *Handbook of Human Factors*. New York: John Wiley & Sons, 1987.

Saunders, Byron W. "The Industrial Engineering Profession." In *Handbook of Industrial Engineering*. Ed. Gavriel Salvendy. New York: John Wiley & Sons, 1982.

Taylor, F. W. *The Principles of Scientific Management*. New York: Harper, 1911.

Thompson, C. Bertrand. *The Taylor System of Scientific Management*. Chicago, IL: A. W. Shaw, 1917.

United Nations Industrial Development Organization. *Industry in the 1980s: Structural Change and Interdependence*. New York: United Nations, 1985.

CHAPTER 2

Problem-Solving Tools

KEY POINTS:

- *Select the project* with Pareto analyses, fish diagrams, Gantt charts, PERT charts, and job/worksite analysis guides.
- *Get and present data* with operation, flow, worker/machine, and gang process charts, and flow diagrams.
- *Develop the ideal method* for worker/machine relationships with synchronous and random servicing and line balancing calculations.

Regardless of how methods work is being used—to design a new work center or to improve one already in operation—both the problem and the factual information related to the problem must be presented in clear, logical form. Just as the machinist uses tools such as micrometers and calipers to facilitate performance, so the methods engineer uses appropriate tools to do a better job in a shorter time. A variety of such problem-solving tools are available, and each has specific applications.

The first five tools are primarily used in the first step of methods analysis, *select the project*. Pareto analysis and fish diagrams evolved from Japanese quality circles of the early 1960s (see Chapter 18) and were quite successful in improving quality and reducing costs in their manufacturing processes. Gantt and PERT charts emerged during the 1940s in response to a need for better project planning and control of complex military projects. They can also be very useful in identifying problems in an industrial setting.

Typically, project selection is based on three considerations: economic (probably the most important), technical, and human. Economic considerations may involve new products, for which standards have not been implemented, or existing products that have a high cost of manufacturing. Problems could be large amounts of scrap or rework, excessive material handling, in terms of either cost or distance, or simply "bottleneck" operations. Technical considerations may include processing techniques that need to be improved, quality control problems due to the method, or product performance problems compared to the competition. Human considerations may involve highly repetitive jobs, leading to work-related musculoskeletal injuries, high-accident-rate jobs, excessively fatiguing jobs, or jobs about which workers constantly complain.

The first four tools are most typically used in the analyst's office. The fifth tool, job/worksite analysis guide, helps identify problems within a particular area, department or worksite and is best developed as part of a physical walk-through and on-site observations. The guide provides a subjective identification of key worker, task, environmental, or administrative factors that may cause potential problems. It also indicates appropriate tools for further, more quantitative evaluations. Use of the job/worksite analysis guide should be a necessary first step before extensive quantitative data are collected on the present method.

The next five tools are used to record the present method, and they comprise the second step of methods analysis, *get and present the data.* Pertinent factual information—such as the production quantity, delivery schedules, operational times, facilities, machine capacities, special materials, and special tools—may have an important bearing on the solution of the problem and such information needs to be recorded. (The data are also useful in the third step of methods analysis, *analyze the data.*)

The final three tools are more useful as a quantitative approach in the fourth step of methods analysis, *develop the ideal method.* Once the facts are presented clearly and accurately, they are examined critically, so that the most practical, economic, and effective method can be defined and installed. They should therefore be used in conjunction with the operational analysis techniques described in Chapter 3. Note that most of the tools from all three groupings can easily be utilized in the operational analysis phase of development.

EXPLORATORY TOOLS

Pareto Analysis

Problem areas can be defined by a technique developed by the economist Vilfredo Pareto to explain the concentration of wealth. In *Pareto analysis*, items of interest are identified and measured on a common scale and are then ordered in ascending order, creating a cumulative distribution. Typically 20 percent of the ranked items account for 80 percent or more of the total activity; consequently, the technique is sometimes called the *80-20 rule*. For example, 80 percent of the total inventory is found in only 20 percent of the inventory items, or 20 percent of the jobs account for approximately 80 percent of the accidents (Figure 2-1), or 20 percent of the jobs account for 80 percent of the Workers' Compensation costs. Conceptually, the methods analyst concentrates the greatest effort on the few jobs that produce most of the problems. In many cases, the Pareto distribution can be transformed to a straight line using a log–normal transformation, from which further quantitative analyses can be performed (Herron, 1976).

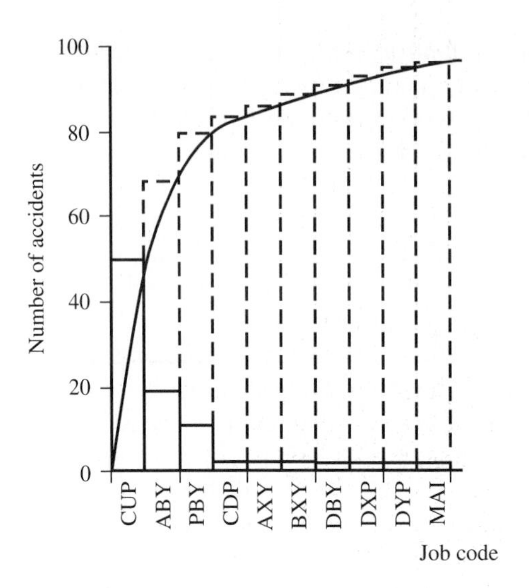

FIGURE 2–1
Pareto distribution of industrial accidents [20% of job codes (CUP and ABY) cause approximately 80% of accidents].

Fish Diagrams

Fish diagrams, also known as *cause-and-effect diagrams*, were developed by Ishikawa in the early 1950s while working on a quality control project for Kawasaki Steel Company. The method consists of defining an occurrence of a typically undesirable event or problem, that is, the *effect*, as the "fish head" and then identifying contributing factors, that is, the *causes*, as "fish bones" attached to a backbone and the fish head. The principal causes are typically subdivided into four or five major categories: the human, machines, methods, materials, environmental, administrative, etc., each of which is further subdivided into subcauses. The process is continued until all possible causes are listed. A good diagram will have several levels of bones and will provide a very good overview of a problem and its contributing factors. The factors are then critically analyzed in terms of their probable contribution to the overall problem. Hopefully, this process will also tend to identify potential solutions. An example of a fish diagram used to identify problems in achieving 100 percent efficiency in grinding operations is shown in Figure 2-2.

Fish diagrams have worked quite successfully in Japanese quality circles, where input is expected from all levels of workers and managers. Such diagrams may prove to be less successful in U.S. industry, where the cooperation between labor and management may be less effective in producing the desired solutions and outcomes (Cole, 1979).

Gantt Chart

The Gantt chart was probably the first project planning and control technique to emerge during the 1940s in response to the need to manage complex defense projects and systems better. A Gantt chart simply shows the anticipated completion

FIGURE 2–2
Fish diagram for operator health complaints on cut-off operation.

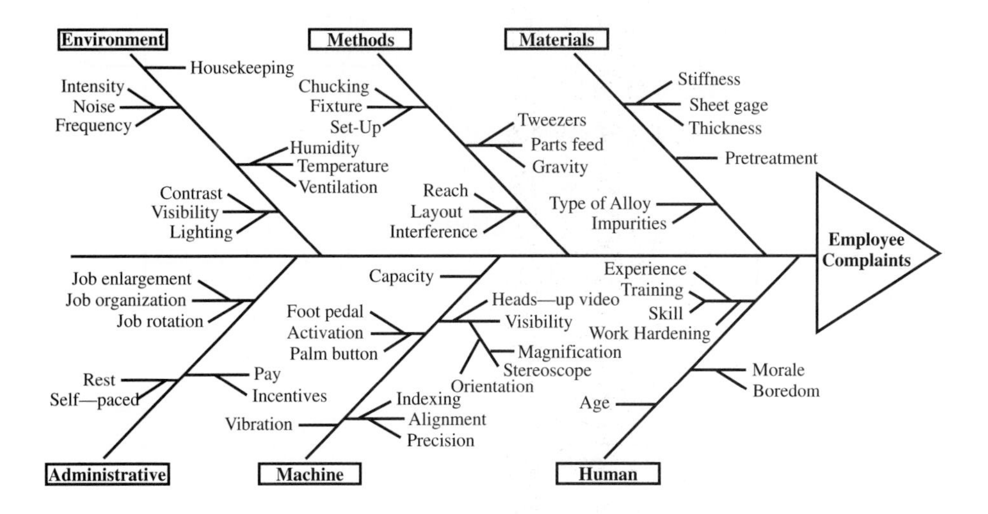

times for various project activities as bars plotted against time on the horizontal axis (Figure 2–3a). Actual completion times are shown by shading the bars appropriately. If a vertical line is drawn through a given date, you can easily determine which project components are ahead of or behind schedule. For example, in Figure 2–3a, by mid-April, mock-up work is behind schedule in two phases. A Gantt chart forces the project planner to develop a plan ahead of time, and provides a quick snapshot of the progress of the project at any given point in time. Unfortunately, it doesn't always completely describe the interaction between different project activities. More analytical techniques, such as PERT charts, are required for that purpose.

The Gantt chart can also be utilized for sequencing machine activity on the plant floor. The machine-based chart can include repair or maintenance activity by crossing out the time period in which the planned downtime will occur. For example, in Figure 2–3b, at approximately 10:00 P.M. Wednesday, mill work is behind schedule, while production on the punch is ahead of schedule.

PERT Charting

PERT stands for Program Evaluation and Review Technique. A *PERT chart,* also referred to as a network diagram or critical path, is a planning and control tool that graphically portrays the optimum way to attain some predetermined objective, generally in terms of time. This technique was employed by the U.S. military in the design of a such processes as the development of the Polaris missile and the operation of control systems in nuclear-powered submarines. Methods analysts usually use PERT charting to improve scheduling through cost reduction or customer satisfaction.

FIGURE 2–3
Example of (a) project-based Gantt chart, and (b) machine- or process-based Gantt chart

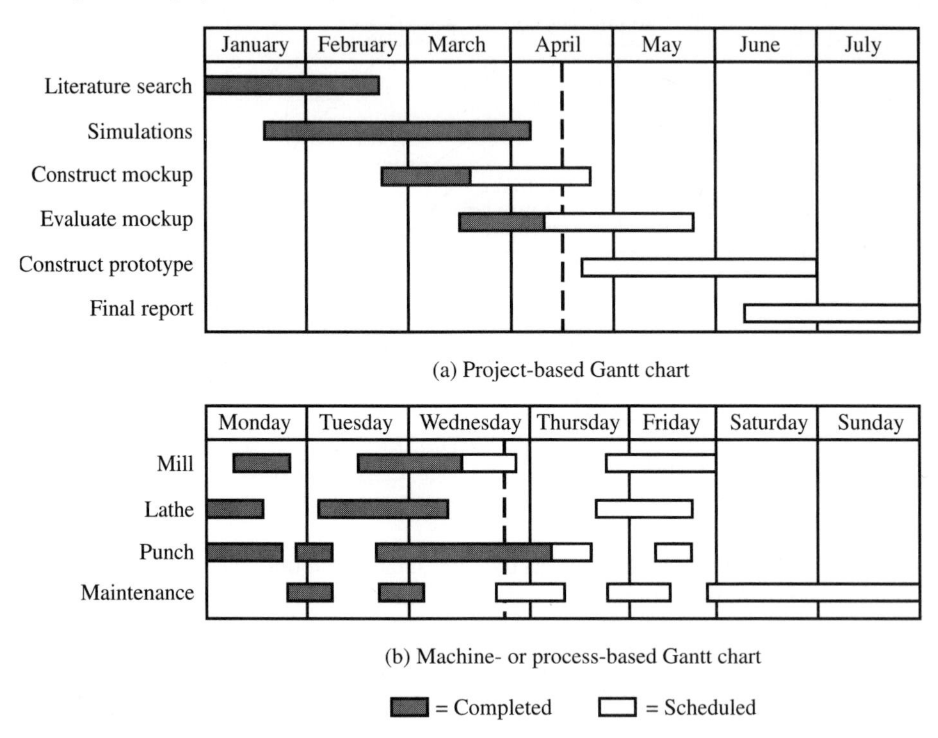

In using PERT for scheduling, analysts generally provide two or three time estimates for each activity. For example, if three time estimates are used, they are based on the following questions:

1. How much time is required to complete a specific activity if everything works out ideally (optimistic estimate)?
2. Under average conditions, what would be the most likely duration of this activity?
3. What is the time required to complete this activity if almost everything goes wrong (pessimistic estimate)?

With these estimates, the analyst can develop a probability distribution of the time required to perform the activity. (See Table 2–1).

On the PERT chart, events (represented by nodes) are positions in time that show the start and completion of a particular operation or group of operations. Each operation or group of operations in a department is defined as an activity and is called an arc. Each arc has an attached number representing the time (days, weeks, months) needed to complete the activity. Activities that utilize no time or cost, yet are necessary to maintain a correct sequence, are called dummy activities and are shown as dotted lines.

The minimum time needed to complete the entire project corresponds to the longest path from the initial node to the final node. In Figure 2–4, the minimum

TABLE 2–1
Cost and Time Values to Perform a Variety of Activities under Normal and Emergency Conditions

	Normal		Emergency	
Activities	Weeks	Dollars	Weeks	Dollars
A	4	4,000	2	6,000
B	2	1,200	1	2,500
C	3	3,600	2	4,800
D	1	1,000	0.5	1,800
E	5	6,000	3	8,000
F	4	3,200	3	5,000
G	3	3,000	2	5,000
H	0	0	0	0
I	6	7,200	4	8,400
J	2	1,600	1	2,000
K	5	3,000	3	4,000
L	3	3,000	2	4,000
M	4	1,600	3	2,000
N	1	700	1	700
O	4	4,400	2	6,000
P	2	1,600	1	2,400

time needed to complete the project is the longest path from node 1 to node 12. The longest path is also defined as the critical path. While there is always one such path through any project, more than one path can reflect the minimum time needed to complete the project.

Activities not on the critical path have a certain time flexibility. This time flexibility, or freedom, is referred to as "float." The amount of float is computed by subtracting the normal time from the time available. In other words, the float is the amount of time that a noncritical activity can be lengthened without delaying the project's completion date. This implies that when the intent is to reduce the project completion time, it is better to concentrate on activities that lie on the critical path, rather than those on other pathways.

Figure 2–4 illustrates an elementary network and its critical path. This path, identified by a heavy line, would involve a time duration of 27 weeks. Several methods can be used to shorten the project's duration, and the cost of the various alternatives can be estimated. For example, assume that the following cost table has been developed and that a linear relation exists between the time and the cost per week. The cost of various time alternatives can be readily computed:

27-week schedule–normal duration of project .... cost = $22,500
26-week schedule–the least expensive way to gain one week would be to reduce activity M or J by one week for an additional cost of $400 .... cost = $22,900
25-week schedule–the least expensive way to gain two weeks would be to reduce activities M and J by one week each for an additional cost of $800 .... cost = $23,300

FIGURE 2–4
Network showing critical path (heavy line).
Code numbers within nodes signify events. Connecting lines with directional arrows indicate operations that are dependent on prerequisite operations. Time values on the connecting lines represent normal duration in weeks. Hexagonals associated with events show the earliest event time. Dotted circles associated with events show the latest event time.

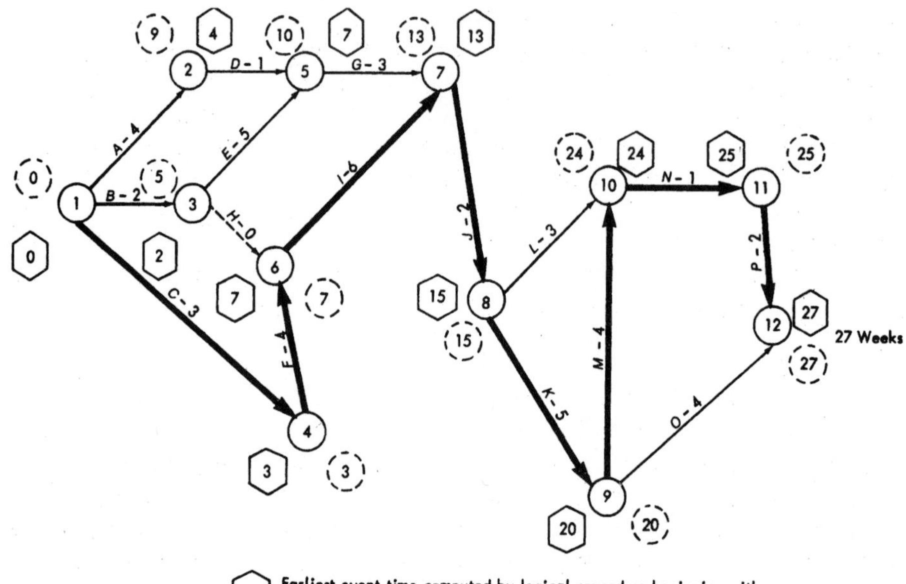

24-week schedule–the least expensive way to gain three weeks would be to reduce activities M, J, and K by one week each for an additional cost of $1,300 . . . . cost = $23,800

23-week schedule–the least expensive way to gain four weeks would be to reduce activities M and J by one week each and activity K by two weeks for an additional cost of $1,800 . . cost = $24,300

22-week schedule–the least expensive way to gain five weeks would be to reduce activities M and J by one week each, activity K by two weeks, and activity I by one week for an additional cost of $2,400 . . . . cost = $24,900

21-week schedule–the least expensive way to gain six weeks would be to reduce activities M and J by one week each and activities K and I by two weeks each for an additional cost of $3,000 . . . . cost = $25,500

20-week schedule–the least expensive way to gain seven weeks would be to reduce activities M, J, and P by one week each and activities K and I by two weeks each for an additional cost of $3,800 . . . . cost = $26,300

19-week schedule–the least expensive way to gain eight weeks would be to reduce activities M, J, P, and C by one week each and activities K and I by two weeks each for an additional cost of $5,000 .... cost = $27,500
(Note that a second critical path is now developed through nodes 1, 3, 5, and 7.)

18-week schedule–the least expensive way to gain nine weeks would be to reduce activities M, J, P, C, E, and F by one week each and activities K and I by two weeks each for an additional cost of $7,800 .... cost = $30,300
(Note that by shortening the time to 18 weeks, we develop a second critical path.)

Job/Worksite Analysis Guide

The *job/worksite analysis guide* (see Figure 2–5) identifies problems within a particular area, department, or worksite. Before collecting quantitative data, the analyst first walks through the area and observes the worker, the task, the workplace, and the surrounding working environment. In addition, the analyst identifies any administrative factors that may affect the worker's behavior or performance. All of these factors provide an overall perspective of the situation and help guide the analyst in using other, more quantitative tools for collecting and analyzing data. The example in Figure 2–5 shows the application of the job/worksite analysis guide to a hot-end operation in a television manufacturing facility. Key concerns include the lifting of heavy loads, heat stress, and noise exposure.

RECORDING AND ANALYSIS TOOLS

Operation Process Chart

The *operation process chart* shows the chronological sequence of all operations, inspections, time allowances, and materials used in a manufacturing or business process, from the arrival of raw material to the packaging of the finished product. The chart depicts the entrance of all components and subassemblies to the main assembly. Just as a blueprint displays such design details as fits, tolerances, and specifications, the operation process chart gives manufacturing and business details at a glance.

Two symbols are used in constructing the operation process chart: a small circle, usually 3/8 inch in diameter, which denotes an operation; and a small square, usually 3/8 inch on a side, which denotes an inspection. An operation takes place when a part being studied is intentionally transformed or when it is being studied or planned prior to productive work being performed on it. Some analysts prefer to separate manual operations from those for paperwork. Manual operations are usually related to direct labor, while information analyses are frequently a portion of

FIGURE 2-5
Job/worksite analysis guide for a hot-end job in a television manufacturing facility.

Job/Worksite Analysis Guide

Job/Worksite: HOT END	Analyst: AF Date: 1-27-
Description: INSERTING STEM TO FUNNEL	
Worker Factors	
Name: **Age:** 42 **Gender:** (M) F **Height:** 6' **Weight:** 180	
Motivation: High Medium (Low)	**Job Satisfaction:** High Medium (Low)
Education Level: Some HS (HS) College	**Fitness Level:** High (Medium) Low
Personal Protective Equipment: (Safety Glasses) Hard Hat Safety Shoes (Ear Plugs) Other GLOVES SLEEVES	
Task Factors	***Refer To:***
What happens? How do parts flow in/out? FUNNEL FROM BELT TO INSERTING MACHINE, THEN SEALER, THEN BACK TO BELT	*Flow Process Charts*
What kinds of motions are involved? REPETITIVE LIFTING, WALKING, GRASPING	*Video Analysis, Principles of Motion Economy*
Are there any jigs/fixtures? Automation? YES, POSITION FUNNEL, YES, FOR BASIC PROCESS, NO FOR HANDLING	
Are any tools being used? NO	*Tool Evaluation Checklist*
Is the workplace laid out well? Any long reaches? NO - EXCESSIVE WALKING & REACHING	*Workstation Evaluation Checklist*
Are there awkward finger/wrist motions? How frequent? NO	*CTD Risk Index*
Is there any lifting? YES, HEAVY GLASS FUNNELS	*NIOSH Lifting Analysis, UM2D Model*
Is the worker fatigued? Physical workload? YES, YES	*Heart Rate Analysis, Work-rest Allowances*
Is there any decision making? Mental workload? MINIMAL	
How long is each cycle? What is the standard time? ~ 1 1/2 MIN	*Time Study, MTM-2 Checklist*
Work Environment Factors	***Work Environment Checklist***
Is the illumination acceptable? Is there glare? YES NO	*IESNA Recommended Values*
Is the noise level acceptable? NO - EARPLUGS REQUIRED	*OSHA Levels*
Is there heat stress? YES!	*WBGT*
Is there vibration? NO	*ISO Standards*
Administrative Factors	Remarks: LOOK AT POSITIONING BELT & MACHINES CLOSER, VERY HOT!
Are there wage incentives? NO	
Is there job rotation? Job enlargement? YES NO	
Is training or work hardening provided? YES	
What are the overall management policies? —	

indirect or expense costs. An inspection takes place when the part is being examined to determine its conformity to a standard. Note that some analysts prefer to outline only the operations, calling the result an *outline process chart.*

Before beginning the actual construction of the operation process chart, analysts identify the chart with a title, Operation Process Chart, and other information,

FIGURE 2–6
Flowcharting conventions.

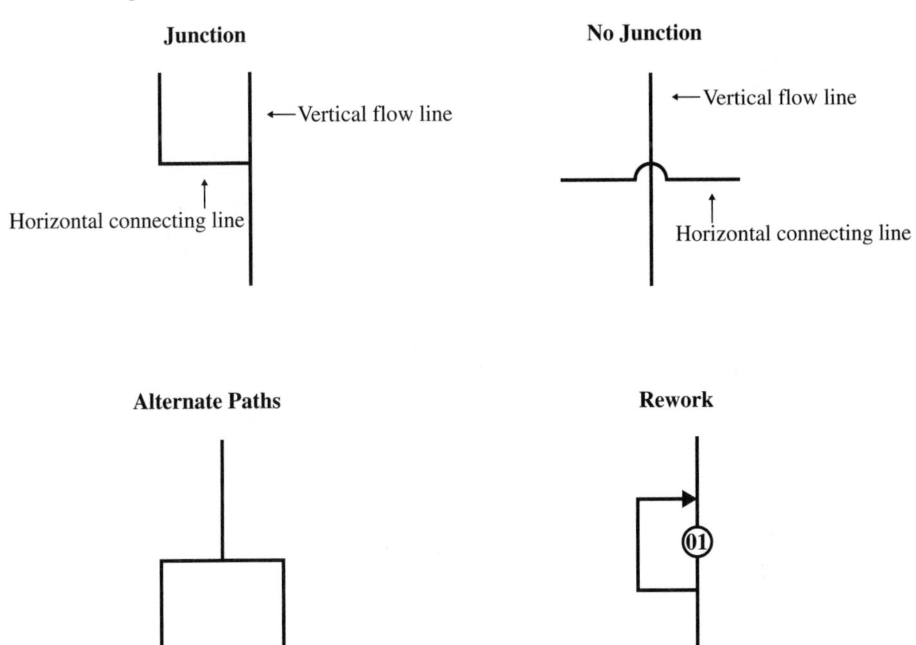

such as the part number, drawing number, process description, present or proposed method, date, and name of the person doing the charting. Additional information may include such items as the chart number, plant, building, and department.

Vertical lines indicate the general flow of the process as work is accomplished, while horizontal lines feeding into the vertical flow lines indicate material, either purchased or worked on during the process. Parts are shown as entering a vertical line for assembly or leaving a vertical line for disassembly. Materials that are disassembled or extracted are represented by horizontal material lines drawn to the right of the vertical flow line, while assembly materials are shown as horizontal lines drawn to the left of the vertical flow line.

In general, the operation process chart is constructed such that vertical flow lines and horizontal material lines do not cross. If it becomes necessary to cross a vertical and a horizontal line, use conventional practice to show that no juncture occurs; that is, draw a small semicircle in the horizontal line at the point where the vertical line crosses it (see Figure 2–6).

Time values, based either on estimates or actual measurements, are assigned to each operation and inspection. A typical completed operation process chart illustrating the manufacture of telephone stands is shown in Figure 2–7.

The completed operation process chart helps analysts visualize the present method, with all its details, so that new and better procedures may be devised. It shows analysts what effect a change on a given operation will have on the preceding and subsequent operations. It is not unusual to realize a 30 percent reduction in

FIGURE 2–7
Operation process chart illustrating manufacture of telephone stands.

OPERATION PROCESS CHART

Manufacturing Type 2834421 Telephone Stands--Present Method
Part 2834421 Dwg. No. SK2834421
Charted By B.W.N. 4-12-

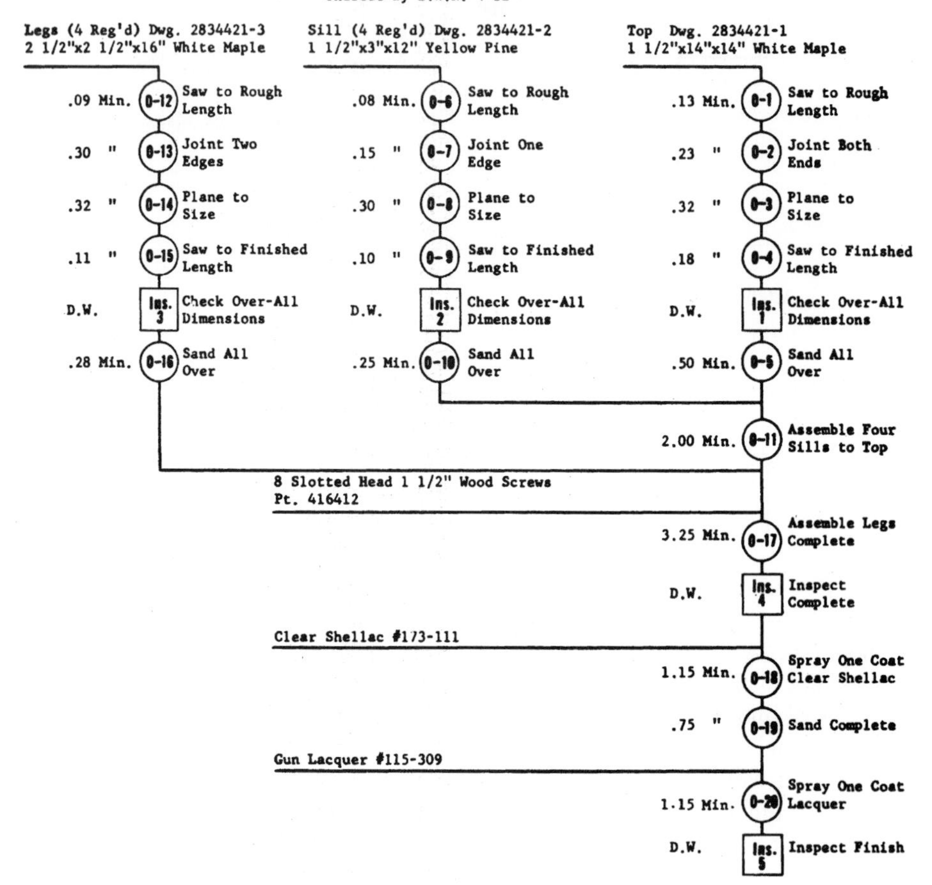

SUMMARY:

Event	Number	Time
Operations	20	17.58 minutes
Inspections	5	Day work

performance time utilizing the principles of operations analysis in conjunction with the operation process chart. The construction of the chart inevitably suggests possibilities for improvement.

The operation process chart indicates the general flow of all components in a product, and since each step is shown in its proper chronological sequence, the chart in itself is an ideal plant layout. Consequently, methods analysts, plant layout engineers, and persons in related fields find this tool extremely helpful in developing new layouts and improving existing ones.

The operation process chart is also an aid in promoting and explaining a proposed method. Since it gives so much information so clearly, it provides an ideal comparison between two competing solutions. This important tool:

1. Identifies all operations, inspections, materials, moves, storages, and delays involved in making a part or completing a process.
2. Shows all events in correct sequence.
3. Clearly shows the relationship between parts and fabrication complexity.
4. Distinguishes between produced and purchased parts.
5. Provides information on the number of employees utilized and the time required to perform each operation and inspection.

Flow Process Chart

In general, the *flow process chart* contains considerably more detail than the operation process chart. Consequently, it is not usually applied to entire assemblies. Rather, it is used primarily for each component of an assembly or a system to effect the maximum savings in manufacturing or in the procedures applicable to a particular component or sequence of work. The flow process chart is especially valuable in recording nonproduction hidden costs, such as distances traveled, delays, and temporary storages. Once these nonproduction periods are highlighted, analysts can take steps to minimize them and hence their costs.

In addition to recording operations and inspections, flow process charts show all the moves and storage delays encountered by an item as it goes through the plant. Flow process charts therefore need several symbols in addition to the operation and inspection symbols used in operation process charts. A small arrow signifies transportation, which can be defined as moving an object from one place to another, except when the movement takes place during the normal course of an operation or inspection. A large capital D indicates a delay, which occurs when a part is not immediately permitted to be processed at the next workstation. An equilateral triangle standing on its vertex signifies a storage, which occurs when a part is held and protected against unauthorized removal. These five symbols (see Figure 2–8) are the standard set of process chart symbols (ASME, 1972). Several other nonstandard symbols may sometimes be utilized for clerical or paperwork operations and for combined operations, as shown in Figure 2–9.

Two types of flowcharts are currently in general use: product or material (see Figure 2–10, preparation of direct mail advertising), and operative or person (see Figure 2–11, service personnel inspecting LUX field units). The product chart provides the details of the events involving a product or a material, and the operative flowchart details how a person performs an operational sequence.

Like the operation process chart, the flow process chart is identified by a title, Flow Process Chart, and accompanying information that usually includes the part

FIGURE 2–8
The ASME standard set of process chart symbols.

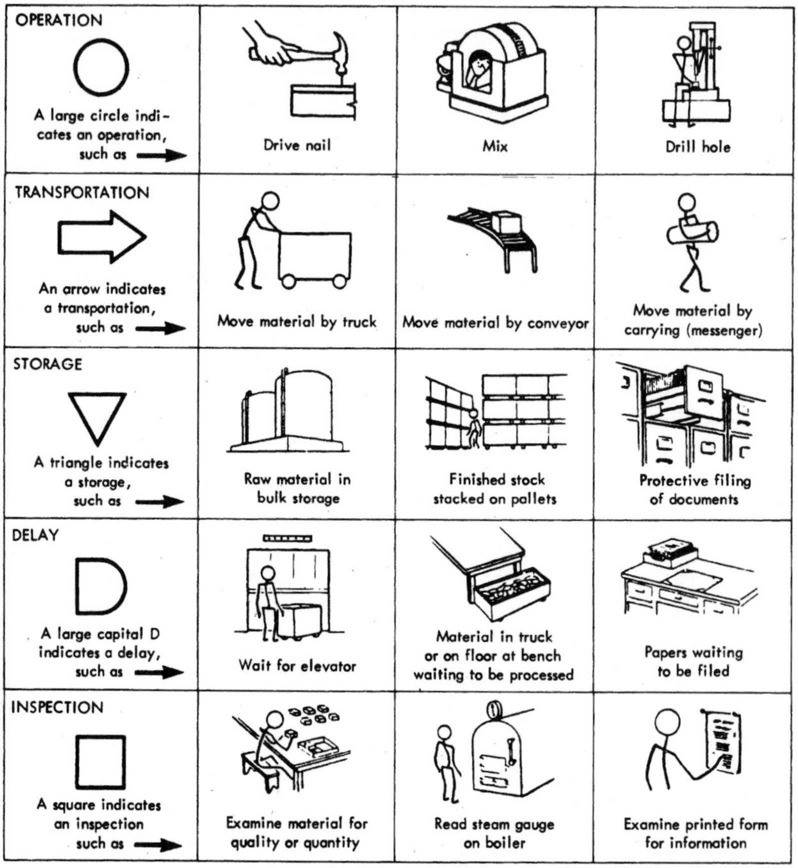

OPERATION ○ A large circle indicates an operation, such as →	Drive nail	Mix	Drill hole
TRANSPORTATION ⇨ An arrow indicates a transportation, such as →	Move material by truck	Move material by conveyor	Move material by carrying (messenger)
STORAGE ▽ A triangle indicates a storage, such as →	Raw material in bulk storage	Finished stock stacked on pallets	Protective filing of documents
DELAY D A large capital D indicates a delay, such as →	Wait for elevator	Material in truck or on floor at bench waiting to be processed	Papers waiting to be filed
INSPECTION □ A square indicates an inspection such as →	Examine material for quality or quantity	Read steam gauge on boiler	Examine printed form for information

number, drawing number, process description, present or proposed method, date, and name of the person doing the charting. Additional data that may be valuable for completely identifying the job being charted includes: the plant, building, or department; chart number; quantity; and cost.

For each event in the process, the analyst writes a description of the event, circles the appropriate process chart symbol, and indicates the times for processes or delays and distances for transportations. The analyst then connects succeeding event symbols with a vertical line. The right-hand column provides space for the analyst to enter comments or make recommendations for potential changes.

To determine the distance moved, the analyst need not measure each move accurately with a tape or a 6-foot rule. A sufficiently correct figure usually results by counting the number of columns that the material moves past and then multiplying

FIGURE 2–9
Nonstandard process chart symbols.

A record was created.

Information was added to a record.

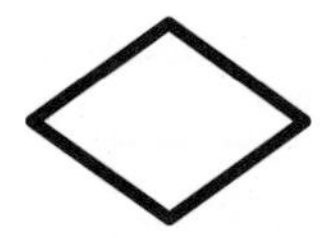
A decision was made.

An inspection was performed in conjunction with an operation.

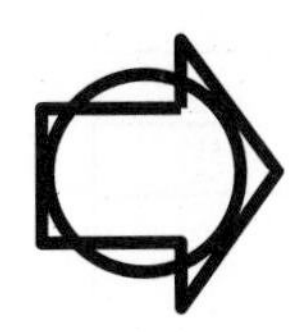
An operation and transportation took place simultaneously.

this number, less one, by the span. Moves of 5 feet or less are usually not recorded; however, they may be if the analyst feels that they materially affect the overall cost of the method being plotted.

All delay and storage times must be included on the chart. However, it is not sufficient to simply indicate that a delay or storage takes place. The longer a part stays in storage or is delayed, the more cost it accumulates and the longer the customer must wait for delivery. It is therefore important to know how much time a part spends at each delay or storage. The most economical method of determining the duration of delays and storages is to mark several parts with chalk indicating the exact time they went into a storage or were delayed. Then check the section periodically to see when the marked parts are brought back into production. By taking a number of cases, recording the elapsed time, and then averaging the results, analysts can obtain sufficiently accurate time values.

FIGURE 2–10

Flow process chart (material) for preparation of direct mail advertising.

Flow Process Chart Page 1 of 1

Location: Dorben Ad Agency		Summary			
Activity: Preparing Direct Mail Ads		**Event**	**Present**	**Proposed**	**Savings**
Date: 1-26-98		Operation	4		
Operator: J.S.	Analyst: A.F.	Transport	4		
Circle appropriate Method and Type:		Delay	4		
Method: *Present* *Proposed*		Inspection	0		
Type: *Worker* *Material* *Machine*		Storage	2		
Remarks:		Time (min)			
		Distance (ft)	340		
		Cost			

Event Description	Symbol	Time (In Minutes)	Distance (In Feet)	Method Recommendation
stock room	○ ⇨ D □ ▽			
to collating room	○ ⇨ D □ ▽		100	
collating rack by type	○ ⇨ D □ ▽			
collate 4 sheets	○ ⇨ D □ ▽			
stack	○ ⇨ D □ ▽			
to folding room	○ ⇨ D □ ▽		20	
jog, fold, crease	○ ⇨ D □ ▽			
stack	○ ⇨ D □ ▽			
to angle stapler	○ ⇨ D □ ▽		20	
staple	○ ⇨ D □ ▽			
stack	○ ⇨ D □ ▽			
to mail room	○ ⇨ D □ ▽		200	
addressing	○ ⇨ D □ ▽			
mailbag	○ ⇨ D □ ▽			
	○ ⇨ D □ ▽			
	○ ⇨ D □ ▽			
	○ ⇨ D □ ▽			
	○ ⇨ D □ ▽			
	○ ⇨ D □ ▽			

The flow process chart, like the operation process chart, is not an end in and of itself; it is merely a means to an end. This tool facilitates the elimination or reduction of the hidden costs of a component. Since the flow chart clearly shows all transportations, delays, and storages, the information it provides can lead to a reduction of both the quantity and duration of these elements. Also, since distances are recorded on the flow process chart, the chart is exceptionally valuable in showing how the layout of a plant can be improved. These techniques are described in greater detail in Chapter 3.

FIGURE 2–11

Flow process chart (worker) for field inspection of LUX.

Flow Process Chart Page 1 of 1

		Summary			
Location: Dorben Co.		Event	Present	Proposed	Savings
Activity: Field Inspection of LUX		Operation	7		
Date: 4-17-97		Transport	6		
Operator: T.Smith	Analyst: R. Ruhf	Delay	2		
Circle appropriate Method and Type: **Method:** *Present* (circled) *Proposed*		Inspection	6		
Type: *Worker* (circled) *Material* *Machine*		Storage	0		
Remarks:		Time (min)	32.60		
		Distance (ft)	375		
		Cost			

Event Description	Symbol	Time (In Minutes)	Distance (In Feet)	Method Recommendation
Leave vehicle, walk to front door, ring bell.	○ ⇨ D □ ▽	1.00	75	Call home in advance to reduce waiting delays.
Wait, enter home.	○ ⇨ D □ ▽			
Walk to field reservoir.	○ ⇨ D □ ▽	.25	25	
Disconnect field reservoir from unit.	○ ⇨ D □ ▽	.35		
Inspect for dents, cracks in shroud, cracked glass or missing hardware.	○ ⇨ D □ ▽	1.25		This can be done while walking back to vehicle.
Clean unit with approved cleaner and disinfectant.	○ ⇨ D □ ▽	2.25		This can be done more effectively at vehicle.
Return to vehicle with empty tank.	○ ⇨ D □ ▽	1.00	75	
Unlock vehicle, place empty tank in fixture and connect hardware.	○ ⇨ D □ ▽	1.75		
Open valve; begin fill.	○ ⇨ D □ ▽	.25		
Wait for tank to fill.	○ ⇨ D □ ▽	12.00		Clean unit while being filled.
Check humidifier for proper function.	○ ⇨ D □ ▽	.5		Eliminate. No need to do this twice.
Check pressure (indicator).	○ ⇨ D □ ▽	.2		
Check reservoir contents (indicator).	○ ⇨ D □ ▽	.2		
Return to patient with filled tank.	○ ⇨ D □ ▽	1.10	100	
Hook up filled tank.	○ ⇨ D □ ▽	1.00		
Check humidifier for proper function.	○ ⇨ D □ ▽	.75		
Wait for patient to remove nasal cannula or face mask.	○ ⇨ D □ ▽	2.00		
Install new nasal cannula or face mask.	○ ⇨ D □ ▽	2.50		
Check flows with patient.	○ ⇨ D □ ▽	2.25		
Affix a dated, intialed inspection sticker.	○ ⇨ D □ ▽	1.00		Perform this while unit being filled.
Return to vehicle.	○ ⇨ D □ ▽	1.00	100	

Flow Diagram

Although the flow process chart gives most of the pertinent information related to a manufacturing process, it does not show a pictorial plan of the flow of work. Sometimes this information is helpful in developing a new method. For example, before a transportation can be shortened, the analyst needs to see or visualize where room can be made to add a facility so that the transportation distance can be shortened. Likewise, it is helpful to visualize potential temporary and permanent storage areas, inspection stations, and work points.

The best way to provide this information is to take an existing drawing of the plant areas involved, and then sketch in the flow lines indicating the movement of the material from one activity to the next. A pictorial representation of the layout of floors and buildings, showing the locations of all activities on the flow process chart, is a *flow diagram*. When constructing a flow diagram, analysts identify each activity by symbols and numbers corresponding to those appearing on the flow process chart. The direction of flow is indicated by placing small arrows periodically along the flow lines. Different colors can be used to indicate flow lines for more than one part.

Figure 2–12 illustrates a flow diagram made in conjunction with a flow process chart to improve the production of the Garand (M1) rifle at Springfield Armory. This pictorial representation, together with the flow process chart, resulted in savings that increased production from 500 rifle barrels per shift to 3,600—with the same number of employees. Figure 2–13 illustrates the flow diagram of the revised layout.

FIGURE 2–12
Flow diagram of the old layout of a group of operations on the Garand rifle. *(Shaded section of plant represents the total floor space needed for the revised layout [Figure 2–13]. This represented a 40 percent savings in floor space.)*

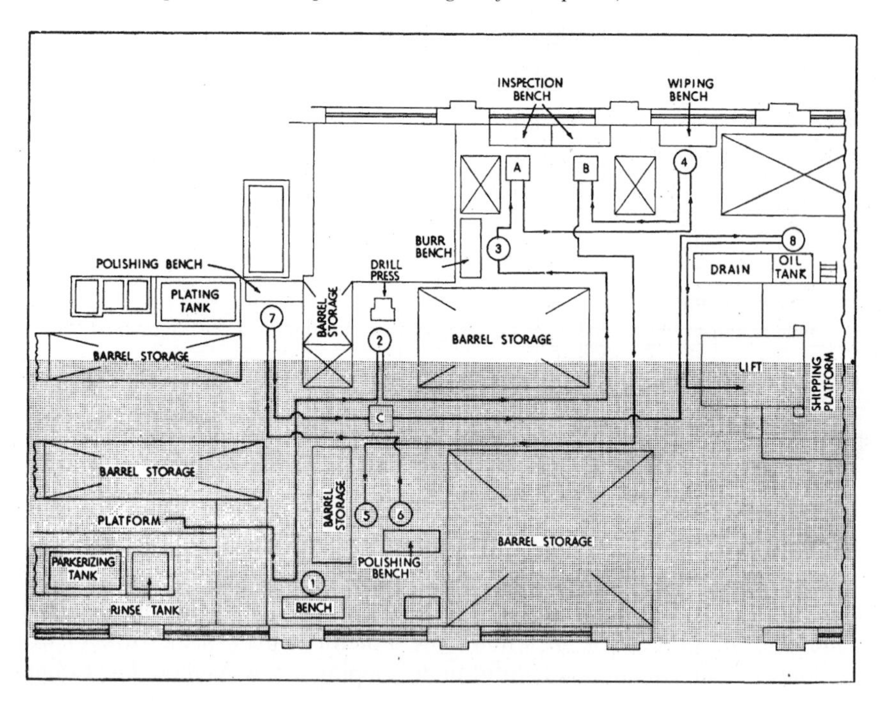

The flow diagram is a helpful supplement to the flow process chart because it indicates backtracking and possible traffic congestion areas, and it facilitates developing an ideal plant layout.

Worker and Machine Process Charts

The *worker and machine process chart* is used to study, analyze, and improve one workstation at a time. The chart shows the exact time relationship between the working cycle of the person and the operating cycle of the machine. These facts can lead to a fuller utilization of both worker and machine time, and a better balance of the work cycle.

Many machine tools are either completely automatic (the automatic screw machine) or partially automatic (the turret lathe). With these types of facilities, the operator is often idle for a portion of the cycle. The utilization of this idle time can increase operator earnings and improve production efficiency.

The practice of having one employee operate more than one machine is known as *machine coupling.* Because organized labor may resist this concept, the best way to sell machine coupling is to demonstrate the opportunity for added earnings. Since machine coupling increases the percentage of "effort time" during the operating cycle, greater incentive earnings are possible if a company is on an incentive wage payment plan. Also, higher base rates result when machine coupling is practiced, since the operator has greater responsibility and can exercise more mental and physical effort.

When constructing the worker and machine process chart, the analyst must first identify the chart with a title such as Worker and Machine Process Chart. Additional identifying information would include: part number, drawing number, operation description, present or proposed method, date, and name of the person doing the charting.

Since workers and machine charts are always drawn to scale, the analyst selects a distance in inches to conform with a unit of time such that the chart can be neatly arranged. The longer the cycle time of the operation being charted, the shorter the

FIGURE 2–13
Flow diagram of the revised layout of a group of operations on the Garand rifle.

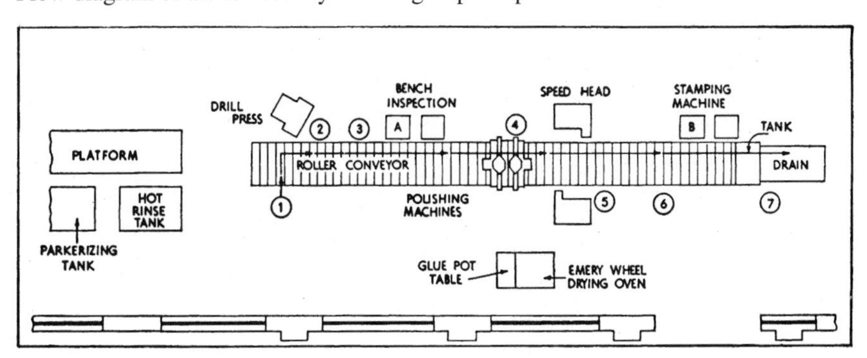

distance per decimal minute of time. Once exact values have been established for the distance, in inches per unit of time, the chart is begun. The left side shows the operations and time for the worker, and the right shows the working time and the idle time of the machine or machines. A solid line drawn vertically represents the employee's working time. A break in the vertical work–time line signifies idle time. Likewise, a solid vertical line under each machine heading indicates machine operating time, and a break in the vertical machine line designates idle machine time. A dotted line under the machine column indicates loading and unloading machine time, during which the machine is neither idle nor productive (see Figure 2–14).

FIGURE 2–14
Worker and machine process chart for milling machine operation.

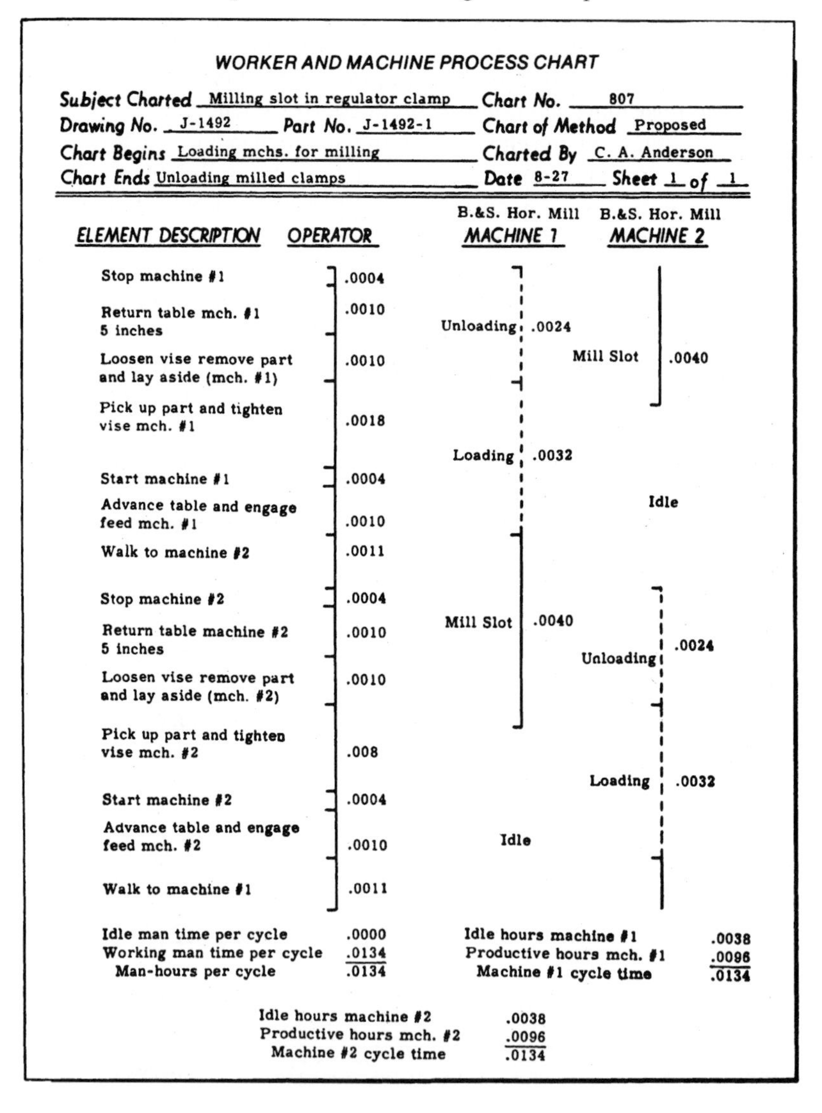

WORKER AND MACHINE PROCESS CHART

Subject Charted: Milling slot in regulator clamp — Chart No. 807
Drawing No. J-1492 — Part No. J-1492-1 — Chart of Method: Proposed
Chart Begins: Loading mchs. for milling — Charted By: C. A. Anderson
Chart Ends: Unloading milled clamps — Date 8-27 — Sheet 1 of 1

ELEMENT DESCRIPTION	OPERATOR	B.&S. Hor. Mill MACHINE 1	B.&S. Hor. Mill MACHINE 2
Stop machine #1	.0004	Unloading .0024	Mill Slot .0040
Return table mch. #1 5 inches	.0010		
Loosen vise remove part and lay aside (mch. #1)	.0010		
Pick up part and tighten vise mch. #1	.0018	Loading .0032	Idle
Start machine #1	.0004		
Advance table and engage feed mch. #1	.0010		
Walk to machine #2	.0011	Mill Slot .0040	
Stop machine #2	.0004		Unloading .0024
Return table machine #2 5 inches	.0010		
Loosen vise remove part and lay aside (mch. #2)	.0010		
Pick up part and tighten vise mch. #2	.008	Idle	Loading .0032
Start machine #2	.0004		
Advance table and engage feed mch. #2	.0010		
Walk to machine #1	.0011		

Idle man time per cycle	.0000
Working man time per cycle	.0134
Man-hours per cycle	.0134

Idle hours machine #1	.0038
Productive hours mch. #1	.0096
Machine #1 cycle time	.0134

Idle hours machine #2	.0038
Productive hours mch. #2	.0096
Machine #2 cycle time	.0134

The analyst charts all elements of occupied and idle time for both the worker and the machine through the termination of the cycle. The bottom of the chart shows the employee's total working time and total idle time, as well as the total working time and idle time of each machine. The productive time plus the idle time of the worker must equal the productive time plus the idle time of each machine the worker operates.

Accurate elemental time values are necessary before the worker and machine chart can be constructed. These time values should represent standard times that include an acceptable allowance for fatigue, unavoidable delays, and personal delays (see Chapter 10 for more details). The analyst should never use overall stop-watch readings in the construction of the chart.

The completed worker and machine process chart clearly shows the areas in which both idle machine time and worker time occur. These areas are generally a good place to start in effecting improvements. However, the analyst must also compare the cost of the idle machine with that of the idle worker. It is only when total cost is considered that the analyst can safely recommend one method over another. Economical considerations are presented in the next section.

Gang Process Charts

The *gang process chart* is, in a sense, an adaptation of the worker and machine chart. A worker and machine process chart helps determine the most economical number of machines one worker can operate. However, several processes and facilities are of such magnitude that instead of one worker operating several machines, several workers are needed to operate one machine effectively. The gang process chart shows the exact relationship between the idle and operating cycle of the machine and the idle and operating times per cycle of the workers who service that machine. This chart reveals the possibilities for improvement by reducing both idle operator time and idle machine time.

Figure 2–15 illustrates a gang process chart for a process in which a large number of idle work-hours exist, up to 18.4 hours per 8-hour shift. The chart also shows that the company is employing two more operators than are needed. By relocating some of the controls of the process, the company was able to reassign the elements of work so that four, rather than six, workers could effectively operate the extrusion press. A better operation of the same process is shown on the gang process chart in Figure 2–16. The savings of 16 hours per shift was easily developed through the use of this chart.

QUANTITATIVE TOOLS, WORKER AND MACHINE RELATIONSHIPS

Although the worker and machine process chart can illustrate the number of facilities that can be assigned to an operator, this can often be computed in much less time through the development of a mathematical model. A worker and machine

FIGURE 2–15
Gang process chart of the present method of operation of a hydraulic extrusion process.

GANG PROCESS CHART OF PRESENT METHOD

HYDRAULIC EXTRUSION PRESS   DEPT. 11   BELLEFONTE PA. PLANT

CHARTED BY B.W.N.   4-15-   CHART NO. G-85

Worker	Operation	Time
MACHINE	Elevate Billet	.07
	Position Billet	.08
	Position Dummy	.04
	Build Pressure	.05
	Extrude	.45
	Unlock Die	.06
	Loosen & Push Out Shell	.10
	Withdraw Ram & Lock Die in Head	.15
PRESS OPERATOR	Elevate Billet	.07
	Position Billet	.08
	Position Dummy	.04
	Build Pressure	.05
	Extrude	.45
	Unlock Die	.06
	Loosen & Push Out Shell	.10
	Withdraw Ram & Lock Die in Head	.15
ASSISTANT PRESS OPERATOR	Grease Die & Position Back in Die Head	.12
	Idle Time	.68
	Run Head & Shell Out	.11
	Shear Rod from Shell	.04
	Pull Die Off End of Rod	.05
FURNACE MAN	Rearrange Billets in Furnace	.20
	Idle Time	.51
	Open Furnace Door & Remove Billet	.19
	Ram Billet from Furnace & Close Furnace Door	.10
DUMMY KNOCKER	Position Shell on Small Press	.10
	Press Dummy Out of Shell	.12
	Dispose of Shell	.18
	Dispose of Dummy and Lay Aside Tongs	.12
	Idle	.43
	Grab Tongs & Move to Position	.05
ASSISTANT DUMMY KNOCKER	Move Away from Small Press and Lay Aside Tongs	.12
	Idle Time	.68
	Guide Shell from Shear to Small Press	.20
PULL-OUT MAN	Pull Rod toward Cooling Rack	.20
	Walk Back toward Press	.15
	Grab Rod with Tongs and Pull Out	.45
	Straighten Rod End with Mallet	.11
	Hold Rod while Die Removed at Press	.09

	Machine	Press Operator	Assistant Press Operator	Furnace Man	Dummy Knocker	Assistant Dummy Knocker	Pull-Out Man
WORKING TIME	1.00 MIN.	1.00 MIN.	.32 MIN.	.49 MIN.	.57 MIN.	.32 MIN.	1.00 MIN.
IDLE TIME	0 "	0 "	.68 "	.51 "	.43 "	.68 "	0 "

IDLE TIME = 2.30 MAN-MINUTES PER CYCLE = 18.4 MAN-HOURS PER EIGHT-HOUR DAY

FIGURE 2–16
Gang process chart of the proposed method of operation of a hydraulic extrusion process.

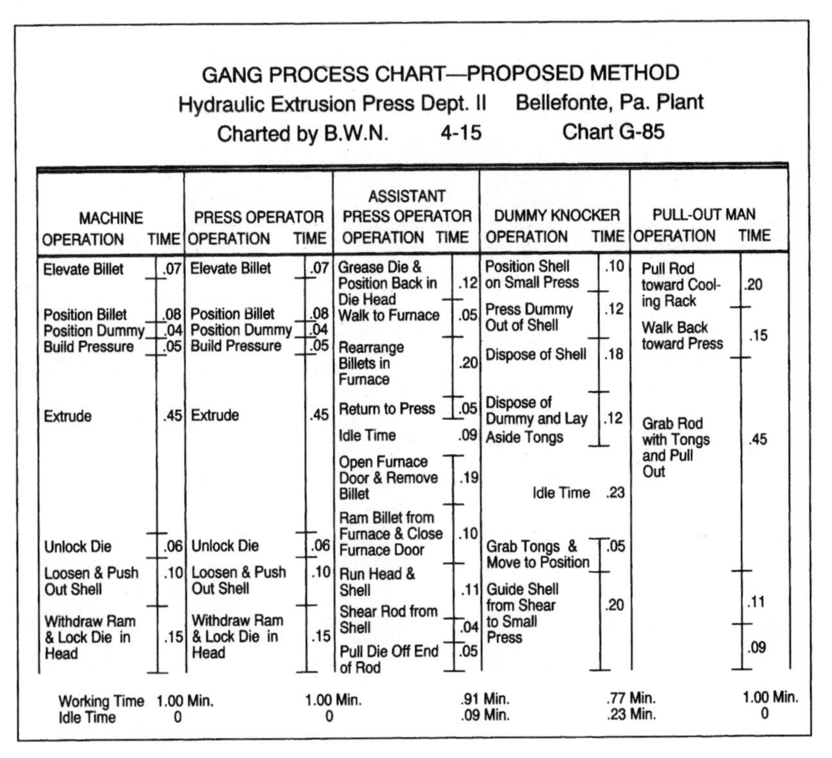

relationship is usually one of three types: (1) synchronous servicing, (2) completely random servicing, and (3) a combination of synchronous and random servicing.

Synchronous Servicing

Assigning more than one machine to an operator seldom results in the ideal case where both the worker and the machine are occupied during the whole cycle. Such ideal cases are referred to as *synchronous servicing,* and the number of machines to be assigned can be computed as:

$$N = \frac{l + m}{l}$$

where: N = Number of machines the operator is assigned.
l = Total operator loading and unloading (servicing) time per machine.
m = Total machine running time (automatic power feed).

For example, if the total operator servicing time is one minute, while the cycle time of the machine is four minutes, synchronous servicing would result in the

FIGURE 2–17
Synchronous servicing assignment for one operator and five machines.

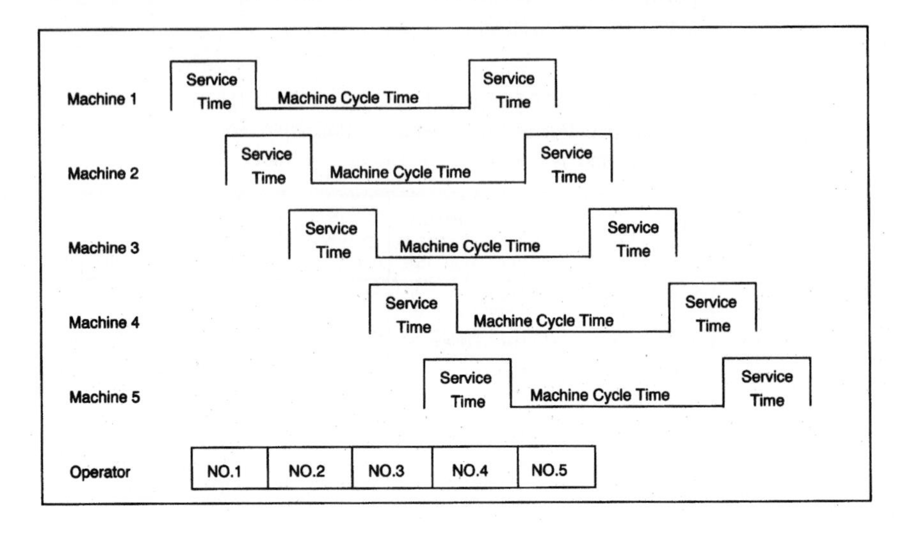

assignment of $(1 + 4)/1 = 5$ machines. Graphically, this assignment would appear as shown in Figure 2–17.

If the number of machines in this example is increased, machine interference takes place, and we have a situation in which one or more of the facilities sit idle for a portion of the work cycle. If the number of machines is reduced to some figure less than five, then the operator is idle for a portion of the cycle. In such cases, the minimum total cost per piece usually represents the criterion for optimum operation. To establish the best method, the analyst must consider the cost of each idle machine and the hourly rate of the operator. Quantitative techniques can determine the best arrangement. The procedure is first to estimate the number of machines that the operator should be assigned under realistic conditions by establishing the lowest whole number from the following equation:

$$N_1 \leq \frac{l + m}{l + w}$$

where: N_1 = Lowest whole number.
w = Total worker time (not directly interacting with the machine, typically walking time to the next machine).

The cycle time with the operator servicing N_1 machines is $\ell + m$, since in this case, the operator is not busy the whole cycle, yet the facilities are occupied during the entire cycle.

Using N_1, we can compute the total expected cost (T.E.C.) as follows:

$$\text{T.E.C.}_{N_1} = \frac{K_1(l + m) + N_1 K_2(l + m)}{N_1}$$

$$= \frac{(l + m)(K_1 + N_1 K_2)}{N_1}$$

where: T.E.C. = Total expected cost of production per cycle from one machine.
K_1 = Operator rate, in dollars per unit of time.
K_2 = Cost of machine, in dollars per unit of time.

After this cost is computed, a cost should be calculated with $N_1 + 1$ machines assigned to the operator. In this case, the cycle time is governed by the working cycle of the operator, since there is some idle machine time. The cycle time is now $(N_1 + 1)(\ell + w)$. Let $N_2 = N_1 + 1$. Then the total expected cost with N_2 facilities is:

$$\text{T.E.C.}_{N_2} = \frac{(K_1)(N_2)(l + w) + (K_2)(N_2)\,(N_2)(l + w)}{(N_2)}$$
$$= [(l + w)][K_1 + K_2\,(N_2)]$$

The number of machines assigned depends on whether N_1 or N_2 gives the lowest total expected cost per piece.

Random Servicing

Completely *random servicing* situations are those cases in which it is not known when a facility needs to be serviced or how long servicing takes. Mean values are usually known or can be determined; with these averages, the laws of probability can provide a useful tool in determining the number of machines to assign a single operator.

The successive terms of the binomial expansion give a useful approximation of the probability of 0, 1, 2, 3, . . ., n machines down (where n is relatively small), assuming that each machine is down at random times during the day and that the probability of down time is p and the probability of runtime is $q = (1 - p)$. Each term of the binomial expansion can be expressed as a probability of M (out of N) machines down:

$$\text{P (M of N)} = \frac{\text{N!}}{\text{M! (N} - \text{M)!}}\, \text{p}^{\text{M}}\, \text{q}^{\text{N} - \text{M}}$$

As an example, let us determine the minimum proportion of machine time lost for various numbers of turret lathes assigned to an operator where the average machine runs unattended 60 percent of the time. Operator attention time (machine is down or requires servicing) at irregular intervals is 40 percent on average. The analyst estimates that three turret lathes should be assigned per operator on this class of work. Under this arrangement, the probabilities of M (out of N) machines down would be:

Machines down (M)	Probability
0	$\frac{3!}{0!\,(3-3)!}\,.4^0\,.6^3 = (1)(1)(.216) = .216$
1	$\frac{3!}{1!\,(3-1)!}\,.4^1\,.6^2 = (3)(.4)(.36) = .432$
2	$\frac{3!}{2!\,(3-2)!}\,.4^2\,.6^1 = (3)(.16)(.6) = .288$
3	$\frac{3!}{3!\,(3-3)!}\,.4^3\,.6^0 = (1)(.064)(1) = .064$

Using this approach, the proportion of time that some machines are down may be determined, and the resulting lost time of one operator per three machines may be readily computed. In this example, we have:

No. of machines down	Probability	Machine hours lost per 8-hour day
0	0.216	0
1	0.432	0*
2	0.288	(0.288)(8) = 2.304
3	0.064	(2)(0.064)(8) = 1.024
	1.000	3.328

*Since only one machine is down at a time, the operator can be attending the down machine.

$$\text{Proportion of machine time lost} = \frac{3.328}{24.0} = 13.9 \text{ percent}$$

Similar computations can be made for more or less machine assignments to determine the assignment resulting in the least machine downtime. The most satisfactory assignment is usually the arrangement showing the least total expected cost per piece, while the total expected cost per piece for a given arrangement is computed by the expression:

$$\text{T.E.C.} = \frac{K_1 + NK_2}{R}$$

where: K_1 = Hourly rate of the operator.
K_2 = Hourly rate of the machine.
N = Number of machines assigned.
R = Rate of production, pieces from N machines per hour.

The pieces per hour from N machines is computed with the mean machine time required per piece, the average machine servicing time per piece, and the expected down or lost time per hour.

For example, under a five-machine assignment to one operator, an analyst determined that the machining time per piece was 0.82 hours, the machine serving time per piece was 0.17 hours, and the machine downtime was an average of 0.11 hours per machine per hour. Thus, each machine was available for production work only 0.89 hours each hour. The average time required to produce one piece per machine would be:

$$\frac{0.82 + 0.17}{0.89} = 1.11$$

Therefore, the five machines would produce 4.5 pieces per hour. With an operator hourly rate of $12 and a machine hourly rate of $22, we have a total expected cost per piece of:

$$\frac{\$12.00 + 5(\$22.00)}{4.5} = \$27.11$$

Combinations of synchronous and random servicing are perhaps the most common type of worker and machine relationships. Here the servicing time is constant, but the machine downtime is random. Winding, coning, and quilling operations used in the textile industry are characteristic of this type of worker and machine relationship. As in the former examples, algebra and probability can establish the mathematical model that points to a realistic solution.

Line Balancing

The problem of determining the ideal number of workers to be assigned to a production line is analogous to that of determining the number of workers to be assigned to a workstation; the gang process chart solves both problems. Perhaps the most elementary *line balancing* situation, yet one that is very often encountered, is one in which several operators, each performing consecutive operations, work as a unit. In such a situation, the rate of production is dependent on the slowest operator. For example, we may have a line of five operators assembling bonded rubber mountings prior to the curing process. The specific work assignments might be as follows: Operator 1, 0.52 minutes; Operator 2, 0.48 minutes; Operator 3, 0.65 minutes; Operator 4, 0.41 minutes; Operator 5, 0.55 minutes. Operator 3 establishes the pace, as is evidenced by the following:

Operator	Standard minutes to perform operation	Wait time based on slowest operator	Allowed standard minutes
1	0.52	0.13	0.65
2	0.48	0.17	0.65
3	0.65	—	0.65
4	0.41	0.24	0.65
5	0.55	0.10	0.65
Totals	2.61		3.25

The efficiency of this line can be computed as the ratio of the total actual standard minutes to the total allowed standard minutes, or:

$$E = \frac{\sum_{1}^{5} S.M.}{\sum_{1}^{5} A.M.} \times 100 = \frac{2.61}{3.25} \times 100 = 80 \text{ percent}$$

where: E = Efficiency.
$S.M.$ = Standard minutes per operation.
$A.M.$ = Allowed standard minutes per operation.

Details on standard times will be covered later in Chapter 8.

Some analysts prefer to consider percent idle time (%Idle):

$$\%\text{Idle} = 100 - E = 20\%$$

In a real-life situation similar to this example, the opportunity for significant savings exists. If an analyst can save 0.10 minute on Operator 3, the net savings per cycle is not 0.10 minutes, but 0.10×5 or 0.50 minutes.

Only in the most unusual situations would a line be perfectly balanced; that is, the standard minutes to perform an operation would be identical for each member of the team. The "standard minutes to perform an operation" is not really a standard. It is only a standard to the individual who established it. Thus, in our example, where Operator 3 has a standard time of 0.65 minutes to perform the first operation, a different work measurement analyst might have allowed as little as 0.61 minutes, or as much as 0.69 minutes. The range of standards established by different work measurement analysts on the same operation might be even greater than the range suggested. The point is that whether the issued standard is 0.61, 0.65, or 0.69, the typical conscientious operator should have little difficulty in meeting the standard. In fact, the operator will probably better the standard in view of the performance of the operators on the line with less work content in their assignments. Those operators who have a wait time based on the output of the slowest operator are seldom observed as actually waiting. Instead, they reduce the tempo of their movements to utilize the number of standard minutes established by the slowest operator.

The number of operators needed for the required rate of production is equal to:

$$N = R \times \Sigma AM = R \times \frac{\Sigma SM}{E}$$

where: N = Number of operators needed in the line.
R = Desired rate of production.

For example, assume that we have a new design for which we are establishing an assembly line. Eight distinct operations are involved. The line must produce 700 units per day (or 700/480 = 1.458 units per minute), and since it is desirable to minimize storage, we do not want to produce many more than 700 units per day. The eight operations involve the following standard minutes based on existing standard data: Operation 1, 1.25 minutes; Operation 2, 1.38 minutes; Operation 3, 2.58 minutes; Operation 4, 3.84 minutes; Operation 5, 1.27 minutes; Operation 6, 1.29 minutes; Operation 7, 2.48 minutes; and Operation 8, 1.28 minutes. To plan this assembly line for the most economical setup, we estimate the number of operators required for a given level of efficiency (ideally, 100%), as follows:

$$\begin{aligned} N &= 1.458 \times (1.25 + 1.38 + 2.58 + 3.84 + 1.27 + 1.29 + 2.48 + 1.28)/1.00 \\ &= 22.4 \end{aligned}$$

For a more realistic 95 percent efficiency, the number of operators becomes 22.4/.95 = 23.6.

Since it is impossible to have six-tenths of an operator, you would endeavor to set up the line utilizing 24 operators.

Next, we estimate the number of operators to be utilized at each of the eight specific operations. Since 700 units of work are required a day, it will be necessary to produce one unit in about 0.685 minutes (480/700). We estimate the number of operators needed on each operation by dividing the number of minutes allowable to produce one piece into the standard minutes for each operation, as follows:

Operation	Standard minutes	Standard minutes / Minutes/unit	No. of operators
Operation 1	1.25	1.83	2
Operation 2	1.38	2.02	2
Operation 3	2.58	3.77	4
Operation 4	3.84	5.62	6
Operation 5	1.27	1.86	2
Operation 6	1.29	1.88	2
Operation 7	2.48	3.62	4
Operation 8	1.28	1.87	2
Total	15.37		24

To identify the slowest operation, we divide the estimated number of operators into the standard minutes for each of the eight operations. The results are shown in the following table.

Operation 1	1.25/2 = 0.625
Operation 2	1.38/2 = 0.690
Operation 3	2.58/4 = 0.645
Operation 4	3.84/6 = 0.640
Operation 5	1.27/2 = 0.635
Operation 6	1.29/2 = 0.645
Operation 7	2.48/4 = 0.620
Operation 8	1.28/2 = 0.640

Thus, Operation 2 determines the output from the line. In this case, it is:

$$\frac{2 \text{ workers} \times 60 \text{ min.}}{1.38 \text{ standard minutes}} = 87 \text{ pieces per hour, or } 696 \text{ pieces per day}$$

If this rate of production is inadequate, we would need to increase the rate of production of Operator 2. This can be accomplished by:

1. Working one or both of the operators at the second operation overtime, thus accumulating a small inventory at this workstation.
2. Utilizing the services of a third part-time worker at the workstation of Operation 2.
3. Reassigning some of the work of Operation 2 to Operation 1 or Operation 3. (It would be preferable to assign more work to Operation 1.)
4. Improving the method at Operation 2 to diminish the cycle time of this operation.

In the preceding example, given a cycle time and operation times, an analyst can determine the number of operators needed for each operation to meet a desired production schedule. The production line work assignment problem can also be minimizing the number of workstations, given the desired cycle time; or, given the number of workstations, assigning work elements to the workstations, within the restrictions established, to minimize the cycle time.

An important strategy in assembly line balancing is work element sharing. Two or more operators whose work cycle includes some idle time might share the work of another station, to make the entire line more efficient. For example, Figure 2–18 shows an assembly line involving six workstations. Station 1 has three work elements—A, B, and C—for a total of 45 seconds. Note that work elements, B, D, and E cannot begin until A is completed and that B, D, and E can occur in any order. It may be possible to share element H between stations 2 and 4, with only a 1-second increase in cycle time (from 45 seconds to 46 seconds), while saving 30 seconds per assembled unit. We should note that element sharing may result in an increase in material handling, since parts may have to be delivered to more than one location. In addition, element sharing may necessitate added costs for duplicate tooling.

A second possibility for improving the balance of an assembly line involves dividing a work element. Referring again to Figure 2-18, it may be possible to divide element H, rather than have half of the parts go to station 2 and the other half to station 4.

Many times, it is not economical to divide an element. An example would be driving home eight machine screws with a power screwdriver. Once the operator has located the part in a fixture, gained control of the power tool, and brought the tool to the work, it would usually be more advantageous to drive home all eight screws, rather than only a portion of them, leaving the rest for a different operator. Whenever elements can be divided, workstations may be better balanced as a result of the division.

A different assembly sequence may also produce more favorable results. Product design generally dictates the assembly sequence. However, alternatives should not be overlooked. Balanced assembly lines are not only less costly, but they assist in maintaining worker morale, since little differential exists in the work content of the different workers in such lines.

FIGURE 2–18
Assembly line involving six workstations.

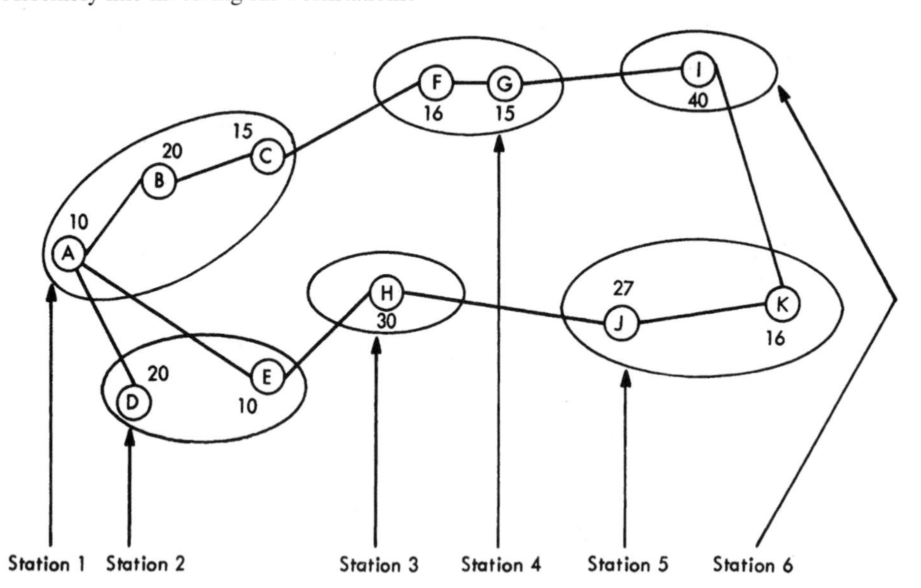

The following procedure for solving an assembly line balancing problem is based on General Electric's Assembly Line Balancing. The method assumes the following:

1. Operators are not able to move from one workstation to another to help maintain a uniform workload.
2. The work elements that have been established are of such magnitude that further division would substantially decrease the efficiency of performing the work element. (Once established, the work elements should be identified with a code.)

The first step in the solution of the problem is to determine the sequence of individual work elements. The fewer the restrictions on the order in which the work elements can be done, the greater the probability that a favorable balance in the work assignments will be achieved. To determine the sequence of the work elements, the analyst determines the answer to the following question: "What other work elements, if any, must be completed before this work element can be started?" This question is applied to each element to establish a precedence chart for the production line under study (see Figure 2–19). Functional design, available production

FIGURE 2–19
Partially completed precedence chart.
Note that work elements 002 and 003 may be done in any sequence with respect to any of the other work elements and that 032 cannot be started until 005, 006, 008, and 009 have been completed. Note also that after 004 has been finished, we may start either 033, 017, 021, 007, 008, or 009.

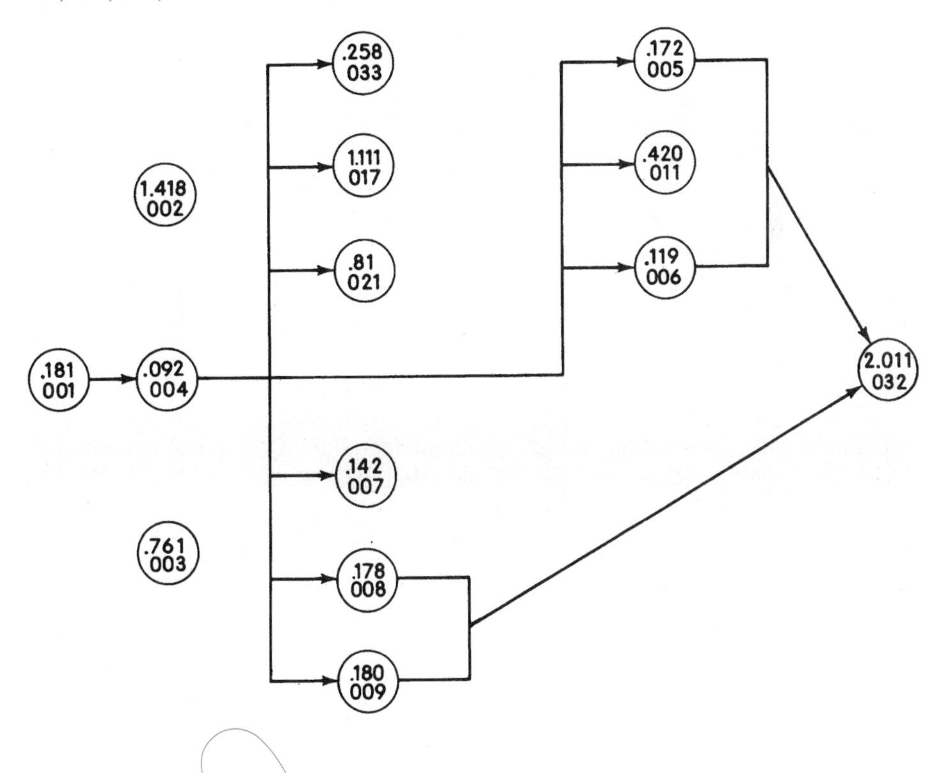

methods, floor space, and so on, can all introduce constraints with respect to work element sequence.

A second consideration in the production line work assignment problem is zoning restraints. A zone represents a subdivision that may or may not be physically separated or identified from other zones in the system. Confining certain work elements to a given zone may be justified, to congregate similar jobs, working conditions, or pay rates. Or zoning restraints may help to identify physically specific stages of a component, such as keeping it in a certain position while performing certain work elements. As an example, all work elements related to one side of a component might be performed in a certain zone before the component is allowed to be turned over.

Obviously, the more zoning restraints placed on the system, the fewer the combinational possibilities available for investigation. The analyst begins by making a sketch of the system and coding the applicable zones. Within each zone, the work elements that may be done in that area are shown. The analyst then estimates the production rate, using the expression:

$$\text{Production per day} = \frac{\text{Working min/day}}{\text{Cycle time of system (min/unit)}}$$

where the cycle time of the system is the standard time of the limiting zone or station.

Next, the precedence graph is established:

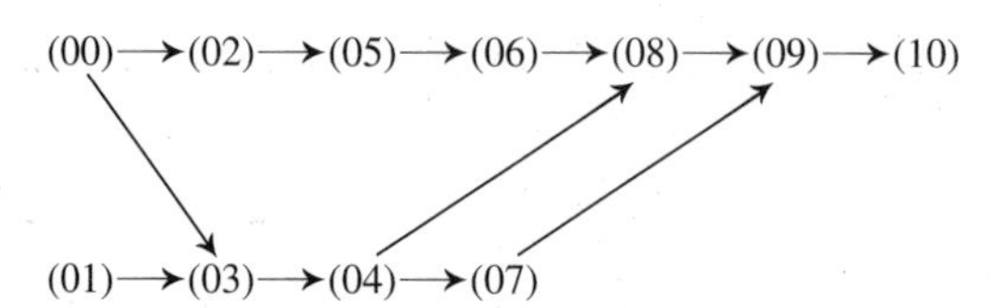

This precedence graph shows that work unit (00) must be completed before (02), (03), (05), (06), (04), (07), (08), (09), and (10); and work unit (01) must be completed before (03), (04), (07), (08), (09), and (10). Either (00) or (01) can be done first, or they can be done concurrently. Also, work unit (03) cannot be started until work units (00) and (01) are completed, and so on.

To describe these relationships, the precedence matrix illustrated in Figure 2–20 is established. Here the numeral 1 signifies a "must precede" relationship. For example, work unit (00) must precede work units (02), (03), (04), (05), (06), (07), (08), (09), and (10). Also, work unit (09) must precede only work unit (10).

Now, a "positional weight" must be computed for each work unit. This is done by computing the summation of each work unit and all the work units that must follow it. Thus, the "positional weight" for work unit (00) would be:

$$\begin{aligned} &\Sigma\ 00, 02, 03, 04, 05, 06, 07, 08, 09, 10 \\ &= 0.46 + 0.25 + 0.22 + 1.10 + 0.87 + 0.28 + 0.72 + 1.32 + 0.49 + 0.55 \\ &= 6.26. \end{aligned}$$

Listing the positional weights in decreasing order of magnitude gives the following:

FIGURE 2–20
A precedence matrix as used by a digital computer for a line balancing problem.

Estimated work unit time (minutes)	*Work unit*	*Work unit*										
		00	*01*	*02*	*03*	*04*	*05*	*06*	*07*	*08*	*09*	*10*
0.46	*00*			1	1	1	1	1	1	1	1	1
0.35	*01*				1	1			1	1	1	1
0.25	*02*						1	1		1	1	1
0.22	*03*					1			1	1	1	1
1.10	*04*								1	1	1	1
0.87	*05*							1		1	1	1
0.28	*06*									1	1	1
0.72	*07*										1	1
1.32	*08*										1	1
0.49	*09*											1
0.55	*10*											
6.61												

Unsorted work elements	Sorted work elements	Positional weight	Immediate predecessors
00	00	6.26	—
01	01	4.75	—
02	03	4.40	(00), (01)
03	04	4.18	(03)
04	02	3.76	(00)
05	05	3.56	(02)
06	06	2.64	(05)
07	08	2.36	(04), (06)
08	07	1.76	(04)
09	09	1.04	(07), (08)
10	10	0.55	(09)

Work elements must then be assigned to various workstations. This process is based on the positional weights (i.e., those work elements with the highest positional weights are assigned first) and the cycle time of the system. The work element with the highest positional weight is assigned to the first workstation. The unassigned time for this workstation is determined by subtracting the sum of the assigned work element times from the estimated cycle time. If there is adequate unassigned time, the work element with the next highest positional weight is assigned, provided that the work elements in the "immediate predecessors" column have already been assigned. Once a workstation's allotted time has been filled, the analyst moves on to the next workstation, and the procedure continues until all of the work elements have been assigned.

As an example, assume that the required production per 450-minute shift is 300 units. The cycle time of the system is 450/300 = 1.50 minutes, and the final balanced line is shown in Table 2–2.

Under the arrangement illustrated, with six workstations, we have a cycle time of 1.32 minutes (workstation 4). This arrangement produces 450/1.32 = 341 units, which more than meets the daily requirement of 300.

However, with six workstations, we also have considerable idle time. The idle time per cycle is:

$$\sum_{1}^{6} 0.04 + 0.22 + 0.17 + 0 + 0.11 + 0.77 = 1.31 \text{ min}$$

For more favorable balancing, the problem can be solved for cycle times of less than 1.50 minutes. This may result in more operators and more production per day,

TABLE 2–2
Balanced Assembly Line

Work					Station time		
Station	Element	Positional weight	Immediate predecessors	Work element time	Cumulative	Unassigned	Remarks*
1	00	6.26	—	0.46	.46	1.04	—
1	01	4.75	—	0.35	.81	0.69	—
1	03	4.40	(00), (01)	0.22	1.03	0.47	—
1	~~04~~	~~4.18~~	~~(03)~~	~~1.10~~	2.13	—	N.A.
1	02	3.76	(00)	0.25	1.28	0.22	—
1	~~05~~	~~3.56~~	~~(02)~~	~~0.87~~	2.05	—	N.A.
2	04	4.18	(03)	1.10	1.10	0.40	—
2	~~05~~	~~3.56~~	~~(02)~~	~~0.87~~	1.97	—	N.A.
3	05	3.56	(02)	0.87	.87	0.63	—
3	06	2.64	(05)	0.28	1.15	0.35	—
3	~~08~~	~~2.36~~	~~(04), (06)~~	~~1.32~~	2.47	—	N.A.
4	08	2.36	(04), (06)	1.32	1.32	0.18	—
4	~~07~~	~~1.76~~	~~(04)~~	~~0.72~~	2.04	—	N.A.
5	07	1.76	(04)	0.72	.72	0.78	—
5	09	1.04	(07), (08)	0.49	1.21	0.29	—
5	~~10~~	~~.55~~	~~(09)~~	~~0.55~~	1.76	—	N.A.
6	10	.55	(09)	0.55	.55	0.95	—

*N.A. means not acceptable.

which may have to be stored. Another possibility involves operating the line under a more efficient balancing for a limited number of hours per day.

A variety of commercially available software packages, such as STORM, eliminate the drudgery of the calculations and perform these steps automatically.

SUMMARY

The various charts presented in this chapter are valuable tools for presenting and solving problems. Just as several types of tools are available for a particular job, so several chart designs can help solve an engineering problem. However, for a specific solution, one chart usually has advantages over another. Analysts should understand the specific functions of each process chart and choose the correct one for solving a specific problem. By understanding the specific functions of the various charts, the analyst can select the appropriate one for improving operations.

Pareto analyses and fish diagrams are used to select a critical operation and to identify the root causes and contributing factors leading to the problem. Gantt and PERT charts are project scheduling tools. The Gantt chart provides only a good overview, and the PERT chart quantifies the interactions between different activities. The job/worksite analysis guide is primarily used on a physical walk-through to identify key worker, task, environmental and administrative factors that may cause potential problems. The operation process chart provides a good overview of the relationships between different operations and inspections on assemblies involving several components. The flow process chart provides more detail for the analysis of manufacturing operations, to find hidden or indirect costs, such as delay time, storage costs, and material handling costs. The flow diagram is a useful supplement to the flow process chart in developing plant layouts. The worker/machine and gang process charts show machines or facilities in conjunction with the operator or operators, and are used to analyze idle operator time and idle machine time. Synchronous and random servicing calculations and line balancing techniques are used to develop more efficient operations through quantitative methods.

These 13 tools are very important for methods analysts. The charts are valuable descriptive and communicative aids for understanding a process and its related activities. Their correct use can aid in presenting and solving the problem, and in selling and installing the solution. Quantitative techniques can determine the optimum arrangement of operators and machines. Analysts should be acquainted with sufficient algebra and probability theory to develop a mathematical model that provides the best solution to the machine or facility problem. Thus, they are effective in presenting improved methods to management, training employees in the prescribed method, and focusing pertinent details, in conjunction with plant layout work.

QUESTIONS

1. What does the operation process chart show?
2. What symbols are used in constructing the operation process chart?
3. How does the operation process chart show materials introduced into the general flow?
4. How does the flow process chart differ from the operation process chart?
5. What is the principal purpose of the flow process chart?
6. What symbols are used in constructing the flow process chart?
7. Why is it necessary to construct process charts from direct observation, as opposed to information obtained from the foreman?
8. In the construction of the flow process chart, what method can be used to estimate distances moved?
9. How can delay times be determined in the construction of the flow process chart? Storage times?
10. When would you advocate using the flow diagram?
11. How can the flow of several different products be shown on the flow diagram?
12. What two flowchart symbols are used exclusively in the study of paperwork?
13. What are the limitations of the operation and flow process charts and the flow diagram?
14. Explain how PERT charting can save a company money.
15. Explain how you would obtain "optimistic" and "pessimistic" times to be used in connection with a PERT chart.
16. When is it advisable to construct a worker and machine process chart?
17. What is machine coupling?
18. In what way does an operator benefit through machine coupling?
19. Explain how you would sell machine coupling to union officials strongly opposed to the technique.
20. How does the gang process chart differ from the worker and machine process chart?
21. In a process plant, which of the following process charts has the greatest application: worker and machine, gang, operation, flow? Why?
22. What is the difference between synchronous and random servicing?

PROBLEMS

1. Based on the following "emergency" cost table, what would be the minimum time to complete the project described by Figure 2–4, whose normal costs are shown in Table 2–1? What would be the added cost to complete the project within this time period?

	Emergency Schedule	
	Weeks	Dollars
A	2	$7,000
B	1	$2,500
C	2	$5,000
D	0.5	$2,000
E	4	$6,000
F	3	$5,000
G	2	$6,000
H	0	0
I	4	$7,600
J	1	$2,200
K	4	$4,500
L	2	$2,200
M	3	$3,000
N	1	$1,000
O	2	$6,000
P	1	$3,000

2. In the Dorben Company's automatic screw machine department, five machines are assigned to each operator. On a given job, the machining time per piece is 0.164 hours, the machine serving time is 0.038 hours, and the average machine downtime is 0.12 hours per machine per hour. With an operator rate of $12.80 per hour and a machine rate of $14 per hour, calculate the expected cost per unit of output. Exclude material cost.

3. In the Dorben Company, a worker is assigned to operate three similar facilities. Each of these facilities is down at random times during the day. A work sampling study indicates that, on average, the machines operate unattended 60 percent of the time. Operator attention time at irregular intervals averages 40 percent. This arrangement results in the loss of about 14 percent of the available machine time due to machine interference. If the machine rate is $20 per hour and the operator rate is $12 per hour, what would be the most favorable number of machines (from an economic standpoint) that should be operated by one operator?

4. The analyst in the Dorben Company wishes to assign a number of similar facilities to an operator, based on minimizing the cost per unit of output. A detailed study of the facilities reveals the following:

   Loading machine standard time = 0.34 minutes
   Unloading machine standard time = 0.26 minutes
   Walk time between two machines = 0.06 minutes
   Operator rate = $12.00 per hour
   Machine rate (both idle and working) = $18.00 per hour
   Power feed time = 1.48 minutes

   How many of these machines should be assigned to each operator?

5. A study reveals that a group of three semiautomatic machines assigned to one operator operates unattended 70 percent of the time. Operator service time at irregular intervals averages 30 percent of the time on these three machines. What would be the estimated machine hours lost per 8-hour day because of lack of an operator?

6. Based upon the following data, develop your recommended allocation of work and the number of work stations.

Work unit	Estimated work unit time in minutes
0	0.76
1	1.24
2	0.84
3	2.07
4	1.47
5	2.40
6	0.62
7	2.16
8	4.75
9	0.65
10	1.45

The minimum required production per day is 90 assemblies. The following precedence matrix was developed by the analyst.

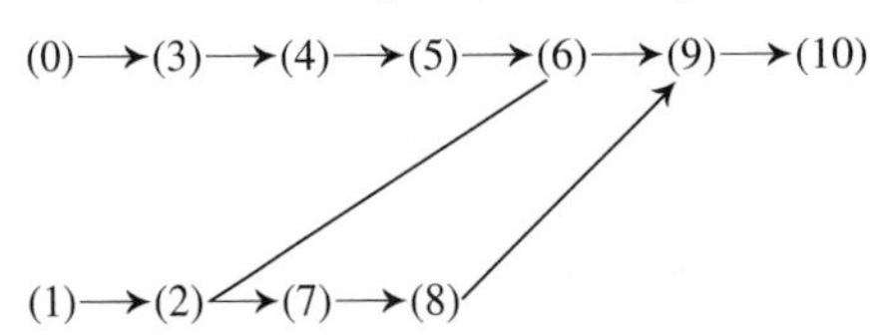

7. How many machines should be assigned to an operator when:
a. Loading and unloading time on one machine is 1.41 minutes.
b. Walking time to the next facility is 0.08 minutes.
c. Machine time (power feed) is 4.34 minutes.
d. Operator rate is $13.20 per hour.
e. Machine rate is $18.00 per hour.

8. What proportion of machine time would be lost in operating four machines when each machine operates unattended 70 percent of the time and the operator attention time at irregular intervals averages 30 percent? Is this the best arrangement for minimizing the proportion of machine time lost?

9. In an assembly process involving six distinct operations, it is necessary to produce 250 units per 8-hour day. The measured operation times are as follows:
a. 7.56 minutes.
b. 4.25 minutes.
c. 12.11 minutes.
d. 1.58 minutes.
e. 3.72 minutes.
f. 8.44 minutes.

How many operators would be required at 80 percent efficiency? How many operators will be utilized at each of the six operations?

REFERENCES

ASME. *ASME Standard—Operation and Flow Process Charts, ANSI Y15.3-1974.* New York: American Society of Mechanical Engineers, 1974.

Baker, Kenneth R. *Elements of Sequencing and Scheduling.* Hannover, NJ: K.R. Baker, 1995.

Buffa, E. S., and W. H. Taubert. *Production-Inventory Systems: Planning and Control.* 3rd ed. Homewood, IL: Richard D. Irwin, 1979.

Cole, R. *Work, Mobility, and Participation: A Comparative Study of American and Japanese Industry.* Berkeley, CA: University of California Press, 1979.

Herron, D. "Industrial Engineering Applications of ABC Curves." *AIIE Transactions,* 8, no.2 (June 1976), pp. 210–218.

Johnson, L. A., and Douglas C. Montgomery. *Operations Research in Production Planning, Scheduling, and Inventory Control.* New York: John Wiley & Sons, 1974.

Moodie, C. "Assembly Line Balancing." In *Handbook of Industrial Engineering,* ed. Gavriel Salvendy. New York: John Wiley & Sons, 1982.

Stecke, K. "Machine Interference." In *Handbook of Industrial Engineering,* ed. Gavriel Salvendy. New York: John Wiley & Sons, 1982.

Takeji, K. "Charting Techniques." In *Handbook of Industrial Engineering,* ed. Gavriel Salvendy. New York: John Wiley & Sons, 1982.

SELECTED SOFTWARE

STORM, Quantitative Modeling for Decision Support, P.O. Box 21196, Cleveland, OH, 44121-0196.

CHAPTER 3

Operation Analysis

KEY POINTS:

- Use operation analysis to *improve the method.*
- Focus on the purpose of operation by asking *Why.*
- Focus on design, materials, tolerances, processes, and tools by asking *How.*
- Focus on the operator and work design by asking *Who.*
- Focus on the layout of the work by asking *Where.*
- Focus on the sequence of manufacture by asking *When.*
- Always try to *simplify* by eliminating, combining, and rearranging operations.

Methods analysts use operation analysis to study all productive and nonproductive elements of an operation, to increase productivity per unit of time and reduce unit costs while maintaining or improving quality. Operation analysis is just as effective in planning new work centers as it is in improving those already in operation. By using the questioning approach on all facets of the workstation, dependent tooling, and product design, analysts can develop an efficient work center.

The improvement of existing operations is a continuing process in industry, and this chapter deals primarily with that process. These principles are equally valid and important in planning new work centers. Operation analysis immediately follows obtaining and presenting facts using a variety of flow process charting tools. Operation analysis is the third methods step, the one in which analysis takes place and the various components of the proposed method crystallize.

Experience has proven that practically all operations can be improved if sufficient study is given to them. Since the systematic analysis procedure is equally effective for both large and small industries, in job shops as well as in mass production, operation analysis is applicable to all areas of manufacturing, business, and government. When properly utilized, it develops a better method of doing the work by simplifying operational procedures and material handling, and by utilizing equipment more effectively. Thus, firms are able to: increase output and reduce unit cost; ensure quality and reduce defective workmanship; and facilitate operator enthusiasm by improving working conditions, minimizing operator fatigue, and permitting higher operator earnings.

THE NINE PRIMARY OPERATION ANALYSIS APPROACHES

Probably one of the most common attitudes of management is that its problems are unique. Consequently, it feels that any new method will be impractical. Actually, all work, whether clerical, maintenance, office, machine, assembly, or general labor, productive or nonproductive, is much the same; it consists of combinations of 17 basic elements, as defined by the Gilbreths (see Chapter 4). For example, the elements of work in driving a car are much like those required to operate a turret lathe; the basic motions employed in dealing a bridge hand are almost identical with certain manual inspection and machine loading elements. The fact that all work is similar in many respects verifies the principle that if methods can be improved in one plant, opportunities exist for methods improvements in all plants.

As outlined in Chapter 1, the use of a systematic procedure is invaluable in producing real savings. The first step is to gather all information related to the anticipated volume of the work. To determine how much time and effort should be devoted to improving the present method or planning the new job, analysts evaluate the expected volume, the chance of repeat business, the life of the job, the chance for design changes, and the labor content of the job. If the job promises to be quite active, a more detailed study is justified.

Once the quantity, job life, and labor content are estimated, operations analysts then collect all factual manufacturing information. This information includes all operations, the facilities used to perform the operations, and operational times; all moves or transportations, the facilities used for transportation, and transportation distances; all inspections, inspection facilities, and inspection times; all storages, storage facilities, and storage times; all vendor operations together with their prices; all drawings, and quality and design specifications. After all this information affecting quality and cost is gathered, it must be presented in a form suitable for study. One of the most effective ways of doing this is through the various charting tools presented in Chapter 2. The analyst should review each operation and inspection presented graphically on these charts and should ask a number of questions, the most important of which is "Why":

1. "Why is this operation necessary?"
2. "Why is this operation performed in this manner?"
3. "Why are these tolerances this close?"
4. "Why has this material been specified?"
5. "Why has this class of operator been assigned to do the work?"

The question "Why" immediately suggests other questions, including "How" "Who" "Where" and "When." Thus, analysts might ask:

1. "How can the operation be performed better?"
2. "Who can best perform the operation?"
3. "Where could the operation be performed at a lower cost or improved quality?"
4. "When should the operation be performed to yield the least amount of material handling?"

For example, in the operation process chart shown in Figure 2–7, analysts might ask the questions listed in Table 3–1 to determine the practicability of the methods improvements indicated. Answering these questions helps initiate the elimination, combination, and simplification of the operations.

In obtaining the answers to such questions, analysts become aware of other questions that may lead to improvement. Ideas seem to generate more ideas, and experienced analysts usually arrive at several improvement possibilities. Analysts must keep an open mind, so that previous disappointments do not discourage the trial of new ideas.

When the nine primary approaches to operation analysis are used in studying each individual operation, attention focuses on the items most likely to produce improvement. However, not all of these approaches are applicable to each activity on the flowchart; yet, usually more than one should be considered. The recommended analysis method takes each step in the present method and analyzes it, considering all the key points, with a specific improvement approach clearly in mind. Follow this same procedure on the succeeding operations, inspections, moves, storages, and so on. After each element has been analyzed, consider the entire product as a whole, and reconsider all points of analysis, looking for overall improvement possibilities. Unlimited opportunities for methods improvement usually appear in every plant. To develop maximum savings, carefully study the individual and collective operations as outlined. Wherever this procedure has been followed by competent engineers and related analysts, beneficial results have been realized.

1. OPERATION PURPOSE

This is probably the most important of the nine points of operation analysis. The best way to simplify an operation is to devise some way to get the same or better results at no additional cost. An analyst's cardinal rule is to try to *eliminate* or *combine* an operation before trying to improve it. In our experience, as much as 25 percent of the operations being performed by American industry can be eliminated if sufficient study is given to the design and process.

Far too much unnecessary work is done today. In many instances, the task or the process should not be simplified or improved, but eliminated entirely. Eliminating an activity saves money on the installation of an improved method, and there is no interruption or delay because no improved method is being developed, tested, and installed. Operators need not be trained on the new method, and resistance to change is minimized when an unnecessary task or activity is eliminated. With respect to paperwork, before a form is developed for information transfer,

TABLE 3–1
Questions to ask in the Manufacture of Telephone Stands

	Question	Method Improvement
1.	Can fixed lengths of 1 1/2″ x 14″ white maple may be purchased at no extra square footage cost?	Eliminate waste ends from lengths that are not multiples of 14″.
2.	Can purchased maple boards be secured with edges smooth and parallel?	Eliminate jointing of ends (operation 2).
3.	Can boards be purchased to thickness size and have at least one side planed smooth? If so, how much extra will this cost?	Eliminate planing to size.
4.	Why cannot two boards be stacked and sawed into 14” sections simultaneously?	Reduce time of 0.18 (operation 4).
5.	What percentage of rejects do we have at the first inspection station?	If the percentage is low, perhaps this inspection can be eliminated.
6.	Why should the top of the table be sanded all over?	Eliminate sanding of one side of top and reduce time (operation 5).
7.	Can fixed lengths of 1 1/2″ x 3″ yellow pine be purchased at no extra square footage cost?	Eliminate waste ends from lengths that are not multiples of 12″.
8.	Can purchased yellow pine boards be secured with edges smooth and parallel?	Eliminate jointing of one edge.
9.	Can sill boards be purchased to thickness size and have one side planed smooth? If so, how much extra will this cost?	Eliminate planing to size.
10.	Why cannot two or more boards be stacked and sawed into 14” sections simultaneously?	Reduce time of 0.10 (operation 9).
11.	What percentage of rejects do we have at the first inspection of the sills?	If the percentage is low, perhaps this inspection can be eliminated.
12.	Why is it necessary to sand the sills all over?	Eliminate some sanding and reduce time (operation 10).
13.	Can fixed lengths of 2 1/2″ x 2 1/2″ white maple be purchased at no extra square footage cost?	Eliminate waste ends from lengths that are not multiples of 16″.
14.	Can a smaller size than 2 1/2″ x 2 1/2″ be used?	Reduce material cost.
15.	Can purchased white maple boards be secured with edges smooth and parallel?	Eliminate jointing of edges.
16.	Can leg boards be purchased to thickness size and have sides planed smooth? If so, how much extra will this cost?	Eliminate planing to size.
17.	Why cannot two or more boards be stacked and sawed into 14″ sections simultaneously?	Reduce time (operation 15).
18.	What percentage of rejects do we have at the first inspection of the legs?	If the percentage is low, perhaps this inspection can be eliminated.
19.	Why is it necessary to sand the legs all over?	Eliminate some sanding and reduce time (operation 16).
20.	Could a fixture facilitate assembly of the sills to the top?	Reduce assembly time (operation 11).
21.	Can a sampling inspection be used on the first inspection of the assembly?	Reduce inspection time (operation 4).
22.	Is it necessary to sand after one coat of shellac?	Eliminate operation 19.

analysts should ask, "Is the form really needed?" The advance of today's computer-controlled systems should reduce the generation of forms and paperwork.

Unnecessary operations frequently result from improper planning when the job is first set up. Once a standard routine is established, it is difficult to change, even if such a change would eliminate a portion of the work and make the job easier. When new jobs are planned, the planner may include an extra operation if there is any possibility that the product would be rejected without that extra work. For example, in turning a steel shaft, if there is some question whether to take two or three cuts to maintain a 40-microinch finish, the planner invariably specifies three cuts, even though proper maintenance of the cutting tools, supplemented by ideal feeds and speeds, would allow the job to be done with two cuts.

Unnecessary operations often develop because of the improper performance of a previous operation. A second operation must be done to "touch up" or make acceptable the work done by the first operation. In one plant, for example, armatures were previously spray painted in a fixture, making it impossible to cover the bottom of the armature with paint because the fixture shielded the bottom from the spray blast. It was therefore necessary to touch up the armature bottoms after spray painting. A study of the job resulted in a redesigned fixture that held the armature and still allowed complete coverage. In addition, the new fixture permitted seven armatures to be spray painted simultaneously, while the old method called for spray painting one at a time. Thus, by considering that an unnecessary operation may have developed because of the improper performance of a previous operation, the analyst was able to eliminate the touch-up operation (see Figure 3–1).

As another example, in the manufacture of large gears, it was necessary to introduce a hand-scraping and lapping operation to remove waves in the teeth after they had been hobbed. An investigation disclosed that contraction and expansion, brought about by temperature changes in the course of the day, were responsible for the waviness in the teeth's surfaces. By enclosing the whole unit and installing an air-conditioning system within the enclosure, the company was able to maintain the proper temperature during the whole day. The waviness disappeared immediately, and it was no longer necessary to continue the hand-scraping and lapping operations.

To eliminate an operation, analysts should ask and answer the following question: "Can an outside supplier perform the operation more economically?" In one example, ball bearings purchased from an outside vendor had to be packed in grease prior to assembly. A study of bearing vendors revealed that "sealed-for-life" bearings could be purchased from another supplier at lower cost.

The examples given in this section highlight the need to establish the purpose of each operation before endeavoring to improve the operation. Once the necessity of the operation has been determined, the remaining nine steps to operation analysis should help to determine how it can be improved.

2. PART DESIGN

Methods engineers are often inclined to feel that once a design has been accepted, their only recourse is to plan its economical manufacture. While introducing even

FIGURE 3–1
A. Painted armature as removed from the old fixture and as removed from the improved fixture.

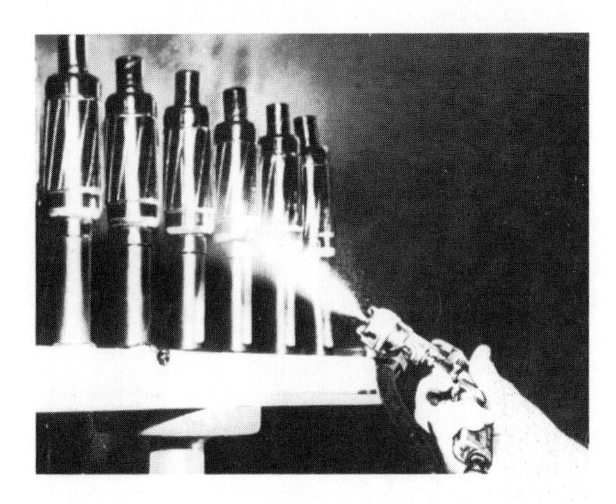

B. Armature in spray-painting fixture allowing complete coverage of the armature bottom.

a slight design change may be difficult, a good methods analyst should still review every design for possible improvements. Designs can be changed, and if improvement is the result, and the activity of the job is significant, then the change should be made.

To improve the design, analysts should keep in mind the following pointers for lower cost designs on each component and each subassembly:

1. Reduce the number of parts by simplifying the design.
2. Reduce the number of operations and the length of travel in manufacturing by joining the parts better and by making the machining and assembly easier.
3. Utilize a better material.
4. Liberalize tolerances and rely on key operations for accuracy, rather than on series of closely held limits.
5. Design for manufacturability and assembly.

The General Electric Company summarized the ideas for developing minimum cost designs in Table 3–2.

TABLE 3–2

Methods for Minimum Cost Design

Castings

1. Eliminate dry sand (baked-sand) cores.
2. Minimize depth to obtain flatter castings.
3. Use minimum weight consistent with sufficient thickness to cast without chilling.
4. Choose simple forms.
5. Symmetrical forms produce uniform shrinkage.
6. Liberal radii—no sharp corners.
7. If surfaces are to be accurate with relation to each other, they should be in the same part of the pattern, if possible.
8. Locate parting lines so that they will not affect looks and utility, and need not be ground smooth.
9. Specify multiple patterns instead of single ones.
10. Metal patterns are preferable to wood.
11. Permanent molds instead of metal patterns.

Moldings

1. Eliminate inserts from parts.
2. Design molds with smallest number of parts.
3. Use simple shapes.
4. Locate flash lines so that the flash does not need to be filed and polished.
5. Minimum weight.

Punchings

1. Punched parts instead of molded, cast, machined, or fabricated parts.
2. "Nestable" punchings to economize on material.
3. Holes requiring accurate relation to each other to be made by the same die.
4. Design to use coil stock.
5. Punchings designed to have minimum sheared length and maximum die strength with fewest die moves.

Formed parts

1. Drawn parts instead of spun, welded, or forged parts.
2. Shallow draws if possible.
3. Liberal radii on corners.
4. Bent parts instead of drawn.
5. Parts formed of strip or wire instead of punched from sheet.

Fabricated parts

1. Self-tapping screws instead of standard screws.
2. Drive pins instead of standard screws.
3. Rivets instead of screws.
4. Hollow rivets instead of solid rivets.
5. Spot or projection welding instead of riveting.
6. Welding instead of brazing or soldering.
7. Use die castings or molded parts instead of fabricated construction requiring several parts.

Machined parts

1. Use rotary machining processes instead of shaping methods.
2. Use automatic or semiautomatic machining instead of hand-operated.
3. Reduce the number of shoulders.
4. Omit finishes where possible.
5. Use rough finish when satisfactory.
6. Dimension drawings from same point as used by factory in measuring and inspecting.
7. Use centerless grinding instead of between-center grinding.
8. Avoid tapers and formed contours.
9. Allow a radius or undercut at shoulders.

Screw machine parts

1. Eliminate second operation.
2. Use cold-rolled stock.
3. Design for header instead of screw machine.
4. Use rolled threads instead of cut threads.

Welded parts

1. Fabricated construction instead of castings or forgings.
2. Minimum sizes of welds.
3. Welds made in flat position rather than vertical or overhead.
4. Eliminate chamfering edges before welding.
5. Use "burnouts" (torch-cut contours) instead of machined contours.
6. Lay out parts to cut to best advantage from standard rectangular plates and avoid scrap.
7. Use intermittent instead of continuous weld.
8. Design for circular or straight-line welding to use automatic machines.

Treatments and finishes

1. Reduce baking time to minimum.
2. Use air drying instead of baking.
3. Use fewer or thinner coats.
4. Eliminate treatments and finishes entirely.

Assemblies

1. Make assemblies simple.
2. Make assemblies progressive.
3. Make only one assembly and eliminate trial assemblies.
4. Make component parts RIGHT in the first place so that fitting and adjusting will not be required in assembly.

   This means that drawings must be correct, with proper tolerances, and that parts must be made according to drawings.

General

1. Reduce number of parts
2. Reduce number of operations.
3. Reduce length of travel in manufacturing.

Source: Adapted from *American Machinist,* reference sheets, 12th ed., New York: McGraw-Hill Publishing Co.

The following examples of methods improvement resulted from considering a better material or process in an effort to improve the design.

1. Conduit boxes were originally built of cast iron. The improved design, making a stronger, neater, lighter, and less expensive conduit box, was fabricated from sheet steel.
2. A brass cam switch used in control equipment was originally made as a brass die casting. The design was slightly altered so that the less expensive process of extruding could be utilized. The extruded sections were then cut to the desired length to produce the cam switch (see Figure 3–2).
3. Design simplification through the better joining of parts was used in assembling terminal clips to their mating conductors. The original practice required turning up the end of the clip to form a socket. The socket was filled with solder, and the wire conductor was then tinned, inserted into the solder-filled socket, and held there until the solder solidified. The altered design called for resistance welding the clip to the wire conductor, eliminating both the forming and dipping operations.
4. A motor cover thumbscrew was originally made of three components: head, pin, and screw. The components were assembled by joining the head to the screw with the pin. A significantly less costly thumbscrew was developed by

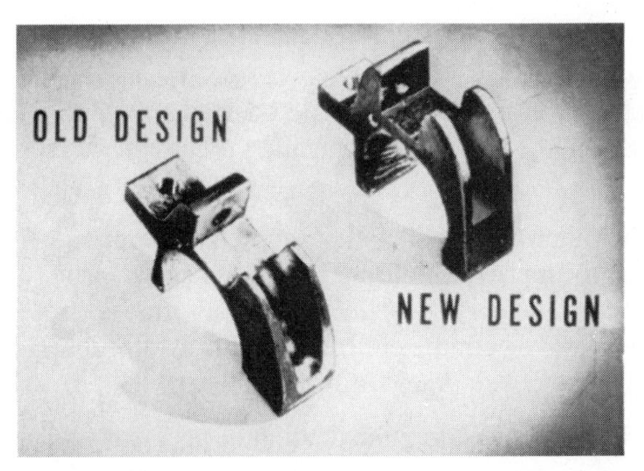

FIGURE 3–2
A. Redesigned brass cam switch allowing part to be made from extrusion.

B. New design shown cut to length from section of brass extrusion.

FIGURE 3–3
A. Old thumbscrew in three parts.

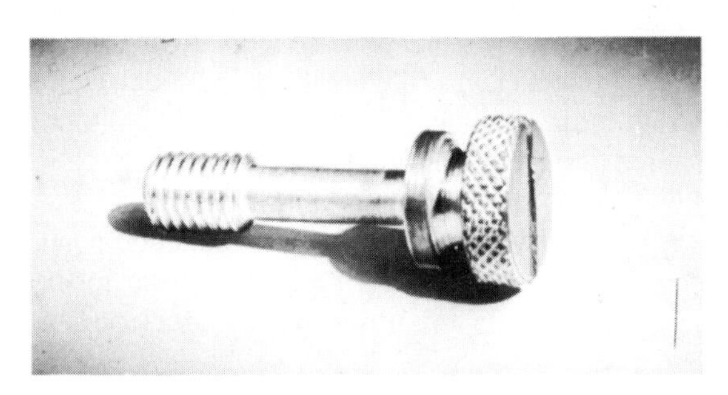

B. Improved one-piece design of motor cover thumbscrew.

redesigning the part for an automatic screw machine that was able to turn out the part complete with no secondary operations. A simplified design resulted in a less expensive part that still met all the service and operating requirements (see Figure 3–3).

Just as opportunities exist to improve productivity through better product design, similar opportunities exist to improve the design of forms (whether hard copy or electronic) used throughout an industry or business. Once a form is proved necessary, it should be studied to improve both the collection and flow of information. The following criteria apply to the development of forms:

1. Maintain simplicity in the form design, keeping the amount of necessary input information at a minimum.
2. Provide ample space for each bit of information, allowing for different input methods (writing, typewriter, word processor).
3. Sequence the information input in a logical pattern.
4. Color code the form to facilitate distribution and routing.
5. Provide adequate margins to accommodate standard filing facilities and procedures.
6. Confine computer forms to one page.

3.
TOLERANCES AND SPECIFICATIONS

The third of the nine points of operation analysis concerns tolerances and specifications that relate to the quality of the product, that is, its ability to satisfy given needs. While tolerances and specifications are always considered when reviewing the design, this is usually not sufficient; they should be considered independently of the other approaches to operation analysis.

Designers may have a tendency to incorporate specifications that are more rigid than necessary when developing a product. This can be due to a lack of knowledge about cost and the thought that it is necessary to specify closer tolerances and specifications than are actually needed to have the manufacturing departments produce to the actual required tolerance range.

Methods analysts should be well versed in the details of cost and should be fully aware of what unnecessarily close tolerances and/or rejects can do to the selling price. Figure 3–4 illustrates the relationship between the cost and the machining tolerance. If designers are being needlessly tight in establishing tolerances and specifications, management should embark on a training program clearly presenting the economies of specifications. Also, consideration should be given to the extra cost of products because of scrap and/or rejects. Today, there is only one way that a company can be competitive: all parts in every product must be produced to the precise dimensions given on the drawings. Developing quality products in a manner that actually reduces costs is a major tenet of the approach to quality instituted by Taguchi (1986). This approach involves combining engineering and statistical methods to achieve improvements in cost and quality by optimizing product design and manufacturing methods.

One manufacturer's drawings called for a 0.0005-inch tolerance on a shoulder ring for a DC motor shaft. The original specifications called for a 1.8105 to 1.8110-inch tolerance on the inside diameter. This close tolerance was deemed necessary because the shoulder ring was shrunk onto the motor shaft. Investigation revealed that a 0.003–inch tolerance was adequate for the shrink fit. The drawing was immediately changed to specify a 1.809 to 1.812-inch inside diameter. This change meant that a reaming operation was eliminated because someone questioned the absolute necessity of a close tolerance.

Analysts must also be alert for too liberal, as well as too restrictive, specifications. Closing up a tolerance often facilitates an assembly operation or some other subsequent step. This may be economically sound, even though it may increase the time required to perform a preassembly operation. In this connection, analysts should recognize that the overall tolerance is equal to the square root of the sum of the squares of the individual tolerances comprising the overall tolerance.

Analysts should also take into consideration the ideal inspection procedure. Inspection is a verification of quantity, quality, dimensions, and performance. Such inspections can usually be performed by a variety of techniques: spot inspection, lot-by-lot inspection, or 100 percent inspection. Spot inspection is a periodic check to assure that established standards are being realized. For example, a nonprecision blanking and piercing operation set up on a punch press should have a spot inspection to assure the maintenance of size and the absence of burrs. As the die begins to wear or as deficiencies in the material being worked begin to show up, the spot inspection would catch the trouble in time to make the necessary changes, without generating an appreciable number of rejects.

Lot-by-lot inspection is a sampling procedure in which a sample is examined to determine the quality of the production run or lot. The size of the sample depends on

FIGURE 3–4
Approximate relationship between cost and machining tolerance.

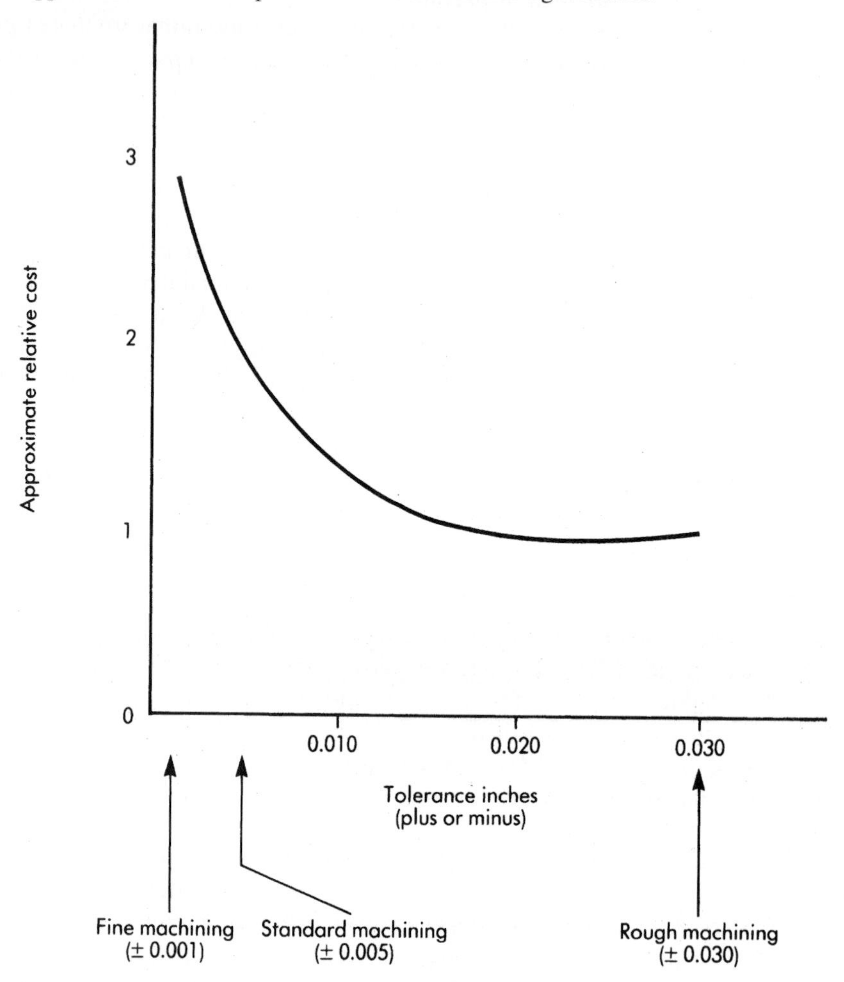

the allowable percentage of defective unity and the size of the production lot being checked. A 100 percent inspection involves inspecting every unit of production and rejecting the defective units. However, experience has shown that this type of inspection does not assure a perfect product. The monotony of screening tends to create fatigue, thus lowering operator attention. The inspector may pass some defective parts, or reject good parts. Because a perfect product is not assured under 100 percent inspection, acceptable quality may be realized by the considerably more economical methods of lot-by-lot or spot inspection.

For example, in one shop, a certain automatic polishing operation had a normal rejection quantity of 1 percent. Subjecting each lot of polished goods to 100 percent inspection would have been quite expensive. Management therefore decided, at an appreciable saving, to consider 1 percent the allowable percentage defective, even though this quantity of defective material would go through to plating and finishing, only to be thrown out in the final inspection before shipment.

By investigating tolerances and specifications and taking action when desirable, the company can reduce the costs of inspection, minimize scrap, diminish repair costs, and keep quality high.

4. MATERIAL

One of the first questions an engineer considers when designing a new product is, "What material shall be used?" Since choosing the correct material may be difficult because of the great variety available, it is often more practical to incorporate a better and more economical material into an existing design.

Methods analysts should consider the following possibilities for the direct and indirect materials utilized in a process:

1. Finding a less expensive material.
2. Finding materials that are easier to process.
3. Using materials more economically.
4. Using salvage materials.
5. Using supplies and tools more economically.
6. Standardizing materials.
7. Finding the best vendor from the standpoint of price and vendor stocking.

Finding a Less Expensive Material

Industry is continually developing new processes for producing and refining materials. Monthly publications summarize the approximate cost per pound of steel sheets, bars, and plates, and the cost of cast iron, cast steel, cast aluminum, cast bronze, thermoplastic and thermosetting resins, and other basic materials. These costs can be used as anchor points from which to judge the application of new materials. A material that was not competitive in price yesterday may be very competitive today.

One company used Micarta spacer bars between the windings of transformer coils. Separating the windings permitted the circulation of air between the windings. An investigation revealed that glass tubing could be substituted for the Micarta bars at a considerable savings. The glass tubing was less expensive, and it met service requirements better because the glass could withstand higher temperatures. Furthermore, the hollow tubing permitted more air circulation than did the solid Micarta bars.

Another company also used a less expensive material that still met service requirements in the production of distribution transformers. Originally, a porcelain plate separated and held the wire leads coming out of the transformers. The company found that a fullerboard plate stood up just as well in service, yet was considerably less expensive.

Today, many plastics are competing effectively with metals and wood. For example, Figure 3–5 illustrates a change in material in the manufacture of gasoline mechanical pump computer wheels; a 13–cent-per-unit savings was made, as well

FIGURE 3–5
Gasoline mechanical pump computer wheel. (*Courtesy of:* Veeder Root Company Subsidiary of Danaher Co.)
At left is the stamped and coined gear that was converted to a plastic gear, shown at right. The steel pawl shown assembled to the body (upper right) was converted to a Delrin plastic (bottom). The assembly is shown in the center.

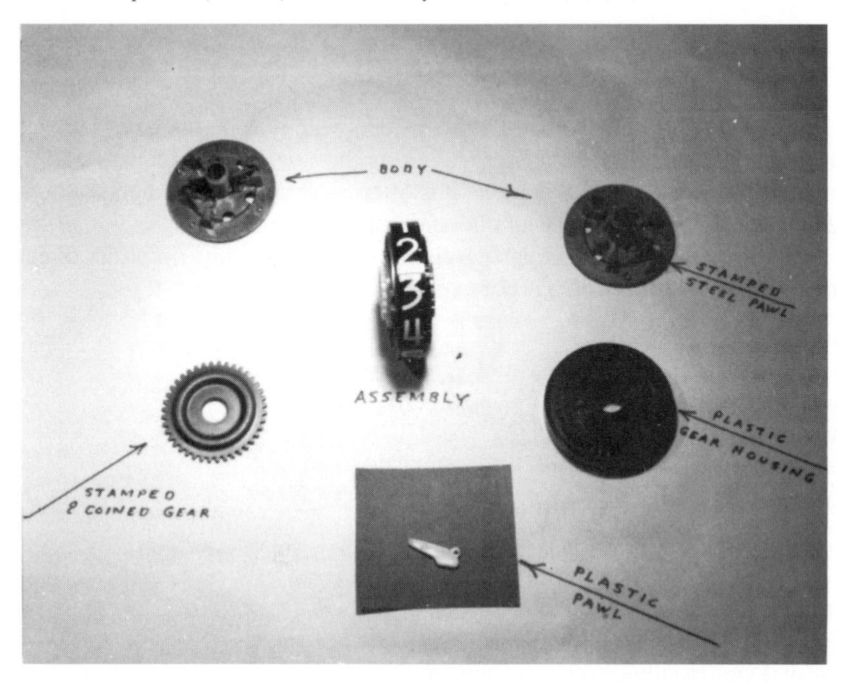

as a $10,000-per-year savings in tool maintenance, by converting two parts of the assembly from steel to plastic. The pawl had originally been made of sheet steel that had to be blanked, tumbled, and ground to tolerance. The new pawl was made by injection molding the plastic Delrin. Similarly, the stamped and coined steel gear was changed to a redesigned injection-molded thermoplastic. The new design of the mechanical pump computer wheel assembly was not only 13 cents less expensive to produce, but it proved to have greater reliability and be maintenance free, giving a longer service life.

Methods analysts should remember that items such as valves, relays, air cylinders, transformers, pipe fittings, bearings, couplings, chains, hinges, hardware, and motors can usually be purchased at less cost than they can be manufactured.

Finding a Material that is Easier to Process

Some materials are usually more readily processed than others. Referring to handbook data on the physical properties usually helps analysts discern which material will react most favorably to the processes to which it must be subjected in its conversion from raw material to finished product. For example, machinability varies inversely with hardness, and hardness usually varies directly with strength.

Today the most versatile material is reinforced composites. Resin transfer molding can produce more complex parts advantageously from the standpoint of quality and production rate than most other metal and plastic forming procedures. Thus, by specifying a plastic made of reinforcing carbon fibers and epoxy, the analyst can substitute a composite for a metal part, at both a quality and a cost advantage.

Using Material More Economically

The possibility of using material more economically is a fertile field for analysis. If the ratio of scrap material to that actually going into the product is high, then greater utilization should be examined. For example, if the material put into a plastic compression mold is preweighed, it may be possible to use only the exact amount required to fill the cavity; excessive flash can also be eliminated.

In another example, the production of stampings from sheet metal, if the skeleton seems to contain an undue amount of scrap material, the analyst would consider going to the next higher standard width of material and utilizing a multiple die. If a multiple die is used, the cuts should be carefully arranged to assure maximum utilization of material. Figure 3–6 illustrates how the careful nesting of parts permits the maximum utilization of flat stock.

Many world-class manufacturers are finding it not only desirable, but absolutely necessary, to take weight out of existing designs. For example, the average 1997 automobile must lose about 1200 pounds to meet an 80-mile-per-gallon fuel efficiency rating for the next generation of automobiles. This goal will require functional designers and methods analysts to reengineer many automobile components. For example, we can expect to see the cladding of stainless steel to high-strength aluminum to replace chrome plated steel bumpers, as well as a much greater use of plastics and structural composites to replace ferrous components. Similar weight reduction is taking place on many other well-known products, such as washing machines, video cameras, VCRs, suitcases, TV sets, etc.

Today, powder coating is a proven technology that is replacing many other methods of metal finishing. Coating powders are finely divided particles of organic polymers (acrylic, epoxy, polyester, or blends) that usually contain pigments, fillers and additives. Powder coating is the application of a suitable formulation to a substrate, which are then fused into a continuous film by the application of heat, forming a protective and decorative finish. In view of current environmental regulations affecting traditional metal finishing operations, such as electroplating and wet painting, powder coating offers a safer and cleaner environment. The methodology can also provide a durable, attractive, cost-effective finish for metal surfaces used in many commercial products, such as wire shelving, control boxes, trailer hitches, water meters, handrails, boat racks, office partitions, snow shovels, etc.

FIGURE 3–6
Method of torch cutting heavy gear case side plates (note nesting of the point and the heel for most effective use of plate). (*Source:* General Electric Co.)

Using Salvage Materials

Materials can often be salvaged, rather than sold as scrap. Byproducts from an unworked portion or scrap section can sometimes offer real possibilities for savings. For example, one manufacturer of stainless steel cooling cabinets had 4- to 8-inch-wide sections left as cuttings on the shear. An analysis identified electric light switchplate covers as a possible byproduct. Another manufacturer, after salvaging the steel insert from defective bonded rubber ringer rolls, was able to utilize the hollow, cylindrical rubber rolls as bumpers for protecting moored motorboats and sailboats.

If it is not possible to develop a byproduct, then scrap materials should be separated to obtain top scrap prices. Separate bins should be provided for tool steel, steel, brass, copper, and aluminum. Chip-haulers and floor sweepers should specifically be instructed to keep the scrap segregated. For electric light bulbs, for example, the brass socket would be stored in one area, and after the glass bulb is broken and disposed of, the tungsten filament is removed and stored separately for greatest residual value. Many companies save wooden boxes from incoming shipments, and then saw the boards to standard lengths for use in making smaller boxes for outgoing shipments. This practice is usually economical, and it is now being followed by many large industries, as well as by service maintenance centers.

There are also a few interesting examples from the food industry. A manufacturer of tofu processes the beans, centrifuges out the edible protein material, and leaves behind tons of waste fiber. Rather than paying to haul it away to a landfill, the manufacturer gave it away to local farmers for hog feed, as long as they came and picked it up. Similarly, meatpackers utilize everything from a cow: hides, bones, even blood, all except the "moo."

Using Supplies and Tools Fully

Management should encourage full use of all shop supplies. One manufacturer of dairy equipment introduced the policy that no new welding rod was to be distributed to workers without the return of old tips under 2 inches long. The cost of welding rods was reduced immediately by more than 15 percent. Brazing or welding is usually the most economical way to repair expensive cutting tools, such as broaches, special form tools, and milling cutters. If it has been company practice to discard broken tools of this nature, the analyst should investigate the potential savings of a tool salvage program.

Analysts can also find a use for the unworn portions of grinding wheels, emery disks, and so forth. Also, items such as gloves and rags should not be discarded simply because they are soiled. Storing dirty items and then laundering them is less expensive than replacing them. Methods analysts can make a real contribution to a company by simply minimizing waste, which today claims about one-fifth of our material.

Standardizing Materials

Methods analysts should always be alert to the possibility of standardizing materials. They must minimize the sizes, shapes, grades, and so on of each material utilized in the production and assembly processes. The typical economies resulting from reductions in the sizes and grades of the materials employed include the following:

- Purchase orders are used for larger amounts, which are almost always less expensive per unit.
- Inventories are smaller, since less material must be maintained as a reserve.
- Fewer entries need to be made in storage records.
- Fewer invoices need to be paid.
- Fewer spaces are needed to house materials in the storeroom.
- Sampling inspection reduces the total number of parts inspected.
- Fewer price quotations and purchase orders are needed.

The standardization of materials, like other methods improvement techniques, is a continuing process. It requires the continual cooperation of the design, production planning, and purchasing departments.

Finding the Best Vendor

For the vast majority of materials, supplies, and parts, numerous suppliers will quote different prices, quality levels, delivery times, and the willingness to hold inventories. It is usually the responsibility of the purchasing department to locate the most favorable supplier. However, the best supplier last year may not be the best one now. The methods analyst should encourage the purchasing department to rebid the highest-cost materials, supplies, and parts to obtain better prices and superior quality and to increase vendor stocking, where the vendors agree to hold inventories for their customers. It is not unusual for methods analysts to achieve a 10 percent reduction in the cost of materials and a 15 percent reduction in inventories by regularly pursuing this approach through their purchasing departments.

Perhaps the most important reason for continued Japanese success in the manufacturing sector is the *keiretsu*. This is a form of business and manufacturing organization that links businesses together. It can be thought of as a web of interlocking relationships among manufacturers—often between a large manufacturer and its principal suppliers. Thus, in Japan such companies as Hitachi and Toyota and other international competitors are able to acquire parts for their products from regular suppliers who produce to the quality called for and are continually looking for improvement so as to provide better prices for the firms in their network. Alert purchasing departments are often able to create relationships with suppliers comparable to the so-called production keiretsu.

5. MANUFACTURE SEQUENCE AND PROCESS

As manufacturing technology in the 21st century eliminates labor-intensive manufacturing in favor of capital intensive procedures, the methods engineer will focus on multiaxis and multifunctioning machining and assembly. Modern equipment is capable of cutting at higher speeds on more accurate, rigid and flexible machines that utilize both advanced controls and tool materials. Programming functions permit in-process and post-process gaging for tool sensing and compensation, resulting in dependable quality control.

The methods engineer must understand that the time utilized by the manufacturing process is divided into three steps: inventory control and planning, setup operations, and in-process manufacturing. Furthermore, it is not unusual to find that these procedures, in aggregate, are only about 30 percent efficient from the standpoint of process improvement.

To improve the manufacturing process, the analyst should consider the following: (1) rearranging the operations; (2) mechanizing manual operations; (3) utilizing more efficient facilities on mechanical operations; (4) operating mechanical facilities more efficiently; (5) manufacturing near the net shape; and (6) using robots.

Rearranging Operations

Rearranging operations often results in savings. As an example, the flange of a motor conduit box required four drilled holes, one in each corner. Also, the base had to be smooth and flat. Originally, the operator began by grinding the base, then drilling the four holes using a drill jig. The drilling operation threw up burrs, which then had to be removed in another step. By rearranging the operation so that the holes were drilled first and the base then ground, analysts eliminated the deburring operation. The basegrinding operation automatically removed the burrs.

Combining operations usually reduces costs. For example, a manufacturer fabricated the fan motor support and the outlet box of its electric fans. After painting the parts separately, operators then riveted them together. By having the outlet box riveted to the fan motor support prior to painting, analysts effected an appreciable time savings for the painting operation.

In another example, the market for aluminum cylinder head castings is growing, and foundries are finding it cost effective to go from the steel-mold casting process to the lost foam process. Lost foam is an investment casting procedure that uses an expendable pattern of polystyrene foam surrounded by a thin ceramic shell. Steel-mold castings require considerable subsequent machining. In comparison, the lost foam process reduces the amount of machining and also eliminates the sand disposal costs usually associated with investment casting.

Before changing any operation, however, the analyst must consider possible detrimental effects on subsequent operations down the line. Reducing the cost of one operation could result in higher costs for other operations. For example, a change recommended in the manufacture of AC field coils resulted in higher costs and was therefore not practical. The field coils were made of heavy copper bands, which were formed and then insulated with mica tape. The mica tape was hand wrapped on the already coiled parts. The company decided to machine wrap the copper bands prior to coiling. This did not prove practical, as the forming of the coils cracked the mica tape, necessitating time-consuming repairs prior to product acceptance (see Figure 3–7).

Mechanizing Manual Operations

Today, any practicing methods analyst should consider using special-purpose and automatic equipment and tooling, especially if production quantities are large. Notable among industry's latest offerings are program controlled, numerically controlled (NC), and computer controlled (CNC) machining and other equipment. These afford substantial savings in labor cost, as well as the following advantages: reduced work-in-process inventory, less parts damage due to handling, less scrap, reduced floor space, and reduced production throughput time. For example, one company experienced a 40 percent direct labor savings in the machining of a precision stainless steel component (see Figure 3–8). Prior to the acquisition of the CNC equipment, three machines with a total of eight cutting tools were used to produce

FIGURE 3–7
Machine-wrapped copper bands prove unsatisfactory in manufacture of AC field coils.
A. Method of forming heavy copper bands.

B. Hand-wrapped field coils.

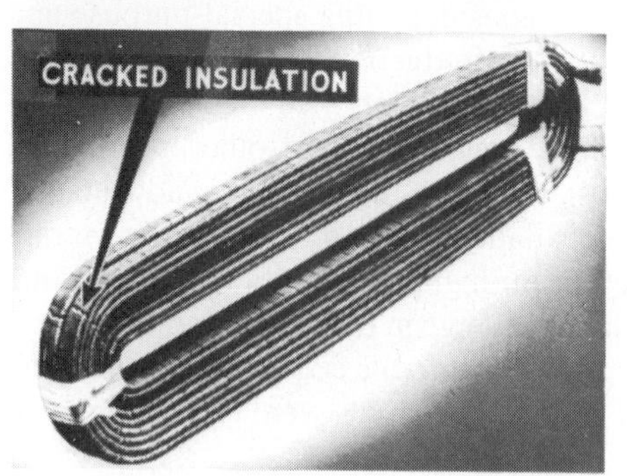

C. Cracked insulation in coils machine wrapped prior to forming.

FIGURE 3–8
Machining of a pulse dampner top plate. (*Courtesy of:* Scientific Systems Incorporated.)
Figure 3–A illustrates the complete machining of a pulse dampner top plate. This precision stainless steel part is being machined on a CNC Bridgeport machining center using a CAM program. Seven tools are utilized from the 24-position carousel. A 40 percent savings resulted from changing the machining of this part from three separate operations on three different machines to this CNC equipment. (See Figures, 3–8B, 3–8C, and 3–8D).

A.

B.

C.

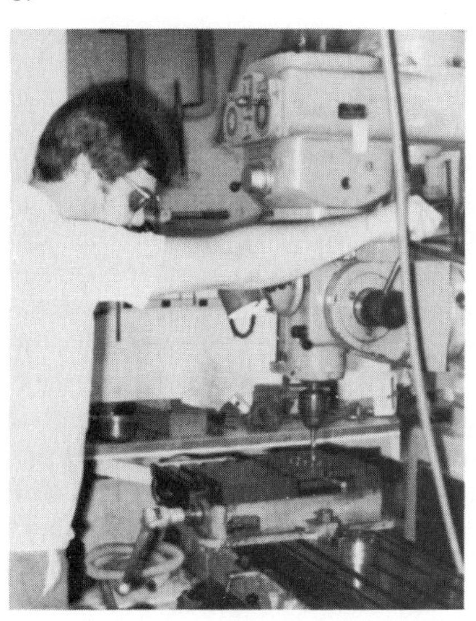

D.

the part. The change in method realized a 60 percent savings in floor space, a reduction in rejects, and a lower in-process inventory.

Other automatic equipment includes: automatic screw machines; multiple-spindle drilling, boring, and tapping machines; index-table machine tools; automatic casting equipment combining automatic sand-mold making, pouring, shakeout, and grinding; and automatic painting and plating finishing equipment. The use of power assembly tools, such as power nut- and screwdrivers, electric or air hammers, and mechanical feeders, is often more economical than the use of hand tools.

To illustrate, a company that produces specialty windows was using manual methods to press rails over both ends of plate window glass that had been covered with a synthetic rubber wrap. The plates of glass were held in position by two pads that were pneumatically squeezed together. The operator would pick up a rail and position it over the end of the window glass and then pick up a mallet and hammer the rail into position over the glass. The operation was slow and it resulted in considerable operator work-related musculoskeletal disorders. Furthermore, scrap was high because of glass breakage due to pounding the rails over the glass. A new facility was designed that pneumatically squeezed rails onto the window glass over the synthetic rubber wrap. Operators enthusiastically accepted the new facility because the work was much easier to perform; health problems disappeared, productivity increased, and glass breakage dropped to near zero.

The application of mechanization applies not only to process operations, but also to paperwork. For example, bar coding applications can be invaluable to the operations analyst. Bar coding can rapidly and accurately enter a variety of data. Computers can then manipulate the data for some desired objective, such as counting and controlling inventory, routing specific items to or through a process, or identifying the state of completion and the operator currently working on each item in a work-in-process.

Utilizing More Efficient Mechanical Facilities

If an operation is done mechanically, there is always the possibility of a more efficient means of mechanization. At one company, for example, turbine blade roots were machined by using three separate milling operations. Both the cycle time and the costs were high. When external broaching was introduced, all three surfaces could be finished at once, for considerable time and cost savings. Another company overlooked the possibility of utilizing a press operation. This process is one of the fastest for forming and sizing processes. A stamped bracket had four holes that were drilled after the bracket was formed. By using a die designed to pierce the holes, the work could be performed in a fraction of the drilling time.

Work mechanization applies to more than just manual work. For example, one company in the food industry was checking the weight of various product lines with a balance. This equipment required the operator to note the weight visually, record the weight on a form, and subsequently perform several calculations. A methods engineering study resulted in the introduction of a statistical weight control system. Under the improved method, the operator weighs the product on a digital scale programmed to accept the product within a certain weight range. As the product is weighed, the weight information is transferred to a personal computer that compiles the information and prints the desired report.

Operating Mechanical Facilities More Efficiently

A good slogan for methods analysts is, "Design for two at a time." Usually multiple-die operation in presswork is more economical than single-stage operation. Again, multiple cavities in diecasting, molding, and similar processes are viable options when there is sufficient volume. On machine operations, analysts should be sure that proper feeds and speeds are used. They should investigate the grinding of cutting tools for maximum performance. They should check to see whether the cutting tools are properly mounted, whether the right lubricant is being used, and whether the machine tool is in good condition and is adequately maintained. Many machine tools are operated at a fraction of their possible output. Endeavoring to operate mechanical facilities more efficiently nearly always pays dividends.

Manufacturing Near the Net Shape

Using a manufacturing process that produces components closer to the final shape can maximize material use, reduce scrap, minimize secondary processing such as final machining and finishing, and permit manufacturing with more environmentally friendly materials. For example, forming parts with powder metals (PM) instead of conventional casting or forging often provides the manufacture of near-net shapes for many components, resulting in dramatic economic savings, as well as functional advantages. In the case of forged PM connecting rods, it has been reported that they have reduced the reciprocating mass of competing alternatives, resulting in less noise and vibration, as well as major cost economies.

Considering the Use of Robots

For cost and productivity reasons, it is advantageous today to consider the use of robots in many manufacturing areas (see Figure 3–9). For example, assembly areas include work that typically has a high direct labor cost, in some cases accounting for as much as half of the manufacturing cost of a product. The principal advantage of integrating a modern robot in the assembly process is its inherent flexibility. It can assemble multiple products on a single system and can be reprogrammed to handle various tasks with part variations. In addition, robotic assembly can provide consistently repeatable quality with predictable product output.

A robot's typical life is approximately 10 years. If well maintained and if used for moving small payloads, the life can be extended to up to 15 years. Consequently, a robot's depreciation cost can be relatively low. Also, if a given robot's size and configuration are appropriate, it can be used in a variety of operations. For example,

FIGURE 3–9
Illustration of a few common industrial robot applications.
One welding robot is shown (a), but typically a number of robots would be used along an automotive assembly line. In a diecasting application (b), a robot unloads diecasting machines, performs quench operations, and loads material into a press. The production machining line is used for producing cam housings (c). The assembly line (d) uses a combination of robots, parts feeders, and human operators.

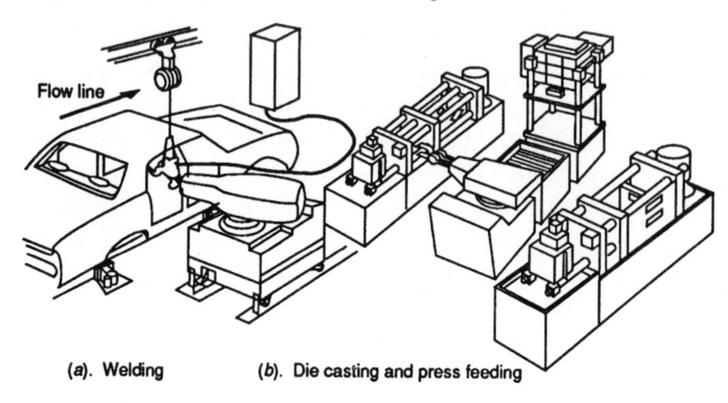

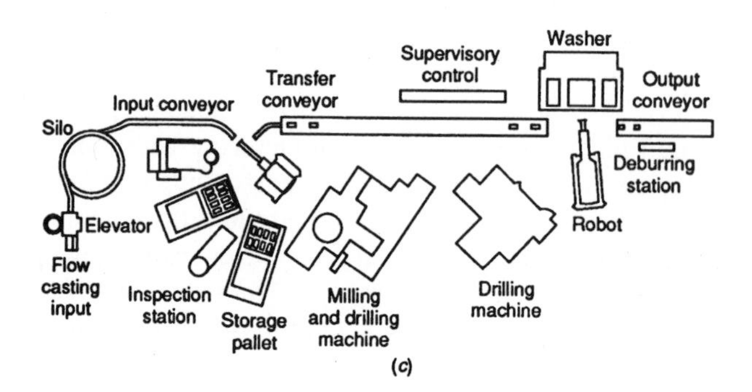

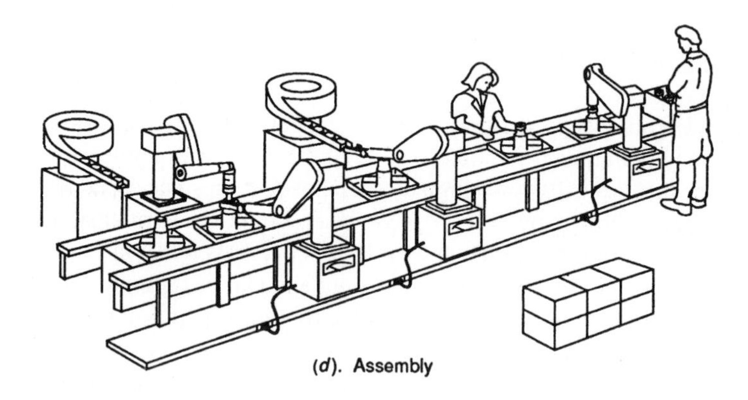

a robot could be used to: load a die-casting facility, load a quenching tank, load and unload a board drop-hammer forging operation, load a plate glass washing operation, and so on. In theory, a robot of the correct size and configuration can be programmed to do any job.

In addition to productivity advantages, robots also offer safety advantages. They can be used in work centers where there is danger to the worker because of the nature of the process (see Chapter 6). For example, in the die-casting process, there can be considerable danger due to hot metal splashing when the molten metal is injected into the die cavity. One of the original applications for robots was die casting. In one company, a five-axis robot developed by Unimation, Inc., serves a 600-ton microprocessor-controlled die-casting machine. In the operation, the robot moves into position when the die opens, grasps the casting by its slug, and clears it from the cavity. At the same time, it initiates automatic die-lubrication sprays. The robot displays the casting to infrared scanners, then signals the die-casting machine to accept another shot. The casting is deposited by the robot on an output station for trimming. Here an operator, remote from the die-casting machine, safely trims the casting preparatory to subsequent operations.

Japanese motor vehicle manufacturers have placed particular emphasis on the use of robots in welding. For example, at Nissan Motors, 95 percent of the welds on vehicles are made by robots; and Mitsubishi Motors reported that about 70 percent of its welding is performed by robots. In these companies, robot downtime averages less than 1 percent.

6. SETUP AND TOOLS

One of the most important elements of all forms of work holders, tools, and setups is economics. The amount of tooling up that proves most advantageous depends on: (1) the production quantity, (2) repeat business, (3) labor, (4) delivery requirements, and (5) required capital.

The most prevalent mistake of planners and toolmakers is to tie up money in fixtures that may show a large savings when in use, but are seldom used. For example, a savings of 10 percent in direct labor cost on a job in constant use would probably justify greater expense in tools than an 80 or 90 percent savings on a small job that appears on the production schedule only a few times a year. (This is an example of Pareto analysis, from Chapter 2.) The economic advantage of lower labor costs is the controlling factor in determining the tooling; consequently, jigs and fixtures may be desirable, even when only small quantities are involved. Other considerations, such as improved interchangeability, increased accuracy, or labor trouble reduction, may provide the dominant reasons for elaborate tooling, although this is usually not the case. An example of the trade-off between fixturing and tooling costs is discussed in Chapter 7 in the section on break-even charts.

Once the needed amount of tooling has been determined (or if tooling already exists, once the ideal amount needed has been determined), specific considerations for producing the most favorable designs should be evaluated. These are outlined in the Setup and Tooling Evaluation Checklist shown in Figure 3–10.

FIGURE 3–10
Setup and tooling evaluation checklist.

Fixtures	**Yes**	**No**
1. Can the same fixture be used for other products?	❑	❑
2. Can the fixture be made similar to some other that has been used advantageously? If so, how can it be improved?	❑	❑
3. Can any stock hardware be used for making the fixture?	❑	❑
4. Can the output be increased by placing more than one part in the fixture?	❑	❑
5. Can the chips be readily removed from the fixture?	❑	❑
6. Are the clamps on the fixture strong enough to prevent them from buckling when they are tightened down on the work?	❑	❑
7. Can regular wrenches be used with the fixture?	❑	❑
8. Can special milling cutters, arbors, or collars be avoided?	❑	❑
9. Are indexing arrangements accurate for rotary type fixtures?	❑	❑
10. Can the fixture be used on a standard rotary indexing head?	❑	❑
11. Can the fixture be made to handle more than one operation?	❑	❑
12. Is the work as close to the miller's table as possible?	❑	❑
13. Can the work be gaged in the fixture? Can a snap gage be used?	❑	❑
14. Can jack pins be used to help support the work while it is being milled?	❑	❑
15. Have springs been placed under all clamps?	❑	❑
16. Are all steel contact points, clamps, etc., hardened?	❑	❑
17. Are the simplest class of jigs being utilized?	❑	❑
18. Is a double or triple thread used on the screw that holds the work in the jig, so that it will take fewer turns to get the screw out of the way, to remove the part more quickly?	❑	❑
19. Can the toolmaker make the jig?	❑	❑
20. Are the legs on the jig long enough to allow the drill, the reamer, or the pilot of the reamer to pass through the part a reasonable distance without striking the table of the drill press?	❑	❑
21. Is the jig light enough, to handle easily?	❑	❑
22. Is the jig identified with both a location number for storing and a part number that identifies the part or parts that the jig helps produce?	❑	❑
23. Is the work adequately supported so that the clamping force will not bend or distort it?	❑	❑
Parts	**Yes**	**No**
1. Has the part undergone any previous operations? If so, can any of these surfaces be used to locate or master from?	❑	❑
2. Can the part be quickly placed in the fixture?	❑	❑
3. Can the part be quickly removed from the fixture?	❑	❑
4. Is the part held firmly so that it cannot work loose, spring, or chatter while the cut is being made? (The cut should be against the solid part of the fixture and not against the clamp.)		
5. Can the part be milled in a standard vise, thus doing away with an expensive fixture?	❑	❑
6. If the part is to be milled at an angle, can the fixture be simplified by using a standard adjustable milling angle?	❑	❑
7. Can lugs be cast on the part to be machined, for better grip?	❑	❑
8. Have notes been made on the drawing, or all loose parts stamped, indicating the jig they were made for, so that lost or misplaced parts can be returned to the jig when found?	❑	❑
9. Are all necessary corners rounded?	❑	❑
Drills	**Yes**	**No**
1. Is the thrust of the drill taken up by the fixture?	❑	❑
2. Can jack pins or screws be used to support the work while it is being drilled?	❑	❑
3. Are the drill bushings short enough so that extension drills are not necessary?	❑	❑
4. Are all clamps located in such a way as to resist or help resist the pressure of the drill?	❑	❑
5. Has the drill press the necessary speeds for drilling and reaming all holes?	❑	❑
6. Can drill-press tapping attachments be avoided?	❑	❑
7. Drilling and reaming several small holes and only one large one in the jig is not practical since quicker results can be obtained by drilling the small holes on a small drill press, while having only one large one would require the jig to be used on a large machine:		
a. Is it cheaper to drill the large hole in another jig?	❑	❑
b. Will the result of doing so be accurate enough?	❑	❑
Others	**Yes**	**No**
1. Can a gage be designed, or hardened pins added, to help the operator set the milling cutters or check up on the work?	❑	❑
2. Is there plenty of clearance for the arbor collars to pass over the work without striking?	❑	❑

Setup ties in very closely with tooling, because tooling invariably determines the setup and teardown time. When we speak of setup time, we usually include such items as arriving on the job; procuring instructions, drawings, tools, and material; preparing workstations so that production can begin in the prescribed manner (setting up tools; adjusting stops; setting feeds, speeds, and cut depth; and so on); tearing down the setup; and returning tools to the crib.

Setup operations are especially important in the job shop where production runs tend to be small. Even if this type of shop has modern facilities and puts forth a high effort, it may still have difficulty meeting the competition if setups are too long because of poor planning and inefficient tooling. When the ratio of setup time to production-run time is high, a methods analyst can usually develop several possibilities for setup and tool improvement. One notable option is a group technology system.

The essence of group technology is the classification of the various components of a company's products, so that parts similar in shape and processing sequence are identified numerically. Parts belonging to the same family group, such as rings, sleeves, discs, and collars, are scheduled for production over the same time interval on a general-purpose line arranged in the optimal operational sequence. Since both the size and shape of the parts in a given family vary considerably, the line is usually equipped with universal-type, quick-acting jigs and fixtures.

The resulting line can mean greater output, less setup time, greater machine utilization, less material handling, shorter cycle time, and better cost improvement. The design and development of universal-type jigs and fixtures means that less equipment is required, and hidden costs such as tool storage and obsolescence, are reduced.

As an example, Figure 3–11 illustrates a system grouping subdivided into nine classes of parts. Note the similarity of parts within each vertical column. If we were machining a shaft with external threads and a partial bore at one end, the part would be identified as Class 206.

To develop better methods, analysts should investigate the setup and tools to: (1) reduce setup time by better planning, methods, and production control; (2) utilize the full capacity of the machine; and (3) introduce more efficient tooling.

FIGURE 3–11
Subdivision of a system grouping for group technology.

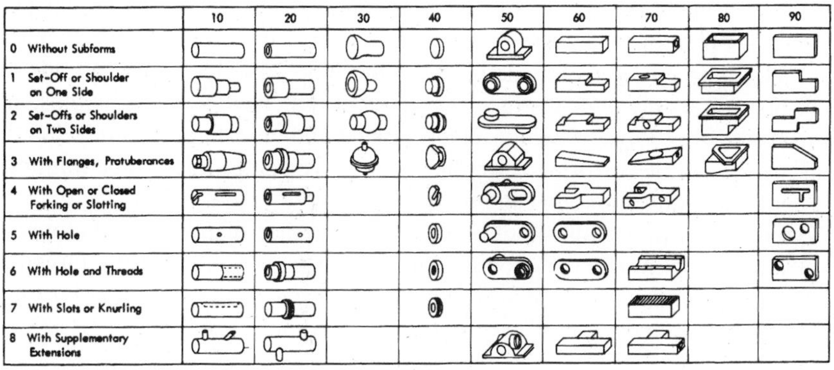

	10	20	30	40	50	60	70	80	90
0 Without Subforms									
1 Set-Off or Shoulder on One Side									
2 Set-Offs or Shoulders on Two Sides									
3 With Flanges, Protuberances									
4 With Open or Closed Forking or Slotting									
5 With Hole									
6 With Hole and Threads									
7 With Slots or Knurling									
8 With Supplementary Extensions									

Reduce Setup Time

Just-in-time (JIT) techniques, which have become popular in recent years, emphasize decreasing the setup times to the minimum by simplifying or eliminating them. The SMED (single minute exchange of die) System of the Toyota Production System (Shingo, 1981) is a good example of this approach. A significant portion of setup time can often be eliminated by assuring that: raw materials are within specifications, tools are sharp, and fixtures are available and in good condition. Producing in smaller lots can often prove cost effective. Smaller lot sizes can lead to smaller inventories, with reduced carrying costs and shelf-life problems, such as contamination, corrosion, deterioration, obsolescence, theft, etc. The analyst must understand that decreasing the lot size will result in an increase in total setup costs for the same total production quantity over a given period. Several points that should be considered in reducing setup time are as follows:

1. Work that can be done while the equipment is running should be done at that time. For example, presetting tools for numerical control (NC) equipment can be done while the machine is running.
2. Use the most efficient clamping. Usually, quick-acting clamps that employ cam action, levers, wedges, and so on, are much faster, provide adequate force, and are usually a good alternative to threaded fasteners. When threaded fasteners must be used (for clamping force), "C" washers or slotted holes can be used so that nuts and bolts do not have to be removed from the machine and can be reused, reducing the setup time on the next job.
3. Eliminate machine base adjustment. Redesigning part fixtures and using preset tooling may eliminate the need for spacers or guide-block adjustments to the table position.
4. Use templates or block gages to make quick adjustments to machine stops.

The time spent in requisitioning tools and materials, preparing the workstation for actual production, cleaning up the workstation, and returning the tools to the tool crib is usually included in setup time. This time is often difficult to control, and the work usually performed least efficiently. Effective production control can often reduce this time. Making the dispatch section responsible for seeing that the tools, gages, instructions, and materials are provided at the correct time, and that the tools are returned to their respective cribs after the job has been completed, eliminates the need for the operator to leave the work area. The operator then only has to perform the actual setting up and tearing down of the machine. The clerical and routine function of providing drawings, instructions, and tools can be performed by those more familiar with this type of work. Thus, large numbers of requisitions for these requirements can be performed simultaneously, and setup time can be minimized. Here again, group technology can be advantageous.

Duplicate cutting tools should be available, rather than having the operators sharpen their tools. When the operators get new tools, the dull ones are turned in to the tool crib attendant and replaced with sharp ones. Tool sharpening becomes a separate function, and the tools can be standardized more readily.

To minimize downtime, each operator should have a constant backlog of work. The operators should always know what the next work assignment is. A technique frequently used to keep the workload apparent to the operator, supervisor, and superintendent is a board over each production facility, with three wire clips or pockets to receive work orders. The first clip contains all work orders scheduled ahead; the second clip holds the orders currently being worked on; and the last clip holds the completed orders. When issuing work orders, the dispatcher places them in the work-ahead station. At the same time, the dispatcher picks up all completed job tickets from the work-completed station and delivers them to the scheduling department for recording. This system assures the operators of continuous loads and makes it unnecessary for them to go to the supervisor for their next work assignments.

Making a record of difficult, recurring setups can save considerable setup time when repeat business is received. Perhaps the simplest and yet most effective way to compile a record of a setup is to take a photograph of the setup once it is complete. The photograph should either be stapled to and filed with the production operation card, or placed in a plastic envelope and attached to the tooling prior to storage in the tool crib.

Utilize the Full Capacity of the Machine

A careful review of many jobs often reveals possibilities for utilizing a greater share of the machine's capacity. For example, a milling setup for a toggle lever was changed so that the six faces were milled simultaneously by five cutters. The old setup required that the job be done in three steps, which meant that the part had to be placed in a separate fixture three different times. The new setup reduced the total machining time and increased the accuracy of the relationship between the six machined faces.

Analysts should also consider positioning one part while another is being machined. This opportunity exists on many milling machine jobs where it is possible to conventional mill on one stroke of the table and climb mill on the return stroke. While the operator is loading a fixture at one end of the machine table, a similar fixture is holding a piece being machined by power feed. As the table of the machine returns, the operator removes the first piece from the machine and reloads the fixture. While this internal work is taking place, the machine is cutting the piece in the second fixture.

In view of the ever-increasing cost of energy, it is important to utilize the most economical equipment to do the job. Several years ago, the cost of energy was such an insignificant proportion of total cost that little attention was given to utilizing the full capacity of machines. There are literally thousands of operations where only a fraction of machine capacity is utilized, with a resulting waste of electric power. In the metal trades industry today, the cost of power is over 2.5 percent of total cost, with strong indications that the present cost of power will increase by at least 50 percent in the next decade. It is highly probable that careful planning to utilize a larger proportion of the capacity of a machine to do the work can effect a 50 percent

EXAMPLE 3–1
Savings Through the Use of Energy-Efficient Motor

An analyst considers replacing a 25 HP motor that is 10 years old with a new energy-efficient motor. The motor will operate an estimated 6,000 hours per year at 91 percent of full load. The estimated annual savings for this motor, based on a 3 percent improvement in efficiency and a cost of power of \$0.05/kwh, is:

$$\text{Power cost/yr (standard motor)} = \text{HP} \times 0.746 \times \text{F} \times \text{H} \times \text{C} \times 1/\text{E}$$

where: HP = Horsepower of motor
0.746 = conversion factor (horsepower to watts)
F = Percent of full load
H = Annual hours of operation
E = Efficiency
C = Cost of power

$$\text{Power cost/yr (standard motor)} = 25 \times 0.746 \times 0.91 \times 6000 \times 0.05/0.88 = \$5{,}786$$

$$\text{Power cost/yr (energy-efficient motor)} = 25 \times 0.746 \times 0.91 \times 6000 \times 0.05/0.91 = \$5{,}595$$

This indicates a \$191 power savings per year. The analyst needs to compare this annual savings with the extra cost of the energy-efficient motor. If the extra cost can be saved within a three-year period, the analyst should proceed.

savings in power usage in many of our plants. Typically, for most motors, if the percent of the rated full load is increased from 25 percent to 50 percent, as much as an 11 percent increase in efficiency could be realized.

Similarly energy-efficient AC motors provide between 2 and 4 percent more efficient operation than standard motors. Also, energy-efficient motors run cooler than standard motors, thus giving longer service life. Energy-efficient AC motors work well on continuous installations, such as compressors, pumps, fans, and blowers. Standard motors with a lower initial cost are usually more cost effective for intermittent use situations.

Introduce More Efficient Tooling

Just as new processing techniques are continually being developed, new and more efficient tooling should be considered. Coated cutting tools have dramatically improved the critical wear-resistance/breakage-resistance combination. For example, TiC-coated tools have provided a 50 to 100 percent increase in speed over uncoated carbide where each have the same breakage resistance. Advantages include: harder surfaces, thus reducing abrasive wear; excellent adhesion to the substrates; low coefficient of friction with most work piece materials; chemical inertness; and resistance to elevated temperatures.

Carbide tools are usually more cost effective than high-speed steel tools on many jobs. For example, one company realized a 60 percent savings by changing the milling operation of a magnesium casting. Originally, the base was milled complete in two operations, using high-speed steel milling cutters. An analysis resulted in the employment of three carbide-tipped fly cutters mounted in a special holder to mill parts complete. Faster feeds and speeds were possible, and surface finish was not impaired.

Savings can often be achieved by altering tool geometries. Each setup has different requirements that can be achieved only by designing an engineered system that optimizes the feed range for chip control, cutting forces, and edge strength. For example, single-sided low-force geometries may be designed to provide both good chip control and force reduction. In this case, high positive rake angles are grouped to reduce the chip thickness ratio, providing a low cutting force and cutting temperature.

While introducing more efficient tooling, the analyst should develop better methods for holding the work. The work must be held so that it can be positioned and removed quickly. Figure 3–12 illustrates a hinge support assembly made up of three components held together with three rivets (see A). In the original method, operators loaded these components into a fixture by hand and then activated the riveter by palm pushbuttons (see B). The new method (see C) introduced a dial indexing table with automatic stations to load the roller, assemble three rivets, and eject the completed assembly. The loading of the metal parts was still a manual operation, but there was a 280 percent increase in productivity. Another benefit of the new method was an increase in quality. The machine was designed to check automatically for the presence of all parts in the assembly, thus assuring that the completed assembly was correct.

Finally, management should provide proper hand tools. Many hand tools, such as screwdrivers, are designed for a specific task. A screwdriver that is efficient under one set of conditions may be very inefficient under another set. Analysts should specify that the most effective hand tools be used. More details on the selection of hand tools will be provided in Chapter 5.

7. MATERIAL HANDLING

Material handling includes motion, time, place, quantity, and space. First, material handling must ensure that parts, raw materials, in-process materials, finished products, and supplies are moved periodically from location to location. Second, since each operation requires materials and supplies at a particular time, material handling assures that no production process or customer is hampered by either the early or late arrival of materials. Third, material handling must ensure that materials are delivered to the correct place. Fourth, material handling must ensure that materials are delivered at each location without damage and in the proper quantity. Finally, material handling must consider storage space, both temporary and dormant.

FIGURE 3–12
Use of indexed tooling to increase productivity. (*Source:* General Electric Co., Louisville, Kentucky.)

A. Parts assembled with rivets.

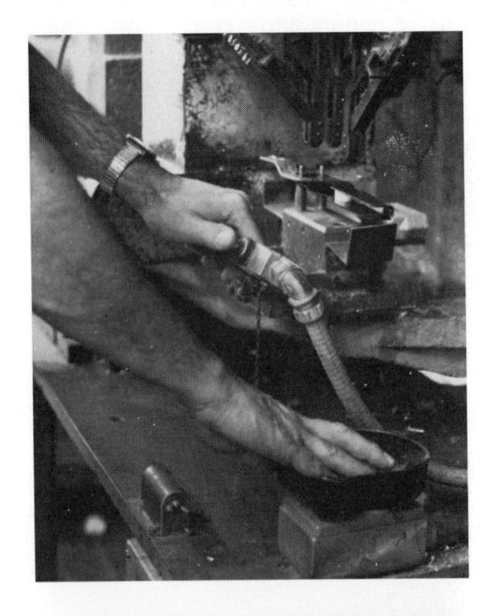

B. Components manually loaded into fixture and riveter activated with pushbuttons.

C. Improved methods of holding work and assembly—280 percent improvement in productivity.

A study conducted by the Material Handling Institute revealed that between 30 and 85 percent of the cost of bringing a product to market is associated with material handling. Axiomatically, the best handled part is the least manually handled part. Whether the distances of the moves are large or small, these moves should be scrutinized. The following six points should be considered for reducing the time spent in handling material: (1) reduce the time spent in picking up material; (2) use mechanized or automated equipment; (3) make better use of existing handling facilities; (4) handle material with greater care; and (5) consider the application of bar coding for inventory and related applications.

A good example of the application of these six points is the evolution of warehousing; the former storage center has become an automated distribution center. Today, the automated warehouse uses computer control for material movement, as well as information flow through data processing. In this type of automated warehouse, receiving, transporting, storing, retrieving, and controlling inventory are treated as an integrated function.

Reduce the Time Spent in Picking Up Material

Material handling is often thought of as only transportation, neglecting consideration of positioning at the workstation, which is equally important. Since it is often overlooked, workstation positioning of material may offer even greater opportunities for savings than does transportation. Reducing the time spent in picking up material minimizes tiring, costly manual handling at the machine or the workplace. It gives the operator a chance to do the job faster with less fatigue and greater safety.

For example, consider eliminating loose piling on the floor. Perhaps the material can be stacked directly on pallets or skids after being processed at the workstation. This can result in a substantial reduction of terminal transportation time (the time that material handling equipment stands idle while loading and unloading take place). Usually some type of conveyor or mechanical fingers can bring material to the workstation, thus reducing or eliminating the time needed to pick up the material. Plants can also install gravity conveyors, in conjunction with the automatic removal of finished parts, thus minimizing material handling at the workstation. Figure 3–13 shows examples of typical handling equipment.

Interfaces between different types of handling and storage equipment should be studied to develop more efficient arrangements. For example, the sketch in Figure 3–14 shows the order picking arrangements, depicting how materials can be removed from the reserve or staging storage either by a man-aboard order picking vehicle (left), or manually (right). A lift truck can be used to replenish pallet racks. After the required items are removed from the flow rack, they are sent by conveyor to order accumulation and packaging operations.

Use Mechanical Equipment

Mechanizing the handling of material usually reduces labor costs, reduces materials damage, improves safety, alleviates fatigue, and increases production. However,

FIGURE 3–13
Typical handling equipment used in industry today. (*Source:* The Material Handling Institute.)

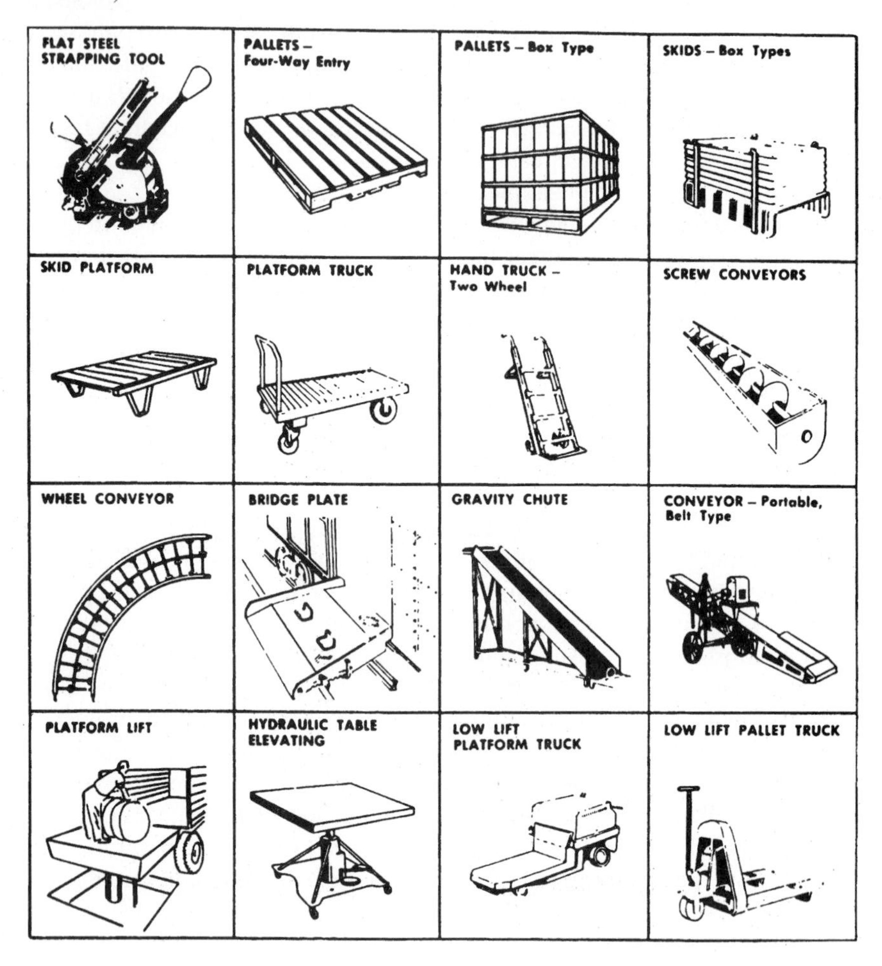

care must be exercised in selecting the proper equipment and methods. Equipment standardization is important because it simplifies operator training, allows equipment interchangeability, and requires fewer repair parts.

The savings possible through the mechanization of material handling equipment are typified by the following examples. At the outset of the IBM 360 program, to build a panel, the operator would go to the storage crib, select the correct cards required for a specific panel based on its "plug" list, return to the workbench, and then proceed to insert the circuit cards into the panel in accordance with the plug list. The improved method utilizes two automated, vertical storage machines, each with 10 carriers and four pullout drawers per carrier (see Figure 3–15). The carriers move up and around in a system that is a compressed version of a Ferris wheel. With 20 possible stop positions on call, the unit always selects the closest route—either for-

FIGURE 3–13 *(concluded)*
Typical handling equipment used in industry today. (*Source:* The Material Handling Institute.)

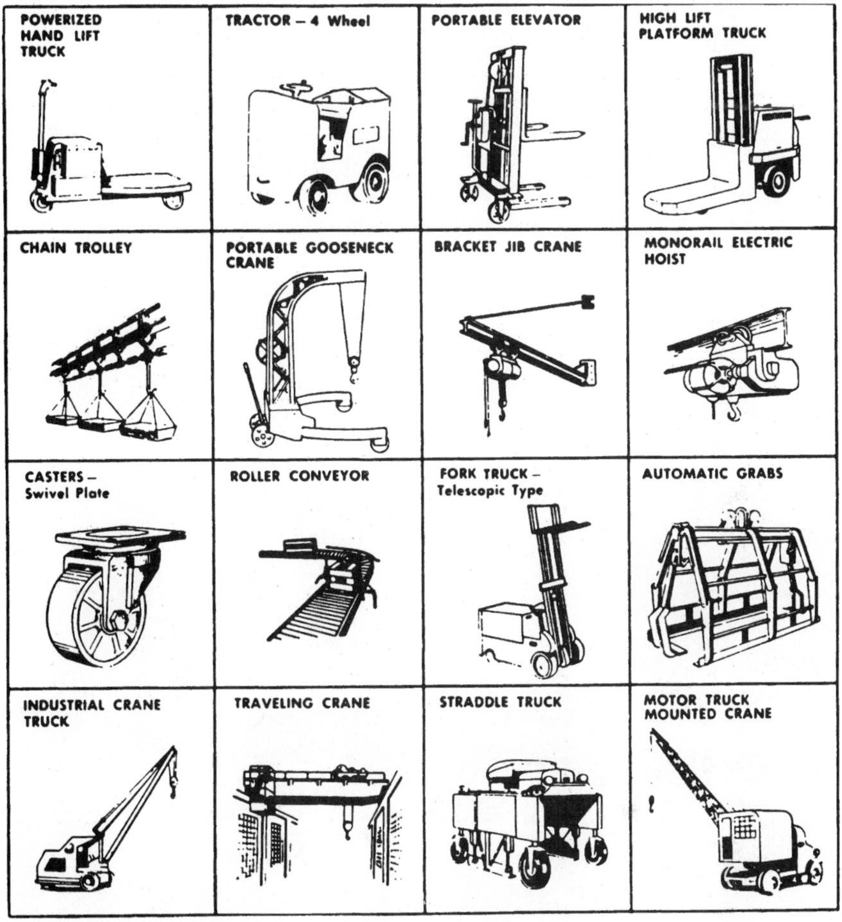

ward or backward—to bring the proper drawers to the opening in the shortest possible time. From a seated position, the operator dials the correct stop, pulls open the drawer to expose the needed cards, withdraws the proper card, and places it in the panel. The improved method has reduced the required storage area by approximately 50 percent, improved workstation layout, and substantially reduced populating errors by minimizing operator handling, decisionmaking, and fatigue.

Often, an Automated Guided Vehicle (AGV) can replace a driver. AGVs are successfully used in a variety of applications, such as mail delivery. Typically, these vehicles are not programmed; rather, they follow a magnetic or optical guide for a planned route. Stops are made at specific locations for a predetermined period, giving an employee adequate time for unloading and loading. By pressing a "hold" button and then pressing a "start" button at the conclusion of the loading/unloading

FIGURE 3–14
Schematic of efficient warehousing operations.

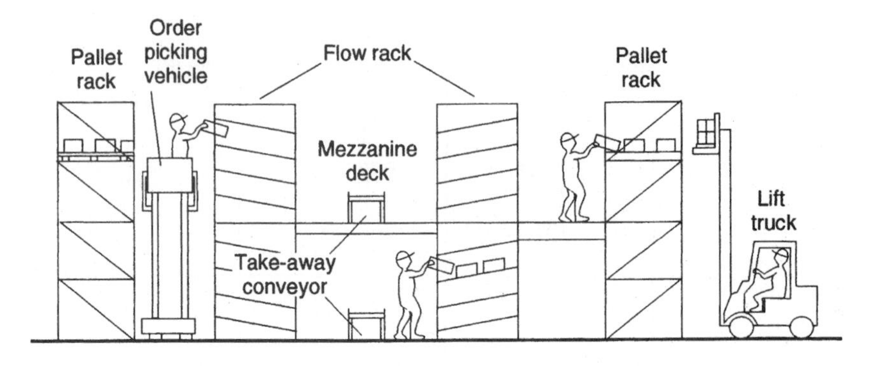

FIGURE 3–15
Work area of vertical storage machine used in the assembly of computer panels.

FIGURE 3–16
Hydraulic lift table used to minimize manual lifting. (*Courtesy:* Lee Engineering Co., Pawtucket, RI.)

operation, the operator can lengthen the dwell period at each stop. AGVs can be programmed to go to any location over more than one path. They are equipped with sensing and control instrumentation to avoid collisions with other vehicles. Also, when such guide path equipment is used, material handling costs vary little with distance.

Mechanization is also useful for manual materials handling, such as palletizing. There are a variety of devices under the generic label of lift tables, which eliminate most of the lifting required of the operator. Some lift tables are spring loaded, which, when set with a proper spring stiffness, will adjust automatically to the optimal height for the operator as boxes are placed on a pallet on top of the lift. (See Chapter 4 for a discussion on the determination of optimal lifting heights.) Others are pneumatic (see Figure 3–16) and can be easily adjusted with a control, so that lifting is eliminated and material can be slid from one surface to another. Some tilt for easier access into bins, while others rotate, facilitating palletizing. In general, lift tables are probably the least expensive engineering control measure used in conjunction with the NIOSH[1] lifting guidelines (see Chapter 4).

Make Better Use of Existing Handling Facilities

To ensure the greatest return from material handling equipment, that equipment must be used effectively. Thus, both the methods and the equipment should be sufficiently flexible that a variety of material handling tasks can be accomplished

[1]NIOSH is the National Institute of Occupational Safety and Health.

EXAMPLE 3–2
Maximum Net Load that can Safely be Handled by Fork Trucks

Start by computing the torque rating by multiplying the distance from the center of the front axle to the center of the load (see Figure 3–17):

$$\text{Load} = \text{Torque rating}/\text{B}$$

where B is distance C + D, with D = A/2.

If the distance C from the center of the front axle to the front end of the fork truck is 18 inches, and the length of the pallet A is 60 inches, then the maximum gross weight that a 200,000 inch-pounds fork truck should handle would be:

$$\text{L} = \frac{200{,}000}{18 + 60/2} = 4{,}167 \text{ lbs}$$

By planning the pallet size to make full use of the equipment, the company can realize a greater return from the material handling equipment.

FIGURE 3–17
Typical forklift truck.

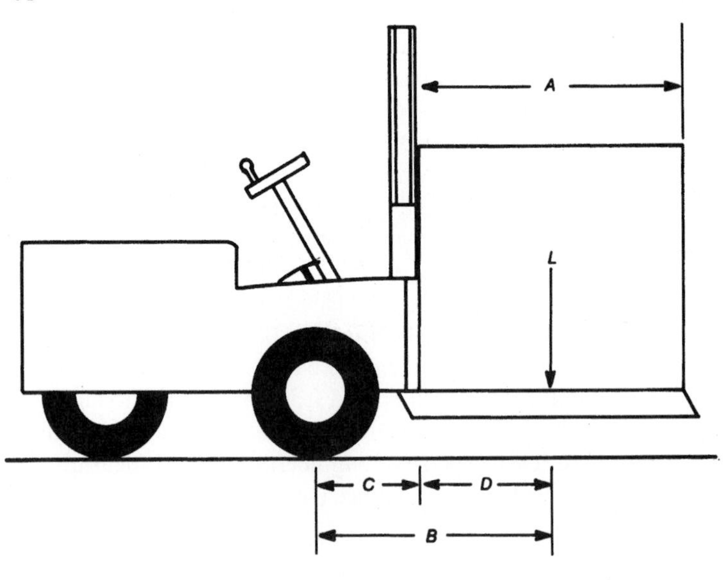

under variable conditions. Palletizing material in temporary and permanent storage allows greater quantities to be transported faster than storing material without the use of pallets, saving up to 65 percent in labor costs. Sometimes, material can be handled in larger or more convenient units by designing special racks. When this is done, the compartments, hooks, pins, or supports for holding the work should be in multiples of 10 for ease of counting during processing and final inspection. If any material handling equipment is used only part of the time, consider the possibility of putting it to use a greater share of the time. By relocating production facilities or adapting material handling equipment to diversified areas of work, companies may achieve greater utilization.

Handle Material with Greater Care

Industrial surveys indicate that approximately 40 percent of plant accidents happen during material handling operations. Of these, 25 percent are caused by lifting and shifting material. By exercising greater care in handling material, and using mechanical mechanisms wherever possible for material handling, employees can reduce fatigue and accidents. Records prove that the safe factory is also an efficient factory. Safety guards at points of power transmission, safe operating practices, good lighting, and good housekeeping are essential to making material handling equipment safer. Workers should install and operate all material handling equipment in a manner compatible with existing safety codes.

Better handling also reduces product damage. If the number of reject parts is at all significant in the handling of parts between workstations, then this area should be investigated. Usually, parts damage during handling can be minimized if specially designed racks or trays are fabricated to hold the parts immediately after processing. For example, one manufacturer of aircraft engine parts incurred a sizable number of damaged external threads on one component that was stored in metal tote pans after the completion of each operation. When two-wheeled hand trucks moved the filled tote pans to the next workstation, the machined forgings bumped against one another and against the sides of the metal pan to such an extent that they became badly damaged. Someone investigated the cause of the rejects and suggested making wooden racks with individual compartments to support the machined forgings. This prevented the parts from bumping against one another or the metal tote pan, thereby significantly reducing the number of damaged parts. Production runs were also more easily controlled because of the faster counting of parts and rejects.

In a city hospital, a portable mechanized lift permitted much greater use of a Hubbard tank for physical therapy treatment (see Figure 3–18). With this controllable material handling equipment, patients could be comfortably immersed in the tank in either a sitting or a prone position.

Consider Bar Coding for Inventory and Related Applications

The majority of technical people have some familiarity with bar coding and bar code scanning. Bar coding has shortened queues at grocery and department store checkout lines. The black bars and white spaces represent digits that uniquely identify both the item and the manufacturer. Once this Universal Product Code (UPC) is scanned by a reader at the checkout counter, the decoded data are sent to a computer that records timely information on labor productivity, inventory status, and sales.

FIGURE 3–18

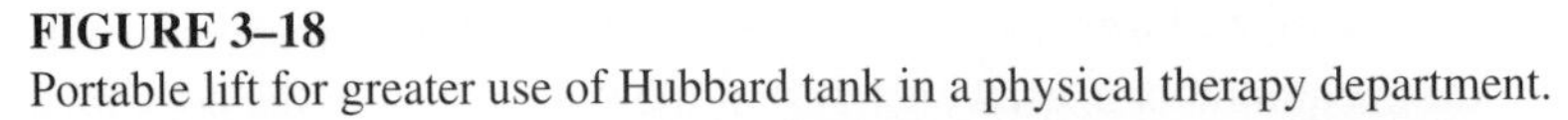

Portable lift for greater use of Hubbard tank in a physical therapy department.

The following five reasons justify the use of bar coding for inventory and related applications:

1. *Accuracy.* Typically representative performance is less than one error in 3.4 million characters. This compares favorably with the 2 to 5 percent error that is characteristic of keyboard data entry.
2. *Performance.* A bar code scanner enters data three to four times faster than typical keyboard entry.
3. *Acceptance.* Most employees enjoy using the scanning wand. Inevitably, they prefer using a wand to keyboard entry.
4. *Low cost.* Since bar codes are printed on packages and containers, the cost of adding this identification is extremely low.
5. *Portability.* An operator can carry a bar code scanner into any area of the plant to determine inventories, order status, etc.

Bar coding is useful for receiving, warehousing, job tracking, labor reporting, tool crib control, shipping, failure reporting, quality assurance, tracking, production control, and scheduling. For example, the typical storage bin label provides the following information: part description, size, packing quantity, department number, storage number, basic stock level, and order point. Considerable time can be saved by using a scanning wand to gather these data for inventory reordering.

Some practical applications reported by Accu-Sort Systems, Inc., include automatically controlling conveyor systems, diverting material to the location where it is needed; and providing material handlers with clear, concise instructions about where to take materials, automatically verifying that the proper material is handled. If bar coding is incorporated into programmable controllers and automatic packaging equipment, on-line real-time verification of packing labels with container contents can be used to avoid costly product recalls.

Summary: Material Handling

Analysts should always be looking for ways to eliminate inefficient material handling. To assist the methods analyst in this endeavor, the Materials Handling Institute (1998) has developed 10 principles of material handling. These are:

1. *Planning principle.* All material handling should be the result of a deliberate plan in which the needs, performance objectives, and functional specifications of the proposed methods are completely defined at the outset.
2. *Standardization principle.* Material handling methods, equipment, controls, and software should be standardized within the limits of achieving overall performance objectives and without sacrificing needed flexibility, modularity, and throughput.
3. *Work principle.* Material handling work should be minimized without sacrificing productivity or the level of service required of the operation.
4. *Ergonomic principle.* Human capabilities and limitations must be recognized and respected in the design of material handling tasks and equipment, to ensure safe and effective operations.
5. *Unit load principle.* Unit loads shall be appropriately sized and configured in a way that achieves the material flow and inventory objectives at each stage in the supply chain.
6. *Space utilization principle.* Effective and efficient use must be made of all available space.
7. *System principle.* Material movement and storage activities should be fully integrated to form a coordinated, operational system that spans receiving, inspection, storage, production, assembly, packaging, unitizing, order selection, shipping, transportation, and returns handling.
8. *Automation principle.* Material handling operations should be mechanized and/or automated where feasible, to improve operational efficiency, increase responsiveness, improve consistency and predictability, decrease operating costs, and eliminate repetitive or potentially unsafe manual labor.
9. *Environmental principle.* Environmental impacts and energy consumption are criteria to be considered when designing or selecting alternative equipment and material handling systems.
10. *Life-cycle-cost principle.* A thorough economic analysis should account for the entire life cycle of all material handling equipment and resulting systems.

To reiterate, the predominant principle is that the less a material is handled, the better it is handled.

8.
PLANT LAYOUT

The principal objective of effective plant layout is to develop a production system that permits the manufacture of the desired number of products with the desired quality at the least cost. Physical layout is an important element of an entire production system that embraces operation cards, inventory control, material handling, scheduling, routing, and dispatching. All of these elements must be carefully integrated to fulfill the stated objective. Although it is difficult and costly to make changes in arrangements that already exist, analysts should critically review every portion of every layout. Poor plant layouts result in major costs. Unfortunately, most of these costs are hidden and, consequently, cannot be readily exposed. The indirect labor expense of long moves, backtracking, delays, and work stoppages due to bottlenecks are characteristic of a plant with an antiquated and costly layout.

Layout Types

Is there one type of layout that tends to be the best? The answer is no. A given layout can be best in one set of conditions and yet poor in a different set of conditions. In general, all plant layouts represent one or a combination of two basic layouts: *product* or *straight-line layouts* and *process* or *functional layouts.* In the straight-line layout, the machinery is located such that the flow from one operation to the next is minimized for any product class. In an organization that utilizes this technique, it would not be unusual to see a surface grinder located between a milling machine and a turret lathe, with an assembly bench and plating tanks in the immediate area. This type of layout is quite popular for certain mass-production manufacture, because material handling costs are lower than for process grouping.

Product layout has some distinct disadvantages. Since a broad variety of occupations are represented in a relatively small area, employee dissatisfaction can escalate. This is especially true when different opportunities carry a significant money rate differential. Because unlike facilities are grouped together, operator training can be more cumbersome, especially if an experienced employee is not available in the immediate area to train a new operator. The problem of finding competent supervisors is also exacerbated, due to the variety of facilities and jobs that must be supervised. Then, too, this type of layout invariably necessitates a larger initial investment because duplicate service lines are required, such as air, water, gas, oil, and power. Another disadvantage of product grouping is the fact that this arrangement tends to appear disorderly and chaotic. With these conditions, it is often difficult to promote good housekeeping. In general, however, the disadvantages of product grouping are more than offset by the advantages, if production requirements are substantial.

Process layout is the grouping of similar facilities. Thus, all turret lathes would be grouped in one section, department, or building. Milling machines, drill presses, and punch presses would also be grouped in their respective sections. This type of arrangement gives a general appearance of neatness and orderliness, and tends to promote good housekeeping. Another advantage of functional layout is the ease with which a new operator can be trained. Surrounded by experienced employees operating similar machines, the new worker has a greater opportunity to learn from them. The problem of finding competent supervisors is lessened, because the job demands are not as great. Since these supervisors need only be familiar with one general type or class of facilities, their backgrounds do not have to be as extensive as those of supervisors in shops using product grouping. Also, if production quantities of similar products are limited and there are frequent "job" or special orders, a process layout is more satisfactory.

The disadvantage of process grouping is the possibility that long moves and backtracking will be needed on jobs that require a series of operations on diversified machines. For example, if the operation card of a job specifies a sequence of drill, turn, mill, ream, and grind, the movement of the material from one section to the next could prove extremely costly. Another major disadvantage of process grouping is the large volume of paperwork required to issue orders and control production between sections.

Travel Charts

Before designing a new layout or correcting an old one, analysts must accumulate the facts that may influence that layout. *Travel* or *from-to charts* can be helpful in diagnosing problems related to the arrangement of departments and service areas, as well as the location of equipment within a given sector of the plant. The travel chart is a matrix that presents the magnitude of material handling that takes place between two facilities per time period. The unit identifying the amount of handling may be whatever seems most appropriate to the analyst. It can be pounds, tons, handling frequency, and so on. Figure 3–19 illustrates a very elementary travel chart.

Muther's Systematic Layout Planning

A systematic approach to plant layout developed by Muther (1973) is termed Systematic Layout Planning (SLP). The goal of SLP is to locate two areas with high frequency and logical relationships close to one another using a straightforward six-step procedure:

1. **Chart relationships.** In the first step, the relationships between different areas are established and then charted on a special form called the *relationship chart* (or rel chart for short; see Figure 3–20). A relationship is the relative degree of closeness, desired or required, among different activities, areas, departments, rooms, etc., as determined from quantitative flow information (volume, time, cost, routing) from a from-to chart, or more qualitatively from functional

FIGURE 3–19
The travel chart, a useful tool in solving material handling and plant layout problems related to process-type layouts.

		To							
		No. 4 W. & S. Turret Lathe	Delta 17" Drill Press	2-Spindle L. & G. Drill	No. 2 Cinn. Hor. Mill	No. 3B. & S. Verticle Mill	Niagara 100Ton Press	No. 2 Cinn. Centerless	No. 3 Excello Thd. Grinder
From	No. 4 W. & S. Turret Lathe		20	45	80	32	4	6	2
	Delta 17" Drill Press			6	8	4	22	2	3
	2-Spindle L. & G. Drill				22	14	18	4	4
	No. 2 Cinn. Hor. Mill	120				10	5	4	2
	No. 3B. & S. Verticle Mill						6	3	1
	Niagara 100Ton Press		60	12	2			0	1
	No. 2 Cinn. Centerless		15						15
	No. 3 Excello Thd. Grinder				15	8			

interactions or subjective information. For example, although painting may be the logical step between finishing and final inspection and packing, the toxic materials and hazardous or flammable conditions may require that the paint area be completely separated from the other areas. The relationship ratings range in value from 4 to −1, based on the vowels that semantically define the relationship, as shown in Table 3–3.

2. **Space requirements.** In the second step, space requirements are established in terms of square footage. These values can be calculated based on production requirements, extrapolated from existing areas, projected for future expansion, or fixed by legal standards, such as the ADA or architectural standards. In addition to square footage, the kind and shape of the area being laid out, or the location with respect to required utilities, may also be very important.
3. **Activity relationships diagram.** In the third step, a visual representation of the different activities is drawn. The analyst starts with the absolutely important relationships (A's), using four short, parallel lines to join the two areas. The ana-

FIGURE 3–20

Relationship Chart for Dorben Consulting.

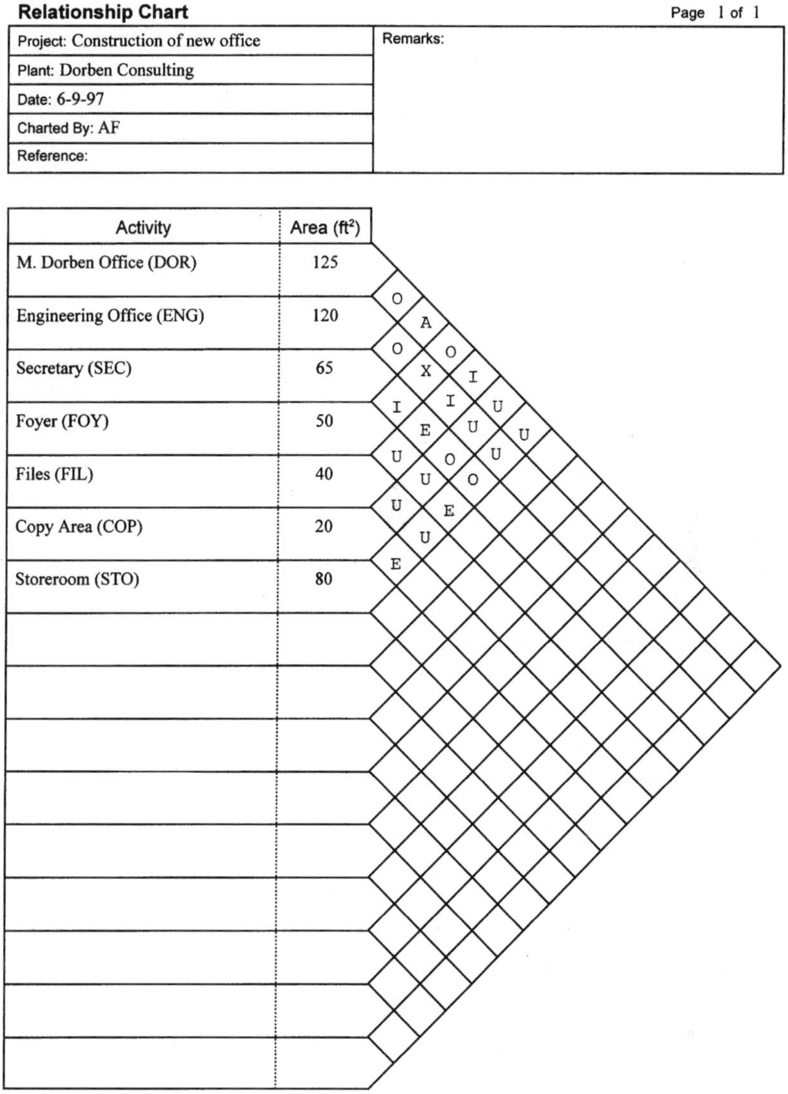

Relationship Chart Page 1 of 1

Project: Construction of new office	Remarks:
Plant: Dorben Consulting	
Date: 6-9-97	
Charted By: AF	
Reference:	

Activity	Area (ft²)
M. Dorben Office (DOR)	125
Engineering Office (ENG)	120
Secretary (SEC)	65
Foyer (FOY)	50
Files (FIL)	40
Copy Area (COP)	20
Storeroom (STO)	80

	DOR	ENG	SEC	FOY	FIL	COP	STO
DOR		O	A	O	I	U	U
ENG			O	X	I	U	U
SEC				I	E	O	O
FOY					U	U	E
FIL						U	U
COP							E

lyst then proceeds to the E's, using three parallel lines approximately double the length of the A lines. The analyst continues this procedure for the I's, O's, etc., progressively increasing the length of the lines, while attempting to avoid crossing or tangling the lines. For undesirable relationships, the two areas are placed as far apart as possible and a squiggly line (representing a spring) is drawn between them. (Some analysts may also define an extremely undesirable relationship with a −2 value and a double squiggly line.)

TABLE 3–3
SLP Relationship Ratings

Relationship	Closeness Rating	Value	Diagram Lines	Color
Absolutely necessary	A	4		Red
Especially important	E	3		Yellow
Important	I	2		Green
Ordinary	O	1		Blue
Unimportant	U	0		
Not desirable	X	−1		Brown

4. **Space relationship layout.** Next, a spatial representation is created by scaling the areas in terms of relative size. Once the analyst is satisfied with the layout, the areas are compressed into a floor plan. This is typically not as easy as it sounds, and the analyst may want to utilize templates. In addition, modifications may be made to layout based on material handling requirements (e.g., a shipping or receiving department would necessarily be located on an exterior wall), storage facilities (perhaps similar exterior access requirements), personnel requirements (a cafeteria or restroom located close by), building features (crane activities in a high bay area; forklift operations on the ground floor), and utilities.
5. **Alternative arrangements evaluation.** With numerous possible layouts, it would not be unusual to find that several appear to be equally likely possibilities. In that case, the analyst will need to evaluate the different alternatives to determine the best solution. First, the analyst will need to identify factors deemed important: for example, future expansion capability, flexibility, flow efficiency, material handling effectiveness, safety, supervision ease, appearance or aesthetics, etc. Second, the relative importance of these factors will need to be established through a system of weights, such as a 0 to 10 basis. Next, each alternative is rated for satisfying each factor. Muther (1973) suggests the same 4 to −1 scale: 4 being almost perfect; 3, especially good; 2, important; 1, ordinary result; 0, unimportant; and −1, not acceptable. Each rating is then multiplied by the weight. The products for each alternative are then summed, with the largest value indicating the best solution. At this point, scale models providing a third dimension to the layouts may be helpful to an analyst endeavoring to sell a contemplated layout to a top executive who has neither the time nor the familiarity to grasp all the details of a two-dimensional layout.
6. **Selected layout and installation.** The final step is to implement the new method.

Computer-Aided Layout

Commercially available software can help analysts develop realistic layouts rapidly and inexpensively. The Computerized Relative Allocation Facilities (CRAFT) program is one that has been used extensively. An activity center could be a department or work center within a department. Any one activity center can be identified as

EXAMPLE 3–3
Plant Layout of Dorben Consulting Using SLP.

The Dorben Consulting group would like to lay out a new office area. There are seven activity areas: M. Dorben's office, engineering office (occupied by two engineers), secretarial area, foyer and waiting area for visitors, file area, copy area, and storeroom. The activity relationships are subjectively assessed by M. Dorben to be as shown in the rel chart in Figure 3–20. The chart also indicates space allotments for each area, ranging from a low of 20 square ft. for the copy area to 125 square ft. for M. Dorben's office. For example, the relationship between M. Dorben and the secretary is deemed absolutely important (A), while the relationship between the engineering area and the foyer is deemed not desirable (X), so that the engineers are not disrupted in their work by visitors.

A relatively good first attempt at an activity relationship diagram yields Figure 3–21. Adding in the relative size of each area yields the space relationship chart in Figure 3–22. Compressing the areas yields the final floor plan in Figure 3–23.

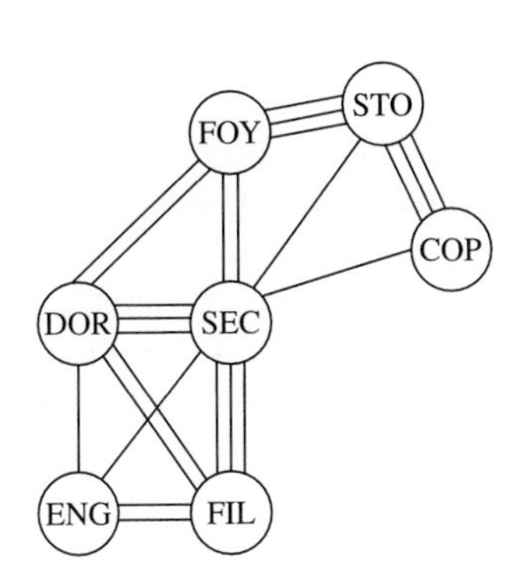

FIGURE 3–21
Activity relationship diagram for Dorben Consulting.

FIGURE 3–22
Space relationship layout for Dorben Consulting.

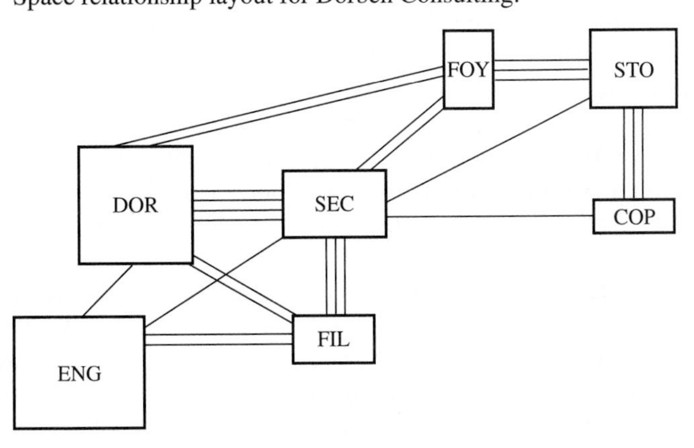

EXAMPLE 3–3 ***(continued)***

FIGURE 3–23
Floor Plan for Dorben Consulting.

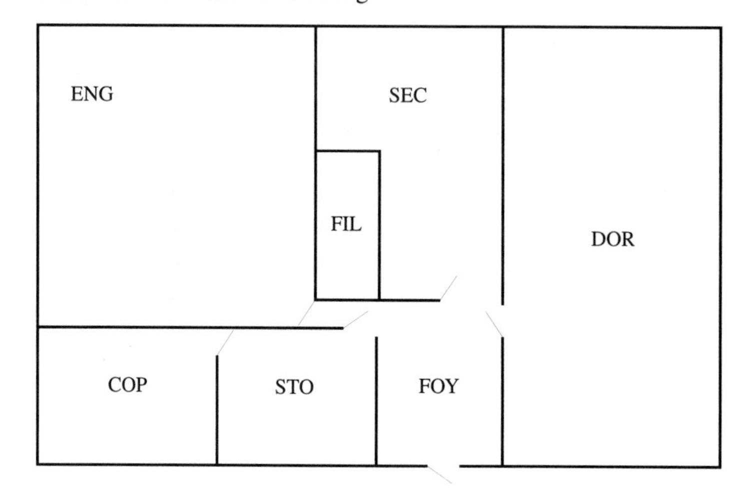

Since Dorben's office and the engineering area are practically the same size, they could easily be interchanged, leaving two alternative layouts. These are evaluated (Figure 3–24) on the basis of personnel isolation (which is very important to M. Dorben, yielding a high weight of 8), supplies movement, visitor reception, and flexibility. The big difference in the layouts is the closeness of the engineering area to the foyer. Thus, Alternative B (shown in Figure 3–23) at 68 points, compared to 60 points for Alternative A, turns out to be the preferred layout.

fixed, freezing it and allowing freedom of movement in those that can be readily moved. For example, it is often desirable to freeze such activity centers as elevators, restrooms, and stairways. Input data include fixed work center numbers and locations, material handling costs, interactivity center flow, and a block layout representation. The governing heuristic algorithm asks, What change in material handling costs would result if work centers were exchanged? Once the answer is stored, the computer proceeds in an iterative manner until it converges on a good solution. CRAFT calculates the distance matrix as the rectangular distances from the department centroids.

Another program currently available is CORELAP. The input requirements for CORELAP are the number of departments, the departmental areas, the departmental relationships, and the weights for these relationships. CORELAP constructs layouts by locating the departments, using rectangular areas. The objective is to provide a layout with "high-ranking" departments close together.

EXAMPLE 3–3 ***(concluded)***

FIGURE 3–24
Evaluating Alternatives for Dorben Consulting.

Evaluating Alternatives Page 1 of 1

Plant: Dorben Consulting	Alternatives	A		B		C		D		E		
Project: New Office Construction		Dorben office facing West		Dorben office facing East								
Date: 6-9-97												
Analyst: AF												
Factor/Consideration	**Wt.**	**Ratings and Weighted Ratings**										**Comments**
		A		B		C		D		E		
Isolation of personnel	8	1	8	3	24							
Movement of supplies	4	3	12	3	12							
Visitor reception	4	4	16	4	16							
Flexibility	8	3	24	2	16							
Totals		60		68								

Remarks:
Alternative B, with Dorben's office facing East and engineers' office facing West, lessens disruptions of the engineers' work due to visitors.

ALDEP, still another available program, constructs plant layouts by randomly selecting a department and locating it in a given layout. The relationship chart is then scanned, and a department that has a high closeness rating is introduced into the layout. This process continues until the program places all departments. ALDEP then computes a score for the layout, and repeats the process a specific number of times. The program also has the ability to provide multifloor layouts.

All of these plant layout programs were originally developed for large mainframe computers. With the advent of personal computers, the algorithms have been incorporated into PC programs, as have other algorithms. One such program, SPIRAL, attempts to optimize the adjacency relationship by summing the positive relationships and deducting the negative relationships for adjacent areas. This is essentially a quantified Muther's approach and is described in more detail in Goetschalckx (1992). For example, entering the data for the Dorben Consulting example yields a slightly different layout, as shown in Figure 3–25.

Note that there is a tendency to generate long, narrow rooms, to minimize the distance between room centers. This is an especially big problem with CRAFT, ALDEP, etc. SPIRAL at least attempts to modify this tendency by adding a shape penalty. Also, there is a tendency for many of these programs (i.e., those that are *improvement* programs, such as CRAFT, that build upon an initial layout) to reach a local minimum and not attain the optimum layout. This problem can be circumvented by starting with alternate layouts. This is less a problem with *construction* programs, such as SPIRAL, which generate a solution from scratch. A more powerful and perhaps more useful program is FactoryCAD, which inputs existing AutoCAD files of floor plans and creates very detailed layouts suitable for architectural planning.

9. WORK DESIGN

Because of the recent regulatory (i.e., OSHA) and health (i.e., rising medical and Workers' Compensation costs) concerns, work design techniques will be covered in detail in separate chapters. Chapter 4 addresses manual work and the principles of motion economy, Chapter 5 addresses ergonomic principles of workplace and tool design, and Chapter 6 covers working and environmental conditions.

SUMMARY

The 9 primary approaches to operation analysis represent a systematic approach to analyzing the facts presented on the operation and flow process charts. Regardless of the nature of the work, whether continuous or intermittent, process or job shop, soft or hard goods, when systematic operation analysis is applied by competent personnel, real savings result. These principles are just as applicable to the planning of new work as to the improvement of work already in production. While increased output and improved quality are the primary outcomes of operation analysis, it also distributes the benefits of improved production to all workers and helps develop better working conditions and methods. The result is that the worker can do more work at the plant, can do a good job, and can still enjoy life.

A systematic method for remembering and applying the 9 operation analyses is offered by a checklist of pertinent questions, as shown in Figure 3–26. In the figure,

FIGURE 3–25
SPIRAL input files;
(a) DORBEN.DAT,
(b) DORBEN.DEP, and
(c) Resulting layout for the DORBEN Consulting example.

a)

[project_name]	DORBEN
[number_of_departments]	7
[department_file_name]	DORBEN.DEP
[building_width]	25
[building_depth]	20
[seed]	12345
[tolerance]	0.00010
[time_limit]	120
[number_of_iterations]	20
[report_level]	2
[max_shape_ratio]	2.50
[shape_penalty]	500.00

b)

```
DOR   0   0   125   0   0   GREEN    Dorben
ENG   0   0   120   0   0   BLUE     Engineers
SEC   0   0    65   0   0   RED      Secretary
FOY   0   0    50   0   0   YELLOW   Foyer
FIL   0   0    40   0   0   BROWN    Files
COP   0   0    20   0   0   GRAY     Copy
STO   0   0    80   0   0   BLACK    Storeroom
DOR   ENG   1
DOR   SEC   4
DOR   FOY   1
DOR   FIL   2
ENG   SEC   1
ENG   FOY  -1
ENG   FIL   2
SEC   FOY   2
SEC   FIL   3
SEC   COP   1
SEC   STO   1
FOY   STO   3
COP   STO   3
OUT   OUT   0
```

c)

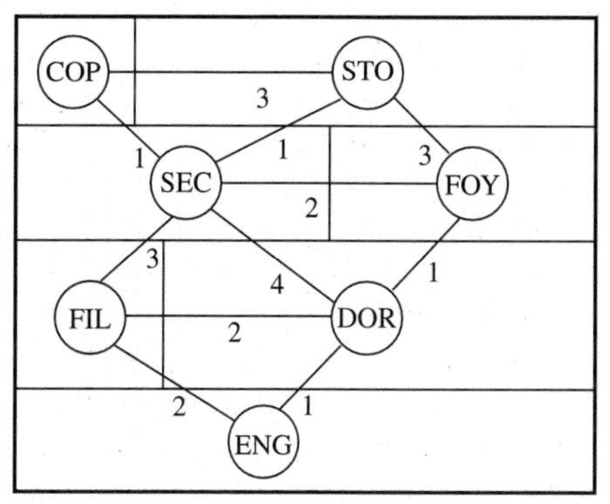

FIGURE 3–26
Operations analysis checklist for manufacture of blanket control knob shaft.

Date 9/15 Dept. 11 Dwg. 18-4612 Sub. 2
Mould — Die — Style — Item 2
Pattern — Ins. Spec. C L. Spec. — Sub. —
Part Description Blanket control knob shaft

Operation Turn, groove, drill, tap, knurl, thread, cut-off Operator Blazer

DETERMINE AND DESCRIBE

1. **PURPOSE OF OPERATION**
To form contours of 3/8"S.A.E. 1112 rod on automatic screw machine to achieve drawing specifications.

2. **COMPLETE LIST OF ALL OPERATIONS PERFORMED ON PART**

No.	Description	Work Sta.	Dept.
1.	Turn, groove, drill, tap, knurl, thread, cut-off	B. & S.	11
2.	Burr	Bench	12
3.	Inspect 1%	Bench	18
4.			
5.			
6.			
7.			
8.			
9.			
10.			

3. **INSPECTION REQUIREMENTS**
a—Of previous oprn.
b—Of this oprn. Yes. Perhaps S.Q.C. will reduce amount of inspection.

DETAILS OF ANALYSIS

Can purpose be accomplished better otherwise? *Yes—by die casting*

Can oprn. being analysed
be eliminated? *No*
be combined with another? *No*
be performed during idle period of another? *Yes, by machine coupling*
Is sequence of oprns. best possible? *Yes*

Should oprn. be done in another dept. to save cost or handling? *Perhaps can be purchased outside at a savings.*

Are tolerance, allowance, finish and other requirements necessary?

c—Of next opra.

	too costly? suitable to purpose?
4. MATERIAL Zinc base die cast metal would be less expensive. Cutting compounds and other supply materials	Consider size, suitability, straightness, and condition. Can cheaper material be substituted?
5. MATERIAL HANDLING a—Brought by 4 wheel truck to automatics b—Removed by hand 2 wheel trucks c—Handled at work station by	Should crane, gravity conveyors, totepans, or special trucks be used? Consider layout with respect to distance moved. *Perhaps gravity to burring station.*
6. SET-UP (Accompany description with sketches if necessary) This is satisfactory as being done. a—Tool Equipment Present Suggestions Redesign part to be made as zinc base die casting rather than S.A.E. 1112 screw machine part.	How are dwgs. and tools secured? Can set-up be improved? Trial pieces. Machine Adjustments. **Tools** Suitable? Provided? Ratchet Tools Power Tools Spl. Purpose Tools Jigs, Vises Special Clamps Fixtures Multiple Duplicate

Methods Engineering Council
Form No. 101

ANALYSIS SHEET

FIGURE 3–26 *(concluded)*

	RECOMMENDED ACTION
7. CONSIDER THE FOLLOWING POSSIBILITIES.	
1. Install gravity delivery chutes.	____
2. Use drop delivery	yes, to accumulate for tumbling.
3. Compare methods if more than one operator is working on same job.	
4. Provide correct chair for operator.	____
5. Improve jigs or fixtures by providing ejectors, quick-acting clamps, etc.	____
6. Use foot operated mechanisms.	____
7. Arrange for two handed operation.	____
8. Arrange tools and parts within normal working area.	____
9. Change layout to eliminate back tracking and to permit coupling of machines.	____
10. Utilize all improvements developed for other jobs.	
8. WORKING CONDITIONS.	Light O.K.
Generally satisfactory.	Heat O.K.
	Ventilation, Fumes O.K.
	Drinking Fountains O.K.
	Wash Rooms O.K.
a—Other Conditions	Safety Aspects O.K.
	Design of Part O.K.
	Clerical Work Required (to fill out time cards, etc.) O.K.
	Probability of Delays O.K.
	Probable Mfg. Quantities O.K.
9. METHOD (Accompany with sketches or Process Charts if necessary.)	Arrangement of Work Area
	Placement of
a—Before Analysis and Motion Study.	Tools.
	Materials.
	Supplies.

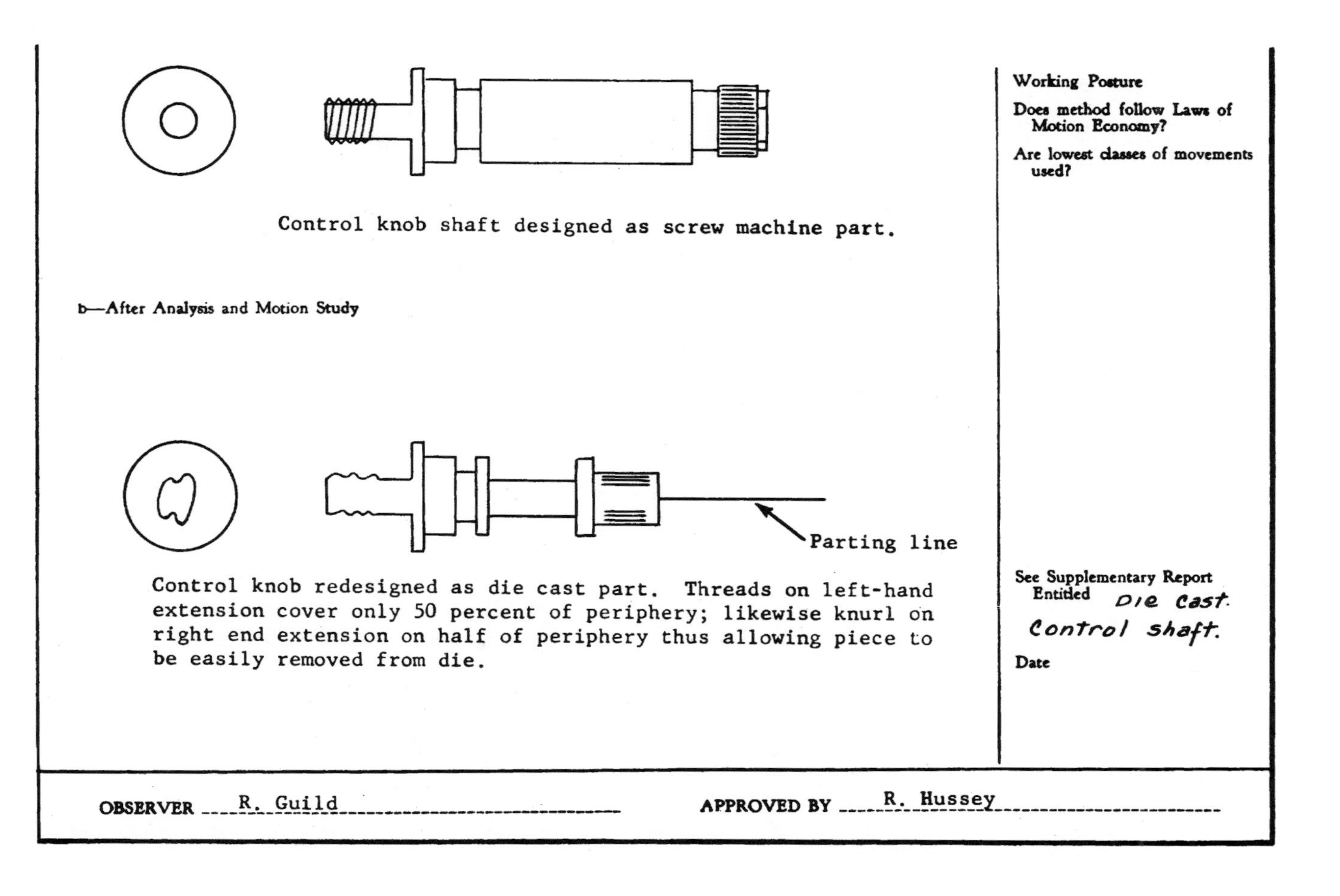
Control knob shaft designed as screw machine part.

b—After Analysis and Motion Study

Control knob redesigned as die cast part. Threads on left-hand extension cover only 50 percent of periphery; likewise knurl on right end extension on half of periphery thus allowing piece to be easily removed from die.

Working Posture

Does method follow Laws of Motion Economy?

Are lowest classes of movements used?

See Supplementary Report Entitled *Die Cast Control shaft.*

Date

OBSERVER R. Guild APPROVED BY R. Hussey

the checklist demonstrates how its use resulted in a cost reduction on an electric blanket control knob shaft. Redesigning the shaft so that it could be economically produced as a die casting rather than a screw machine part reduced factory cost from $68.75 per thousand pieces to $17.19 per thousand pieces.

Analysts should obtain the answers to all questions on the check sheet related to all steps appearing on the flowchart. This procedure invariably leads to efficient ways of performing the work. As ideas develop, they are immediately recorded so they won't be forgotten, and sketches are included at this time. Analysts may be surprised by the number of inefficiencies revealed and should have little trouble in compiling a number of improvement possibilities. One improvement usually leads to another. To be successful in this type of work, analysts must have open minds and creative ability. The check sheet is also useful in providing methods training to factory foremen and superintendents. Thought-provoking questions, when intelligently used, help factory supervisors to develop constructive ideas. The check sheet serves as an outline, which can be referred to by the discussion leader handling the methods training.

QUESTIONS

1. Explain how design simplification can be applied to the manufacturing process.

2. How is operation analysis related to methods engineering?

3. How do unnecessary operations develop in an industry?

4. What has been the impact of the computer in connection with paperwork?

5. What four thoughts should analysts keep in mind to improve design?

6. What is meant by "tight" tolerances?

7. Explain why it may be desirable to "tighten up" tolerances and specifications.

8. What is meant by lot-by-lot inspection?

9. When is an elaborate quality control procedure not justified?

10. What six points should be considered when endeavoring to reduce material cost?

11. How does a changing labor and equipment situation affect the cost of purchased components?

12. What six points should be remembered when designing forms?

13. Explain how rearranging operations can result in savings.

14. What process is usually considered the fastest for forming and sizing operations?

15. How should the analyst investigate the setup and tools to develop better methods?

16. Give some applications of bar coding for the improvement of productivity.

17. When would you recommend the use of energy-efficient motors?

18. What are the two general types of plant layout? Explain each in detail.

19. What is the best way to test a proposed layout?

20. Which question should the analyst ask when studying work performed at a specific workstation?

21. Explain the advantages of using a checklist.

22. In connection with automated guided vehicles, why do costs vary little with distance?

23. On what does the extent of tooling depend?

24. How can planning and production control affect setup time?

25. How can a material best be handled?

26. How is the travel chart related to Muther's SLP?

27. Why does the travel chart have more application in process layout than in product layout?

28. Explain the fundamental purpose of group technology.

29. Explain how the conservation of welding rods can result in 20 percent material savings.

30. Identify several automobile components that have been converted from metal to plastic in recent years.

31. Where would you find application for a hydraulic elevating table?

32. What is the difference between a skid and a pallet?

33. When would you recommend using three-dimensional models in layout work?

PROBLEMS

1. How much more efficient is a typical 50 HP energy-efficient motor than a standard 50 HP motor?

2. The finish tolerance on the shaft in Figure 3–4 was changed from 0.004 inches to 0.008 inches. How much cost improvement resulted from this change?

3. What overall tolerance would be applied to three components making up the overall dimension if: component one had a tolerance of 0.002 inches; component two, 0.004 inches; and component three, 0.005 inches?

4. A ceramic material is being considered as a possible mold material in conjunction with the die casting of 60–40 brass. A cylinder of the material 8 inches in diameter and 10 inches long was used to obtain a stress–strain relationship in compression. The material failed under a load of 265,000 pounds and a total strain of 0.012 inches. What was the material's fracture strength, percent contraction at fracture, and modulus of toughness?

5. The Dorben Company is designing a cast-iron part whose strength T is a known function of the carbon content C, where: $T = 2C^2 + 3/4\ C - C^3 + k$. To maximize strength, what carbon content should be specified?

6. To make a given part interchangeable, it was necessary to reduce the tolerance on the outside diameter from ± 0.010 to ± 0.005 at a resulting cost increase of 50 percent of the turning operation. The turning operation represented 20 percent of the total cost. Making the part interchangeable meant that the volume of this part could be increased by 30 percent. The increase in volume would permit production at 90 percent of the former cost. Should the methods engineer proceed with the tolerance change? Explain.

7. The analyst in the Dorben Company is considering replacing five 50 HP motors with five 50 HP energy-efficient motors. These motors will all be operated seven days per week and three shifts per day, at an estimated 85 percent of full load. If the cost of electric power is $0.06 per KWH, how much can the company afford to pay for the five energy-efficient motors?

8. The methods analyst in the Dorben Company is considering the installation of a state-of-the-art solid-state electronic energy conservation system that will tune and balance the whole fluorescent system in the plant, including lamps, ballast, and power supply. The energy conservation system will regulate the voltage and the current on an ongoing basis, to establish and hold the fluorescent system at optimum performance, and will protect the fluorescent system with special safety circuits. Based on estimates received from the supplier, the new energy conservation system will save 30 percent of the lighting energy cost, 50 percent of the lamp replacement cost, and all of the present ballast replacement cost. If the cost of the installed energy conservation system is $15,000, how many months will it take for the investment to pay for itself, based on the following data:

   Lighting energy costs:
   Number of fixtures—445.
   Average KWH per fixture—0.187.
   Cost per KWH—$0.085.
   Annual operating hours—4,440.

   Lamp replacement cost:
   Number of fixtures—445.
   Average lamps per fixture—2.25.
   Average lamp life—2.5 years.
   Cost of lamp installed—$6.50.

FIGURE 3–27
Information for Problem 3.9.

Activity	Area (sq.ft)
A	160
B	160
C	240
D	160
E	80

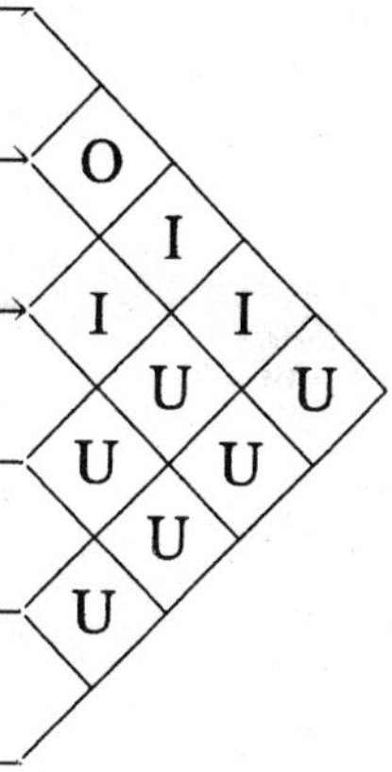

O
I
I I
U U
U U
U
U

Ballast replacement cost:

Number of fixtures—445.
Average ballasts per fixture—1.1.
Average ballast life—4.5 years.
Cost of ballast installed—$40.00.

9. The Dorben Group suite consists of five rooms, with areas and relationships as shown in Figure 3–27. Obtain an optimal layout using Muther's SLP and SPIRAL. Compare and contrast the resulting layouts.

REFERENCES

Bralla, James G. *Handbook of Product Design for Manufacturing.* New York: McGraw-Hill, 1986.

Buffa, Elwood S. *Modern Production Operations Management.* 6th ed. New York: John Wiley & Sons, 1980.

Chang, Ning San. "Pattern Recognition and Bar Code Technology." In *Handbook of Industrial Engineering,* 2nd ed. Ed. Gavriel Salvendy. New York: John Wiley & Sons, 1992.

Chang, Tien-Chien, Richard A. Wysk, and Wang Hsu-Pin. *Computer Aided Manufacturing.* Englewood Cliffs, NJ: Prentice Hall, 1991.

DeMarle, David J., and M. Larry Shillito. "Value Engineering." In *Handbook of Industrial Engineering,* 2nd ed. Ed. Gavriel Salvendy. New York: John Wiley & Sons, 1992.

Drury, Colin G. "Inspection Performance." In *Handbook of Industrial Engineering,* 2nd ed. Ed. Gavriel Salvendy. New York: John Wiley & Sons, 1992.

Fallon, Carlos. *Value Analysis to Improve Productivity.* New York: John Wiley & Sons, 1971.

Francis, Richard L., and John A. White. *Facility Layout and Location: An Analytical Approach.* Englewood Cliffs, NJ: Prentice-Hall, 1974.

Goetschalckx, M. "An Interactive Layout Heuristic Based on Hexagonal Adjacency Graphs." *European Journal of Operations Research,* 63, no.2 (December 1992), pp. 304–321.

Hyer, Nancy Lea., ed. *Capabilities of Group Technology.* Dearborn, MI: Society of Manufacturing Engineers, 1987.

Kadota, Takeji, and Shigeyasu Sakamoto. "Methods Analysis and Design." In *Handbook of Industrial Engineering,* 2nd ed. Ed. Gavriel Salvendy. New York: John Wiley & Sons, 1992.

Konz, Stephan. *Facility Design.* New York: John Wiley & Sons, 1985.

Material Handling Institute. *The Ten Principles of Material Handling.* Charlotte, NC, 1998.

Mudge, Arthur E. *Value Engineering: A Systematic Approach.* New York: McGraw-Hill, 1971.

Muther, R. *Systematic Layout Planning,* 2nd ed. New York: Van Nostrand Reinhold, 1973.

Niebel, Benjamin W., and C. Richard Liu. "Designing for Manufacturing." In *Handbook of Industrial Engineering,* 2nd ed. Ed. Gavriel Salvendy. New York: John Wiley & Sons, 1992.

Niebel, Benjamin W. "Designing for Manufacturing." In *Handbook of Industrial Engineering.* Ed. Gavriel Salvendy. New York: John Wiley & Sons, 1982.

Nof, Shimon Y. "Industrial Robotics." In *Handbook of Industrial Engineering,* 2nd ed. Ed. Gavriel Salvendy. New York: John Wiley & Sons, 1992.

Shingo, S. *Study of Toyota Production System.* Tokyo, Japan: Japan Management Assoc. (1981), pp. 167–182.

Sims, Ralph E. "Material Handling Systems." In *Handbook of Industrial Engineering,* 2nd ed. Ed. Gavriel Salvendy. New York: John Wiley & Sons, 1992.

Smith, Michael J. "Design for Health and Safety." In *Handbook of Industrial Engineering,* 2nd ed. Ed. Gavriel Salvendy. New York: John Wiley & Sons, 1992.

Society of Manufacturing Engineers. *Applying Industrial Bar Coding.* Dearborn, MI: Society of Manufacturing Engineers, 1985.

Spur, Gunter. "Numerical Control Machines." In *Handbook of Industrial Engineering,* 2nd ed. Ed. Gavriel Salvendy. New York: John Wiley & Sons, 1992.

Taguchi, Genichi. *Introduction to Quality Engineering.* Tokyo, Japan: Asian Productivity Organization, 1986.

Trucks, H. E. *Designing for Economical Production.* Detroit, MI: Society of Manufacturing Engineers, 1974.

Tompkins, J. "Plant Layout." In *Handbook of Industrial Engineering.* Ed. Gavriel Salvendy. New York: John Wiley & Sons, 1982.

Wemmerlov, Urban, and Nancy Lea Hyer. "Group Technology." In *Handbook of Industrial Engineering,* 2nd ed. Ed. Gavriel Salvendy. New York: John Wiley & Sons, 1992.

Wick, Charles, and Raymond F. Veilleux. *Quality Control and Assembly, 4.* Detroit, MI: Society of Manufacturing Engineers, 1987.

SELECTED SOFTWARE

ALDEP, IBM Corporation, program order no. 360D-15.0.004.
CORRELAP, Engineering Management Associates, Boston, MA.
CRAFT, IBM share library No. SDA 3391.

FactoryCAD, *Tutorial and Reference Manual,* Cimtechnologies Corp., 2501 North Loop Dr. Suite 700, Iowa State University Research Park, Ames, IA, 50010, 1995.
SPIRAL, *User's Manual,* 4031 Bradbury Dr., Marietta, GA, 30062, 1994.

SELECTED VIDEOTAPES

Automated Inspection/Non-Destructive Testing. Manufacturing Insights Videotape Series. 1/29 VHS VT281-1368 & 3/49 U-Matic VT281U-1368. Dearborn, MI: Society of Manufacturing Engineers, 1988.
Design for Manufacturing. Manufacturing Insights Videotape Series. 1/29 VHS VT396-1368 & 3/49 U-Matic VT396U-1368. Dearborn, MI: Society of Manufacturing Engineers, 1990.
Layout Improvements for Just-In-Time. Manufacturing Insights Videotape Series. 1/29 VHS VT393–1368 & 3/49 U-Matic VT393U-1368. Dearborn, MI: Society of Manufacturing Engineers, 1990.
Total Quality Management. Manufacturing Insights Videotape Series. 1/29 VHS VT395-1368 & 3/49 U-Matic VT395U-1368. Dearborn, MI: Society of Manufacturing Engineers, 1990.
Automated Material Handling. Manufacturing Insights Videotape Series. 1/29 VHS: VT251-1368 & 3/49 U-Matic: VT251U-1368. Dearborn, MI: Society of Manufacturing Engineers, 1986.
Cutting Tools. Manufacturing Insights Videotape Series. 1/29 VHS: VT249-1368 & 3/49 U-Matic: VT249U-1368. Dearborn, MI: Society of Manufacturing Engineers, 1986.
Flexible Small Lot Production for Just-In-Time. Manufacturing Insights Videotape Series. 1/29 VHS: VT415-1368 & 3/49 U-Matic VT415U-1368. Dearborn, MI: Society of Manufacturing Engineers, 1991.
Programmable Controllers. Manufacturing Insights Videotape Series. 1/29 VHS: VT254-1368 & 3/49 U-Matic VT254U-1368. Dearborn, MI: Society of Manufacturing Engineers, 1987.
Setup Reduction for Just-In-Time. Manufacturing Insights Videotape Series. 1/29 VHS: VT392-1368 & 3/49 U-Matic VT392U-1368. Dearborn, MI: Society of Manufacturing Engineers, 1990.

CHAPTER 4

Manual Work Design

KEY POINTS:

- Design work according to human capabilities and limitations.
- For manipulative tasks:
  - Use dynamic motions rather than static holds.
  - Keep the strength requirement below 15 percent of maximum.
  - Avoid extreme ranges of motion.
  - Use the smallest muscles for speed and precision.
  - Use the largest muscles for strength.
- For lifting and other heavy manual work:
  - Keep workloads below one-third of the maximum work capacity.
  - Minimize horizontal load distances.
  - Avoid twisting.
  - Use frequent, short work/rest cycles.

The design of manual work was introduced by the Gilbreths through motion study and the principles of motion economy, and later was further developed by Barnes (1980). The principles have traditionally been broken down into three basic subdivisions: (1) the use of the human body; (2) the arrangement and conditions of the workplace; and (3) the design of tools and equipment. More importantly, although developed empirically, the principles are in fact based on established anatomical, biomechanical, and physiological principles of the human body. They form the scientific basis for ergonomics and work design. Accordingly, some theoretical background will be presented so that the principles of motion economy can be understood better rather than merely being accepted as memorized rules. Furthermore, the traditional principles of motion economy have been considerably expanded and are now called the principles of and guidelines for work design. This chapter presents the principles related to the human body and the guidelines for the design of work as related to physical activity. Chapter 5 covers those principles related to the design of workstations, tools and equipment. Chapter 6 presents guidelines for the design of the work environment.

THE MUSCULOSKELETAL SYSTEM

The human body is able to produce movements because of a complex system of muscles and bones, termed the *musculoskeletal system.* The muscles are attached to the bones on either side of a joint (see Figure 4–1), so that one or several muscles, termed *agonists*, act as the prime activators of motion. Other muscles, termed *antagonists,* counteract the agonists and oppose the motion. For elbow *flexion*, which is a decrease in the internal joint angle, the biceps or brachioradialis form the agonist, while the triceps form the antagonist. However, on elbow *extension*, which is an increase in the joint angle, the triceps become the agonist, while the biceps become the antagonist.

There are three types of muscles in the human body: skeletal or striated muscles, attached to the bones; cardiac muscle, found in the heart; and smooth muscle, found in the internal organs and the walls of the blood vessels. Only the skeletal muscles (of which there are approximately 500 in the body) will be discussed here, because of their relevance to motion.

Each muscle is made up of a large number of muscle fibers, approximately 0.004 inches (0.1 mm) in diameter and ranging in length from 0.2 to 5.5 inches (5–140 mm), depending on the size of the muscle. These fibers are typically bound together in bundles by connective tissue, which extends to the end of the muscle and assists in firmly attaching the muscle and muscle fibers to the bone (see Figure 4–2.) These bundles are penetrated by tiny blood vessels that carry oxygen and nutrients to the muscle fibers, as well as by small nerve endings that carry electrical impulses from the spinal cord and brain.

FIGURE 4–1
The musculoskeletal system of the arm.

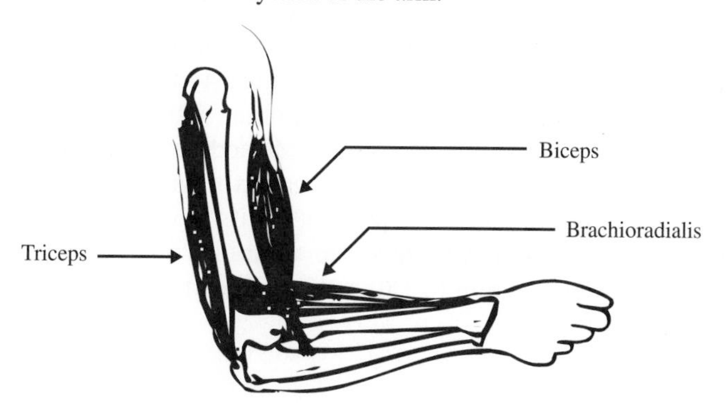

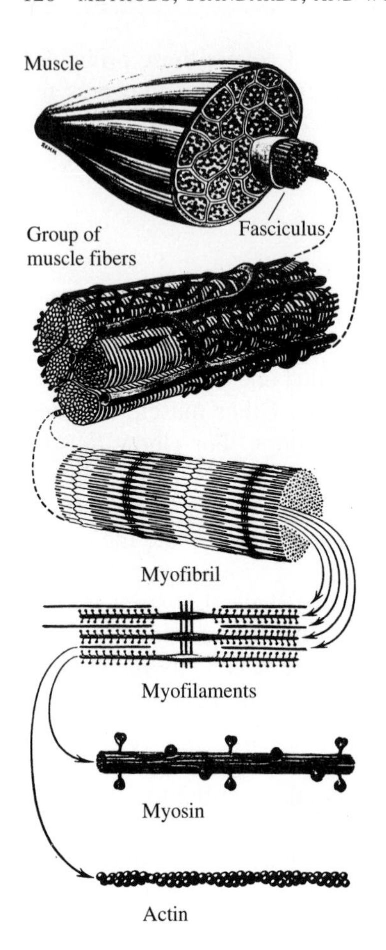

FIGURE 4–2
The structure of muscle. (*From:* Gray's Anatomy, 1973, by permission of W. B. Saunders Co., London.)

Each muscle fiber is further subdivided into smaller *myofibrils* and ultimately into the protein filaments that provide the contractile mechanism. There are two types of filaments: thick filaments, comprised of long proteins with molecular heads, called myosin; and thin filaments, comprised of globular proteins, called actin. The two types of filaments are interlaced, giving rise to the striated appearance and alternate name, as shown in Figure 4–3. This allows the muscle to contract as the filaments slide over one another, which occurs as molecular bridges or bonds are formed, broken, and reformed between the myosin heads and actin globules. This *sliding filament theory* explains how the muscle length can change from approximately 50 percent of its *resting length* (the neutral uncontracted length at approximately the midpoint in the normal range of motion) at complete contraction to 180 percent of its resting length at complete extension (see Figure 4–3).

FIGURE 4–3
Force–length relationship of skeletal muscle. (*From:* Winter, 1979, p. 114. Reprinted by permission of John Wiley & Sons, Inc.)

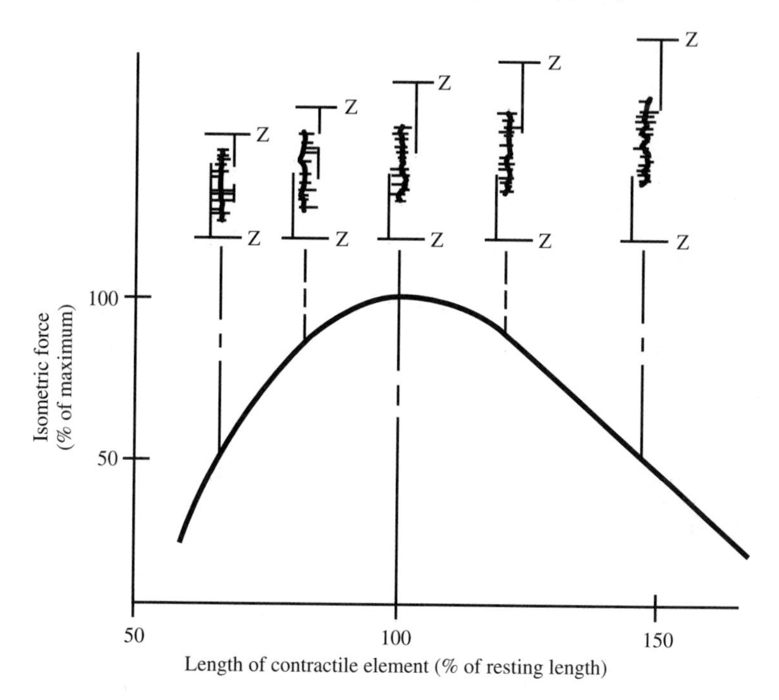

PRINCIPLES OF WORK DESIGN: MOTION ECONOMY

Achieve the Maximum Muscle Strength at the Midrange of Motion

The first principle of motion economy derives from the inverted-U-shaped property of muscle contraction shown in Figure 4–3. At the resting length, optimal bonding occurs between the thick and thin filaments. In the stretched state, there is minimal overlap or bonding between the thick and thin filaments, resulting in considerably decreased (almost zero) muscle force. Similarly, in the completely contracted state, interference occurs between the opposing thin filaments, again preventing optimum bonding and decreasing muscle force. This muscle property is typically termed the *force–length relationship.* Therefore, a task requiring considerable muscle force should be performed at the optimum position. For example, the neutral or straight position will provide the strongest grip strength for wrist motions. For elbow flexion, the strongest position would be with the elbow bent somewhat beyond the 90-degree position. For plantar flexion (i.e., depressing a pedal), again the optimum position is slightly beyond 90 degrees. A rough rule of

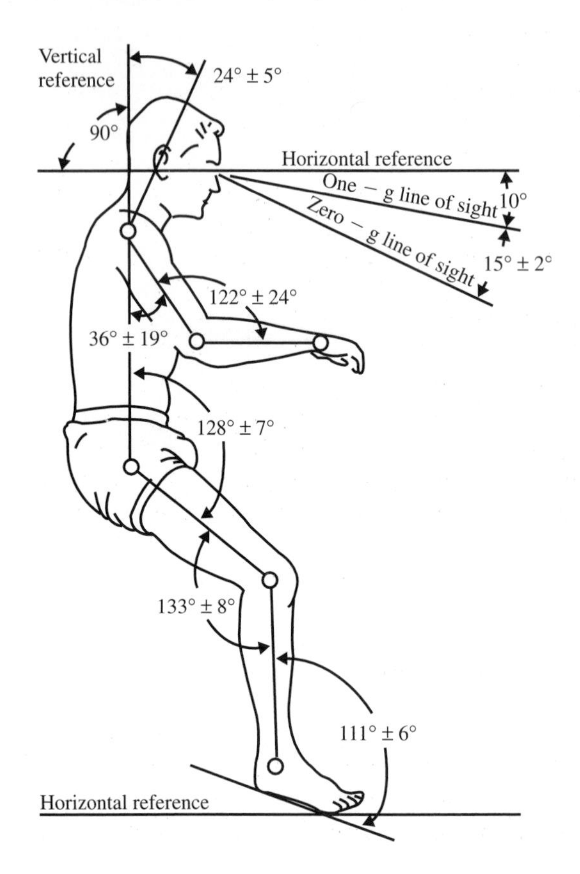

FIGURE 4–4
Typical relaxed posture assumed by people in weightless conditions. (*From:* Thornton, 1978, Fig. 16.)

thumb for finding the midrange of motion is to consider the posture assumed by an astronaut in weightless conditions when both agonist and antagonist muscles surrounding the joint are most relaxed and the limb attains a neutral position (see Figure 4–4).

Achieve the Maximum Muscle Strength with Slow Movements

The second principle of motion economy is based on another property of the sliding filament theory and muscle contraction. The faster the molecular bonds are formed, broken, and reformed, the less effective is the bonding and the less muscular force is produced. This is a pronounced nonlinear effect (see Figure 4–5) with maximum muscle force being produced with no externally measurable shortening (i.e., zero velocity or a static contraction), and minimal muscle force being produced at the maximum velocity of muscle shortening. The force is only sufficient to move the mass of that body segment. This muscle property is known as the *force–velocity relationship* and is especially important with respect to heavy manual work.

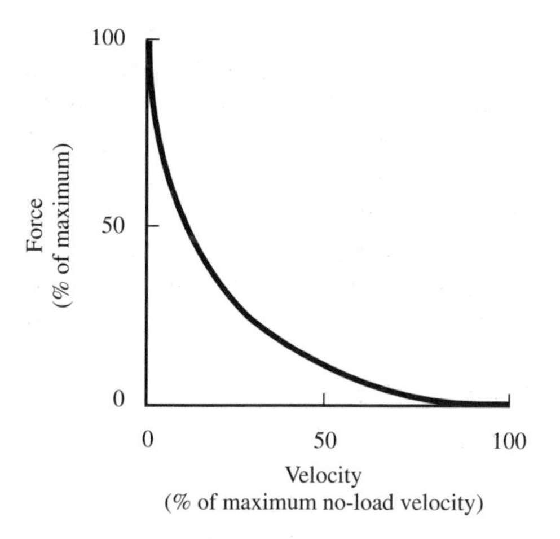

FIGURE 4–5
Force–velocity relationship of skeletal muscle.

Use Momentum to Assist Workers Wherever Possible; Minimize It If It Is Counteracted by Muscular Effort

There is a tradeoff between the second and third principles. Faster movements produce higher momentum and higher impact forces in the case of blows. Downward motions are more effective than upward motions, because of the assistance from gravity. To make full use of the momentum built up, workstations should allow operators to release a finished part into a delivery area while their hands are on their way to get component parts or tools to begin the next work cycle.

Design Tasks to Optimize Human Strength Capability

Human strength capability depends on three major task factors: (1) the type of strength; (2) the muscle or joint motion being utilized; and (3) posture. There are three types of muscle exertions, defined primarily by the way the strength of the exertion is measured. Muscular exertions resulting in body motions result from dynamic strength. These are sometimes termed *isotonic* contractions, because the load and body segments lifted nominally maintain a constant external force on the muscle. (However, the internal force produced by the muscle varies due to the geometry of the effective moment arms.) Because of the many variables involved in such contractions, some variables necessarily need to be constrained to obtain a measurable strength. Thus, dynamic strength measurements have typically been made using constant-velocity (*isokinetic*) dynamometers, such as the Cybex or the Mini-Gym (Freivalds and Fotouhi, 1987). In the case where the body motion is restrained, an *isometric* or static strength is obtained. As seen in Figure 4–5, an isometric strength is necessarily greater than a dynamic strength because of the more efficient bonding in the slower sliding muscle filaments. Some representative isometric muscle strengths for various postures are given in

TABLE 4–1

A. Static Muscle Strength Moment Data (ft-lbs) for 25 Men and 22 Women Employed in Manual Jobs in Industry

		Male (%ile)			Female (%ile)		
Muscle Function	**Joint Angles**	**5**	**50**	**95**	**5**	**50**	**95**
Elbow flexion	90° Included to arm (arm at side)	31	57	82	12	30	41
Elbow extension	70° Included to arm (arm at side)	23	34	49	7	20	28
Medial humeral (shoulder) rotation	90° Vertical shoulder (abducted)	21	38	61	7	15	24
Lateral humeral (shoulder) rotation	5° Vertical shoulder (at side)	17	24	38	10	14	21
Shoulder horizontal flexion	90° Vertical shoulder (abducted)	32	68	89	9	30	44
Shoulder horizontal extension	90° Vertical shoulder (abducted)	32	49	76	14	24	42
Shoulder vertical adduction	90° Vertical shoulder (abducted)	26	49	85	10	22	40
Shoulder vertical abduction	90° Vertical shoulder (abducted)	32	52	75	11	27	42
Ankle extension (plantar flexion)	90° Included to shank	51	93	175	29	60	97
Knee extension	120° Included to thigh (seated)	62	124	235	38	78	162
Knee flexion	135° Included to thigh (seated)	43	74	116	16	46	77
Hip extension	100° Included to torso (seated)	69	140	309	28	72	133
Hip flexion	110° Included to torso (seated)	87	137	252	42	93	131
Torso extension	100° Included to thigh (seated)	121	173	371	52	136	257
Torso flexion	100° Included to thigh (seated)	66	106	159	36	55	119
Torso lateral flexion	Sitting erect	70	117	193	37	69	120

B. Static Muscle Strength Moment Data (Nm) for 25 Men and 22 Women Employed in Manual Jobs in Industry

		Male (%ile)			Female (%ile)		
Muscle Function	**Joint Angles**	**5**	**50**	**95**	**5**	**50**	**95**
Elbow flexion	90° Included to arm (arm at side)	42	77	111	16	41	55
Elbow extension	70° Included to arm (arm at side)	31	46	67	9	27	39
Medial humeral (shoulder) rotation	90° Vertical shoulder (abducted)	28	52	83	9	21	33
Lateral humeral (shoulder) rotation	5° Vertical shoulder (at side)	23	33	51	13	19	28
Shoulder horizontal flexion	90° Vertical shoulder (abducted)	44	92	119	12	40	60
Shoulder horizontal extension	90° Vertical shoulder (abducted)	43	67	103	19	33	57
Shoulder vertical adduction	90° Vertical shoulder (abducted)	35	67	115	13	30	54
Shoulder vertical abduction	90° Vertical shoulder (abducted)	43	71	101	15	37	57
Ankle extension (plantar flexion)	90° Included to shank	69	126	237	31	81	131
Knee extension	120° Included to thigh (seated)	84	168	318	52	106	219
Knee flexion	135° Included to thigh (seated)	58	100	157	22	62	104
Hip extension	100° Included to torso (seated)	94	190	419	38	97	180
Hip flexion	110° Included to torso (seated)	118	185	342	57	126	177
Torso extension	100° Included to thigh (seated)	164	234	503	71	184	348
Torso flexion	100° Included to thigh (seated)	89	143	216	49	75	161
Torso lateral flexion	Sitting erect	95	159	261	50	94	162

(*From:* Chaffin and Anderson, 1991). (Reprinted by permission of John Wiley & Sons, Inc).

FIGURE 4–6
Static strength positions and results for 443 males, 108 females. (Chaffin et al., 1977)

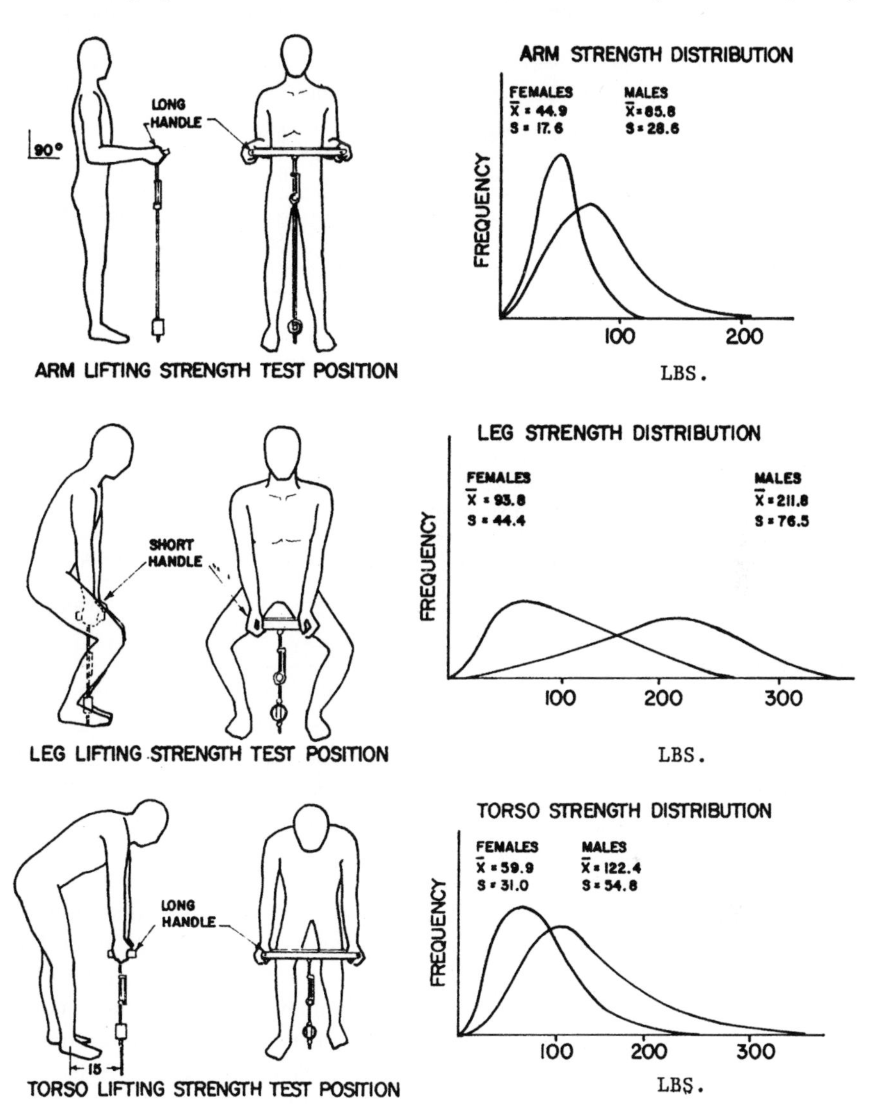

Table 4–1, and representative lifting strengths for 551 industrial workers in different postures are shown in Figure 4–6.

Most industrial tasks typically involve some movement; therefore, completely isometric contractions are relatively rare. Most typically, the movement range is somewhat limited, and the dynamic contraction is not a true isokinetic contraction, but a set of quasi-static contractions. Thus, dynamic strengths are very much task and condition dependent, and little is published regarding dynamic strength data.

TABLE 4–2
Maximum Weights (in lbs and kg) Acceptable to Average Males and Females Lifting Compact Boxes [14 inches (34 cm) wide] with Handles (Adapted from Snook and Ciriello, 1991)

	1 lift per 0.5 min				1 lift per 1 min				1 lift per 30 min			
	Males		Females		Males		Females		Males		Females	
Task	lbs	kg	lbs	kg	lbs	kg	lbs	kg	lbs	kg	lbs	kg
Floor to knuckle height	42	19	26	12	66	30	31	14	84	38	37	17
Knuckle to shoulder height	42	19	20	9	55	25	29	13	64	29	33	15
Shoulder to arm reach	37	17	18	8	51	23	24	11	59	27	29	13

For lowering increase values by 6%.
For boxes without handles, decrease values by 15%.
Increasing the box size (away from body) to 30 inches (75 cm) decrease values by 16%.

TABLE 4–3
Push Forces (in lbs and kg) at Waist Height Acceptable to Average Males and Females (I = Initial, S = Sustained) (Adapted from Snook and Ciriello, 1991)

	1 lift per minute								1 lift per 30 minutes							
	Males				Females				Males				Females			
	I		S		I		S		I		S		I		S	
Distance Pushed in Feet (Meters)	lbs	kg	lbs	kg	lbs	kg	lbs	kg	lbs	kg	lbs	kg	lbs	kg	lbs	kg
150 (45)	51	23	26	12	40	18	22	10	66	30	42	19	51	23	26	12
50 (15)	77	35	42	19	44	20	29	13	84	38	51	23	53	24	33	15
7 (2)	95	43	62	28	55	25	40	18	99	45	75	34	66	30	46	21

For push forces at shoulder heights or knuckle/knee heights, decrease values by 11%.

Finally, a third type of muscle strength capability, *psychophysical* strength, has been defined for those situations in which the strength demands are required for an extended period of time. A static strength capability is not necessarily representative of what would be repetitively possible over an 8-hour shift. Typically, the maximum acceptable load (determined by adjusting the load lifted or force exerted until the subject feels that the load or force would be acceptable on a repetitive basis for the given time period) is 40 to 50 percent less than a one-time static exertion. Extensive tables have been compiled for psychophysical strengths of various frequencies and postures (Snook and Ciriello, 1991). A summary of these values is provided in Tables 4–2, 4–3, and 4–4.

Use Large Muscles for Tasks Requiring Strength

Muscle strength is directly proportional to the size of the muscle, as defined by the cross-sectional area (specifically, 87 psi [60 N/cm^2] for both males and females)

TABLE 4–4
Pull Forces (in lbs and kg) at Waist Height Acceptable to Average Males and Females (I = Initial, S = Sustained) (Adapted from Snook and Ciriello, 1991)

	1 lift per minute								1 lift per 30 minutes							
	Males				Females				Males				Females			
Distance Pulled	I		S		I		S		I		S		I		S	
in Feet (Meters)	lbs	kg	lbs	kg	lbs	kg	lbs	kg	lbs	kg	lbs	kg	lbs	kg	lbs	kg
150 (45)	37	17	26	12	40	18	24	11	48	22	42	19	48	22	26	12
50 (15)	57	26	42	19	42	19	26	12	62	28	51	23	51	23	33	15
7 (2)	68	31	57	26	55	25	35	16	73	33	70	32	66	30	44	20

For pull forces at knuckle/knee heights, increases values by 75%.
For pull forces at shoulder heights, decreases values by 15%.

(Ikai and Fukunaga, 1968). For example, leg and trunk muscles should be used in heavy load lifting, rather than weaker arm muscles. The posture factor, although somewhat confounded by geometrical changes in the muscle moment or lever arm, is related to the resting length of the muscle fibers being roughly mid-range of motion for most joints as stated in the first principal of motion economy.

Stay Below 15 Percent of Maximum Voluntary Force

Muscle fatigue is a very important but little utilized criterion in designing tasks appropriately for the human operator. The human body and muscle tissue rely primarily on two types of energy sources, *aerobic* and *anaerobic* (see later section on Manual Work). Since the anaerobic metabolism can supply energy for only a very small period of time, the oxygen supplied to the muscle fibers via peripheral blood flow becomes critical in determining how long the muscle contractions will last. Unfortunately, the harder the muscle fibers contract, the more the interlaced arterioles and capillaries are compressed (see Figure 4–2), the more the blood flow and oxygen supplies are restricted, and the faster the muscle fatigues. The result is the endurance curve in Figure 4–7. The relationship is very nonlinear, ranging from a very short endurance time of approximately six seconds at a maximal contraction, at which point the muscle force rapidly drops off, to a rather indefinite endurance time at approximately 15 percent of a maximal contraction.

This relationship can be modeled by:

$$T = 1.2/(f - 0.15)^{0.618} - 1.21$$

where: T = endurance time (min)
f = required force, expressed as a fraction of maximum isometric strength

For example, a worker would be able to sustain a force level of 50 percent of maximum strength for only about one minute:

$$T = 1.2/(0.5 - 0.15)^{0.618} - 1.21 = 1.09 \text{ min}$$

FIGURE 4–7
Static muscle endurance–exertion level relationship with ±1 SD ranges depicted. (*From:* Chaffin and Andersson, 1991) (Reprinted by permission of John Wiley & Sons, Inc)

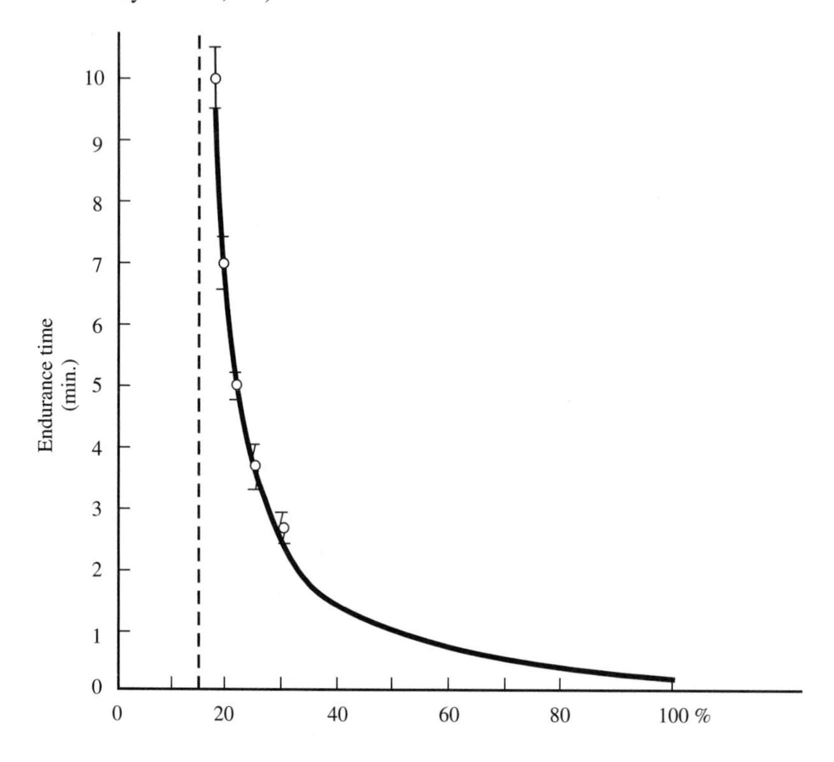

The indefinite asymptote is due to early researchers stopping their experimentation without reaching complete muscle fatigue. Later researchers suggested reducing this level of acceptable static force levels from 15 percent to below 10 percent, perhaps even 5 percent (Jonsson, 1978). The amount of rest needed to recover from a static hold will be presented as a set of relaxation allowances that depend on the force exerted and the holding time (see Chapter 10).

Use Short, Frequent, Intermittent Work–Rest Cycles

Whether performing repeated static contractions (such as holding a load in elbow flexion) or a series of dynamic work elements (such as cranking with the arms or legs), work and recovery should be apportioned in short, frequent cycles. This is due primarily to a fast initial recovery period, which then tends to level off with increasing time. Thus, most of the benefit is gained in a relatively short period. A much higher percentage of maximum strength can be maintained if the strength is exerted

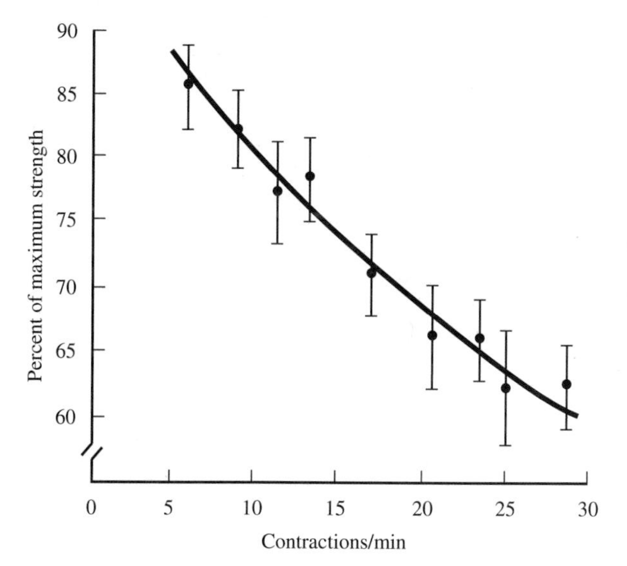

FIGURE 4–8
Percent of maximum isometric strength that can be maintained in a steady state during rhythmic contractions.
Points are averages for finger muscles, hand muscles, arm muscles, and leg muscles, combined. Vertical lines denote ± standard error.
(*From:* Åstrand and Rodahl, 1986)

as a series of repetitive contractions rather than one sustained static contraction (see Figure 4–8). However, if the person is driven to complete muscle (or whole body) fatigue, full recovery will take a fairly long time, perhaps several hours.

Design Tasks So That Most Workers Can Do Them

As can be seen in Figure 4–6, for a given muscle group, there is a considerable range of strength in the normal, healthy adult population, with the strongest being five to eight times stronger than the weakest. These large ranges are due to individual factors that affect strength performance: gender, age, handedness, and fitness/training. Gender accounts for the largest variation in muscle strength, with average female strength ranging from 35 to 85 percent of average male strength, with an average effect of 66 percent (see Figure 4–9).The difference is greatest for upper extremity strengths and smallest for lower extremity strengths. However, this effect is primarily due to average body size (i.e., total muscle mass) and not strictly to gender; the average female is considerably smaller and lighter than the average male. Furthermore, with the wide distribution for a given muscle strength, there are many females who are stronger than many males.

In terms of age, muscle strength appears to peak in the mid 20s and then decreases linearly by 20 to 25 percent by the mid 60s (see Figure 4–9). The decrease in strength is due to reduced muscle mass and a loss of muscle fibers. However, whether this loss is due to physiological changes of aging or just a gradual reduction in activity levels is not well known. It has definitely been shown that by starting a strength training program, a person can increase strength by 30 percent in the first several weeks, with maximum increases approaching 100 percent (Åstrand and Rodahl, 1986). In terms of handedness, the nondominant hand typically produces

FIGURE 4–9
Changes in maximal isometric strength with age in women and men. (*From:* Åstrand and Rodahl, 1986)

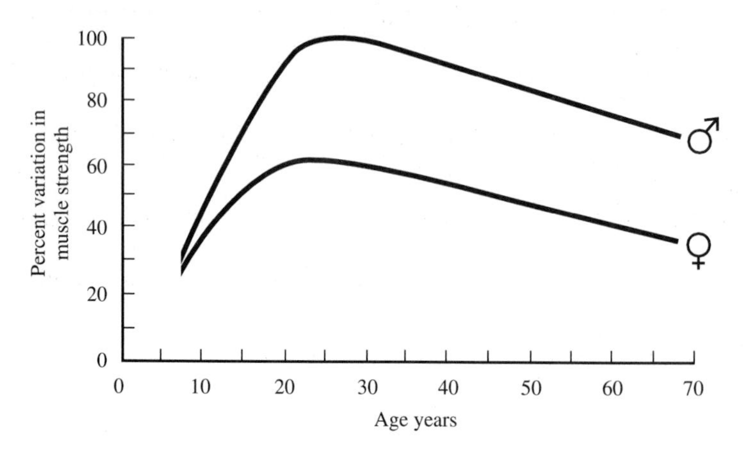

about 90 percent of the dominant hand's grip strength, with the effect being less pronounced in lefties, probably because they have been forced to adapt to a right-handed world (Miller and Freivalds, 1987). In any case, it is best to design tools and machines such that they can be used by either hand, to avoid placing any individual at a strength disadvantage.

Use Low Force for Precise Movements or Fine Motor Control

Muscle contractions are initiated by neural innervation from the brain and spinal cord, which together comprise the central nervous system. A typical motor neuron, or nerve cell leading to the muscle from the central nervous system, may innervate or have connections with several hundred muscle fibers. The innervation ratio of the number of fibers per neuron ranges from less than 10 in the small muscles of the eye to over 1,000 for the large calf muscles, and can vary considerably even within the same muscle. Such functional arrangement is called a *motor unit* and has important implications in movement control. Once a neuron is stimulated, the electrical potential is transferred simultaneously to all of the muscle fibers innervated by that neuron and the motor unit acts as one contractile or motor control unit. Also, the central nervous system tends to recruit these motor units selectively by increasing size as higher muscle forces are needed (Figure 4–10). The initial motor units recruited are small in size, with few muscle fibers and low produced forces. However, since these are small and low in tension, the change in force production from one to two or more motor units recruited is very gradual, and very fine precision in motor control can be produced. Near the end of motor recruitment, the total muscle force is high, and each additional motor unit recruited becomes a large increment in force, with little sensitivity in terms of precision or control. This muscle property is sometimes termed the *size principle*.

FIGURE 4–10
Muscle recruitment demonstrating size principle.

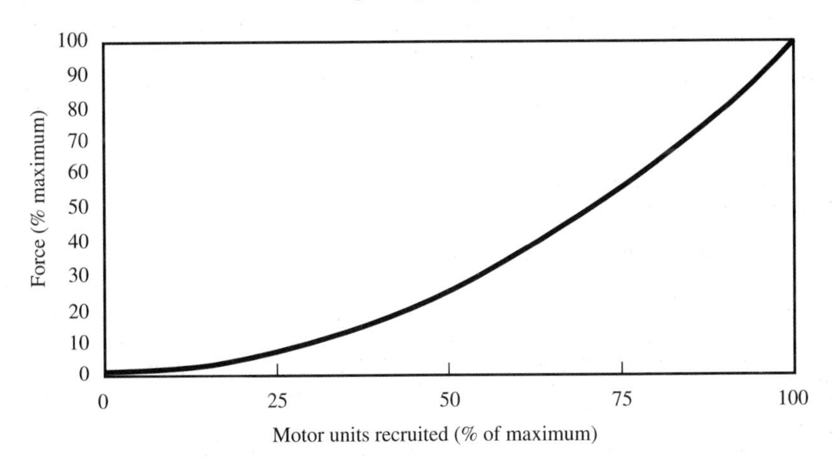

The electrical activity of muscles, termed *electromyograms* (EMG), is a useful measure of local muscle activity. Such activity is measured by placing recording electrodes on the skin surface over the muscles of interest, then modifying and processing to the amplitude and frequency of the signal. For amplitude analysis, the signal is typically rectified and smoothed (with a resistor–capacitor circuit). The result has a reasonably linear relationship to the muscle force exerted (Bouisset, 1973). The frequency approach involves digitizing the signal and performing a Fast Fourier Transform analysis to yield a frequency spectrum. As the muscle begins to fatigue, muscle activity shifts from high frequency (>60 Hz) to lower frequencies (<60 Hz) (Chaffin, 1969). Also, the EMG amplitude tends to increase with fatigue, for a given level of exertion.

Do Not Attempt Precise Movements or Fine Control Immediately after Heavy Work

This is a corollary to the previous principle of motion economy. The small motor units tend to be used continually during normal motion, and although more resistant to fatigue than the large motor units, they can still experience fatigue. A typical example where this principle is violated occurs when operators load their workstations before their shift or replenish parts during a shift. Lifting heavy parts containers requires the recruitment of the small motor units, as well as the later larger motor units, to generate the necessary muscle forces. During lifting and restocking, some of the motor units will fatigue and others will be recruited to compensate for the fatigued ones. Once the operator has restocked the bins and returned to more precise assembly work, some of the motor units, including the smaller precision ones, will not be available for use. The larger motor units recruited to replace the fatigued ones will provide larger increments in force and less precise motor control. After several minutes, the motor units will have recovered and will be available, but in the meantime, the quality and speed of the assembly work may suffer. One solution would be to provide less-skilled laborers to restock the bins on a regular basis.

Use Ballistic Movements for Speed

Through spinal reflexes, cross innervation of agonists and antagonists always occurs. This minimizes any unnecessary conflict between the muscles, as well as the consequent excess energy expenditure. Typically, in a short (less than 200 msec), gross, voluntary motion, the agonist is activated and the antagonist is inhibited (termed *reciprocal inhibition*), to reduce counterproductive muscle contractions. On the other hand, for precise movements, feedback control from both sets of muscles is utilized, increasing motion time. This is sometimes referred to as the *speed–accuracy trade-off.*

Begin and End Motions with Both Hands Simultaneously

When the right hand is working in the normal area to the right of the body and the left hand is working in the normal area to the left of the body, a feeling of balance tends to induce a rhythm in the operator's performance, which leads to maximum productivity. The left hand, in right-handed people, can be just as effective as the right hand and it should be used. A right-handed boxer learns to jab just as effectively with his left hand as with his right hand. A speed typist is just as proficient with one hand as the other. In a large number of instances, workstations can be designed to do "two at a time." Using dual fixtures to hold two components, both hands can work at the same time, making symmetrical moves in opposite directions (see Figure 4–11). A corollary to this principle is that both hands should not be idle at the same time, except during rest periods. (This principle was the one followed by Frank Gilbreth in shaving with both hands simultaneously.)

Move the Hands Symmetrically and Simultaneously To and From the Center of the Body

It is natural for the hands to move in symmetrical patterns. Deviations from symmetry in a two-handed workstation result in slow, awkward movements of the operator. The difficulty of patting the stomach with the left hand while rubbing the top of the head with the right hand is familiar to many. Another experiment that can readily illustrate the difficulty of performing nonsymmetrical operations is to try to draw a circle with the left hand while trying to draw a square with the right hand. Figure 4–12 illustrates an ideal workstation that allows the operator to assemble two products by going through a series of symmetrical, simultaneous motions away from and toward the center of the body.

FIGURE 4–11
Workstation and fixture designed to allow productive work with both hands. (General Electric Co.)

Use the Natural Rhythms of the Body

The spinal reflexes that excite or inhibit muscles also lead to natural rhythms in the motion of body segments. These can be logically compared to second-order mass–spring–dashpot systems, with the body segments providing mass and the muscle having internal resistance and damping. The natural frequency of the system

FIGURE 4–12
An ideal workstation that permits the operator to assemble two products by going through a series of symmetrical motions made simultaneously away from and toward the center of the body. General Electric Co.

will depend on all three parameters, but the segment mass will have the greatest effect. This natural frequency is essential to the smooth and automatic performance of a task. Drillis (1963) has studied a variety of very common manual tasks and has suggested optimum work tempos, as follows:

Filing metal—	60–78 strokes per minute
Chiseling—	60 strokes per minute
Arm cranking—	35 revolutions per minute
Leg cranking—	60–72 revolutions per minute
Shoveling—	14–17 tosses per minute

Use Continuous Curved Motions

Because of the nature of body segment linkages (typically approximating pin joints), it is easier for the human to produce curved motions, that is, to pivot around a joint. Straight-line motions involving sudden and sharp changes in direction

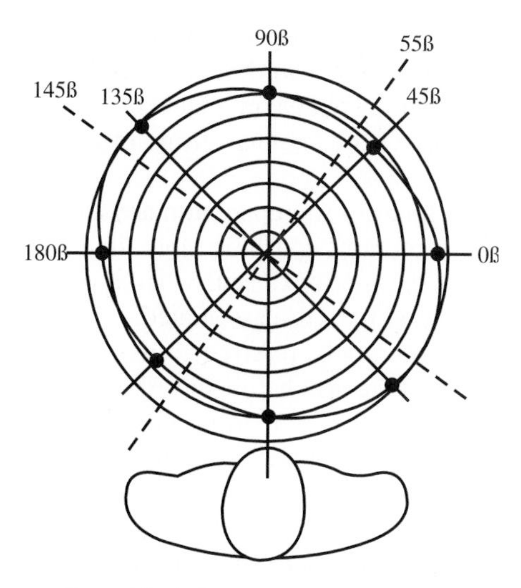

Concentric circles represent
equal time intervals.

FIGURE 4–13
Forearm motion is best while pivoting on elbow (*From:* Sanders and McCormick, 1993) (Reproduced with permission of the McGraw-Hill Companies.)

require more time and are less accurate. This law is very easily demonstrated by moving either hand in a rectangular pattern, and then moving it in a circular pattern of about the same magnitude. The greater amount of time required to make the abrupt 90-degree directional changes is quite apparent. To make a directional change, the hand must decelerate, change direction, and accelerate until it is time to decelerate again for the next directional change. Continuous curved motions do not require deceleration, and are consequently performed faster per unit of distance. This is demonstrated very nicely in Figure 4–13, with subjects who made positioning movements with the right hand in eight directions in a horizontal plane from a center starting point. Motion from the lower left to the upper right (pivoting about the elbow) required 20 percent less time than the perpendicular motion from the lower right to upper left (additional awkward shoulder and arm line movements).

Use the Lowest Practical Classification of Movement

Understanding the classifications of motions plays a major role in using this fundamental law of motion economy appropriately in methods studies. The classifications are as follows:

1. Finger motions are the fastest of the five motion classes and are readily recognized, for they are made by moving the finger or fingers while the remainder of the arm is kept stationary. Typical finger motions are running a nut down on a stud, depressing the keys of a typewriter, or grasping a small part. There is usually a significant difference in the time required to perform finger motions with the various fingers. In most cases, the index finger moves considerably faster than the other fingers.

Note that repetitive finger motions can result in fatigue and work-related musculoskeletal disorders, such as cumulative trauma disorders. Therefore, finger forces should be kept low by using bar switches in place of trigger switches, and by designing tasks that involve high manual effort to higher motion classifications.

2. Finger and wrist motions are made while the forearm and upper arm are stationary. In the majority of cases, finger and wrist motions consume more time than strictly finger motions. Typical finger and wrist motions occur when a part is positioned in a jig or fixture, or when two mating parts are assembled.
3. Finger, wrist, and lower arm motions are commonly referred to as *forearm motions* and include those movements made by the arm below the elbow while the upper arm is stationary. Since the forearm includes a strong muscle, such motions are usually considered efficient because they are not fatiguing. However, repetitive work involving force with the arms extended can induce soreness. This can usually be relieved by designing the workstation so that the elbows can be kept at 90 degrees while work is being done. The time required by a given operator to make forearm motions depends on the distance moved and the amount of resistance overcome during the movement. Workstations should be designed so that these third-class motions, rather than fourth-class motions, are used to perform transport motions, to minimize cycle times.
4. Finger, wrist, lower arm, and upper arm motions, commonly known as *fourth-class* or *shoulder motions,* are probably used more than any other motion class. The fourth-class motion for a given distance takes considerably more time than the three classes previously described. Fourth-class motions are required to perform transport motions for parts that cannot be reached without extending the arm. The time required to perform fourth-class motions depends primarily on the distance of the move and the resistance to the move. To reduce static loading of shoulder motions, tools should be designed so that the elbow is not elevated while the work is being performed. For example, by using a socket wrench instead of an open-end wrench, the operator can approach the nut to be assembled from an angle without having to lift the elbow.
5. Fifth-class motions include body motions, which are the most time-consuming. Body motions include movements of the ankle, knee, and thigh, as well as movements of the trunk.

First-class motions require the least amount of effort and time, while fifth-class motions are considered the least efficient. Therefore, always utilize the lowest practicable motion classification to perform the work properly. This will involve careful consideration of the location of tools and materials, so that ideal motion patterns can be arranged.

This classification of movement was shown experimentally by Langolf, Chaffin, and Foulke (1976), in a series of positioning movements to and from targets, known as Fitts' Tapping Task (Fitts, 1954). Mathematically, the task can be expressed as:

$$MT = a + b \times \log_2(2D/W)$$

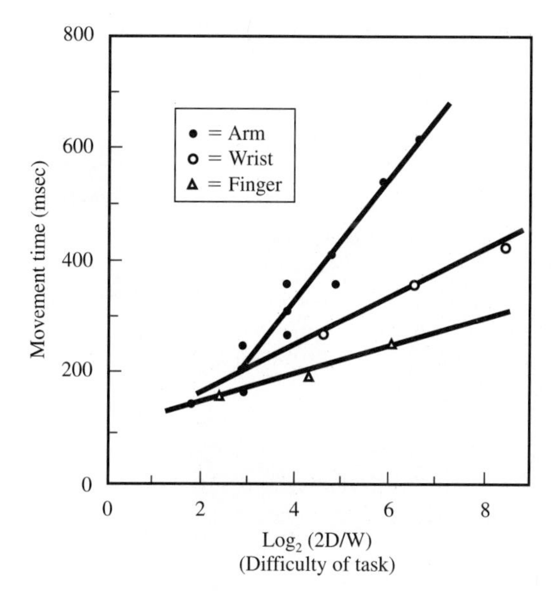

FIGURE 4–14
Classifications of movements (*From:* Sanders and McCormick, 1993) (Reproduced with permission of the McGraw-Hill Companies.)

where: MT = Movement time (msec)
D = Distance between centers of targets
W = Width of target
a = Intercept of the regression line
b = Slope of the regression line

The movement time increases with the difficulty of the task (see Figure 4–14), but also increases with higher levels of classification; that is, the slope for the arm (b = 105 msec) is steeper than for the wrist (b = 45 msec), which in turn is steeper than for the finger (b = 26 msec). The effect is due simply to the added time required for the central nervous system to process additional joints, motor units, and receptors.

Work with Both Hands and Feet Simultaneously

Since the major part of work cycles is performed by the hands, it is economical to relieve the hands of work that can be done by the feet, but only if this work is performed while the hands are occupied. Since the hands are more skillful than the feet, it would be folly to have the feet perform elements while the hands are idle. Foot pedal devices that allow clamping, parts ejection, or feeding can often be arranged, freeing the hands for other, more useful work and consequently reducing the cycle time (see Figure 4–15). When the hands are moving, the feet should not be moving, since the simultaneous movement of the hands and feet is difficult. However, the feet can be applying pressure to something, such as a foot pedal. Also, the operator should be seated, as it is difficult to operate a foot pedal while standing, which would mean maintaining the full body weight on the other foot.

FIGURE 4–15
This foot-operated press permits the operator's hands to procure parts in preparation for the next cycle, while the press is in operation. General Electric Co.

Minimize Eye Fixations

Although eye fixations or eye movements cannot be eliminated for most work, the location of the primary visual targets should be optimized with respect to the human operator. The normal line of sight is roughly about 15 degrees below the horizontal, and the primary visual field is roughly defined as a cone ±15 degrees in arc centered on the line of sight. The implication is that within this area, minimal eye movements are needed and eye fatigue is minimized.

Summary: Motion Economy Principles

The principles of motion economy are based on an elementary understanding of human physiology and should be very useful in applying methods analysis with the human operator in mind. However, the analyst need not be an expert in human anatomy and physiology to be able to apply these principles. In fact, for most task analysis purposes, it may be sufficient to use the Motion Economy Checklist, which summarizes most of the principles in a questionnaire format (see Figure 4–16).

FIGURE 4–16
Motion Economy Checklist.

Suboperations	**Yes**	**No**
1. Can a suboperation be eliminated?	❑	❑
a. As unnecessary?	❑	❑
b. By a change in the order of the work?	❑	❑
c. By a change of tools or equipment?	❑	❑
d. By a change in layout of the workplace?	❑	❑
e. By combining tools?	❑	❑
f. By a slight change of material?	❑	❑
g. By a slight change in product?	❑	❑
h. By a quick-acting clamp on the jigs or fixtures?	❑	❑
2. Can a suboperation be made easier?	❑	❑
a. By better tools?	❑	❑
b. By changing leverages?	❑	❑
c. By changing positions of controls or tools?	❑	❑
d. By better material containers?	❑	❑
e. By using inertia where possible?	❑	❑
f. By lessening visual requirements?	❑	❑
g. By better workplace heights?	❑	❑

Movements	**Yes**	**No**
1. Can a movement be eliminated?	❑	❑
a. As unnecessary?	❑	❑
b. By a change in the order of work?	❑	❑
c. By combining tools?	❑	❑
d. By a change in tools or equipment?	❑	❑
e. By a drop disposal of finished material?	❑	❑
2. Can a movement be made easier?	❑	❑
a. By a change in layout, shortening distances?	❑	❑
b. By changing the direction of movements?	❑	❑
c. By using different muscles?	❑	❑
Use the first muscle group that is strong enough for the task:		
(1) Finger?	❑	❑
(2) Wrist?	❑	❑
(3) Forearm?	❑	❑
(4) Upper arm?	❑	❑
(5) Trunk?	❑	❑
d. By making movements continuous rather than jerky?	❑	❑

Holds	**Yes**	**No**
1. Can a hold be eliminated? (Holding is extremely fatiguing.)	❑	❑
a. As unnecessary?	❑	❑
b. By a simple holding device or fixture?	❑	❑
2. Can a hold be made easier?	❑	❑
a. By shortening its duration?	❑	❑
b. By using stronger muscle groups, such as the legs with foot-operated vises?	❑	❑

Delays	**Yes**	**No**
1. Can a delay be eliminated or shortened?	❑	❑
a. As unnecessary?	❑	❑
b. By a change in the work each body member does?	❑	❑
c. By balancing the work between the body members?	❑	❑
d. By working simultaneously on two items?	❑	❑
e. By alternating the work, each hand doing the same job, but out of phase?	❑	❑

Cycles	**Yes**	**No**
1. Can the cycle be rearranged so that more of the handwork is done during running time?	❑	❑
a. By automatic feed?	❑	❑
b. By automatic supply of material?	❑	❑
c. By change of man and machine phase relationship?	❑	❑
d. By automatic power cutoff at completion of cut or in case of tool or material failure?	❑	❑

Machine Time	**Yes**	**No**
1. Can the machine time be shortened?	❑	❑
a. By better tools?	❑	❑
b. By combined tools?	❑	❑
c. By higher feeds or speeds?	❑	❑

MOTION STUDY

Motion study is the careful analysis of body motions employed in doing a job. The purpose of motion study is to eliminate or reduce ineffective movements, and facilitate and speed effective movements. Through motion study, in conjunction with the principles of motion economy, the job is redesigned to be more effective and to produce a higher rate of output. The Gilbreths pioneered the study of manual motion and developed basic laws of motion economy that are still considered fundamental. They were also responsible for the development of detailed motion picture studies, known as *micromotion studies*, which have proved invaluable in studying highly repetitive manual operations. Motion study, in the broad sense, covers both studies that are performed as a simple visual analysis and studies that utilize more expensive equipment. Traditionally, motion picture film cameras were utilized, but today the videotape camera is used exclusively, because of the easy ability to rewind and replay sections, the freeze-frame capability in 4-head videotape cassette recorders (VCR), and the elimination of the need for film development. In view of its much higher cost, micromotion is usually practical only on extremely active jobs with high repetitiveness.

The two types of studies may be compared to viewing a part under a magnifying glass versus viewing it under a microscope. The added detail revealed by the microscope is needed only on the most productive jobs. Traditionally, micromotion studies were recorded on a *simultaneous motion (simo) chart*, while motion studies are recorded on a *two-hand process chart.* A true simo chart is hardly used today, but the term is sometimes applied to a two-hand process chart.

Basic Motions

As part of motion analysis, the Gilbreths concluded that all work, whether productive or nonproductive, is done by using combinations of 17 basic motions that they called *therbligs* (Gilbreth spelled backward). The therbligs can be either effective or ineffective. Effective therbligs directly advance the progress of the work. They can frequently be shortened, but typically cannot be completely eliminated. Ineffective therbligs do not advance the progress of the work and should be eliminated by applying the principles of motion economy. The 17 therbligs, along with their symbols and definitions, are shown in Table 4–5.

The Two-Hand Process Chart

The *two-hand process chart*, sometimes referred to as an *operator process chart* is a motion study tool. This chart shows all movements and delays made by the right and left hands, and the relationships between the basic divisions of accomplishment as performed by the hands. The purpose of the two-hand process chart is to present a given operation in sufficient detail that the operation can be analyzed and improved. Usually, it is not practical to make a detailed study through the two-hand

TABLE 4–5
Gilbreths' Therbligs

Effective Therbligs
(Directly advance progress of work. May be shortened but difficult to eliminate completely.)

Therblig	Symbol	Description
Reach	RE	Motion of empty hand to or from object; time depends on distance moved; usually preceded by Release and followed by Grasp.
Move	M	Movement of loaded hand; time depends on distance, weight, and type of move; usually preceded by Grasp and followed by Release or Position.
Grasp	G	Closing fingers around an object; begins as the fingers contact the object and ends when control has been gained; depends on type of grasp; usually preceded by Reach and followed by Move.
Release	RL	Relinquishing control of object, typically the shortest of the therbligs.
Pre-Position	PP	Positioning object in predetermined location for later use; usually occurs in conjunction with Move, as in orienting a pen for writing.
Use	U	Manipulating tool for intended use; easily detected, as it advances the progress of work.
Assemble	A	Bringing two mating parts together; usually preceded by Position or Move; followed by Release.
Disassemble	DA	Opposite of assemble, separating mating parts; usually preceded by Grasp and followed by Move or Release.

Ineffective Therbligs
(Do not advance progress of work. Should be eliminated if possible.)

Therblig	Symbol	Description
Search	S	Eyes or hands groping for object; begins as the eyes move in to locate an object.
Select	SE	Choosing one item from several; usually follows Search.
Position	P	Orienting object during work, usually preceded by Move and followed by Release (as opposed to *during* for Pre-Position).
Inspect	I	Comparing object with standard, typically with sight, but could also be with the other senses.
Plan	PL	Pausing to determine next action; usually detected as a hesitation preceding Motion.
Unavoidable Delay	UD	Beyond the operator's control due to the nature of the operation, e.g., left hand waiting while right hand completes a longer Reach.
Avoidable Delay	D	Operator solely responsible for idle time, e.g., coughing.
Rest to Overcome Fatigue	R	Appears periodically, not every cycle, depends on the physical workload.
Hold	H	One hand supports object while other does useful work.

process chart unless a highly repetitive manual operation is involved. Through the motion analysis of the two-hand process chart, inefficient motion patterns can be identified, and violations of the principles of motion economy can readily be observed. This chart facilitates changing a method so that a balanced two-handed operation can be achieved and ineffective motions can be either reduced or eliminated. The result is a smoother, more rhythmic cycle that keeps both delays and operator fatigue to a minimum.

As usual, the analyst heads the chart Two-Hand Process Chart, and adds all necessary identifying information, including the part number, drawing number, operation or process description, present or proposed method, date, and name of the person doing the charting. Immediately below the identifying information, the analyst sketches the workstation, drawn to scale. The sketch materially aids in presenting the method under study. Figure 4–17 shows a typical two-hand process chart for a cable-clamp assembly, with the times for each therblig obtained from stopwatch timing.

After completely identifying the operation and making a sketch of the workstation, showing the dimensional relationships, the analyst begins constructing the two-hand process chart. This chart is also drawn to scale. The analyst observes the duration of the cycle, to determine the amount of time to be represented by each 1/4 inch of vertical space on the chart. For example, if the operation has a cycle time of 0.70 minutes and there are 7 vertical inches of available charting space, then each 1/4 inch of chart space would equal 0.025 minutes.

It is usually less confusing to chart the activities of one hand completely, and then chart all the basic divisions of accomplishment performed by the other hand. Although there is no fixed rule regarding what part of the work cycle to use as a starting point, it is usually best to start plotting immediately after the release of the finished part. If this release is done with the right hand, the next movement that would normally take place would be the first motion shown on the two-hand process chart. For the right hand, this would probably be "reach for new part." If the analyst observes that the "reach" element takes about 0.025 minutes to perform, this duration is indicated by a horizontal line drawn across the right-hand side of the paper 1/4 inch from the top. Under the "Symbols" column, "RE" (for 'reach') is written, indicating that an effective motion has been accomplished. Immediately to the right of the symbol, a brief description of the event, such as "Reach 20 inches for a 1/2-inch nut" is entered. Immediately below, the next basic division is shown, and so on until the cycle is completed. The analyst then proceeds to plot the left hand activities during the cycle. While plotting the left hand's activities, the analyst verifies that the end points of the therbligs actually occur at the same point, as indicated—a check for overall plotting errors.

All elements must be large enough to be measured, since it is not possible in most instances to time individual therbligs. For example, in Figure 4–17 the first element performed by the left hand was "Get U-bolt." This element comprised the therbligs "reach" and "grasp.' It would not have been possible to time either of these therbligs individually. Only through the use of a motion picture camera or videotape can time values as short as these be measured. The observer in Figure 4–17 was using a decimal minute watch. By observing one element at a time, the observer broke the elements into fractional minute periods.

FIGURE 4–17
Two-hand process chart for assembly of cable clamps

Two-Hand Process Chart Page 1 of 1

Operation: Assemble Cable Clamps	Part: SK-112	Summary	Left Hand	Right Hand
Operator Name and No.: J.B. #1157		Effective Time:	2.9	12.2
Analyst: G. Thuering	Date: 6-11-98	Ineffective Time:	11.4	2.1
Method (circle choice): Present (circled) Proposed		Cycle Time =	14.30 sec.	

Sketch:

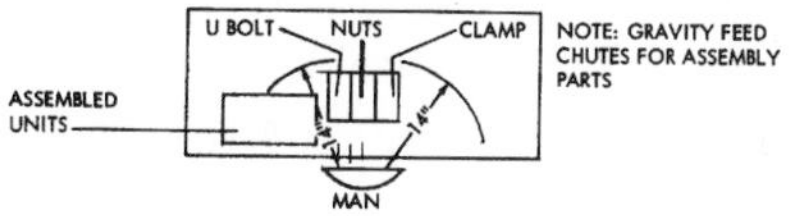

U BOLT, NUTS, CLAMP, ASSEMBLED UNITS, MAN

NOTE: GRAVITY FEED CHUTES FOR ASSEMBLY PARTS

Left Hand Description	Symbol	Time	Time	Symbol	Right Hand Description
Get U-Bolt (10")	RE G	1.00	1.00	RE G	Get Cable Clamp (10")
Place U-Bolt (10")	M P RL	1.20	1.20	M P RL	Place Cable Clamp (10")
Hold U-Bolt	H	11.00	1.00	RE G	Get First Nut (9")
			1.20	M P	Place First Nut (9")
			3.40	U RL	Run Down First Nut
			1.00	RE G	Get Second Nut (9")
			1.20	M P	Place Second Nut (9")
			3.40	U RL	Run Down Second Nut
Dispose of Assembly	M RL	1.10	0.90	UD	Wait

After the activities of both the right and the left hand have been charted, the analyst creates a summary at the bottom of the sheet, indicating the cycle time, pieces per cycle, and time per piece. Once the two-hand process chart has been completed for an existing method, the analyst can determine what improvements can be introduced. Several important corollaries to the principles of motion economy should be applied at this point. They include:

1. Establish the best sequences of therbligs.
2. Investigate any substantial variation in the time required for a given therblig and determine the cause.
3. Examine and analyze hesitations, to determine and then eliminate their causes.
4. As a goal, aim for cycles and portions of cycles completed in the least amount of time. Study deviations from these minimum times to determine the causes.

In the example, the "delays" and "holds" are good places to begin. For example, in Figure 4–17, the left hand acted as a holding device for almost the entire cycle. This would suggest the development of a fixture to hold the U-bolt. Further considerations to achieve balanced motions of both hands would suggest that when the fixture holds the U-bolts, the left hand and the right hand would each completely assemble a cable clamp. Additional study of this chart might result in the introduction of an automatic ejector and gravity chute, to eliminate the final cycle element, "dispose of assembly." The use of the Therblig Analysis Checklist (see Figure 4–18) may also be helpful in this analysis.

In summary, the two-hand process chart is an effective tool to:

1. Balance the motions of both hands and reduce fatigue.
2. Reduce or eliminate nonproductive motions.
3. Shorten the duration of productive motions.
4. Train new operators in the ideal method.
5. Sell the proposed method.

MANUAL WORK AND DESIGN GUIDELINES

Although automation has significantly reduced the demands for human power in the modern industrial environment, muscular strength still remains an essential part of many occupations, particularly those involving manual materials handling (MMH) or manual work. In these activities, overexertion from moving heavy loads can highly stress the musculoskeletal system, resulting in nearly a third of all occupational injuries. The low back alone accounts for almost a quarter of these injuries and a quarter of the annual workers' compensation costs (National Safety Council, 1997). Back injuries are especially detrimental because they often result in permanent disorders, with considerable discomfort and limitations for the employee as well as a large expense for the employer (an average case involving surgery may exceed $60,000 in direct costs).

Energy Expenditure and Workload Guidelines

Energy is required for the muscle contraction process. The molecule called *adenosine triphosphate* (ATP) is the immediate energy source, which physically interacts with the protein cross bridging as one of the ATP's high-energy phosphate bonds is broken. This source is very limited, lasting only several seconds, and the ATP must

FIGURE 4–18
Therblig analysis checklist.

Reach and Move	Yes	No
1. Can either of these therbligs be eliminated?	❑	❑
2. Can distances be shortened to advantage?	❑	❑
3. Are the best means (conveyors, tongs, tweezers) being used?	❑	❑
4. Is the correct body member (fingers, wrist, forearm, shoulder) being used?	❑	❑
5. Can a gravity chute be employed?	❑	❑
6. Can transports be effected through mechanization and foot-operated devices?	❑	❑
7. Will time be reduced by transporting in larger units?	❑	❑
8. Is time increased because of the nature of the material being moved or because of a subsequent delicate positioning?	❑	❑
9. Can abrupt changes in direction be eliminated?	❑	❑

Grasp	Yes	No
1. Would it be advisable for the operator to grasp more than one part or object at a time?	❑	❑
2. Can a contact grasp be used rather than a pickup grasp?	❑	❑
3. In other words, can objects be slid instead of carried?	❑	❑
4. Will a lip on the front of bins simplify grasping small parts?	❑	❑
5. Can tools or parts be pre-positioned for easy grasp?	❑	❑
6. Can a vacuum, magnet, rubber fingertip, or other device be used to advantage?	❑	❑
7. Can a conveyor be used?	❑	❑
8. Has the jig been designed so that operators may grasp the part easily when removing it?	❑	❑
9. Can the previous operator pre-position the tool or the work, simplifying grasp for the next operator?	❑	❑
10. Can tools be pre-positioned on a swinging bracket?	❑	❑
11. Can the work table surface be covered with a layer of sponge material so that the fingers can enclose small parts more easily?	❑	❑

Release	Yes	No
1. Can the release be made in transit?	❑	❑
2. Can a mechanical ejector be used?	❑	❑
3. Are the bins that contain the part after its release the proper size and design?	❑	❑
4. At the end of the therblig release, are the hands in the most advantageous position for the next therblig?	❑	❑
5. Can multiple units be released?	❑	❑

Pre-Position	Yes	No
1. Can a holding device at the workstation keep tools in the proper positions and the handles in upright positions?	❑	❑
2. Can tools be suspended?	❑	❑
3. Can a guide be used?	❑	❑
4. Can a magazine feed be used?	❑	❑
5. Can a stacking device be used?	❑	❑
6. Can a rotating fixture be used?	❑	❑

Use	Yes	No
1. Can a jig or fixture be used?	❑	❑
2. Does the activity justify mechanized or automated equipment?	❑	❑
3. Would it be practical to make the assembly in multiple units?	❑	❑
4. Can a more efficient tool be used?	❑	❑
5. Can stops be used?	❑	❑
6. Is the tool being operated at the most efficient feeds and speeds?	❑	❑
7. Should a power tool be employed?	❑	❑

(continued)

immediately be replenished from another molecule *creatine phosphate (CP)*. The CP source is also limited, less than one minute of duration (see Figure 4–19), and must ultimately be regenerated from the metabolism of the basic foods we eat: carbohydrates, fats, and proteins. This metabolism can occur in two different modes: *aerobic*, requiring oxygen, and *anaerobic*, not using oxygen. Aerobic metabolism is

FIGURE 4–18 ***(concluded)***
Therblig analysis checklist.

Search	**Yes**	**No**
1. Are articles properly identified?	❑	❑
2. Perhaps labels or color could be utilized?	❑	❑
3. Can transparent containers be used?	❑	❑
4. Will a better layout of the workstation eliminate searching?	❑	❑
5. Is proper lighting being used?	❑	❑
6. Can tools and parts be pre-positioned?	❑	❑

Select	**Yes**	**No**
1. Are common parts interchangeable?	❑	❑
2. Can tools be standardized?	❑	❑
3. Are parts and materials stored in the same bin?	❑	❑
4. Can parts be pre-positioned in a rack or tray?	❑	❑

Position	**Yes**	**No**
1. Can such devices as a guide, funnel, bushing, stop, swinging bracket, locating pin, recess, key, pilot, or chamfer be used?	❑	❑
2. Can tolerances be changed?	❑	❑
3. Can the hole be counterbored or countersunk?	❑	❑
4. Can a template be used?	❑	❑
5. Can the elimination of burrs decrease the problem of positioning?	❑	❑
6. Can the article be pointed to act as a pilot?	❑	❑

Inspect	**Yes**	**No**
1. Can inspection be eliminated or combined with another operation or therblig?	❑	❑
2. Can multiple gages or tests be used?	❑	❑
3. Will inspection time be reduced by increasing the illumination?	❑	❑
4. Are the articles being inspected at the correct distance from the worker's eyes?	❑	❑
5. Will a grazing light accentuate defects and facilitate inspection?	❑	❑
6. Would an electric eye be useful?	❑	❑
7. Does the volume justify automatic electronic inspection?	❑	❑
8. Would a magnifying glass facilitate the inspection of small parts?	❑	❑
9. Is the best inspection method being used?	❑	❑
10. Has consideration been given to polarized light, template gages, sound tests, performance tests, and so on?	❑	❑

Rest to Overcome Fatigue	**Yes**	**No**
1. Is the best order-of-muscles classification being used?	❑	❑
2. Are temperature, humidity, ventilation, noise, light, and other working conditions satisfactory?	❑	❑
3. Are benches of the proper height?	❑	❑
4. Can the operator alternately sit and stand while performing work?	❑	❑
5. Does the operator have a comfortable chair of the right height?	❑	❑
6. Are mechanical means being used for heavy loads?	❑	❑
7. Is the operator aware of his or her average intake requirements in calories per day?	❑	❑

Hold	**Yes**	**No**
1. Can a mechanical jig, such as a vise, pin, hook, rack, clip, or vacuum, be used?	❑	❑
2. Can friction be used?	❑	❑
3. Can a magnetic device be used?	❑	❑
4. Should a twin holding fixture be used?	❑	❑

much more efficient, generating 38 ATPs for each *glucose* molecule (basic unit of carbohydrates), but it is relatively slow. Anaerobic metabolism is very inefficient, producing only 2 ATPs for each glucose molecule, but it is much quicker. Also, the glucose molecule is only partially broken down into two lactate molecules, which in the watery environment of the body, forms *lactic acid*, a direct correlate of fatigue. Thus, during the first few minutes of heavy work, the ATP and CP energy sources

FIGURE 4–19
Sources of energy during the first few minutes of moderately heavy work.
High-energy phosphate stores (ATP and CP) provide most of the energy during the first seconds of work. Anaerobic glycolysis supplies less and less of the energy required as the duration of work increases, and aerobic metabolism takes over. (*From:* Sanders and McCormick, 1993) (Reproduced with permission of the McGraw-Hill Companies.)

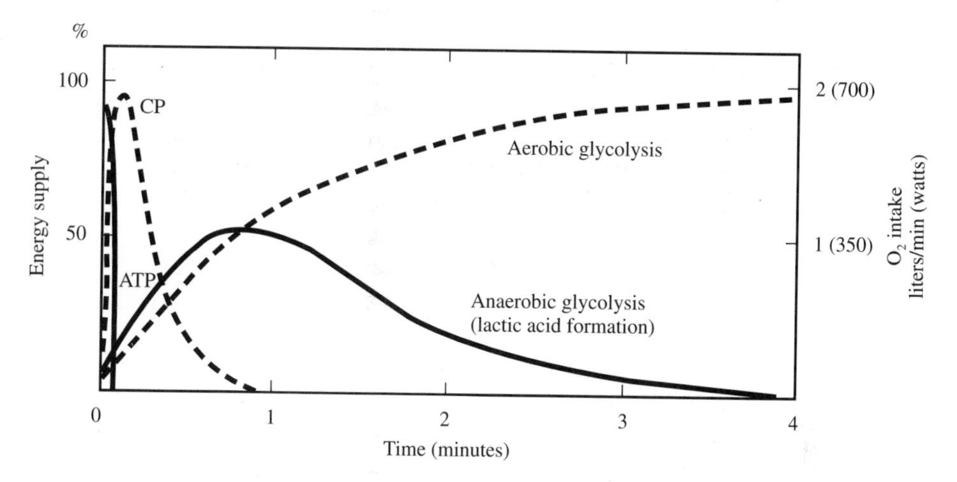

are used up very quickly, and anaerobic metabolism must be utilized to regenerate the ATP stores. Eventually, as the worker reaches steady state, the aerobic metabolism catches up and maintains the energy output, as the anaerobic metabolism slows down. By warming up and starting heavy work slowly, the worker can minimize the amount of anaerobic metabolism and the concurrent buildup of lactic acid associated with feelings of fatigue. This delay of full aerobic metabolism is termed *oxygen deficit* and must eventually be repaid by *the oxygen debt* of a cooling down period, which is always larger than the oxygen deficit.

The energy expended on a task can be estimated by assuming that most of the energy is produced through aerobic metabolism and measuring the amount of oxygen consumed by the worker. The amount of inspired air is measured with a flow meter and assumed to contain 21 percent oxygen. However, not all of this oxygen is utilized by the body; therefore, the expired oxygen must also be measured. Typically, the volume of air inspired and expired is the same, and only the percent of expired oxygen must be found, using an oxygen meter. A conversion factor is included for a typical diet in which 4.9 kcal (19.6 BTU) of energy are produced for each liter of oxygen used in metabolism.

$$\dot{E}(\text{kcal/min}) = 4.9 * V\,(0.21 - E_{O_2})$$

where: $\dot{E}$ = energy expenditure (kcal/min).
V = volume of air inspired (liters/min).
E_{O_2} = fraction of oxygen (O_2) in expired air (roughly 0.16–0.18).

The energy expended on a task varies by the type of task being performed, the posture maintained during the task, and the type of load carriage. Energy

FIGURE 4–20
Examples of energy costs of various types of human activity. Energy costs are given in kilocalories per minute. (*From:* Sanders & McCormick, 1993) (Reproduced with permission of the McGraw-Hill Companies.)

expenditure data on several hundred different types of tasks have been collected, with the most common summarized in Figure 4–20. For manual materials handling, the manner in which the load is carried is most critical, with lowest energy costs for balanced loads held closest to the center of gravity of the body, which has the largest muscle groups. For example, a backpack supported by the trunk muscles is less demanding than holding an equal weight in two suitcases, one in each arm. Although balanced, the latter situation places the load far from the center of gravity and on the smaller arm muscles. Posture also plays an important role, with the least amount of energy expenditure for supported postures. Thus, a posture with the trunk bent over, with no arm support, will expend twice the energy than a kneeling posture with the hands on the floor supporting the trunk.

A 5.33 kcal/min (21.3 BTU/min) limit for acceptable energy expenditure for an 8-hour work day has been proposed by Bink (1962). This number corresponds to *one-third the maximum energy expenditure* of the average American male [for females, it would be $1/3 * 12 = 4$ kcal/min (16 BTU/min)]. If the overall workload is exceedingly high (i.e., exceeds the recommended limits), aerobic metabolism may not be sufficient to provide all of the energy requirements, and the worker may rely on greater amounts of anaerobic metabolism, resulting in fatigue and the buildup of lactic acid. Sufficient recovery must then be provided to allow the body to recover from fatigue and recycle the lactic acid. One guideline for rest allocation was developed by Murrell (1965):

$$R = (W - 5.33)/(W - 1.33)$$

where: R = time required for rest, as percent of total time.
W = average energy expenditure during work (kcal/min).

The value of 1.33 kcal/min (5.3 BTU/min) is the energy expenditure during rest. Consider a strenuous task of shoveling coal into a hopper, which has an energy expenditure of 9.33 kcal/min (37.3 BTU/min). Entering W = 9.33 into the equation yields R = 0.5. Therefore, to provide adequate time for recovery from fatigue, the worker would need to spend roughly one-half of an 8-hour shift, or 4 hours, resting.

The manner in which rest is also allocated is also important. It serves no purpose to have the laborer work for 4 hours straight at a rate of 9.33 kcal/min (37.3 BTU/min), suffer from extreme fatigue, and then rest for 4 hours. In general, the duration of the work cycle is the primary determinant of fatigue buildup. With heavy work, blood flow tends to be occluded, further accelerating the use of anaerobic pathways. In addition, the recovery process tends to be exponential, with later times providing minimal incremental benefits. Therefore, short bursts (approximately 1/2–1 minute) of heavy work interspersed with short rest periods provide maximum benefit. During the 1/2–1 minute periods, the immediate energy sources of ATP and CP get used up, but they can also be quickly replenished. Once lactic acid builds up during longer work periods, it becomes more difficult to remove. Micropauses of 1–3 seconds duration are also useful for flushing any occluded blood vessels, and active breaks, during which the worker alternates hands or uses other muscles, serve to relieve the fatigued muscles. Also, it is best for workers to decide when to take the breaks, whenever they feel the need for rest (self-paced), as opposed to prescribed (or machine-paced) breaks. In summary, the use of *frequent, short work/rest cycles* is highly recommended.

Heart Rate Guidelines

Unfortunately, the measurement of oxygen consumption and the computation of energy expenditure is both costly and cumbersome in an industrial work situation. The equipment costs several thousand dollars and interferes with the worker performing the job. An alternative indirect measure of energy expenditure is the heart rate level. Since the heart pumps the blood carrying oxygen to the working muscles, the higher the required energy expenditure, the higher the corresponding heart rate (Figure 4–21). The instrumentation needed to measure heart rate is inexpensive (less than $100 for a visual readout, several hundred dollars for a PC interface) and relatively nonintrusive (worn regularly by athletes to monitor performance). On the other hand, the analyst must be careful, since heart rate measurement is most appropriate for dynamic work involving the large muscles of the body at fairly high levels (40 percent of maximum) and can vary considerably between individuals, depending on their fitness levels and age. In addition, heart rate can be confounded

FIGURE 4–21
Linear increase in heart rate with physical workload, as measured by energy expenditure.

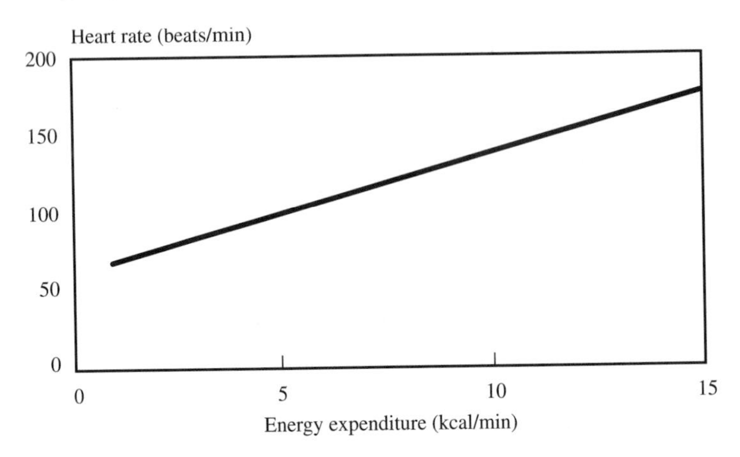

by other stressors including heat, humidity, emotional levels, and mental stress. Limiting these external influences will result in a better estimate of physical workload. However, if the desired goal is to obtain the overall stress on the worker on the job, this may not be necessary.

A methodology for interpreting heart rate has been proposed by various German researchers (cited in Grandjean, 1988). The average working heart rate is compared to the resting prework heart rate, with 40 beats/minute being proposed as an acceptable increase. This increase corresponds very nicely with the recommended working energy expenditure limits. The average increase in heart rate per increase in energy expenditure for dynamic work (i.e., the slope in Figure 4–21) is 10 beats/min per 1 kcal/min. Thus, a 5.33 kcal/min workload (4 kcal/min above the resting level of 1.33 kcal/min) produces a 40 beat/min increase in heart rate, which is the limit for an acceptable workload. This value also corresponds closely to the heart rate recovery index presented by Brouha (1967).

The average heart rate is measured in two time periods during recovery after the cessation of work (see Figure 4–22): (1) between 1/2–1 minutes after cessation and (2) between 2 1/2–3 minutes after cessation. Acceptable heart rate recovery (and therefore acceptable work load) occurs if the first reading does not exceed 110 beats/min and the difference between the two readings is at least 20 beats. Given a typical resting heart rate of 72 beats/min, the addition of the acceptable increment of 40 beats/min yields a working heart rate of 112 beats/min, which corresponds closely to Brouha's first criterion.

As a final note on heart rate, it is very important to observe the course of heart rate during the working hours. An increase in heart rate during steady-state work (see Figure 4–22), termed *heart rate creep*, indicates an increasing buildup of fatigue and insufficient recovery during rest pauses (Brouha, 1967). This fatigue most likely results from the physical workload, but could also result from heat and

FIGURE 4–22
Heart rate for two different work loads. The two marked time periods are used in Brouha's criteria.

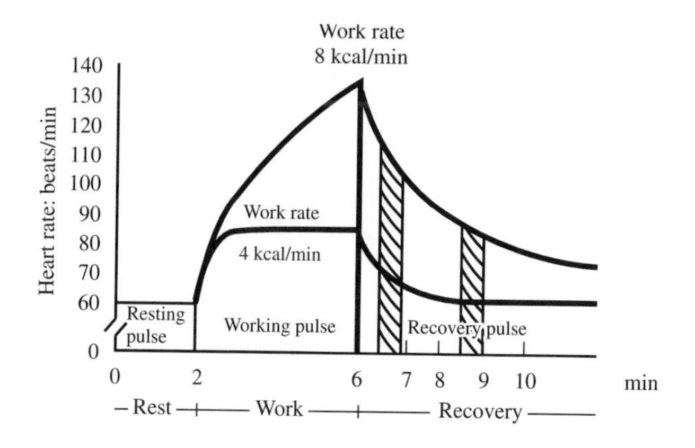

mental stress, and a greater proportion of static rather than dynamic work. In any case, heart rate creep should be avoided by providing additional rest.

Subjective Ratings of Perceived Exertion

An even simpler approach to estimating workload and the stress on the worker is the use of *subjective ratings of perceived exertion.* These can replace the expensive and relatively cumbersome equipment required for physiological measurements with the simplicity of verbal ratings. Borg (1967) developed the most popular scale for assessing the perceived exertion during dynamic whole-body activities—the so called *Borg Rating of Perceived Exertion (RPE) Scale.* The scale is constructed such that the ratings 6 through 20 correspond directly to the heart rate (divided by 10) expected for that level of exertion (Table 4–6). Verbal anchors are provided to assist the worker in performing the ratings. Therefore, to insure an acceptable heart rate recovery, based on the previous heart rate guidelines, the Borg scale should probably not exceed a rating of 13.

Note that the ratings, being subjective, can be affected by previous experience and the individual's level of motivation. Therefore, the ratings should be used with caution and should perhaps be normalized to each individual's maximum rating.

Low Back Compressive Forces

The adult human spine, or vertebral column, is an S-shaped assembly of 25 separate bones (*vertebrae*) divided into four major regions: 7 *cervical* vertebrae in the neck, 12 *thoracic* vertebrae in the upper back, 5 *lumbar* vertebrae in the low back, and the sacrum in the pelvic area (Figure 4–23). The bones have a roughly cylindrical body, with several bony processes emanating from the rear, which serve as attachments

TABLE 4–6
Borg's (1967) RPE Scale with Verbal Anchors

Rating	Verbal Anchor
6	No exertion at all
7	Extremely light
8	
9	Very light
10	
11	Light
12	
13	Somewhat hard
14	
15	Hard
16	
17	Very hard
18	
19	Extremely hard
20	Maximal exertion

FIGURE 4–23
Anatomy of the Human Spine. (*From:* Rowe, 1983)

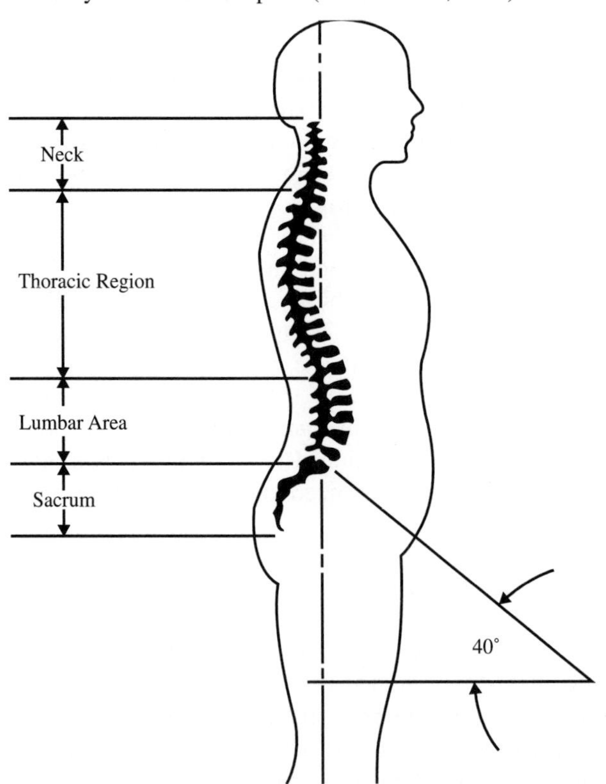

FIGURE 4–24
Anatomy of a vertebra and the process of disk degeneration.
(a) Normal state: (1) body of vertebra, (2) spinous process, serves as muscle attachment point, (3) intervertebral disk, (4) spinal cord, (5) nerve root.
(b) Narrowing of the disk space, allowing the nerve root to be pinched.
(c) Herniated disk, allowing the gel material to extrude and impinge upon the nerve root.
(Adapted from Rowe, 1983)

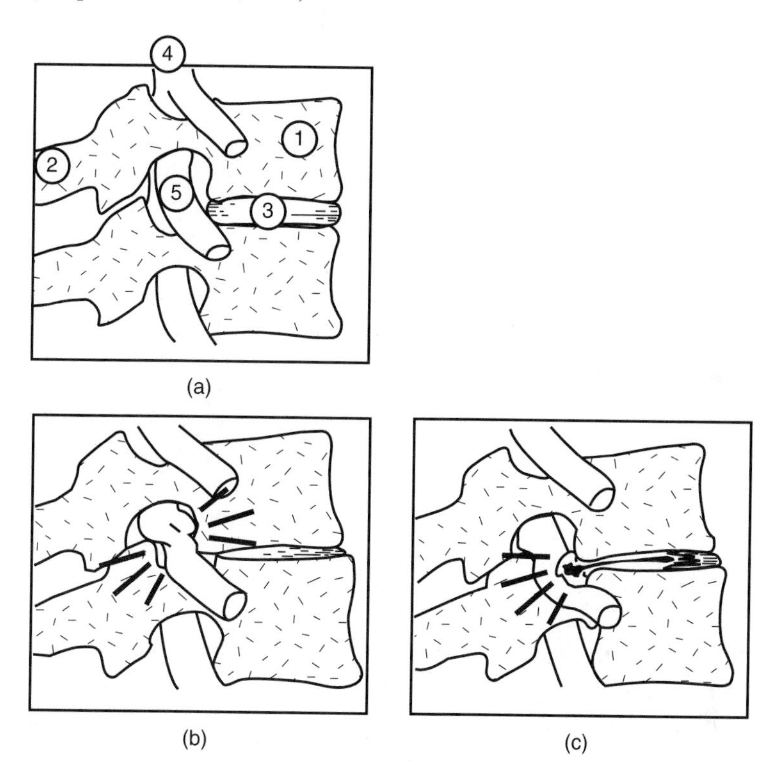

for the back muscles, the *erector spinae*. Through the center of each vertebrae is an opening that contains and protects the spinal cord as it travels from the brain to the end of the vertebral column (Figure 4–24) At various points along the way, spinal nerve roots separate from the spinal cord and pass between the vertebral bones out to the extremities, heart, organs, etc.

The vertebral bones are separated by softer tissue, the vertebral *disks*. These serve as joints, allowing a large range of motion in the spine, although most trunk flexion occurs in the two lowest joints, the one on the border between the lowest lumbar vertebra and the sacrum (termed the L_5/S_1 disk, where the numbering of vertebrae is top down by region), and the next one up (L_4/L_5 disk). The disks also act as cushions between the vertebral bones, and, along with the S-shaped spine, they help to protect the head and brain from the jarring impacts of walking, running, jumping, etc. The disks are composed of a gel-like center surrounded by onion-like layers of fibers, separated from the bone by a cartilage end plate. Considerable movement of fluid occurs between the gel center and the surrounding tissue, depending on the pressure on the disk. Consequently, the length of the vertebral column (measured by change in overall stature) can change by as much as 1/2–1 inch (1.3-2.5 cm) over the course of a work day and is sometimes used as an independent measure of an individual's physical workload. (Interestingly, astronauts in space, removed from the effects of gravity, can be as much as 2 inches taller.)

Unfortunately, due to the combined effects of aging and heavy manual work exposure (these effects are hard to separate), the disks can weaken with time. Some of the enclosing fiber can become frayed, or the cartilage end plate can suffer microfractures, releasing some of the gelatinous material, reducing inner pressures, and allowing the center to start drying up. Correspondingly, the disk space narrows, allowing the vertebral bones to come closer together and eventually even touch, causing irritation and pain. Even worse, the nerve roots are impinged upon, leading to pain and sensory and motor impairments. As the fibers lose integrity, the vertebral bones may shift, causing uneven pressure on the disks and even more pain. In more catastrophic cases, termed *disk herniation,* or more commonly, a *slipped disk*, the fiber casings can actually rupture, allowing large amounts of the gel substance to extrude and impinge upon the nerve roots even more (Figure 4–24c).

The causes for low-back problems are not always easy to identify. As with most occupational diseases, both job and individual factors are at play. The latter may include a genetic predisposition toward weaker connective tissues, disks, ligaments, etc., and personal lifestyle conditions, such as smoking, obesity, etc., over which the industrial engineer has very little control. Changes can only be made with the job factors. Although epidemiological data is easily confounded with survivor population effects or individual compensatory mechanisms, it can be shown statistically that heavy work leads to an increase in low-back problems. Heavy work includes more than just the frequent lifting of large loads; it also encompasses the static maintenance of forward-bending trunk postures for long periods of time. Long periods of immobility, even in sitting postures, and whole-body vibration are also contributing factors. Therefore, scientists have associated the buildup of high disk pressures with eventual disk failures and have resorted to biomechanical calculations or estimations of disk compressive forces from intra-abdominal or direct intradisk pressure measurements, neither of which are practical for industry.

A crude but useful analogy (Figure 4–25) considers a freebody diagram of the L_5/S_1 disk (where most of trunk flexion and disk herniation occurs) and models the components as a first-class lever, with the center of the disk acting as the fulcrum. The load acting through a moment arm determined by the distance from the center of the hands to the center of the disc creates a clockwise moment, while erector spinae muscle is modeled as a force acting downwards through a very small moment arm [approximately 2 inches (5 cm)], creating a counter-clockwise moment barely sufficient to maintain equilibrium. Thus, the two moments must be equal, allowing calculation of the internal force of the erector spinae muscle:

$$2 \times F_M = 30 \times 50$$

where: F_M = muscle force.

FIGURE 4–25
Back compressive forces modeled as a first class lever. (*From:* Sanders and McCormick, 1993)

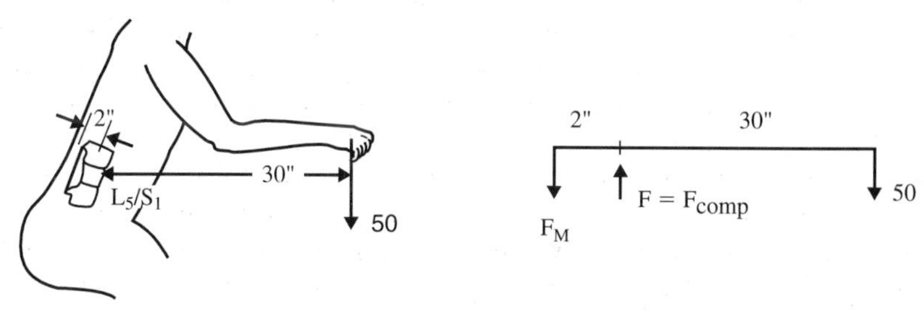

Then, F_M equals 1,500/2 or 750 lbs (341 kg). Solving for the total compressive force (F_{COMP}) exerted on the disk yields:

$$F_{COMP} = F_M + 50 = 800$$

This disk compressive force of 800 lbs (364 kg) is a considerable load, which may cause injury in certain individuals.

Note that this simple analogy neglects the offset alignment of the disks, weights of the body segments, multiple action points of erector spinae components, and other factors, and probably underpredicts the extremely high compressive forces typically obtained in the low back area. More accurate values for various loads and horizontal distances are presented in Figure 4–26. Due to the considerable individual variation in force levels resulting in disk failures, NIOSH (1981) has recommended that a compressive force of 770 lbs (350 kg) be considered the danger threshold.

The hand calculation of such compressive forces through biomechanical modeling is exceedingly time consuming and has led to the development of various computerized biomechanical models. The best known of these are the 2-D and 3-D Static Strength Prediction Models and the Vision 3000 system, which also allows the analyst to capture postures of interest directly from videotape. The marked joint data are linked directly to the 2-D Static Strength Prediction Model for further analysis. Interestingly, this latter system can be compared directly to the now rather neglected chronocyclograph analysis techniques of the Gilbreths (see Chapter 1), allowing the analyst to study motion patterns in greater detail. This feature is quite useful in determining the frequency and range of motion for highly repetitive hand motions, for evaluating the risk of work-related musculoskeletal disorders (see Chapter 5).

Note that although disk herniation may be the most severe of low back injuries, there are other problems, such as soft tissue injuries involving ligaments, muscles, and tendons. These are probably more common, resulting in the backache that most people associate with manual work. Such pain, although uncomfortable, will probably recede over the course of several days with moderate rest. Physicians are currently recommending moderate daily activity to accelerate recovery, rather than the traditional complete bed rest. In addition, researchers are incorporating the soft tissue components in ever more complex back models.

FIGURE 4–26
Effect of weight of load and horizontal distance between the load center of gravity and the L_5/S_1 disc on the predicted compressive force on the L_5/S_1 disc. (*From:* Sanders and McCormick, 1993)

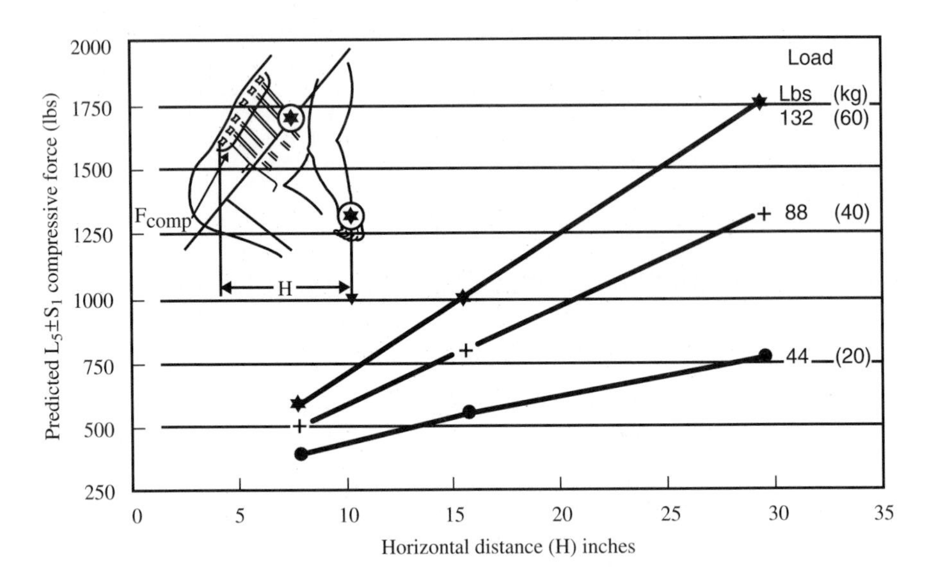

NIOSH Lifting Guidelines

Recognizing and attempting to control the growing problem of work-related back injuries, the National Institute for Occupational Safety and Health (NIOSH) issued what is commonly referred to as the *NIOSH lifting guidelines* (NIOSH, 1981). The guidelines were developed by a distinguished group of scientists and engineers who considered the biomechanical, physiological, psychophysical, and epidemiological bases for an overexertion injury resulting from job demands that exceed a worker's capacity. The NIOSH lifting guidelines have been accepted by the scientific community, recommended by the National Safety Council, and used extensively by OSHA in its workplace inspections. Although these are only guidelines, OSHA will issue citations based on these through the General Duty Clause.

The guidelines were recently updated with the latest epidemiological data and additional features to increase their versatility. This resulted in the *Revised NIOSH Lifting Equation* (Waters et al., 1994) and the Recommended Weight Limit (RWL). The RWL is based on the concept of an optimum weight, with adjustments for various factors related to task variables. The RWL is meant to be a load that can be handled by almost everyone, that is:

1. The 770-lb (350 kg) compression force on the L_5/S_1 disc, created by the RWL, can be tolerated by most young, healthy workers.
2. Over 75 percent of women and over 99 percent of men have the strength capability to lift a load described by the RWL.

3. Maximum resulting energy expenditures of 4.7 kcal/min (18.8 BTU/min) will not exceed recommended limits.

Once the RWL is exceeded, musculoskeletal injury incidences and severity rates increase considerably. The formulation for RWL is based on a maximum load that can be handled in an optimum posture. As the posture deviates from the optimum, adjustments for various task factors, in the form of multipliers, decrease the acceptable load.

$$RWL = LC * HM * VM * DM * AM * FM * CM$$

where: LC = Load constant = 51 lbs
HM = Horizontal multiplier = 10/H
VM = Vertical multiplier = $1 - .0075|V - 30|$
DM = Distance multiplier = $0.82 + 1.8/D$
AM = Asymmetry multiplier = $1 - 0.0032*A$
FM = Frequency multiplier from Table 4–7
CM = Coupling multiplier from Table 4–8
H = Horizontal location of the load cg forward of the midpoint between the ankles, $10 \leq H \leq 25$ inches
V = Vertical location of the load cg, $0 \leq V \leq 70$ inches
D = Vertical travel distance between the origin and destination of the lift, $10 \leq D \leq 70$ inches
A = Angle of asymmetry between the hands and feet (degrees), $0° \leq A \leq 135°$

More simply:

$$RWL\ (lbs) = 51 * (10/H) * (1 - .0075|V - 30|) * (.82 + 1.8/D) * (1 - .0032*A) * FM * CM$$

Note that these multipliers range from a minimum value of zero for extreme postures to a maximum value of 1 for an optimal posture or condition. Table 4–7 provides frequency multipliers for three different work durations and for frequencies varying from 0.2/min to 15/min. Work duration is divided into three categories:

1. Short duration. One hour or less followed by a recovery time equal to 1.2 times the work time. (Thus, even though an individual works for three 1-hour periods, as long as these work periods are interspersed with recovery times of 1.2 hours, the overall work will still be considered of short duration.)
2. Moderate duration. Between 1 and 2 hours of work, followed by a recovery period of at least 0.3 times the work time.
3. Long duration. Anything longer than 2 hours but less than 8 hours.

The coupling multiplier depends on the nature of the hand-to-object interface. In general, a good interface or grip will reduce the maximum grasp forces required and increase the acceptable weight for lifting. On the other hand, a poor interface will require large grasp forces and decrease the acceptable weight. For the revised NIOSH guidelines, three classes of couplings are used: good, fair, and poor.

A good coupling is obtained if the container is of optimal design, such as boxes and crates with well-defined handles or hand-hold cutouts. An optimal container has

TABLE 4–7
Frequency Multiplier Table (FM)

Frequency Lifts/min (F)‡	Work Duration ≤ 1 Hour		>1 but ≤ 2 Hours		>2 but ≤ 8 Hours	
	V < 30†	V ≥ 30	V < 30	V ≥ 30	V < 30	V ≥ 30
≤ 0.2	1.00	1.00	0.95	0.95	0.85	0.85
0.5	0.97	0.97	0.92	0.92	0.81	0.81
1	0.94	0.94	0.88	0.88	0.75	0.75
2	0.91	0.91	0.84	0.84	0.65	0.65
3	0.88	0.88	0.79	0.79	0.55	0.55
4	0.84	0.84	0.72	0.72	0.45	0.45
5	0.80	0.80	0.60	0.60	0.35	0.35
6	0.75	0.75	0.50	0.50	0.27	0.27
7	0.70	0.70	0.42	0.42	0.22	0.22
8	0.60	0.60	0.35	0.35	0.18	0.18
9	0.52	0.52	0.30	0.30	0.00	0.15
10	0.45	0.45	0.26	0.26	0.00	0.13
11	0.41	0.41	0.00	0.23	0.00	0.00
12	0.37	0.37	0.00	0.21	0.00	0.00
13	0.00	0.34	0.00	0.00	0.00	0.00
14	0.00	0.31	0.00	0.00	0.00	0.00
15	0.00	0.28	0.00	0.00	0.00	0.00
>15	0.00	0.00	0.00	0.00	0.00	0.00

†Values of V are in inches.
‡For lifting less frequently than once per 5 minutes, set F = .2 lifts/minute.

TABLE 4–8
Coupling Multiplier

Coupling Type	Coupling Multiplier V< 30 inches (75 cm)	V ≥ 30 inches (75 cm)
Good	1.00	1.00
Fair	0.95	1.00
Poor	0.90	0.90

a smooth, nonslip texture, is < 16 inches (40 cm) in the horizontal direction and < 12 inches (30 cm) high. An optimal handle is cylindrical, with a smooth, nonslip surface, 0.75–1.5 inches (1.9–3.8 cm) in diameter, > 4.5 inches (11.3 cm) long, and with 2 inches (5 cm) clearance. For loose parts or irregular objects that are not found in containers, a good coupling would consist of a comfortable grip in which the hand can comfortably wrap around the object without any large wrist deviations (typically, small parts in a *power grip*).

A fair coupling results from less than optimal interfaces due to less than optimal handles or hand-hold cutouts. For containers of optimal design but with no handles or cutouts, or for loose parts, a fair coupling results if the hand cannot wrap all the way around but is flexed to only 90 degrees. This would typically apply to most industrial packaging boxes.

A poor coupling results from containers of less than optimal design with no handles or hand-hold cutouts, or from loose parts that are bulky or hard to handle. Any container with rough or slippery surfaces, sharp edges, an asymmetric center of gravity, or unstable contents, or one that requires gloves would result in a poor coupling, by definition. To assist in the classification of couplings, the decision tree shown in Figure 4–27 may be useful.

The multipliers for each variable act as simple design tools for fairly straightforward job redesign. For example, if HM = 0.4, 60 percent of the potential lifting capability is lost due to a large horizontal distance. Therefore, the horizontal distance should be reduced as much as possible.

NIOSH also devised a Lifting Index (LI) to provide a simple estimate of the hazard level of lifting a given load, with values exceeding 1.0 deemed to be hazardous. Also, the LI is useful in prioritizing jobs for ergonomic redesign.

$$LI = \text{Load Weight}/RWL$$

In terms of controlling the hazard, NIOSH recommends engineering controls, physical changes or a job and workplace redesign rather than administrative controls consisting of specialized selection and training of workers. Most common changes include: avoiding high and low locations, using lift and tilt tables, using handles or specialized containers for handling loads, and reducing the horizontal distance by cutting out work surfaces and bringing loads closer to the body.

Multitask Lifting Guidelines

For jobs with a variety of lifting tasks, the overall physical/metabolic load is increased compared to the single lifting task. This is reflected in a decreased RWL and an increased LI, and there is a special procedure to handle such situations. The concept is a Composite Lifting Index (CLI), which represents the collective demands of the job. The CLI equals the largest Single Task Lifting Index (STLI), and it increases incrementally CLI for each subsequent task. The multitask procedure is as follows:

1. Compute a Single Task RWL (STRWL) for each task.
2. Compute a Frequency Independent RWL (FIRWL) for each task by setting FM = 1.
3. Compute a Single Task LI (STLI) by dividing the load by STRWL.
4. Compute a Frequency Independent LI (FILI) by dividing the load by FIRWL.
5. Compute the CLI for the overall job by rank-ordering the tasks according to decreasing physical stress, that is, the STLI for each task. The CLI is then:

$$CLI = STLI_1 + \Sigma\Delta LI$$

where: $\Sigma\Delta LI = FILI_2 * (1/FM_{1,2} - 1/FM_1) + FILI_3 * (1/FM_{1,2,3} - 1/FM_{1,2}) + \text{etc.}$

FIGURE 4–27
Decision tree for coupling quality

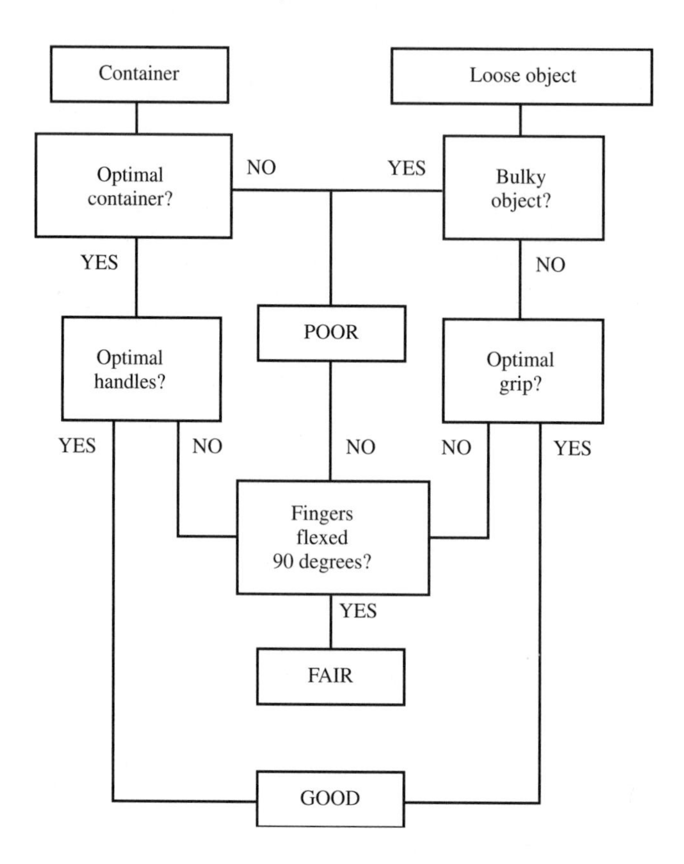

Consider the three-task lifting job shown in Table 4–9. The multitask lifting analysis is as follows:

1. The task with the greatest Lifting Index is new Task #1 (old Task #2) with STLI = 1.6.
2. The sum of the frequencies for new Tasks #1 and #2 is 1 + 2 = 3.
3. The sum of the frequencies for new Tasks #1, #2, and #3 is 1 + 2 + 4 = 7.
4. From the Table 4–7, the new frequency multipliers are: $FM_1 = 0.94$, $FM_{1,2} = 0.88$, and $FM_{1,2,3} = 0.70$
5. The combined lifting index is therefore:

$$CLI = 1.6 + 1 * (1/.88 - 1/0.94) + 0.67 * (1/0.7 - 1/0.88) = 1.60 + 0.07 + .20 = 1.90$$

This procedure is facilitated by the NIOSH Multitask Job Analysis Worksheet (see Figure 4–29). However, once the number of tasks exceeds three or four, it

EXAMPLE 4–1.
NIOSH Analysis of Lifting a Box Into the Trunk of a Car.

Before recent automotive design changes, it was not unusual to have to lean forward and extend the arms while placing an object into the trunk of a car (Figure 4–28). Assume the occupant lifts a 30-lb box from the ground into a trunk. Being lazy, the occupant simply twists 90 degrees to pick up the box from the ground level (V = 0) at a short horizontal distance (H ~ 10 in). The vertical travel distance is the difference between the vertical location of the box at the destination (assume the bottom of the trunk is 25 inches from the ground) and the vertical location of the box at the origin (V = 0), yielding D = 25. Assume that this is a one-time lift; therefore, FM = 1. Also, assume that the box is fairly small and compact, but has no handles. Thus, coupling is fair with CM = 0.95. This yields the following calculation for the origin:

$$\begin{aligned} RWL_{ORG} &= 51 * (10/10) * (1 - 0.0075|0 - 30|) * (0.82 + 1.8/25) * \\ &\quad (1 - 0.0032 * 90) * (1) * (0.95) \\ &= 51 * (1) * (0.775) * (0.892) * (0.712) * (1) * (0.95) = 23.8 \end{aligned}$$

Assuming a larger reach (H=25 inches) into the trunk because of the bumper and high trunk lip, no twisting, the distance traveled remaining the same and the coupling remaining fair, the calculation at the destination is:

$$\begin{aligned} RWL_{DEST} &= 51 * (10/25) * (1 - 0.0075|25 - 30|) * (0.82 + 1.8/25) * \\ &\quad (1 - 0.0032 * 0) * (1) * (0.95) \\ &= 51 * (0.4) * (0.963) * (0.892) * (1) * (1) * (0.95) = 16.6 \end{aligned}$$

and:

$$LI = 30/16.6 = 1.8$$

FIGURE 4–28
Posture for trunk loading example

(continued)

EXAMPLE 4–1. ***(concluded)***
NIOSH Analysis of Lifting a Box Into the Trunk of a Car.

Thus, using the worst case approach, only 16.6 lbs could be lifted safely by most individuals, and the 30-lb box would create a hazard almost twice the acceptable level. The biggest reduction in capability is the horizontal distance at the destination, due to the trunk design. Decreasing the horizontal distance to 10 inches would increase the H factor to 10/10=1 and increase RWL to 33.4 lbs. For most newer cars, this has been accomplished by the auto manufacturers by opening the front part of the trunk such that once the load is lifted to the lower lip, minimum horizontal lifting is needed; the load can be simply pushed forward. However, the limiting case is now the origin, which can be improved by moving the feet and eliminating the twist, increasing RWL to 33.4 lbs. Note that a two-step analysis is necessary if the occupant lifts the load from ground level to the lip of the trunk and then lowers the box into the trunk. This lift is also improved in newer model cars because of a decrease in the vertical height of the lip, which also decreases the distance lifted.

TABLE 4–9
A Sample Three-Task Lifting Job Characteristics:

Task Number	1	2	3
Load weight (L)	20	30	10
Task frequency (F)	2	1	4
FIRWL	20	20	15
FM	0.91	0.94	0.84
STRWL	18.2	18.8	12.6
FILI	1.0	1.5	0.67
STLI	1.1	1.6	0.8
New task number	2	1	3

becomes very time consuming to calculate CLI by hand. A variety of software programs are now available to assist the user in this effort. One of the better known was devised by Garg (1994), one of the developers of the original NIOSH Lifting Guidelines. Of course, the best solution overall is to avoid manual materials handling and use mechanical assist devices or completely automated material handling systems (see Chapter 3).

General Guidelines: Manual Lifting

Although no one optimal lifting technique is suitable for all individuals or task conditions, several guidelines are generally appropriate overall (see Figure 4–30). First plan the lift by evaluating the size and shape of the load, determining whether

FIGURE 4–29
Multitask job analysis worksheet

MULTI-TASK JOB ANALYSIS WORKSHEET

DEPARTMENT ______________ JOB DESCRIPTION
JOB TITLE ______________ ______________
ANALYST'S NAME ______________ ______________
DATE ______________ ______________

STEP 1. Measure and Record Task Variable Data

Task No.	Object Weight (lbs)		Hand Location (in) Origin		Hand Location (in) Dest.		Vertical Distance (in)	Asymmetry Angle (degs) Origin	Asymmetry Angle (degs) Dest.	Frequency Rate lifts/min	Duration Hrs	Coupling
	L (Avg.)	L (Max.)	H	V	H	V	D	A	A	F		C

STEP 2. Compute multipliers and FIRWL, STRWL, FILI, and STLI for Each Task

Task No.	LC x	HM x	VM x	DM x	AM x	CM	FIRWL x	FM	STRWL	FILI = L/FIRWL	STLI = L/STRWL	New Task No.	F
	51												
	51												
	51												
	51												
	51												

STEP 3. Compute the Composite Lifting Index for the Job (After renumbering tasks)

CLI =	$STLI_1$ +	$\Delta FILI_2$ +	$\Delta FILI_3$ +	$\Delta FILI_4$ +	$\Delta FILI_5$	
		$FILI_2(1/FM_{1,2} - 1/FM_1)$	$FILI_3(1/FM_{1,2,3} - 1/FM_{1,2})$	$FILI_4(1/FM_{1,2,3,4} - 1/FM_{1,2,3})$	$FILI_5(1/FM_{1,2,3,4,5} - 1/FM_{1,2,3,4})$	
CLI =						

assistance is needed, and ascertaining what worksite conditions may interfere with the lift. Second, determine the best lifting technique. In general, a squat lift, keeping the back relatively straight and lifting with the bent knees, is the safest in terms of low-back compressive forces. However, bulky loads may interfere with the knees, and a stoop lift, in which the individual bends over and then extends the back, may be required. Third, spread the feet apart, both sideways and fore–aft, to maintain a good balance and stable posture. Fourth, secure a good grip on the load. These last two are especially important in avoiding sudden twisting and jerking movements, both of which are extremely detrimental to the low back. Fifth, hold the load close to the body to minimize the horizontal moment arm created by the load and the resulting moment on the low back.

Avoiding twisting and jerky motions is critical. The first produces an asymmetrical orientation of the discs, leading to increased disc pressures, while the second generates additional accelerative forces on the back. One nonintuitive method for discouraging twisting in workers is actually to increase the horizontal distance between the origin and the destination. This will force the worker to take a step and, in so doing, turn the whole body, rather than twisting the trunk. Carrying uneven loads in both arms or an entire load in only one arm, generates similar asymmetric disk orientations and should also be avoided.

The General Posture and Task Evaluation Checklist (see Figure 4–31) can be very useful in reminding the analyst of the basic principles of good work design.

FIGURE 4–30
Safe lifting procedure. (Available through S.H. Rodgers, Ph.D. P.O. Box 23446, Rochester, N.Y. 14692)

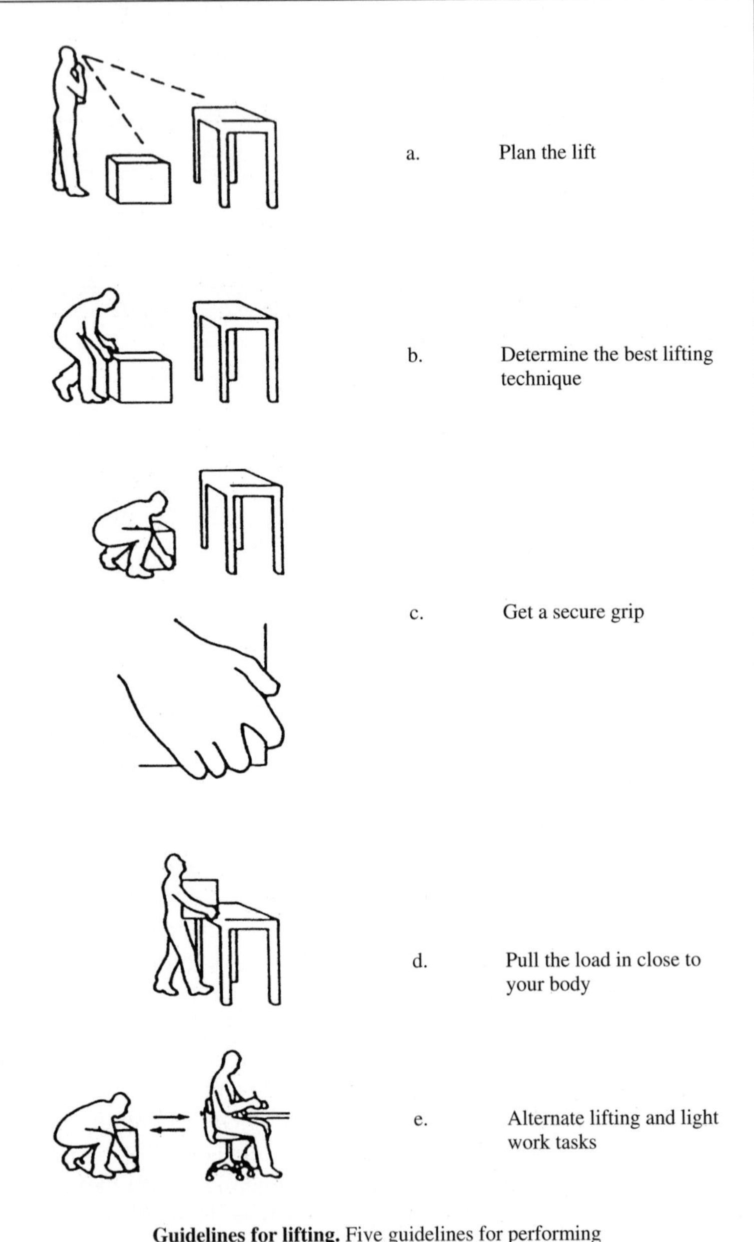

Guidelines for lifting. Five guidelines for performing manual lifting tasks are illustrated. See the text for further discussion.

FIGURE 4–31
General posture and task evaluation checklist.

General Posture Evaluation	Yes	No
1. Are the joints maintained in a neutral position (most are straight, elbow is at 90°)?	❑	❑
2. Is the work or load held close to the body?	❑	❑
3. Are forward bending postures avoided?	❑	❑
4. Are twisting postures of the trunk avoided?	❑	❑
5. Are sudden movements or jerks avoided?	❑	❑
6. Are static postures avoided? i.e., Are there changes in posture?	❑	❑
7. Are excessive reaches avoided?	❑	❑
8. Are the hands utilized in front of the body?	❑	❑

Task Evaluation	Yes	No
1. Are static muscle exertions avoided?	❑	❑
a. Are repetitive static exertions limited to < 15% of maximum strength?	❑	❑
b. Are durations of static exertion limited to several seconds?	❑	❑
2. Are pinch grips used only for low-force precision tasks?	❑	❑
3. Are large muscle groups and power grips utilized for tasks requiring force?	❑	❑
4. Is momentum utilized to assist the operator?	❑	❑
5. Are curved motions pivoting around the lowest-order joints utilized?	❑	❑
6. Are materials and tools placed within the normal working area?	❑	❑
7. Are gravity bins and drop deliveries utilized?	❑	❑
8. Are tasks carried out below shoulder level and above knuckle height?	❑	❑
9. Are lifts performed slowly with knees bent?	❑	❑
10. Are mechanical assists or additional help utilized for loads exceeding 50 pounds?	❑	❑
11. Is the workload low enough that the heart rate is steady and below 110 beats/min?	❑	❑
12. Are frequent, short rest breaks provided?	❑	❑

Back Belts

A cautionary note should be given regarding back belts. Although commonly found on many workers and automatically prescribed in some companies, back belts are not the ultimate panacea and must be regarded with caution. Back belts originated from early studies of weightlifting, showing that for extreme loads, belts relieved 15–30 percent of low-back compressive forces, as estimated from back electromyograms (Morris et al., 1961). However, these studies were performed on trained weightlifters, lifting much larger loads, in a completely sagittal plane. Industrial workers lift much lighter loads, producing a much lower effect. Twisting because of misaligned muscles probably reduces this effect even further. There is also anecdotal data of the "superman" effect—industrial workers with back belts selecting heavier loads than those without back belts, and some workers having coronary incidents from the 10–15 mm Hg increase in blood pressure due to the abdominal compression.

Finally, a longitudinal study of airline baggage handlers (Ridell et al., 1992) concluded there was no significant difference in back injuries between workers using back belts and "control" workers without back belts. Surprisingly, a smaller group of workers, who for one reason or another (e.g., discomfort, heat, etc.) quit wearing the belts but continued in the study, had significantly higher injuries. This may be attributed to atrophy of the abdominal muscles, which should naturally provide an internal back belt but were weakened due to decreased stress. A positive approach may be to encourage workers to strengthen these abdominal muscles through abdominal crunches (modified sit-ups), regular exercise, and body weight reduction. Back belts should only be used with proper training and only after engineering controls have been attempted.

SUMMARY

Chapter 4 introduces some of the theoretical concepts of the human musculoskeletal and physiological systems as a means of providing a framework for a better understanding of the principles of motion economy and work design. These principles are presented as a set of rules to be utilized in redesigning manual assembly work as part of the motion study. Hopefully, with a better understanding of the functioning of the human body, the analyst will view these rules as less arbitrary. These same concepts will be elaborated on in Chapter 5, for the discussions on the design of the workplace, tools, and equipment.

QUESTIONS

1. What structural components are found in muscles? What do these components have to do with muscle performance?
2. Explain the elements of static and dynamic muscle performance with the sliding filament theory.
3. Describe the different types of muscle fibers and relate their properties to muscle performance.
4. Why does a change in the number of active motor units not result in a proportional change in muscle tension?
5. What does EMG measure? How is EMG interpreted?
6. Explain why workstation designers should endeavor to have operators perform work elements without lifting their elbows.
7. What viewing distance would you recommend for a seated operator working at a computer terminal?
8. Define and give examples of the 17 fundamental motions, or therbligs.
9. How may the basic motion "search" be eliminated from the work cycle?
10. What basic motion generally precedes "reach"?
11. What three variables affect the time for the basic motion "move"?

12. How does the analyst determine when the operator is performing the element "inspect"?

13. Explain the difference between avoidable and unavoidable delays.

14. Which of the 17 therbligs are classed as effective and usually cannot be removed from the work cycle?

15. Why should fixed locations be provided at the workstation for all tools and materials?

16. Which of the five classes of motions is used most by industrial workers?

17. Why is it desirable to have the feet working only when the hands are occupied?

18. In a motion study, why is it inadvisable to analyze both hands simultaneously?

19. What task factors increase the index of difficulty in a Fitts Tapping Task?

20. What factors affect back compressive forces during lifting?

21. What factors influence the measurement of isometric muscle strength?

22. Why do psychophysical, dynamic, and static strength capabilities differ?

23. What methods can be used to estimate the energy requirements of a job?

24. What factors change the energy expended for a given job?

25. How does work capability vary with gender and age?

26. What limits endurance in a whole-body manual task?

PROBLEMS

1. What would be the effective visual surface area (in square inches) of an average employee whose workstation was centered 26 inches from the centerpoint between the eyes?

2. In the packing department, a worker stands sideways between the end of a conveyor and a pallet. The surface of the conveyor is 40 inches from the floor and the top of the pallet is 6 inches from the floor. As a box moves to the end of the conveyor, the worker twists 90 degrees to pick up the box, then twists 180 degrees in the opposite direction and sets the box down on the pallet. Each box is 12 inches on a side and weighs 25 lbs. Assume the worker moves five boxes per minute for an 8-hour shift. Using the NIOSH lifting equation, calculate RWL and LI. Redesign the task to improve it. What is the RWL and LI now?

3. For problem #2 calculate the low-back compressive forces incurred in the performance of this job, using the University of Michigan 2-Dimensional Static Strength Prediction Model.

4. A 95th percentile male is holding a 20-lb load in his outstretched arm in 90-degree abduction. What is the reactive torque at the shoulder?

5. A worker is shoveling sand at a rate of 8 kcal/min. How much rest does he need during an 8-hour shift? How should the rest be allocated?

6. A current problem in the U.S. Army is the neck/shoulder fatigue experienced by helicopter pilots. To be able to fly missions at night, the pilots wear night vision goggles, which are attached to the front of the helmet. Unfortunately, these are fairly heavy, causing a large downward torque of the head. This torque must be counteracted by the neck muscles, which then fatigue. To alleviate this problem, many pilots have started attaching random lead weights to the back of the helmet. Find the appropriate weight that would best balance the head and minimize neck fatigue. Assumptions: (a) cg of goggles is 8 inches in front of neck pivot point; (b) goggles weigh 2 lbs.; (c) max. volitional neck torque is 480 in lbs.; (d) cg of lead weight is 5 inches behind neck pivot point, (e) bare helmet weighs 4 lbs.; and (f) cg of helmet is 0.5 inches in front of neck pivot.

7. The laborer on a palletizing operation has been complaining about fatigue and the lack of rest. You measure his heart rate and find it to be 130 beats/min and slowly increasing during work. When he sits down, his heart rates dropped to 125 beats/min by the end of the first minute of rest and 120 beats/min by the end of the third minute. What do you conclude?

8. A grievance has been filed by the union at Dorben Co. regarding the final inspection station, in which the operator slightly lifts a 20-lb assembly, examines all sides, and, if acceptable, sets it back down on the conveyor that takes the assembly to the packing station. On average, the inspector examines five assemblies per minute, at an energy expenditure level of 6 kcal/min. The conveyor is 40 inches off the floor and the assembly is roughly 20 inches from the inspector while being inspected. Evaluate the job with respect to the NIOSH lifting guidelines and metabolic energy expenditure considerations. Indicate whether the job exceeds allowable limits. If it does, calculate how many assemblies the inspector may inspect per minute without exceeding acceptable guidelines.

9. The Dorben Foundry utilizes an overhead crane to move iron from a scrap pile to the furnace. The crane operator uses various levers to control the direction of movement of the crane, as well as three pedals to accelerate (Pedal C), decelerate (Pedal B), and completely disengage (Pedal A) the crane. Assume that the operator always returns his foot to the rest position (Pedal D) after activating a pedal. Calculate the average index of difficulty for the crane operation task, assuming that: all pedals are used equally; each pedal is 2 inches wide; and each pedal is separated from the adjacent pedal by a 2-inch gap.

REFERENCES

Åstrand, P. O., and K. Rodahl. *Textbook of Work Physiology*. New York: McGraw-Hill, 1986.

Barnes, Ralph M. *Motion and Time Study: Design and Measurement of Work*. 7th ed. New York: John Wiley & Sons, 1980.

Bink, B. "The Physical Working Capacity in Relation to Working Time and Age." *Ergonomics*, 5, no.1 (January 1962), pp. 25–28.

Borg, G., and H. Linderholm. "Perceived Exertion and Pulse Rate During Graded Exercise in Various Age Groups." *Acta Medica Scandinavica*, Suppl. 472, (1967), pp. 194–206.

Bouisset, S. "EMG and Muscle Force in Normal Motor Activities." In *New Developments in EMG and Clinical Neurophysiology*. Ed. J. E. Desmedt. Basel, Switzerland: S. Karger, 1973.

Brouha, L. *Physiology in Industry*. New York: Pergamon Press, 1967.

Chaffin, D. B. "Electromyography—A Method of Measuring Local Muscle Fatigue." *The Journal of Methods-Time Measurement*, 14, (1969), pp. 29–36.

Chaffin, D. B., and G. B. J. Andersson. *Occupational Biomechanics*. New York: John Wiley & Sons, 1991.

Chaffin, D. B., G. D. Herrin, W. M. Keyserling, and J. A. Foulke. *Preemployment Strength Testing*. NIOSH Publication 77-163. Cincinnati, OH: National Institute for Occupational Safety and Health, 1977.

Clark, Daniel O., and Guy C. Close. "Motion Study." In *Handbook of Industrial Engineering*. Ed. Gavriel Salvendy. New York: John Wiley & Sons, 1982.

Drillis, R. "Folk Norms and Biomechanics." *Human Factors,* 5 (October 1963), pp. 427–441.

Dul, J., and B. Weerdmeester. *Ergonomics for Beginners*. London: Taylor & Francis, 1993.

Eastman Kodak Co., Human Factors Section. *Ergonomic Design for People at Work*. New York: Van Nostrand Reinhold, 1983.

Fitts, P. "The Information Capacity of the Human Motor System in Controlling the Amplitude of Movement." *Journal of Experimental Psychology,* 47, no. 6 (June 1954), pp. 381–391.

Freivalds, A., and D. M. Fotouhi. "Comparison of Dynamic Strength as Measured by the Cybex and Mini-Gym Isokinetic Dynamometers." *International Journal of Industrial Ergonomics*, 1, no. 3 (May 1987), pp. 189–208.

Grandjean, E. *Fitting the Task to the Man*. New York: Taylor & Francis, 1988.

Gray, H. *Gray's Anatomy*. 35th ed. Ed. R. Warrick and P. Williams. Philadelphia: W.B. Saunders, 1973.

Ikai, M., and T. Fukunaga. "Calculation of Muscle Strength per Unit Cross-Sectional Area of Human Muscle by Means of Ultrasonic Measurement." *Internationale Zeitschrift für angewandte Physiologie einschließlich Arbeitsphysiologie,* 26, (1968), pp. 26–32.

Jones, R., N. Morgan, E. Campbell, R. Edwards, and D. Robertson. *Clinical Exercise Testing*. Philadelphia: W.B. Saunders, 1975, p. 17.

Jonsson, B. "Kinesiology." In *Contemporary Clinical Neurophysiology (EEG Sup. 34).* New York: Elsevier-North-Holland, 1978.

Konz, Stephen. *Work Design.* 4th ed. Columbus, OH: Grid, 1995.

Langolf, G., D. G. Chaffin, and J. A. Foulke. "An Investigation of Fitt's Law Using a Wide Range of Movement Amplitudes." *Journal of Motor Behavior*, 8, no.2 (June 1976), pp. 113–128.

Miller, G. D., and A. Freivalds. "Gender and Handedness in Grip Strength—a Double Whammy for Females." *Proceedings of the Human Factors Society,* 31, (1987), pp. 906–910.

Morris, J. M., D. B. Lucas, and B. Bressler. "Role of the Trunk in Stability of the Spine." *Journal of Bone and Joint Surgery*, 43-A, no. 3 (April 1961), pp. 327–351.

Mundel, M. E., and D. L. Danner. *Motion and Time Study*. 7th ed. Englewood Cliffs, NJ: Prentice Hall, 1994.

Murrell, K.F.H. *Human Performance in Industry*. New York: Reinhold Publishing, 1965.

National Safety Council. *Accident Facts*. Chicago, IL: National Safety Council, 1997.
NIOSH. *Work Practices Guide for Manual Lifting*. Pub. No. TR 81-122, Cincinnati, OH: National Institute for Occupational Safety and Health, 1981.
Ridell, C. R., J. J. Congleton, R. D. Huchingson, and J. T. Montgomery. "An Evaluation of a Weightlifting Belt and Back Injury Prevention Training Class for Airline Baggage Handlers." *Applied Ergonomics*, 23, no. 5 (October 1992), pp. 319–329.
Rodgers, S. H. *Working with Backache*, Fairport, NY: Perinton Press, 1983.
Rowe, M. L. *Backache at Work*. Fairport, NY: Perinton Press, 1983.
Sanders, M. S., and E. J. McCormick. *Human Factors in Engineering and Design*, New York: McGraw-Hill, 1993.
Snook, S. H. and V. M. Ciriello. "The Design of Manual Handling Tasks: Revised Tables of Maximum Acceptable Weights and Forces." *Ergonomics*, 34, no. 9 (September 1991), pp. 1197–1213.
Thorton, W. "Anthropometric Changes in Weightlessness." In *Anthropometric Source Book, 1*, ed. Anthropology Research Project, Webb Associates. NASA RP1024. Houston, TX: National Aeronautics and Space Administration, 1978.
Waters, T. R., V. Putz-Anderson, and A. Garg. *Revised NIOSH Lifting Equation*, Pub. No. 94-110, Cincinnati, OH: National Institute for Occupational Safety and Health, 1994.
Winter, D. A. *Biomechanics of Human Movement*. New York: John Wiley & Sons, 1979.

SELECTED SOFTWARE

2D Static Strength Prediction Program, Ann Arbor, MI: University of Michigan Software, 1990.
3D Static Strength Prediction Program, Ann Arbor, MI: University of Michigan Software, 1990.
Energy Expenditure Prediction Program, Ann Arbor, MI: University of Michigan Software, 1989.
ErgoTRACK (NIOSH Lifting Equation) Carrboro, NC: Performance Track Software, 1992.
The Revised NIOSH Guide Program for Manual Lifting. Brown Deer, WI: A. Garg, 1994.
Vision 3000, Joliet, IL: Promatek, 1993.

CHAPTER 5

Workplace, Equipment, and Tool Design

KEY POINTS:

- Fit the workplace to the operator.
- Provide adjustability.
- Maintain neutral postures (joints in midrange).
- Minimize repetitions.
- Use power grips when force is required.
- Use pinch grips for precision, and not force.

Designing the workplace, tools, equipment, and work environment to fit the human operator is called *ergonomics*. Rather than devoting a lot of space to the underlying theory of the physiology, capabilities, and limitations of the human, this chapter presents the principles of work design and appropriate checklists to facilitate the use of these design principles. With each design principle, a brief explanation of its origin or relationship to the human is provided. This approach will better assist the methods analyst in designing the workplace, equipment, and tools to meet the simultaneous goals of (1) increased production and efficiency of the operation, and (2) decreased injury rates for the human operator.

ANTHROPOMETRY AND DESIGN

The primary guideline is to design the workplace to accommodate most individuals with regard to structural size of the human body. The science of measuring the human body is termed *anthropometry* and typically utilizes a variety of caliper-like devices to measure structural dimensions, for example, stature, forearm length, etc. Practically speaking, however, few ergonomists or engineers collect their own data,

EXAMPLE 5–1
Probability Distributions and Percentiles

A kth *percentile* is defined as a value such that k percent of the data values (plotted in ascending order) are at or below this value and $100 - k$ percent of the data values are at or above this value. A histogram plot of U.S. adult male statures shows a bell-shaped curve, termed a *normal distribution*, with a median value of 68.3 inches (see Figure 5-1). This is also the 50th percentile value, for example, half of all males are shorter than 68.3 inches, while half are taller. The 5th percentile male is only 63.7 inches tall, while a 95th percentile male is 72.8 inches tall. The proof is as follows.

Typically, in a statistical approach, the approximately bell-shaped curve is normalized by the transformation:

$$z = (x - \bar{x})/\sigma$$

where: $\bar{x}$ = Mean
σ = Standard deviation (measure of dispersion)

to form a standard normal distribution (also termed a z distribution; see Figure 5–2).

FIGURE 5–1
Normal distribution of U.S. adult male statures

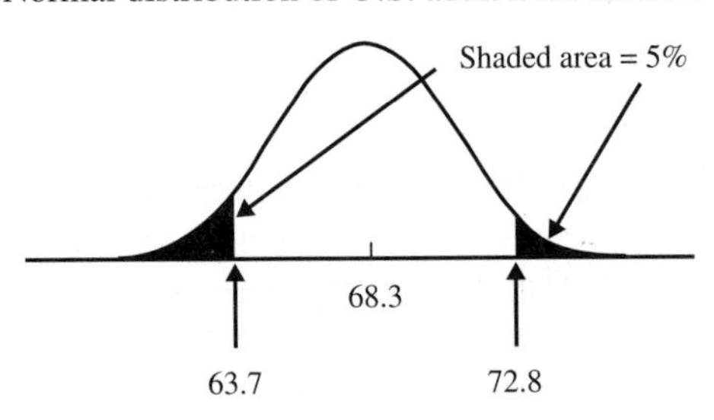

FIGURE 5–2
Standard normal distribution of male weights

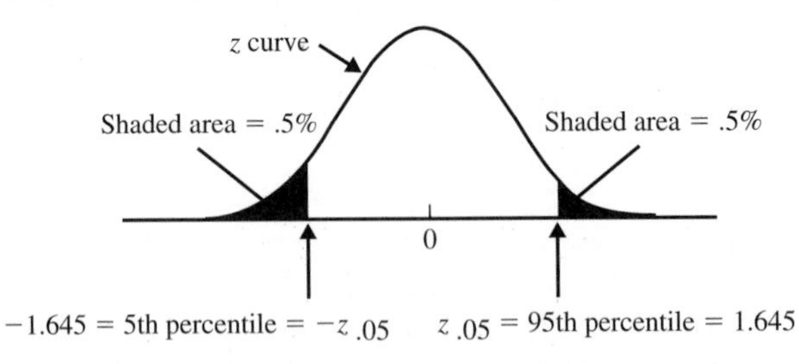

EXAMPLE 5–1
(concluded)

Once normalized, any approximately bell-shaped population distribution will have the same statistical properties. This allows easy calculation of any percentile value desired, using the appropriate k and z values, as follows:

k^{th} percentile	10 or 90	5 or 95	2.5 or 97.5	1 or 99
z value	±1.28	±1.645	±1.96	±2.33

$$k^{th} \text{ percentile} = \bar{x} \pm z\sigma$$

Given that the mean stature for males in the United States is 68.2 inches (173 cm), while the standard deviation is 2.71 inches (6.9 cm) (Webb Associates, 1978), the 95th percentile male stature is calculated as:

$$68.2 + 1.645\ (2.71) = 72.66 \text{ inches}$$

while the 5th percentile male stature is calculated as:

$$68.2 - 1.645\ (2.71) = 63.74 \text{ inches}$$

Note that the calculated value of 72.66 inches is not exactly equal to the actual value of 72.8 inches. This is because the U.S. male height distribution is not completely normal. It is skewed at one end.

because of the wealth of data that has already been collected and tabulated. Close to 1000 different body dimensions, for close to 100 different population types, for example, U.S. army soldiers, National Health Examination Survey males and females, Dutch civilians, Japanese civilians, etc., are available in the *Anthropometric Source Book* (Webb Associates, 1978). Useful dimensions that apply to the particular postures needed for workplace design for U. S. males and females are given in (Table 5-1). Some of these data have been incorporated into mannequins that can be manipulated, or nomograms that provide acceptable workplace dimensions based on the stature of the operator (Diffrient et al. 1978).

Finally, for computer assisted layouts, computerized human models, such as COMBIMAN, Transom Jack, and MannequinPro, provide easy size adjustments and even indicate limitations in ranges of motion or visibility. The first two are more powerful, but require SGI or RISC-type workstations, while MannequinPro can be loaded on personal computers.

Design for Extremes

Designing for most individuals is an approach that involves the use of one of three different specific design principles, as determined by the type of design

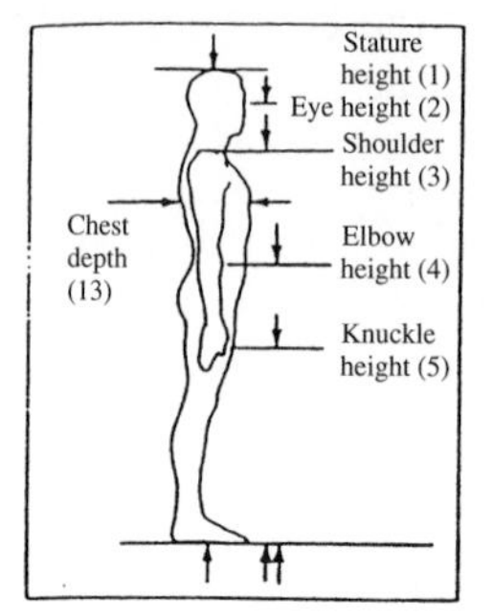

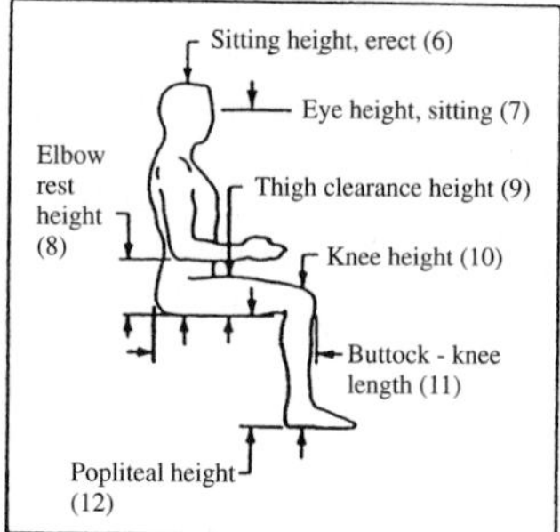

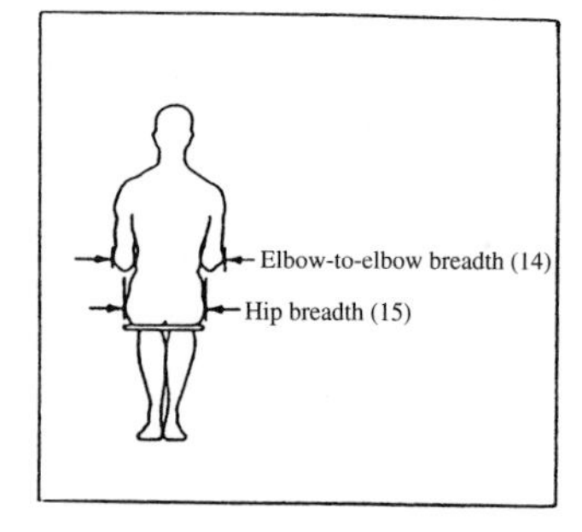

TABLE 5–1
Selected Body Dimensions and Weights of U.S. Adult Civilians*

Body dimension	Sex	Dimension, In			Dimension, cm		
		5th	50th	95th	5th	50th	95th
1. Stature (height)	Male	63.7	68.3	72.6	161.8	173.6	184.4
	Female	58.9	63.2	67.4	149.5	160.5	171.3
2. Eye height	Male	59.5	63.9	68.0	151.1	162.4	172.7
	Female	54.4	58.6	62.7	138.3	148.9	159.3
3. Shoulder height	Male	52.1	56.2	60.0	132.3	142.8	152.4
	Female	47.7	51.6	55.9	121.1	131.1	141.9
4. Elbow height	Male	39.4	43.3	46.9	100.0	109.9	119.0
	Female	36.9	39.8	42.8	93.6	101.2	108.8
5. Knuckle height	Male	27.5	29.7	31.7	69.8	75.4	80.4
	Female	25.3	27.6	29.9	64.3	70.2	75.9
6. Height, sitting	Male	33.1	35.7	38.1	84.2	90.6	96.7
	Female	30.9	33.5	35.7	78.6	85.0	90.7
7. Eye height, sitting	Male	28.6	30.9	33.2	72.6	78.6	84.4
	Female	26.6	28.9	30.9	67.5	73.3	78.5
8. Elbow rest height, sitting	Male	7.5	9.6	11.6	19.0	24.3	29.4
	Female	7.1	9.2	11.1	18.1	23.3	28.1
9. Thigh clearance height	Male	4.5	5.7	7.0	11.4	14.4	17.7
	Female	4.2	5.4	6.9	10.6	13.7	17.5
10. Knee height, sitting	Male	19.4	21.4	23.3	49.3	54.3	59.3
	Female	17.8	19.6	21.5	45.2	49.8	54.5
11. Buttock-knee distance, sitting	Male	21.3	23.4	25.3	54.0	59.4	64.2
	Female	20.4	22.4	24.6	51.8	56.9	62.5
12. Popliteal height, sitting	Male	15.4	17.4	19.2	39.2	44.2	48.8
	Female	14.0	15.7	17.4	35.5	39.8	44.3
13. Chest depth	Male	8.4	9.5	10.9	21.4	24.2	27.6
	Female	8.4	9.5	11.7	21.4	24.2	29.7
14. Elbow-elbow breadth	Male	13.8	16.4	19.9	35.0	41.7	50.6
	Female	12.4	15.1	19.3	31.5	38.4	49.1
15. Hip breadth, sitting	Male	12.1	13.9	16.0	30.8	35.4	40.6
	Female	12.3	14.3	17.2	31.2	36.4	43.7
X. Weight (lbs and kg)	Male	123.6	162.8	213.6	56.2	74.0	97.1
	Female	101.6	134.4	197.8	46.2	61.1	89.9

*(*From:* Sanders & McCormick, 1993) (Reproduced with permission of the McGraw-Hill Companies.)

problem. *Design for extremes* implies that a specific design feature is a limiting factor in determining either the maximum or minimum value of a population variable that will be accommodated. For example, clearances, such as a doorway or an entry opening into a storage tank, should be designed for the maximum individual, that is, a 95th percentile male stature or shoulder width. Then, 95 percent of all males and almost all females will be able to enter the opening. Obviously, for doorways, space is not at a premium, and the opening can be designed to accommodate even larger individuals. On the other hand, added space in military aircraft or submarines is expensive, and these areas are therefore designed to accommodate only a certain (smaller) range of individuals. Reaches, for such things as a brake pedal or control knob, are designed for the minimum individual, that is, a 5th percentile female leg or arm length. Then, 95 percent of all females and practically all males will have a longer reach and will be able to activate the pedal or control.

Design for Adjustability

Design for adjustability is typically used for equipment or facilities than can be adjusted to fit a wider range of individuals. Chairs, tables, desks, vehicle seats, steering columns, and tool supports are devices that are typically adjusted to accommodate the worker population ranging from 5th percentile females to 95th percentile males. Obviously, designing for adjustability is the preferred method of design, but there is a trade-off with the cost of implementation. (Specific adjustment ranges for seat design are given later in Table 5–2.)

Design for the Average

Design for the average is the cheapest but least preferred approach. Even though there is no individual with all average dimensions, there are certain situations where it would be impractical or too costly to include adjustability for all features. For example, most industrial machine tools are too large and too heavy to include height adjustability for the operator. Designing the operating height at the 50th percentile of the elbow height for the combined female and male populations (roughly the average of the male and female 50th percentile values), means that most individuals will not be unduly inconvenienced. However, the exceptionally tall male or very short female may experience some postural discomfort.

Finally, the industrial designer should also consider the legal ramifications of design work. Due to the passage of the Americans with Disabilities Act of 1990 (see Chapter 7), reasonable effort must be made to accommodate individuals with all abilities. Special accessibility guidelines (U.S. Department of Justice, 1991) have been issued regarding parking lots, entryways into buildings, assembly areas, hallways, ramps, elevators, doors, water fountains, lavatories, restaurant or cafeteria facilities, alarms, telephones, etc.

EXAMPLE 5–2
Designing Seating in a Large Training Room

The following example will show the step-by-step procedures utilized in a typical design problem—arranging seating in an industrial training room such that most individuals will have an unobstructed view of the speaker and screen (see Figure 5–3).

1. Determine the body dimensions critical to the design—sitting height, erect; and eye height, sitting.
2. Define the population being served—U.S. adult males and females.
3. Select a design principle and the percentage of the population to be accommodated—Designing for extremes and accommodating 95 percent of the population. The key principle is to allow a 5th percentile female sitting behind a 95th percentile male to have an unimpeded line of sight.
4. Find appropriate anthropometric values from Table 5-1. The 5th percentile female seated eye height is 26.6 inches (67.6 cm), while the 95th percentile male erect sitting height is 38.1 inches (96.8 cm). Thus, for the small female to see over the large male, a rise height of 11.5 inches (29.2 cm) is necessary between the two rows. This would be a very large rise height, which would create a very steep slope. Typically, therefore, the seats are staggered, so that the individual in the back is looking over the head of an individual two rows in front, decreasing the rise height by one-half.
5. Add allowances and test. Many anthropometric measurements have been made on nude human bodies. Therefore, allowances for heavy clothing, hats, or shoes may be necessary. For example, if all the trainees will be wearing hard hats, an additional 2–3 inches might be needed for the rise height. It would be much more practical to remove the hard hats in the training room.

FIGURE 5–3
Seating design in a large training room.

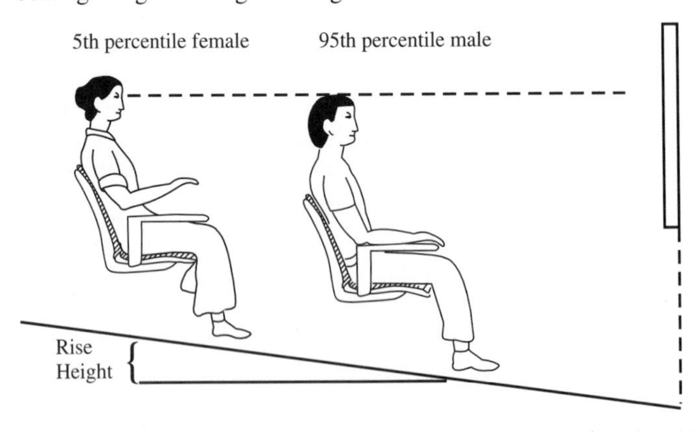

It is also very useful, if practical and cost effective, to build a full-scale mockup of the equipment or facility being designed and then have the users evaluate the mockup. Anthropometric measurements are typically made in standardized postures. In real life, people slouch or have relaxed postures, changing the effective dimensions and the ultimate design. Many costly errors have been identified during production, because of the lack of mockup evaluations. In Example 5–2, the final design actually accommodates a larger percentage of the population than anticipated, for example, the rise height was larger than necessary. The true design should have used the body dimensions for a combined male and female population. However, such combined data is rarely available. The data can be created through statistical techniques, but the general design approach is sufficient for most industrial applications.

PRINCIPLES OF WORK DESIGN: THE WORKPLACE

Determine Work Surface Height by Elbow Height

The work surface height (whether the worker is seated or standing) should be determined by a comfortable working posture for the operator. Typically, this means that the upper arms are hanging down naturally and the elbows are flexed at 90 degrees so that the forearms are parallel to the ground (see Figure 5–4). The elbow height becomes the proper operation or work surface height. If the work surface is too high, the upper arms are abducted, leading to shoulder fatigue. If the work surface is too low, the neck or back is flexed forward, leading to back fatigue.

FIGURE 5–4
Graphic aid for determining correct worksurface height. (*From:* Putz-Anderson, 1988)

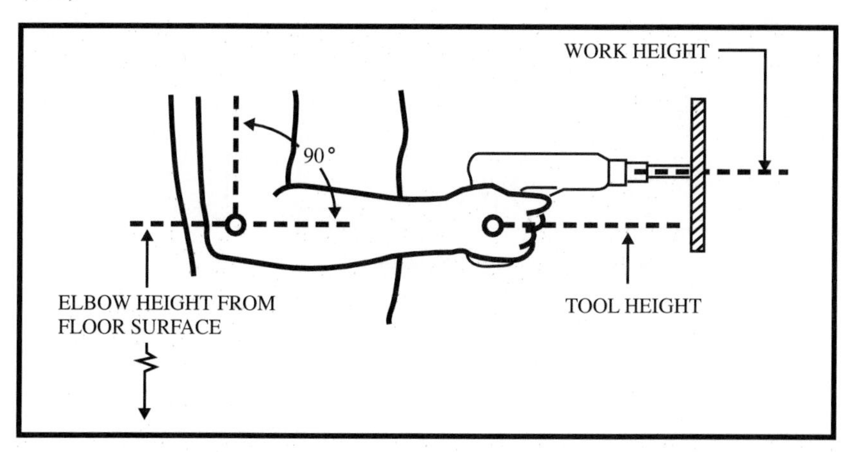

Adjust the Work Surface Height Based on the Task Being Performed

There are exceptions to the first principle. For rough assembly involving the lifting of heavy parts, it is more advantageous to lower the work surface by as much as 8 inches (20 cm), to take advantage of the stronger trunk muscles (see Figure 5–5). For fine assembly involving minute visual details, it is more advantageous to raise the work surface by up to 8 inches (20 cm), to bring the details closer to the optimum line of sight of 15 degrees (Principle from Chapter 4). Another, perhaps better, alternative is to slant the work surface approximately 15 degrees; then, both principles can be satisfied. However, rounded parts then have a tendency to roll off the surface.

These principles also apply to a seated workstation. A majority of tasks, such as writing or light assembly, are best performed at the resting-elbow height. If the job requires the perception of fine detail, it may be necessary to raise the work to bring it closer to the eyes. Seated workstations should be provided with adjustable chairs

FIGURE 5–5
Recommended standing workplace dimensions.

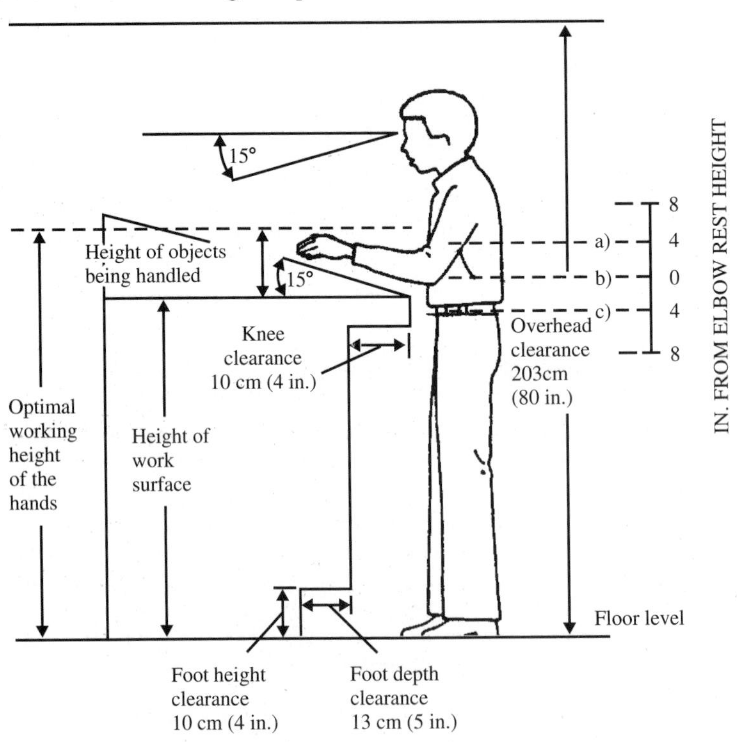

and adjustable footrests (see Figure 5–6). Ideally, after the operator is comfortably seated with both feet on the floor, the work surface is positioned at the appropriate elbow height to accommodate the operation. Thus, the workstation also needs to be adjustable. Short operators whose feet do not reach the floor, even after adjusting the chair, should utilize a footrest to provide support for the feet.

Provide a Comfortable Chair for the Seated Operator

The seated posture is important from the standpoint of reducing both the stress on the feet and the overall energy expenditure. Because comfort is a very individual response, strict principles for good seating are somewhat difficult to define. Furthermore, few chairs will comfortably adapt to the many possible seating postures (see Figure 5–7). However, several general principles hold true for all seats. When standing erect, the lumbar portion of the spine (the small of the back, approximately at the belt level) curves naturally inward, which is termed *lordosis*. However, as you sit down, the pelvis rotates backward, flattening the lordotic curve and increasing the pressure on the disks in the vertebral column (see Figure 5–8). Therefore, it is very important to provide *lumbar support* in the form of an outward bulge in the seat back, or even a simple lumbar pad placed at the belt level.

Another approach to preventing flattening of the lordotic curve is to reduce the pelvic rotation by maintaining a large angle between the torso and thighs, via a forward-tilting seat (kneeling posture in Figure 5–7). The theory is that this is a shape maintained by astronauts in the weightless environment of space (see Figure 4–4). The disadvantage of this type of seat is that it may put additional stress on the knees. The addition of a pommel to the forward-sloping seat, forming a saddle-like seat,

FIGURE 5–6
Adjustable Chair. (Specific seat parameter values found in Table 5–2)

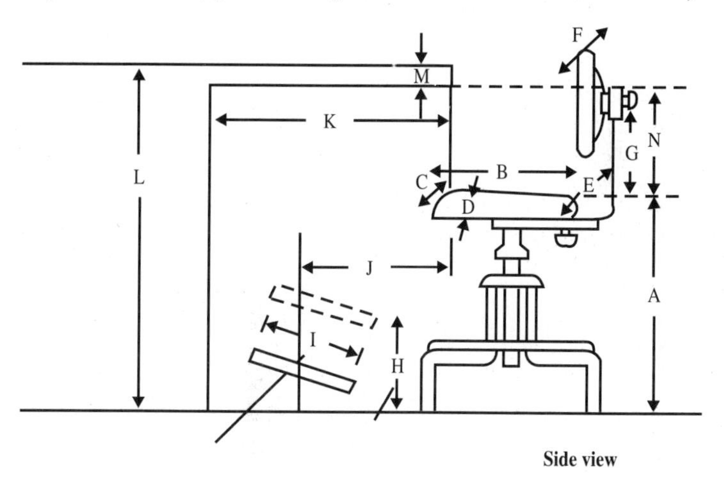

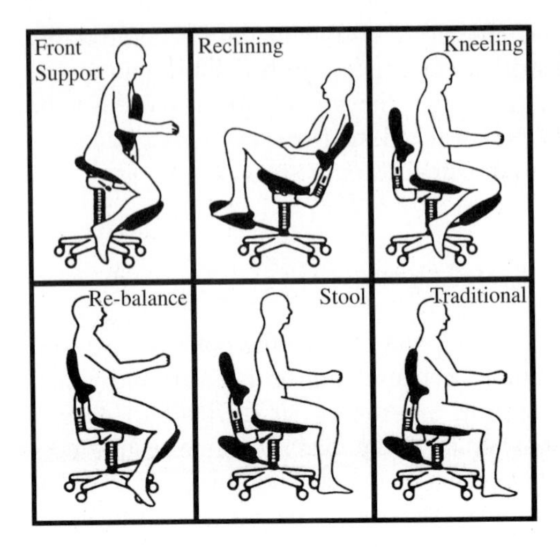

FIGURE 5–7
Six basic seating postures. (*From:* Serber, 1990. Reprinted with permission of the Human Factors and Ergonomics Society. All rights reserved.)

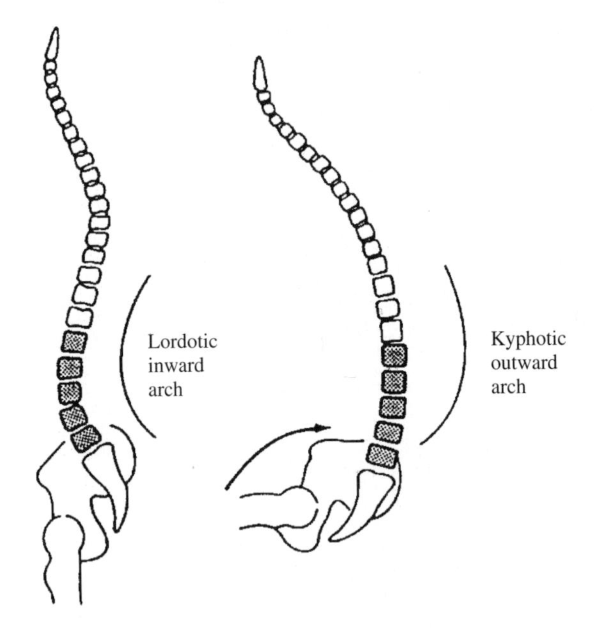

FIGURE 5–8
Posture of the spine when (a) standing and (b) sitting. Lumbar portion of spine is lordotic when standing and kyphotic when sitting. The shaded vertebrae are the lumbar portion of the spine. (*From:* Sanders and McCormick, 1993) (Reproduced with permission of the McGraw-Hill Companies.)

may be a better overall approach, as it eliminates the need for knee supports and still allows for back support (see Figure 5–9).

Provide Adjustability in the Seat

A second consideration is the reduction of disk pressure, which can increase considerably with a forward tilt of the trunk. Reclining the back rest from the vertical also has a dramatic effect in reducing disk pressures (Andersson et al. 1974).

FIGURE 5–9
Saddle-seat. (*Courtesy:* Neutral Posture Ergonomics, Inc., Bryan, TX)

Unfortunately, there is a tradeoff. With increasing angles, it becomes more difficult to look down and perform productive work.

Another factor is the need to provide easy adjustability for specific seat parameters. Seat height is most critical, with ideal height being determined by the person's popliteal height which is defined in the figure accompanying (Table 5–1). A seat that is too high will uncomfortably compress the underside of the thighs. A seat that is too low will raise the knees uncomfortably high and decrease trunk angle, again increasing disk pressure. Specific recommendations for seat height and other seat parameters (shown in Figure 5–6) are given in Table 5–2.

In addition, armrests for shoulder and arm support and footrests for shorter individuals are recommended. Casters assist in movement and ingress/egress from workstations. However, there may be situations where a stationary chair is desired. In general, the chair should be slightly contoured, slightly cushioned, and covered in a breathable fabric to prevent moisture buildup. Overly soft cushioning restricts posture and may restrict circulation in the legs. An overall optimum working posture and workstation is shown in Figure 5-10.

Encourage Postural Flexibility

The work station height should be adjustable so that the work can be performed efficiently either standing or sitting. The human body is not designed for long periods of

TABLE 5–2
Recommended Seat Adjustment Ranges

Seat parameter	Design Value [in inches (cm) unless specified]	Comments
A–Seat height	16-20.5 (40-52)	Too high–compresses thighs; too low–disk pressure increases
B–Seat depth	15-17 (38-43)	Too long–cuts popliteal region, use waterfall contour
C–Seat width	≥18.2 (≥46.2)	Wider seats recommended for overweight individuals
D–Seat pan angle	-10°-+10°	Downward tilting requires more friction in the fabric
E–Seat back to pan angle	>90°	>105° preferred, but requires workstation modifications
F–Seat back width	12 (30.5)	Measured in the lumbar region
G–Lumbar support	6-9 (15-23)	Vertical height from seat pan to center of lumbar support
H–Foot rest height	1-9 (2.5-23)	
I–Foot rest depth	12 (30.5)	
J–Foot rest distance	16.5 (42)	
K–Leg clearance	26 (66)	
L–Work surface height	~32 (~81)	Determined by elbow rest height
M–Worksurface thickness	<2 (<5)	Maximum value
N–Thigh clearance	>8 (>20)	Minimum value

[A–G from ANSI (1988); H–M from Eastman Kodak (1983)]

FIGURE 5–10
Properly adjusted workstation.

Arms: When operator's hands are on keyboard, upper arm and forearm should form right angle; hands should be lined up with forearm; if hands are angled up from the wrist, try using attached to front of keyboard; optional arm rests should be adjustable.

Telephone: Cradling telephone receiver between head and shoulder can cause muscle strain; headset allows head, neck to remain straight while keeping hands free.

Document holder: Same height and distance from user as the screen, so eyes can remain focused as they look from one to the other.

Backrest: Adjustable for occasional variations; shape should match contour of lower back, providing even pressure and support.

Posture: Sit all the way back into chair for proper back support; back, neck should be as comfortably straight ahead; knees should be slightly lower than hips; do not cross legs or shift weight to one side; give joints, muscles a chance to relax; periodically, get up and walk around.

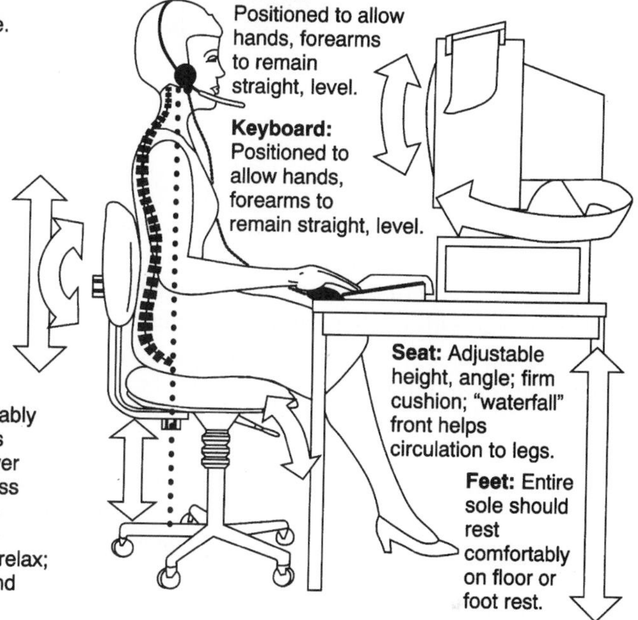

Desk: Thin work surface to allow leg room and posture adjustments; adjustable surface height preferable; table should be large enough for books, files, telephone while permitting different positions of screen, keyboard, mouse pad.

Avoiding eye strain:
1. Getting glasses that improve focus on screen; measure distance before visiting eye doctor.
2. Try to position screen or lamps so that lighting is indirect; do not have light shining directly at screen or into eyes.
3. Use a glare-reducing screen.
4. Periodically rest eyes by looking into the distance.

sitting. The disks between the vertebrae do not have a separate blood supply, and they rely on pressure changes resulting from movement to receive nutrients and remove wastes. Postural rigidity also reduces blood flow to the muscles and induces muscle fatigue and cramping. An alternate compromise is to provide a sit/stand stool so that the operator can change postures easily. Two key features for a sit/stand stool are: height adjustability, and a large base of support so that the stool does not tip, preferably long enough that the feet can rest on and counterbalance it (see Figure 5–11).

FIGURE 5–11
Industrial sit/stand stools. (*Courtesy:* Biofit, Waterville, OH)

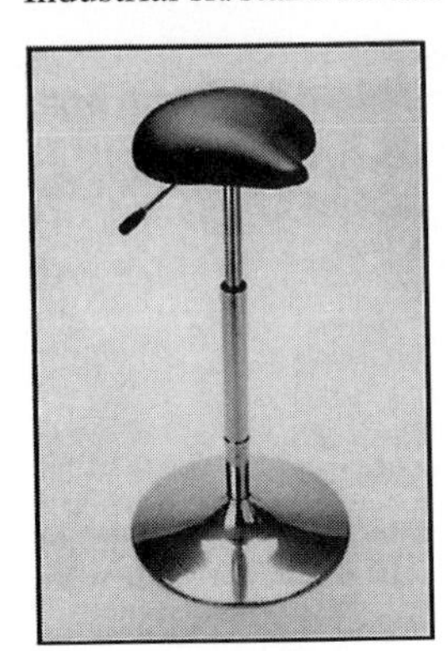

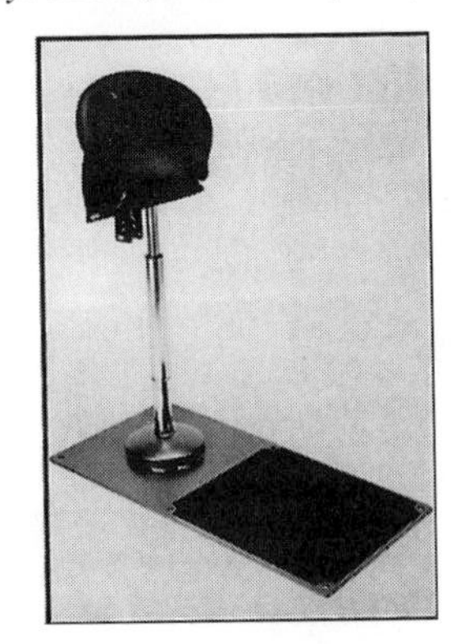

Provide Antifatigue Mats for a Standing Operator

Standing for extending periods of time on a cement floor is fatiguing. The operators should be provided with resilient antifatigue mats. The mats allow small muscle contractions in the legs, forcing the blood to move and keeping it from tending to pool in the lower extremities.

Locate All Tools and Materials Within the Normal Working Area

In every motion, a distance is involved. The greater the distance, the larger the muscular effort, control, and time. It is therefore important to minimize distances. The normal working area in the horizontal plane of the right hand includes the area circumscribed by the arm below the elbow when it is moved in an arc pivoted at the elbow (see Figure 5–12). This area represents the most convenient zone within which motions may be made by that hand with a normal expenditure of energy. The normal area of the left hand may be similarly established. Since movements are made in the third dimension, as well as in the horizontal plane, the normal working area also applies to the vertical plane. The normal area relative to height for the right hand includes the area circumscribed by the lower arm in an upright position hinged at the elbow moving in an arc. There is a similar normal area in the vertical plane (see Figure 5–13).

Fix Locations for All Tools and Materials to Permit the Best Sequence

In driving an automobile, we are all familiar with the short time required to apply the foot brake. The reason is obvious: since the brake pedal is in a fixed location, no time is required to decide where the brake is located. The body responds instinc-

FIGURE 5–12
Normal and maximum working areas in the horizontal plane for women (for men, multiply by 1.09).

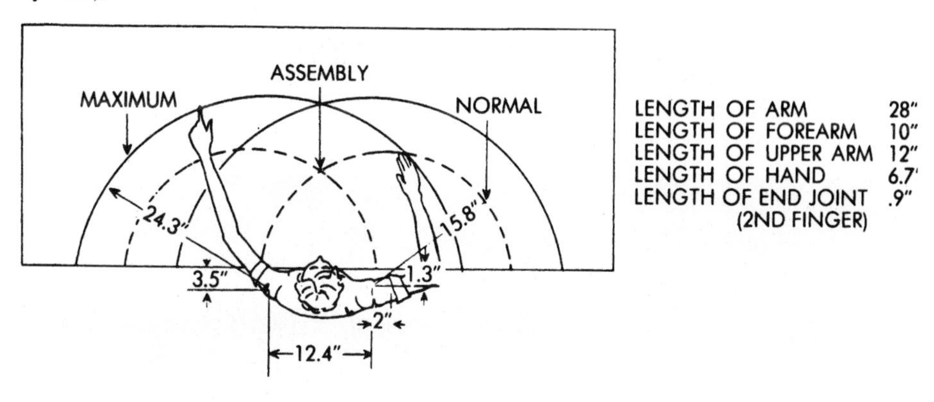

FIGURE 5–13
Normal and maximum working areas in the vertical plane for women (for men, multiply by 1.09).

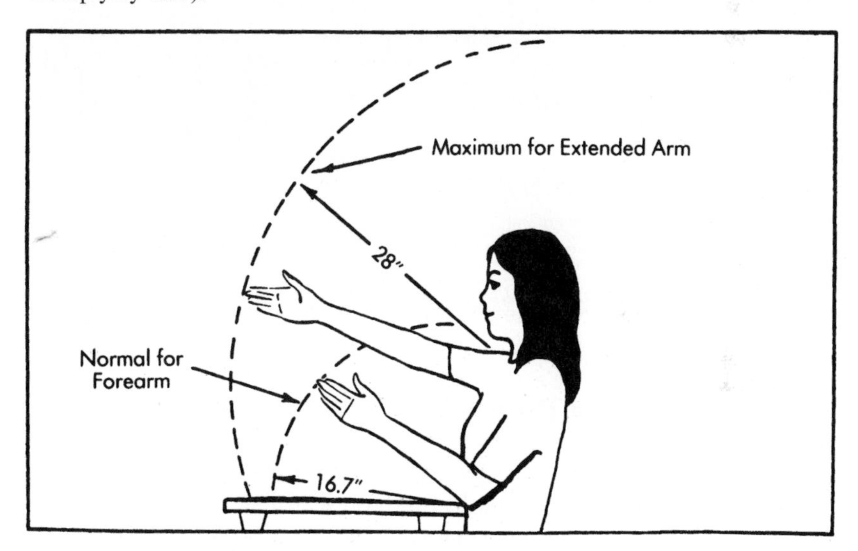

tively and applies pressure to the area where the driver knows the foot pedal is located. If the location of the brake foot pedal varied, the driver would need considerably more time to brake the car. Similarly, providing fixed locations for all tools and materials at the workstation eliminates, or at least minimizes, the short hesitations required to search for and select the objects needed to do the work. These are the ineffective "search" and "select" therbligs discussed in Chapter 4 (see Figure 5–14).

FIGURE 5–14
Fixed locations for all materials and tools in this workstation minimize search and select hesitations. *The operation of subassembly of crankshafts utilizes an indexing table. At the rear of the index table is an induction heating coil used to heat the thrust coolers sitting atop the crankshaft. The heated part drops down onto the top of the cast iron bushing where, on cooling, it shrinks on the shaft and provides the proper clearance required by the pump end bushing.* (*Source:* General Electric Co.)

Use Gravity Bins and Drop Delivery to Reduce Reach and Move Times

The time required to perform both of the transport therbligs, "reach" and "move," is directly proportional to the distance that the hands must move in performing these therbligs. Utilizing gravity bins, components can be continuously brought to the normal work area, thus eliminating long reaches to get these supplies (see Figure 5–15). Likewise, gravity chutes allow the disposal of completed parts within the normal area, eliminating the necessity for long moves to do so. Sometimes, ejectors can remove finished products automatically. Gravity chutes make a clean work area possible, as finished material is carried away from the work area, rather than stacked up all around it. A bin raised off the work surface (so that the hand can partially slide underneath) will also decrease the time required to perform this task by approximately 10-15%.

FIGURE 5–15
A workstation utilizing gravity bins and a belt conveyor to reduce reach and move times. (*Source:* Alden Systems Co.)
The conveyor in the background carries other parts past this particular workstation. The operator is feeding the conveyor from under the platform by merely dropping assembled parts onto the feeder belt.

Arrange Tools, Controls, and Other Components Optimally, to Minimize Motions

The optimum arrangement depends on many characteristics, both human (strength, reach, sensory) and task (loads, repetition, orientation). Obviously, all factors cannot be optimized. The designer must set priorities and make trade-offs in the layout of the workplace. However, certain basic principles should be followed. First, the designer must consider the general location of components relative to other components, using the *importance* and *frequency-of-use* principles. The most important, as determined by overall goals or objectives, or most frequently used components should be placed in the most convenient locations. For example, an emergency stop button should be placed in a readily visible, reachable, or convenient position. Similarly, a regularly used activation button, or the most often used fasteners, should be within easy reach of the operator.

Once the general location has been determined for a group of components, that is, the most frequently used parts for assembly, the principles of *functionality* and *sequence of use* must be considered. Functionality refers to the grouping of components by similar function, for example, all fasteners in one area, all gaskets and rubber components in another area, etc. Since many products are assembled in a strict sequence, cycle after cycle, it is very important to place the components or subassemblies in the order that they are assembled, since this will have a very large effect on reducing wasteful motions. The designer should also consider using Muther's Simplified Layout Planning (see Chapter 3) or other types of adjacency layout diagramming techniques, to develop a quantitative or relative comparison of the various layouts of components on a worksurface. The relationships between components can be modified from traditional data on the flow from one area to another, and should include visual links (eye movements), auditory links (voice communications or signals), tactile and control motions, etc.

These principles of work design for workstations are summarized in the Workstation Evaluation Checklist (see Figure 5–16). The analyst may find this useful in evaluating existing workstations or implementing new workstations.

PRINCIPLES OF WORK DESIGN: MACHINES AND EQUIPMENT

Take Multiple Cuts Whenever Possible by Combining Two or More Tools in One, or by Arranging Simultaneous Cuts from Both Feeding Devices

Advanced production planning for the most efficient manufacture includes taking multiple cuts with combination tools and simultaneous cuts with different tools. Of course, the type of work to be processed and the number of parts to be produced determine the desirability of combining cuts, such as cuts from both the square turret and the hexagon turret.

Locate All Control Devices for Best Operator Accessibility and Strength Capability

Many of our machine tools and other devices are mechanically perfect, yet incapable of effective operation, because the facility designer overlooked various human factors. Handwheels, cranks, and levers should be of such a size and placed in such positions that operators can manipulate them with maximum proficiency and minimum fatigue. Frequently used controls should be positioned between elbow and shoulder height. Seated operators can apply maximum force to levers located at elbow level; standing operators, to levers located at shoulder height. Handwheel and crank diameters depend on the torque to be expended and the mounting position. The maximum diameters of handgrips depend on the forces to be exerted. For example, for a 10- to 15-pound (4.5–6 kg) force, the diameter should be no less than 0.25 inch (0.6 cm) and preferably larger; for 15 to 25 pounds

FIGURE 5–16
Workstation evaluation checklist.

Sitting Workstation	Yes	No
1. Is the chair easily adjustable according to the following features:	❑	❑
a. Is the seat height adjustable from 15 to 22 inches?	❑	❑
b. Is the seat width a minimum of 18 inches?	❑	❑
c. Is the seat depth 15 to 16 inches?	❑	❑
d. Can the seat be sloped ±10° from horizontal?	❑	❑
e. Is a back rest with lumbar support provided?	❑	❑
f. Is the back rest a minimum of 8 × 12 inches in size?	❑	❑
g. Can the back rest be moved 7 to 10 inches above the seat?	❑	❑
h. Can the back rest be moved 12 to 17 inches from the front of the seat?	❑	❑
i. Does the chair have five legs for support?	❑	❑
j. Are casters and swivel capability provided for mobile tasks?	❑	❑
k. Is the chair covering breathable?	❑	❑
l. Is a footrest (large, stable and adjustable in height and slope) provided?	❑	❑
2. Has the chair been adjusted properly?	❑	❑
a. Is the seat height adjusted to the popliteal height with the feet flat on the floor?	❑	❑
b. Is there approximately a 90° angle between the trunk and thigh?	❑	❑
c. Is the lumbar area of the back support in the small of the back (~ belt line)?	❑	❑
d. Is there sufficient legroom (i.e., to the back of the workstation)?	❑	❑
3. Is the workstation surface adjustable?	❑	❑
a. Is the workstation surface roughly at elbow rest height?	❑	❑
b. Is the surface lowered 2 to 4 inches for heavy assembly?	❑	❑
c. Is the surface raised 2 to 4 inches (or tilted) for detailed assembly or visually intensive tasks?	❑	❑
d. Is there sufficient thigh room (i.e., from the bottom of the worksurface)?	❑	❑
4. Is sitting alternated with standing or walking?	❑	❑

Computer Workstation	Yes	No
1. Has the chair been adjusted first, then keyboard and mouse, finally the monitor?	❑	❑
2. Is the keyboard as low as possible (without hitting the legs)?	❑	❑
a. Are the shoulders relaxed, upper arms hanging down comfortably and forearms are below horizontal (i.e., elbow angle >90°)?	❑	❑
b. Is a keyboard shelf utilized (i.e., lower than a normal 28 inch writing surface)?	❑	❑
c. Is the keyboard sloped downward so as to maintain a neutral wrist position?	❑	❑
d. Is the mouse positioned next to the keyboard at the same height?	❑	❑
e. Are armrests (adjustable in height at least 5 inches) provided?	❑	❑
f. If no armrests, are wrist rests provided?	❑	❑
3. Is the monitor positioned 16 to 30 inches (roughly arm's length) from the eyes?	❑	❑
a. Is the top of the screen slightly below eye level?	❑	❑
b. Is the bottom of the screen roughly 30° down from horizontal eye level?	❑	❑
c. Is the monitor positioned at a 90° angle to windows to minimize glare?	❑	❑
d. Can the windows be covered with curtains or blinds to reduce bright light?	❑	❑
e. Is the monitor tilted to minimize ceiling light reflections?	❑	❑
f. If glare still exists, is an antiglare filter utilized?	❑	❑
g. Is a document holder utilized for data transfer from papers?	❑	❑
h. Is the main visual task (monitor or documents) placed directly in front?	❑	❑

Standing Workstation	Yes	No
1. Is the workstation surface adjustable?	❑	❑
a. Is the workstation surface roughly at elbow rest height?	❑	❑
b. Is the surface lowered 4 to 8 inches for heavy assembly?	❑	❑
c. Is the surface raised 4 to 8 inches (or tilted) for detailed assembly or visually intensive tasks?	❑	❑
2. Is there sufficient legroom?	❑	❑
3. Is a sit/stand stool (adjustable in height) provided?	❑	❑
4. Is standing alternated with sitting?	❑	❑

(6.8-11.4 kg) , a minimum of 0.5 inch (1.3 cm) should be used; and for 25 or more pounds (11.4 kg), a minimum of 0.75 inch (1.9 cm). However, diameters should not exceed 1.5 inches (3.8 cm), and the grip length should be at least 4 inches (10 cm), to accommodate the breadth of the hand.

Guidelines for crank and handwheel radii are: light loads, radii of 3 to 5 inches (7.6–12.7 cm); medium to heavy loads, radii of 4 to 7 inches (10.2–17.8 cm); very heavy loads, radii of more than 8 inches (20 cm) but not in excess of 20 inches (51 cm). Knob diameters of 0.5 to 2 inches (1.3–5.1 cm) are usually satisfactory. The diameters of knobs should be increased as greater torques are needed.

Use a Fixture Instead of the Hand as a Holding Device

If either hand is used as a holding device during the processing of a part, then the hand is not performing useful work. Invariably, a fixture can be designed to hold the work satisfactorily, thus allowing both hands to do useful work. Fixtures not only save time in processing parts, but permit better quality in that the work can be held more accurately and firmly. Many times, foot-operated mechanisms allow both hands to perform productive work.

An example will help clarify the principle of using a fixture, as opposed to the hands, for holding work. A company that produced specialty windows needed to remove a 0.75-inch-wide strip of protective paper from around all four edges of both sides of Lexan panels. An operator would pick up a single sheet of Lexan and bring it to the work area. The operator would then pick up a pencil and square and mark the four corners of the Lexan panel. The pencil and square would be laid aside and a template would be picked up and located on the pencil marks. The operator would then strip the protective paper from around the periphery of the Lexan panels. The standard time developed by MTM-1 was 1.063 min per piece.

A simple wood fixture was developed to hold three Lexan panels while each was stripped of the 0.75-inch wide protection paper around the periphery. With the fixture method, the worker picked up three Lexan sheets and located them in the fixture (see Figure 5–17). The protective paper was stripped, the sheets were turned

FIGURE 5–17
Fixture for stripping 3/4" wide protection paper around periphery of Lexan sheets

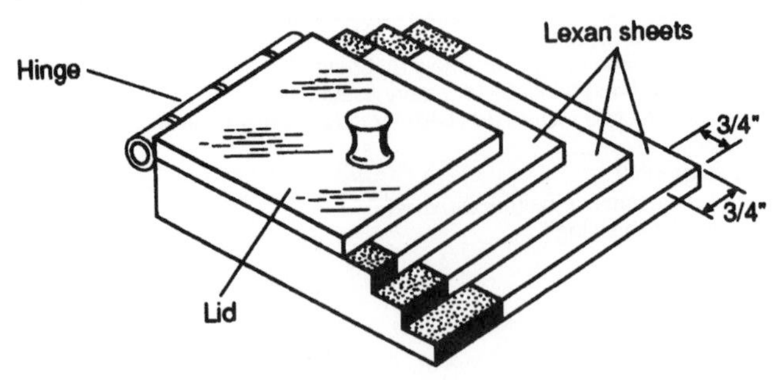

180 degrees, and the protective paper was removed from the remaining two sides. The improved method resulted in a standard of 0.46 min per panel, or a savings of 0.603 min of direct labor per panel.

Use Indicator Lights to Get the Attention of the Worker

Indicator or warning lights are probably the most commonly used visual display. Several basic requirements should be incorporated into their use. They should indicate what is wrong and what action the operator should take. Generally, only one warning light should be used with a given system. Other lights that identify the cause and the action to be taken, and operate with the single warning light, may be located in less central positions. The warning light should remain on until the condition that caused it to come on has been remedied. The warning light should be red or yellow and of sufficient size and intensity to be noticed immediately. A good rule is to make the light twice the size (at least 1 degree of visual arc) and brightness of other panel indicators, and place it not more than 30 degrees off the operator's expected line of sight.

One flashing light attracts attention quickly; several lose most of this ability. A flashing light should flash four times per second. Immediately after the operator takes action, the flashing should stop, but the light should remain on until the improper condition has been completely remedied.

Display Information Appropriately

Table 5–3 shows the advantages and disadvantages of using moving pointers, moving scales, and counters. Operator errors in reading display information increase as the density of information per unit display area increases, and as the operator time for reading and responding to the display decreases. Coding improves the readability of the display and the operator's viewing efficiency. The best three coding methods are: color, alphanumerics (letters and digits), and shapes (geometric figures). All three techniques require little space and allow easy identification, though operators may require some training in their interpretation.

In Western culture, red universally indicates stop, as characterized by traffic control. Red also usually symbolizes danger. On the other hand, green is interpreted as proceed or go ahead. Yellow is usually thought of as caution. It is widely used in connection with hunting sportswear, as well as in traffic lights to convey caution. A recommended coding of simple indicator lights is shown in Table 5–4.

Alphanumeric coding provides many more combinations than color coding. However, from an efficiency standpoint, it parallels color coding. For highly effective alphanumeric coding, consider the stroke width of the numerals and letters, the width–height ratio, and the type form, or font. Based on a viewing distance of up to 20 inches (51 cm), under a range of illuminating conditions, the letter or numeral height should be at least 0.09 inches (0.23 cm), and the stroke width at least 0.02 inches (0.05 cm), yielding a width–height ratio of 1:5. This creates a visual angle

TABLE 5–3
Comparison of Pointers, Scales, and Counters

Indicator	Service rendered			
	Quantitative reading	Qualitative reading	Setting	Tracking
Moving pointer	Fair	Good (changes are easily detected)	Good (easily discernible relation between setting knob and pointer)	Good (pointer position is easily controlled and monitored)
Moving scale	Fair	Poor (may be difficult to identify direction and magnitude)	Fair (may be difficult to identify relation between setting and motion)	Fair (may have ambiguous relationship to manual-control motion)
Counter	Good (minimum time to read and results in minimum error)	Poor (position change may not indicate qualitative change)	Good (accurate method to monitor numerical setting)	Poor (not readily monitored)

TABLE 5–4
Recommended Coding of Indicator Lights

Diameter	State	Color			
		Red	Yellow	Green	White
12.5 mm	Steady	Failure; Stop action; Malfunction	Delay; Inspect	Circuit energized; Go ahead; Ready; Producing	Functional; In position; Normal (on)
25 mm or larger	Steady	System or subsystem in stop action	Caution	System or subsystem in go-ahead state	
25 mm or larger	Flashing	Emergency condition			

(see Chapter 6) of 12 arc–minutes, as required by ANSI/HFS 100 (1988). Use a broader stroke with dark letters on a bright background, and a narrower stroke with bright letters on a dark background. The font refers to the available type styles, such as Gothic, Futura, and Tempo. In general, for a few words capital (uppercase) letters grab one's attention quicker than lowercase letters. Consequently, use uppercase letters with a width to height ratio of about 3:5.

Use Acoustic Signals for Warnings

In some instances, it is better to use auditory signals than visual presentations. For example, auditory signals are usually more efficient if the worker's job necessitates continual movement about the plant or business, or if the person receiving the signal is in a work area where it would be difficult to see a visual signal, such as a dark or excessively bright area. Also, short, simple messages are usually handled better by auditory means.

The human auditory system is continuously alert. It can detect the sources of different signals without orienting the body, as is usually necessary with visual signals. Since hearing is omnidirectional, and since reaction times to sounds are shorter than for visual indications, auditory messages are especially desirable for warning signals. Of course, only acoustic means are satisfactory for speech.

There are also cases where auditory signals should not be considered as an alternative to visual signals, but rather as an addition to them. In cases where the visual system of the operator may already be overburdened, it may be more efficient to add an auditory system. A modulated signal (either on/off or changing in pitch and/or loudness) is especially effective for warnings.

Use Shape, Texture, and Size Coding for Tactual Identification

Shape coding, using two- or three-dimensional geometric configurations, permits both tactual and visual identification. It is especially useful under low-light conditions, or in situations where redundant or double-quality identification is desirable, thus helping to minimize errors. Shape coding, along with variations in texture, permits a relatively large number of discriminable shapes. An especially useful set of known shapes and textures that are seldom confused are shown in Figure 5–18. However, as the number of shapes increases, discrimination can be difficult and slow if the operator must identify controls without vision. If the operator is obliged to wear gloves, then shape coding is only desirable for visual discrimination, or for the tactual discrimination of only two to four shapes.

Size coding, analogous to shape coding, permits both tactual and visual identification of controls. Size coding is used principally where the controls cannot be seen by the operators. Of course, as is the case with shape coding, size coding permits redundant coding, since controls can be discriminated both tactually and visually. In general, try to limit the size categories to three or four, with at least a 0.5–inch size difference between controls. Operational coding, requiring a unique movement (e.g., putting the gearshift into reverse) is especially useful for critical controls that shouldn't be activated inadvertently.

Use Proper Control Size, Displacement, and Resistance

In their work assignments, workers continually use various types and designs of controls. The three parameters that have a major impact on performance are: control size, control–response ratio, and control resistance when engaged. A control that is

FIGURE 5–18
Examples of knob designs for three classes of use that are seldom confused by touch. *The diameter or length of these controls should be between 0.5 and 4.0 in (1.3 to 10 cm), except for class C, where 0.75 in (1.9 cm) is the minimum suggested. The height should be between 0.5 and 1 in (1.3 to 2.5 cm). Illustration of some of the knob designs used in study of tactual discrimination of surface textures. Smooth: A; fluted: B (6 troughs), C (9), D (18); and knurled; E (full rectangular), F (half rectangular), G (quarter rectangular), H (full diamond, I (half diamond), and J (quarter diamond). (From:* Sanders and McCormick, 1993) (Reproduced with permission of the McGraw-Hill Companies)

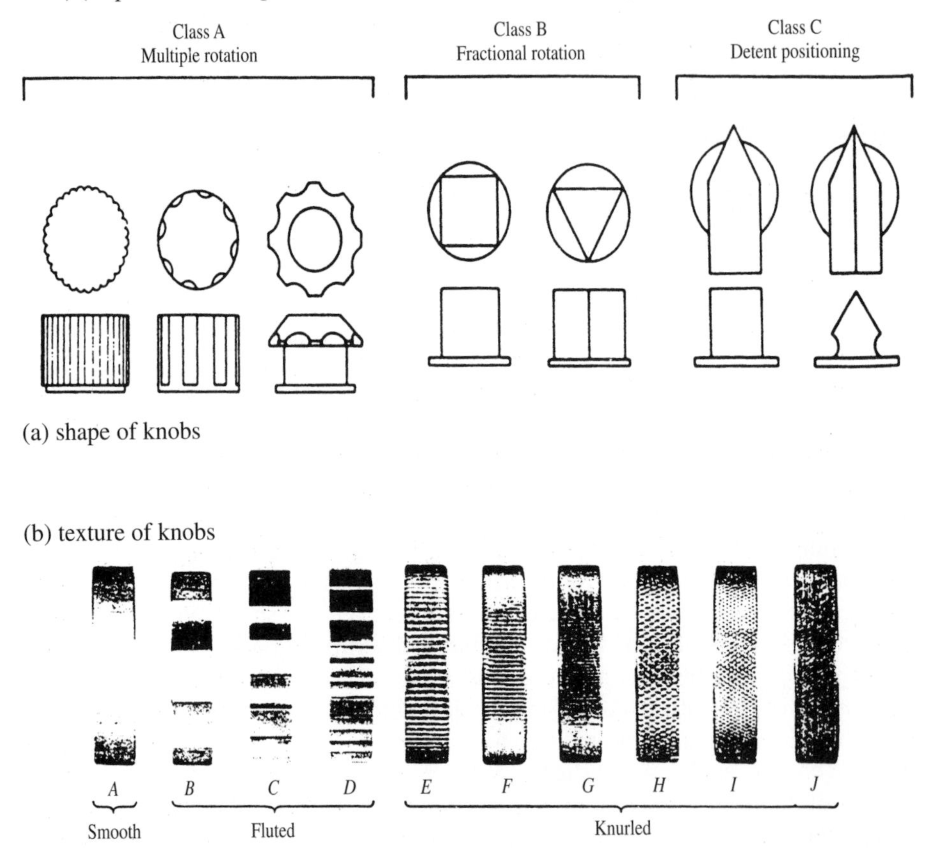

either too small or too large cannot be activated efficiently. Tables 5–5, 5–6, and 5–7 provide helpful design information about minimum and maximum dimensions for various control mechanisms.

The *control–response (CR) ratio* is defined as the amount of movement in a control divided by the amount of movement in the response (see Figure 5–19). A low CR ratio indicates high sensitivity, such as in the coarse adjustment of a micrometer. A high CR ratio means low sensitivity, such as the fine adjustment on a micrometer. Overall control movement depends on the combination of the primary travel time to reach the approximate target setting and the secondary adjust time to reach the exact target setting accurately. The optimum CR ratio that mini-

TABLE 5–5
Control Size Criteria

Control		Dimension	Control size Minimum (mm)	Control size Maximum (mm)
Pushbutton	Fingertip	Diameter	13	*
	Thumb/palm	Diameter	19	*
	Foot	Diameter	8	*
Toggle switch		Tip diameter	3	25
		Lever arm length	13	50
Rotary selector		Length	25	*
		Width	*	25
		Depth	16	*
Continuous adjustment knob	Finger/thumb	Depth	13	25
		Diameter	10	100
	Hand/palm	Depth	19	*
		Diameter	38	75
Cranks	For rate	Radius	13	113
	For force	Radius	13	500
Handwheel		Diameter	175	525
		Rim thickness	19	50
Thumbwheel		Diameter	38	*
		Width	*	*
		Protrusion from surface	3	*
Lever handle	Finger	Diameter	13	75
	Hand	Diameter	38	75
Crank handle		Grasp area	75	*
Pedal		Length	88	†
		Width	25	†
Valve handle		Diameter	75 inches per inch of valve size	

*No limit set by operator performance.
†Dependent on space available.

mizes this total movement time depends on the type of control and task conditions (see Figure 5-20). Note that there is also a *range effect*—the tendency to overshoot short distances and undershoot long distances.

Control resistance is important in terms of providing feedback to the operator. Ideally, it can be of two types: pure displacement with no resistance, or pure force with no displacement. The first has the advantage of being less fatiguing, while the second has the *deadman* feature, that is, the control returns to zero upon release. Real-life controls are typically spring-loaded, incorporating the features of both.

TABLE 5–6
Control Displacement Criteria

Control	Condition	Displacement Minimum	Displacement Maximum
Pushbutton	Thumb/fingertip operation	3 mm	25 mm
	Foot Normal	13 mm	—
	Heavy boot	25 mm	—
	Ankle flexion only	—	63 mm
	Leg movement	—	100 mm
Toggle switch	Between adjacent positions	30°	—
	Total	—	120°
Rotary selector	Between adjacent detents: Visual	15°	—
	Nonvisual	30°	—
	For facilitating performance	—	40°
	When special engineering is required	—	90°
Continuous adjustment knob	Determined by desired control/display ratio (mm of control movement for each mm of display movement)		
Crank	Determined by desired control/display ratio		
Handwheel	Determined by desired control/display ratio		90°-120°†
Thumbwheel	Determined by number of positions		
Lever handle	Fore-aft movement	*	350 mm
	Lateral movement	*	950 mm
Pedals	Normal	13 mm	—
	Heavy boot	25 mm	—
	Ankle flexion (raising)	—	63 mm
	Leg movement	—	175 mm

* None established.
†Provided optimum control/display ratio is not hindered.

Faulty control aspects include: high initial static friction, excessive viscous damping, and *deadspace*, that is, control movement with no response. All three impair tracking and use performance. However, the first two are sometimes incorporated purposely to prevent inadvertent activation of the control (Sanders and McCormick, 1993).

Insure Proper Compatibility Between Controls and Displays

Compatibility is defined as the relationship between controls and displays that is consistent with human expectations. Basic principles include: *affordance*, the per-

TABLE 5–7
Control Resistance Criteria

Control	Condition	Resistance Minimum (kg)	Resistance Maximum (kg)
Push button	Fingertip	0.17	1.14
	Foot: Normally off control	1.82	9.10
	Rested on control	4.55	9.10
Toggle switch	Finger operation	0.17	1.14
Rotary selector	Torque	1 cm-kg	7 cm-kg
Continuous adjustment knob	Torque: Fingertip <1-in dia	*	0.3 cm-kg
	Fingertip >1-in dia	*	0.4 cm-kg
Crank	Rapid, steady turning: <3-in radius	0.91	2.28
	5-8 in radius	2.28	4.55
	Precise settings	1.14	3.64
Handwheel†	Precision operation: <3-in radius	*	*
	5–8 in radius	1.14	3.64
	Resistance at rim: One-hand	2.28	13.64
	Two-hand	2.28	22.73
Thumbwheel	Torque	1 cm-kg	3 cm-kg
Lever handle	Finger grasp	0.34	1.14
	Hand grasp: One-hand	0.91	—
	Two-hands	1.82	—
	Fore-aft: Along median plane:		
	One-hand—10 in forward SRP§	—	13.64
	—16-24 in forward SRP	—	22.73
	Two-hand—10-19 in forward SRP	—	45
	Lateral:		
	One-hand—10-19 in forward SRP	—	9.09
	Two-hand—10-19 in forward SRP	—	22.73
Pedal	Foot: Normally off control	1.82	—
	Rested on control	4.55	—
	Ankle flexion only	—	4.55
	Leg movement	—	80

*Not established.
†For valve handles/wheels: 25 ± cm-kg of torque/cm of valve size (8 cm-kg of torque/cm of handle diameter).
§SRP = Seat reference point.

ceived property results in the desired action; *mapping*; and *feedback*, so that the operator knows that the function has been accomplished. For example, good affordance is a door with a handle that pulls open or a door with a plate that pushes open. Spatial mapping is provided on well-designed stoves. Movement compatibility is provided by direct drive action, scale readings that increase from left to right, and clockwise movements that increase settings. For circular displays, the best compatibility is accomplished with a fixed scale and moving pointer display. For vertical or horizontal displays, *Warrick's principle*, which says that points closest on the display and control move in the same direction, provides the best compatibility (see Figure 5–19). For controls and displays in different planes, a clockwise movement for increases and the right-hand screw rule (the display advances in the direction of

FIGURE 5–19
Generalized illustrations of low and high control–response ratios (C/R ratios) for lever and rotary controls. The C/R ratio is a function of the linkage between the control and the display. (*From:* Sanders and McCormick, 1993) (Reproduced with permission of the McGraw-Hill Companies.)

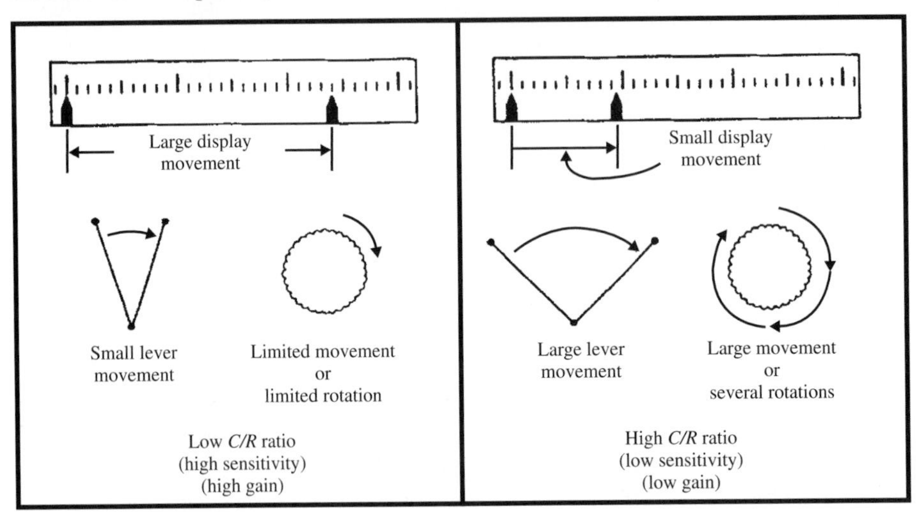

FIGURE 5–20
Relationship between C/R ratio and movement time (travel time and adjust time). (*From:* Sanders and McCormick, 1993) (Reproduced with permission of the McGraw-Hill Companies.)
The specific C/R ratios are not meaningful out of their original context, so are omitted here. These data, however, very typically depict the nature of the relationships, especially for knob controls.

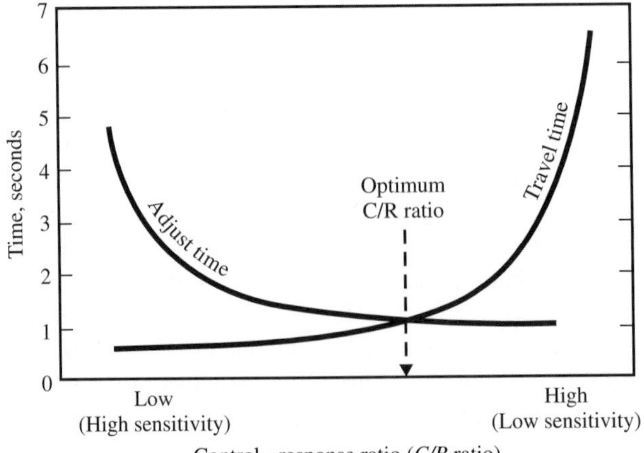

motion of a right-handed screw or control) are most compatible. For stick controls of a direct drive, the best approach is up results in up movement. (Sanders and McCormick, 1993).

The principles of work design for machines and equipment are summarized in the Machine Evaluation Checklist (Figure 5–21). The analyst may find this useful in evaluating and designing machines or other equipment.

CUMULATIVE TRAUMA DISORDERS

The cost of work-related musculoskeletal disorders such as cumulative trauma disorders (CTDs) in U.S. industry, although not all due to improper work design, is quite high. Data from the National Safety Council (1997) suggest that 15–20 percent of workers in key industries (meatpacking, poultry processing, auto assembly, and garment manufacturing) are at potential risk for CTD, and 61 percent of all occupational injuries are associated with repetitive actions. The worst industry is manufacturing, while the worst occupational title is butchering, with 222 CTD claims per 100,000 workers. With such high rates, and with average medical costs of $30,000 per case, NIOSH and OSHA have focused on the reduction of incidence rates for work-related musculoskeletal disorders as one of their main objectives.

Cumulative trauma disorders (sometimes called *repetitive motion injuries,* or work-related musculoskeletal disorders) are injuries to the musculoskeletal system that develop gradually as a result of repeated microtrauma due to poor design and the excessive use of hand tools and other equipment. Because of the slow onset and relatively mild nature of the trauma, the condition is often ignored until the symptoms become chronic and more severe injury occurs. These problems are a collection of a variety of problems, including repetitive motion disorders, carpal tunnel syndrome, tendinitis, ganglionitis, tenosynovitis, and bursitis, with these terms sometimes being used interchangeably.

Four major work-related factors seem to lead to the development of CTD: (1) the use of excessive force during normal motions; (2) awkward or extreme joint motions; (3) high repetitions of the same movement; and (4) the lack of sufficient rest allowing the traumatized joint to recover. The most common symptoms associated with CTD include: pain, joint movement restriction, and soft tissue swelling. In the early stages, there may be few visible signs; however, if the nerves are affected, sensory responses and motor control may be impaired. If left untreated, CTD can result in permanent disability.

The human hand is a complex structure of bones, arteries, nerves, ligaments, and tendons. The fingers are controlled by the extensor carpi and flexor carpi muscles in the forearm. The muscles are connected to the fingers by tendons, which pass through a channel in the wrist, formed by the bones of the back of the hand on one side and the transverse carpal ligament on the other. Through this channel, called the carpal tunnel, also pass various arteries and nerves (see Figure 5–22). The bones of the wrist connect to two long bones in the forearm, the ulna and the radius. The radius connects to the thumb side of the wrist and the ulna connects to

FIGURE 5–21
Machine evaluation checklist.

Machine Efficiency and Safety	Yes	No
1. Are multiple or simultaneous cuts possible?	❑	❑
2. Are handles, wheels, and levers readily accessible?	❑	❑
3. Are handles, wheels, and levers designed for best mechanical advantage?	❑	❑
a. Are knobs at least 0.5-2 inches in diameter, with larger sizes for greater torque?	❑	❑
b. Are cranks and handwheels a minimum of 3-5 inches in diameter for low loads?	❑	❑
c. Are cranks and handwheels more than 8 inches in diameter for heavy loads?	❑	❑
4. Are fixtures used to, avoid holding with the hand?	❑	❑
5. Are guards or interlocks used, to prevent unintended entry?	❑	❑

Design of General Controls	Yes	No
1. Are different colors used for different controls?	❑	❑
2. Are controls clearly labeled?	❑	❑
3. Are shape and texture coding used for tactual identification?	❑	❑
a. Are no more than seven unique codes being utilized?	❑	❑
4. Is size coding used for tactual identification?	❑	❑
a. Are no more than three unique codes being utilized?	❑	❑
b. Are size differences greater than 0.5 inch?	❑	❑

Design of Emergency Controls	Yes	No
1. Are power-on controls designed to prevent accidental activation?	❑	❑
2. Do activation controls require a unique or dual action motion?	❑	❑
3. Are power-on buttons recessed?	❑	❑
4. Are activation controls colored green?	❑	❑
5. Are deadman controls utilized for continually activated controls?	❑	❑
6. Are emergency controls designed for quick activation?	❑	❑
7. Are stop buttons protruding?	❑	❑
8. Are emergency controls large and easy to activate?	❑	❑
9. Are emergency controls easily reachable?	❑	❑
10. Are emergency controls visible and colored red?	❑	❑
11. Are emergency controls placed away from other normally used controls?	❑	❑

Control Placement	Yes	No
1. Are primary controls placed in front of the operator at elbow height?	❑	❑
a. Are frequency-of-use and importance principles used to identify primary controls?	❑	❑
2. Are secondary controls placed next to primary controls, but still within reach?	❑	❑
3. Is twisting avoided in reaching for controls?	❑	❑
4. Are controls located in the proper sequence of operation?	❑	❑
5. Are mutually related controls grouped together?	❑	❑
6. Are hand operated controls separated by at least 2 inches?	❑	❑
7. Are three or less foot pedals utilized?	❑	❑
8. Are foot pedals located at floor level, to avoid raising the leg?	❑	❑
9. Is a sit/stand stool provided for extended foot pedal operation?	❑	❑

Display Design	Yes	No
1. Are displays located on the visual cone of sight (horizontal to 30° down)?	❑	❑
2. Are indicator lights used to attract the operator's attention?	❑	❑
3. Are acoustic signals used for critical warnings?	❑	❑
4. Are movable pointers used to indicate trends?	❑	❑
5. Are counters provided for accurate readings?	❑	❑
6. Are displays grouped so as to accentuate an abnormal display?	❑	❑
7. Are mutually related displays grouped together?	❑	❑

(continued)

FIGURE 5–21 ***(concluded)***
Machine evaluation checklist.

Control-Display Compatibility	Yes	No
1. Is affordance (perceived property results in desired action) used?	❑	❑
2. Is feedback utilized to indicate completion of action?	❑	❑
3. Does the control and display have a direct-drive relationship?	❑	❑
4. Does the display reading increase from left to right?	❑	❑
5. Do clockwise motions increase settings?	❑	❑
6. Do clockwise motions close valves?	❑	❑
7. For stick controls, does upward or backward motion produce upward motion?	❑	❑
8. For controls out of plane, does the right-hand rule apply?	❑	❑

Label Design	Yes	No
1. Is clear and concise wording used?	❑	❑
2. Do the letters subtend at least 12 arcminutes of visual angle?	❑	❑
3. Are dark letters used on a white background?	❑	❑
4. Are uppercase letters used for only a few words?	❑	❑
5. Are symbols (preferably simple) used only if clearly understood?	❑	❑

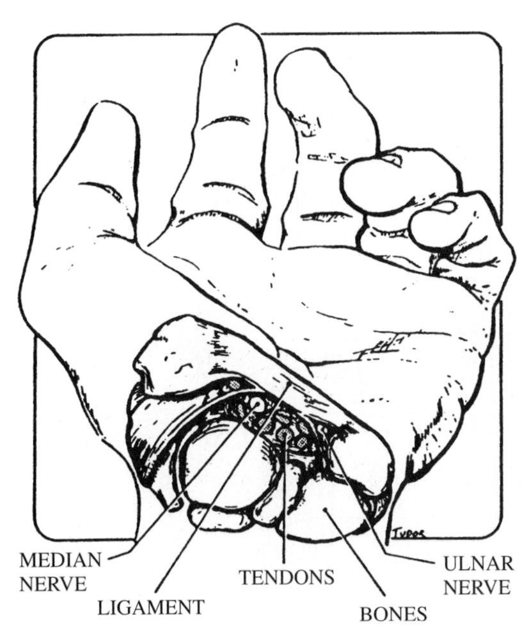

FIGURE 5–22
A pictorial view of the carpal tunnel
(*From:* Putz-Anderson, 1988)

the little-finger side of the wrist. The orientation of the wrist joint allows movement in two planes, at 90 degrees to each other (see Figure 5–23). The first permits *palmar flexion* and *dorsiflexion* (or *extension*). The second movement plane permits *ulnar* and *radial deviation*. Also, rotation of the forearm can result in *pronation* with the palm down or *supination* with the palm up.

Tenosynovitis, one of the more common CTDs, is the inflammation of the tendon sheaths due to overuse or unaccustomed use of improperly designed tools. If the inflammation spreads to the tendons, it becomes *tendinitis*. It is often experienced by trainees exposed to large ulnar deviations, coupled with supination of the

FIGURE 5–23
Positions of the hand and arm. (*From:* Putz-Anderson, 1988)

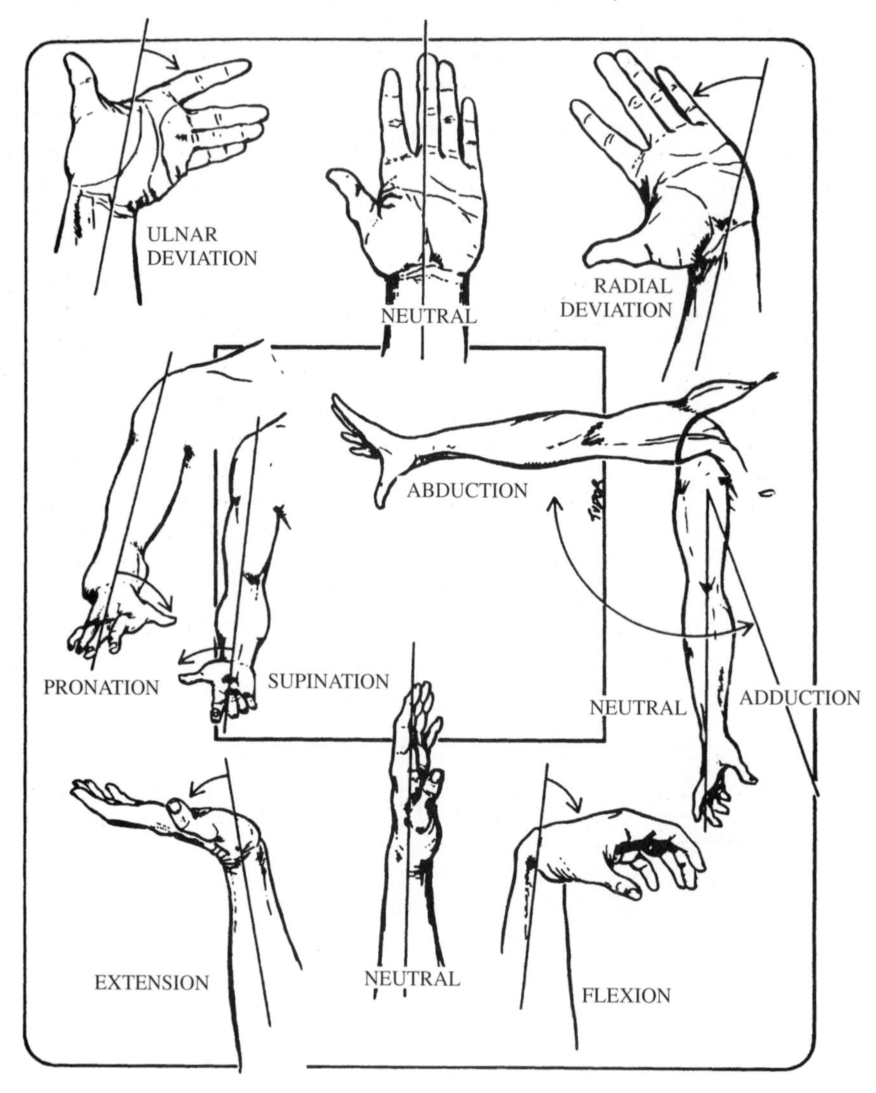

wrist. Repetitive motions and impact shocks may further aggravate the condition. *Carpal tunnel syndrome* is a disorder of the hand caused by injury of the median nerve inside the wrist. Repetitive flexion and extension of the wrist under stress may cause inflammation of the tendon sheaths. The sheaths, sensing increased friction, secrete more fluid to lubricate the sheaths and facilitate tendon movement. The resulting buildup of fluid in the carpal tunnel increases pressure, which in turn compresses the median nerve. Symptoms include impaired or lost nervous function in the first $3^{1}/_{2}$ digits, manifesting as numbness, tingling, pain, and loss of dexterity. Again, proper tool design is very important for avoiding these extreme wrist posi-

FIGURE 5–24
Body discomfort chart (Adapted from Corlett and Bishop, 1970)

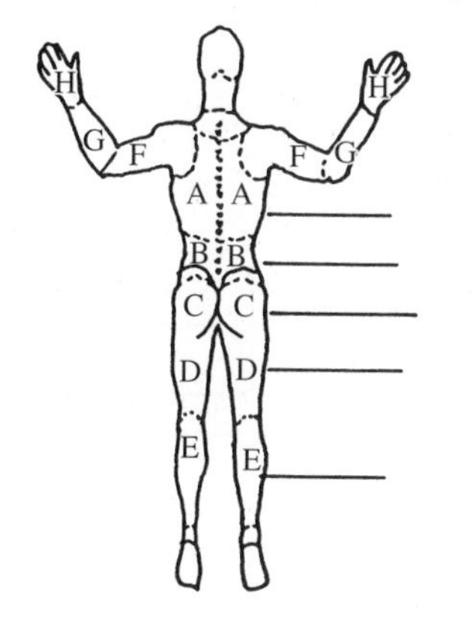

0	Nothing at all	
.5	Extremely weak	(just noticeable)
1	Very weak	
2	Weak	(light)
3	Moderate	
4		
5	Strong	(heavy)
6		
7	Very strong	
8		
9		
10	Extremely strong	(almost max)
•	Maximal	

tions. Extreme radial deviations of the wrist result in pressure between the head of the radius and the adjoining part of the humerus, resulting in tennis elbow, a form of tendinitis. Similarly, simultaneous extension of the wrist, concurrent with full pronation, is equally stressful on the elbow.

Trigger finger is a form of tendinitis resulting from a work situation in which the distal phalanx of the index finger must be bent and flexed against resistance before more proximal phalanges are flexed. Excessive isometric forces impress a groove on the bone, or the tendon enlarges due to inflammation. When the tendon moves within the sheath, it may jerk or snap with an audible click. *White finger* results from excessive vibration from power tools, inducing the constriction of arterioles within the digits. The resulting lack of blood flow appears as a blanching of the skin, with a corresponding loss of motor control. A similar effect can occur as a result of exposure to cold and is termed *Raynaud's syndrome*. A very good introduction to these and other CTDs can be found in Putz-Anderson (1988).

Not all incidences are traumatic. Short-term fatigue and discomfort have also been shown to result from poor handle and work orientation in hammering, improper tool shape and work height in work with screwdrivers, etc. Typically, a poor tool grip design leads to the exertion of higher grip forces and to extreme wrist deviations, resulting in more fatigue (Freivalds, 1996).

To evaluate the level of CTD problems in a plant, the methods analyst or ergonomist typically starts out by surveying the workers to determine their health and discomfort at work. One common tool utilized for this purpose is the *body discomfort map* (Corlett and Bishop, 1970; see Figure 5–24) in which the worker rates the level of pain or body discomfort for various parts of the body, on a scale from 0 (nothing at all) to 10 (almost maximum). The rating scale is based on Borg's (1990) *category ratio scale* (*CR-10*) with verbal anchors shown in Figure 5–24.

A more quantitative approach is the novel CTD risk analysis procedure that sums risk values for all three major causative factors into one risk score (see Figure 5–25; Seth, Weston and Freivalds, 1998). A frequency factor is determined by the

FIGURE 5–25
CTD risk index

CTD Risk Index

Job Title: CUT-OFF	VCR Counter No.: 2371	Date: 1-26-
Job Description: CUT-OFF KNIFE BLADES	Department: KNIFE	Analyst: AF

Cycle Time (in seconds; obtain from videotape. from videotape)			① 5
# Cycle/Day = $\frac{(480 - Lunch - Breaks) \times 60}{CycleTime}$ = $\frac{(480-30-20) \times 60}{5}$		②a 5160	③ Larger of ②a or ②b: 5160
# Parts / Day (if known)	—	②b —	
# Handmotions / Cycle			④ 3
# Handmotions / Day (③ x ④)			⑤ 15,480
		Frequency Factor (Divide ⑤ by 10,000) =	1.55

(Circle appropriate condition)	Points 0	1	2	3
Working Posture	Sit	Stand		
Hand Posture 1: Pulp Pinch	No	Yes		
Hand Posture 2: Lateral Pinch	No	Yes		
Hand Posture 3: Palm Pinch	No	Yes		
Hand Posture 4: Finger Press	No	Yes		
Hand Posture 5: Power Grip	Yes	No		
Type of Reach	Horizontal	Up/Down		
Hand Deviation 1: Flexion	No	Yes		
Hand Deviation 2: Extension	No	Yes		
Hand Deviation 3: Radial Dev.	No	Yes		
Hand Deviation 4: Ulnar Dev.	No	Yes		
Forearm Rotation	Neutral	In/Out		
Elbow Angle	=90°	≠90°		
Shoulder Abduction	0	<45°	<90°	>90°
Shoulder Flexion	0	<90°	<180°	>180°
Back Angle	0	<45°	<90°	>90°
Balance	Yes	No		
		Total the Points for the Circled Conditions		⑥ 8
		Posture Factor (Divide ⑥ by 10) =		.80

Grip or Pinch Force Used on Task	⑦ 30 lbs.	⑨Divide⑦ by⑧:
Max Grip or Pinch Force	⑧ 100 lbs.	.30
	Force Factor (Divide ⑨ by .15) =	2.00

(Circle appropriate condition)	Points 0	1	2	3
Sharp Edge	No	Yes		
Glove	No	Yes		
Vibration	No	Yes		
Type of Action	Dynamic	Intermittent	Static	
Temperature	Warm	Cold		
		Total the Points for the Circled Conditions		⑩ 1
		Miscellaneous Factor (Divide ⑩ by 3) =		.33

CTD Risk Index = .3 x (Frequency + Posture + Force Factors) + .1 x (Miscellaneous Factor)

CTD Risk Index = .3 x (1.55 + .80 + 2.00 **) + .1 x (** .33 **)** = 1.34

number of damaging wrist motions, and then scaled by a threshold value of 10,000. A posture factor is determined from the degree of deviation from the neutral posture for major upper extremity motions. A force factor is determined from the relative percentage of maximum muscle exertion required for the task, and then scaled by 15 percent, the maximum allowed for extended static contractions (see Chapter 4). A final miscellaneous factor incorporates a variety of conditions that may have a role in CTD causation, such as vibration, temperature, etc. They are weighted appropriately and then summed to yield a final CTD risk index. For relatively safe conditions, the index should be less than one (similar to the NIOSH Lifting Index, Chapter 4).

One example (see Figure 5–25) analyzes the CTD stress incurred on a highly repetitive cut-off operation described in greater detail in Example 7.1 of Chapter 7. Both the frequency factor of 1.55 and the force factor of 2.00 exceed the safety threshold of 1.0, leading to a total risk value of 1.34, which also exceeds 1.0. Thus, the most cost effective approach is to decrease the frequency by eliminating or combining unnecessary motions (which may or may not be possible) and decrease the force component by modifying the grasp utilized (the basis for methods change #5 in Example 7.1 of Chapter 7).

The CTD index has proven to be quite successful at identifying injurious jobs, but it works much better on a relative basis, rather than an absolute basis, for example, rank ordering critical jobs. Note that the CTD risk index also serves as both a useful checklist for identifying poor postures, as well as a design tool for selecting key conditions to redesign.

PRINCIPLES OF WORK DESIGN: TOOLS

Use a Power Grip for Tasks Requiring Force and Pinch Grips for Tasks Requiring Precision

Prehension of the hand can basically be defined as variations of grip between two extremes: a power grip and a pinch grip. In a power grip, the cylindrical handle of the tool, whose axis is more or less perpendicular to the forearm, is held in a clamp formed by the partly flexed fingers and the palm. Opposing pressure is applied by the thumb, which slightly overlaps the middle finger (see Figure 5–26). The line of action of the force can vary with: (1) the force parallel to the forearm, as in sawing; (2) the force at an angle to the forearm, as in hammering; and (3) the force acting on a moment arm, creating torque about the forearm, as in using a screwdriver. As the name implies, the power grip is used for power or for holding heavy objects. However, the more the fingers or the thumb deviate from the cylindrical grip, the less force is produced and the greater the precision that can be provided. For example, in holding a light hammer as in tacking, the thumb may deviate from opposing the fingers to aligning with the handle. If the index finger also deviates to the tool axis, as in holding a knife for a precise cut, then a pinch grip is approached, with the blade being pinched between the thumb and index finger. This grip is sometimes called an internal precision grip (Konz, 1995). A hook grip, used for holding a box or a handle, is an incomplete power grip in which the thumb counterforce is not applied, thereby considerably reducing the available grip force.

The pinch grip is used for control or precision. In a pinch grip, the item is held between the distal ends of one or more fingers and the opposing thumb (the thumb is sometimes omitted). The relative position of the thumb and fingers determines

FIGURE 5–26
Types of grip.

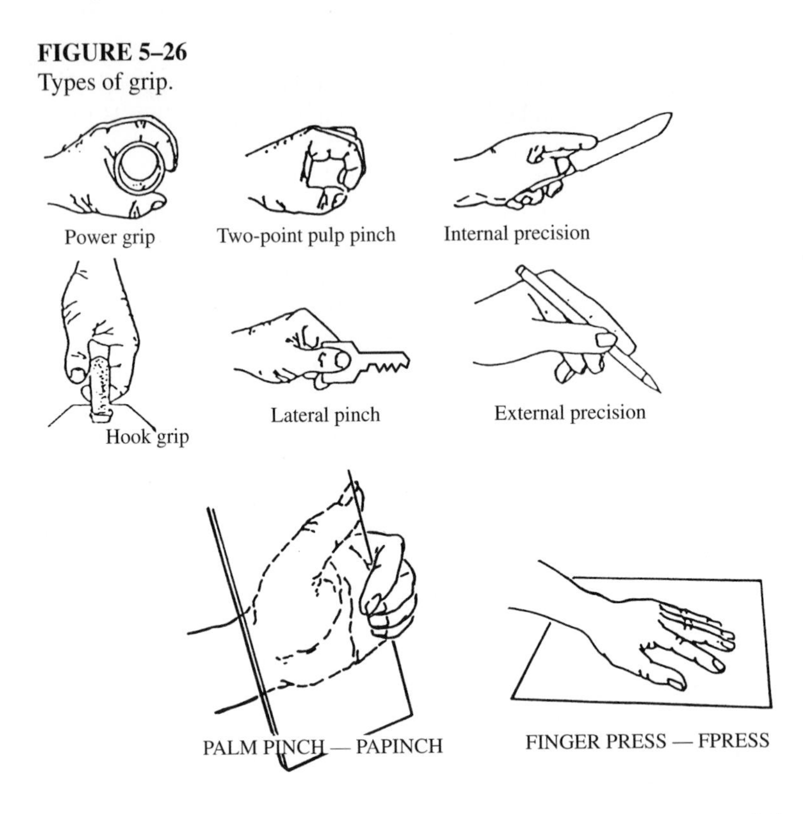

how much force can be applied and provides a sensory surface for receiving the feedback necessary to give the precision needed. There are four basic types of pinch grips, with many variations: (see Figure 5–26) (1) lateral pinch, thumb opposes the side of the index finger; (2) two- and three-point tip (or pulp) pinches, in which the tip (or palmar pad) of the thumb opposes to the tips (or palmar pads) of one or more fingers (for a relatively small cylindrical object, the three digits act as a chuck, resulting in a chuck grip); (3) palm pinch, the fingers oppose the palm of the hand without the thumb participating, as in glass windshield handling; and (4) finger press, the thumbs and fingers press against a surface, as in garment workers pushing cloth into a sewing machine. One specialized grip is an external precision or writing grip, which is a combination of a lateral pinch with the middle finger and a two-point pinch to hold the writing implement (Konz, 1995).

Complete gradation and naming of grips can be found in Kroemer (1986). Note the significantly decreased strength capability of the various pinch grips as compared to the power grip (see Table 5–8). Large forces should never be applied with pinch grips.

Avoid Prolonged Static Muscle Loading

When tools are used in situations in which the arms must be elevated or the tools must be held for extended periods, muscles of the shoulders, arms, and hands may be

TABLE 5–8
Relative Strengths for Different Types of Grips (Adapted from An, et al., 1986)

	Male		Female		
Grip	**lbs**	**kg**	**lbs**	**kg**	**Mean % of Power Grip**
Power	89.9	40.9	51.2	23.3	100
Tip Pinch	14.6	6.6	10.1	4.6	17.5
Pulp Pinch	13.7	6.2	9.7	4.4	16.6
Lateral Pinch	24.5	11.1	17.1	7.8	29.5

statically loaded, resulting in fatigue, reduced work capacity, and soreness. Abduction of the shoulder, with corresponding elevation of the elbow, will occur if work must be done with a pistol-grip tool on a horizontal workplace. An in-line or straight tool reduces the need to raise the arm and also permits a neutral wrist posture. Prolonged work with arms extended, as in assembly tasks done with force, can produce soreness in the forearm. Rearranging the workplace so as to keep the elbows at 90 degrees eliminates most of the problem (see Figure 5–5). Similarly, continuously holding an activation switch can result in fatigue of the fingers and reduced flexibility.

Perform Twisting Motions with the Elbows Bent

When the elbow is extended, tendons and muscles in the arm stretch out and provide low force capability. When the elbow is bent 90 degrees or less, the biceps brachii has a good mechanical advantage and can contribute to forearm rotation.

Maintain a Straight Wrist

As the wrist is moved from its neutral position, a loss of grip strength occurs. Starting from a neutral wrist position, pronation decreases grip strength by 12 percent, flexion/extension by 25 percent, and radial/ulnar deviation by 15 percent (see Figure 5–27). Furthermore, awkward hand positions may result in soreness of the wrist, loss of grip, and, if sustained for extended periods of time, the occurrence of carpal tunnel syndrome. To reduce this problem, the workplace or tools should be redesigned to allow for a straight wrist, for example, lower work surface and edges of containers, and tilt jigs toward the user. Similarly, the tool handle should reflect the axis of grasp, which is about 78 degrees from horizontal, and should be oriented such that the eventual tool axis is in line with the index finger; examples are bent plier handles and a pistol-grip knife (see Figure 5–28).

Avoid Tissue Compression

Often, in the operation of hand tools, considerable force is applied by the hand. Such actions can concentrate considerable compressive force on the palm of the

FIGURE 5–27
Grip strength as a function of wrist and forearm position. (*From:* Sanders and McCormick, 1993) (Reproduced with permission of the McGraw-Hill Companies.)
Grip strength is the average maximal grip sustained for 3 s, expressed as a percentage of neutral supinated grip strength. Based on data from Terrell and Purswell, 1976, Table 1.

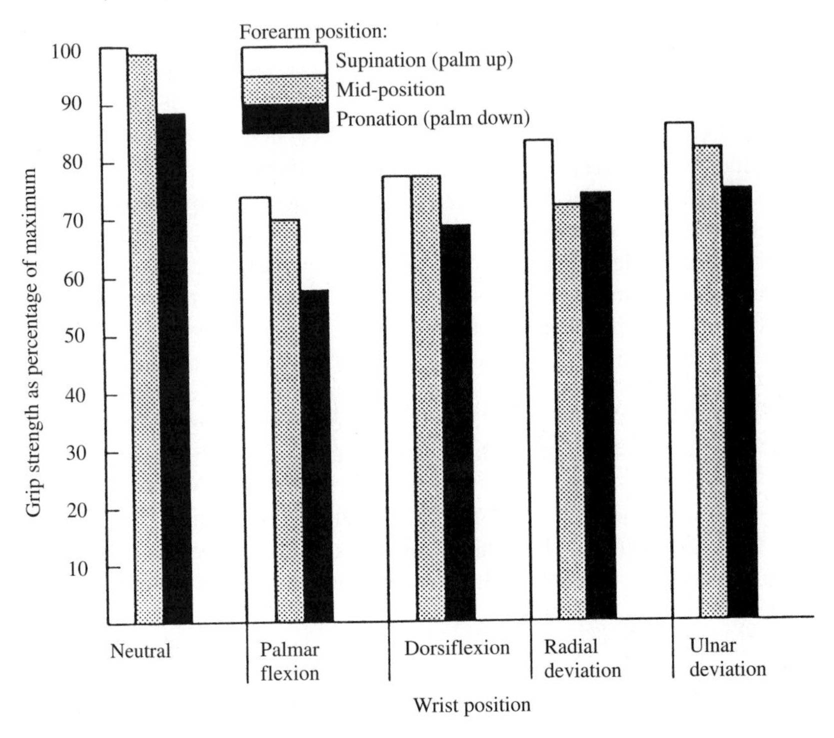

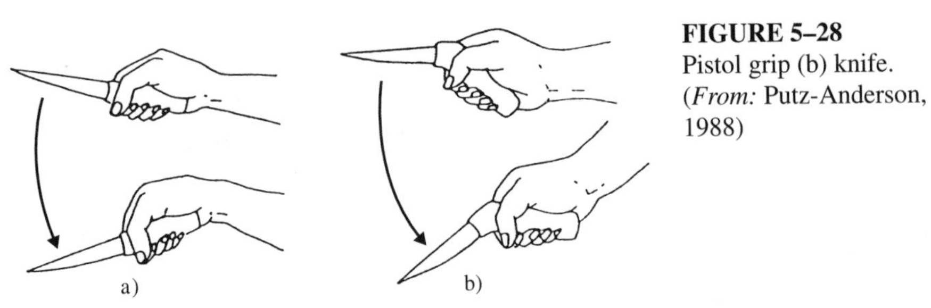

FIGURE 5–28
Pistol grip (b) knife. (*From:* Putz-Anderson, 1988)

hand or the fingers, resulting in ischemia, which is the obstruction of blood flow to the tissues and eventual numbness and tingling of the fingers. Handles should be designed with large contact surfaces, to distribute the force over a larger area (see Figure 5–29) or to direct it to less-sensitive areas, such as the tissue between the thumb and index finger. Similarly, finger grooves or recesses in tool handles should be avoided. Since hands vary considerably in size, such grooves would accommodate only a fraction of the population.

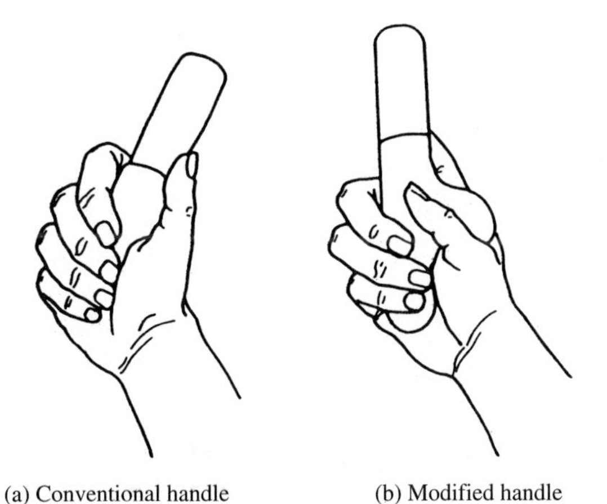

FIGURE 5–29
A conventional paint scraper that presses on the ulnar artery, and a modified handle that rests on the tough tissues between thumb and index finger, thus preventing pressure on the critical areas of the hand. Note that the handle extends beyond the base of the palm. (*From:* Sanders and McCormick, 1993) (Reproduced with permission of the McGraw-Hill Companies.)

Design Tools So That They Can Be Used by Either Hand and by Most Individuals

Alternating hands allows the reduction of local muscle fatigue. However, in many situations, this is not possible, as the tool use is one-handed. Furthermore, if the tool is designated for the user's preferred hand, which for 90 percent of the population is the right hand, then 10 percent are left out. Good examples of right-handed tools that cannot be used by a left-handed person include a power drill with side handle on the left side only, a circular saw, and a serrated knife leveled on one side only. Typically, right-handed males show a 12-percent strength decrement in the left hand, while right-handed females show a 7-percent strength decrement. Surprisingly, both left-handed males and females had nearly equal strengths in both hands. One conclusion is that left-handed subjects are forced to adapt to a right-handed world (Miller and Freivalds, 1976).

Female grip strength typically ranges from 50–67 percent of male strength (Pheasant and Scriven, 1983), for example, the average male can be expected to exert approximately 110 pounds (50 kg), while the average female can be expected to exert approximately 60 pounds (27.3 kg). Females have a twofold disadvantage: an average lower grip strength, and an average smaller grip span. The best solution is to provide a variety of tool sizes.

Avoid Repetitive Finger Action

If the index finger is used excessively for operating triggers, symptoms of trigger finger develop. Trigger forces should be kept low, preferably below 2 pounds (0.9 kg) (Eastman Kodak, 1983), to reduce the load on the index finger. Two or three finger-operated controls are preferable (see Figure 5–30); finger strip controls or a power grip bar are even better, because they require the use of more and stronger fingers. Absolute finger flexion strengths and their relative contributions to grip are shown in Table 5–9.

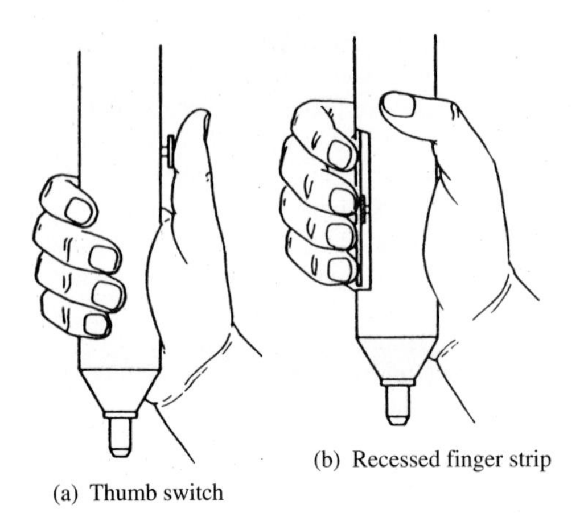

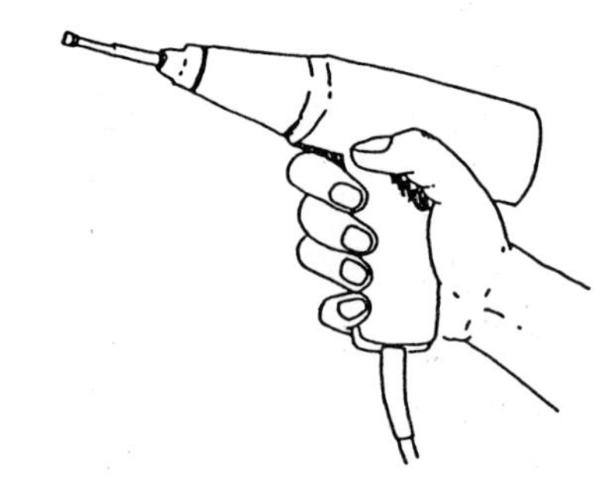

FIGURE 5–30
Thumb-operated and finger-strip-operated pneumatic tool. (Adapted from Sanders and McCormick, 1993) (Reproduced with permission of the McGraw-Hill Companies.)
Thumb operation results in overextension of the thumb. Finger-strip control allows all the fingers to share the load and the thumb to grip and guide the tool.

TABLE 5–9
Maximal Static Finger Flexion Forces [Adapted from Hertzberg (1973)]

Digit	Max Force lbs	Max Force kg	% Force (of thumb)	% Contribution to Power Grip
Thumb	16	7.3	100	—
Index	13	5.9	81	29
Middle	14	6.4	88	31
Ring	11	5.0	69	24
Little	7	3.2	44	16

For a two-handled tool, a spring-loaded return saves the fingers from having to return the tool to its starting position. In addition, the high number of repetitions must be reduced. Although critical levels of repetitions are not known, NIOSH (1989) found high rates of muscle–tendon disorders in workers exceeding 10,000 motions per day.

Use the Strongest Working Fingers: The Middle Finger and the Thumb

Although the index finger is usually the finger that is capable of moving the fastest, it is not the strongest finger (see Table 5–9). Where a relatively heavy load is involved, it is usually more efficient to use the middle finger, or a combination of the middle finger and the index finger.

Design Handle Diameters for Power Grips in the Range of 1.5–2 Inches

Power grips around a cylindrical object should entirely surround the circumference of the cylinder, with the fingers and thumb barely touching. For most individuals, this would entail a handle diameter of 1.5–2 inches (3.8–5.1 cm), resulting in minimum EMG activity, minimum grip endurance deterioration, and maximum thrust forces. In general, the upper end of the range is best for maximum torque, and the lower end is best for dexterity and speed. The handle diameter for precision grips should be approximately 0.5 inches (1.3 cm) (Freivalds, 1996).

Design Handle Lengths to Be a Minimum of 4 Inches

For both handles and cutouts, there should be enough space to allow for all four fingers. Hand breadth across the metacarpals ranges from 2.8 inches (7.1 cm) for a 5th percentile female to 3.8 inches (9.7 cm) for a 95th percentile male (Garrett, 1971). Thus, 4 inches (10 cm) may be a reasonable minimum, but 5 inches (12.5 cm) may be more comfortable. If the grip is enclosed, or if gloves are used, even larger openings are recommended. For an external precision grip, the tool shaft must be long enough to be supported at the base of the first finger or thumb. For an internal precision grip, the tool should extend past the palm, but not far enough to hit the wrist (Konz, 1995).

Design a 3-Inch Grip Span for Two-Handled Tools

Grip strength and the resulting stress on finger flexor tendons vary with the size of the object being grasped. On a dynamometer with handles angled inward, a maximum grip strength is achieved at about 3–3.2 inches (7.68.1 cm) (Chaffin and Andersson, 1991). At distances different from the optimum, the percent grip strength decreases (see Figure 5–31), as defined by:

$$\%\text{Grip strength} = 100 - 0.28*S - 65.8*S^2$$

where S is the given grip span minus the optimum grip span (3 inches for females and 3.2 inches for males). For dynamometers with parallel sides, this optimum span decreases to 1.8–2 inches (4.5-5 cm) (Pheasant and Scriven, 1983). Because of the large variation in individual strength capacities, and the need to accommodate most of the working population (i.e., the 5th percentile female), maximal grip

FIGURE 5–31
Grip strength capability for various population distributions as a function of grip span (*From:* Greenberg & Chaffin, 1976)

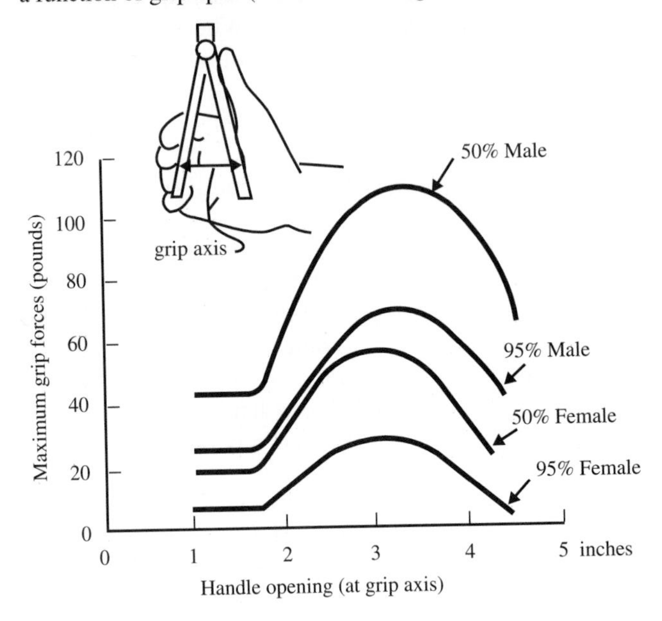

requirements should be limited to less than 20 pounds. A similar effect is found for pinch strength (see Figure 5–32). However, the overall pinch force is at a much more reduced force level (approximately 20 percent of power grip) and the optimum pinch span (for a 4-point pulp pinch) ranges from 0.5–2 inches (1.3–5.1 cm) and then drops sharply for larger spans.

Design Appropriately Shaped Handles

For a power grip, design for maximum surface contact, to minimize unit pressure of the hand. Typically, a tool with a circular cross section is thought to give the largest torque. However, the shape may be dependent on the type of task and the motions involved (Cochran and Riley, 1986). For example, the maximum pull force and the best thrusting actions are actually obtained with a triangular cross section. For a rolling type of manipulation, the triangular shape is slowest. A rectangular shape (with corners rounded) with width to height ratios of from 1:1.25 to 1:1.5 appears to be a good compromise. A further advantage of a rectangular cross section is that the tool does not roll when placed on a table. Also, the handles should not have the shape of a true cylinder, except for a hook grip. For screwdriver-type tools, the handle end should be rounded to prevent undue pressure at the palm; for hammer-type tools, the handle may have some flattening curving, to indicate the end of the handle.

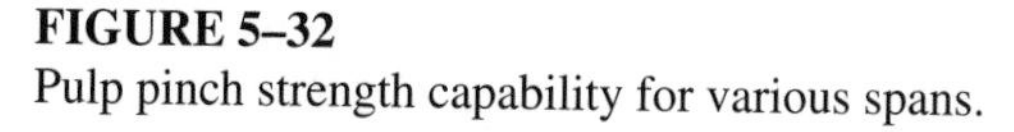

FIGURE 5–32
Pulp pinch strength capability for various spans.

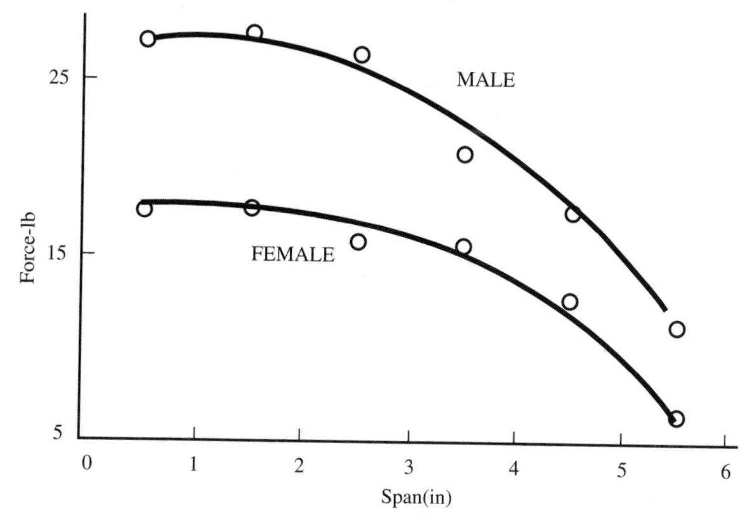

In a departure from the circular, cylindrically shaped handles, Bullinger and Solf (1979) proposed a more radical design using a hexagonal cross section, shaped as two truncated cones joined at the largest ends. Such a shape fits the contours of the palm and thumb best, in both precision and power grips, and it yielded the highest torques in comparison with more conventional handles. A similar dual-truncated conical shape was also developed for a file handle. In this case, the heavily rounded square-shaped cross section was found to be markedly superior to more conventional shapes.

A final note on shape is that T-handles yield a much higher torque (up to 50 percent more) than straight screwdriver handles. The slanting of the T-handle generates even larger torques by allowing the wrist to remain straight (Saran, 1973).

Design Grip Surface to Be Compressible and Nonconductive

For centuries, wood was the material of choice for tool handles. Wood is readily available and easily worked. It has good resistance to shock and thermal and electrical conductivity, and has good frictional qualities, even when wet. Since wooden handles can break and stain with grease and oil, there has recently been a shift to plastic and even metal. However, metal should be covered with rubber or leather, to reduce shock and electrical conductivity and increase friction (Fraser, 1980). Such compressible materials also dampen vibration and allow a better distribution of pressure, reducing fatigue and hand tenderness (Fellows and Freivalds, 1991). However, the grip material should not be too soft, otherwise sharp objects, such as metal chips, will get embedded in the grip and make it difficult to use. The grip surface area should be maximized, to ensure pressure distribution over as large an area as possible. Excessive localized pressure may cause pain sufficient to interrupt the work.

The frictional characteristics of the tool surface vary with the pressure exerted by the hand, the smoothness and porosity of the surface, and the type of contamination (Bobjer et al., 1993). Sweat increases the coefficient of friction, while oil and fat reduce it. Adhesive tape and suede provide good friction when moisture is present. The type of surface pattern, as defined by the ratio of ridge area to groove area, shows some interesting characteristics. When the hand is clean or sweaty, the maximum frictions are obtained with high ratios (maximizing the hand–surface contact area); when the hand is contaminated, maximum frictions are obtained with low ratios (maximizing the capacity to channel contaminants away).

Keep the Weight of the Tool Below 5 Pounds

The weight of the hand tool will determine how long it can be held or used and how precisely it can be manipulated. For tools held in one hand with the elbow at 90 degrees for extended periods of time, Greenberg and Chaffin (1976) recommend loads of no more than 5 pounds (2.3 kg). In addition, the tool should be well balanced, with the center of gravity as close as possible to the center of gravity of the hand (unless the purpose of the tool is to transfer force, as in a hammer). Thus, the hand or arm muscles do not need to oppose any torque development by an unbalanced tool. Heavy tools used to absorb impact or vibration should be mounted on telescoping arms or tool balancers to reduce the effort required by the operator. For precision operations, tool weights greater than 1 pound are not recommended, unless a counterbalanced system is used.

Use Gloves Judiciously

Gloves are often used with hand tools for safety and comfort. Safety gloves are seldom bulky, but gloves worn in subfreezing climates can be very heavy and can interfere with grasping ability. Wearing woolen or leather gloves may add 0.2 inches (0.5 cm) to the hand thickness and 0.3 inches (0.8 cm) to the hand breadth at the thumb, while heavy mittens add 1 inch (2.5 cm) and 1.6 inches (4.0 cm), respectively (Damon et al., 1966). More importantly, gloves reduce grip strength 10–20 percent (Hertzberg, 1973), torque production, and manual dexterity performance times. Neoprene gloves slow performance times by 12.5 percent over barehanded performance, terry cloth by 36 percent, leather by 45 percent, and PVC by 64 percent (Weidman, 1974). A trade-off between increased safety and decreased performance with gloves must be considered.

Use Power Tools Such as Nut- and Screwdrivers Instead of Manual Tools

Power hand tools not only perform work faster than manual tools, but also do the work with considerably less operator fatigue. Greater uniformity of product can be expected when power hand tools are used. For example, a power nut driver can drive nuts consistently to a predetermined tightness in inch–pounds, while a manual nut driver cannot be expected to maintain a constant driving pressure due to operator fatigue.

There is, however another tradeoff. Powered hand tools produce vibration, which can induce white finger syndrome, the primary symptom of which is a reduction in blood flow to the fingers and hand due to *vasoconstriction* of the blood vessels. As a result, there is a sensory feedback loss and decreased performance, and the condition may contribute to the development of carpal tunnel syndrome, especially in jobs with a combination of forceful and repetitive exertions. It is generally recommended that vibrations in the critical range of 40–130 Hz or a slightly larger (but safer) range of 2–200 Hz (Lundstrom and Johansson, 1986) be avoided. The exposure to vibration can be minimized through a reduction in the driving force, the use of specially designed vibration damping handles (Andersson, 1990) or vibration absorbing gloves, and better maintenance to decrease misalignments or unbalanced shafts.

Use the Proper Configuration and Orientation of Power Tools

In a power drill or other power tools, the major function of the operator is to hold, stabilize, and monitor the tool against a workpiece, while the tools perform the main effort of the job. Although the operator may at times need to shift or orient the tool, the main function for the operator is effectively to grasp and hold the tool. A hand drill is comprised of a head, body, and handle, with all three ideally being in line. The line of action is the line from the extended index finger, which means that in the ideal drill, the head is off-center with respect to the central axis of the body.

Handle configuration is also important, with the choices being pistol-grip, in-line, or right-angle. As a rule of thumb, in-line and right-angle are best for tightening downward on a horizontal surface, while pistol-grips are best for tightening on a vertical surface, with the aim being to obtain a standing posture with a straight back, upper arms hanging down, and the wrist straight (see Figure 5–33). For the pistol grip, this results in the handle being at an angle of approximately 78 degrees with the horizontal (Fraser, 1980).

Another important factor is the center of gravity. If it is too far forward in the body of the tool, a turning moment is created, which must be overcome by the muscles of the hand and forearm. This requires muscular effort additional to that required for holding, positioning, and pushing the drill into the workpiece. The primary handle is placed directly under the center of gravity, so that the body juts out behind the handle, as well as in front. For heavy drills, a secondary supportive handle may be needed, either to the side or preferably below the tool, so that the supporting arm can be tucked in against the body, rather than being abducted.

Choose a Power Tool with the Proper Characteristics

Power tools, such as nutrunners used to tighten nuts, are commercially available in a variety of handle configurations, spindle diameters, speeds, weights, shut-off mechanisms, and torque outputs. The torque is transferred from the motor to the spindle through a variety of mechanisms, so that the power (often compressed air)

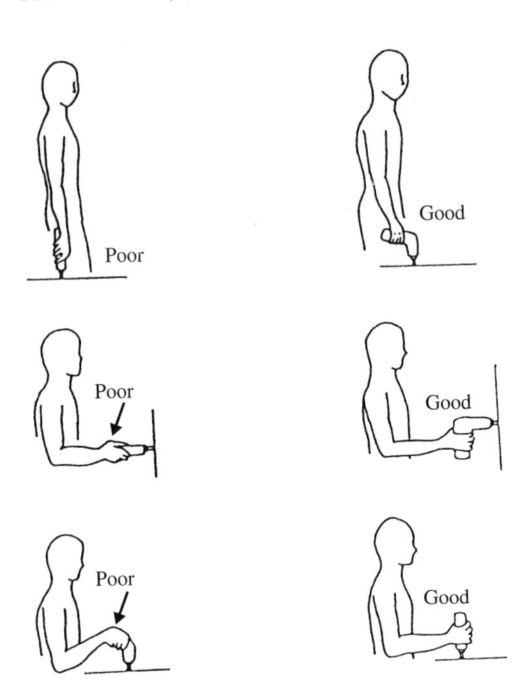

FIGURE 5–33
Proper orientation of power tools in the workplace. (*From:* Armstrong, 1993)

can be shut-off quickly once the nut or other fastener is tight. The simplest and cheapest mechanism is a direct-drive, which is under the operator's control, but because of the long time to release the trigger once the nut is tightened, direct-drive transfers a very large reaction torque to the operator's arm. Mechanical friction clutches will allow the spindle to slip, reducing some of this reaction torque. A better mechanism for reducing the reaction torque is the air-flow shut-off, which automatically senses when to cut off the air supply as the nut is tightened. A still faster mechanism is an automatic mechanical clutch shut-off. The most recent mechanisms include the hydraulic pulse system, in which the rotational energy from the motor is transferred over a pulse unit containing an oil cushion (filtering off the high-frequency pulses, as well as noise), and a similar electrical pulse system, both of which reduce the reaction torque to a large extent (Freivalds and Eklund, 1993).

Variations of torque delivered to the nut depends on several conditions, including: properties of the tool; the operator; properties of the joint, for example, the combination of the fastener and the material being fastened (ranging from soft, in which the materials have elastic properties, such as body panels, to hard, in which there are two stiff surfaces, such as pulleys on a crankshaft); stability of the air supply; etc. The torque experienced by the user (the reaction torque) depends on these factors plus the torque shut-off system. In general, using electrical tools at lower than normal rpm levels, or underpowering pneumatic tools, results in larger reaction torques and more stressful ratings. Pulse-type tools produce the lowest reaction torques, perhaps because the short pulses "chop up" the reaction torque. Other potential problems include: noise from the pneumatic mechanism reaching levels as high as 95 dB(A); vibration levels exceeding 132 dB(V); and dust or oil fumes emanating from the exhaust (Freivalds and Eklund, 1993).

Use Reaction Bars and Tool Balancers for Power Tools

Reaction torque bars should be provided if the torque exceeds: 53 in–lbs (6 Nm) for in-line tools used for a downward action; 106 in–lbs (12 Nm) for pistol-grip tools used in a horizontal mode; and 444 in–lbs (50 Nm) for right-angled tools used in a downward or upward motion (Mital and Kilbom, 1992).

This information is summarized in an evaluative checklist for tools (see Figure 5–34). If the tool does not conform to the recommendations and desired features, it should be redesigned or replaced.

FIGURE 5–34
Tool evaluation checklist.

Basic Principles	Yes	No
1. Does the tool perform the desired function effectively?	❑	❑
2. Does the tool match the size and strength of the operator?	❑	❑
3. Can the tool be used without undue fatigue?	❑	❑
4. Does the tool provide sensory feedback?	❑	❑
5. Are the tool capital and maintenance costs reasonable?	❑	❑

Anatomical Concerns	Yes	No
1. If force is required, can the tool be grasped in a power grip (i.e., handshake)?	❑	❑
2. Can the tool be used without shoulder abduction?	❑	❑
3. Can the tool be used with a 90° elbow angle (i.e., forearms horizontal)?	❑	❑
4. Can the tool be used with the wrist straight?	❑	❑
5. Does the tool handle have large contact surfaces to distribute forces?	❑	❑
6. Can the tool be used comfortably by a 5th percentile female operator?	❑	❑
7. Can the tool be used in either hand?	❑	❑

Handles and Grips	Yes	No
1. For power uses, is the tool grip 1.5 - 2 inches in diameter?	❑	❑
a. Can the handle be grasped with the thumb and fingers slightly overlapped?	❑	❑
2. For precision tasks, is the tool grip 5/16 - 5/8 inches in diameter?	❑	❑
3. Is the grip cross section circular?	❑	❑
4. Is the grip length at least 4 inches (5 inches if gloves are worn)?	❑	❑
5. Is the grip surface finely textured and slightly compressible?	❑	❑
6. Is the handle nonconductive and stain free?	❑	❑
7. For power uses, does the tool have a pistol grip angled at 78°?	❑	❑
8. Can a two-handled tool be operated with less than 20 pounds grip force?	❑	❑
9. Is the span of the tool handles between $2\frac{3}{4}$ - $3\frac{1}{4}$ inches?	❑	❑

Power Tool Considerations	Yes	No
1. Are trigger activation forces less than 1 pound?	❑	❑
2. For repetitive use, is a finger strip trigger present?	❑	❑
3. Are less than 10,000 triggering actions required per shift?	❑	❑
4. Is a reaction bar provided for torques exceeding ...		
a. 50 inch-pounds for in-line tools?	❑	❑
b. 100 inch-pounds for pistol-grip tools?	❑	❑
c. 400 inch-pounds for right-angled tools?	❑	❑
5. Does the tool create less than 85 dBA for a full day of noise exposure?	❑	❑
6. Does the tool vibrate?	❑	❑
a. Are the vibrations outside the 2 - 200 Hz range?	❑	❑

Miscellaneous and General Considerations	Yes	No
1. For general use, is the weight of the tool less than 5 pounds?	❑	❑
2. For precision tasks, is the weight of the tool less than 1 pound?	❑	❑
3. For extended use, is the tool suspended?	❑	❑
4. Is the tool balanced (i.e., center of gravity on the grip axis)?	❑	❑
5. Can the tool be used without gloves?	❑	❑
6. Does the tool have stops to limit closure and prevent pinching?	❑	❑
7. Does the tool have smooth and rounded edges?	❑	❑

SUMMARY

Many factors significantly impact both the productivity and the well-being of the operator at the workstation. Sound ergonomics technology applies to both the equipment being used and the general conditions surrounding the work area. For the equipment point and the workstation environment, adequate flexibility should be provided so that variations in employee height, reach, strength, reflex time, and so on can be accommodated. A workbench that is 32 inches high (81 cm) may be just right for a 75-inch (191-cm) tall worker, but would definitely be too high for a 66-inch (167.6-cm) tall employee. Adjustable-height workstations and chairs are desirable to accommodate the full range of workers, based on plus or minus two standard deviations from the norm. The better able we are to provide a flexible work center to accommodate the total range of the work force, the more satisfactory will be the productivity results and worker satisfaction.

Just as there are significant variations in height and size in the workforce, there are equal or greater variations in visual capacity, hearing ability, feeling ability, and manual dexterity. The vast majority of workstations can be improved. Applying ergonomic considerations along with methods engineering will lead to more efficient competitive work environments that will improve the well-being of the workers, the quality of the product, the labor turnover of the business, and the prestige of the organization.

QUESTIONS

1. What are the principal objectives of operation analysis and motion study?
2. For which situation is an auditory display appropriate?
3. For which situation is a visual display appropriate?
4. What display features would be most effective for a search task?
5. In designing controls and displays, to what does Warrick's principle refer?
6. What key design features should be included in a warning?
7. What are the three most important task factors leading to cumulative trauma disorders?
8. What is the most important factor leading to white finger?
9. What is trigger finger?
10. What is the optimum line of sight?
11. List three principles for arranging components on a panel.

12. What is the range effect?

13. List the three principles for effective control–display compatibility.

14. What is operational coding?

15. What is the main disadvantage of tactile controls?

16. What is "control movement without system response" known as?

17. If the CR ratio is increased from 1.0 to 4.0, what happens to travel time, adjust time, and total time?

18. Describe the progression of the disease state for carpal tunnel syndrome.

19. What is lordosis and how does it relate to a lumbar pad?

PROBLEMS

1. Because of the Challenger disaster, NASA has decided to include a personal escape capability (i.e., a launch compartment) for each space shuttle astronaut. Because space is at a premium, proper anthropometric design is crucial. Also, because of budget restrictions, the design is to be nonadjustable, for example, the same design must fit all present and future astronauts, both males and females. For each launch compartment feature, indicate the body feature used in the design, the design principle used, and the actual value (in inches) to be used in its construction.

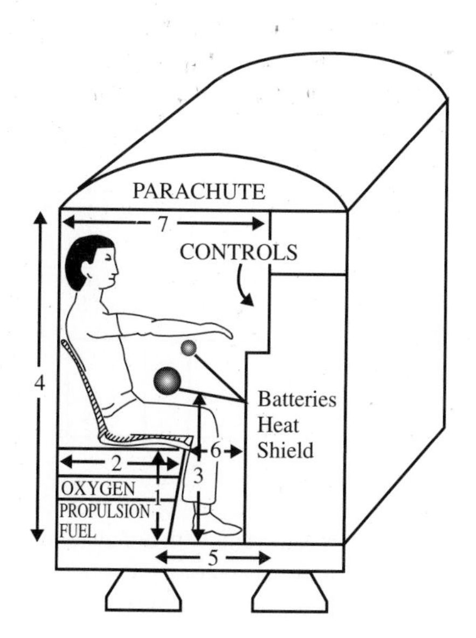

Launch Compartment Feature	Body Feature	Design Principle	Actual Value
1. Height of seat			
2. Seat depth			
3. Height of joystick			
4. Height of compartment			
5. Depth of foot area			
6. Depth of leg area			
7. Depth of chamber			
8. Width of compartment			
9. Weight limit			

2. You are asked to design a control/display panel for the NASA escape launch. After the initial escape, propulsion is to be used to decelerate against the earth's gravitational field. The parachute can only be released within a given, narrow altitude range. Arrange the seven displays/controls, using the same sized dials as shown on the following control panel. Explain the logic for your arrangement.

Control/Display	% of Viewing Time	Importance	# Times Used
Launch release	1	Critical	1
Propulsion fuel level	20	Very important	10
Air speed indicator	15	Important	5
Oxygen pressure	1	Unimportant	2
Electric power level	2	Important	3
Altitude indicator	60	Critical	50
Parachute release	1	Critical	1

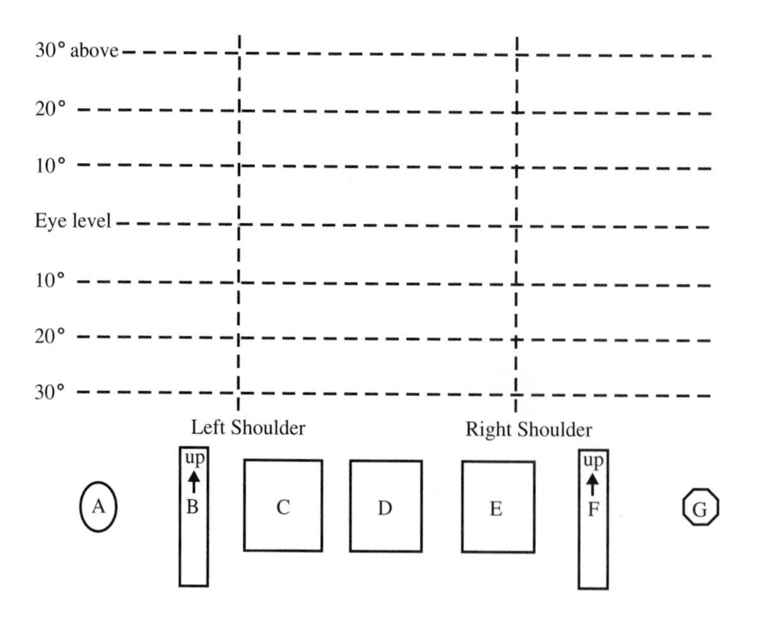

3. The Dorben Foundry uses an overhead crane with a magnetic head to load scrap iron into the blast furnace. The crane operator uses various levers to control the three degrees of freedom needed for the crane and its magnetic head. A control is used to activate/deactivate the magnetic pull of the head. The operator is above the operation and looking down much of the time. Operators frequently complain of back pain. Information on various commercially available lever controls is as follows:

Lever	Throw Distance (in)	Crane Movement (ft)	Time to Target (sec)
A	20	20	1.2
B	20	10	2.2
C	20	80	1.8
D	20	40	1.2

a. Design an appropriate control system for the crane operator. Indicate the number of controls needed, their location (especially in reference to the operator's line of sight), their direction of movement, and their type of feedback.
b. Indicate an appropriate control–response ratio for these controls.
c. What other factors may be important in designing these controls?
d. Draw a side and front view showing the location of the operator with respect to the controls.

REFERENCES

An, K., L. Askew, and E. Chao. "Biomechanics and Functional Assessment of Upper Extremities." In *Trends in Ergonomics/Human Factors III.* Ed. W. Karwowski. Amsterdam: Elsevier, 1986, pp. 573–580.

Andersson, G. B. J., R. Örtengren, A. Nachemson, and G. Elfström. "Lumbar Disc Pressure and Myoelectric Back Muscle Activity During Sitting." *Scandinavian Journal of Rehabilitation Medicine,* 6, no. 3 (1974), pp. 104–133.

Andersson, E. R. "Design and Testing of a Vibration Attenuating Handle." *International Journal of Industrial Ergonomics*, 6, no. 2 (September 1990), pp. 119–125.

ANSI (American National Standards Institute). *ANSI Standard for Human Factors Engineering of Visual Display Terminal Workstations.* ANSI/HFS 100-1988. Santa Monica, CA: Human Factors Society, 1988.

Armstrong, T. J. *Ergonomics Guide to Carpal Tunnel Syndrome.* Fairfax, VA: American Industrial Hygiene Association, 1983.

Bobjer, O., S. E. Johansson, and S. Piguet. "Friction Between Hand and Handle. Effects of Oil and Lard on Textured and Non-textured Surfaces; Perception of Discomfort." *Applied Ergonomics*, 24, no.3 (June 1993), pp. 190–202.

Borg, G. "Psychophysical Scaling with Applications in Physical Work and the Perception of Exertion." *Scandinavian Journal of Work Environment and Health*, 16, Supplement 1 (1990), pp. 55–58.

Bullinger, H. J., and J. J. Solf. *Ergononomische Arbeitsmittel-gestaltung, II - Handgeführte Werkzeuge - Fallstudien.* Dortmund, Germany: Bundesanstalt für Arbeitsschutz und Unfallforschung, 1979.

Chaffin, D. B., and G. Andersson. *Occupational Biomechanics.* New York: John Wiley & Sons, 1991, pp. 355–368.

Cochran, D. J., and M. W. Riley. "An Evaluation of Knife Handle Guarding." *Human Factors*, 28, no.3 (June 1986), pp. 295–301.

Congleton, J. J. *The Design and Evaluation of the Neutral Posture Chair.* (Doctoral Dissertation). Lubbock, TX: Texas Tech University, 1983.

Corlett, E. N., and R. A. Bishop. "A Technique for Assessing Postural Discomfort." *Ergonomics*, 19, no. 2 (March 1976), pp. 175–182.

Damon, A., H. W. Stoudt, and R. A. McFarland. *The Human Body in Equipment Design.* Cambridge, MA: Harvard University Press, 1966.

Diffrient, N., A. R. Tilley, and J. C. Bardagjy. *Humanscale, H.* New York: Dreyfuss Associates, 1978.

Eastman Kodak Co. *Ergonomic Design for People at Work.* Belmont, CA: Lifetime Learning Pub., 1983.

Fellows, G. L., and A. Freivalds. "Ergonomics Evaluation of a Foam Rubber Grip for Tool Handles." *Applied Ergonomics*, 22, no. 4 (August 1991), pp. 225–230.

Fraser, T. M. *Ergonomic Principles in the Design of Hand Tools.* Geneva, Switzerland: International Labor Office, 1980.

Freivalds, A. "Tool Evaluation and Design." In *Occupational Ergonomics.* Ed. A. Bhattacharya and J. D. McGlothlin. New York: Marcel Dekker, 1996, pp. 303–327.

Freivalds, A., and J. Eklund. "Reaction Torques and Operator Stress While Using Powered Nutrunners." *Applied Ergonomics*, 24, no. 3 (June 1993), pp. 158–164.

Garrett, J. "The Adult Human Hand: Some Anthropometric and Biomechanical Considerations." *Human Factors*, 13, no. 2 (April 1971), pp. 117–131.

Greenberg, L., and D. B. Chaffin. *Workers and Their Tools.* Midland, MI: Pendell Press, 1976.

Hertzberg, H. "Engineering Anthropometry." In *Human Engineering Guide to Equipment Design.* Ed. H. Van Cott, and R. Kincaid. Washington, DC: U.S. Government Printing Office, 1973, pp. 467–584.

Konz, S. *Work Design.* Scottsdale, AZ: Publishing Horizons, 1995.

Kroemer, K. H. E. "Coupling the Hand with the Handle: an Improved Notation of Touch, Grip and Grasp." *Human Factors*, 28, no. 3 (June 1986), pp. 337–339.

Lundstrom, R., and R. S. Johannson. "Acute Impairment of the Sensitivity of Skin Mechanoreceptive Units Caused by Vibration Exposure of the Hand." *Ergonomics*, 29, no. 5 (May 1986), pp. 687–698.

Miller, G., and A. Freivalds. "Gender and Handedness in Grip Strength." *Proceedings of the Human Factors Society 31st Annual Meeting.* Santa Monica, CA, 1987, pp. 906–909.

Mital, A., and Å. Kilbom. "Design, Selection and Use of Hand Tools to Alleviate Trauma of the Upper Extremities." *International Journal of Industrial Ergonomics*, 10, no. 1 (January 1992), pp. 1–21.

National Safety Council. *Accident Facts.* Chicago, IL: National Safety Council, 1993.

NIOSH, *Health hazard evaluation-Eagle Convex Glass, Co.* HETA-89-137-2005. Cincinnati, OH: National Institute for Occupational Safety and Health, 1989.

Osinski, C., and A. Freivalds. "Pinching Forces as a Function of Pinch Span." Submitted to *Applied Ergonomics,* 1998.

Pheasant, S. T., and S. J. Scriven. " Sex Differences in Strength, Some Implications for the Design of Handtools." In *Proceedings of the Ergonomics Society.* Ed. K. Coombes. London: Taylor & Francis, 1983, pp. 9–13.

Putz-Anderson, V. *Cumulative Trauma Disorders.* London: Taylor & Francis, 1988.

Sanders, M. S., and E. J. McCormick. *Human Factors in Engineering and Design.* New York: McGraw-Hill, 1993.

Saran, C. "Biomechanical Evaluation of T-handles for a Pronation Supination Task." *Journal of Occupational Medicine*, 15, no. 9 (September 1973), pp. 712–716.

Serber, H. "New Developments in the Science of Seating." *Human Factors Bulletin*, 33, no. 2 (February 1990), pp. 1–3.

Terrell, R., and J. Purswell. "The Influence of Forearm and Wrist Orientation on Static Grip Strength as a Design Criterion for Hand Tools." *Proceedings of the Human Factors Society 20th Annual Meeting*, Santa Monica, CA, 1976, pp. 28–32.

Tichauer, E. R. "Some Aspects of Stress on Forearm and Hand in Industry." *Journal of Occupational Medicine*, 8, no. 2 (February 1966), pp. 63–71.

U.S. Department of Justice. *Americans with Disabilities Act Handbook.* EEOC-BK-19. Washington, DC: U.S. Government Printing Office, 1991.

Webb Associates. *Anthropometric Source Book.* II, Pub. 1024. Washington, DC: National Aeronautics and Space Administration, 1978.

Weidman, B. *Effect of Safety Gloves on Simulated Work Tasks.* AD 738981. Springfield, VA: National Technical Information Service, 1970.

Seth, V., R. Weston, and A. Freivalds. "Development of a Cumulative Trauma Disorder Risk Assessment Model." *International Journal of Industrial Ergonomics*, (to appear 1998).

SELECTED SOFTWARE

COMBIMAN, User's Guide for COMBIMAN, CSERIAC. Dayton, OH: Wright-Patterson AFB, 1992.

Transom Jack, Ann Arbor MI: Transom Technologies Inc., 1998.

MannequinPro. Melville, NY: HUMANCAD Systems, Inc., 1998.

CHAPTER 6

Work Environment Design

KEY POINTS:

- Provide both general and task lighting—avoid glare.
- Control noise at the source.
- Control heat stress with radiation shielding and ventilation.
- Provide both overall air movement and local ventilation for hot areas.
- Dampen tool handles and seats to reduce vibration exposure.
- Avoid radiation exposure.
- Use rapid, forward-rotating shifts, if shiftwork can't be avoided.
- Promote good housekeeping and general safety.
- Sponsor a well-formulated OSHA ergonomics program.

Methods analysts should provide good, safe, comfortable working conditions for the operator. Experience has conclusively proven that plants with good working conditions outproduce those with poor conditions. The economic return from investment in an improved working environment is usually significant. In addition to increasing production, ideal working conditions improve the safety record, reduce absenteeism, tardiness, and labor turnover, raise employee morale, and improve public relations. If these reasons are not sufficient to encourage a company to improve the working environment for its workers, then there is always the "threat" of an OSHA inspection, citation, and fine. The acceptable levels for working conditions and the recommended control measures for problem areas are covered under the OSHA act and presented in greater detail in this chapter.

ILLUMINATION

Theory

Many concepts, terms, and units relate to the measurement of illumination. The basic theory applies to a point source of light (such as a candle) of a given *luminous intensity*, measured in *candelas* (cd) (see Figure 6–1). Light emanates spherically in all directions from the source. The amount of light striking a surface, or a section of this sphere, is termed *illumination* or *illuminance* and is measured in *foot-candles* (fc). The amount of illumination striking a surface drops off as the square of the distance (d) in feet from the source to the surface:

$$\text{illuminance} = \text{intensity}/d^2$$

Some of that light is absorbed and some of it is reflected (for translucent materials, some is also transmitted), which allows humans to "see" that object and provides a perception of brightness. The amount reflected is termed *luminance* and is measured in *foot-Lamberts* (fL). It is determined by the reflective properties of the surface, known as *reflectance*:

$$\text{luminance} = \text{illuminance} \times \text{reflectance}$$

FIGURE 6–1
Illustration of the distribution of light from a light source following the inverse-square law. (*From:* Sanders and McCormick, 1993.) (Reprinted with permission from of the McGraw-Hill Companies.)

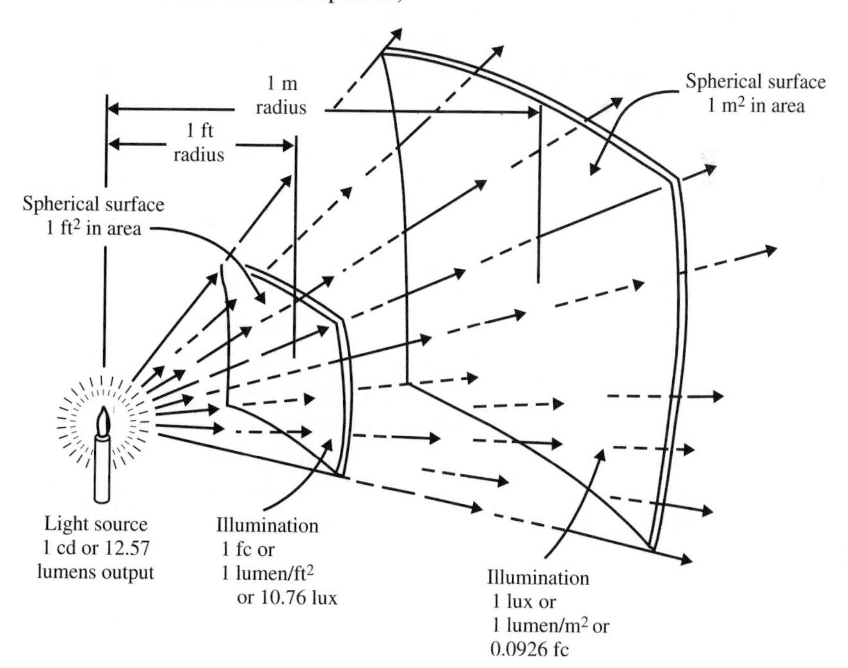

TABLE 6–1
Reflectances of Typical Paint and Wood Finishes

Color or finish	Percent of reflected light	Color or finish	Percent of reflected light
White	85	Medium blue	35
Light cream	75	Dark gray	30
Light gray	75	Dark red	13
Light yellow	75	Dark brown	10
Light buff	70	Dark blue	8
Light green	65	Dark green	7
Light blue	55	Maple	42
Medium yellow	65	Satinwood	34
Medium buff	63	Walnut	16
Medium gray	55	Mahogany	12
Medium green	52		

Reflectance is a unitless proportion and ranges from 0 to 100 percent. High-quality white paper has a reflectance of about 90 percent, newsprint and concrete around 55 percent, cardboard 30 percent, and matte black paint 5 percent. The reflectances for various color paints or finishes are presented in Table 6–1.

Visibility

The clarity with which the human sees something is usually referred to as *visibility*. The three critical factors of visibility are: *visual angle*, *contrast*, and most importantly, *illuminance*. Visual angle is the angle subtended at the eye by the target, and contrast is the difference in luminance between a visual target and its background. Visual angle is usually defined in arc minutes (1/60 of a degree) for small targets by:

$$\text{visual angle (arc min)} = 3438 \times h/d$$

where h is the height of the target or critical detail (or stroke width for printed matter), and d is the distance from of the target to the eye (in the same units as h).

Contrast can be defined in several ways. A typical one is:

$$\text{contrast} = (L_{max} - L_{min})/L_{max}$$

where L is luminance. Contrast, then, is related to the difference in maximum and minimum luminances of the target and background. Note that contrast is unitless.

Other less important factors for visibility are: exposure time, target motion, age, known location, and training, which will not be included here.

The relationship between these three critical factors was quantified by Blackwell (1959) in a series of experiments that led to the development of the Illuminating Engineering Society of North America (IESNA, 1995) standards for illumination. Although the Blackwell curves (see Figure 6–2) as such are not often used today, they show the tradeoff between the size of the object, the amount of illumination (in this case, measured as luminance reflected from the target), and the

FIGURE 6–2
Smoothed threshold contrast curves for disks of diameter (d).
(*Adapted from:* Blackwell, 1959)

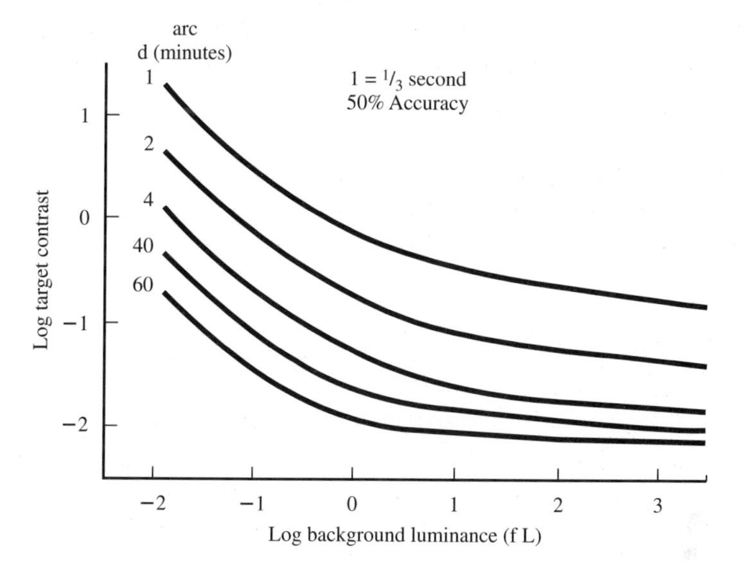

contrast between the target and background. Thus, although increasing the amount of illumination is the simplest approach to improving task visibility, it can also be improved by increasing the contrast or increasing the size of the target.

Illuminance

Recognizing the complexity of extending the point source theory to real light sources (which can be anything but a point source) and some of the uncertainties and constraints of Blackwell's (1959) laboratory setting, the IESNA adopted a much simpler approach for determining minimum levels of illumination (IESNA, 1995). The first step is to identify the general type of activity to be performed and classify it into one of nine categories, shown in Table 6–2. A more extensive list of specific tasks for this process can be found in IESNA (1995). Note that categories A, B, and C do not involve specific visual tasks. For each category, there is a range of illuminances (low, middle, high). The appropriate value is selected by calculating a weighting factor (−1, 0, +1) based on three task and worker characteristics, shown in Table 6–3. These weights are then summed to obtain the total weighing factor. Note that since categories A, B, and C do not involve visual tasks, the speed/accuracy characteristic is not utilized for these categories, and overall room surfaces are utilized in place of task background. If the total sum of the two or three weighting factors is −2 or −3, the low value of the three illuminances is used; if −1, 0, or +1, the middle value is used; and if +2 or +3, the high value is used.

TABLE 6–2
Recommended Illumination Levels for Use in Interior Lighting Design

Category	Range of Illuminance (fc)	Type of Activity	Reference Area
A	2-3-5	Public areas with dark surroundings.	
B	5-7.5-10	Simple orientation for short temporary visits.	General lighting throughout room or area.
C	10-15-20	Working spaces where visual tasks are performed only occasionally.	
D	20-30-50	Performance of visual tasks of high contrast or large size, e.g., reading printed material, typed originals, handwriting in ink and xerography; rough bench and machine work; ordinary inspection; rough assembly.	
E	50-75-100	Performance of visual tasks of medium contrast or small size, e.g., reading medium-pencil handwriting, poorly printed or reproduced material; medium bench and machine work; difficult inspection; medium assembly.	Illuminance on task.
F	100-150-200	Performance of visual tasks of low contrast or very small size, e.g., reading handwriting in hard pencil on poor-quality paper and very poorly reproduced material; highly difficult inspection.	
G	200-300-500	Performance of visual tasks of low contrast and very small size over a prolonged period, e.g., fine assembly; very difficult inspection; fine bench and machine work; extra fine assembly.	
H	500-750-1000	Performance of very prolonged and exacting visual tasks, e.g., the most difficult inspection; extra fine bench and machine work; extra fine assembly.	Illuminance on task via a combination of general and supplementary local lighting.
I	1000-1500-2000	Performance of very special visual tasks of extremely low contrast and small size, e.g., surgical procedures.	

(Adapted from IESNA, 1995)

In practice, illumination is typically measured with a light meter (similar to one found on cameras, but in different units), while luminance is measured with a photometer (typically, a separate attachment to the light meter). Reflectance is usually calculated as the ratio between the luminance of the target surface and the lumi-

TABLE 6–3
Weighting Factors to be Considered in Selecting Specific Illumination Levels Within Each Category of Table 6–2

	Weight		
Task and Worker Characteristics	**−1**	**0**	**+1**
Age	< 40	40-55	> 55
Reflectance of task/surface background	> 70%	30-70%	< 30%
Speed and accuracy (only for categories D - I)	Not important	Important	Critical

(Adapted from IESNA, 1995)

nance of a standard surface of known reflectance (e.g., a Kodak neutral test card of reflectance = 0.9) placed at the same position on the target surface. The reflectance of the target is then:

$$\text{reflectance} = 0.9 \times L_{target}/L_{standard}$$

Light Sources

After determining the illumination requirements for the area under study, analysts select appropriate artificial light sources. Two important parameters related to artificial lighting are *efficiency* (light output per unit energy; typically, lumens/watt) and *color rendering*. Efficiency is particularly important, since it is related to cost; efficient light sources reduce energy consumption. Color rendering relates to the closeness with which the perceived colors of the object being observed match the perceived colors of the same object when illuminated by standard light sources. The more efficient light sources (high- and low-pressure sodium) have only fair to poor color rendering characteristics and consequently may not be suitable for certain inspection operations where color discrimination is necessary. Table 6–4 provides efficiency and color rendering information for the principal types of artificial light. Typical industrial lighting sources, that is, luminaires, are shown in Figure 6–3.

Light Distribution

Luminaires for general lighting are classified in accordance with the percentage of total light output emitted above and below the horizontal (see Figure 6–4). *Indirect lighting* illuminates the ceiling, which in turn reflects light downward. Thus, the ceilings should be the brightest surface in the room (see Figure 6–5), with reflectances above 80 percent. The other areas of the room should reflect lower and lower percentages of the light as one moves downward from the ceiling until the floor is reached, which should reflect no more than 20–40 percent of the light, to avoid glare. To avoid excessive luminance, the luminaires should be evenly distributed across the ceiling.

EXAMPLE 6–1
Calculation of Required Illumination

Consider workers of all ages performing an important, medium-difficulty assembly on a dingy metal workstation with a reflectance of 35 percent. The appropriate weights would be: age = +1, reflectance = 0, and accuracy = 0. The total weight of +1 implies that the middle value of category E is utilized with a required illumination of 75 fc.

TABLE 6–4
Artificial Light Sources

Type	Efficiency (lm/W)	Color rendering	Comments
Incandescent	17-23	Good	A commonly used light source, but the least efficient. Lamp cost is low. Lamp life is typically less than one year.
Fluorescent	50-80	Fair to good	Efficiency and color rendering vary considerably with type of lamp: cool white, warm white, deluxe cool white. Significant energy cost reductions are possible with new energy-saving lamps and ballasts. Lamp life is typically 5–8 years.
Mercury	50-55	Very poor to fair	A very long lamp life (9–12 years), but efficiency drops off substantially with age.
Metal halide	80-90	Fair to moderate	Color rendering is adequate for many applications. Lamp life is typically 1–3 years.
High-pressure sodium	85-125	Fair	Very efficient light source. Lamp life is 3–6 years at average burning rates, up to 12 hours per day.
Low-pressure sodium	100-180	Poor	The most efficient light source. Lamp life is 4–5 years at average burning rate of 12 hours per day. Mainly used for roadways and warehouse lighting.

The efficiency (column 2), in lumens per watt (lm/W), and color rendering (column 3) of six frequently used light sources (column 1) are indicated. Lamp life and other features are given in column 4. Color rendering is a measure of how colors appear under any of these artificial light sources compared with their color under a standard light source. Higher values for efficiency indicate better energy conservation.

Adapted from Lum-i-neering Associates, 1979; Ross and Baruzzini, Inc. 1975; courtesy Human Factors Section, Eastman Kodak Co.

FIGURE 6–3
Types of industrial ceiling-mounted luminaires. (a,c) downlighting, (b,d) diffuse, (e) damp location, (f) high bay, (g) low bay. (*From:* IESNA, 1995)

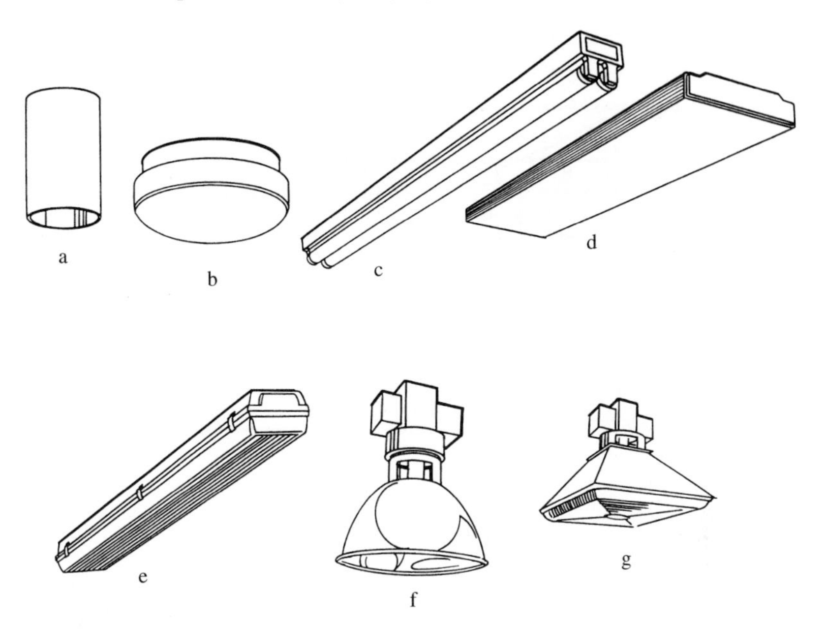

FIGURE 6–4
Luminaires for general lighting are classified in accordance with the percentage of total light output emitted above and below the horizontal. Three of the classifications are: (a) direct lighting, (b) indirect lighting, and (c) direct–indirect lighting. (*From:* IESNA, 1995)

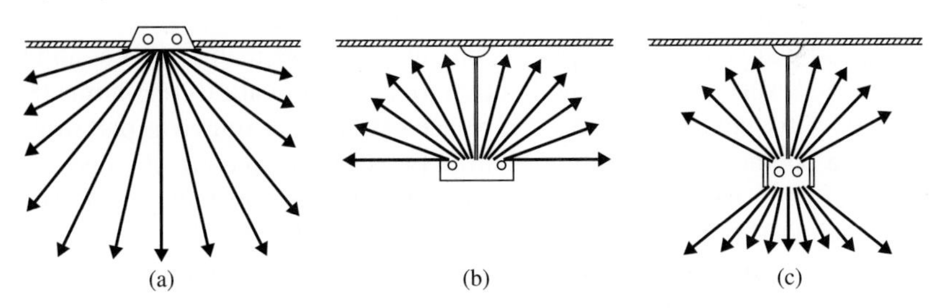

Direct lighting de-emphasizes the ceiling surface and places more of the light on the work surfaces and the floor. Direct–indirect lighting is a combination of both. This distribution of lighting is important, as IESNA (1995) recommends that the ratio of luminances of any adjacent areas in the visual field should not exceed 3/1. The purpose of this is to avoid glare and problems in adaptation.

FIGURE 6–5
Reflectances recommended for room and furniture surfaces in offices. (*From:* IESNA, 1995)

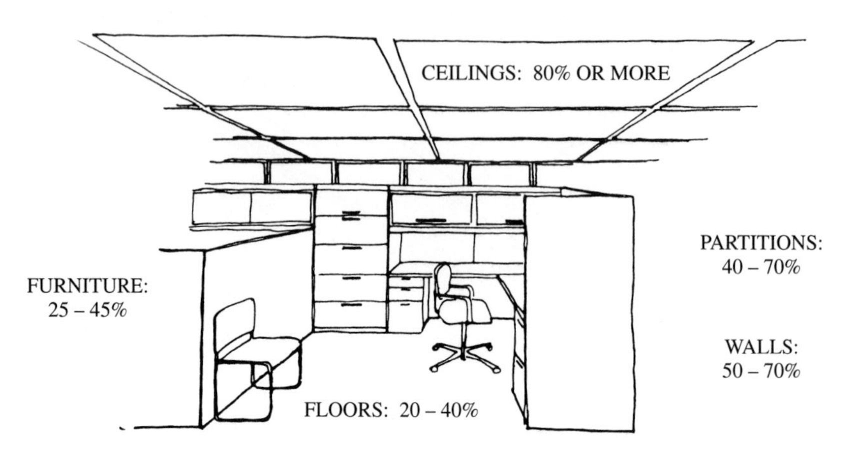

Glare

Glare is the excessive brightness in the field of vision. This excessive light is scattered in the cornea, lens, and even corrective lenses (Freivalds, Harpster, and Heckman, 1983), decreasing visibility so that additional time is required for the eyes to adapt from light to darker conditions. Also, unfortunately, the eyes tend to be drawn directly to the brightest light source, which is known as *phototropism.* Glare can be either direct, as caused by light sources directly in the field of view, or indirect, as reflected from a surface in the field of view. Direct glare can be reduced by using more luminaires with lower intensities, using baffles or diffusers on luminaires, placing the work surface perpendicular to the light source, and increasing overall background lighting so as to decrease the contrast.

Reflected glare can be reduced by using nonglossy or matte surfaces and reorienting the work surface or task, in addition to the modifications recommended for direct glare. Also, polarizing filters can be used at the light source as part of glasses worn by the operator. A special problem is the stroboscopic effect caused by the reflections from moving parts or machinery. Avoiding polished mirror-like surfaces is important here. For example, the mirror-like qualities of the glass screen on computer monitors is a problem in office areas. Repositioning the monitor or using a screen filter is helpful. Typically, most jobs will require supplementary task lighting. This can be provided in a variety of forms, depending on the nature of the task (see Figure 6–6).

Color

Both color and texture have psychological effects on people. For example, yellow is the accepted color of butter; therefore, margarine must be made yellow to appeal to

FIGURE 6–6
Examples of placement of supplementary luminaires: (a) Luminaire located to prevent veiling reflections and reflected glare; reflected light does not coincide with angle of view. (b) Reflected light coincides with angle of view. (c) Low-angle (grazing) lighting to emphasize surface irregularities. (d) Large-area surface source and pattern are reflected toward the eye. (e) Transillumination from diffuse source. (*From:* IESNA, 1995)

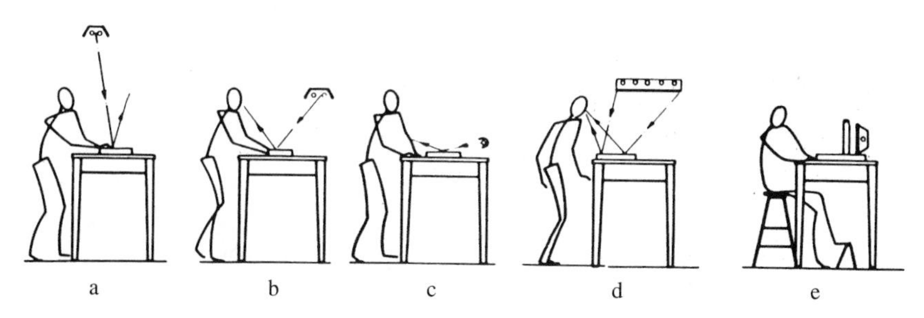

the appetite. Steak is another example. Cooked in 45 seconds on an electronic grill, it does not appeal to customers because it lacks a seared, brown, "appetizing" surface. A special attachment had to be designed to sear the steak. In a third example, employees in an air-conditioned Midwestern plant complained of feeling cold, although the temperature was maintained at 72°F (22.2°C). When the white walls of the plant were repainted in a warm coral color, all complaints ceased. In another instance, workers in a factory complained that boxes were too heavy, until the plant engineer had the old boxes repainted a light green. The next day, several workers said to the supervisor, "Say, those new lightweight boxes sure make a difference."

Perhaps the most important use of color is to improve the environmental conditions of the workers by providing more visual comfort. Analysts use colors to reduce sharp contrasts, increase reflectance, highlight hazards, and call attention to features of the work environment.

Sales are also affected or conditioned by colors. People recognize a company's products instantly by the pattern of colors used on packages, trademarks, letterheads, trucks, and buildings. Some research has indicated that color preferences are influenced by nationality, location, and climate. Sales of a product formerly made in one color increased when several colors suited to the differences in customer demands were supplied. Table 6–5 illustrates the typical emotional effects and psychological significances of the principal colors.

NOISE

Theory

From the analyst's point of view, noise is any unwanted sound. Sound waves originate from the vibration of some object, which in turns sets up a succession of

TABLE 6–5
The Emotional and Psychological Significance of the Principal Colors

Color	Characteristics
Yellow	Has the highest visibility of any color under practically all lighting conditions. It tends to instill a feeling of freshness and dryness. It can give the sensation of wealth and glory, yet can also suggest cowardice and sickness.
Orange	Tends to combine the high visibility of yellow and the vitality and intensity characteristic of red. It attracts more attention than any other color in the spectrum. It gives a feeling of warmth, and frequently has a stimulating or cheering effect.
Red	A high-visibility color having intensity and vitality. It is the physical color associated with blood. It suggests heat, stimulation, and action.
Blue	A low-visibility color. It tends to lead the mind to thoughtfulness and deliberation. It tends to be a soothing color, although it can promote a depressed mood.
Green	A low-visibility color. It imparts a feeling of restfulness, coolness, and stability.
Purple and violet	Low-visibility colors. They are associated with pain, passion, suffering, heroism, and so on. They tend to bring a feeling of fragility, limpness, and dullness.

compression and expansion waves through the transporting medium (air, water, and so on). Thus, sound can be transmitted not only through air and liquids, but also through solids, such as machine tool structures. We know that the velocity of sound waves in air is approximately 1,100 ft/sec (340 m/sec). In viscoelastic materials, such as lead and putty, sound energy is dissipated rapidly as viscous friction.

Sound can be defined in terms of the frequencies that determine its tone and quality, along with the amplitudes that determine its intensity. Frequencies audible to the human ear range from approximately 20 to 20,000 cycles per second, commonly called Hertz and abbreviated Hz. The fundamental equation of wave propagation is:

$$c = f\lambda$$

where: c = Sound velocity (1,100 ft/sec)
f = Frequency in Hz
λ = Wave length in ft

Note that as the wave length increases, the frequency decreases. Methods analysts measure sound intensity with a sound level meter; the unit of sound intensity is the *decibel* (dB). The greater the amplitude of the sound waves, the greater the sound pressure, measured on the decibel scale.

Measurement

Because of the very large range of sound intensities encountered in the normal human environment, the decibel scale has been chosen. In effect, it is the logarithmic

FIGURE 6–7
Decibel values of typical sounds (dBA).

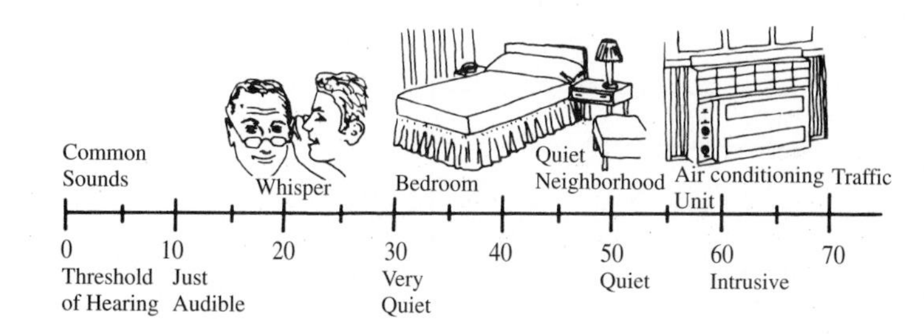

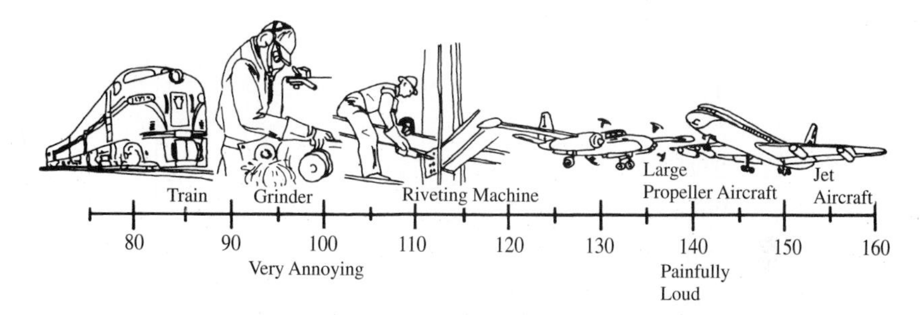

ratio of the actual sound intensity to the sound intensity at the threshold of hearing of a young person. Thus, the sound pressure level (L) in decibels (dB) is given by:

$$L = 20 \log_{10} P_{rms}/P_{ref}$$

where: P_{rms} = Root-mean-square sound pressure in microbars
P_{ref} = Sound pressure at the threshold of hearing of a young person at 1,000 Hz (0.0002 microbars)

Since sound pressure levels are logarithmic quantities, the effect of the coexistence of two or more sound sources in one location requires that a logarithmic addition be performed as follows:

$$L_{TOT} = 10 \log_{10} (10^{L_1/10} + 10^{L_2/10} + ..)$$

where: L_{TOT} is the total noise
L_1 and L_2 are the two noise sources

The A-weighted sound level used in Figure 6–7 is the most widely accepted measure of environmental noise. The A weighting recognizes that from both the psychological and physiological points of view, the low frequencies (50–500 Hz) are far less annoying and harmful than sounds in the critical frequency range of 1,000–4,000 Hz. Above frequencies of 10,000 Hz, hearing acuity (and therefore noise effects) again drops off (see Figure 6–8). The appropriate electronic network is built into sound level meters to attenuate low and high frequencies, so that the sound level meter can read in dBA units directly, to correspond to the effect on the average human ear.

FIGURE 6–8
Equivalent sound level contours.

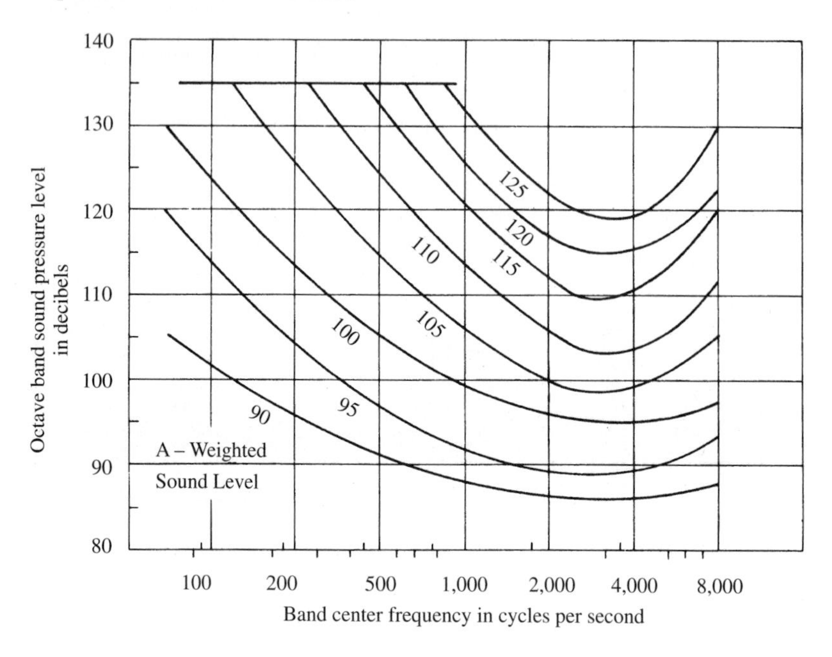

Hearing Loss

The chances of damage to the ear, resulting in "nerve" deafness, increase as the frequency approaches the 2,400 to 4,800 Hz range. This loss of hearing is a result of a loss of receptors in the inner ear, which then fail to transmit the sound waves further to the brain. Also, as the exposure time increases, especially where higher intensities are involved, there will eventually be an impairment in hearing. Nerve deafness is due most commonly to excess exposure to occupational noise. Individuals vary widely in their susceptibility to noise-induced deafness.

In general, noise is classified as either broadband noise or meaningful noise. Broadband noise is made up of frequencies covering a significant part of the sound spectrum. This type of noise can be either continuous or intermittent. Meaningful noise represents distracting information that impacts the worker's efficiency. In long-term situations, broadband noise can result in deafness; in day-to-day operations, it can result in reduced worker efficiency and ineffective communication.

Continuous broadband noise is typical of such industries as the textile industry and an automatic screw machine shop, where the noise level does not deviate significantly during the entire working day. Intermittent broadband noise is characteristic of a drop forge plant and a lumber mill. When a person is exposed to noise that exceeds the damage level, the initial effect is likely to be a temporary hearing loss from which there is complete recovery within a few hours after leaving the work environment. If repeated exposure continues over a long period, irreversible hearing

TABLE 6–6
Permissible Noise Exposures

Duration per day (hours)	Sound level (dA)
8	90
6	92
4	95
3	97
2	100
1.5	102
1	105
0.5	110
0.25 or less	115

Note: When the daily noise exposure is composed of two or more periods of noise exposure of different levels, their combined effect should be considered rather than the individual effects of each. If the sum of the following fractions $C_1/T_1 + C_2/T_2 \ldots C_n/T_n$ exceeds unity, then the mixed exposure should be considered to exceed the limit value. C_n indicates the total time of exposure at a specified noise level and T_n equals the total time of exposure permitted during the work day.

Exposure to impulsive or impact noise should not exceed 140 dB peak sound pressure level.

damage can result. The effects of excessive noise depend on the total energy that the ear has received during the work period. Thus, reducing the time of exposure to excessive noise during the work shift reduces the probability of permanent hearing impairment.

Both broadband and meaningful noise have proved to be sufficiently distracting and annoying to result in decreased productivity and increased employee fatigue. However, federal legislation was enacted primarily because of the possibility of permanent hearing damage due to occupational noise exposure. The OSHA (1970) limits for permissible noise exposure are contained in Table 6–6.

When noise levels are determined by *octave band analysis* (a special filter attachment to the sound-level meter that decomposes the noise into component frequencies), the equivalent A-weighted sound level may be determined as follows: Plot the octave-band sound pressure levels on the graph in Figure 6–8 and note the A-weighted sound level corresponding to the point of highest penetration into the sound level contours. This is the dBA value to be used in further calculations.

Noise Dose

OSHA uses the concept of *noise dose*, with the exposure to any sound level above 80 dBA causing the listener to incur a partial dose. If the total daily exposure consists of several partial exposures to different noise levels, then the several partial doses are added to obtain a combined exposure:

$$D = 100 \times (C_1/T_1 + C_2/T_2 + .. + C_n/T_n) \leq 100$$

where: D = Noise dose
C = Time spent at specified noise level (hours)
T = Time permitted at specified noise level (hours) (see Table 6–6)

EXAMPLE 6–2
Calculation of OSHA Noise Dose

A worker is exposed to 95 dBA for 3 hours and 90 dBA for 5 hours. Although each partial dose is separately permissible, the combined dose is not:

$$D = 100 \times (3/4 + 5/8) = 137.5 > 100$$

The total exposure to various noise levels cannot exceed a 100-percent dose.

Thus, 90 dBA is the maximum permissible level for an 8-hour day, and any sound level above 90 dBA will require some noise abatement. All sound levels between 80-130 dBA must be included in the noise dose computations (although continuous levels above 115 dBA are not allowed at all). Since Table 6–6 provides only certain key times, a computational formula can be used for intermediate noise levels:

$$T = 8/2^{(L - 90)/5}$$

where: L = noise level (dBA)

The noise dose can also be converted to an *8-hour time weighted average* (TWA) sound level. This is the sound level that would produce a given noise dose if a worker were exposed to that sound level continuously over a 8-hour workday. The TWA is defined by:

$$\text{TWA} = 16.61 \times \log_{10}(D/100) + 90$$

Thus, in the last example, a 139.3 percent dose would yield a TWA of:

$$\text{TWA} = 16.61 \times \log_{10}(139.3/100) + 90 = 92.39 \text{ dB}$$

Today, OSHA also requires a mandatory hearing conservation program, including exposure monitoring, audiometric testing, and training, for all employees who have occupational noise exposures equal to or exceeding TWA of 85 dB. Although noise levels below 85 dB may not cause hearing loss, they contribute to distraction and annoyance, resulting in poor worker performance. For example, typical office noises, although not loud, can make it difficult to concentrate, resulting in low productivity in design and other creative work. Also, the effectiveness of telephone and face-to-face communications can be considerably distracted by noise levels less than 85 dB.

Performance Effects

Generally, performance decrements are most often observed in difficult tasks that place high demands on perceptual, information processing, and short-term memory capacities. Surprisingly, noise may have no effect, or may even improve perfor-

EXAMPLE 6–3
Calculation of OSHA Noise Dose with Additional Exposures

A worker is exposed to one hour at 80 dBA, four hours at 90 dBA, and three hours at 96 dBA. The worker is permitted 32 hours for the first exposure, 8 hours for the second exposure, and:

$$T = 8/2^{(96\text{-}90)/5} = 3.48$$

hours for the third exposure. The total noise dose becomes:

$$D = 100 \times (1/32 + 4/8 + 3/3.48) = 139.3$$

Thus, for this worker, the 8-hour noise exposure dosage exceeds OSHA requirements, and either the noise must be abated or the worker must be provided with a rest allowance (see Chapter 10) to comply with OSHA requirements.

mance, on simple routine tasks. Without the noise source, the person's attention may wander due to boredom.

Annoyance is even more complicated and is fraught with emotional issues. Acoustic factors, such as intensity, frequency, duration, fluctuations in level, and spectral composition, play a major role, as do nonacoustic factors, such as past noise experience, activity, personality, noise occurrence predictability, time of day and year, and type of locale. There are approximately a dozen different methods for evaluating annoyance aspects (Sanders and McCormick, 1993). However, most of these measures involve community type issues with noise levels in the 60–70 dBA range, much lower than could reasonably be applied in an industrial situation.

Noise Control

Management can control the noise level in three ways. The best, and usually the most difficult, is to reduce the noise level at its source. However, it would be very difficult to redesign such equipment as pneumatic hammers, steam forging presses, board drop hammers, and woodworking planers and joiners, such that the efficiency of the equipment would be maintained while the noise level is being brought into a tolerable range. In some instances, however, more quietly operating facilities may be substituted for those operating at a high noise level. For example, a hydraulic riveter may be substituted for a pneumatic riveter, an electrically operated apparatus for a steam operated apparatus, and an elastomer-lined tumble barrel for an unlined barrel. Low-frequency noise at the source is effectively controlled at the source using rubber mounts and better alignment and maintenance of the equipment.

If the noise cannot be controlled at its source, then analysts should investigate the opportunity to isolate the equipment responsible for the noise; that is, control the noise that emanates from a machine by housing all or a substantial portion of the facility in an insulating enclosure. This has frequently been done in connection with power presses having automatic feeds. Ambient noise can frequently be reduced by isolating the noise source from the remainder of the structure, thus preventing a sounding board effect. This can be done by mounting the facility on a shear-type elastomer, thus damping the telegraphing of noise.

In situations where enclosing the facility would not interfere with operation and accessibility, the following steps can assure the most satisfactory enclosure design:

1. Clearly establish the design goals and determine the acoustical performance required of the enclosure. Establish octave-band criteria at 3 ft (1 m) from the major machine surfaces.
2. Make actual measurements of the octave-band noise levels of the equipment to be enclosed, at the locations recommended in step 1.
3. Determine the accumulation of noise and then the net noise level when multiple facilities are being used.
4. Determine the spectral attenuation of each enclosure. This is the difference between the design criteria determined in step 1 and the net noise level determined in step 3.
5. Select the acoustical panels and wall configuration for the enclosure. Table 6–7 provides several materials that are popular for relatively small enclosures. A viscoelastic damping material should be applied if any of these materials (with the exception of lead) is used. This can provide an additional attenuation of 3 to 5 dB.

If the noise cannot be reduced at its source, and if the noise-making source cannot be acoustically isolated, then acoustic absorption can provide beneficial results. The purpose of installing acoustical materials on walls, ceilings, and floors is to reduce reverberation. Figure 6–9 illustrates the amount of noise reduction typically possible through such acoustical treatment.

Note that some sounds are desirable in a work environment. For example, background music has been used in factories for many years to improve the work environment. The majority of production and indirect workers (maintenance, shipping, receiving, etc.) enjoy listening to music while they work. When introducing music to a work environment, first consult the employees to determine the type of music to be played and the schedule for playing. Intervals of 20 to 30 minutes on and 20 to 30 minutes off have worked out well.

Hearing Protection

The personnel in the area can wear hearing protection, though in most cases, OSHA accepts this as only a temporary solution. Personal protective equipment can include various types of earplugs, some of which are able to attenuate noises in all frequencies up to sound pressure levels of 110 dB or more. Also available are earmuffs that attenuate noises to 125 dB above 600 Hz and up to 115 dB below this frequency. Earplug effectiveness is measured quantitatively by a *noise reduction rating* (NRR), which is marked on the packaging. The equivalent noise exposure for the listener is

TABLE 6–7

Octave Band Noise Reduction of Single-Layer Materials Commonly Used for Enclosures

Octave band center frequency	125	250	500	1,000	2,000	4,000
16-gauge steel	15	23	31	31	35	41
7-mm steel	25	38	41	45	41	48
7-mm plywood 0.32 kg/0.1 square meter	11	15	20	24	29	30
3/4-in. plywood 0.9 kg/0.1 square meter	19	24	27	30	33	35
14-mm gypsum board 1 kg/0.1 square meter	14	20	30	35	38	37
7-mm fiberglass 0.23 kg/0.1 square meter	5	15	23	24	32	33
0.2 mm lead 0.45 kg/0.1 square meter	19	19	24	28	33	38
0.4-mm lead 0.9 kg/0.1 square meter	23	24	29	33	40	43

FIGURE 6–9

Illustrations of the possible effects of some noise control measures. (*From:* Sanders and McCormick, 1993.) (Reprinted with permission from of the McGraw-Hill Companies.)

The lines on the graph show the possible reductions in noise (from the original level) that might be expected by vibration insulation, a; an enclosure of acoustic absorbing material, b; a rigid, sealed enclosure, c; a single combined enclosure plus vibration insulation, a + b + c; and a double combined enclosure plus vibration insulation, a + 2b + 2c.

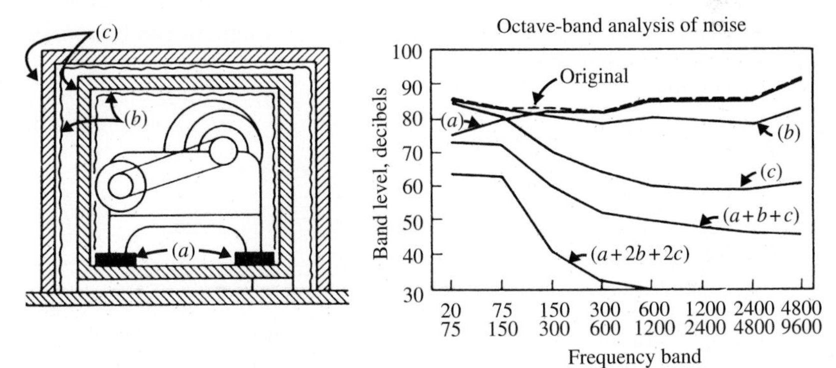

equal to the TWA plus 7 minus the NRR (OSHA, 1983). In general, insert-type (e.g., expandable foam) devices provide better protection than muff-type devices. A combination of an insert device and a muff device can yield NRR values as high as

30. Note that this is a laboratory value obtained under ideal conditions. Typically, in a real-world setting, with hair, beards, eyeglasses, improper fit, etc., the NRR value is going to be considerably lower, perhaps by as much as 10 (Sanders and McCormick, 1993).

TEMPERATURE

Most workers are exposed to excessive heat at one time or another. In many situations, artificially hot climates are created by the demands of the particular industry. Miners are subjected to hot working conditions due to the increase of temperature with depth, as well as a lack of ventilation. Textile workers are subjected to the hot, humid conditions needed for weaving cloth. Steel, coke, aluminum, etc., workers are subject to intense radiative loads from open hearth furnaces and refractory ovens. Such conditions, while present for only a limited part of the day, may exceed the climatic stress found in the most extreme, naturally occurring climates.

Theory

The human is typically modeled as a cylinder with a shell, corresponding to the skin, surface tissues, and limbs, and with a core, corresponding to the deeper tissues of the trunk and head. Core temperatures exhibit a narrow range around a normal value of 98.6° F (37° C). At values between 100–102° F (37.8–38.9° C), physiological performance drops sharply. At temperatures above 105° F (40.6° C), the sweating mechanism may fail, resulting in a rapid rise in core temperature and eventual death. The shell tissues of the body, on the other hand, can vary over a much wider range of temperatures without serious loss of efficiency, and can act as a buffer to protect core temperatures. Clothing, if worn, acts as a second shell to insulate the core temperature further.

The heat exchanges between the body and its environment can be represented by the following heat balance equation:

$$S = M \pm C \pm R - E$$

where: M = Heat gain of metabolism
C = Heat gained (or lost) due to convection
R = Heat gained (or lost) due to radiation
E = Heat lost through evaporation of sweat
S = Heat storage (or loss) of the body

For thermal neutrality, S must be zero. If the summation of the various heat exchanges across the body result in a heat gain, the resulting heat will be stored in the tissues of the body, with a concomitant increase in core temperature and a potential heat stress problem.

A thermal comfort zone, for areas where eight hours of sedentary or light work is done, has been defined as the range of temperatures of from 66–79° F

FIGURE 6–10
The thermal comfort zone. (*Courtesy:* Eastman Kodak Co.)

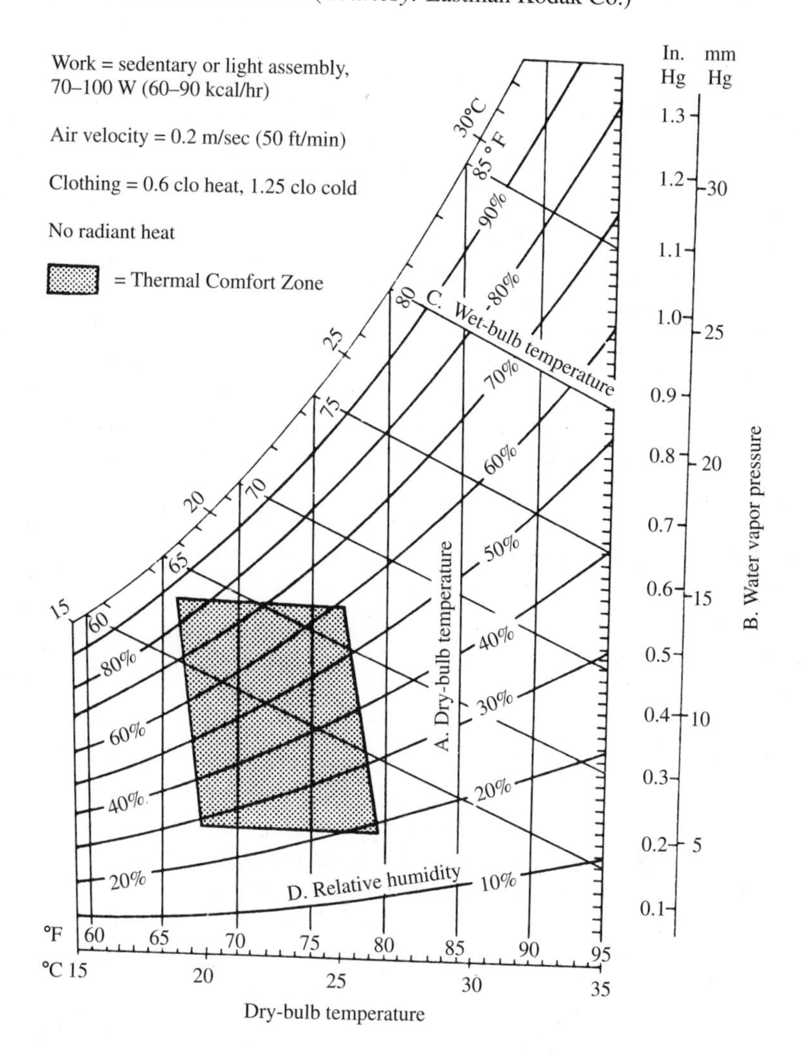

(18.9–26.1° C), with a relative humidity ranging from 20 to 80 percent (see Figure 6–10). Of course, the workload, clothing, and radiant heat load all affect the individual's sense of comfort within the comfort zone.

Heat Stress: WBGT

Many attempts have been made to combine into one index the physiological manifestations of these heat exchanges with environmental measurements. Such attempts have centered around designing instruments intended to simulate the human body, or devising formulas and models based on theoretical or empirical data to estimate the environmental stresses or the resulting physiological strains. In the simplest form, an index consists of the dominant factor, such as the dry bulb temperature, which is used by most people in temperate zones.

Probably the most commonly used index in industry today establishes heat exposure limits and work–rest cycles based on the *wet bulb globe temperature*, or *WBGT* (Yaglou and Minard, 1957), and the metabolic load. In slightly different forms it is recommended by ACGIH (1985), NIOSH (1986) and ASHRAE (1991). For outdoors with a solar load, the WBGT is defined as:

$$\text{WBGT} = 0.7\ \text{NWB} + 0.2\ \text{GT} + 0.1\ \text{DB}$$

and indoors or outdoors with no solar load, the WBGT is:

$$\text{WBGT} = 0.7\ \text{NWB} + .03\ \text{GT}$$

where: NWB = *natural wet bulb temperature* (measure of evaporative cooling, using a thermometer with a wet wick and natural air movement)
GT = *globe temperature* (measure of radiative load, using a thermometer in a 6–in-diameter black copper sphere)
DB = *dry bulb temperature* (basic ambient temperature; thermometer shielded from radiation)

Note that NWB is different from a psychometric wet bulb, which uses maximum air velocity and is used in conjunction with DB to establish relative humidity and thermal comfort zones.

Once the WBGT is measured (commercially available instruments provide instantaneous weighted readings), it is used along with the metabolic load of the workers to establish the amount of time an unacclimatized worker and an acclimatized worker are allowed to work under the given conditions (see Figure 6–11). These limits are based on the individual's core temperature having increased by approximately 1.8° F (1° C) as calculated by the heat balance equation. The 1.8° F increase has been established by NIOSH (1986) as the upper acceptable limit for heat storage in the body. The appropriate amount of rest is assumed to be under the same conditions. Obviously, if the worker rests in a more comfortable area, less rest time would be needed.

Control Methods

Heat stress can be reduced by implementing either engineering controls, that is, modifying the environment, or by administrative controls. Modifying the environment follows directly from the heat balance equation. If the metabolic load is a significant contributor to heat storage, the workload should be reduced by mechanization of the operation. Working more slowly will also decrease the workload, but will have the negative effect of decreasing productivity. The radiative load can be decreased by controlling the heat at the source: insulating hot equipment, providing drains for hot water, maintaining tight joints where steam may escape, and using local exhaust ventilation to disperse heated air rising from a hot process. Radiation

EXAMPLE 6–4
Calculation of WBGT and Heat Stress Level

Consider an unacclimatized worker palletizing a skid at 400 kcal/hr (1,600 BTU/hr) with a thermal load of WBGT = 77° F (25° C). This individual would be able to work for 45 minutes and would then need to rest. At this point, the worker must rest for at least 15 minutes in the same environment or a shorter time in a less stressful environment.

can also be intercepted before it reaches the operator, via radiation shielding: sheets of reflective material, such as aluminum or foil-covered plasterboard, or metal chain curtains, wire mesh screens, or tempered glass, if visibility is required. Reflective garments, protective clothing, or even long-sleeved clothing will also help in reducing the radiative load.

FIGURE 6–11
Recommended heat stress levels based on metabolic heat (1-h time-weighted average), acclimatization, and work-rest cycle. (*From:* NIOSH, 1986, Figs. 1 and 2)
Temperature limits are 1-h time-weighted average WGBT. RAL = recommended alert limit for unacclimatized workers. REL = recommended exposure limit for acclimatized workers.

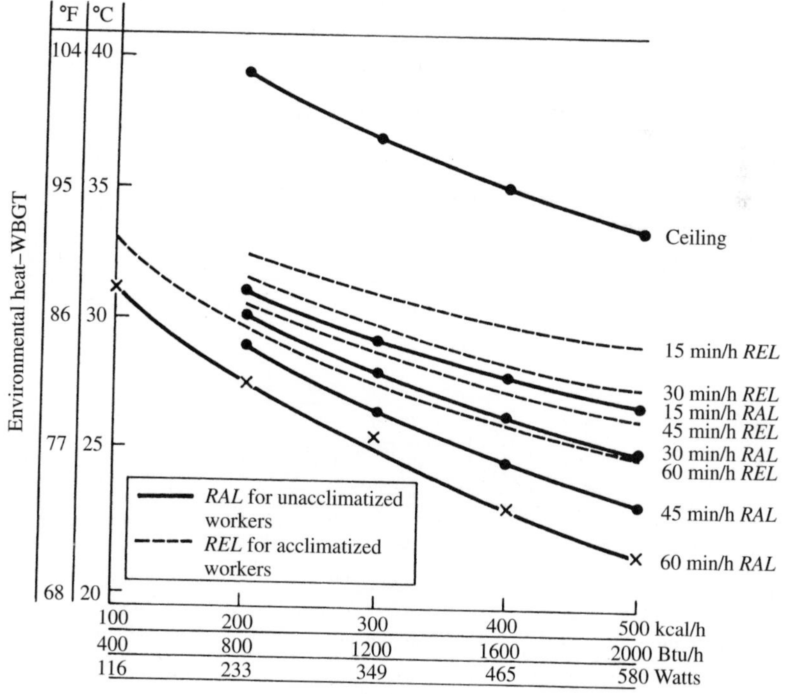

Convective heat loss from the operator can be increased by increasing air movement through ventilation, as long as the dry bulb temperature is less than skin temperature, which is typically around 95° F (35° C) in such environments. Convection is more effective over bare skin; however, bare skin also absorbs more radiation. Thus, there is a trade-off between convection and radiation. Evaporative heat loss from the operator can be improved by again increasing air movement and decreasing the ambient water vapor pressure, using dehumidifiers or air conditioning. Unfortunately, the latter approach, although creating a very pleasant environment, is quite costly and is often not practical for the typical production facility.

Administrative measures, though less effective, include: modifying work schedules to decrease the metabolic load, using work/rest schedules per Figure 6–11, acclimatizing workers (this may take close to two weeks, and the effect is lost over a similar time period), rotating workers into and out of the hot environment, and using cooling vests. The cheapest vests utilize ice frozen in small, plastic packets placed into numerous pockets in the vest (Kamon et al., 1986).

Cold Stress

The most commonly used cold stress index is the *wind chill index*. It describes the rate of heat loss by radiation and convection as a function of ambient temperature and wind velocity (Siple and Passel, 1945). Typically, the wind chill index is not used directly, but is converted to an *equivalent wind chill temperature*. This is the ambient temperature that, in calm conditions, would produce the same wind chill index as the actual combination of air temperature and wind velocity (Table 6–8,

TABLE 6–8
Equivalent Wind Chill Temperatures (°F) of Cold Environments Under Calm Conditions

	Actual Thermometer Reading (°F)							
Wind Speed (mph)	**40**	**30**	**20**	**10**	**0**	**−10**	**−20**	**−30**
0	40	30	20	10	0	−10	−20	−30
5	37	27	16	6	−5	−15	−26	−36
10	28	16	3	−9	−21	−34	−46	−58
15	22	9	−5	−18	−32	−45	−59	−72
20	18	4	−11	−25	−39	−53	−68	−82
30	13	−3	−18	−33	−48	−64	−79	−94
40	10	−6	−22	−38	−53	−69	−85	−101

Little Danger: Exposed, dry flesh won't freeze for 5 hours.	**Increasing Danger:** Exposed flesh may freeze within one minute.	**Great Danger:** Flesh may freeze within 30 seconds.

(*From:* ASHRAE, 1993)

ASHRAE, 1993). For the operator to maintain thermal balance under such low temperature conditions, there must be a close relationship between the worker's physical activity (heat production) and the insulation provided by protective clothing (see Figure 6–12). Here, *clo,* represents the insulation needed to maintain comfort for a person sitting where the relative humidity is 50 percent, the air movement is 20 feet/minute, and the dry bulb temperature is 70° F (21.1° C). A light business suit is approximately equivalent to one *clo* of insulation.

Probably the most critical effects for industrial workers exposed to outdoor conditions are decreased tactile sensitivity and manual dexterity due to vasodilation and decreased blood flow to the hands. Manual performance may decrease as much as 50 percent as the hand skin temperature drops from 65° F to 45° F (18.3° C to 7.2° C) (Lockhart, Kiess, and Clegg, 1975). Auxiliary heaters, hand warmers,

FIGURE 6–12

Prediction of the total insulation required as a function of the ambient temperature (M = Heat production in cal/m²/hr). (*From:* Redrawn from Belding, H. S. Index for evaluating heat stress in terms of physiological strains. Heat, Pipe, and Air Cond., 27, 129.)

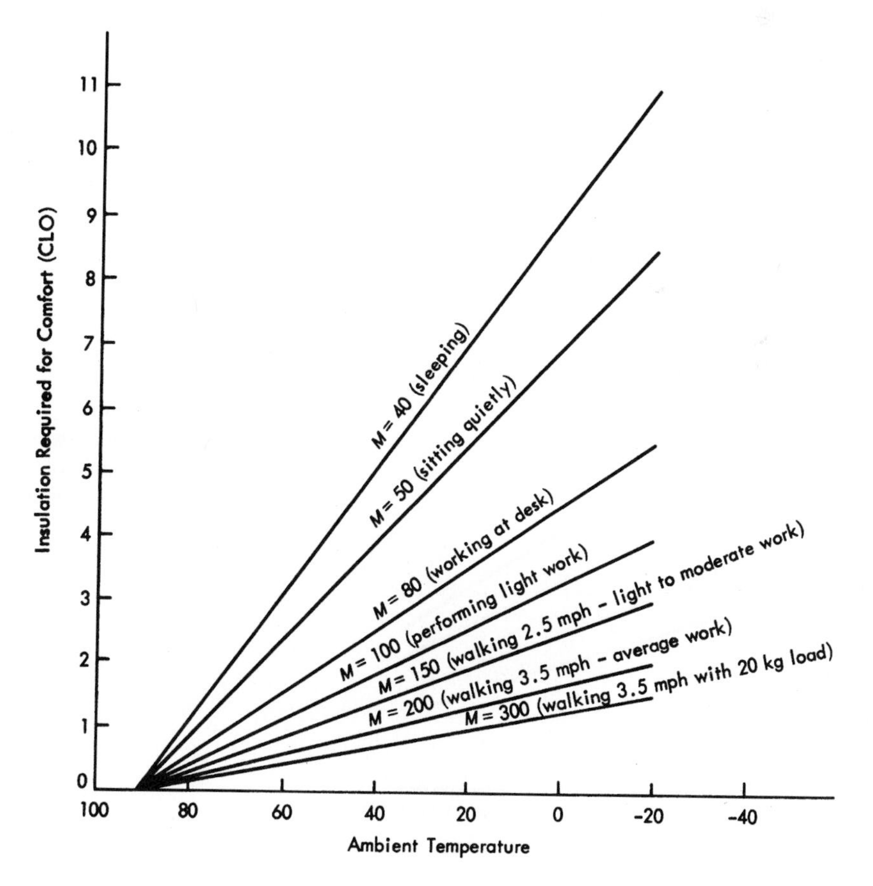

and gloves are potential solutions to the problem. Unfortunately, as indicated in Chapter 5, gloves can impair manual performance and decrease grip strength. A compromise that protects the hands and minimally affects performance may be fingerless gloves (Riley and Cochran, 1984).

VENTILATION

If a room has people, machinery, or activities in it, the air in the room will deteriorate due to the release of odors, the release of heat, the formation of water vapor, the production of carbon dioxide, and the production of toxic vapors. Ventilation must be provided to dilute these contaminants, exhaust the stale air, and supply fresh air. This can be done in one or more of three approaches: general, local, or spot. General or displacement ventilation is delivered at the 8–12 feet (2.4–3.6 m) level and displaces the warm air rising from the equipment, lights, and workers. Recommended guidelines for fresh air requirements, based on the room volume per person, are shown in Figure 6–13 (Yaglou et al., 1936). A rough rule of thumb is 300 ft^3 (8.5 m^3) of fresh air per person per hour.

In a building with only a few work areas, it would be impractical to ventilate the whole building. In that case, local ventilation can be provided at a lower level, or perhaps in an enclosed area, such as a ventilated control booth or crane cab. Note that fan velocity drops rapidly with increasing distance from the fan (see Figure 6–14) and is directionality very critical. Acceptable air velocities at the worker are specified in Table 6–9 (ASHRAE, 1991). A rough rule of thumb is that at a distance of 30 fan diameters, the fan velocity drops to zero (Konz, 1995). Finally, in areas with localized heat sources, such as refractory ovens, spot cooling with a direct high-velocity airstream at the worker will increase convective and evaporative cooling.

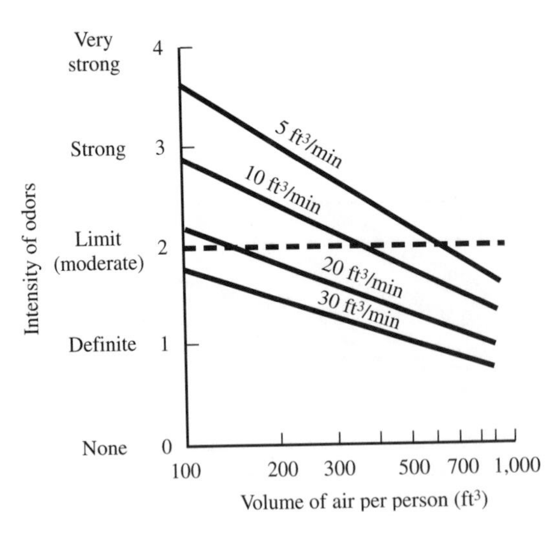

FIGURE 6–13
Guidelines for ventilation requirements for sedentary workers given the available volume of room air. (*Adapted from:* Yaglou et al. 1936)

FIGURE 6–14
Air velocity versus distance for fan placement. (*From:* Konz, 1995)

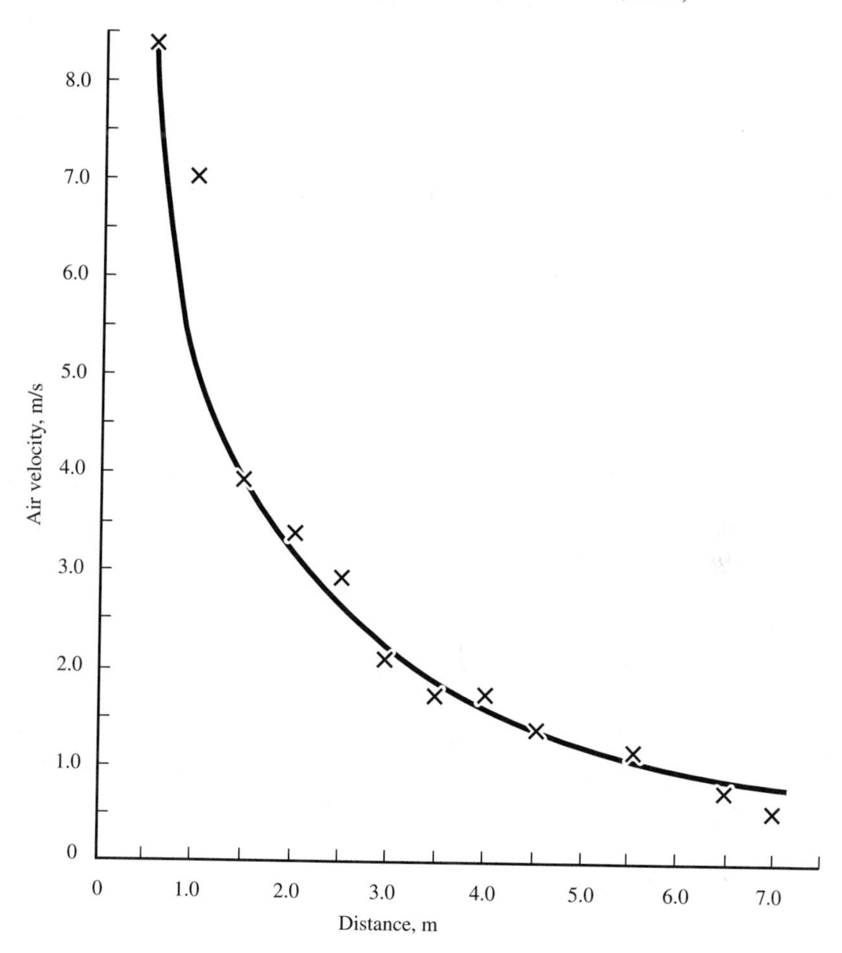

TABLE 6–9
Acceptable Air Motion at the Worker

Exposure	Air Velocity (fpm)
Continuous	
Air-conditioned space	50 to 75
Fixed workstation, general ventilation, or spot cooling	
Sitting	75 to 125
Standing	100 to 200
Intermittent, spot cooling, or relief stations	
Light heat loads and activity	1000 to 2000
Moderate heat loads and activity	2000 to 3000
High heat loads and activity	3000 to 4000

(Reprinted with permission from ASHRAE, 1991.)

VIBRATION

Vibration can cause detrimental effects on human performance. Vibrations of high amplitude and low frequency have especially undesirable effects on body organs and tissue. The parameters of vibration are frequency, amplitude, velocity, acceleration, and jerk. For sinusoidal vibrations, amplitude and its derivations with respect to time are:

$$\text{Amplitude } (s) = \text{Maximum displacement from static position (in)}$$

$$\text{Maximum velocity } \frac{ds}{dt} = 2\pi(s)(f) \text{ in/sec}$$

$$\text{Maximum acceleration } \frac{d^2s}{dt^2} = 4\pi^2(s)(f^2) \text{ in/sec}^2$$

$$\text{Maximum jerk } \frac{d^3s}{dt^3} = 8\pi^3(s)(f^3) \text{ in/sec}^3$$

where: f = Frequency
s = Displacement amplitude

Displacement and maximum acceleration are the principal parameters used to characterize the intensity of vibration.

There are three classifications of vibration exposure:

1. Circumstances in which the whole or a major portion of the body surface is affected; for example, when high-intensity sound in air or water excites vibration.
2. Cases in which vibrations are transmitted to the body through a supporting area; for example, through the buttocks of a person driving a truck, or through the feet of a person standing by a shakeout facility in a foundry.
3. Instances in which vibrations are applied to a localized body area; for example, to the hand when holding and operating a power tool.

Every mechanical system can be modeled using a mass, spring, and dashpot, which, in combination, result in the system having its own *natural frequency*. The nearer the vibration comes to this frequency, the greater the effect on that system. In fact, if the forced vibrations induce larger amplitude vibrations in the system, then the system is in *resonance*. This can have dramatic effects, for example, large winds causing a bridge in Washington to oscillate and eventually collapse, or soldiers breaking step in crossing bridges. For a sitting person, the critical resonant frequencies are given in Table 6–10.

On the other hand, oscillations in the body, or any system, tend to be dampened. Thus, in a standing posture, the muscles of the legs heavily dampen vibrations. Frequencies above 35 Hz are especially dampened. Amplitudes of oscillations induced in the fingers will reduce 50% in the hands, 66% in the elbows, and 90% in the shoulders.

The human tolerance for vibration decreases as the exposure time increases. Thus, the tolerable acceleration level increases with decreasing exposure time.

TABLE 6–10
Resonant Frequencies for Different Body Parts

Frequency (Hz)	Body Part Affected
3–4	Cervical vertebrae
4	Lumbar vertebrae (key for forklift and truck operators)
5	Shoulder girdle
20–30	Between head and shoulder
>30	Fingers, hands, and arms (key for power tool operators)
60–90	Eyeballs (key for pilots and astronauts)

The limits for whole-body vibration have been developed by both the International Standards Organization (ISO) and the American National Standards Institute (ANSI) (ASA, 1980) for transportation and industrial applications. The standards specify limits in terms of acceleration, frequency, and time duration (Figure 6–15). The plotted lines show fatigue/performance limits. For comfort limits, the acceleration values are divided by 3.15; for safety limits, the values are multiplied by 2. Unfortunately, no limits have been developed for the hands and upper extremities.

Low-frequency (0.2–0.7 Hz), high-amplitude vibrations are the principal cause of motion sickness in sea and air travel. Workers experience fatigue much more rapidly when they are exposed to vibrations in the range of 1 to 250 Hz. Early symptoms of vibration fatigue are headache, vision problems, loss of appetite, and loss of interest. Later problems include motor control impairments, disc degeneration, bone atrophy, and arthritis. Vibrations experienced in this range are often characteristic of the trucking industry. The vertical vibrations of many rubber-tired trucks when traveling at typical speeds over ordinary roads range from 3 to about 7 Hz, which are exactly in the critical range for resonances in the human trunk.

Power tools with frequencies between 40–300 Hz tend to occlude blood flow and affect nerves, resulting in the "*white fingers*" *syndrome*. The problem is exacerbated in cold conditions, with the additional problem of cold-induced occlusion of blood flow, or *Raynaud's syndrome*. Better dampened tools, the exchange of detachable handles with special vibration-absorbing handles, and the wearing of gloves, especially those padded with vibration-absorbing gel, will help reduce the problem.

Management can protect employees against vibration in several ways. The applied forces responsible for initiating the vibration may be reduced by modifying the speed, feed, or motion, and by properly maintaining the equipment, balancing and/or replacing worn parts. Analysts can place equipment on antivibration mountings (springs, shear type elastomers, compression pads) or alter workers' body positions to lessen the disturbing vibratory forces. They can also reduce the time workers are exposed to the vibration by alternating work assignments within a group of employees. Lastly, they can introduce supports that cushion the body and thus dampen higher amplitude vibrations. Seat suspension systems involving hydraulic shock absorbers, coil or leaf springs, rubber shear-type mountings, or torsion bars may be used. In standing operations, a soft, elastomer floor mat usually proves helpful.

FIGURE 6–15
The fatigue-decreased proficiency boundary for vertical vibration contained in ISO 2631 and ANSI S3.18-1979. (*From:* Sanders and McCormick, 1993.) (Reprinted with permission from of the McGraw-Hill Companies)
To obtain the boundary for reduced comfort, subtract 10dB (i.e., divide each value by 3.15); to obtain the boundary for safe physiological exposure, add 6 dB (i.e., multiply each value by 2.0).

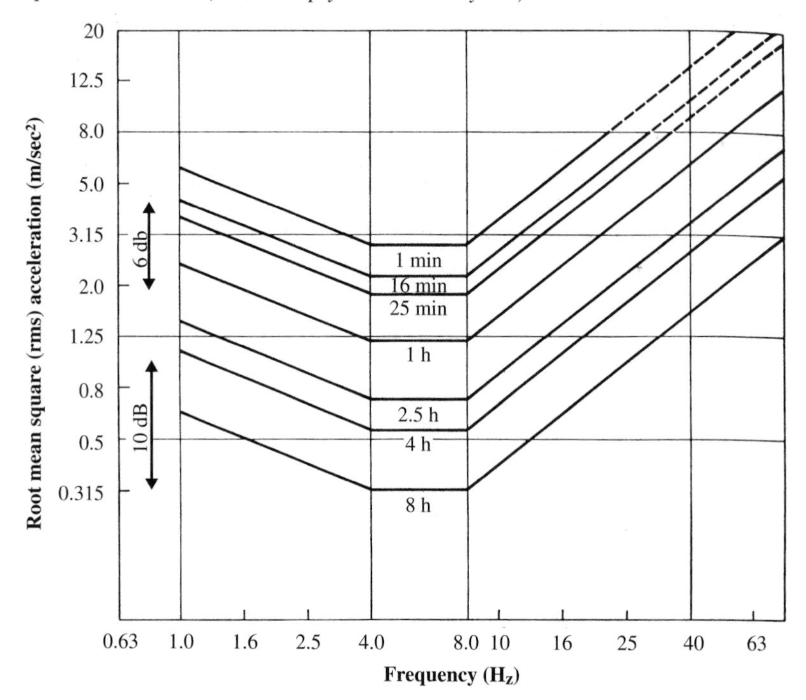

RADIATION

Although all types of ionizing radiation can damage tissue, beta and alpha radiation are so easy to shield that most attention today is given to gamma ray, X ray, and neutron radiation. High-energy electron beams impinging on metal in vacuum equipment can produce very penetrating X rays that may require much more shielding than the electron beam itself.

The absorbed dose is the amount of energy imparted by ionizing radiation to a given mass of material. The unit of absorbed dose is the *rad*, which is equivalent to the absorption of 0.01 joules per kilogram (100 ergs per gram). The dose equivalent is a way of correcting for the differences in the biological effect of various types of ionizing radiation on humans. The unit of dose equivalent is the *rem*, which produces a biological effect essentially the same as that of one rad of absorbed dose of X or gamma radiation. The *Roentgen* (R) is a unit of exposure that measures the amount of ionization produced in air by X or gamma radiation. Tissue located at a point where the exposure is one Roentgen receives an absorbed dose of approximately one rad.

Very large doses of ionizing radiation—100 rads or more—received over a short time span by the entire body can cause radiation sickness. An absorbed dose of about 400 rads to the whole body would be fatal to approximately one-half of adults. Small doses received over a longer period of time may increase the probability of contracting various types of cancers or other diseases. The overall risk of a fatal cancer from a radiation dose equivalent of one rem is about 10^{-4}; that is, a person receiving a dose equivalent of one rem has about 1 chance in 10,000 of dying from a cancer produced by the radiation. The risk can also be expressed as the expectation of one fatal cancer in a group of 10,000 persons, if each person receives a dose equivalent of one rem.

Persons working in areas where access is controlled for the purpose of radiation protection are generally limited to a dose equivalent of five rem per year. The limit in uncontrolled areas is usually the same. Working within these limits should have no significant effect on the health of the individuals involved. All persons are exposed to radiation from naturally occurring radioisotopes in the body, cosmic radiation, and radiation emitted from the earth and building materials. The dose equivalent from natural background sources is about 0.1 rem (100 millirem) per year.

SHIFTWORK AND WORKING HOURS

Shiftwork

Shiftwork, defined as working other than daytime hours, is becoming an ever-increasing problem for industry. Traditionally, the need for continuous services from police, fire, and medical personnel, or for continuous operations in the chemical or pharmaceutical industries, has required the use of shiftwork. More recently, however, the economics of manufacturing, that is, the capitalization or payback of ever more expensive automated machinery is also increasing the demand for shiftwork. Similarly, just-in-time production and seasonal demands for products (i.e., decreased inventory space) has also required more shiftwork.

The problem with shiftwork is the stress on *circadian rhythms,* which are the roughly 24-hour variations in bodily functions in humans (as well as other organisms). The length of the cycle varies from 22 to 25 hours, but is kept synchronized into a 24-hour cycle by various timekeepers, such as the daily light–dark changes, social contacts, work, and clock time. The most marked cyclic changes occur in sleep, core temperature, heart rate, blood pressure, and task performance, such as critical tracking capability (see Figure 6–16). Typically, bodily functions and performance start increasing upon awakening, to a peak in midafternoon, then steadily decline to a low point in the middle of the night. There may also be a dip after midday, typically known as the *post-lunch dip*. Thus, individuals who are asked to work on night shift will exhibit a marked degradation in performance, from truck drivers falling asleep at the wheel to gas inspectors reading meters (Grandjean, 1988).

It could be assumed that night workers would adapt to night work because of the change in work patterns. Unfortunately, the other social interactions still play a

FIGURE 6–16
Examples of circadian rhythms. (*From:* Freivalds, 1983)

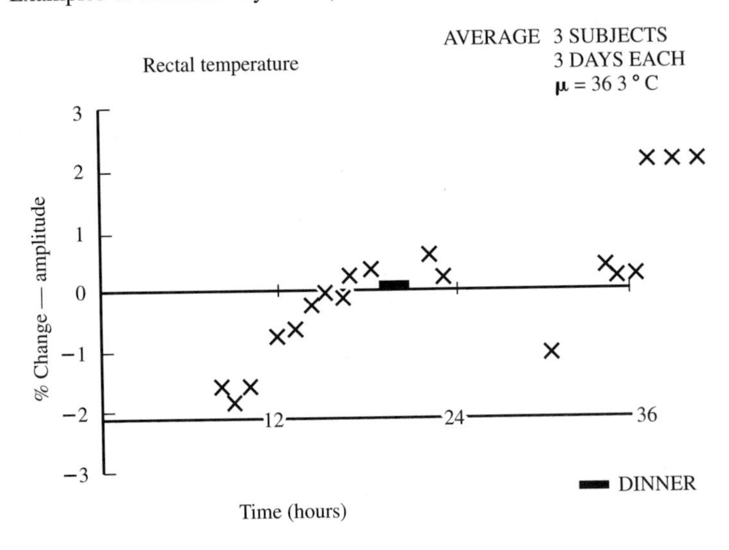

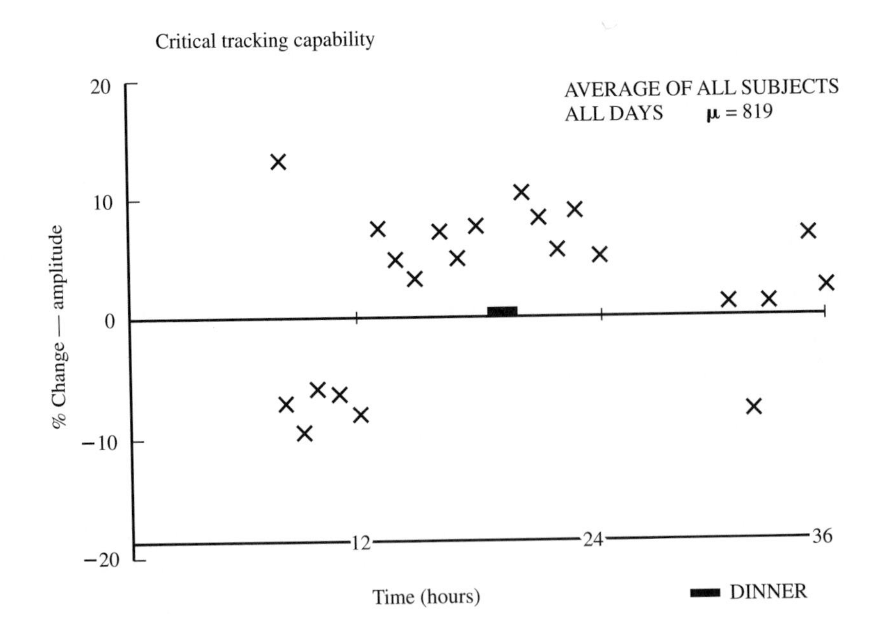

very important role, and the circadian rhythm never truly shifts (as it would for individuals traveling to the other side of the globe for extensive time periods) and may even grow, which some researchers consider to be a worse scenario. Thus, night workers also experience health problems, such as appetite loss, digestive problems, ulcers, and increased sickness rates. The problems become even worse as the worker becomes older.

There are many ways to organize shiftwork. Typically, a three-shift system has an early (E) shift from 8 AM to 4 PM, a late-afternoon shift (L) from 4 PM to 12 PM (midnight), and a night (N) shift from 12 PM to 8 AM. In the simplest case, because of short-term increased production demands, a company may go from just an early shift to both an early and a late-afternoon shift. Usually, because of seniority, the early shift is requested by older, established workers, while new hirees start on the late afternoon shift. The rotation of the two shifts on a weekly basis does not cause major physiological problems, since the sleep pattern is not disrupted. However, the social patterns can be considerably disrupted.

Progressing to a third, night shift becomes more problematic. Since there is difficulty adjusting to a new circadian rhythm, even over the course of several weeks, most researchers advocate a *rapid rotation*, with shift changes every two or three days. This maintains the quality of sleep as well as possible, and does not disrupt family life and social contacts for extended time periods. The weekly rotation typically found in the United States is perhaps the worst scenario, because the workers never truly adjust to any one shift.

One rapid-rotation shiftwork system for a five-day production system (i.e., weekends off) is given in Table 6–11. However, in many companies, the night shift is mainly a maintenance shift with limited production. In that case, a full crew is not needed, and it may be simpler to rotate only the early and late-afternoon shifts, and operate a smaller, fixed night shift, which can be staffed primarily by volunteers who can better adapt to that shift.

For continuous round-the-clock operations, a rapid-rotation seven-day shift system is needed. Two plans commonly used in Europe are the 2-2-2 system, with no more than two days on any one shift (see Table 6–12), and the 2-2-3 system, with no more than three days on any one shift (see Table 6–13). There are tradeoffs with each of these systems. The 2-2-2 system provides a free weekend only once in eight weeks. The 2-2-3 system provide a free three-day weekend once in four weeks, but requires workers to work seven days straight, which is not appealing. A basic

TABLE 6–11
Eight-Hour Shift Rotation (Weekends Off)

Week	M	T	W	Th	F	S	Su
1	E	E	L	L	L	—	—
2	N	N	E	E	E	—	—
3	L	L	N	N	N	—	—

E = early
L = late afternoon
N = night

TABLE 6–12
The 2-2-2 Shift Rotation (8-Hour Continuous)

Week	M	T	W	Th	F	S	Su
1	E	E	L	L	N	N	—
2	—	E	E	L	L	N	N
3	—	—	E	E	L	L	N
4	N	—	—	E	E	L	L
5	N	N	—	—	E	E	L
6	L	N	N	—	—	E	E
7	L	L	N	N	—	—	E
8	E	L	L	N	N	—	—

E = early
L = late afternoon
N = night

TABLE 6–13
The 2-2-3 Shift Rotation (8-Hour Continuous)

Week	M	T	W	Th	F	S	Su
1	E	E	L	L	N	N	N
2	—	—	E	E	L	L	L
3	N	N	—	—	E	E	E
4	L	L	N	N	—	—	—

E = early
L = late afternoon
N = night

problem in both systems is that with 8-hour shifts, a total of 42 hours per week is worked. Alternative systems with more crews and shorter hours may be required (Kodak, 1986).

Another possible approach is to schedule 12-hour shifts. Under these systems, workers either work 12-hour day (D) shifts or 12-hour night (N) shifts, on either a regular three days on and three days off schedule (see Table 6–14), or a more complicated two or three days on or off, with every other weekend free (see Table 6–15). There are several advantages in that there are longer rest periods between work days, and at least one-half of the rest days coincide with a weekend. Of course, the obvious disadvantage is having to work extended days or essentially regular overtime (see next section).

More complicated systems exist with reduced hours (40 or less) per week. These can be studied in further detail in Kodak (1986) or Schwarzenau, et al., (1986).

In summary, definite health and accident risks are associated with shiftwork. However, if shiftwork is unavoidable, due to manufacturing process considerations, the following recommendations should be considered:

1. Avoid shiftwork for workers older than 50.
2. Use rapid rotations as opposed to weekly or monthly cycles.

TABLE 6–14
Twelve-Hour Shift Rotation (3 Days On, 3 Days Off)

Week	M	T	W	Th	F	S	Su
1	D	D	D	—	—	—	N
2	N	N	—	—	—	D	D
3	D	—	—	—	N	N	N
4	—	—	—	D	D	D	—
5	—	—	N	N	N	—	—
6	—	D	D	D	—	—	—

D = day
N = night

TABLE 6–15
Twelve-Hour Shift Rotation (Every Other Weekend Off)

Week	M	T	W	Th	F	S	Su
1	D	—	—	N	N	—	—
2	—	D	D	—	—	N	N
3	N	—	—	D	D	—	—
4	—	N	N	—	—	D	D

D = day
N = night

3. Schedule as few night shifts (three or less) in succession as possible.
4. Use forward rotation of shifts if possible (e.g., E-L-N or D-N)
5. Limit the total number of working shifts in succession to seven or less.
6. Include some free weekends, with at least two successive full days off.
7. Schedule rest days after night shifts.
8. Keep the plans simple, predictable, and equitable for all workers.

Overtime

Many studies have shown that changes in the length of the working day or week have a direct effect on work output, Unfortunately, the result is not typically the direct proportionality expected. Note that in Figure 6–17, the theoretical daily performance is linear (Line 1), but, in practice, is more S-shaped (Curve 2). For example, there is an initial set-up or preparatory period with little productivity (Area A), a gradual warming up, a steeper section with greater than the theoretical productivity (Area B), and graduate leveling off as the end of the shift approaches. In an eight-hour shift, the two-areas, sub-par (Area A) and excess productivity (Area B), are equal, whereas in longer than eight-hour shifts for heavy manual work (Curve 3), the sub-par productivity is greater than the excess productivity, especially with additional sub-par performance (Area C) in the last few hours (Lehmann, 1953).

The results of an old British survey (cited in Grandjean, 1988) found that *shortening* the working day resulted in a *higher* hourly output, with fewer rest pauses

FIGURE 6–17
Productivity as a function of working hours. (*Adapted from:* Lehmann, 1953)

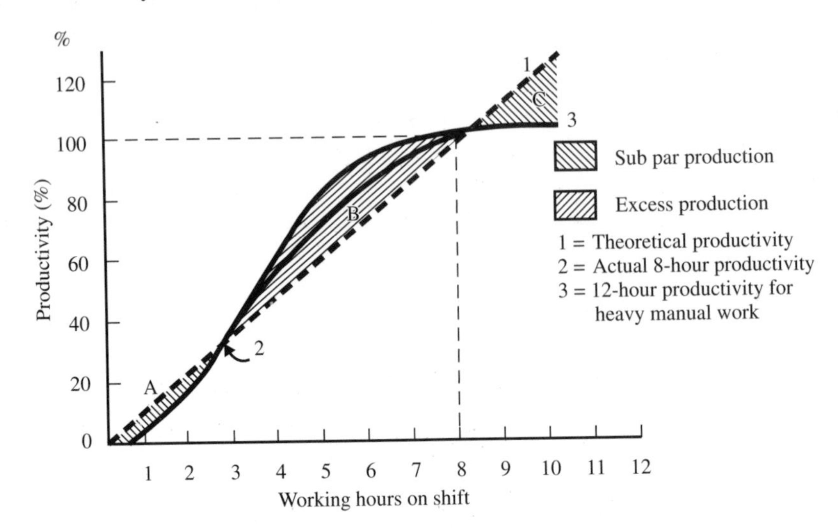

taken. This change in the working performance required at least several days (sometimes longer) before steady state was reached. Conversely, making the working day longer, that is, assigning overtime, causes productivity to fall, sometimes to the point that the total output over the course of the shift actually drops, even though the total hours worked are longer (see Curve 3 in Figure 6–17). Therefore, any expected benefit from increased hours is typically offset by decreased productivity. This effect depends on the level of the physical workload: the more strenuous the work, the greater the decrease in productivity, with the worker using more rest to pace himself or herself.

More recent data (cited in Kodak, 1986) indicate that the expected increase in output is approximately 10 percent for each 25 percent increase in hours worked. This definitely does not justify the time-and-a-half pay expended for overtime work. This discussion presumes a daywork pay scale (see Chapter 17). With an incentive scheme throughout the extended hours, the fall in productivity may not be so great. Similarly, if the work is machine paced, productivity is tied to that machine pace. However, the operator may reach unacceptable fatigue levels, and additional rest per appropriate allowances (see Chapter 10) may be needed. A secondary effect of overtime is that excessive or continuous overtime is accompanied by increased accident rates and sick leave (Grandjean, 1988).

Scheduling overtime on a regular basis cannot be recommended. However, overtime may be necessary for transient short periods, to maintain production or alleviate temporary labor shortages. In such cases, the following guidelines should be followed:

1. Avoid overtime for heavy manual work.
2. Reevaluate machine-paced work for appropriate rest periods or lowered rates.
3. For continuous or long periods of overtime, rotate the work among several workers, or examine alternate shift systems.

4. In choosing between extending a series of work days by one or two hours versus extending the work week by one day, most workers will opt for the former, to avoid losing a weekend day with the family (Kodak, 1986).

Compressed Work Week

A compressed work week implies that 40 hours are performed in fewer than five days. Typically, this occurs in the form of four 10-hour days, three 12-hour days, or four 9-hour days with a half day on Friday. From the management perspective, this concept offers several advantages: reduced absenteeism, relatively less time spent on coffee or lunch breaks, and reduced start-up and shut-down costs (relative to operating time). For example, heat treating, forging, and melting facilities require a significant amount of time, up to 15 percent of the 8-hour working day or more, to bring the facility and material up to the required temperature before production can begin. By going to a 10-hour day, the operation can gain an additional two hours of production time with no additional set-up time. Here, the economic savings from the longer work day can be significant. Workers also gain from increased leisure time, reduced commuter time (relative to working time), and lower commuting costs.

On the other hand, based on the discussions on overtime, a compressed work week essentially amounts to continuous overtime. Although, the total hours worked are less, the hours worked on a given day are more. Therefore, many of the same disadvantages of overtime would apply to a compressed work week (Kodak, 1986). Other objections to the 10-hour day, 4-day week stem from members of management who state that they are obliged to be on the job not only 10 hours for 4 days, but at least 8 hours on the fifth day.

Alternative Work Schedules

With the greater influx of women, especially mothers with school-age children, single parents, older workers, and dual-career-family workers into the workforce, and with the increased concerns for the cost and time of commuting and the value of quality of life, alternate work schedules are needed. One such schedule is *flextime*, where the starting and stopping times are established by the workers, within limits set up by management. Various plans of this nature currently exist. Some require employees to work at least eight hours a day, some require a specified number of hours in a week or a month, while others require all workers to be on site four or five middle hours of the shift.

There are many advantages to these plans for both employees and management. Employees can work the morning or evening hours most conducive to their circadian rhythms, they can better handle family needs or emergencies, and can take care of personal business during business hours without requiring special leave from work. Management gains from reduced tardiness and sick leave. Even the surrounding community gains from decreased traffic congestion and better use of recreational and service facilities. On the other hand, flextime may have limited use in manufacturing, machine-paced, and continuous-process operations, because of problems in scheduling and coordinating the labor force. However, in situations where work groups (see Chapter 18) are utilized, flextime may still be possible (Kodak, 1986).

Part-time employment and job sharing may be especially useful to single parents with children or to retirees seeking to supplement their retirement incomes. Both groups can provide considerable talent and services to a company, but may be limited by circumstances from performing traditional 8-hour shifts. While there may be problems regarding benefits or other fixed employee costs, these may be handled on a prorated or other creative basis.

SAFETY

General Housekeeping

One of the objectives of any progressive management team is to provide a safe and healthful workplace for the employees. This requires controlling the physical environment of the business or operation. Most injuries are the result of accidents caused by an unsafe condition, an unsafe act, or a combination of the two. The unsafe condition is related to the physical environment, which involves the equipment used and all of the physical conditions surrounding the workplace. For example, hazards can stem from lack of guarding or inadequate guarding for the equipment, the location of machines, the condition of storage areas, or the condition of the building.

General safety considerations related to the building include adequate floor-loading capacity. This is especially important in storage areas, where overloading has caused many serious accidents every year. The danger signs of overloading include cracks in walls or ceilings, excessive vibration, and the displacement of structural members.

Aisles, stairs, and other walkways should be investigated periodically to assure that they are free of obstacles, are not uneven, and are not covered with oil or other material that could lead to slips and falls. In many old buildings, stairs should be inspected, since they are the cause of numerous lost-time accidents. Stairs should have a slope of 30 to 35 degrees, with tread widths of approximately 9.5 in (24 cm). Riser heights should not exceed 8 in (20 cm). All stairways should be equipped with handrails, should have at least 10 foot-candles (100 lux) of illumination, and be painted in light colors.

There should be at least two exits on all floors of a building, and the size of the exits should be in accordance with the Life Safety Code of the National Fire Protection Association. This code considers the occupancy and relative fire hazard that the exit area is servicing. Adequate fire protection should be incorporated, based on both OSHA standards and specific local regulations. The building should contain adequate fire extinguishers, sprinkler systems, and standpipe and hose.

Aisles should be plainly marked and straight, with well-rounded corners or diagonals at turn points. If aisles are to accommodate vehicle travel, they should be at least 3 feet wider than twice the width of the broadest vehicle. (When traffic is only one way, then 2 feet wider than the broadest vehicle is adequate.) In general, aisles should have at least 5 foot-candles (50 lux) of illumination. The initial installation of sufficient fixtures does not assure adequate illumination. Only a continuing maintenance effort can ensure the periodic cleaning of fixtures and replacement of blown lights. Color should be used throughout, to identify hazardous conditions. The color recommendations shown in Table 6–16 comply with OSHA standards.

Most machine tools can be satisfactorily guarded to minimize the probability of a worker being injured while operating the machine. The problem is that many machines are not guarded. In these instances, immediate action should be taken to see that a guard is provided and that it is workable and routinely used. An alternate approach is to provide a two-button operation, such as that shown for the press operations in Figure 6–18. Note that each person has two-hand buttons spread well apart (36 in, or 91 cm), so the operator must have both hands in a safe position when the press starts. Activated palm buttons are the number one safety device for protecting an operator while loading and unloading fabrication presses.

There are, of course, exceptions, such as a jointer or a circular cutoff saw, for which the process does not lend itself to foolproof guarding. In such cases, partial guarding is easily attainable, but complete guarding is excessively expensive or impossible, because it interferes with the operator's manipulations. Several alternatives should be considered. Sometimes, the process can be automated,

TABLE 6–16
Color Recommendations

Color	Used for	Examples
Red	Fire protection equipment, danger, and as a stop signal	Fire alarm boxes, location of fire extinguishers and fire hose, sprinkler piping, safety cans for flammables, danger signs, emergency stop buttons
Orange	Dangerous parts of machines, other hazards	Inside of movable guards, safety starting buttons, edges of exposed parts of moving equipment
Yellow	Designating caution, physical hazards	Construction and material handling equipment, corner markings, edges of platforms, pits, stair treads, projections. Black stripes or checks may be used in conjunction with yellow
Green	Safety	Location of first-aid equipment, gas masks, safety deluge showers
Blue	Designating caution against starting or using equipment	Warning flags at starting point of machines, electrical controls, valves about tanks and boilers
Purple	Radiation hazards	Container for radioactive materials or sources
Black and white	Traffic and housekeeping markings	Location of aisles, direction signs, clear floor areas around emergency equipment

FIGURE 6–18
Two-button press operation.

completely freeing the operator from the "nip point." In other instances, a robot manipulator can be used in place of an operator, or the method can be planned and the operator trained to use manual feeders or devices to keep hands and other portions of the body away from danger points.

A quality control and maintenance system should be incorporated in the tool room and the tool cribs, so that only reliable tools in good working condition are released to workers. Examples of unsafe tools that should not be released to operators include: power tools with broken insulation, electrically driven power tools lacking grounding plugs or wires, poorly sharpened tools, hammers with mushroomed heads, cracked grinding wheels, grinding wheels without guards, and tools with split handles or sprung jaws.

There are also potentially dangerous materials to be considered. A large segment of the business and manufacturing industries uses hazardous chemicals. As a matter of company policy, the composition of every chemical compound used should be ascertained, its hazards determined, and control measures established to protect employees. The long-term health effects of many materials are still unknown. As hazards are identified by OSHA and other groups, companies are initiating new procedures to deal with those substances.

Materials known to cause health and/or safety problems fall into one of three categories: corrosive materials, toxic or irritant materials, and flammable materials.

Corrosive materials include a variety of acids and caustics that can burn and destroy human tissue upon contact. The chemical action of corrosive materials can take place by direct contact with the skin, or through the inhalation of fumes or vapors. To avoid the potential danger resulting from the use of corrosive materials, consider the following measures:

1. Be sure that the material handling methods are completely foolproof.
2. Avoid any spilling or spattering, especially during initial delivery processes.
3. Be sure that operators exposed to corrosive materials have and are using correctly designed personal protective equipment and waste disposal procedures.
4. Ensure that the dispensary or the first-aid area is equipped with the necessary emergency provisions, including deluge showers and eye baths.

Toxic or irritating materials include gases, liquids, or solids that poison the body or disrupt normal processes by ingestion, absorption through the skin, or inhalation. To control toxic materials, use the following methods:

1. Completely isolate the process from workers.
2. Provide adequate exhaust ventilation.
3. Provide workers with reliable personal protective equipment.
4. Substitute a nontoxic or nonirritating material, wherever possible.

Flammable materials and strong oxidizing agents present fire and explosion hazards. The spontaneous ignition of combustible materials can take place when there is insufficient ventilation to remove the heat from a process of slow oxidation. To prevent such fires, combustible materials need to be stored in a well-ventilated, cool, dry area. Small quantities should be stored in covered metal containers.

Some combustible dusts, such as sawdust, are not ordinarily known to be explosive. However, explosions can occur when such dusts, flammable vapors, or gases are present in the air in large enough concentrations to ignite. For both gases and dusts, there are limiting concentrations in air below which and above which explosions do not occur. For light dusts, the generally accepted lower explosive limit is 0.015 ounce per cubic foot, and for heavy dusts, 0.5 ounce per cubic foot. Vapors and gases have a wider range over which an explosion is liable to take place. Concentrations of 0.5 percent by volume in air are frequently listed as lower limits. An increase in temperature depresses the lower limit.

To avoid explosions, prevent ignition by providing adequate ventilation–exhaust systems. Adequate control of the manufacturing processes minimizes the generation of dust and the liberation of gases and vapors. Gases and vapors may be removed from gas streams by absorption in liquids or solids, adsorption on solids, condensation, and catalytic combustion and incineration. In absorption, the gas or vapor becomes distributed in the collecting liquid or solid. Absorption equipment includes absorption towers, such as bubble-cap plate columns, packed towers, spray towers, and wet-cell washers. The adsorption of gases and vapors uses a variety of solid adsorbents with an affinity for certain substances. Charcoal, for example, adsorbs many different substances, including benzene, carbon tetrachloride, chloroform, nitrous oxide, and acetaldehyde.

Personal Protective Equipment

Because of the nature of some operations and due to economic considerations, changing methods, equipment, or tools may not eliminate certain hazards. When this is the case, operators can often be fully protected by personal protective equipment. Such equipment would include goggles, face shields, helmets, aprons, jackets, trousers, leggings, gloves, shoes, and respiratory equipment.

To ensure that operating personnel are conscientiously using protective equipment, companies should furnish it to employees either at cost or at no expense. The policy of having the company absorb the cost of personal protective equipment is becoming more and more common. Innumerable cases can be cited where personal protective equipment has saved an eye, a hand, a foot, or a life. For example, one steel company reported that 20 fatalities were prevented in one year by the enforced wearing of company-provided helmets. A Northwest lumber company reported that within a 20-day period, the use of protective hats prevented six serious head injuries.

A study recently conducted by the U.S. Bureau of Labor Statistics showed that three out of five workers who suffered eye injuries or chemical burns were not wearing eye protection at the time of the accident. In 75 percent of those cases, the workers thought protective eyewear was not required in that particular situation. Eye protection equipment is not expensive and should always be available. Employees must be taught the importance of utilizing the protective equipment specified and should develop the attitude that they will not deviate from the prescribed use of such equipment. Compliance should be a condition of employment.

OCCUPATIONAL SAFETY AND HEALTH ADMINISTRATION (OSHA)

OSHA Act

The Occupational Safety and Health Act of 1970 was passed by Congress "to assure so far as possible every working man and woman in the Nation safe and healthful working conditions and to preserve our human resources." Under the act, the Occupational Safety and Health Administration (OSHA) was created to:

1. Encourage employers and employees to reduce workplace hazards and to implement new or improve existing safety and health programs.
2. Establish "separate but dependent responsibilities and rights" for employers and employees for the achievement of better safety and health conditions.
3. Maintain a reporting and recordkeeping system to monitor job-related injuries and illnesses.
4. Develop mandatory job safety and health standards and enforce them effectively.
5. Provide for the development, analysis, evaluation, and approval of state occupational safety and health programs.

Since the act can intimately affect the design of the workplace, methods analysts should be knowledgeable regarding the details of this act. The general-duty clause of the act states that each employer "must furnish a place of employment which is free from recognized hazards that cause or are likely to cause death or serious physical harm to employees." Furthermore, the act brings out that it is the employers' responsibility to become familiar with standards applicable to their establishments and to ensure that employees have and use personal protective gear and equipment for safety.

OSHA standards fall into four categories: general industry, maritime, construction, and agriculture. All OSHA standards are published in the *Federal Register*, which is available in most public libraries, in a separate book of regulations (OSHA, 1997), and on the Web (http://www.osha.gov/). OSHA can begin standards-setting procedures on its own initiative or on the basis of petitions from the Secretary of Health and Human Services (HHS), the National Institute for Occupational Safety and Health (NIOSH), state and local governments, nationally recognized standards-producing organizations such as the ASME, and employer or labor representatives. Of these groups, NIOSH, an agency of HHS, is quite active in making recommendations for standards. It conducts research on various safety and health problems and provides considerable technical assistance to OSHA. Especially important is the investigation of toxic substances by NIOSH, and its development of criteria for the use of such substances in the workplace.

OSHA also provides free on-site consultation services for employers in all 50 states. This service is available on request, and priority is given to smaller businesses, which are generally less able to afford private sector consultations. These consultants help employers identify hazardous conditions and determine corrective measures.

The Act also requires employers of 11 or more employees to maintain records of occupational injuries and illnesses as they occur. An occupational injury is defined as "any injury such as a cut, fracture, sprain or amputation which results from a work-related accident or from exposure involving a single incident in the work environment." An occupational illness is "any abnormal condition or disorder, other than one resulting from an occupational injury, caused by exposure to environmental factors associated with employment." Occupational illnesses include acute and chronic illnesses that may be caused by inhalation, absorption, ingestion, or direct contact with toxic substances or harmful agents. Specifically, they must be recorded if the result is death, loss of one or more workdays, restriction in motion or ability to do the work that had been done, loss of consciousness, transfer to another job, or medical treatment other than first aid.

Workplace Inspections

To enforce its standards, OSHA is authorized to conduct workplace inspections. Consequently, every establishment covered by the Act is subject to inspection by OSHA compliance safety and health officers. The Act states that "upon presenting appropriate credentials to the owner, operator, or agent in charge," an OSHA compliance officer is authorized to enter without delay any factory or workplace in order to inspect all pertinent conditions, equipment, and materials therein and to question the employer, operator, or employees.

OSHA inspections, with few exceptions, are concluded without advance notice. In fact, alerting an employer in advance of an OSHA inspection can bring a fine of up to $1,000 and/or a 6–month jail term. Special circumstances under which OSHA may give notice of inspection to an employer include those where:

1. Imminently dangerous situations exist that require correction as soon as possible.
2. Inspections necessitate special preparation or must take place after regular business hours.
3. Prior notice ensures that the employer and employee representatives or other personnel will be present.
4. The OSHA area director determines that advance notice would produce a more thorough or more effective inspection.

Upon inspection, if an imminently dangerous situation is found, the compliance officer asks the employer to abate the hazard voluntarily and to remove endangered employees from exposure. Notice of the imminent danger must also be posted. Before the OSHA inspector leaves the workplace, he or she will advise all affected employees of the hazard.

At the time of the inspection, the employer is asked to select an employer representative to accompany the compliance officer during the inspection. An authorized employee representative is also given the opportunity to attend the opening conference and to accompany the compliance officer during the inspection. In those plants with a union, the union ordinarily designates the employee representative to accompany the compliance officer. Under no circumstances may the employer select the employee representative for the inspection. The Act does not require an employee representative for each inspection; however, where there is no authorized employee representative, the compliance officer must consult with a reasonable number of employees concerning safety and health matters in the workplace.

After the inspection tour, a closing conference is held between the compliance officer and the employer or the employer representative. Subsequently, the compliance officer reports the findings to the OSHA office, and the area director determines what citations, if any, will be issued, and what penalties, if any, will be proposed.

Citations

Citations inform the employer and employees of the regulations and standards alleged to have been violated, and the proposed time set for their abatement. The employer will receive citations and notices of proposed penalties by certified mail. The employer must post a copy of each citation at or near the place where a violation has occurred, for three days or until the violation is abated, whichever is longer.

The compliance officer has the authority to issue citations at the worksite, following the closing conference. To do so, he or she must first discuss each apparent violation with the area director and must receive approval to issue the citations.

The six types of violations that may be cited, and the penalties that may be imposed, are:

1. *De minimis (no penalty)*. This type of violation has no immediate relationship to safety or health, e.g., number of toilets.
2. *Nonserious violation*. This type of violation has a direct relationship to job safety and health, but probably would not cause death or serious physical harm. A proposed penalty of up to $7,000 for each violation is discretionary. A penalty for a nonserious violation may be decreased considerably depending on the employer's good faith (demonstrated efforts to comply with the Act), history of previous violations, and size of business.
3. *Serious violation*. This is a violation in which there is substantial probability that death or serious harm could result, stemming from a hazard about which the employer knew or should have known. A mandatory penalty of up to $7,000 is assessed for each violation.
4. *Willful violation*. This is a violation that the employer intentionally and knowingly commits. The employer either knows that his or her actions constitute a violation, or is aware that a hazardous condition exists and has made no reasonable effort to eliminate it. Penalties of up to $70,000 may be proposed for each willful violation. If an employer is convicted of a willful violation that has resulted in the death of an employee, there may also be imprisonment for up to six months. A second conviction doubles these maximum penalties.
5. *Repeated violation*. A repeated violation occurs when a violation of any standard, regulation, rule, or order is reinspected and another violation of the previously cited section is found. A citation for a repeated violation is not necessarily issued for violations involving the same piece of equipment or location. If, on reinspection, a violation of the previously cited standard, regulation, rule, or order is found, but it involves another piece of equipment and/or a different location in the establishment or worksite, it may be considered a repeated violation. Each repeated violation can bring a fine of up to $70,000.
6. *Imminent danger*. This is a situation in which there is reasonable certainty that a danger exists that can be expected to cause death or serious physical harm either immediately or before the danger can be eliminated through normal enforcement procedures. An imminent danger violation may result in a cessation of the operation or even complete plant shutdown.

Other violations for which citations and proposed penalties may be issued are as follows:

1. Falsifying records, reports, or applications, on conviction, can bring a fine of $10,000 and six months in jail.
2. Violating the posting requirements can bring a civil penalty of up to $7,000.
3. Failing to correct a violation can bring a civil penalty of up to $7,000 for each day the violation continues beyond the prescribed abatement date.

OSHA ERGONOMICS PROGRAM

In 1990, the high incidences and severity of work-related musculoskeletal disorders found in the meatpacking industry led OSHA to develop ergonomics guidelines to be used in protecting meatpackers from these hazards (OSHA, 1990). The publication and dissemination of these guidelines was meant to be a first step in assisting the meatpacking industry in implementing a comprehensive safety and health program that would include ergonomics. Although the guidelines were initially meant to be advisory in nature, they were eventually to be developed into a new industry-wide ergonomics standards. The guidelines were meant to provide information so that the employers could determine if they have ergonomics-type problems, identify the nature and location of those problems, and implement measures to reduce or eliminate them.

The ergonomics program for meatpacking plants is divided into five sections: (1) management commitment and employee involvement; (2) worksite analysis; (3) recommended hazard prevention and controls; (4) medical management; and (5) training and education. Detailed examples tailored for the meatpacking industry are also provided.

Commitment and Involvement

Commitment and involvement are essential elements in any sound safety and health program. Commitment by management is especially important in providing both the motivating force and the necessary resources to solve the problems. Similarly, employee involvement is necessary to maintain and continue the program. An effective program should have a team approach, with top management as the team leader, using the following principles:

1. A written program for job safety, health, and ergonomics, with clear goals and objectives to meet these goals, endorsed and advocated by the highest levels of management.
2. A personal concern for employee health and safety, emphasizing the elimination of ergonomics hazards.
3. A policy that places the same emphasis on health and safety as on production.
4. Assignment and communication of the responsibility of the ergonomics program to the appropriate managers, supervisors, and employees.
5. A program ensuring accountability from these managers, supervisors, and employees for carrying out these responsibilities.
6. Implementation of a regular review and evaluation of the ergonomics program. This might include: trend analyses of injury data, employee surveys, "before and after" evaluations of workplace changes, logs of job improvements, etc.

Employees can be involved via the following:

1. A complaint or suggestion procedure for voicing their concerns to management without fear or reprisal.

2. A procedure for prompt and accurate recording of the first signs of work-related musculoskeletal disorders, so that prompt controls and treatment can be implemented.
3. Ergonomics committees that receive reports of, analyze, and correct ergonomics problems.
4. Ergonomics teams with the required skills to identify and analyze jobs for ergonomics stress.

An effective ergonomics program includes four major program elements: worksite analysis, hazard control, medical management, and training and education.

Worksite Analysis

Worksite analysis identifies existing hazards and conditions, as well as operations and workplaces where such hazards may develop. The analysis also includes a detailed tracking and statistical analysis of injury and illness records, to identify patterns of work-related musculoskeletal disorder development. The first step in implementing the analysis program should be a review and analysis of injury and illness records, for example, medical records, insurance records, and OSHA-200 logs. Incidence rates (IR) of work-related musculoskeletal disorders should be calculated by reporting the number of incidences per 100 full-time workers per year per facility:

$$IR = 200{,}000 \times I/H$$

where: I = Number of injuries in a given time period
H = Employee hours worked in that same time period

Similarly, severity rates (SR) of the number of lost-time days can be calculated and tracked:

$$SR = 200{,}000 \times LT/H$$

where: LT = number of lost-time days

The second step in worksite analysis is to conduct baseline screening surveys to identify jobs that put employees at risk of developing work-related musculoskeletal disorders. The survey is typically performed with a questionnaire to identify potential ergonomics risk factors in the job process, workplace, or work method, as well as the location and severity of the potential musculoskeletal problems for the individual worker.

The essential third step of the program is a physical worksite analysis. This would involve a physical walkthrough of the plant, using ergonomics or work design checklists presented in previous chapters. Videotaping is recommended for this analysis phase. Slow-motion features allow more accurate analysis of the postures, angles, and demands of the job, as well as how individual workers perform their tasks.

The fourth step in worksite analysis is periodic reviews. These may uncover previously missed risk factors or design deficiencies. This review process should allow employees to notify management about conditions that appear to be ergonomic hazards, and utilize their experience to assist in controlling the problems. Trends of injuries and illnesses should be calculated and examined at regular intervals, as a quantitative check on the effectiveness of the ergonomics program.

Hazard Prevention and Controls

Hazard control involves the same engineering controls, work practice controls, personal protective equipment, and administrative controls as discussed throughout this book. Engineering controls, where feasible, are the OSHA preferred method of control.

Medical Management

Proper medical management, including the early identification of signs and the effective treatment of symptoms, is necessary to reduce the risk of developing work-related musculoskeletal disorders. A physician or occupational nurse with experience in musculoskeletal disorders should supervise the program. They should conduct periodic, systematic workplace walkthroughs to remain knowledgeable about the jobs, identify potential light-duty jobs, and maintain close contact with the employees. This information will allow the health providers to recommend assignments of recovering workers to restricted-duty jobs with minimal ergonomic stress on the injured muscle/tendon groups.

Health care providers should participate in the training and education of all employees, including supervisors, on different types of work-related musculoskeletal disorders, means of prevention, causes, early symptoms, and treatments. This demonstration will assist in the early detection of work-related musculoskeletal disorders prior to the development of more severe conditions. Employees should be encouraged to report early signs and symptoms of work-related musculoskeletal disorders, for timely treatment without fear of retribution by management. Written protocols for health surveillance, evaluation, and treatment will assist in maintaining properly controlled procedures.

Training and Education

Training and education are critical components of an ergonomics program for employees potentially exposed to ergonomics hazards. Training allows managers, supervisors, and employees to understand the ergonomics problems associated with their jobs, as well as the prevention, control and medical consequences of those problems.

1. General training on work-related musculoskeletal disorder risk factors, symptoms, and hazards associated with the job should be given annually to those employees who are potentially exposed.

2. Job-specific training on tools, knives, guards, safety, and proper lifting should be given to new employees prior to being placed on a full-time job.
3. Supervisors should be trained to recognize the early signs of work-related musculoskeletal disorders and hazardous work practices.
4. Managers should be trained to be aware of their health and safety responsibilities.
5. Engineers should be trained in the prevention and correction of ergonomics hazards through workplace redesign.

A rough-draft version of the guidelines for general industry, as a precursor to an ergonomics standard, was released in 1990. It contained primarily the same information as found in the guidelines for the meatpacking industry. However, there was considerable negative reaction from industry, and with the Republicans gaining control of Congress in 1992, the ergonomics standard was effectively shelved for the time being. It is currently (1998) being re-examined.

SUMMARY

A proper working environment is important not only from the standpoint of increasing productivity and improving the physical health and safety of the workers, but also for promoting worker morale and consequent reductions in worker absenteeism and labor turnover. Although many of these factors may seem intangible or of marginal effect, controlled scientific studies have shown the positive benefits of improved illumination, decreased noise and heat stress, better ventilation, etc.

Visibility is directly dependent on the illumination provided, but is also affected by the visual angle of the target viewed and the contrast of the target with the background. Consequently, improvement in task visibility can be accomplished through various means and does not always depend on increasing the light source.

Extended exposures to loud noise, although not directly affecting productivity, can cause hearing loss and is definitely annoying. The control of noise (and vibration) is simplest at the source and, typically, becomes more costly further away. Although using hearing protection may seem the simplest, it requires the expense of continuous motivation and enforcement.

Similarly, the effect of climate on productivity is quite variable, depending on individual motivation. A comfortable climate is a function of the amount and velocity of air exchange, the temperature, and the humidity. For hot areas, the climate is controlled most easily through adequate ventilation to remove pollutants and improve the evaporation of sweat. (Air conditioning is more effective, but is also more expensive.) For cold climates, adequate clothing is the primary control.

Shiftwork should utilize short, rapid, forward-rotating schedules with limited overtime.

Overall safety should focus on unsafe conditions, good housekeeping, proper guarding, OSHA regulations, employee involvement, and management commitment.

To assist the methods analyst in utilizing the various factors discussed in Chapter 6, they have been summarized in the Work Environment Checklist shown in Figure 6–19.

FIGURE 6–19
Work environment checklist.

Illumination	**Yes**	**No**
1. Is the illumination sufficient for the task, per IESNA recommendations?	❑	❑
a. To increase illumination, are more luminaires provided, rather than increasing the wattage of existing ones?	❑	❑
2. Is there general lighting as well as supplementary lighting?	❑	❑
3. Are the workplace and lighting arranged so as to avoid glare?	❑	❑
a. Are direct luminaires placed away from the field of vision?	❑	❑
b. Do the luminaires have baffles or diffusers?	❑	❑
c. Are work surfaces laid out perpendicular to the luminaires?	❑	❑
d. Are surfaces matted or nonglossy?	❑	❑
4. If necessary, are screen filters available for computer monitors?	❑	❑

Thermal conditions — Heat	**Yes**	**No**
1. Is the worker within the thermal comfort zone?	❑	❑
a. If not within the thermal comfort zone, has the WBGT of the working environment been measured?		
2. Are the thermal conditions within ASHRAE guidelines?	❑	❑
a. If not within guidelines, is sufficient recovery time provided?	❑	❑
3. Are procedures in place for control of potential heat stress conditions?	❑	❑
a. Is the escape of heat controlled at the source?	❑	❑
b. Are radiation shields in place?	❑	❑
c. Is ventilation provided?	❑	❑
d. Is the air dehumidified?	❑	❑
e. Is air-conditioning provided?	❑	❑

Thermal conditions — Cold	**Yes**	**No**
1. Is the worker adequately clothed for the equivalent wind chill temperature?	❑	❑
2. Are auxiliary heaters provided?	❑	❑
3. Are gloves provided?	❑	❑

Ventilation	**Yes**	**No**
1. Are ventilation levels acceptable per guidelines?	❑	❑
a. Is a minimum of 300 ft^3/hr/person provided?	❑	❑
2. If necessary, are local fans provided for workers?	❑	❑
a. Are these fans within a distance of 30 × fan-diameter?	❑	❑
3. For local heat sources, is spot cooling provided?	❑	❑

Noise Levels	**Yes**	**No**
1. Are noise levels below 90 dBA?	❑	❑
a. If the noise levels exceed 90 dBA, is there sufficient rest such that the 8-hr dose is less than 100%?	❑	❑
2. Are noise control measures in place?	❑	❑
a. Is the noise controlled at the source with better maintenance, mufflers, and rubber mounts?	❑	❑
b. Is the noise source isolated?	❑	❑
c. Are acoustical treatments being utilized?	❑	❑
d. As a last resort, are earplugs (or earmuffs) being used properly?	❑	❑

Vibration	**Yes**	**No**
1. Are vibration levels within acceptable ANSI standards?	❑	❑
2. If there is vibration, can the vibration causing sources be eliminated?	❑	❑
3. Have specially dampened seats been installed on vehicles?	❑	❑
4. Have vibration absorbing handles been attached to power tools?	❑	❑
5. Have resilient, fatigue-resistant mats been supplied to standing operators?	❑	❑

QUESTIONS

1. What factors affect the quantity of light needed to perform a task satisfactorily?
2. Explain the color rendering effect of low-pressure sodium lamps.
3. What is the relationship between contrast and visibility?
4. What foot-candle intensity would you recommend 30 inches above the floor in the company washroom?
5. Explain how sales may be influenced by colors.
6. What color has the highest visibility?
7. How is sound energy dissipated in viscoelastic materials?
8. A frequency of 2,000 Hz would have approximately what wavelength in meters?
9. What would be the approximate decibel value of a grinder being used to grind a high-carbon steel?
10. Distinguish between broadband noise and meaningful noise.
11. Would you advocate background music at the workstation? What results would you anticipate?
12. According to the present OSHA law, how many continuous hours per day of a 100 dBA sound level would be permissible?
13. What three classifications have been identified from the standpoint of exposure to vibration?
14. In what ways can workers be protected from vibration?
15. What is meant by the environmental temperature?
16. Explain what is meant by the thermal comfort zone.
17. What is the maximum rise in body temperature that analysts should allow?
18. How would you go about estimating the maximum length of time that a worker should be exposed to a particular heat environment?
19. What is WBGT?
20. What is the WBGT with a dry-bulb temperature of 80° F, a wet-bulb temperature of 70° F, and a globe temperature of 100° F?

21. Which type of radiation is given the most attention by the safety engineer?

22. What is meant by absorbed dose of radiation? What is the unit of absorbed dose?

23. What is meant by the rem?

24. What steps would you take to increase the amount of light in the following assembly department by about 15 percent? The department currently uses fluorescent fixtures, and the walls and ceiling are painted a medium green. The assembly benches are a dark brown.

25. What color combination would you use to attract attention to a new product being displayed?

26. When would you advocate that the company purchase aluminized clothing?

27. Are possible health hazards associated with electron beam machining? With laser beam machining? Explain.

28. Explain the impact of noise levels below 85 dBA on office work.

29. What environmental factors affect heat stress? How can each be measured?

30. How would you determine if a job places an excessive heat load on the worker?

PROBLEMS

1. A work area has a reflectivity of 60 percent, based on the color combinations of the workstations and the immediate environment. The seeing task of the assembly work could be classified as difficult. What would be your recommended illumination?

2. What is the combined noise level of two sounds of 86 and 96 decibels?

3. In the Dorben Company, an industrial engineer designed a workstation where the seeing task was difficult because of the size of the components going into the assembly. The desired brightness was 100 foot-Lamberts, and the workstation was painted a medium green with a reflectance of 50 percent. What illumination in foot-candles would be required at this workstation to provide the desired brightness? Estimate the required illumination if you repainted the workstation with a light cream paint.

4. In the Dorben Company, an industrial engineer was assigned to alter the work methods in the press department to meet OSHA standards relative to permissible noise exposures. He found that the sound level averaged 100 db and that the standard deviation was 10 db. The 20 operators in this department wore earplugs provided by Dorben. Also, the power output from the public-address system was altered from 30 watts to 20 watts. The deadening of the sound level by the earplugs was estimated to be 20 percent effective.

What improvement resulted? Do you feel that this department is now in compliance with the law for 99 percent of the employees? Explain.

5. In the Dorben Co., an all-day study revealed the following noise sources: 0.5 hrs, 100 dBA; 1 hr, less than 80 dBA; 3.5 hrs, 90 dBA; 3 hrs, 92 dBA. Is this company in compliance? What is the dose exposure? What is the TWA noise level?

6. In problem #5, consider that the last exposure is in the press room, which currently has five presses operating. Assuming that Dorben Co. can eliminate some of the presses and transfer production to the remaining presses, how many presses should Dorben eliminate so as not to exceed 100% dose exposure for the workers?

7. What is the illumination on a surface 6 inches from a 2-candela source?

8. What is the luminance of a surface having a 50 percent reflectance and 4 foot-candle illumination?

9. What is the contrast created by black text (reflectance = 10%) on white paper (reflectance = 90%)?

10. How much louder is an 80 dB noise than a 60 dB noise?

11. What is the increase in dBs of a noise that doubles in intensity?

12. A supervisor is sitting at her desk illuminated by a 180-candela source 3 feet above it. She is writing with green ink (reflectance = 30%) on a yellow note pad (reflectance = 60%). What is the illumination on the note pad? Is that sufficient? If not, what amount of illumination is needed? What is the contrast of the writing task? What is the luminance of the notepad?

REFERENCES

ACGIH. *Threshold Limit Values for Chemical Substances and Physical Agents in the Work Environment*, Cincinnati, OH: American Conference of Government Industrial Hygienists, 1985.

ASA. *American National Standard: Guide for the Evaluation of Human Exposure to Whole Body Vibration (ANSI S3.18-1979)*. New York: Acoustical Society of America, 1980.

ASHRAE. *Handbook, Heating, Ventilation and Air Conditioning Applications. (Ch. 25)*. Atlanta, GA: American Society of Heating, Refrigeration and Air Conditioning Engineers, 1991.

ASHRAE. *Handbook, Fundamentals. (Ch. 8)*. Atlanta, GA: American Society of Heating, Refrigeration and Air Conditioning Engineers, 1993.

Blackwell, H. R. "Development and Use of a Quantitative Method for Specification of Interior Illumination Levels on the Basis of Performance Data." *Illuminating Engineer*, 54 (June 1959), pp. 317–353.

Eastman Kodak Co. *Ergonomic Design for People at Work*. New York: Van Nostrand Reinhold, 1983.

Freivalds, A., D. B. Chaffin, and G. D. Langolf. "Quantification of Human Performance Circadian Rhythms." *Journal of the American Industrial Hygiene Association,* 44, no. 9 (September 1983), pp. 643–648.

Grandjean, E. *Fitting the Task to Man.* 4th ed. London: Taylor & Francis, 1988.

IESNA. *Lighting Handbook.* 8th ed. Ed. M. S. Rea. New York: Illuminating Engineering Society of North America, 1995, pp. 459–478.

Kamon, E., W. L. Kenney, N. S. Deno, K. J. Soto, and A. J. Carpenter. "Readdressing Personal Cooling with Ice." *Journal of the American Industrial Hygiene Association,* 47, no. 5 (May 1986), pp. 293–298.

Konz, S. *Work Design.* Scottsdale, AZ: Publishing Horizons, 1995.

Lehmann, G., *Praktische Arbeitsphysiologie.* Stuttgart: G. Thieme, 1953.

Lockhart, J. M., H. O. Kiess, and T. J. Clegg. "Effect of Rate and Level of Lowered Finger-surface Temperature on Manual Performance." *Journal of Applied Psychology,* 60, no. 1 (February 1975), pp. 106–113.

NIOSH. *Criteria for a Recommended Standard ... Occupational Exposure to Hot Environments, Revised Criteria.* Washington, DC: National Institute for Occupational Safety and Health, Superintendent of Documents, 1986.

OSHA. *Code of Federal Regulations—Labor. (29 CFR 1910).* Washington, DC: Office of the Federal Register, 1997.

OSHA. *Ergonomics Program Management Guidelines for Meatpacking Plants.* OSHA 3123. Washington, DC: The Bureau of National Affairs, Inc., 1990.

Riley, M. W., and D. J. Cochran. "Partial Gloves and Reduced Temperature." In *Proceedings of the Human Factors Society 28th Annual Meeting.* Santa Monica, CA: Human Factors and Ergonomics Society, 1984, pp. 179–182.

Sanders, M. S., and E. J. McCormick. *Human Factors in Engineering and Design.* 7th ed. New York: McGraw-Hill, 1993.

Schwarzenau, P., P. Knauth, E. Kiessvetter, W. Brockmann, and J. Rutenfranz. "Algorithms for the Computerized Construction of Shift Systems Which Meet Ergonomic Criteria." *Applied Ergonomics,* 17, no. 3 (September 1986), pp. 169–176.

Sipple, P. A. and C. G. Passel. "Movement of Dry Atmospheric Cooling in Subfreezing Temperatures." *Proceedings of the American Philosophical Society,* 89 (1945), pp. 177–199.

Yaglou, C. P., E. C. Riley, and D. I. Coggins. "Ventilation Requirements." *American Society of Heating, Refrigeration and Air Conditioning Engineers Transactions,* 42 (1936), pp. 133–158.

Yaglou, C. P., and D. Minard. "Control of Heat Casualties at Military Training Centers." *AMA Archives of Industrial Health,* 16 (1957), pp. 302–316.

CHAPTER 7

Proposed Method Implementation

KEY POINTS:

- Decide among alternate methods, using value engineering, cost-benefit analysis, cross-over charts, and economic analyses.
- *Sell* the new method; people are resistant to change.
- Establish sound base rates using reliable job evaluations.
- Accommodate workers of all abilities.

Presenting and installing the proposed method is the fifth step in the systematic development of a work center to produce a product or perform a service. However, the analyst must first choose which proposed method to present. Several alternative methods may be feasible, some more effective than others, some more costly than others. A variety of decisionmaking tools are presented in this chapter to assist the analyst in selecting the best alternative. Obviously, many factors may comprise the definition of "best," and these tools will help the analyst in weighing these factors appropriately.

Selling the proposed method is the next, and probably the most important, element in the presentation procedure. This step is as important as any of the preceding steps, since a method not sold is usually not installed. No matter how thorough the data gathering and analysis and the ingenuity of the proposed method, the value of the project is zero unless it is installed.

Humans naturally resent the attempts of others to influence their thinking. When someone approaches with a new idea, the instinctive reaction is to put up a defense against it and resist any changes. We feel that we must protect our own individuality, preserve the sanctity of our own ego. All of us are just egotistical enough to convince ourselves that our ideas are better than those of anyone else. It is natural for us to react in this manner, even if the new idea is to our own advantage. If the idea has merit, there is a tendency to resent it because we did not think of it first.

The presentation of the proposed method should include the decisionmaking that went into choosing the final design and should emphasize the savings that would be achieved with it. Savings in material (both direct and indirect) and labor (both direct and indirect) should highlight an analyst's report.

The second most important part of the presentation is the quality and reliability improvement possible by installing the improved method. Progressive managers recognize that the key to the continued health of any product or service producer is the ability to supply quality materials in a timely and consistent manner.

The third most important part of the presentation is the recovery of capital investment. Once the proposed method has been properly presented and sold, it can be installed. Installation, like presentation, requires sales ability. During installation, the analyst must continue selling the proposed method to engineers and technicians on their own level, to subordinate executives and supervisors, and to labor and organized labor representatives.

DECISIONMAKING TOOLS

Decision Tables

Decision tables are a structured approach for taking out the subjectivity in making decisions, that is, determining which of several alternative methods changes should be implemented. The tables essentially consist of condition–action statements, similar to if–then statements in computer programs. "If" the right condition or combination of conditions exist, "then" specified actions are taken. Thus, the tables can unambiguously describe complex, multirule, multivariable decision systems.

Such decision tables, also known as *hazard action tables*, are frequently utilized in safety programs to specify certain actions for given hazard conditions. The hazard may be identified by two different variables: frequency, how often the accident is likely to occur; and severity, how severe the loss will be. Frequency may be categorized as extremely remote, remote, reasonably probable, and highly probable, while severity may have levels of negligible, marginal, critical, and catastrophic. This results in a hazard action table with five plans, as shown in Table 7–1.

Consider the right column marked with an asterisk. The analyst would conclude that a condition that is highly probable and would result in a catastrophe, with possible death and severe injury to personnel, should immediately be eliminated by shutting down the operation (Gausch, 1972). Obviously, this is a simplified example that can be envisioned mentally. However, if one has 20 states for each of two different variables, then there are 400 categories of conditions, which cannot be remembered easily. Overall, decision tables emphasize making better-quality decisions through better decision analysis techniques and less time pressure, that is, the action plans can be worked out ahead of time, rather than having to deal with instantaneous pressures, possibly resulting in errors.

TABLE 7–1
Hazard Action Table

	Severity			
Frequency	**Negligible**	**Marginal**	**Critical**	**Catastrophic**
Extremely remote				
Remote				
Reasonably probable				(*)
Probable				
Actions				
Forget it				
Long-range study				
Correct (1 year)				
Correct (90 days)				
Correct (30 days)				
Shutdown				

From: Heinrich, Petersen and Roos, 1980.

Value Engineering

A simple way to expand the evaluation of alternatives is to apply numbers and form a payoff matrix. This is often termed *value engineering* (Gausch, 1974). Each solution may have different values with respect to the desired benefits. A weight is determined for each benefit (0 to 10 is a reasonable range) and then a value (0–4, with 4 being best) is assigned to reflect how well each solution produces the desired benefit. The assigned value is multiplied by the appropriate weight and the products are summed for the final score. The highest sum is the most appropriate solution.

Note that benefits will have different relative weights for different companies, different departments within a company, or even different points in time for the same department. Also note that the Evaluating Alternatives step of Muther's Simplified Layout Planning (see Chapter 3) is a form of value engineering.

Cost-Benefit Analysis

A more quantitative approach to deciding between different alternatives is a *cost-benefit analysis.* This approach requires five steps:

1. Determine what is changed due to better design, that is, increased productivity, better quality, decreased injuries, etc.
2. Quantify these changes (benefits) into monetary units.
3. Determine the cost required to implement the changes.
4. Divide the cost by the benefit for each alternative, to create a ratio.
5. The smallest ratio determines the desired alternative.

Step 2 is probably the most difficult to assess and quantify. It is not always possible to assign dollar values; sometimes it may be percent changes, injuries, or other values. Example 7–1 may help in understanding all three decisionmaking tools. Other examples of cost-benefit analyses as related to less definable benefits, such as health and safety issues, are presented in Brown (1976).

Crossover Charts

Crossover (or *break-even*) *charts* are very useful in deciding which of two alternative methods changes to implement. One may use general-purpose equipment with low capital costs but higher setup costs, while the other may use special equipment at a higher capital cost but with lower setup costs. At some production quantity, the two methods are equal and this is the *crossover point*. This relates to the most prevalent mistake made by planners. Large amounts of money are tied up in fixtures that show large savings while in use, but they are seldom used. For example, a savings of 10 percent in direct labor costs on a job in constant use would probably justify greater expense in tools than an 80 or 90 percent savings on a small job that appears on the production schedule only a few times a year (a good example of the Pareto analysis in Chapter 2).

The economic advantage of lower labor costs is the controlling factor in determining the tooling; consequently, jigs and fixtures may be desirable even where only small quantities are involved. Other considerations, such as improved interchangeability, increased accuracy, or reducing labor troubles, may provide the dominant reason for elaborate tooling, although this is usually not the case.

Multiple Criteria Decisionmaking

Decisionmaking in the presence of multiple, often conflicting, criteria can be approached by a relatively new process called *multiple criteria decisionmaking* (MCDM), developed by Saaty (1980). For example, assume that an analyst has four alternatives to consider (a_1, a_2, a_3, a_4), which would be applied to four possible states of the product or market (S_1, S_2, S_3, S_4). Also assume that this analyst estimates the following outcomes for the various alternatives and states of the market:

	States of product or market			
Alternatives	S_1	S_2	S_3	S_4
a_1	0.30	0.15	0.10	0.06
a_2	0.10	0.14	0.18	0.20
a_3	0.05	0.12	0.20	0.25
a_4	0.01	0.12	0.35	0.25

If the outcomes represent profits or returns, and the state of the market will be S_2, then the analyst would definitely decide on alternative a_1. If the outcomes represent scrap or some other factor that the analyst wishes to minimize, then

EXAMPLE 7–1
Cut-Off Operation

The Dorben Co. manufactures simple, small, knife blades inserted into a plastic handle. One of the operations in the formation of the blade is the cutting off of knife blades from a thin strip of stainless steel via a foot-pedal-operated press. Using tweezers, the worker procures a rubber nib from a parts bin and inserts it over the blade to protect it. After press activation, the cut-off blade is placed on a holder plate for later assembly into the handle (a good example of the effective therblig *pre-position*!). Because of the small blade size, a stereoscope is used to assist in the operation. The operators have complained about wrist, neck, back, and ankle pain. Possible method changes include: (1) replacement of the mechanical pedal with a foot-operated electric switch, to reduce ankle fatigue; (2) better adjustment of the position of the stereoscope, to reduce neck fatigue; (3) implementation of a video projection system, for heads-up viewing; (4) use of a gravity feed bin for the nibs, to improve productivity; and (5) replacement of the tweezers with a vacuum operated stylus, to both improve productivity and eliminate a potential CTD causing pinch grip.

Assume the productivity improvements shown in Table 7–2, based on an MTM-2 analysis (see Chapter 11) and injury reductions based on the CTD Risk Index (refer to Figure 5–30).

Company policy authorizes methods engineers to proceed, with no further authorization needed, if Condition 1 and either Condition 2 or 3 are met: 1) implementation costs are less than $200 (i.e., petty cash), 2) productivity increases are more than 5 percent, 3) injury risks decrease more than 33 percent. In terms of decision tables, the situation could be structured as shown in Table 7–3.

In terms of value engineering, weights of 6, 4, and 8 can be assigned to the three factors of interest: increased productivity, decreased injury rates, and

TABLE 7–2
Expected Changes in Productivity, Injury Risk Potential, and Cost for Various Method Changes in Cut-Off Operation

Work Design and Methods Changes	Δ Productivity (%)	Δ CTD Risk (%)	Cost ($)
1. Foot operated electric switch	0	−1*	175
2. Adjust stereoscope	0	−2	10
3. Video projection system	+1**	−2	2,000
4. Gravity feed bin	+7	−10	40
5. Vacuum stylus	+1**	−40	200

*The current CTD Risk Index does not address lower extremities. However, there is reason to believe that the lower force for the electric switch will have some beneficial effect.
**Can't be quantified from MTM-2, but some benefit is expected.

EXAMPLE 7–1 ***(continued)***

TABLE 7–3
Decision Table for Cut-Off Operation

Methods Changes	Conditions				Action
	#1	#2	#3	Policy	
1. Electric switch					—
2. Adjust stereoscope					—
3. Video projection system					—
4. Gravity feed bin					Proceed
5. Vacuum stylus					Proceed

TABLE 7–4
Value Engineering Analysis of the Cut-Off Operation

Evaluating Alternatives

Plant: Dorben Co.	Alternatives	A		B		C		D		E		
Project: Cut-Off Operation Date: 6-12-97 Analyst: AF		Electric Switch		Adjust Stereo-scope		Video Projection		Gravity Feed Bin		Vacuum Stylus		
Factor/Consideration	**Wt**	**Ratings and Weighted Ratings**										**Comments**
		A		**B**		**C**		**D**		**E**		
Increase in productivity	6	0	0	0	0	1	6	3	18	1	6	
Decrease in injuries	4	1	4	1	4	1	4	2	8	3	12	
Low cost solution	8	3	24	4	32	1	8	4	32	3	24	
Totals		28		36		18		58		42		

Remarks:

Gravity feed bin is the most justifiable methods change.

low-cost solutions. (see Table 7–4). Each solution is rated from 0 to 4 for each of the factors. The resulting product sums are 28, 36, 18, 58, and 42, and the gravity feed bin change at 58 is clearly the best solution.

EXAMPLE 7–1 ***(concluded)***

For a cost-benefit analysis, anticipated benefits could be quantified by both increases in productivity and decreases in injury rates. Assume that the company profits $645 for each 1 percent increase in productivity over the course of the year (for details on costing, see Chapter 12). Similarly, a decrease in Workers' Compensation and medical costs due to a decrease in CTD injuries can be considered a benefit. The company has averaged one CTD case leading to surgery every five years of operation. Assuming one CTD surgery case costs the company $30,000, the expected loss per year is $6,000. For each 1 percent decrease in risk, the company benefits $60 per year. The increases in productivity and decreases in injury rates can be quantified as shown Table 7–5.

From Table 7–5, it is obvious that for any ratio less than one (methods changes #2, #4, #5, and #6), there are more benefits than the costs required to implement the methods change. On the other hand, methods change #4 is by far the most cost effective. Interestingly, a combination of methods changes #2, #4, and #5 (alternative #6) may be worth considering for the comparatively low total dollar amounts expended.

TABLE 7–5
Cost-Benefit Analysis for Cut-Off Operation

	Benefit ($)				
Method Changes	**Productivity**	**Injury Rates**	**Total**	**Cost ($)**	**Cost Benefit**
1. Electric switch	0	60	60	175	2.92
2. Adjust stereoscope	0	120	120	10	0.08
3. Video projection system	645	120	765	2000	2.61
4. Gravity feed bin	4515	600	5115	40	0.01
5. Vacuum stylus	645	2400	3045	100	0.03
6. Methods changes #2, #4, #5	5160	3120	8280	150	0.02

alternative a_3 would be chosen. (Although a_4 also has an outcome of 0.12, the analyst chooses a_3, since there is less variability in outcome under this alternative than with a_4.) Seldom should decisions be made under an assumed certainty. Usually, some risk is involved in predicting the future state of the market. Assume that the analyst is able to estimate the following probability values associated with each of the four states of the market:

S_1	0.10
S_2	0.70
S_3	0.15
S_4	0.05
	1.00

EXAMPLE 7–2
Crossover Analysis of Fixture and Tooling Costs

The production engineer in a machining department has devised two alternate methods involving different tooling for a job being machined in the shop. Data on the present and proposed methods are shown in Table 7–6. Which method would be more economical in view of the activity? The base pay rate is $9.60 per hour. The estimated activity is 10,000 pieces per year. The fixtures are capitalized and depreciated in five years. A cost analysis reveals that a total unit cost of $0.077, represented by Alternate 2, is the most economical for the quantity anticipated.

A crossover chart (see Figure 7–1) allows the analyst to decide which method to use for given quantity requirements. The old method is the best for quantities up to about 7,700 per year, alternate method #1 is better for quantities between 7,700 and 9,100 per year, and method #2 is best for quantities above 9,100 per year. Note that in this latter approach, tooling costs were absorbed upfront.

TABLE 7–6
Fixture & Tooling Costs

Method	Time value (minutes)	Fixture cost	Tool cost	Average tool life	Unit direct labor cost	Unit fixture cost	Unit tool cost	Unit total cost	Annual
Present method . .	.888 each	None	$ 6	10,000 pieces	$0.142	None	$0.0006	$0.426	$1,426
Alternate 1	.606 each	$300	20	20,000 pieces	0.097	$0.006	0.0010	0.104	1,040
Alternate 2	.363 each	600	35	5,000 pieces	0.058	0.012	0.0070	0.077	770

FIGURE 7–1
A crossover (break-even) chart for fixture & tooling costs.

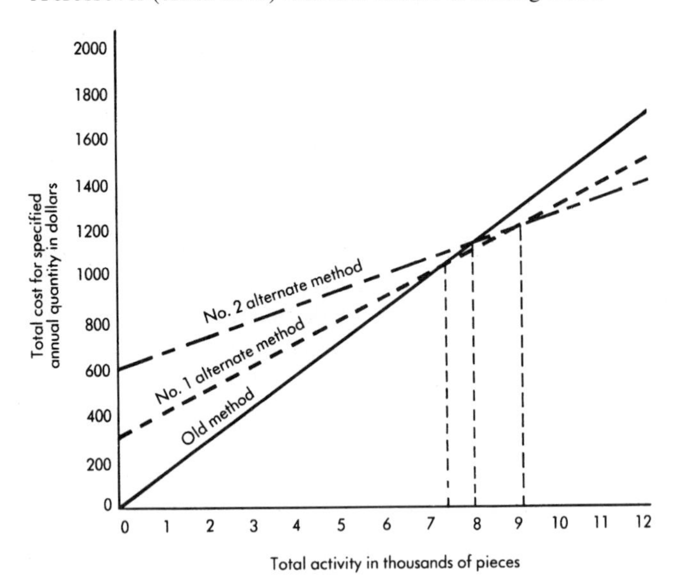

EXAMPLE 7–3
Crossover Analysis of Competing Methods

Insufficient volume may make it impractical to consider many alternative proposals that may offer substantial savings over existing methods. For example, an operation done on a drill press involved a 0.5-inch hole reamed to the tolerance of 0.500 to 0.502 inches. The activity of the job was estimated to be 100,000 pieces. The time study department established a standard of 8.33 hours per thousand to perform the reaming operation, and the reaming fixture cost $2,000. Since a base rate of $7.20 per hour was in effect, the money rate per thousand pieces was $60.

Now, assume that a methods analyst suggests broaching the inside diameter, because the part can be broached at the rate of five hours per thousand. This would be a savings of 3.33 hours per thousand pieces, or a total savings of 333 hours. At the $7.20 base rate, this would mean a direct labor savings of $2,397.60. However, it would not be practical to go ahead with this idea, since the tool cost for broaching is $2,800. Thus, the change would not be sound unless the labor savings could be increased to $2,800 to offset the cost of the new broaching tools.

Since the labor savings in a new broaching setup would be 3.33 × $7.20 per thousand, 116,800 pieces would have to be ordered before the change in tooling would be justified.

$$\frac{\$2{,}800 \times 1{,}000}{\$7.20 \times 3.33} = 116{,}783 \text{ pieces}$$

However, if the broaching method had been used originally instead of the reaming procedure, it would have paid for itself in

$$\frac{\$2{,}800 - \$2{,}000}{\$7.20 \times 3.33/M} = 33{,}367 \text{ pieces}$$

With production requirements of 100,000 pieces, the labor savings would be 3.33 × $7.20 × 66.6 thousand (the difference between 100,000 and 33,400) = $1,596.80 over the present reaming method. Had a motion analysis been made in the planning stage, this savings might have been realized. Figure 7–2 illustrates these relationships with the customary crossover chart.

A logical decisionmaking strategy would be to calculate the expected return under each decision alternative, and then select the largest value to maximize or the smallest value to minimize. Here,

$$E(a) = \sum_{j=1}^{n} P_j C_{ij}$$

$$E(a_1) = 0.153$$

$$E(a_2) = 0.145$$

EXAMPLE 7–3 ***(concluded)***

FIGURE 7–2

Crossover chart illustrating the fixed and variable costs of two competing methods.

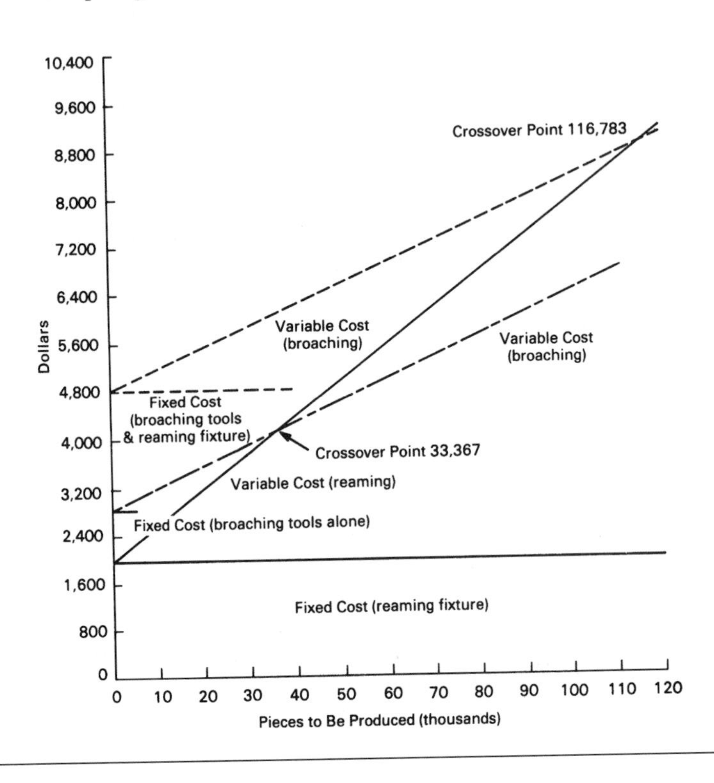

$$E(a_3) = 0.132$$
$$E(a_4) = 0.15$$

Thus, alternative a_1 would be selected to maximize the desired result.

A different decisionmaking strategy would be to consider the state of the market that has the greatest chance of occurring. From the data given, this market state would be S_2, since it carries a probability value of 0.70. The choice, again based on the most probable future, would be alternative a_1, with a 0.15 return.

A third decisionmaking strategy is based on a "level of aspiration." Here, we assign an outcome value (C_{ij}), which represents the consequence of what we are willing to settle for if we are reasonably sure we will get at least this consequence most of the time. This assigned value may be considered to represent a level of aspiration, which we shall denote as A. For each a_j, we then determine the probability that the C_{ij} in connection with each decision alternative is greater than or equal to A. Select the alternative with the greatest $P(C_{ij} \geq A)$.

For example, if assign the consequence value of 0.10 to A, we have the following:

$$
\begin{aligned}
&(C_{ij} \geq 0.10)\\
a_1 &= 0.95\\
a_2 &= 1.00\\
a_3 &= 0.90\\
a_4 &= 0.90
\end{aligned}
$$

Since decision alternative a_2 has the greatest $P(C_{ij} \geq A)$, it would be recommended.

Analysts may be unable to assign probability values to various states of the market with confidence and may therefore want to consider any one of them as being equally likely. A decisionmaking strategy that may be used under these circumstances is based on the *principle of insufficient reason*, since there is no reason to expect that any state is more likely than any other state. Here, we compute the various expected values based on:

$$E(a) = \frac{\sum_{j=1}^{n} C_{ij}}{n}$$

In our example, this would result in:

$$
\begin{aligned}
E(a_1) &= 0.153\\
E(a_2) &= 0.155\\
E(a_3) &= 0.155\\
E(a_4) &= 0.183
\end{aligned}
$$

Based on this choice, alternative four would be proposed.

A second strategy analysts may consider when making decisions under uncertainty is based on the *criterion of pessimism*. When one is pessimistic, one anticipates the worst. Therefore, in a maximization problem, the minimum consequence would be selected for each decision alternative. The analyst compares these minimum values, and selects the alternative that has the maximum of the minimum values. In our example:

Alternative	**Min. C_{ij}**
a_1	0.06
a_2	0.10
a_3	0.05
a_4	0.01

Here, alternative a_2 would be recommended, since its minimum value of 0.10 is a maximum when compared to the minimum values of the other alternatives.

The *plunger criterion* is a third decisionmaking strategy that analysts may want to consider. This criterion is based on an optimistic approach. If one is optimistic, one expects the best, regardless of the alternative chosen. Therefore, in a maximizing problem, the analyst would select the maximum C_{ij} for each alternative, and would then select the alternative with the largest of these maximum values. Thus:

Alternative	Min. C_{ij}
a_1	0.30
a_2	0.20
a_3	0.25
a_4	0.35

Here, decision alternative a_4 would be recommended, because of its maximum value of 0.35.

Most decision makers are neither completely optimistic nor completely pessimistic. Instead, a coefficient of optimism, X, is established, where:

$$0 \leq X \leq 1$$

Then, a Q_i is determined for each alternative, where:

$$Q_i = (X)(\text{Max. } C_{ij}) + (1 - X)(\text{Min. } C_{ij})$$

The alternative recommended is the one associated with the maximum Q_i for maximization, and with the minimum Q_i for minimization.

A final decisionmaking approach based on uncertainty is the *minimax regret criterion*. This criterion involves the calculation of a *regret matrix*. For each alternative, based on a state of the market, analysts calculate a regret value. This regret value is the difference between the payoff actually received and the payoff that could have been received if the decision maker had been able to foresee the state of the market.

To construct the regret matrix, the analyst selects the maximum C_{ij} for each state S_j and then subtracts the C_{ij} value of each alternative associated with that state. In our example, the regret matrix would be:

	States			
Alternative	S_1	S_2	S_3	S_4
a_1	0	0	0.25	0.19
a_2	0.20	0.01	0.17	0.05
a_3	0.25	0.03	0.15	0
a_4	0.29	0.03	0	0

The analyst then selects the alternative associated with the minimum of the maximum regrets (minimax).

Alternative	Min. r_{ij}
a_1	0.25
a_2	0.20
a_3	0.25
a_4	0.29

Based on the minimax regret criterion, a_2 would be selected, in view of its minimum regret of 0.20.

Such decisionmaking is very common in manual materials handling (see Chapter 4), where there is always a trade-off between worker safety and worker productivity. The greater the focus on worker safety, for example, through the reduction in loads and the corresponding biomechanical stresses on the lower back, the worse the productivity of the load handled. To maintain job productivity at the desired level, a reduction of load weight requires an increased task frequency, with corresponding heavier physiological demands. A metabolic evaluation will lead to the conclusion that the infrequent lifting of heavy loads is preferable to the frequent lifting of lighter loads. However, from a biomechanical point of view, the load weight should be minimized, regardless of frequency, resulting in a conflict. This problem was examined by Jung and Freivalds (1991) using MCDM for a critical range of task frequencies of 1 lift/min to 12 lifts/min (see Figure 7–3). For infrequent tasks (< 7/min), the biomechanical stress predominates; for higher frequency tasks (> 7/min), the physiological stress predominates. At about 7 lifts/min, however, both stresses participate equally in determining the overall stress level to the worker. Thus, depending on the alternatives and the effect of each alternative with regard to specific attributes of interest, different solutions can be obtained. Analysts should become familiar with these decisionmaking strategies and should use those that are most appropriate to their organizations.

Economic Decision Tools

The three most frequently used appraisal techniques for determining the desirability of investing in a proposed method are: (1) the return on sales method, (2) the return on investment or payback method, and (3) the discounted cash flow method.

The *return on sales* method involves computing the ratio of (a) the average yearly profit brought about through using the method and (b) the average yearly

FIGURE 7–3
Unacceptability of stress levels used in reconciling the conflicting guidelines. (*From:* Jung and Freivalds, 1991)

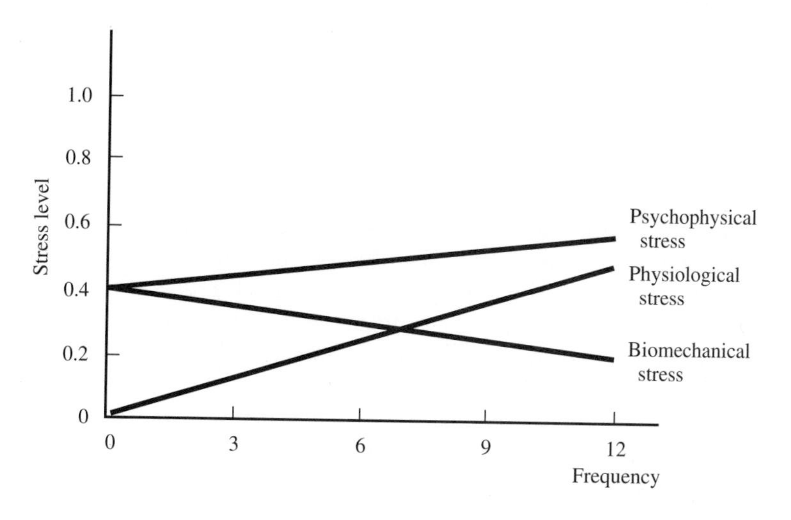

sales or increase in dollar value added to the product, based on the pessimistic estimated life of the product. However, while this ratio provides information on the effectiveness of the method and the resulting sales efforts, it does not consider the original investment required to get started on the method.

The *return on investment* method gives the ratio of (a) the average yearly profit brought about through using the method, based upon the pessimistic estimated life of the product, and (b) the original investment. Of two proposed methods that would result in the same sales and profit potential, management would prefer to use the one requiring the least investment of capital. The reciprocal of the return on investment is often referred to as the *payback* method. This gives the time that it would take to realize a full return on the original investment.

The *discounted cash flow* method computes the ratio of (a) the present worth of cash flow, based on a desired percentage return, and (b) the original investment. This method calculates the rate of flow of money in and through the company and the *time value of money*. The time value of money is important. Because of interest earned, a dollar today is worth more than a dollar at any later date. For example, at 15-percent compound interest, a dollar today is worth \$2.011 five years from now. Expressing it another way, a dollar received five years from now would be worth about 50 cents today. Interest may be thought of as the return obtainable by the productive investment of capital.

The following applies to the present value concept:

Single payment		
– Compound amount factor	(given P, find S)	$S = P(1 + i)^n$
– Present worth factor	(given S, find P)	$P = 1(1 + i)^n$
Uniform series		
– Sinking fund factor	(given S, find R)	$R = i/(1 + i)^n - 1$
– Capital recovery	(given P, find R)	$R = i(1 + i)^n/(1 + i)^n - 1$
– Compound amount factor	(given R, find S)	$S = (1 + i)^n - 1/i$
– Present worth factor	(given R, find P)	$P = (1 + i)^n - 1/I(1 + i)^n$

where: i = Interest rate for a given period.
n = Number of interest periods.
P = Present sum of money (present worth of principal).
S = A sum of money at the end of n periods from the present date; equivalent to P with interest i.
R = The end-of-period payment or receipt in a uniform series continuing for the coming n periods; the entire series equivalent to P at interest i.

An assumed return rate (I) is the basis of the cash flow computation. All cash flows following the initial investment for the new method are estimated and adjusted to their present worth, based on the assumed return rate. The total estimated cash flows for the pessimistic estimated life of the product are then summed up as a profit (or loss) in terms of cash on hand today. Finally, this total is compared to the initial investment.

Estimates of the product demand 10 years hence may deviate considerably from reality. Thus, the element of chance is introduced, and the probabilities of success tend to diminish with the increased length of the payoff period. The results of any

EXAMPLE 7–4
Economic Justification of a Proposed Method

An example should clarify the use of the three methods for appraising the potential of a proposed method.

Investment for proposed method: \$10,000.
Desired return on investment: 10 percent.
Salvage value of jigs, fixtures, and tools: \$500.
Estimated life of the product for which the proposed method will be used: 10 years.

Present worth of cash flow:

$$\begin{array}{ll} (3{,}000)(0.9091) = \$2{,}730 & (3{,}800)(0.5645) = 2{,}140 \\ (3{,}800)(0.8264) = 3{,}140 & (3{,}000)(0.5132) = 1{,}540 \\ (4{,}600)(0.7513) = 3{,}460 & (2{,}200)(0.4665) = 1{,}025 \\ (5{,}400)(0.6830) = 3{,}690 & (1{,}400)(0.4241) = 595 \\ (4{,}600)(0.6209) = 2{,}860 & (\ 500)(0.3855) = \underline{193} \\ & \$21{,}373 \end{array}$$

Salvage value of tools:

$$(500)(0.3855) = \$193$$

Total present worth of anticipated gross profit and tool salvage value: \$21,566. Ratio of present worth to original investment:

$$\frac{21{,}566}{10{,}000} = 2.16$$

TABLE 7–7
Comparison of Economic Justification Methods

End of year	Increase in sales values due to proposed method	Cost of production with proposed method	Gross profit due to proposed method
1	\$ 5,000	\$ 2,000	\$ 3,000
2	6,000	2,200	3,800
3	7,000	2,400	4,600
4	8,000	2,600	5,400
5	7,000	2,400	4,600
6	6,000	2,200	3,800
7	5,000	2,000	3,000
8	4,000	1,800	2,200
9	3,000	1,600	1,400
10	2,000	1,500	500
Totals	\$53,000	\$20,700	\$32,300
Average	\$ 5,300	\$ 2,070	\$ 3,230

Return on sales $= \frac{3{,}230}{5{,}300} = 61\%$

Return on investment $= \frac{3{,}230}{10{,}000} = 32.3\%$

Payback $= 1/0.323 = 3.09$ years

EXAMPLE 7–4 ***(concluded)***

The new method satisfactorily passes all three appraisal methods. A 61-percent return on sales and a 32.3-percent return on capital investment represent very attractive returns. The return of the $10,000 capital investment will take place in 3.09 years and the cash flow analysis reveals that the original investment will be recovered in 4 years while earning 10 percent. During the 10-year anticipated life of the product, $11,566 more than the original investment will be earned.

study are only as valid as the reliability of the input data. Constant follow-up can determine the validity of the assumptions. The analyst should not hesitate to alter decisions if the original data prove to be invalid. Sound financial analysis is intended to facilitate the decisionmaking process, not replace good business judgment.

PROPOSED METHOD PRESENTATION

Analysts make their presentation in both written and oral form. Even if the company does not require a written report, it is good practice to make one for recordkeeping purposes and for future applications. A well-written report is a major step in selling the proposed method, with the summary being the most important section of the report, since that may be the only part read by busy executives. The summary is based on the body of the report; consequently, it is usually not written until after the body has been completed. However, it is presented early in the report, so that those who must evaluate the proposal can obtain the facts quickly before making a decision. The summary should contain three elements: an abstract briefly explaining the nature of the problem; conclusions outlining the results of the analysis; and recommendations setting forth the method proposed and summarizing the estimated savings, quality and reliability improvement, and capital expenditures recovery.

The body of the report includes a section on the nature of the problem, followed by details relative to the data gathering and the analysis methods. This section contains the operation and flow process charts used to present the facts, as well as the worker and machine, gang, and operator process charts used in developing the proposed work center. All sketches, drawings, and specifications showing details of proposed fixtures and tool and machine designs are included. Where robots and/or automation procedures are recommended, type, style, and size details of the proposed facilities are included. The anticipated improvement in quality and reliability, due to consistency and dependability resulting from the mechanization, is emphasized. This section also contains the reasons for the conclusions and recommendations given in the summary. The entire report should be clear, concise, complete, and accurate. It should be prepared so that it can be easily read and studied.

Frequently, analysts are asked to present their methods proposals orally. To help ensure approval of the proposal, the analyst must be prepared to present the benefits and advantages accurately and forcefully. The analyst should provide estimates of the resulting increase in productivity and/or decrease in cost. If quality will be improved or customer service enhanced, this information should also be given. This is a good time to bring out that cycle-time reductions result in better products and services delivered to the market faster.

The analyst should plan the presentation in advance. Data should be available on the advantages of the proposed method, the related costs, the expected savings, and the expected recovery of capital investment. The analyst should list in advance all the information that may be requested, and then be prepared to supply that information, as well as answer objections raised to the proposal. These are usually centered on the initial cost, the time required to adopt the method, and the potential inconvenience while installation is taking place. The analyst should point out that these objections have been carefully considered and plans have been made to cope with them.

INSTALLATION

After a proposed method has been approved, the next step is installation. Too frequently, analysts do not stay close enough to the job during installation. Do not assume that installation will automatically take place according to the proposal. A maintenance person, mechanic, or worker can make a slight change or modification without considering the consequences. This may ultimately mean that the proposed method does not yield the anticipated results.

The analyst should stay with the job during installation, to ensure that all details are carried out in accordance with the proposed plan. The analyst should verify that: the work center being established is equipped with the facilities proposed; the planned working conditions are provided; the tooling is done in accordance with recommendations; and the work is progressing satisfactorily. The analyst must "sell" the new method to the operator, supervisor, setup man, and so on. Then, by the time the installation is complete, these employees should be more accepting of the new method.

Once the new work center has been installed, the analyst must check all aspects to see whether they conform to the specifications established. In particular, the analyst must verify that: the "reach" and "move" distances are the correct length; the tools are correctly sharpened; the mechanisms function soundly; stickiness and sluggishness have been worked out; safety features are operative; material is available in the quantities planned; working conditions associated with the work center are as anticipated; and all parties have been informed of the new method.

Once every aspect of the new method is ready for operation, the supervisor assigns the operator who will be working with the method. The analyst should then stay with the operator as long as necessary to ensure that the operator is familiar with the new assignment. This period may be a matter of a few minutes, several hours, or even several days, depending on the complexity of the assignment and the flexibility and adaptability of the worker.

Once the operator begins to get a feel for the method and can work along systematically, the analyst can proceed with other work. However, the installation phase should not be considered complete until the analyst has checked back several times during the first few days after installation to ensure that the proposed method is working as planned. Also check with line supervisors to assure that they spot check and monitor the new method.

Resistance to Change

It is not unusual for workers to resist changes in the method. Although many want to be, and may think of themselves as, innovators and progressive thinkers, most people are quite comfortable with their present job or workplace, even if it might not be the most comfortable or pleasant. Their fear of change and the impact it might have on their jobs, pay, and security override other concerns (see Maslow's hierarchy of human needs, Chapter 18). Worker reactions to change can be quite obstinate and perplexing, as experienced by Gilbreth in the following classic example. While doing a motion study in a bed manufacturing plant, he noticed a large middle-aged woman ironing bed sheets in an obviously very inefficient and fatiguing manner. To iron each sheet, she picked up a large and heavy iron, sat down, and pressed down hard on the sheet, about 100 times for each sheet. She was fatigued and had back pain. With a few work design changes using counterweights to support the iron (the modern-day equivalent of a tool balancer), he considerably reduced the physical workload. However, the reaction of the woman was completely opposite of that expected. Instead of being the only worker physically capable of doing the job (and receiving praise from her supervisor), she was only one of many. Her status was lost and she was dead set against the change (DeReamer, 1980).

Thus, it is important to "sell" the new method to the operators, supervisors, mechanics, etc. Employees should be notified well in advance about any method changes that will affect them. The resistance to change is directly proportional to the magnitude of the change and the time available to implement the change. Therefore, large changes should be made in small steps. Don't change the whole workstation, chair or stool, and tooling all at once. Start with the chair, maybe the tools next, and then finally the workstation.

Explain the reasons for the change. People resist what they don't understand. Instead of just replacing the worker's pistol-grip tool for a horizontal worksurface with an in-line tool, explain that this tool will weigh less and require less upper arm motion, that is, it will be more comfortable to use.

As a general rule for dealing with emotion, it is better to emphasize positives, for example, "This new tool is going to be much easier to use," and de-emphasize negatives, such as, "That tool is heavy and unsafe." Get the worker to participate directly in the process of the methods change or work design. Workers have a good record of following their own suggestions, and where workers have been involved in the decisionmaking, there has typically been less resistance to the changes. One successful approach is to form worker committees or ergonomic teams (see Chapter 18).

Threats of force, that is, management reprisals for not making the change, may be counterproductive, setting up counter-emotions to resist the change. In addition, people often resist the social aspects of change, rather than the technical aspects. Therefore, if it can be shown that other employees are using the same device, an operator will be much more likely to go along with the change.

The last step in methods engineering, after appropriate standards have been set, is to maintain the method, that is, determine if the anticipated productivity gains are being realized. Here, the industrial engineers must be very careful to determine whether any effect is truly due to the new method.

Hawthorne Effect

This is an often quoted study emphasizing the need for worker involvement in methods changes or production planning leading to increased motivation and productivity. Actually, this was a very poorly designed series of studies from which it is difficult to draw any true scientific effect or conclusion, other than the need to be very careful when assuming that productivity improvements are due strictly to the methods changes. The initial, less well known, study was a joint project between the National Research Council and Western Electric Co. to examine the effects of illumination on productivity. The study was conducted at the very large Hawthorne plant (some 40,000 workers) near Chicago, from 1924 to 1927. In general, management found that employees were reacting to changes in illumination in the way they assumed they were expected to react, that is, when illumination was increased, they were expected to produce more, and they did. When the illumination was decreased, they were expected to produce less and did so. A follow-up study tested this point even further. Light bulbs were changed and workers were allowed to assume that the illumination had been increased. In fact, the light bulbs were replaced with ones of exactly the same wattage. However, the workers commented favorably on the increased illumination and responded with increased productivity. Thus, physiological effects were being confounded by psychological effects (Homans, 1972).

Based on these results, Western Electric decided to conduct a further series of studies on mental attitudes and workers' effectiveness. These are the more famous Hawthorne studies, conducted in conjunction with the Harvard School of Business Administration, from 1927 to 1932 (Mayo, 1960). Six female operators were placed in a separate room and subjected to various experimental conditions: (1) special group incentive for the six workers, as opposed to the more than 100 workers in a given department; (2) inclusion of two 5- or 10-minute rest breaks; (3) shorter workdays; (4) shorter work weeks; and (5) lunches and/or beverages provided by the company. Any changes were discussed with the group of six ahead of time, and any that were seriously objected to were discarded. Also, later, the effects of these changes on the workers were discussed in a rather formal interview process.

With the workers welcoming these opportunities to vent their feelings, these structured interviews degenerated into open-ended gripe sessions. Interestingly, production generally increased during this span of five years (except for minor changes in product or start/stop of vacation periods), regardless of test conditions. In addition, absences and sick days decreased considerably for the six operators, as compared to their coworkers. Western Electric, already known for its concern for worker welfare, attributed this somewhat surprising result to an overall increase in concern for the worker, with a resulting increase in social satisfaction (Pennock, 1929-30).

Unfortunately, these conclusions are oversimplified, as other effects were also confounded in the study. There were large variations in supervisory practices, different and inconsistent measures of productivity were utilized throughout the five years of the study, and significant changes in methods were introduced (Carey, 1972). For example, a drop delivery system was implemented for the experimental group for the purpose of counting the relays produced. However, based on principles of motion economy (Chapter 4), this also tended to increase production rates.

Regardless of the controversy, the Hawthorne studies have three main implications: (1) the basic rule of experimentation, "the act of measuring something changes it," was confirmed; (2) proper human relations can act as a strong motivator (see Chapter 18); and (3) it is very difficult to tease apart confounded factors in an uncontrolled study. Therefore, the analyst should be very careful about jumping to conclusions about the effectiveness of methods changes on productivity. Part of the productivity changes could be due to the improved method, while part could be due to improved morale or motivation of the affected operator. Also, any productivity measurements, even though innocuous on the surface, could have unintended effects, if the workers become aware of them and perform as expected.

JOB ANALYSIS

Closely associated with the installation of the ideal method and the standard time required to perform the operation is the job analysis of the work center and the resulting job evaluation. This is the sixth step in the systematic procedure of applying methods engineering. Every time a method is changed, the job description should be altered to reflect the conditions, duties, and responsibilities of the new method. When a new method is introduced, a job analysis should be done, so that a qualified operator may be assigned to the work center and an appropriate base rate provided. Sound base rates must assure that the money rates are commensurate with the local rates for similar work; they must allow adequate differentials for jobs requiring higher skills and responsibilities; and they must be based on techniques that can be explained and justified.

Appropriate base rates are a result of job evaluation, a technique for equitably determining the relative worth of the different work assignments in an organization. The basis of job evaluation is job analysis, a procedure for making a careful appraisal of each job and then recording the details of the work, so that it can be evaluated fairly by a trained analyst. Before a job description is developed, all aspects of the opportunity should be carefully studied, to ensure that the best methods are being used and that the operator is thoroughly trained in the prescribed methods.

Typically, various job responsibilities, authorities, and poor-decision consequences are included in a job analysis. The analysis should also provide information regarding the machines and tools used in the job, the problem-solving capability required, and the physical and social conditions related to the job. This procedure is similar to that utilized for the Job/Worksite Analysis Guide of Chapter 2, which specifically concentrated on the physical stressors of job and worksite, rather than the mental and cognitive components.

The job description is an essential component of the job analysis. Job descriptions are useful supervisory tools that can aid in the selection, training, and promotion of employees, and in the assessment of work distribution. The job description should identify the job's specific duties and responsibilities, and the minimum requirements of the worker performing the job. The job description should emphasize both doing the job as described and doing the right thing to produce the product or service at the least cost and the highest quality.

A job description should start with an accurate title, and should unambiguously describe what a worker actually does. The worker should be enlisted in accurately defining the job's responsibilities. This procedure can often lead to the development of cost-effective improvements. A combination of personal interviews and questionnaires, along with direct observation, results in a concise definition of each job and the duties that each entails. The mental and physical functions required to perform the work should also be included, and such definitive words as "direct, examine, plan, measure, and operate" should be used. The more accurate the description, the better. Figure 7–4 illustrates an analysis of a clerical job for use in a point job evaluation plan.

JOB EVALUATION

Job evaluation is a procedure by which an organization ranks its jobs in order of their worth or importance. Job evaluation systems became popular during the World War II years. It was the only way that increases in wages could be given, since all wages were frozen by the National War Labor Board. A company had to show that inequities existed in the firm, if wage adjustments were to be initiated. By introducing a job evaluation system, companies found it easy to identify wage and/or salary inequities and obtain permission to provide increases.

The reader should understand that individuals generally believe that others in similar positions work less and are paid more.

The main purpose of any job evaluation plan is to determine the proper compensation for the work performed on each job. A well-conceived job evaluation plan includes the following:

FIGURE 7–4
Job analysis for a shipping and receiving clerk.

JOB TITLE Shipping and Receiving Clerk DEPT. Shipping
MALE X FEMALE ____ DATE ________ TOTAL POINTS 280 CLASS 5

JOB DESCRIPTION

Directs and assists in loading and unloading, counting, and receiving or rejecting purchased parts and supplies, and later delivers to proper departments.

Examines receivals out of line with purchase orders. Maintains file on all purchase orders and/or shipping orders and keeps open orders up-to-date. Maintains daily and weekly shipping reports and monthly inventory reports.

Assists in packing of all foreign and domestic shipments. Makes up request for inspection form on certain materials received and rejection form for all items rejected.

Job requires thorough knowledge of packing, shipping and receiving routine, plant layout, shop supplies, and finished parts. Needs to have a knowledge of simple office routine. Ability to work with other departments, as a service department, and to deal effectively with vendors. Job requires considerable accuracy and dependability. The effects of poor decisions include damaged receivables and shipments, inaccurate inventories, and extra material handling. Considerable lifting of weights up to 100 lbs. is involved. Works in conjunction with two class 4 packers and shippers.

Job Evaluation	Degree	Points
Education	1	15
Experience and Training	2	50
Initiative and Ingenuity	3	50
Analytical Ability	3	50
Personality Requirements	2	30
Supervisory Responsibility	1	25
Responsibility for Loss	1	10
Physical Application	6	25
Mental or Visual Application	1	5
Working Conditions	5	20
		280

1. It provides a basis for explaining to employees why one job is worth more (or less) than another job.
2. It provides a reason to employees whose rates of pay are adjusted because of a change in method.

3. It provides a basis for assigning personnel with specific abilities to certain jobs.
4. It helps determine the criteria for a job when new personnel are employed or promotions are made.
5. It provides assistance in the training of supervisory personnel.
6. It provides a basis for determining where opportunities for methods improvement exist.

Job Evaluation Systems

The majority of job evaluation systems in use today are a variation or combination of four principal systems: the classification method, the point system, the factor comparison method, and the ranking method. The *classification method*, sometimes called the *grade description plan*, consists of a series of definitions designed to differentiate jobs into wage groups. Once the grade levels have been defined, analysts study each job and assign it to the appropriate level, on the basis of the complexity of its duties and responsibilities and its relation to the description of the several levels. The U.S. Civil Service Commission uses this plan extensively.

To use this method of job evaluation, the analyst uses the following procedure:

1. Prepare a grade description scale for each type of job, such as, machine operations, manual operations, skilled (craft) operations, or inspection.
2. Write the grade descriptions for each grade in each scale, using such factors as:
   a. Type of work and complexity of duties.
   b. Education necessary to perform job.
   c. Experience necessary to perform job.
   d. Responsibilities.
   e. Effort demanded.
3. Prepare job descriptions for each job. Classify each job by "slotting" (placing in a specific category) the job description into the proper grade description.

In the *point system*, analysts directly compare all the attributes of a job with the attributes in other jobs, using the following procedure:

1. Establish and define the basic factors common to most jobs, indicating the elements of value in all jobs.
2. Specifically define the degrees of each factor.
3. Establish the points to be accredited to each degree of each factor.
4. Prepare a description of each job.
5. Evaluate each job by determining the degree of each factor contained in it.
6. Sum the points for each factor to get the total points for the job.
7. Convert the job points into a wage rate.

The *factor comparison* method of job evaluation usually has the following elements:

1. Determine the factors establishing the relative worth of all jobs.
2. Establish an evaluation scale that is usually similar to a point scale, except that the units are in terms of money. For example, a $2,000-per-month benchmark

job might attribute $800 to the responsibility factor, $400 to education, $600 to skill, and $200 to experience.
3. Prepare job descriptions.
4. Evaluate key jobs, factor by factor, by ranking each job from the lowest to the highest for each factor.
5. Pay wages on each key job, based on various factors. The money allocation automatically fixes the relationships among jobs for each factor, and therefore establishes the ranks of jobs for each factor.
6. Evaluate other jobs, factor by factor, on the basis of the monetary values assigned to the various factors in the key jobs.
7. Determine a wage by adding up the money value of the various factors.

Both the point system and the factor comparison method are more objective and thorough in their evaluations of the various jobs involved; both plans study the basic factors common to most jobs that influence their relative worth. Of the two plans, the point system is the more commonly used and is generally considered the more accurate method for occupational rating.

The *ranking method* arranges jobs in order of importance, or according to relative worth. Here, the entire job is considered, including: the complexity and degree of difficulty of the duties; the requirements for specific areas of knowledge; the required skills; the amount of experience required; and the level of authority and responsibility assigned to the job. This method became popular in the United States during World War II, because of its simplicity and ease of installation. At that time, the National War Labor Board set up the requirement that all companies working on government contracts must have some type of wage classification system. The ranking method satisfied that requirement. The following steps apply to the ranking method:

1. Prepare job descriptions.
2. Rank jobs (usually, departmentally first) in the order of their relative importance.
3. Determine the class or grade for groups of jobs, using a bracketing process.
4. Establish the wage or wage range for each class or grade.

Generally speaking, the ranking method is less objective than the other techniques. Consequently, it necessitates greater knowledge of all jobs. For this reason, it has not been used extensively in recent years.

Factor Selection

Under the factor comparison method, most companies use five factors. In some point programs, 10 or more factors may be used. However, it is preferable to use a small number of factors. The objective is to use only as many factors as are necessary to provide a clear-cut difference among the jobs. The elements of any job may be classified according to:

1. What job demands the employee meets in the form of physical and mental factors.
2. What the job takes from the employee in the form of physical and mental fatigue.

3. The responsibilities that the job demands.
4. The conditions under which the job is done.

Other factors may include: education, experience, initiative, ingenuity, physical demands, mental and/or visual demands, working conditions, hazards, equipment responsibility, process, materials, products, and the work and safety of others.

These factors are present in varying degrees in all jobs, and any job under consideration falls under one of the several degrees of each factor. The various factors are equally important. To recognize these differences in importance, the analyst assigns weights or points to each degree of each factor, as shown in Table 7–8. Figure 7–5 illustrates a completed job rating, with substantiating data.

Each degree of each factor is carefully defined, so that the degree that characterizes the work situation under study is evident. For example, education may be defined as the requirements for the use of shop mathematics, drawings, measuring instruments, or trade knowledge. First-degree education may require only the ability to read and write, and to add and subtract whole numbers. Second-degree education could be defined as requiring the use of simple arithmetic, such as the addition and subtraction of decimals and fractions, and the ability to read simple drawings and use some measuring instruments, such as calipers and scales. It would be characteristic of two years of high school. Third-degree education may require the use of fairly complicated drawings, advanced shop mathematics, handbook formulas, and a variety of precision measuring instruments, plus some trade knowledge in a specialized field or process. It could be equivalent to four years of high school plus short-term trades training. Fourth-degree education could require the use of complicated drawings and specifications, advanced shop mathematics, and a wide variety of precision measuring instruments, plus broad shop knowledge. It

TABLE 7–8
Points Assigned to Factors and Key to Grades

Factors	1st degree	2nd degree	3rd degree	4th degree	5th degree
Skill					
1. Education	14	28	42	56	70
2. Experience	22	44	66	88	110
3. Initiative and ingenuity	14	28	42	56	70
Effort					
4. Physical demand	10	20	30	40	50
5. Mental and/or visual demand	5	10	15	20	25
Responsibility					
6. Equipment or process	5	10	15	20	25
7. Material or product	5	10	15	20	25
8. Safety of others	5	10	15	20	25
9. Work of others	5	10	15	20	25
Job conditions					
10. Working conditions	10	20	30	40	50
11. Unavoidable hazards	5	10	15	20	25

Source: National Electrical Manufacturers Association.

FIGURE 7–5
Job rating and substantiating form.

JOB RATING - SUBSTANTIATING DATA
DORBEN MFG. CO.
UNIVERSITY PARK, PA.

JOB TITLE: Machinist (General) CODE: 176 DATE: Nov. 12

FACTORS	DEG.	POINTS	BASIS OF RATING
Education	3	42	Requires the use of fairly complicated drawings, advanced shop mathematics, variety of precision instruments, shop trade knowledge. Equivalent to four years of high school or two years of high school plus two to three years of trades training.
Experience	4	88	Three to five years installing, repairing, and maintaining machine tools and other production equipment.
Initiative and Ingenuity	3	42	Rebuild, repair, and maintain a wide variety of medium-size standard automatic and hand-operated machine tools. Diagnose trouble, disassemble machine and fit new parts, such as antifriction and plain bearings, spindles, gears, cams, etc. Manufacture replacement parts as necessary. Involves skilled and accurate machining using a variety of machine tools. Judgment required to diagnose and remedy trouble quickly so as to maintain production.
Physical Demand	2	20	Intermittent physical effort required tearing down, assembling, installing, and maintaining machines.
Mental or Visual Demand	4	20	Concentrated mental and visual attention required. Laying out, setup, machining, checking, inspecting, fitting parts on machines.
Responsibility for Equipment or Process	3	15	Damage seldom over $900. Broken parts of machines. Carelessness in handling gears and intricate parts may cause damage.
Responsibility for Material or Product	2	10	Probable loss due to scrapping of materials or work, seldom over $300.
Responsibility for Safety of Others	3	15	Safety precautions are required to prevent injury to others; fastening work properly to face plates, handling fixtures, etc.
Responsibility for Work of Others	2	10	Responsible for directing one or more helpers a great part of time. Depends on type of work.
Working Conditions	3	30	Somewhat disagreeable conditions due to exposure to oil, grease, and dust.
Unavoidable Hazards	3	15	Exposure to accidents, such as crushed hand or foot, loss of fingers, eye injury from flying particles, possible electric shock, or burns.

REMARKS: Total 307 Points--assign to job class 4.

may be equivalent to four years of high school plus four years of formal trades training. Fifth-degree education may require a basic technical knowledge sufficient to deal with complicated and involved mechanical, electrical, or other engineering problems. Fifth-degree education could be equivalent to four years of technical university training.

Experience appraises the time that an individual with the specified education usually requires to learn to perform the work satisfactorily from the standpoint of both quality and quantity. Here, first degree could involve up to three months; second degree, three months to one year; third degree, one to three years; fourth degree, three to five years; and fifth degree, over five years. In a similar manner, each degree of each factor is identified with a clear definition and with specific examples, when applicable.

Performance Evaluation

Considerable judgment is needed to evaluate each job with respect to the degree of each factor required for the plan. Consequently, it is usually desirable to have a committee perform the evaluation. A separate committee should be appointed for each department of the company or business. A typical committee would include a permanent chairman (usually from industrial relations or industrial engineering), a union representative, the department supervisor, the department steward, and a management representative (usually from industrial relations). The committee should evaluate all jobs for the same factor before proceeding to the next factor. For example, all jobs in the department under study should be evaluated for degree of skill before proceeding to other factors, such as effort, responsibility, and job conditions. Using this pattern, the committee would measure the job, rather than the individual filling the job.

Committee members should assign their degree evaluations independently of the other members. The correlation among different evaluators should be reasonably high, such as 0.85 or higher. The members should then discuss any differences, until there is agreement on the level of the factor.

Job Classification

After all jobs have been evaluated, the points assigned to each job should be tabulated. Next, the number of labor grades within the plant should be determined. This number is a function of the range of points characteristic of the jobs within the plant. Typically, the number of grades runs from 8 (typical of smaller plants and lesser skilled industries) to 15 (typical of larger plants and higher skilled industries) (see Figure 7–6). For example, if the point range of all the jobs within a plant range from 110 to 365, the grades shown in Table 7–9 could be established. Similar ranges are not necessary for the various labor grades. Increasing the point ranges might be desirable for more highly compensated jobs.

The jobs falling within the various labor grades are then reviewed relative to one another, to assure fairness and consistency. For example, it would not be appropriate for a Class A machinist to be in the same grade level as a Class B machinist. Next, hourly rates are assigned to each of the labor grades. These rates are based on area rates for similar work, company policy, and the cost-of-living index.

FIGURE 7–6
Evaluation points and base rate range for nine labor grades.

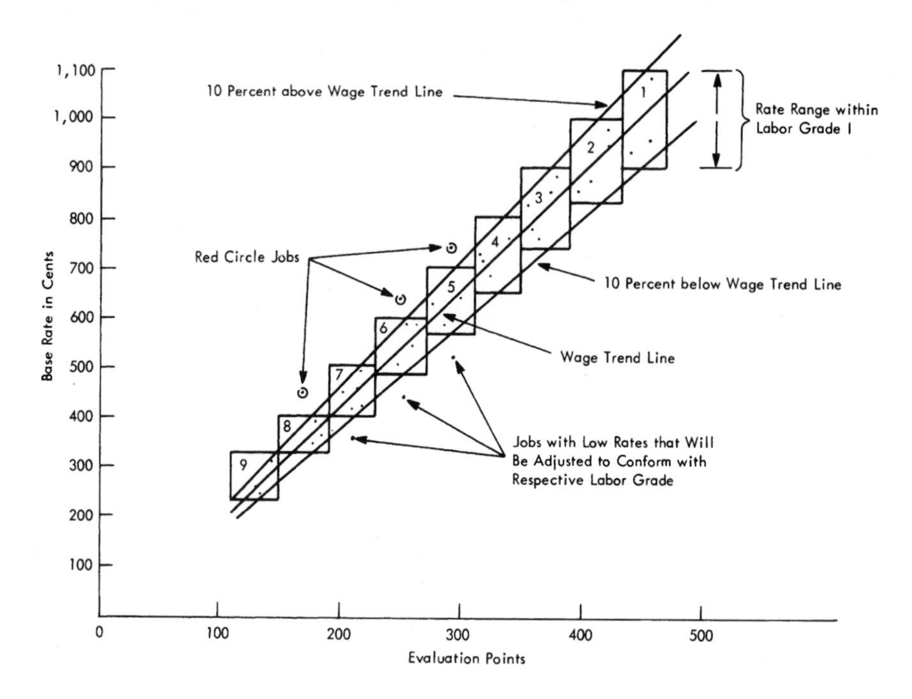

TABLE 7–9
Labor Grades

Grade	Score range (points)	Grade	Score range (points)
12	100–139	6	250–271
11	140–161	5	272–293
10	162–183	4	294–315
9	184–205	3	316–337
8	206–227	2	338–359
7	228–249	1	360 and above

Frequently, analysts establish a rate range for each labor grade. The total performance of each operator determines his or her pay rate within the established range, and total performance refers to quality, quantity, safety, attendance, suggestions, and so on.

Job Evaluation Program Installation

After plotting area rates against the point values of the various jobs, the analyst develops a rate versus point value trend line, which may or may not be a straight line. Regression techniques are helpful in developing this trend line. Several points will be either above or below it. Points significantly above the trend line represent employees whose present rate is higher than that established by the job evaluation plan; points significantly below the trend line represent employees whose present rate is less than that prescribed by the plan.

Employees whose rates are less than that called for by the plan should receive immediate increases to the new rate. Employees whose rates are higher than that called for by the plan (such rates are referred to as *red circle rates*) are not given a rate decrease. However, they are also not given an increase at their next review, unless the cost of living adjustment results in a rate higher than their current pay. Finally, any new employee would be paid the new rate advocated by the job evaluation plan.

Potential Concerns

A point job evaluation system is probably the most favorable approach to bringing both equity and objectivity to the determination of individual compensation. However, the reader should understand that the installation of a point plan can cause several costly and difficult problems.

A point plan tends to emphasize how a job should be done, rather than how it is actually being done. Modern management styles point out that all workers should take the approach that their jobs should always be performed in a manner that proves the best for the company and its customers, not necessarily what is the best for the individual. However, unless the job description is worded carefully, some employees may refuse to perform important work, simply because these tasks are not included in the job description. In addition to job duties, the job description should spell out the desirability of employee development, growth, and superior performance.

A point job evaluation plan can also create unnecessary and undesirable power relationships within the company. Job evaluation point scores inform many in the workforce of the relative scores of the various jobs, providing an obvious pecking order that can interfere with cooperation and group decisionmaking, which are so important in modern management.

Another problem is that individuals soon recognize that a creatively written job description can add enough points that the job will be assigned to the next higher class. Thus, in time, jobs may be written to reflect higher rates than they are really worth. Similarly, employees will recognize that they can increase their job evaluation points by increasing their responsibility. This can often be achieved by adding unnecessary work or adding another clerical or other employee. These additions may be unnecessary and may really add only additional direct and/or overhead costs, in addition to the cost increase of the higher paid job resulting from the point additions.

In some organizations, the points developed by a point job evaluation plan may be inappropriately used for nonpay purposes. For example, they may be used as a basis for privileges, such as parking spaces, shift preferences, mailing lists, meeting invitations, company recreation services usage, and the like.

While point job evaluation plans provide the most favorable approach to equitable wage payment, they also identify comparable worth inequities. For example, a point plan might show that an industrial nurse should be paid $32,000 a year; yet in the community, there is no difficulty in employing nurses at $26,000 a year. If the company paid the higher salary, many of the other salaried people would feel they were underpaid, by comparison. Comparable worth within a geographic area is an important factor. Properly designed point plans should result in companies being able to attract and retain qualified and competent employees and provide internal equity.

One of the principal concerns heard in the courtrooms and legislative hearings deals with the principle of "equal pay for equal work." A point job evaluation system is based on this concept. However, the analyst must realize that there is no inherent worth to any job: it is worth what is provided in the marketplace. If the analyst deviates from the established point plan to provide for a salary based on the marketplace alone, the action will almost always create a new inequity. The analyst should recognize that treating everyone the same is inconsistent with treating individuals equitably in relation to their contribution to the business or industry.

It is also important that a regular follow-up of the plan be done, so that it is adequately maintained. Jobs change, so it is necessary to review all jobs periodically and make adjustments when necessary.

Finally, employees must understand the fairness of the job evaluation plan. George Fry Associates undertook a comprehensive survey of job evaluation practices in over 500 companies. Some of the significant results from this survey appear in Table 7–10. Unfortunately, over half the companies responding to the survey found their employees' understanding of the plan to be poor or nonexistent. If a job evaluation plan is to succeed over the years, the vast majority of the employees should have at least an average understanding of how the plan works.

AMERICANS WITH DISABILITIES ACT

While implementing a new method and performing job evaluations, the analyst must consider the implications of the Americans with Disabilities Act (ADA). The ADA was passed in 1990 to "outlaw discrimination in employment against a qualified individual with a disability." This is an important consideration for all employers with 15 or more employees, as it may entail considerable workplace redesign or other accommodations. The ADA covers such employment practices as recruitment, hiring, promotion, training, pay, layoffs, firing, leave, benefits, and job assignments, the last of which would be the concern of methods analyst. The ADA protects any individual "with a physical or mental impairment that substantially limits a major life activity." Substantial implies something more than minor, while "major life activity" includes hearing, seeing, speaking, breathing, walking, manually feeling or manipulating, learning, or working. Temporary injuries of limited duration are not covered.

TABLE 7–10
Survey of Job Evaluation Plans

	Percent			
Firms using a standard job evaluation plan	66.0			
Plan used:				
Ranking	3.5			
Grade description	1.0			
Factor comparison	10.5			
Point system	85.0			
Average of hourly employees covered	65.0			
Job evaluation function reports to:				
Industrial relations	69.0			
Industrial engineering	16.0			
Results used:				
In employment	87.0			
In employee placement	88.0			
In wage rate bargaining	68.0			
Job evaluation program:				
Is recognized in union contract	83.0			
Is an issue during negotiations	59.0			
Has gone to arbitration (of these cases, management won 74 percent)	27.0			
Understanding and acceptance:	*Good*	*Average*	*Poor*	*None*
By employees	7%	39%	49%	5%
By top-management	50	36	13	1
By middle-management	18	17	5	0
By first-line supervisors	35	52	13	0

The individual with a disability must be qualified to perform the "essential functions" of the job with or without "reasonable accommodations." Essential functions are basic job duties that an employee must be able to perform. They can be determined from the job analysis techniques presented earlier in this chapter. "Reasonable accommodation" is any change or adjustment to a job or work environment that allows the individual to perform the essential functions of the job and enjoy the benefits and privileges that all employees enjoy. Those accommodations could include: physical modification of the tools, equipment, or workstation; job restructuring; modification of work schedules; modification of training materials or policies, etc. The purpose of any modifications would be to make them usable and accessible. One consideration is that any such change will be ergonomically beneficial to all workers. Many of the work design principles (Chapters 4-6) should be useful here also.

A reasonable accommodation is one that does not place an undue hardship on the employer, that is, one that is not unduly costly, extensive, substantial, or disruptive, or that fundamentally alters the nature or operation of the business. Variables that affect cost are the company size, financial resources, and its operational nature or structure. Unfortunately, there is no specific or quantitative definition of the cost factor. Most likely, that will evolve through the legal system as various cases of discrimination are brought forward in the courts. For further information on the definitions, legal aspects, and accessibility modifications guidelines, the methods engineer should consult the ADA (1991).

FOLLOW UP

The eighth and last step in a methods engineering program is Follow Up (see Fig. 7–7). The seventh step, Establish Time Standards, is not strictly part of methods changes and will not be discussed in this section on methods. However, it is a necessary part of any successful work center and will be covered in great detail in Chapters 8-16.

Following up the method, in the short term, includes ensuring that the installation was correct, so that the operators can be trained in the proper work practices and can achieve the desired productivity levels. It also includes economic analyses to verify that the projected savings are truly achieved. If follow up is not done, management may question the need for such changes and may be less willing to support similar methods changes in the future. Finally, it is important to keep everyone sold on the method, so that the operators don't slide back to the old patterns of movement, supervisors do not slack off in the enforcement of new procedures, and management does not waiver in its commitment to the overall program.

Following up the new method is a critical aspect of maintaining a smoothly running and efficient work center. Otherwise, several years later, another methods engineer will be examining the present method and asking the same questions, "why?," "what is the purpose of this operation?," etc., which were asked as part of this operation analysis. Thus, it is very important to close the feedback loop and maintain a continuous improvement cycle, as presented in Figure 7–7.

SUCCESSFUL METHODS IMPLEMENTATIONS

As an example of the effective results of a operation analysis program, one Ohio company realized a 17,496-ton annual savings. By forming a mill section into a ring and welding it, the company could replace an original rough-forged ring weighing 2,198 pounds. The new mill section blank weighed only 740 pounds. The saving of 1,458 pounds of high-grade steel, amounting to twice the weight of the finished piece, was brought about by the simple procedure of reducing the excess material that previously had to be cut away.

In another example an analytic laboratory in a New Jersey plant applied the principles of operation analysis with gratifying results. New workbenches have been laid out in the form of a cross, so that each chemist has an L-shaped worktable. This arrangement allows each chemist to reach any part of the workstation by

FIGURE 7–7
The principal steps in a methods engineering program.

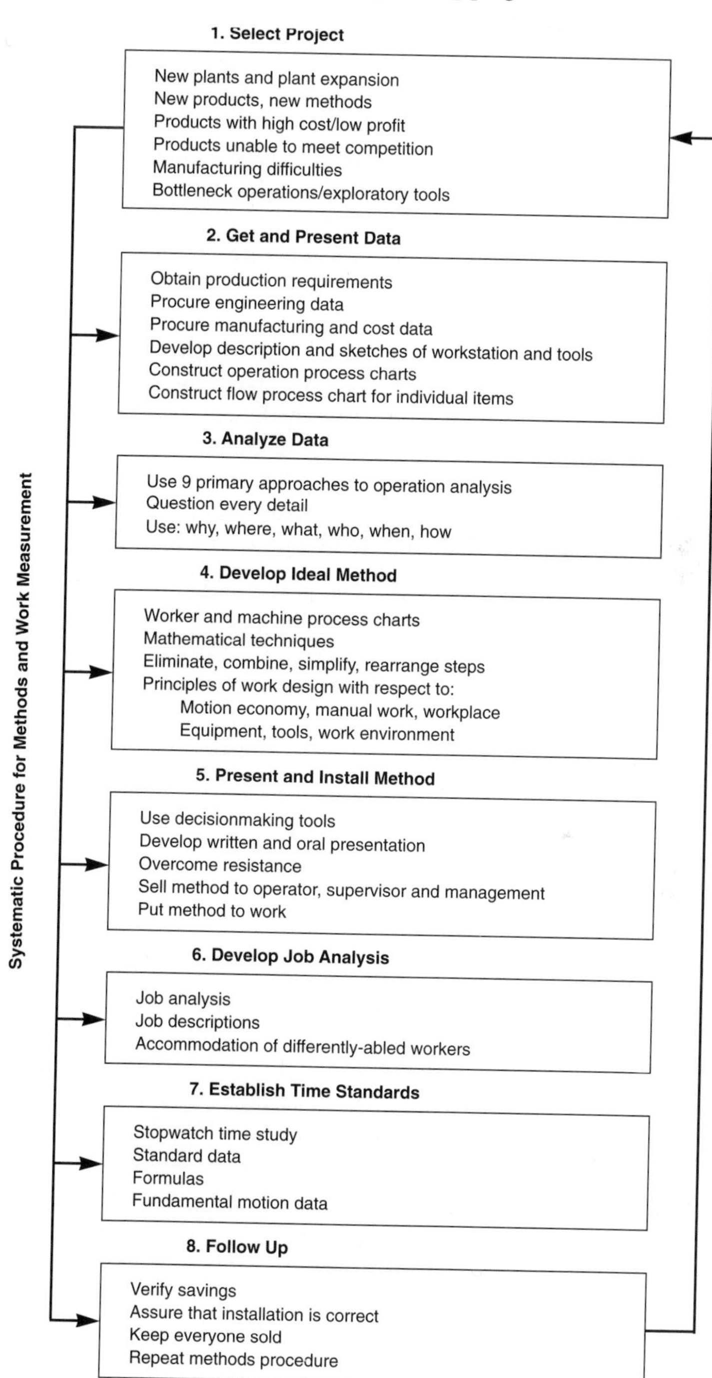

EXAMPLE 7–5
Methods Change for Auto-Starter

A representative case history, which follows operation analysis, is the production of an auto-starter, a device for starting AC motors by reducing the voltage through a transformer. A subassembly of the auto-starter is the arc box. This part sits in the bottom of the auto-starter and acts as a barrier between the contacts so there are no short-circuits. The present design consists of the components shown in Table 7–11.

In assembling these components, the operator places a washer, lock washer, and nut on one end of each rod. Next, the worker inserts the rods through the three holes in the first barrier. Then, the operator places one spacer on each of the rods (three in all) and adds another barrier. Operators repeat this until six barriers are on the rods, separated by the tubing.

It was suggested that the six barriers be made with two slots, one at each end, and that two strips of asbestos be made for supporting the barriers, with six slots in each. These would be slipped together and placed in the bottom of the auto-starter as needed. The manner of assembly would be the same as that used in putting together the separator in an eggbox. A total of 15 suggested improvements occurred after the analysis was completed, with the results shown in Table 7–12.

TABLE 7–11
Auto-Starter Components

Component	Number
Asbestos barriers with three drilled holes	6
Spacers of insulating tubing two inches long	15
Steel rods threaded at both ends	3
Pieces of hardware	18
Pieces total	42

TABLE 7–12
Improvements for Auto-Starter

Old method	New method	Savings
42 parts	8	34
10 workstations	1	9
18 transportations	7	11
7,900 feet of travel	200	7,700
9 storages	4	5
0.45 hours time	0.11 hours time	0.34 hours
$1.55 costs	$0.60	$0.95

taking only one stride. The new workbench has consolidated equipment, thus saving space and eliminating the duplication of facilities. One glassware cabinet services two chemists. A large four-place fume head allows multiple activity in an area that was formerly a bottleneck. All utility outlets are also relocated for maximum efficiency.

Finally, in a third example, in an effort to streamline its organization, a state government division developed an operation analysis program that resulted in an estimated annual savings of more than 50,000 hours. This was brought about by combining, eliminating, and redesigning all paperwork activities; improving the plant layout; and developing paths of authority.

Methods improvement is as effective in office procedures as in production operations. One industrial engineering department of a Pennsylvania company was given the problem of simplifying the paperwork necessary for shipping molded parts manufactured in one of its plants to an outlying plant for assembly. The department developed a new method that reduced the average daily shipment of 45 orders from 552 sheets of paper forms to 50 sheets. The annual savings in paper alone was significant.

Methods improvement should be a part of a continuous improvement program. For example, in the early 1940s, the average savings per member of management in the Procter & Gamble Company was approximately $500. To improve the situation and to maintain an acceptable level of cost reduction, P & G inaugurated a methods program. Manufacturing management attended sessions on the concepts of operation analysis. Also, methods engineers, each of whom had one to five years of company experience, attended a special course. After completion of the operation analysis course, these engineers returned to their respective plants to work as methods specialists. The effect of this training was apparent in a very short time. Annual savings increased to approximately $700 per year per member of factory management. Since the program was a success, management trained additional personnel and extended the program to more plants.

By 1950, the position of the industrial engineer had changed from methods specialist to methods coordinator. They spent approximately two-thirds of their time helping factory supervisors on various projects. Methods coordinators also periodically conducted training courses for both other staff members and members of the line organization. With the active participation of plant management, the rate of savings increased to $2,300 per year per member of factory management by 1950.

Subsequently, teams of four to eight employees, made up equally of line supervisors and staff, were formed to work on selected projects having a high potential savings. More enthusiasm on the part of management resulted. Opportunities for the recognition of good work increased. A friendly rivalry developed between teams for first place in plant standing. Display boards were set up in the front office to show team standings. Factory goals for cost reduction were established. By 1954, company-wide savings amounted to approximately $4,000 per member of management.

These results convinced top management that the cost reduction or profit improvement program should be extended to all plants. To further stimulate the desire of individual plants to make a good showing, P & G circulated a summary sheet comparing the results of the programs in each plant. Annual savings continued to increase each year, and by 1970, they amounted to $27,000 per member of management. This figure is conservative in that no credit is given for savings that continue beyond the first year. Since 1970, the program has spread to involve people in all parts of the business at both the non-management and management levels, making dollars saved per member of management meaningless.

SUMMARY

Increased output and improved quality are the primary outcomes of methods and work design changes, but methods changes also distribute the benefits of improved production to all workers and help develop better working conditions and a safer working environment, so the worker can do more work at the plant, do a good job, and still have enough energy to enjoy life. The examples of effective implementation of methods changes clearly demonstrate the need to follow an orderly approach, as presented in Figure 7–7. The methods engineer should note that it is not sufficient to use sophisticated mathematical algorithms or the latest software tools to develop the ideal method. It is necessary to sell the plan both to management and to the workers themselves. Additional interpersonal techniques and strategies for dealing with people to better sell the method are presented in Chapter 18.

QUESTIONS

1. What are the principal concerns of management with regard to a new method that is relatively costly to install?
2. What is meant by the "cash flow" appraisal technique?
3. What is meant by the payback method? How is it related to the return on investment method?
4. What is the relationship between return on capital investment and the risk associated with the anticipated sales of the product for which a new method will be used?
5. What two specific subjects should be emphasized in writing the job description?
6. Is time a common denominator of labor cost? Why or why not?
7. What is job analysis?
8. Which four methods of job evaluation are being practiced in this country today?
9. Why is it that most people feel that others doing work similar to theirs are paid more?
10. Explain in detail how a point plan works.

11. Which factors influence the relative worth of a job?

12. Why are estimates unsatisfactory for determining direct labor time standards?

13. What is the weakness of using historical records as a means of establishing standards of performance?

14. Which work measurement techniques give valid results when undertaken by competent, trained analysts?

15. Explain why a range of rates, rather than just one rate, should be established for every labor grade.

16. Explain what is meant by total operator performance.

17. What are the principal negative considerations that should be understood prior to the installation of a point job evaluation system?

18. What are the principal benefits of a properly installed job evaluation plan?

19. How is job evaluation dependent on job analysis?

20. Which three considerations constitute a successful job evaluation plan?

PROBLEMS

1. How much capital should be invested in a new method if it is estimated that $5,000 would be saved the first year, $10,000 the second year, and $3,000 the third year? Management expects a 30 percent return on invested capital.

2. You have estimated the life of your design to be three years. You expect that a capital investment of $20,000 will be required to get it into production. You also estimate, based on sales forecasts, that the design will result in an after-tax profit of $12,000 the first year and $16,000 the second year, and a $5,000 loss the third year. Management has asked for an 18 percent return on capital investment. Should the company go ahead with the investment to produce the new design? Explain.

3. In the Dorben Company, a materials handling operation in the warehouse is being done by hand labor. Annual disbursements for this labor and for related expenses (social security, accident insurance, and other fringe benefits) are $8,200. The methods analyst is considering a proposal to build certain equipment to reduce this labor cost. The first cost of this equipment will be $15,000. It is estimated that the equipment will reduce annual disbursements for labor and labor extras to $3,300. Annual payments for power, maintenance, and property taxes plus insurance are estimated to be $400, $1,100, and $300, respectively. The need for this particular operation is anticipated to continue for 10 years. Because the equipment is specially designed for the particular purpose, it will have no salvage value. It is assumed that the annual disbursements for labor, power, and maintenance will be uniform throughout the 10 years. The minimum rate of return before taxes

is 10 percent. Based on an annual cost comparison, should the company proceed with the new material handling equipment?

4. A job evaluation plan based on the point system uses the following factors:
a. Experience: maximum weight 200 points; five grades.
b. Education: maximum weight 100 points; four grades.
c. Effort: maximum weight 100 points; four grades.
d. Responsibility: maximum weight 100 points; four grades.

A floor sweeper is rated as 150 points, and this position carries an hourly rate of $6.50. A class 3 milling machine operator is rated as 320 points, which results in a money rate of $10.00 per hour. What grade of experience would be given to a drill press operator with an $8.50 per hour rate and point ratings of grade 2 education, grade 1 effort, and grade 2 responsibility?

5. A job evaluation plan in the Dorben Company provides for five labor grades, of which grade 5 has the highest base rates and grade 1 the lowest. The linear plan involves a range of 50 to 250 points for skill, 15 to 75 points for effort, 20 to 100 points for responsibility, and 15 to 75 points for job conditions. Each of the four factors has five degrees. Each labor grade has three money rates: low, mean, and high.
a. If the high money rate of labor grade 1 is $8 per hour and the high money rate of labor grade 5 is $20 per hour, what would be the mean money rate of labor grade 3?
b. What degree of skill is required for a labor grade of 4, if second-degree effort, second-degree responsibility, and first-degree job conditions apply?

6. In the Dorben Company, the analyst installed a point job evaluation plan covering all indirect employees in the operating divisions of the plant. Ten factors were used in this plan, and each factor was broken down into five degrees. In the job analysis, the shipping and receiving clerk position was shown as having second-degree initiative and ingenuity, valued at 30 points. The total point value of this job was 250 points. The minimum number of points attainable in the plan was 100, and the maximum was 500.
a. If 10 job classes prevailed, what degree of initiative and ingenuity would be required to elevate the job of shipping and receiving clerk from job class 4 to job class 5?
b. If job class 1 carried a rate of $8 per hour and job class 10 carried a rate of $20 per hour, what rate would job class 7 carry? (Note: Rates are based on the midpoint of job class point ranges.)

REFERENCES

ADA. *Americans with Disabilities Act Handbook.* EEOC-BK-19. Washington, DC: Equal Employment Opportunity Commission and U.S. Dept. of Justice, 1991.

Brown, D. B. *Systems Analysis & Design for Safety.* Englewood Cliffs, NJ: Prentice Hall, 1976.

Carey, A. "The Hawthorne Studies: A Radical Criticism." In *Concepts and Controversy in Organizational Behavior.* Ed. W. R. Nord. Pacific Palisades, CA: Goodyear Publishing Co., 1972.

DeReamer, R. *Modern Safety and Health Technology.* New York: John Wiley & Sons, 1980.

Dunn, J. D., and F. M. Rachel. *Wage and Salary Administration: Total Compensation Systems*. New York: McGraw-Hill, 1971.

Ellig, Bruce R. *Compensation Issues of the Eighties*. Amherst, MA: Human Resource Development Press, Inc., 1988.

Fleischer, G. A. "Economic Risk Analysis." In *Handbook of Industrial Engineering*, 2nd ed. Ed. Gavriel Salvendy. New York: John Wiley & Sons, 1992.

Gausch, J. P. "Safety and Decision-Making Tables." *ASSE Journal*. 17 (November 1972), pp. 33–37.

Gausch, J. P. "Value Engineering and Decision Making." *ASSE Journal*. 19 (May 1974), pp. 14–16.

Grant, E., W. Ireson, and R. S. Leavenworth. *Principles of Engineering Economy*. 6th ed. New York: Ronald Press, 1976.

Heinrich, H. W., D. Petersen, and N. Roos. *Industrial Accident Prevention*. 5th ed. New York: McGraw-Hill, 1980.

Homans, G. "The Western Electric Researches." In *Concepts and Controversy in Organizational Behavior*. Ed. W. R. Nord. Pacific Palisades, CA: Goodyear Publishing Co., 1972.

Jung, E. S., and A. Freivalds. "Multiple Criteria Decision-Making for the Resolution of Conflicting Ergonomic Knowledge in Manual Materials Handling." *Ergonomics*, 34, no. 11 (November 1991), pp. 1351–1356.

Lawler, Edward E. *What's Wrong with Point-Factor Job Evaluation?* Amherst, MA: Human Resource Development Press, Inc., 1988.

Livy, B. *Job Evaluation: A Critical Review*. New York: Halstead, 1973.

Lutz, Raymond P. "Discounted Cash Flow Techniques." In *Handbook of Industrial Engineering*. 2nd ed. Ed. Gavriel Salvendy. New York: John Wiley & Sons, 1992.

Mayo, E. *The Human Problems of an Industrial Civilization*. New York: The Viking Press, 1960.

McCormick, Ernest J. "Job Evaluation." In *Handbook of Industrial Engineering*. Ed. Gavriel Salvendy. New York: John Wiley & Sons, 1982.

Milkovich, George T., Jerry M. Newman, and James T. Brakefield. "Job Evaluation in Organizations." In *Handbook of Industrial Engineering*. 2nd ed. Ed. Gavriel Salvendy. New York: John Wiley & Sons, 1992.

OSHA. *Ergonomics Program Management Guidelines for Meatpacking Plants*. Washington DC: Bureau of National Affairs, 1990.

Otis, Jay, and Richard H. Leukart. *Job Evaluation: A Sound Basis for Wage Administration*. Englewood Cliffs, NJ: Prentice Hall, 1954.

Risner, Howard. *Job Evaluation: Problems and Prospects*. Amherst, MA: Human Resource Development Press, Inc., 1988.

Saaty, T. L. *The Analytic Hierarchy Process*. New York: McGraw-Hill, Inc. 1980.

Salvendy, Gavriel, and Douglas W. Seymour. *Prediction and Development of Industrial Work Performance*. New York: John Wiley & Sons, 1973.

Thuesen, H. G., W. J. Fabrycky, and G. J. Thuesen. *Engineering Economy*. 5th ed. Englewood Cliffs, NJ: Prentice-Hall, 1977.

Wegener, Elaine. *Current Developments in Job Classification and Salary Systems*. Amherst, MA: Human Resource Development Press, Inc., 1988.

CHAPTER 8

Time Study

KEY POINTS:

- Use time study to establish time standards.
- Use both audio and visual breakpoints to divide the operation into elements.
- Use continuous timing to obtain a complete record of times.
- Use snapback timing to avoid clerical errors.
- Perform a time check to confirm the validity of the time study.

The seventh step in the systematic process of developing the efficient work center is the *establishment of time standards*. Three elements help determine time standards: estimates, historical records, and work measurement procedures.

In past years, analysts relied more heavily on estimates as a means of establishing standards. With today's increasing competition from foreign producers, there has been an increasing effort to establish standards based on facts rather than judgment. Experience has shown that no individual can establish consistent and fair standards simply by looking at a job and judging the amount of time required to complete it. Where estimates are used, standards are out of line. Compensating errors sometimes diminish this deviation, but experience shows that over a period of time, estimated values deviate substantially from measured standards. Both historical records and work measurement techniques give much more accurate values than the use of estimates based on judgment alone.

With the historical records method, production standards are based on the records of similar, previously performed jobs. In common practice, the worker punches in on a time clock or data collection hardware every time he or she begins a new job, and then punches out after completing the job. This technique tells how long it actually took to do a job, but not how long it should have taken. Since operators wish to justify their entire working day, some jobs carry personal, unavoidable, and avoidable delay time to a much greater extent than they should, while other jobs do not carry their appropriate share of delay time. Historical records have consistently deviated by as much as 50 percent on the same operation of the same job. Yet, as a basis of determining labor standards, historical records are better than no records at all. Such records give more reliable results than estimates based on judgment alone, but they do not provide sufficiently valid results to assure equitable and competitive labor costs.

Any of the work measurement techniques—stopwatch (electronic or mechanical) time study, fundamental motion data, standard data, time formulas, or work sampling studies—represent a better way to establish fair production standards. All of these techniques are based on facts. All consider each detail of the work and its relation to the normal time required to perform the entire cycle.

Accurately established time standards make it possible to produce more within a given plant, thus increasing the efficiency of the equipment and the operating personnel. Poorly established standards, although better than no standards at all, lead to high costs, labor dissension, and possibly even the failure of the enterprise.

Successful installation of any of the work measurement techniques requires a wholehearted commitment by management, including committing the enthusiasm, time, and financial resources necessary on a continuing basis. As data from the work measurement system become available, they should be used immediately, not shelved for possible future use or, worse yet, simply ignored.

Sound standards have many applications that can mean the difference between the success or failure of a business. Companies should use standards for planning purposes, comparing alternative methods, developing an effective plant layout, determining capacities, purchasing new equipment, balancing the workforce with the available work, controlling production, installing incentives, and instituting standard cost and budgetary controls.

Time study is a technique for establishing an allowed time standard for performing a given task. This technique is based on measuring the work content of the prescribed method, with due allowance for fatigue and for personal and unavoidable delays.

A FAIR DAY'S WORK

Time study is frequently defined as a method of determining a "fair day's work." Practically everyone connected with industry in any way has heard that expression. Yet most people would be unable to define a fair day's work. The intraplant wage rate inequities agreements of the basic steel industries contain the provision that "the fundamental principle of the work and wage relationship is that the employee is entitled to a fair day's pay in return for which the company is entitled to a fair day's work." In these agreements, a fair day's work is defined as the "amount of work that can be produced by a qualified employee when working at a normal pace and effectively utilizing his time where work is not restricted by process limitations." This definition does not clarify what is meant by qualified employees, normal pace, and effective utilization. Although these terms have been defined by the steel industries, a certain amount of flexibility prevails, because firm benchmarks cannot be established on such broad terminology. For example, the term "qualified employee" is defined as "a representative average of those employees who are fully trained and able satisfactorily to perform any and all phases of the work involved, in accordance with the requirements of the job under consideration." This definition leaves some doubt as to the meaning of a "representative average employee."

Then, too, "normal pace" is defined as "the effective rate of performance of a conscientious, self-paced, qualified employee when working neither fast nor slow and giving due consideration to the physical, mental, or visual requirements of the specific job." As an example, the intraplant wage rate inequities agreements specify, "a man walking without load, on smooth, level ground at a rate of three miles per hour." Although the three miles an hour concept tends to tie down what is meant by normal pace, a notable amount of latitude can still prevail if we think of the normal pace on the thousands of different jobs in American industry.

There is also some uncertainty over the definition of "effective utilization." This is explained in the agreements as, "the maintenance of a normal pace while performing essential elements of the job during all portions of the day except that which is required for reasonable rest and personal needs, under circumstances in which the job is not subject to process, equipment or other operating limitations."

In general, a fair day's work is one that is fair to both the company and the employee. This means that the employee should give a full day's work for the time that he or she gets paid, with reasonable allowances for personal delays, unavoidable delays, and fatigue. The worker is expected to operate in the prescribed method at a pace that is neither fast nor slow, but one that may be considered representative of all-day performance by the experienced, cooperative employee.

TIME STUDY REQUIREMENTS

Certain fundamental requirements must be realized before the time study is taken. For example, whether the standard is required on a new job, or on an old job in which the method or part of the method has been altered, the operator should be thoroughly acquainted with the new technique before the operation is studied. Also, the method must be standardized at all points where it is to be used, before the study begins. Unless all details of the method and working conditions have been standardized, the time standards will have little value and will become a continual source of mistrust, grievances, and internal friction.

Analysts should tell the union steward, the department supervisor, and the operator that the job is to be studied. Each of these parties can then make specific advanced plans and take the steps necessary to allow a smooth, coordinated study. The operator should verify that he or she is performing the correct method and should become acquainted with all details of that operation. The supervisor should check the method to make sure that feeds, speeds, cutting tools, lubricants, and so forth conform to standard practice, as established by the methods department. Also, the supervisor should investigate the amount of material available so that no shortages take place during the study. If several operators are available for the study, the supervisor should determine which operator will give the most satisfactory results. The union steward should then make sure that only trained, competent operators are selected, should explain why the study is being taken, and should answer any pertinent questions raised by the operator.

Analyst's Responsibility

All work involves varying degrees of skill, as well as physical and mental effort. There are also differences in the aptitude, physical application, and dexterity of the workers. It is easy for the analyst to observe an employee at work and measure the actual time taken to perform a task. It is considerably more difficult to evaluate all variables and determine the time required for the "normal" operator to perform the task.

Because of the many human interests and reactions associated with the time study technique, it is essential that there be full understanding between the supervisor, employee, union steward, and time study analyst. The time study analyst should ensure that the correct method is being used, accurately record the times taken, honestly evaluate the performance of the operator, and refrain from any operator criticism.

Since time study analysts directly affect the pocketbooks of workers and the profit and loss statements of companies, their work must be completely dependable and accurate. Inaccuracy and poor judgment will not only affect the operator and the company financially, but may also result in loss of confidence by the operator and the union, which may ultimately undo harmonious labor relations that have taken management years to build up. To achieve and maintain good human relations, the time study analyst should always be honest, tactful, pleasing, patient, and enthusiastic, and should always use good judgment. It is imperative that the time study analyst have the best qualifications.

Supervisor's Responsibility

The supervisor should notify the operator in advance that his or her work assignment is to be studied. This clears the way for both the time study analyst and the operator. The operator has the assurance that the supervisor knows a rate is to be established on the job. Therefore, the operator can bring out specific difficulties that he or she feels should be corrected before a standard is set. In addition, the time study analyst should be more at ease knowing that his or her presence is anticipated.

The supervisor should see that the proper method established by the methods department is being utilized, and that the operator selected is competent and has adequate experience on the job. Although the time study analyst should have a practical background in the area of work being studied, analysts can not be expected to know all specifications for all methods and processes. Therefore, the supervisor should verify that the cutting tools are properly ground, the correct lubricant is being used, and a proper selection of feeds, speeds, and cut depths is being made. The supervisor should also make certain that operators use the prescribed method, conscientiously assisting and training all employees in perfecting this method. A supervisor should freely answer any operation-related questions asked by the operator.

Once the time study is completed, the supervisor should sign the original copy, indicating compliance with the study. If a methods change takes place within a department, the supervisor should notify the time study department immediately, so that the standard can be adjusted appropriately. Supervisors who fail in these responsibilities contribute to the establishment of inequitable rates which can result in labor grievances, management pressure, and union dissatisfaction.

Union's Responsibility

Most unions are opposed to work measurement and would prefer to see all standards established by arbitration. However, unions recognize that standards are necessary for the profitable operation of a business, and that management continues to develop such standards using principal work measurement techniques. Furthermore, every union steward knows that poor time standards cause problems for both labor and management.

Through training programs, the union should educate all of its members in the principles, theories, and economic necessity of time study practice. Operators cannot be expected to be enthusiastic about time study if they know nothing about it. This is especially true in view of its background (see Chapter 1).

The union representative should make certain that the time study includes a complete record of the job conditions, that is, work method and workstation layout. The representative should also ascertain that the current job description is accurate and complete.

The union should urge its members to cooperate with the time study analyst and refrain from practices that would place their performance at the low end of the rating scale. Unions that train their members in time study elements, encourage cooperativeness, and stay abreast of management's program benefit by more cooperation at the bargaining table, fewer work stoppages, and more satisfied members.

Operator's Responsibility

Every employee should be sufficiently interested in the welfare of the company to support the practices and procedures inaugurated by management. Operators should give new methods a fair trial. They should cooperate in helping to work out the bugs characteristic of many innovations. Making suggestions for further methods improvements should be an accepted part of each operator's responsibilities. The operator is closer to the job than anyone else, and can make a real contribution to the company by helping to establish ideal methods.

The operator should assist the time study analyst in breaking the job down into elements, thus assuring that all details of the job are specifically covered. The operator should also work at a steady, normal pace while the study is being taken, and should introduce as few foreign elements and extra movements as possible. The worker should use the exact method prescribed, as any action that artificially lengthens the cycle time could result in too liberal standard.

TIME STUDY EQUIPMENT

The minimum equipment required to conduct a time study program includes a stopwatch, time study board, time study forms, and pocket calculator. Videotape equipment can also be very useful.

Stopwatch

Two types of stopwatches are in use today: (1) the traditional decimal minute watch (0.01 minute), and (2) the much more practical electronic stopwatch. The decimal minute watch, shown in Figure 8–1, has 100 divisions on its face, and each division is equal to 0.01 minute, that is, a complete sweep of the long hand requires one minute. The small dial on the watch face has 30 divisions, each of which is equal to one minute. Therefore, for every full revolution of the sweep hand, the small hand moves one division, or one minute. To start this watch, move the side slide toward the crown. Moving the side slide away from the crown stops the watch with the hands in their existing positions. To continue operation of the watch from the point where the hands stopped, move the slide toward the crown. Depressing the crown moves both the sweep hand and the small hand back to zero. Releasing the crown puts the watch back into operation, unless the side slide is moved away from the crown.

Electronic stopwatches cost approximately $50. These watches provide resolution to 0.001 second and an accuracy of ±0.002 percent. They weigh about 4 ounces and are about 4×2×1 inches in size (see Figure 8–2). They permit timing any number of individual elements, while also counting the total elapsed time. Thus, they provide both *continuous* and *snapback timing* (button C), with none of the disadvantages of mechanical watches. To operate the watch, press the top button (button A). Each time the top button is pressed, a numerical readout is presented. Pressing the memory button (button B) causes previous readouts to be retrieved. A slightly fancier version incorporates the watch into an electronic time study board (see Figure 8–3).

With mechanical watches costing $150 and up and electronic watches falling in price, the mechanical watches are quickly disappearing from use.

Computer-Assisted Electronic Stopwatches

The DataMyte 1010 all-solid-state battery-operated data collector, developed by the Electro/General Corporation in 1971, is a practical alternative to both the mechanical and electronic stopwatches. Observational data are keyed in and the data are

FIGURE 8–1
Decimal minute watch. (Meylan Stopwatch Co.)

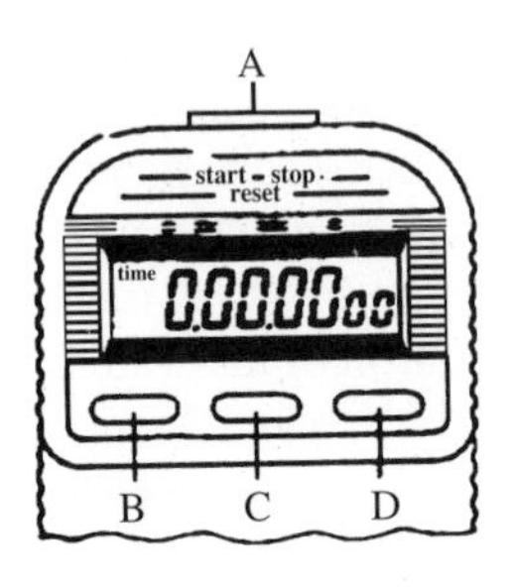

FIGURE 8–2
Electronic stopwatch.

A - Start/Stop
B - Memory retrieval
C - Mode (continuous/snapback)
D - Other functions

recorded in a solid-state memory in computer language. Elapsed time readings are recorded automatically. Input data and elapsed time data may be transmitted directly from the DataMyte to most personal computers through an output cable. The computer prepares printed summaries, eliminating the laborious task of manually computing normal and allowed elemental times and operation standards. This instrument is self-contained and can be carried throughout a plant or organization. The rechargeable battery power provides about 12 hours of continuous operation. Figure 8–4 shows an analyst using the DataMyte. Time studies taken with the DataMyte (see Figure 8–5) and a computer take an estimated 50 to 60 percent of the time needed for a stopwatch and hand calculator.

Another computer-assisted electronic stopwatch that has recently been developed by Faehr Electronic Timers, Inc., is the COMPU-RATE (see Figure 8–6). Manual entries are only required for the element column and the top four lines of the form, allowing the analyst to concentrate on observing the work and operator performance. Studies can be in thousandths of a minute or one-hundred-thousandths

FIGURE 8–3
Digital electronic time-study board. (Meylan Stopwatch Co.)

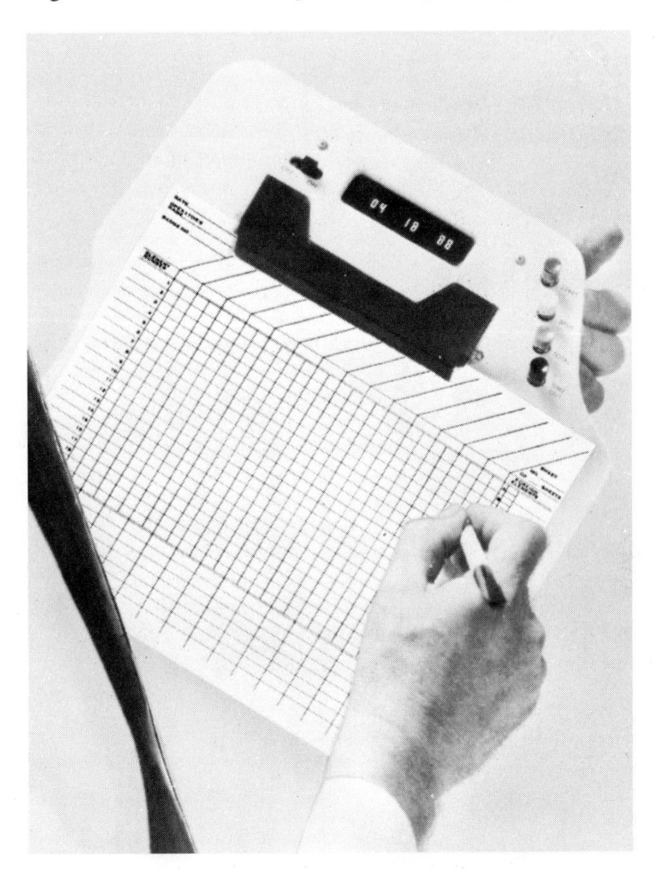

of an hour. The LCD displays can be contrast adjusted to fit the lighting conditions at the workstation. The instrument uses three C-cell rechargeable NiCad batteries, providing about 120 hours of running time before the batteries need to be recharged. The COMPU-RATE software system computes all the typical time study calculations, including mean and median element values, adjustment of mean times to normal time after inputting the performance rating, and allowed times after inputting appropriate allowances. Results are summarized to provide standard times in minutes and/or hours per piece and pieces per hour. An edit function helps in correcting mistakes.

The GageTalker Corporation (formerly, Observational Systems) is marketing the OS-3 Plus Event Recorder. This versatile work measurement recorder is useful for setting and updating standards, for machine downtime studies, and for work sampling (see Chapter 14). The recorder, shown in Figure 8–7, permits the analyst to select the time units most appropriate for the study—0.001 minute, 0.0001 hour, or 0.1 second. After collecting and summarizing input data, the OS-3 interfaces with

FIGURE 8–4
Operator using a DataMyte 1000 Data Collector in conjunction with the DataMyte 1010 Time Study System. (*Courtesy:* DataMyte Corporation)

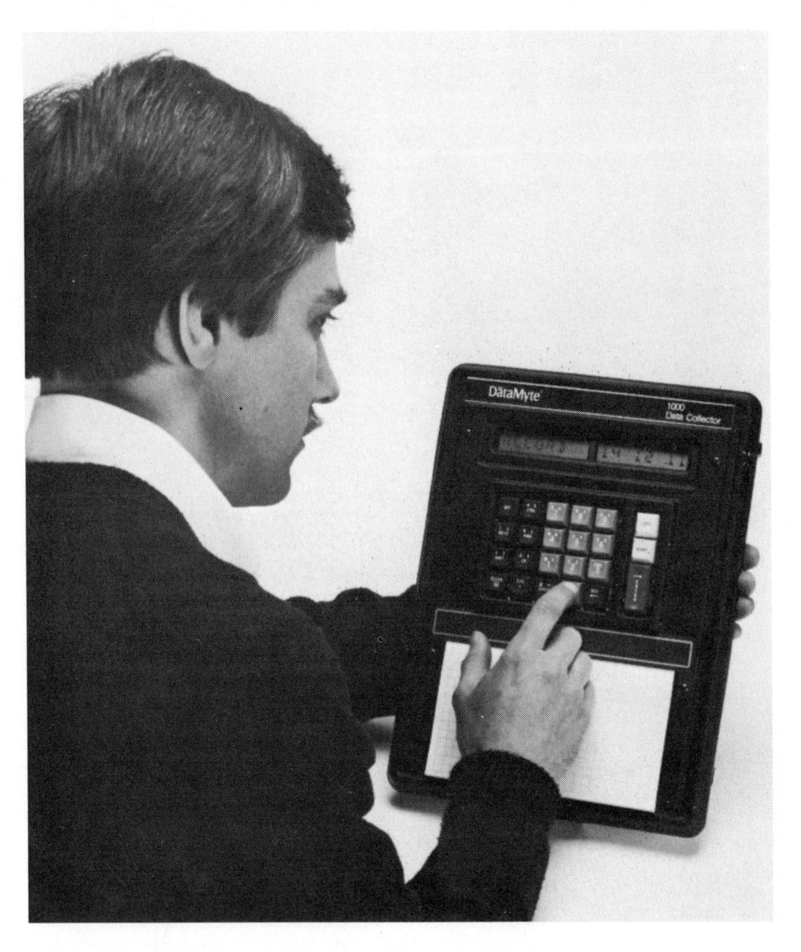

printers to produce hard-copy reports. The total time, frequency, mean times, performance ratings, normal times, allowances, standard times, and pieces per hour are printed. In addition, the standard deviation, the maximum and minimum element values, and frequencies are also available.

Videotape Cameras

Videotape cameras are ideal for recording operators' methods and elapsed time. By taking pictures of the operation and then studying them a frame at a time, analysts can record exact details of the method used and can then assign normal time values. They can also establish standards by projecting the film at the same speed that the

FIGURE 8–5

Application of DataMyte and a computer to the conduct and calculation of a time study.

DATE
STUDY NO T559
SHEET NO./OF /
SHEETS

ELEMENTS		GET PART A		GET PART B		GET WELD GUN		WELD 10 SPOTS		RELEASE GUN DISPOSE TO CONV.	
NUMBER		1		2		3		4		5	
Notes	LINE	T	R	T	R	T	R	T	R	T	R
	1	11	11	13	24	07	31	24	55	05	60
	2	12	72	13	85	08	93	28	121	04	25
	3	10	35	17	52	08	60	22	82	06	88
C →	4	10	332	12	44	11	55	26	81	05	86
	5	13	99	12	411	07	18	25	43	05	48
F →	6	11	59	10	69	08	77	27	523	07	30
F →	7	12	42	11	53	09	62	25	33	08	41
	8	12	53	11	64	10	74	23	97	06	703
H →	9	12	58	13	71	09	80	26	806	04	10
	10										
	11										
	12										
	13										
	14										
	15										
	16										
	17										
	18										
	19										
	20										

	1	2	3	4	5
TOTALS "T"	0103	0112	0077	0226	0050
NO. OBSERVATIONS	9	9	9	9	9
AVERAGE "T"	00114	00124	00086	00251	00056
MINIMUM "T"	0010	0010	0007	0022	0004
MAXIMUM "T"	0013	0017	0011	0028	0008
RATING (SK & EF)	DD				
LEVELING FACTOR	1.00				
LF × AVE. "T"	00114	00124	00086	00251	00056
% ALLOWANCE	0				
TIME ALLOWED	00114	00124	00086	00251	00056

□ DataMyte
○ Computer

[A] Key in ELEMENT NUMBER
[B] Hit ENTER. DataMyte automatically logs in END-POINT READING
(C) Computer calculates the ELEMENT TIME
(D) Computer all of these summaries
[E] These factors can be keyed into DataMyte or entered later at the computer keyboard

FOREIGN ELEMENTS

SYM	R	T	DESCRIPTION
A			
B			
C	322 / 188	134	PARTS DELAY
D			
E	608 / 562	46	MECHANICAL PROBLEM
F	496 / 477	19	MECHANICAL PROBLEM
G			
H	746 / 703	43	PERSONAL DELAY
I			
J			
K			
L			
M			
N			
O			
P			
Q			
R			
S			

SKILL			EFFORT		
A1, A2		SUPER	A1, A2		EXCESSIVE
B1, B2		EXCELLENT	B1, B2		EXCELLENT
C1, C2		GOOD	C1, C2		GOOD
D	✓	AVERAGE	D	✓	AVERAGE
E1, E2		FAIR	E1, E2		FAIR
F1, F2		POOR	F1, F2		POOR

GENERAL RATING FOR STUDY	SKILL	EFFORT	TOTAL
	.00	.00	.00

STUDY STARTED	STUDY FINISHED	OVERALL TIME
10:39 A.M.	10:47 A.M.	.135 Hrs

FIGURE 8–6
Computer-assisted electronic stopwatch. (*Courtesy of:* Faehr Electronic Timers, Inc.)

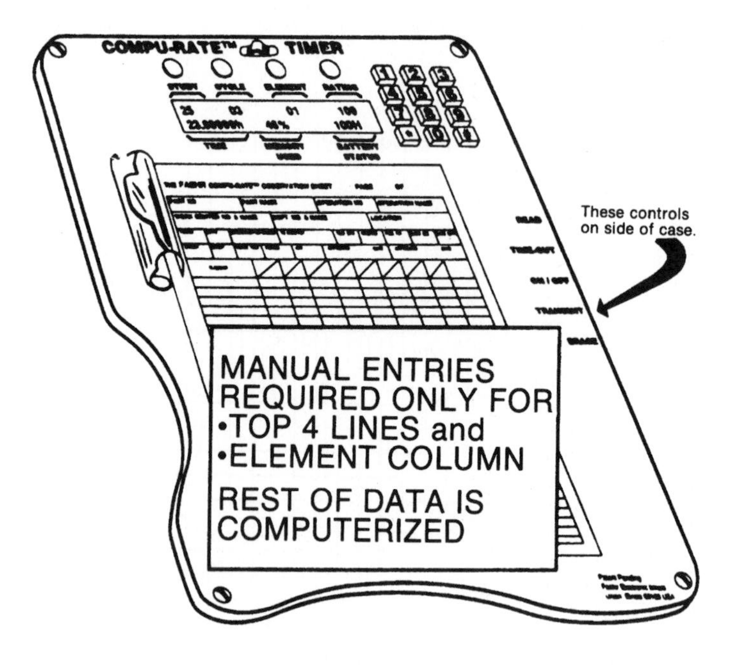

pictures were taken and then performance rating the operator. Because all the facts are there, observing the videotape is a fair and accurate way to rate performance. Then, too, potential methods improvements that would seldom be uncovered with a stopwatch procedure can be revealed through the camera eye. Videotapes are also excellent for training the novice time study analyst, since sections can easily be rewound and repeated until sufficiently mastered.

Time Study Board

When the stopwatch is being used, analysts find it convenient to have a suitable board to hold the time study form and the stopwatch. The board should be light, so as not to tire the arm, and yet strong and sufficiently hard to provide a suitable backing for the time study form. Suitable materials include 1/4-inch plywood or smooth plastic. The board should have both arm and body contacts, for comfortable fit and ease of writing while it is being held. For a right-handed observer, the watch should be mounted in the upper right-hand corner of the board. A spring clip to the left would hold the time study form. Standing in the proper position, the time study analyst can look over the top of the watch to the workstation and follow the operator's movements, while keeping both the watch and the time study form in the immediate field of vision.

FIGURE 8–7
The OS-3 event recorder. (*Courtesy of:* GageTalker Corp.)

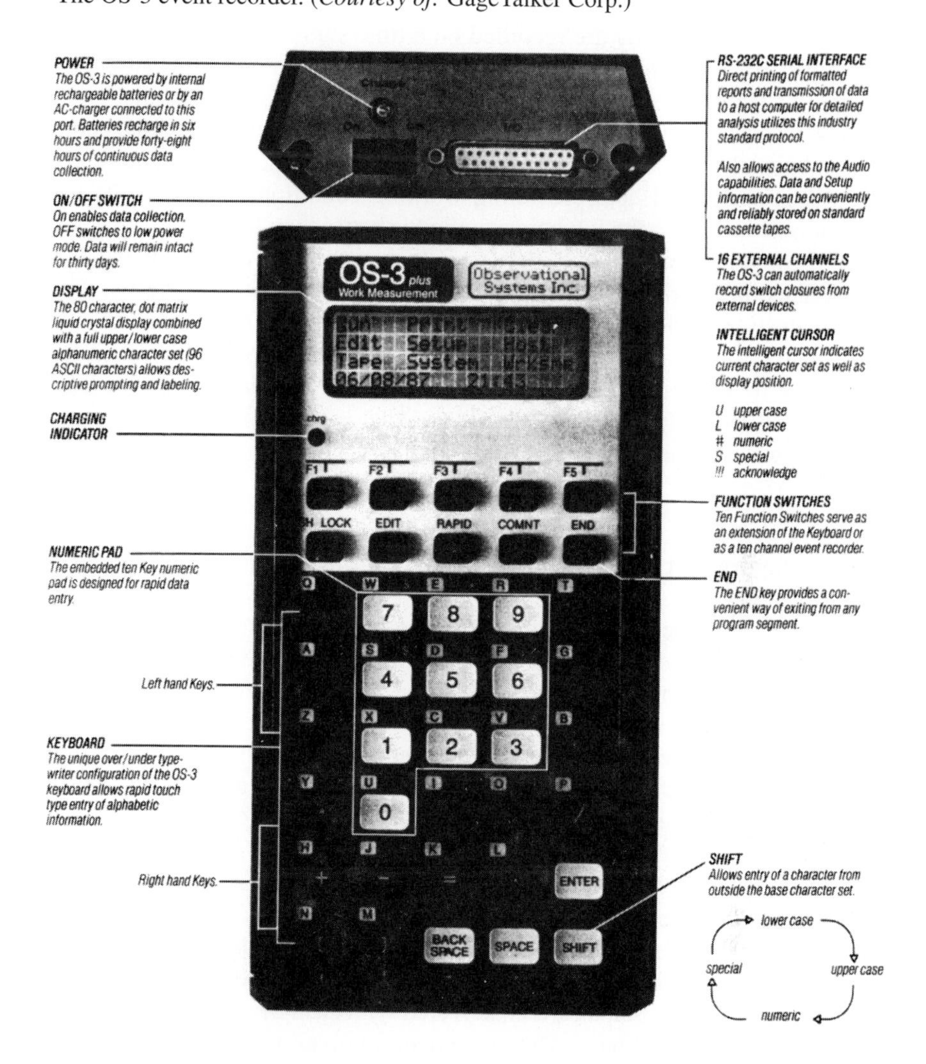

Channel 1. Creative development.
Channel 2. Conference.
Channel 3. Dictation.
Channel 4. Incoming phone calls.
Channel 5. Outgoing phone calls.
Channel 6. Supervision and assigning work.

Time Study Forms

All the details of the study are recorded on a time study form. The form provides space to record all pertinent information concerning the method being studied, tools utilized, etc. The operation being studied is identified by such information as the operator's name and number, operation description and number, machine name and number, special tools used and their respective numbers, department where the operation is performed, and prevailing working conditions. Providing too much information concerning the job being studied is better than too little.

Figure 8–8 illustrates a time study form that has been developed by the authors. It is sufficiently flexible to be used for practically any type of operation. On this form, analysts would record the various elements of the operation horizontally across the top of the sheet, and the cycles studied would be entered vertically, row by row. The four columns under each element are: R for *ratings*; W for *watch time*, that is, watch readout; OT for *observed time*, that is, the differential time between successive watch times; and NT for *normal time.*

Training Equipment

Two inexpensive pieces of equipment that can assist in the training of time study analysts are: the random elapsed time describer, and the metronome. The first device can be programmed (by means of a specially contoured cam) such that successive elements are completed in a known period. The trainee records the durations of the elements as they take place. The trainee is signaled by a lighted bulb and a buzzer at the end of each element. Either the buzzer or the light may be made inoperative, if desired. This exercise provides practice in reading the watch at terminal points and recording the elapsed time.

The second tool is the metronome used by music students. This device can be set to provide a predetermined number of beats per minute, such as 104 beats per minute. This happens to equal the number of cards dealt per minute when dealing at a normal pace (see Chapter 9). By synchronizing the delivery of a card at a four-hand bridge table such that a card is delivered with each beat of the metronome, we can demonstrate normal pace. The movement involves a series of reaches, grasps, moves, and releases and the speed can be easily identified with practice. To illustrate 80-percent performance, the instructor need only set the metronome to 83 beats per minute and then synchronize the card dealing accordingly.

TIME STUDY ELEMENTS

The actual conduct of a time study is both an art and a science. To ensure success, analysts must be able to inspire confidence in, exercise judgment with, and develop a personable approach to everyone with whom they come in contact. In addition,

FIGURE 8–8

Snapback study of a die casting operation (rating elements every cycle).

Time Study Observation Form

Study No.: Z-85	Date: 3-1 -	Page 1 of 1
Operation: DIE CASTING	Operator: B. JONES	Observer: A F

Element No. and Description	1 REMOVE PART FROM DIE, LUBRICATE DIE, INSPECT				2 PLACE PART IN FIXTURE, TRIM ASIDE PART																			
Note / Cycle	R	W	OT	NT	R	W	OT	NT	R	W	OT	NT	R	W	OT	NT	R	W	O	NT	R	W	OT	NT
1	90		30	270	90		23	207																
2	100		27	270	100		21	210																
3	90		31	279	90		23	207																
4	85		35	298	100		20	200																
5	100		28	280	100		20	200																
6	110		25	275	110		18	198																
7	90		31	279	90		24	216																
8	100		28	280	85		24	204																
9	90		32	288	90		23	207																
10	110		26	286	105		19	200																
11																								
12																								
13																								
14																								
15																								
16																								
17																								
18																								

Summary	Element 1	Element 2				
Total OT	2.93	2.15				
Rating	—	—				
Total NT	2.805	2.049				
No. Observations	10	10				
Average NT	.281	.205				
% Allowance	17	17				
Elemental Std. Time	.329	.240				
No. Occurences	1	1				
Standard Time	.329	.240				

Total Standard Time (sum standard time for all elements): .569

Foreign Elements

Sym	W1	W2	OT	Description
A				
B				
C				
D				
E				
F				
G				

Time Check

Finishing Time	3:48.00
Starting Time	3:42.00
Elapsed Time	6.00
TEBS	.60
TEAF	.32
Total Check Time	.92
Effective Time	5.08
Ineffective Time	0
Total Recorded Time	6.00
Unaccounted Time	0
Recording Error %	0

Allowance Summary

Personal Needs	5
Basic Fatigue	4
Variable Fatigue	8
Special	—
Total Allowance %	17

Remarks:

Rating Check

Synthetic Time	%
Observed Time	

their backgrounds and training should prepare them to understand thoroughly and perform the various functions related to the study. These elements include: selecting the operator, analyzing the job and breaking it down into its elements, recording the elapsed elemental values, performance rating the operator, assigning appropriate allowances, and working up the study itself.

Choosing the Operator

The first step in beginning a time study is made through the departmental or line supervisor. After reviewing the job in operation, both the supervisor and the time study analyst should agree that the job is ready to be studied. If more than one operator is performing the work for which the standard is to be established, several things should be considered when selecting which operator to use for the study. In general, an operator who is average or somewhat above average in performance gives a more satisfactory study than a low-skilled or highly superior operator. The average operator usually performs the work consistently and systematically. That operator's pace will tend to be approximately in the normal range (see Chapter 9), thereby making it easier for the time study analyst to apply a correct performance factor.

Of course, the operator should be completely trained in the method, should like the work, and should demonstrate an interest in doing a good job. The operator should be familiar with time study procedures and practices, and should have confidence in both time study methods and the analyst. The operator should also be cooperative enough to follow through willingly with suggestions made by both the supervisor and the time study analyst.

At times, the analyst has no choice of operators, because the operation is performed by only one worker. In these cases, the analyst must be very careful when establishing the performance rating, because the operator may be performing at one of the extreme ends of the rating scale. In one-worker jobs, the method used must be correct and the analyst must approach the operator carefully and tactfully.

The analyst's approach to the selected operator may determine the cooperation received. The analyst should approach the operator in a friendly manner and demonstrate an understanding of the operation to be studied. The operator should have the opportunity to ask questions about the timing technique, method of rating, and the application of allowances. In some instances, the operator may never have been studied before. All questions should be answered frankly and patiently. The operator should be encouraged to offer suggestions, and when the operator does so, the analyst should receive them willingly, thus showing respect for the skill and knowledge of the operator.

The analyst should show interest in the worker's job, and at all times, be fair and straightforward toward the worker. This approach wins over the worker's confidence in the analyst's ability. The resulting respect and goodwill will not only help in establishing a fair standard, but will also facilitate any future work assignments on the production floor.

Recording Significant Information

Analysts should record the machines, hand tools, jigs or fixtures, working conditions, materials, operations, operator name and clock number, department, study date, and observer's name. Space for such detail is provided under *Remarks* on the time study observation form. A sketch of the layout may also be helpful. The more pertinent information is recorded, the more useful the time study becomes over the years. It becomes a resource for establishing standard data (see Chapter 11) and developing formulas (see Chapter 12). It will also be useful for methods improvement, operator evaluation, tool evaluation, and machine performance evaluation.

When machine tools are used, the analyst should specify the name, size, style, capacity, and serial or inventory number, as well as the working conditions. Dies, jigs, gages, and fixtures should be identified by their numbers and with short descriptions. If the working conditions during the study are different from the normal conditions for that job, they will affect the performance of the operator. For example, in a drop forge shop, if a study were taken on an extremely hot day, the working conditions would be poorer than usual, and operator performance would reflect the effect of the intense heat. Consequently, an allowance (see Chapter 10) would be added to the operator's normal time. If the working conditions improve, the allowance can be diminished. Conversely, if the working conditions become poorer, the allowance should be raised.

The operation performed should be very specifically described. For example, "broach 3/8 inch × 3/8 inch keyway in 1-inch bore" is considerably more explicit than "broach keyway." The operator being studied should be identified by name and clock number; there could easily be two John Smiths in one company.

Positioning the Observer

The observer should stand, not sit, a few feet to the rear of the operator, so as not to distract or interfere with the worker. Standing observers are better able to move around and follow the movements of the operator's hands as the operator goes through the work cycle. During the course of the study, the observer should avoid any conversation with the operator, as this could distract the worker or upset the routines.

Dividing the Operation into Elements

For ease of measurement, the operation should be divided into groups of motions known as *elements*. To divide the operation into its individual elements, the analyst should watch the operator for several cycles. However, if the cycle time is over 30 minutes, the analyst can write the description of the elements while taking the study. If possible, the analyst should determine the operational elements before the start of the study. Elements should be broken down into divisions that are as fine as possible and yet not so small that reading accuracy is sacrificed. Elemental divisions of around 0.04 minute are about as fine as can be read consistently by an experienced time study analyst. However, if the preceding and succeeding elements are relatively long, an element as short as 0.02 minute can be readily timed.

To identify endpoints completely and develop consistency in reading the watch from one cycle to the next, consider both sound and sight in the elemental breakdown. For example, the *breakpoints* of elements can be associated with such sounds as: a finished piece hitting the container, a facing tool biting into a casting, a drill breaking through the part being drilled, and a pair of micrometers being laid on a bench.

Each element should be recorded in its proper sequence, including a basic division of work terminated by a distinctive sound or motion. For example, the element "up part to manual chuck and tighten" would include the following basic divisions: reach for part, grasp part, move part, position part, reach for chuck wrench, grasp chuck wrench, move chuck wrench, position chuck wrench, turn chuck wrench, and release chuck wrench. The termination point of this element would be the chuck wrench being dropped on the head of the lathe, as evidenced by the accompanying sound. The element "start machine" could include: reach for lever, grasp lever, move lever, and release lever. The rotation of the machine, with the accompanying sound, would identify the termination point so that readings could be made at exactly the same point in each cycle.

Frequently, different time study analysts in a company adopt a standard elemental breakdown for given classes of facilities, to ensure uniformity in establishing breakpoints. For example, all single-spindle bench-type drill press work may be broken down into standard elements, and all lathe work may be composed of a series of predetermined elements. Having standard elements as a basis for operation breakdown is especially important in the establishment of standard data (see Chapter 11).

Some additional suggestions that may help in breaking elements down are:

1. In general, keep manual and machine elements separate, since machine times are less affected by ratings.
2. Likewise, separate constant elements (those elements for which the time does not deviate within a specified range of work) from variable elements (those elements for which the time does vary within a specified range of work).
3. When an element is repeated, do not include a second description. Instead, in the space provided for the element description, give the identifying number that was used when the element first occurred.

THE STUDY ITSELF

At the start of the study, record the time of day (on a whole minute) from a "master" clock while simultaneously starting the stopwatch. (It is assumed that all data are recorded on the time study form.) This is the *starting time* (① as shown in Figure

TABLE 8–1
Recording Watch Readings for Continuous Timing

Consecutive reading of watch in decimal minutes	Recorded reading
0.08	8
0.25	25
1.32	132
1.35	35
1.41	41
2.01	201
2.10	10
2.15	15
2.71	71
3.05	305
3.17	17
3.25	25

8–11. One of two techniques can be used for recording the elemental times during the study. The *continuous timing* method, as the name implies, allows the stopwatch to run for the entire duration of the study. In this method, the analyst reads the watch at the breakpoint of each element, and the time is allowed to continue. In the *snapback* technique, after the watch is read at the breakpoint of each element, the watch time is returned to zero; as the next element takes place, the time increments from zero.

When recording the watch readings, note only the necessary digits and omit the decimal point, thus giving as much time as possible to observing the performance of the operator. If using a decimal minute watch, if the breakpoint of the first element occurs at 0.08 minute, record only the digit 8 in the W (watch time) column. Other example recordings are shown in Table 8–1.

Snapback Method

The snapback method has both advantages and disadvantages compared to the continuous technique. Some time study analysts use both methods, believing that studies of predominantly long elements are more adapted to snapback readings, while short-cycle studies are better suited to the continuous method.

Since elapsed element values are read directly in the snapback method, no clerical time is needed to make successive subtractions, as for the continuous method. Thus, the readout can be inserted directly in the OT (*observed time*) column. Also, elements performed out of order by the operator can be readily recorded without special notation. In addition, proponents of the snapback method state that delays are not recorded. Also, since elemental values can be compared from one cycle to the next, a decision could be made as to the number of cycles to study. However, it is actually erroneous to use observations of the past few cycles to determine how many additional cycles to study. This practice can lead to studying entirely too small a sample.

Among the disadvantages of the snapback method is that it encourages the removal of individual elements from the operation. These cannot be studied independently, because elemental times depend on the preceding and succeeding elements. Consequently, omitting such factors as delays, foreign elements, and transposed elements, could allow erroneous values in the readings accepted. One of the main objections to the snapback method is the amount of time lost while snapping the hand back to zero. This can be anywhere from 0.0018 to 0.0058 minutes (Lowry, Maynard, and Stegemerten, 1940). However, this has been negated by the use of electronic watches, for which no time is lost in resetting the readout to zero. Also, short elements (0.04 minute and less) are more difficult to time with this method. Finally, the overall time must be verified by summing the elemental watch readings, a process that is more prone to error.

Figure 8–8 illustrates a time study of a die casting operation using the snapback method.

Continuous Method

The continuous method of recording elemental values is superior to the snapback method for several reasons. The most significant is that the resulting study presents a complete record of the entire observation period; as a result, it appeals to the operator and the union. The operator is able to see that no time has been left out of the study, and all delays and foreign elements have been recorded. Since all the facts are clearly presented, this technique of recording times is easier to explain and sell.

The continuous method is also better adapted to measuring and recording very short elements. With practice, a good time study analyst can accurately catch three successive short elements (less than 0.04 minute), if they are followed by an element of about 0.15 minute or longer. This is possible by remembering the watch readings of the breakpoints of the three short elements and then recording their respective values while the fourth, longer element is taking place.

On the other hand, more clerical work is involved in calculating the study if the continuous method is used. Since the watch is read at the breakpoint of each element while the hands of the watch continue their movements, it is necessary to make successive subtractions of the consecutive readings to determine the elapsed elemental times. For example, the following readings might represent the breakpoints of a 10-element study: 4, 14, 19, 121, 25, 52, 61, 76, 211, 16. The elemental values of this cycle would be 4, 10, 5, 102, 4, 27, 9, 15, 35, and 5. Figure 8–9 illustrates a completed time study of the same die casting operation using the continuous method.

DataMyte Data Collector

A typical time study using the DataMyte 1000 Data Collector would be as follows: Assume the work measurement study covers 20 cycles, lasting about 45 minutes, for the spot welding of a special SAE 1112 steel hinge to a cold-rolled-steel tank.

FIGURE 8–9
Continuous timing study of a die-casting operation (rating every cycle).

Time Study Observation Form

Study No.: 2-85	Date: 3-1	Page 1 of 1
Operation: DIE CASTING	Operator: B. JONES	Observer: AF

Element No. and Description		1 REMOVE PART FROM DIE, LUBRICATE DIE, INSPECT				2 PLACE PART IN FIXTURE, TRIM ASIDE PART																			
Note	Cycle	R	W	OT	NT	R	W	OT	NT	R	W	OT	NT	R	W	OT	NT	R	W	O	NT	R	W	OT	NT
	1	90	90	30	270	90	113	23	207																
	2	100	40	27	270	100	61	21	210																
	3	90	92	31	279	90	215	23	207																
	4	85	50	35	298	100	70	20	200																
	5	100	98	28	280	100	318	20	200																
	6	110	43	25	275	110	61	18	198																
	7	90	92	31	279	90	416	24	216																
	8	100	44	28	280	85	68	24	204																
	9	90	500	32	288	90	23	23	207																
	10	110	49	26	286	105	68	19	200																
	11																								
	12																								
	13																								
	14																								
	15																								
	16																								
	17																								
	18																								

Summary						
Total OT	2.93	2.15				
Rating	—	—				
Total NT	2.805	2.049				
No. Observations	10	10				
Average NT	.281	.205				
% Allowance	17	17				
Elemental Std.Time	.329	.205				
No. Occurences	1	1				
Standard Time	.329	.205				

Total Standard Time (sum standard time for all elements): .569

Foreign Elements				
Sym	W1	W2	OT	Description
A				
B				
C				
D				
E				
F				
G				

Time Check		
Finishing Time		3:48,00
Starting Time		3:42,00
Elapsed Time		6.00
TEBS	.60	
TEAF	.32	
Total Check Time	.92	
Effective Time	5.08	
Ineffective Time	0	
Total Recorded Time		6.00
Unaccounted Time		0
Recording Error %		0

Allowance Summary	
Personal Needs	5
Basic Fatigue	4
Variable Fatigue	8
Special	—
Total Allowance %	17
Remarks:	

Rating Check		
Synthetic Time		%
Observed Time		

The analyst, after observing the method, breaks the job down into the following elements:

Element 1: Get and bring tank (on conveyor) to work area.
Element 2: Pick up hinge and place in fixture.
Element 3: Position overhead spot welder and spot weld six spots.
Element 4: Open fixture and move assembled tank 10 feet on conveyor.

During the course of the study, four foreign elements took place:

Element A: Interrupted by supervisor.
Element B: Left workstation to get drink of water.
Element C: Delayed in getting tank on conveyor.
Element D: Sorted out and discarded defective hinge.

In taking the study, the analyst should:

1. Turn on the instrument and set it to input mode 1, data-plus-time.
2. Key in a study identification code of up to 12 characters.
3. Key in a 6-character date.
4. Key in a 4-character begin time.
5. Key in 1 (element 1) key "enter" at end of element.
6. Key in 2 (element 2) key "enter" at end of element.
7. Key in 3 (element 3) key "enter" at end of element.
8. Key in 4 (element 4) key "enter" at end of element.

This process would continue for the 20 cycles. The foreign elements are keyed in as they occur and "enter" is hit at their termination. The elapsed time is logged automatically every time "enter" is hit. If the analyst is assigning an overall performance rating factor, it would be entered at the end of the study. However, performance ratings can be entered at any time—at the end of each element, at the end of each cycle, and so forth. Figure 8–10 illustrates Data Myte output from a food loading study.

Addressing Difficulties

During the time study, analysts may observe variations from the element sequence originally established. Occasionally, analysts may miss specific breakpoints. These difficulties complicate the study; the less often they occur, the easier it is to calculate the study.

When missing a reading, the analyst should immediately indicate an "M" in the W column. In no case should the analyst approximate and endeavor to record the missed value. This practice can destroy the validity of the standard established for the specific element. If the element were then to be used as a source of standard data, appreciable discrepancies in future standards might result. Occasionally, the operator omits an element; this is handled by drawing a horizontal line through the space in the W column. Hopefully, this should only happen infrequently, since it is

FIGURE 8–10
Output from study of loading food items onto a conveyor. (*Courtesy:* Nabisco Brands, Inc.)

```
                    Time Study programs by DataMyte Corporation
                           Model 1031-04    version 07

                              Time Study Summary

Description  TIME STUDY OF SALES BRANCH ORDER PICKER

Study Id: PICKER11
Date: 06/23/86
Begin Time: 1230 (12:30 PM)

********** Deletions - time not used in summary ********************************

                         Time Deleted:    0.00

********** Summary *************************************************************
   obs      raw       rated     min    max     ave      rate              nrm/occ
 1 PICK ONE CASE FROM BIN                                                  0.0771
   211   16.2700   16.2700    0.02   0.34   0.0771   100.0
 2 PICK TWO CASES FROM BIN                                                 0.1158
    19    2.2000    2.2000    0.02   0.34   0.1158   100.0
 3 PICK ONE CASE FROM PALLET                                               0.0805
   147   11.8300   11.8300    0.02   0.40   0.0805   100.0
 4 PICK TWO CASES FROM PALLET                                              0.0778
     9    0.7000    0.7000    0.03   0.21   0.0778   100.0
 5 PICK ONE CASE FROM END OF LINE                                          0.1588
    39    6.8800    6.1920    0.02   0.62   0.1764    90.0
 6 PICK THREE CASES FROM BIN OR PALLET                                     0.0900
     7    0.6300    0.6300    0.05   0.14   0.0900   100.0
 7 PICK FOUR CASES FROM BIN OR PALLET                                      0.0975
     4    0.3900    0.3900    0.04   0.22   0.0975   100.0
 8 PREPARE DISPLAY MATERIAL                                                1.3200
     1    1.3200    1.3200    1.32   1.32   1.3200   100.0
 9 DISCARD DAMAGED CASES                                                   0.2050
     4    0.8200    0.8200    0.11   0.25   0.2050   100.0
10 PAPERWORK/COMMUNICATIONS                                                0.8475
     4    3.3900    3.3900    0.15   2.84   0.8475   100.0
88 IDLE                                                                    3.0067
     3    9.0200    9.0200    0.28   4.52   3.0067   100.0

Totals                                            98.7

Time used:    53.45

Total time:   53.45
```

usually a sign of an inexperienced operator or a lack of standardization in the method. Of course, the operator can inadvertently omit an element, for example, forgetting to "vent cope" in making a bench mold. If elements are omitted repeatedly, the analyst should stop the study and investigate the necessity of performing the omitted elements. This should be done in cooperation with the supervisor and the operator, so that the best method can be established. The observer is expected to be on the alert constantly for better ways to perform the elements; as ideas come to mind, the observer should jot them down in the "note" section of the time study form, for future evaluation.

The observer may also see elements performed out of sequence. This happens fairly frequently when a new or inexperienced employee is studied on a long-cycle job made up of many elements. Avoiding such disturbances is one of the prime reasons for studying a competent, fully trained employee. However, when elements are performed out of order, the analyst should immediately go to the element being performed and draw a horizontal line through the middle of its W space. Directly below this line, the analyst should write the time the operator began the element, and above it, the completion time. This procedure should be repeated for each element performed out of order, as well as for the first element performed back in the normal sequence.

During a time study, the operator may encounter unavoidable delays, such as an interruption by a clerk or supervisor, or tool breakage. The operator may also intentionally cause a change in the order of work by going for a drink of water or stopping to rest. Such interruptions are referred to as "foreign elements."

Foreign elements can occur either at the breakpoint or during the course of an element. The majority of foreign elements, particularly those controlled by the operator, occur at the termination of an element. If a foreign element occurs during an element, it is signified by the letters A, B, C, etc. in the NT column of this element. If the foreign element occurs at the breakpoint, it is recorded in the NT column of the work element that follows the interruption (⑤ in Figure 8–11). The letter A is used to signify the first foreign element, the letter B to signify the second, and so on.

As soon as the foreign element has been properly designated, the analyst should write a short description of it in the lower left-hand corner of the space. The time that the foreign element begins is entered in the W1 block of the foreign element section, and the time it ends is entered in the W2 block. These values can then be subtracted when the time study is calculated, to determine the exact duration of the foreign element. This value is then entered in the OT column of the foreign element section. Figure 8–11 illustrates the correct handling of a foreign element.

Occasionally, a foreign element is of such short duration that it is impossible to record the foreign element in the fashion outlined. Typical examples of this would be dropping a wrench on the floor and quickly picking it up, wiping one's brow with a handkerchief, or turning to speak briefly to the supervisor. In such cases, where the foreign element may be 0.06 minute or less, the most satisfactory method of handling the interruption is to allow it to accumulate in the element and immediately circle the reading, indicating that a "wild" value has been encountered. A short comment should be entered in the "note" section across from the element in which the interruption occurred, justifying the circled number. Cycle 7 in Figure 8–12 illustrates the correct handling of a wild value.

Cycles in Study

Determining how many cycles to study to arrive at an equitable standard is a subject that has caused considerable discussion among time study analysts, as well as union representatives. Since the activity of the job, as well as its cycle time, directly

FIGURE 8–11
Summary of steps in performing and computing a time study.

Time Study Observation Form

Study No.: 1-3	Date: 3-22-	Page 1 of 1
Operation: MACHINING	Operator: J. SMITH	Observer: AF

Element No. and Description		1 FEED BAR TO STOP (3) (4)				2 INDEX, FEED CUTTING TOOL TO BAR				3 TURN 1½" 550 RPM (5)				4 WITHDRAW TOOL AND BAR SET DOWN											
Note	Cycle	R	W	OT	NT	R	W	OT	NT	R	W	OT	NT	R	W	OT	NT	R	W	O	NT	R	W	OT	NT
	1	85		19	162	105		12	126	100		60	(11) 600	90		17	153								
	2	90		22	198	105		13	137	100		60	600	100		16	160								
	3	100		17	170	105		11	116	100		60	600	105		17	179								
	4																								
	5				(10)																				
	6																								
	7																								
	8																								
	9																								
	10																								
	11																								
	12																								
	13																								
	14																								
	15																								
	16																								
	17																								
	18																								

Summary	1	2	3	4		
Total OT	.58	.36	1.80	.50		
Rating (3) →	—	—	—	—		
Total NT	.530	.379	1.800	.492		
No. Observations	3	3	3	3		
Average NT	.177	.126	.600	.164		
% Allowance	10	10	10	10		
Elemental Std. Time	.195	.139	.660	.180		
No. Occurences	1	1	1	1		
Standard Time	.195	.139	.660	.180		

Total Standard Time (sum standard time for all elements): 1.174

Foreign Elements

Sym	W1	W2	OT	Description
A	0 (5)	35	35	CHECK DIMENSIONS
B				
C				
D				
E				
F				
G				

Time Check

Finishing Time	(6) →	9:22.00
Starting Time	(1) →	9:16.00
Elapsed Time	(9) →	6,00
TEBS	(2) →	1.86
TEAF	(7) →	.60
Total Check Time	2.46	← (8)
Effective Time	3.24	← (12)
Ineffective Time	.35	← (13)
Total Recorded Time	(14) →	6.05
Unaccounted Time	(15) →	.05
Recording Error %	(16) →	.8%

Allowance Summary

Personal Needs	5
Basic Fatigue	4
Variable Fatigue	1
Special	—
Total Allowance %	10

Remarks: MACHINE CYCLE (ELEMENT #3) TIME = .60 min

Rating Check

Synthetic Time	%
Observed Time	

TABLE 8–2
Recommended Number of Observation Cycles

Cycle time in minutes	Recommended number of cycles
0.10	200
0.25	100
0.50	60
0.75	40
1.00	30
2.00	20
2.00–5.00	15
5.00–10.00	10
10.00–20.00	8
20.00–40.00	5
40.00–above	3

Source: Information taken from the Time Study Manual of the Erie Works of the General Electric Company, developed under the guidance of Albert E. Shaw, manager of wage administration.

influences the number of cycles that can be studied from an economic standpoint, the analyst cannot be completely governed by sound statistical practice that demands a certain sample size based on the dispersion of individual element readings. The General Electric Company has established Table 8–2 as an approximate guide to the number of cycles to observe.

A more accurate number can be established using statistical methods. Since time study is a sampling procedure, averages of samples ($\overline{x}$) drawn from a normal distribution of observations are distributed normally about the population mean m. The variance about the population mean μ equals σ^2/n, where n equals the sample size and σ^2 equals the population variance. Normal curve theory leads to the following confidence interval:

$$\overline{x} \pm z \frac{\sigma}{\sqrt{n}}$$

The preceding equation assumes that the population standard deviation is known. In general, this is not true, but the population standard deviation may be estimated by the sample standard deviation s, where:

$$s = \sqrt{\frac{\sum_{i=1}^{i=n} (x_i - \overline{x})^2}{n - 1}}$$

However, time studies involve only small samples ($n < 30$) of a population; therefore, a t-distribution must be used. The confidence interval equation is then:

$$\overline{x} \pm t \frac{s}{\sqrt{n}}$$

The $\pm$ term can be considered an error term expressed as a fraction of $\overline{x}$:

$$k\overline{x} = ts/\sqrt{n}$$

EXAMPLE 8–1
Calculation of Required Number of Observations.

A pilot study of 25 readings for a given element showed that $\bar{x} = 0.30$ and $s = 0.09$. A 5-percent probability of error for 24 degrees of freedom (25 minus 1 degree of freedom for estimating one of the parameters) yields $t = 2.064$. (See Table A3-3, Appendix 3, for values of t.) Solving the last equation yields:

$$n = \left\{\frac{0.09 \times 2.064}{0.05 \times 0.30}\right\}^2 = 153.3 \approx 154 \text{ observations}$$

To ensure the required confidence, always round up.

Solving for n yields:

$$n = \left\{\frac{st}{k\bar{x}}\right\}^2$$

It is also possible to solve for n before taking the time study by interpreting historical data of similar elements, or by actually estimating $\bar{x}$ and s from several snap-back readings with the highest variation.

Analysts must decide when and how to observe the recommended number of cycles. If 30 cycles are to be taken, should they be taken as one successive group, as two groups of 15, or as three groups of 10? The mean of three groups of 10 taken at random times during the day probably gives a better estimate of the population mean than one group of 30. A time study is a sampling procedure, and the average of several small samples usually provides more reliable estimates of parent parameters than does one sample of a size equivalent to the total of the small samples.

OPERATOR PERFORMANCE

Since the actual time required to perform each element of the study depends to a high degree on the skill and effort of the operator, it is necessary to adjust upwards to normal the time of the good operator and the time of the poor operator downwards to normal. Therefore, before leaving the workstation, analysts should give a fair and impartial performance rating to the study. On short-cycle, repetitive work, it is customary to apply one rating to the entire study, or an average rating for each element (see Figure 8–12). However, where the elements are long and entail diversified manual movements, it is more practical to evaluate the performance of each element as it occurs. This was done for the die casting study shown in Figures 8–8 and 8–9, in which the elements were over 0.20 minutes in duration. The time study form includes provisions for both the overall rating and the individual element rating.

FIGURE 8–12
Time study with overall rating.

Time Study Observation Form

Study No.: 14	Date: 3/15/	Page 1 of 2
Operation: DIE CASTING	Operator: RAINBOW	Observer: P. ROCHE

Element No. and Description	1 PICK UP & PLACE CASTING IN FIXTURE, DEPRESS 2 PARTS				2 OPEN FIXTURE, GET PART, TURN 90°, PLACE IN 2ND FIXTURE				3 ENGAGE FEED, OPEN FIXTURE, REMOVE PART			
Note / Cycle	R	W	OT	NT	R	W	OT	NT	R	W	OT	NT
1												
2		62	12			78	16			88	10	
3		604	16			21	17			30	9	
4		43	13			59	16			70	11	
5		828	15	Ⓐ		49	21			58	9	
6		71	13			91	20			905	14	
DROPPED CASTING 7		30	(25)			46	16			57	11	
8		70	13			88	18			1002	14	
9		15	13			32	17			40	8	
10		52	12			68	16			78	10	
11		92	14			1112	20			24	12	
12		38	14			56	18			66	10	
13		81	15			1200	19			11	11	
14		25	14			41	16			50	9	
15		63	13			80	17			91	11	
16		1305	14			24	19			34	10	
17		50	16			69	19			83	14	
18												

Element No. and Description	S-1 CLEAN WORKSTATION				S-2 PUNCH-IN				S-3 SET UP STOPS IN FIXTURE				S-4 PUNCH OUT TALLY PRODUCTION			
Cycle	R	W	OT	NT	R	W	OT	NT	R	W	OT	NT	R	W	OT	NT
1		132	132			182	50			415	233			550	135	

Summary	1	2	3	S-1	S-2	S-3	S-4
Total OT	2.07	2.85	1.74	1.32	.50	2.33	1.35
Rating	110	110	110	110	110	110	110
Total NT	2.277	3.135	1.914	1.452	.550	2.563	1.485
No. Observations	15	16	16	1	1	1	1
Average NT	.152	.196	.120	1.452	.550	2.563	1.485
% Allowance	12	12	12	12	12	12	12
Elemental Std. Time	.170	.219	.134	1.626	.616	2.867	1.663
No. Occurences	1	1	1				
Standard Time	.170	.219	.134				

Total Standard Time (sum standard time for all elements): .523

Foreign Elements

Sym	W1	W2	OT	Description
A	670	813	143	TALK TO SUPERVISOR
B				
C				
D				
E				
F				
G				

Rating Check

Synthetic Time	%
Observed Time	

Time Check

Finishing Time	2:11.00
Starting Time	2:25.00
Elapsed Time	14.00
TEBS	0
TEAF	.17
Total Check Time	.17
Effective Time	12.10
Ineffective Time	1.43
Total Recorded Time	14.00
Unaccounted Time	0
Recording Error %	0

Allowance Summary

Personal Needs	5
Basic Fatigue	4
Variable Fatigue	3
Special	–
Total Allowance %	12

Remarks: STANDARD TIME PER PIECE WITHOUT SET UP TIME

In the performance rating or leveling system, the observer evaluates the operator's effectiveness in terms of a "normal" operator performing the same element. The rating value is expressed as a decimal or percentage and is assigned to the observed element, in the R column (③ in Figure 8–11). A "normal" operator is defined as a qualified, thoroughly experienced operator working under customary conditions at the workstation, at a pace neither too fast nor too slow, but representative of average.

The basic principle of performance rating is to adjust the mean observed time (OT) for each element performed during the study to the *normal time* (NT) that would be required by the normal operator to perform the same work:

$$NT = OT \times R/100$$

where R is expressed as a percentage, with 100 percent being standard performance by a normal operator. To do a fair job of rating, the time study analyst must be able to disregard personalities and other varying factors, and consider only the amount of work being done per unit of time, as compared to the amount of work that the normal operator would produce. Chapter 9 more fully explains the performance rating techniques in common use.

ALLOWANCES

No operator can maintain an average pace every minute of the working day. Three classes of interruptions can take place, for which extra time must be provided. The first is personal interruptions, such as trips to the restroom and drinking fountain; the second is fatigue, which can affect even the strongest individual on the lightest work. Finally, there are unavoidable delays, such as tool breakage, supervisor interruptions, slight tool trouble, and material variations, all of which require that some allowance be made. Since the time study is taken over a relatively short period, and since foreign elements should have been removed in determining the normal (leveled) time, an allowance must be added to the normal time to arrive at a fair standard that can reasonably be achieved by an operator. The time required for an average, fully qualified, trained operator, working at a normal pace and exerting average effort, to perform the operation is termed the *standard time* (ST) for that operation. The allowance is typically given as a percentage or a fraction of normal time and is used as a multiplier equal to 1 + allowance:

$$ST = NT + NT \times \text{allowance} = NT \times (1 + \text{allowance})$$

Chapter 10 details the means for arriving at realistic allowance values.

STUDY CALCULATIONS

After properly recording all the necessary information on the time study form, observing an adequate number of cycles, and performance rating the operator, the analyst should record the *finishing time* (⑥ in Figure 8–11) at the same master clock used for the start of the study. For continuous timing, it is very important to verify the final stopwatch reading with the overall elapsed clock reading. These two values should be reasonably close (±2 percent difference). (A sizable discrepancy may mean an error that has occurred, and the time study may need to be repeated.) Finally, the analyst should thank the operator and proceed to the next step, the study computations.

For the continuous method, each watch reading must be subtracted from the preceding reading to get the elapsed time; this value is then recorded in the OT column. Analysts must be especially accurate in this phase, because carelessness at this point can completely destroy the validity of the study. If the elemental performance rating was used, the analyst must multiply the elapsed elemental times by the rating factor and record the result in the NT spaces. Note that since NT is a calculated value, it is typically recorded with three digits.

Elements that have been missed by the observer are signified by an "M" in the W column and are disregarded. Thus, if the operator happened to omit element 7 of cycle 4 in a 30-cycle study, the analyst would have only 29 values of element 7 with which to calculate the mean observed time. The analyst should not only disregard the missed element but also the succeeding one, since the subtracted value in the study would include the time for performing both elements.

To determine the elapsed elemental time on out-of-order elements, it is merely necessary to subtract the appropriate watch times.

For foreign elements, the analyst must deduct the time required for the foreign element from the cycle time of the applicable element. The analyst can obtain the time taken by the foreign element by subtracting the W1 reading in the foreign element section of the time study form from the W2 reading.

After all elapsed times have been calculated and recorded, the analyst should study them carefully for any abnormality. There is no set rule for determining the degree of variation permitted and still keep the value for calculation. If a broad variation on a certain element can be attributed to some influence that was too brief to be handled as a foreign element, yet long enough to affect the time of the element substantially, or if the variation may be attributed to errors in reading the stopwatch, then these values should be immediately circled and excluded from further consideration in working up the study. (For example, in Cycle 7 of element 1 in Figure 8–12, the operator dropped a casting.) However, if wide variations are due to the nature of the work, then it would not be wise to discard any of the values.

Machine elements have little variation from cycle to cycle, while considerably wider variation could be expected in manual elements. When unexplainable time variations occur, the analyst should be quite careful before circling such values. Remember that this is not a performance rating procedure. By arbitrarily discarding high or low values, the analyst may end up with an incorrect standard. A good rule is, "When in doubt, do not discard the value."

If elemental rating is used, then after the elemental elapsed time values have been computed, the analyst should determine the normal elemental time by multiplying each elemental value by its respective performance factor. This normal time is then recorded in the NT columns for each element (in Figure 8–11). Next, the analyst would determine the mean elemental normal value by dividing the total of the times recorded in the NT columns by the number of observations.

After determining all elemental elapsed times, the analyst should check to ensure that no arithmetic or recording errors have been made. One method of checking for such accuracy is to complete the *time check* of the time study form (see Figure 8–11). To do this, though, the analyst needed to have synchronized the starting and stopping of the watch at a master clock, recording the *starting time* (①) and the *finishing time* (⑥) on the form. The analyst then sums three quantities: (1) total observed times, known as *effective time* (⑫ on the form); (2) total foreign elements time, known as *ineffective time* (⑬ on the form); and (3) total of *time elapsed before study* (② on the form) and *time elapsed after study* (⑦ on the form). The time elapsed before study is the readout when the analyst snaps the watch at the start of the first element. The time elapsed after study is the last readout when the analyst snaps the watch at the very end of the study. These last two quantities are sometimes totaled, forming the *check time* (⑧ on the form). The three quantities together equal the *total recorded time* (⑭ on the form). The difference between the finishing and starting times on the master clock equals the actual *elapsed time* (⑨ on the form). Any difference between the total recorded time and the elapsed time is called *unaccounted time* (⑮ on the form). Normally, in a good study, this value would be zero. The unaccounted time divided by the elapsed time is a percentage called the *recording error*. This recording error should be less than 2 percent. If it exceeds 2 percent, the time study should be repeated.

After the normal elemental times have been calculated, the analyst should add the percentage allowance to each element to determine the allowed or standard time. In the time study of Figure 8–12, the normal time for element 1 is multiplied by 1.12, to yield the following elemental standard time for element 1:

$$ST = 0.152 \times (1 + 0.12) = 0.170$$

The nature of the job determines the amount of allowance to be applied, as discussed in Chapter 10. Suffice it to say at this point that 15 percent is the average allowance for manual elements, and 10 percent is the allowance usually applied to machine elements.

In most cases, each element occurs once within each cycle and the *No. Occurrences* is simply 1. In some cases, an element may be repeated within a cycle. In that case, *No. Occurrences* becomes 2 or 3, and the time accrued by that element within the one cycle is doubled or tripled.

The standard times for each element are then summed to obtain the standard time for the entire job, which is recorded in the space labeled *Total Standard Time* on the time study form.

THE STANDARD TIME

The sum of the elemental allowed times gives the standard in minutes per piece, using a decimal minute watch, or hours per piece, using a decimal hour watch. The majority of industrial operations have relatively short cycles (less than 5 minutes); consequently, it is sometimes more convenient to express standards in hours per hundred pieces. For example, the standard on a press operation might be 0.085 hour per hundred pieces. This is a more satisfactory method of expressing the standard than 0.00085 hour per piece or 0.051 minute per piece.

The percent efficiency of the operator can be expressed as:

$$E = 100 \times H_e/H_c$$

where: E = Percent efficiency
H_e = Standard hours earned
H_c = Clock hours on job

Thus, an operator producing 10,000 pieces during the working day would earn 8.5 hours of production, and would perform at an efficiency of 8.5/8 = 106 percent.

Once the allowed time has been computed, the standard is given to the operator in the form of an operation card. The card can either be computer generated or run off on a copier. The operation card serves as the basis for routing, scheduling, instruction, payroll, operator performance, cost, budgeting, and other necessary controls for the effective operation of a business. Figure 8–13 illustrates a typical production operation card.

Temporary Standards

Employees require time to become proficient in any new or different operation. Frequently, time study analysts establish a standard on a relatively new operation on which there is insufficient volume for the operator to reach top efficiency. If the analyst bases operator grading on the usual conception of output (i.e., rating the operator below 100), the resulting standard may seem unduly tight, and the operator will probably be unable to make any incentive earnings (see Chapter 17). On the other hand, if the analyst considers that the job is new and the volume is low, and establishes a liberal standard, then if the size of the order is increased, or if a new order for the same job is received, trouble may occur.

Perhaps the most satisfactory method of handling such situations is the issuance of temporary standards. The analyst establishes the standard by giving consideration to the difficulty of the work assignment and the number of pieces to be produced. Then, by using a learning curve (see Chapter 18) for the work, as well as the existing standard data, the analyst can develop an equitable temporary standard for the work. The resulting standard will be considerably more liberal than if the job involved a large volume. When released to the production floor, the standard is clearly marked "temporary," and will include the maximum quantity for which it applies. When temporary standards are released, they should only be in effect for the duration of the contract, or for 60 days, whichever is shorter. Upon expiration, they should be replaced by permanent standards.

EXAMPLE 8–2
Calculation of Hours Earned and Percent Efficiency.

The standard time for an operation is 11.46 minutes per piece. In an 8-hour shift, the operator would be expected to produce:

$$\frac{8 \text{ hr} \times 60 \text{ min/hr}}{11.46 \text{ min/piece}} = 41.88 \text{ pieces}$$

However, if the operator produced 53 pieces in a given working day, the standard hours earned would be:

$$\mathrm{H_e} = \frac{53 \text{ pieces} \times 11.46 \text{ min/piece}}{60 \text{ min/hr}} = 10.123 \text{ hours}$$

The standard (S_h) expressed in hours per hundred pieces (C) is:

$$S_h = \frac{11.46 \text{ min/piece} \times 100 \text{ pieces/C}}{60 \text{ min/hour}} = 19.1 \text{ hr/C}$$

The standard hours earned would be:

$$H_e = \frac{19.1 \text{ hr/C} \times 53 \text{ pieces}}{100 \text{ pieces/C}} = 10.123 \text{ hr}$$

The operator's efficiency would be:

$$E = 100 \times 10.123/8 = 126.5\%$$

Setup Standards

The elements of work commonly included in setup standards involve all events that take place between completion of the previous job and the start of the present job. The setup standard also includes "teardown" or "put-away" elements, such as punching in on the job, getting tools from the tool crib, getting drawings from the dispatcher, setting up the machine, punching out on the job, removing tools from the machine, returning tools to the crib, and tallying production. Figures 8–12 and 8–14 illustrate a study that involved four setup elements (S-1, S-2, S-3, and S-4).

In establishing setup times, the analyst should use the identical procedure followed in establishing production standards, except that there will be no opportunity to get a series of elemental values for determining the mean times. Also, the analyst cannot observe the operator performing the setup elements in advance; consequently, they are obliged to divide the setup into elements while the study is taking place. However, since setup elements, for the most part, are long duration, there is a reasonable amount of time to break the job down, record the time, and evaluate the performance as the operator proceeds from one work element to the next.

There are two ways of handling setup times distributed by quantity, or allocated by job. In the first method, they would be distributed over a specific manufacturing

FIGURE 8–13

Typical production operation card. (The "F" numbers refer to the fixtures used with the involved operation.)

PRODUCTION OPERATION CARD

DESCRIPTION Shower head face DWG. NO. JB-1102 PART NO. J-1102-1

MADE FROM 2½" diam. 70-30 extruded brass rod

Routing 9-11-12--14-12-18 DATE 9-15

OP. NO.	OPERATION	DEPT.	MACHINE AND SPECIAL TOOLS	SET-UP MINUTES	EACH PC. MINUTES
1	Saw slug	9	J.& L. Air Saw	15 min	.077
2	Forge	11	150 Ton Maxi F-1102	70 min	.234
3	Blank	12	Bliss 72 F-1103	30 min	.061
4	Pickle	14	HCL. Tank	5 min	.007
5	Pierce 6 holes	12	Bliss 74 F-1104	30 min	.075
6	Rough ream and chamfer	12	Delta 17" D.P. F-1105	15 min	.334
7	Drill 13/64" holes	12	Avey D.P. F-1106	15 min	.152
8	Machine stem and face	12	#3 W. & S.	45 min	.648
9	Broach 6 holes	12	Bliss 74½	30 min	.167
10	Inspect	18	F-1109,F-1110,F-1112	Daywork	

FIGURE 8–14

Back of the time study form shown in Figure 8–12.

The study indicates a piece time of 0.523 minute (0.8717 hours/hundred) and a setup time of 6.772 minutes (0.1129 hour).

SKETCH

Parts to be processed; Lock & unlock buttons; Saw; Power actuating; Engage feed; Broach; Finished parts; 16"; 16"; operator

STUDY NO. 14 DATE 3/15/

OPERATION Lug saw & center cut die casting

DEPT. F-114 OPERATOR Rainbow NO. 127

EQUIPMENT Power air saw + piercing broach

MCH. NO. F 114-146

SPECIAL TOOLS, JIGS, FIXTURES, GAGES J-1117-9 & J-1117-10

CONDITIONS Operator standing, clean, average noise, adequate light

MATERIAL Aluminum 380

PART NO. A-1117 DWG. NO. C-1117

PART DESCRIPTION Bracket for mounting assembly

ACT BREAKDOWN

LEFT HAND	RIGHT HAND
Idle	Get casting (16"). Move to fixture, place in fixture,
Press pneumatic locking actuator	Idle
Idle	Press actuating button for power saw
Press unlocking actuator	Idle
Idle	Get part from fixture.
Turn body 60°	Turn body 60°
Lock fixture	Place part in piercing fixture
Idle	Reach to control. Engage feed
Broach 10	Broach 10.
Move to control to open fixture. Engage	Idle
Idle	Get completed part, move 16" to tote box, place in tote box

ELEM. NO.	SMALL TOOL NUMBERS, FEEDS, SPEEDS, DEPTH OF CUT, ETC.	ELEMENTAL TIME	OCC. PER CYCLE	TOTAL TIME ALLOWED
1	J-1117-9, 3600 r.p.m.	0.170	1	0.170
2	J-1117-10, 30'/min	0.222	1	0.222
3		0.131	1	0.131
	Total (each piece)			0.523
S-1		1.626		1.626
S-2		0.616		0.616
S-3		2.867		2.867
S-4		1.663		1.663
	Total (setup)			6.772

EACH PIECE 0.523 min. TOTAL

SET-UP 0.1129 hrs HRS. PER C 0.8717

FOREMAN Eugene Reiter INSPECTOR Jerome Gates

OBSERVER Paul Roche APPROVED BY Dick Henshaw

quantity, such as 1,000 or 10,000 pieces. This method is only satisfactory when the magnitude of the production order is standard. For example, industries that ship from stock and reorder on a basis of minimum–maximum inventories are able to control their production orders to conform to economical lot sizes. In such cases, the setup time can be equitably prorated over the lot size. For example, suppose that the economical lot size of a given item is 1,000 pieces and that reordering is always done on the basis of 1,000 units. If the standard setup time in a given operation is 1.50 hours, then the allowed operation time could be increased by 0.15 hour per 100 pieces to take care of the make-ready and put-away elements.

This method is not at all practical if the size of the order is not controlled. In a plant that requisitions on a job-order basis, that is, releases production orders specifying quantities in accordance with customer requirements, it is impossible to standardize on the size of the work orders. For example, this week an order for 100 units may be issued, and next month an order for 5,000 units of the same part may be needed. In the example, the operator would only be allowed 0.15 hour to set up the machine for the 100-unit order, which would be inadequate. On the 5,000-unit order, however, the operator would be given 7.50 hours, which would be considerably too much time.

It is more practical to establish setup standards as separate allowed times (see Figures 8–12 and 8–14). Then, regardless of the quantity of parts to be produced, a fair standard would prevail. In some concerns, the setup is performed by a person other than the operator who does the job. The advantages of having separate setup personnel are quite obvious. Lower-skilled workers can be utilized as operators when they do not have to set up their own facilities. Setups are more readily standardized and methods changes more easily introduced when the responsibility for setup rests with one individual. Also, when sufficient facilities are available, production can be continuous if the next work assignment is set up while the operator is working on the present job.

Partial Setups

Frequently, it is not necessary to set up a facility completely to perform a given operation, because some of the tools of the previous operation are required in the job being set up. For example, in hand-screw machine or turret lathe setups, careful scheduling of similar work for the same machine allows partial setups from one job to be used for the next. Instead of having to change six tools in the hex turret, it may only be necessary to change two or three. This savings in setup time is one of the principal benefits of a well-formulated group technology program.

Since the sequence of work scheduled for a given machine seldom remains the same, it is difficult to establish partial setup times to cover all possible variations. For example, the standard for a complete setup for a given No. 4 Warner & Swasey turret lathe might be 0.80 hour. However, if this setup is performed after job X, it might take only 0.45 hour; after job Y, it may require 0.57 hour; while following job Z, 0.70 hour may be necessary. The variations possible in partial setup time are so broad that the only practical way to establish their values accurately is to use standard data (see Chapter 11) for each job.

In plants where setup times are less than one hour and production runs are reasonably long, it is common practice to allow operators the full setup time for each job performed. This is advantageous for several reasons: first, if the plant incorporates wage incentives, operators are considerably more satisfied because of higher earnings, and they plan their work to the best possible advantage. This results in more production per unit of time and in lower total costs. Also, considerable time and paperwork are saved by avoiding having to determine a standard for the partial setup operation and its application in all pertinent cases. In fact, this saving tends to approach the extra amount paid to operators resulting from the difference between the time required to make the complete setup and the time required to make the partial setup.

SUMMARY

To summarize, the steps for performing and computing a typical time study are as follows (see Figure 8–11 for numbers corresponding to these steps):

① Synchronize at the master clock and record the starting time.
② Walk to the operation and start the study. The readout at the snap is the time elapsed before the study.
③ Rate operator performance while the element is taking place, and record either the single or the average rating.
④ Snap the watch at the start of the next element. For continuous timing, enter the readout in the W column; for snapback timing, enter the readout in the OT column, as shown.
⑤ For a foreign element, indicate in the appropriate NT column and record the times in the *Foreign Element* section.
⑥ Once all elements have been timed, snap the watch at the master clock. Record the finishing time.
⑦ Record the readout as the time elapsed after the study.
⑧ Add ② and ⑦ to obtain the check time.
⑨ Subtract ① from ⑥ to obtain the elapsed time.
⑩ Calculate the normal time by multiplying the observed time by the rating.
⑪ Sum all observed times and the normal times for each element. Find the average normal time.
⑫ Add all observed time totals to obtain the effective time.
⑬ Add all foreign elements to obtain the ineffective time.
⑭ Add ⑧, ⑫, and ⑬ to obtain the total recorded time.
⑮ Subtract ⑭ from ⑨ to obtain the unaccounted time. Use the absolute value. (The difference could be either negative or positive, and positive numbers are desired.)
⑯ Divide ⑮ by ⑨ to obtain the percent recording error. Hopefully, this value is less than 2 percent!

QUESTIONS

1. How is a fair day's pay determined by wage rate agreements?
2. What benchmark for normal pace is given under these agreements?
3. Why should the supervisor sign the time study?
4. What are the effects of poor time standards?
5. What equipment is needed by the time study analyst?
6. What features of the electronic stopwatch make it attractive to time study analysts?
7. How can the metronome be used as a training tool for performance rating?
8. What considerations should be given to the choice of the operator to be studied?
9. Why is it essential to record complete information on the tools and facility on the time study form?
10. Why are working conditions important in identifying the method being observed?
11. Why would a time study analyst who is hard of hearing have difficulty in performing a time study?
12. Differentiate between constant and variable elements. Why should they be kept separate when dividing the job into elements?
13. What advantages does the continuous method of watch recording offer over the snap-back method?
14. Explain why electronic stopwatches have increased the use of the snapback procedure.
15. Why is the time of day recorded on the time study form?
16. What variations in sequence will the observer occasionally encounter during the course of the time study?
17. Explain what a foreign element is and how foreign elements are handled under the continuous method.
18. What factors enter into the determination of the number of cycles to observe?
19. Why is it necessary to rate the operator?
20. When should individual elements of each cycle be rated?
21. Define a "normal" operator.

22. Why are allowances applied to the normal time?

23. What is the significance of a "circled" elapsed time?

24. What steps are taken in the computation of a time study conducted in accordance with the continuous overall performance rating procedure?

25. How does the walking pace of 3 miles per hour agree with your concept of a normal performance?

26. Define the term "standard time."

27. Why is it usually more convenient to express standards as time per hundred pieces, rather than time per piece?

28. Why would temporary standards be established?

29. Which elements of work are included in the setup standard?

PROBLEMS

1. Based on the Westinghouse guide sheet, how many observations should be taken on an operation for which the annual activity is 750 pieces and the cycle time is estimated at 15 minutes? What would be the number of observations needed, according to the General Electric guide sheet?

2. Take a simple operation that you perform regularly, such as brushing your teeth, shaving, or combing your hair, and estimate the time it takes you to perform that operation. Now measure the time it takes while working at a normal pace. Is your estimate within plus or minus 20 percent of the estimated time?

3. To demonstrate various levels of performance to a group of union stewards, the time study supervisor of the XYZ Company is using the metronome while dealing bridge hands. How many times per minute should the metronome beat to demonstrate the following levels of performance: 60 percent, 75 percent, 100 percent, 125 percent?

4. You used the General Electric guide sheet to determine the number of observations to study. The guide sheet indicated that 10 cycles were required. After taking the study, you used the standard error of the mean to estimate the number of observations needed for a given confidence level. The resulting calculation indicated that 20 cycles should be studied. What would be your procedure? Why?

5. The time study analyst at the Dorben Company developed the following snapback stopwatch readings where elemental performance rating was used. The allowance for this element was assigned a value of 16 percent. What would be the standard time for this element?

Snapback Reading	Performance Factor
28	100
24	115
29	100
32	90
30	95
27	100
38	80
28	100
27	100
26	105

6. What would be the required number of readings if the analyst wanted to be 87 percent confident that the mean observed time was within ± 5 percent of the true mean and the following values were established for an element after 20 cycles were observed: 0.09, 0.08, 0.10, 0.12, 0.09, 0.08, 0.09, 0.12, 0.11, 0.12, 0.09, 0.10, 0.12, 0.10, 0.08, 0.09, 0.10, 0.12, 0.09?

7. The following data resulted from a time study taken on a horizontal milling machine:
Mean manual effort time per cycle: 4.62 minutes.
Mean cutting time (power feed): 3.74 minutes.
Mean performance rating: 115 percent.
Machine allowance (power feed): 10 percent.
Fatigue allowance: 15 percent.
What is the standard time for the operation?

8. A work measurement analyst in the Dorben Company took 10 observations of a high-production job. He performance rated each cycle and then computed the mean normal time for each element. The element with the greatest dispersion had a mean of 0.30 minutes and a standard deviation of 0.03 minutes. If it is desirable to have sampled data within ±5 percent of the true data, how many observations should this time study analyst take of this operation?

9. In the Dorben Company, the work measurement analyst took a detailed time study of the making of shell molds. The third element of this study had the greatest variation in time. After studying nine cycles, the analyst computed the mean and standard deviation of this element, with the following results:

$\bar{x} = 0.42 \quad s = 0.08$

If the analyst wanted to be 90 percent confident that the mean time of the sample was within ±10 percent of the mean of the population, how many total observations should have been taken? Within what percent of the average of the total population is $\bar{x}$ at the 95 percent confidence level, under the measured observations?

10. Based on the data provided in Figure 8–12, what would be the efficiency of an operator who set up the machine and produced an order of 5,000 pieces in a 40-hour workweek?

11. Establish the money rate per hundred pieces from the following data:
Cycle time (averaged measured time): 1.23 minutes.
Base rate: $8.60 per hour.
Pieces per cycle: 4.
Machine time (power feed): 0.52 minutes per cycle.
Allowance: 17 percent on effort time; 12 percent on power feed time.
Element 1 average time: 0.09 minutes.
MTM time for element 1: 132 TMU (One TMU = 0.00001 hour).
Plant uses synthetic performance rating.

12. The following data resulted from a study taken on a horizontal milling machine:
Pieces produced per cycle: 8.
Average measured cycle time: 8.36 minutes.
Average measured effort time per cycle: 4.62 minutes.
Average rapid traverse time: 0.08 minutes.
Average cutting time power feed: 3.66 minutes.
Performance rating: 115 percent.
Allowance (machine time): 10 percent.
Allowance (effort time): 15 percent.
The operator works on the job a full 8-hour day and produces 380 pieces. How many standard hours does the operator earn? What is the operator's efficiency for the 8-hour day?

13. Express the standard of 5.761 minutes in hours per hundred pieces. What would the operator's efficiency be if 92 pieces were completed during a working day? What would the efficiency be if the operator set up the machine (standard for setup = 0.45 hours) and produced 80 pieces during the 8-hour workday?

REFERENCES

Barnes, Ralph M. *Motion and Time Study: Design and Measurement of Work*. 7th ed. New York: John Wiley & Sons, 1980.

Gomberg, William. *A Trade Union Analysis of Time Study*. 2nd ed. Englewood Cliffs, NJ: Prentice Hall, 1955.

Griepentrog, Carl W., and Gilbert Jewell. *Work Measurement: A Guide for Local Union Bargaining Committees and Stewards*. Milwaukee, WI: International Union of Allied Industrial Workers of America, AFL-CIO, 1970.

Lowry, S. M., H. B. Maynard, and G. J. Stegemerten. *Time and Motion Study and Formulas for Wage Incentives*. 3rd ed. New York: McGraw-Hill, 1940.

Mundel, M. E. *Motion and Time Study: Improving Productivity*. 5th ed. Englewood Cliffs, NJ: Prentice Hall, 1978.

Nadler, Gerald. *Work Design: A Systems Concept*. Rev. ed. Homewood, IL: Richard D. Irwin, 1970.

Rotroff, Virgil H. *Work Measurement*. New York: Reinhold Publishing, 1959.

Smith, George, L. *Work Measurement—A Systems Approach*. Columbus, OH: Grid, 1978.

United Auto Workers. *Time Study—Engineering and Education Departments. Is Time Study Scientific?* Publication No. 325. Detroit, MI: Solidarity House, 1972.

CHAPTER 9

Performance Rating

KEY POINTS:

- Use ratings to adjust observed times to those expected with normal performance.
- Speed rating is the fastest and simplest method.
- For a study with long elements, rate each separately.
  - Rate the operator before recording the time.
- For a study with short elements, rate the overall study.
- Practice, practice rating.

During the study, time study analysts carefully observe the performance of the operator. The performance being executed seldom conforms to the exact definition of *normal* or *standard*. Thus, some adjustment must be made to the mean observed time to derive the time required for the normal operator to do the job when working at an average pace. To arrive at the time required by the normal worker, time study analysts must increase the actual time taken by above-standard operators and decrease the time taken by below-standard operators. Only in this manner can they establish a true standard for normal operators.

Performance rating is probably the most important step in the entire work measurement procedure. It is also the step most subject to criticism, since it is based entirely on the experience, training, and judgment of the work measurement analyst. Regardless of whether the rating factor is based on the speed or tempo of the output or on the performance of the operator compared to that of the normal worker, judgment is still the criterion for determining the rating factor. For this reason, analysts must have high personal integrity.

Because of the importance of performance rating, this chapter not only describes the most accepted techniques in use today, but also reviews some of the more important historical methods. An understanding of the historical philosophies and the resulting methods is helpful in developing an appreciation for this very important step.

"NORMAL" PERFORMANCE

There is no universal method of performance rating, and there is no universal concept of "normal" performance. In general, a company engaged in the manufacture of low-cost, highly competitive products has a tighter conception of normal performance than a company producing a line of products protected by patents.

To define normal performance, we use benchmark examples that are familiar to us all. However, method and work requirements must be carefully defined. For example, if the benchmark of dealing 52 cards in 0.50 minute is established, a complete and specific description should be given of the distance of the four hands dealt with respect to the dealer, as well as the technique of grasping, moving, and disposing of the cards. Likewise, if the benchmark of 0.38 minutes is established for walking 100 feet (3 miles per hour or 4.83 km/hr), the definition should specify whether the ground is level, whether a load is being carried, and how heavy the load is. The more clear-cut and specific the definition of normal, the better it is.

The benchmark examples should be supplemented by a clear description of the characteristics of an employee carrying out a normal performance. A representative description of such an employee might be as follows: a worker who is adapted to the work and has attained sufficient experience to perform the job in an efficient manner, with little or no supervision. The worker possesses coordinated mental and physical qualities, enabling him or her to proceed from one element to another without hesitation or delay, in accordance with the principles of motion economy. The worker maintains a good level of efficiency through knowledge and proper use of all tools and equipment related to the job. He or she cooperates and performs at a pace best suited for continuous performance.

Even though personnel departments endeavor to provide only normal or above normal employees for each position within the company, individual differences still exist. These differences can become more pronounced over time. Differences in inherent knowledge, physical capacity, health, trade knowledge, physical dexterity, and training can cause one operator to outperform another consistently and progressively. The variation in the performance of different individuals approximates the ratio of 1 to 2.25. For example, in a random selection of 1,000 employees, the frequency distribution of the output would approximate the normal curve, with less than three cases on average falling outside the three-sigma limits (99.73 percent of the time). If 100 percent should be taken as normal, then based on a ratio of 2.25 to 1 for the slowest operator to the fastest, $x - 3\sigma$ could equal 0.61, and $x + 3\sigma$ would then equal 1.39. This would mean that 68.26 percent of the people would be within a limit of plus or minus one sigma, or between performance rating values of 0.87 and 1.13 (Presgrave, 1957). Graphically, the expected total distribution of the 1,000 people would appear as shown in Figure 9–1.

FIGURE 9–1
Normal distribution curve of the output of 1,000 people selected at random.

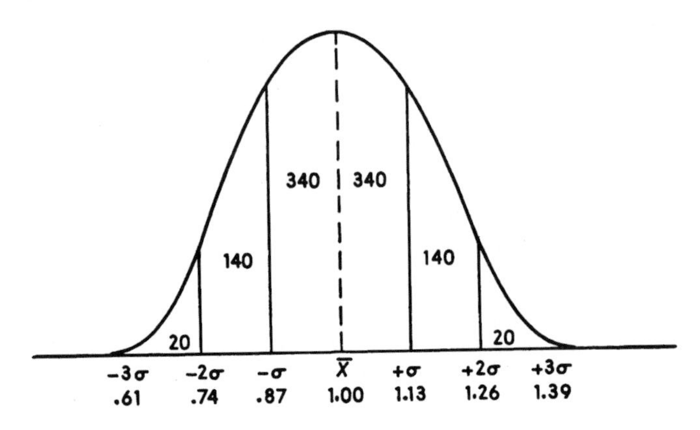

Since employees are usually carefully screened by personnel departments before being assigned to specific jobs, the mean of any selected group may exceed the 100-percent figure representative of a sample taken arbitrarily from the population, and the dispersion of their output may be considerably less than the 2.25-to-1 ratio. In fact, some concerns believe that careful testing programs to select the right person for the job, and intensive training of that employee in the correct method of performance, result in similar output, within close limits, by different operators assigned to the same job. In the majority of cases, however, significant differences in output prevail among those assigned to a given class of work. Therefore, analysts must adjust the performance of the operator being studied to a predetermined concept of normal.

SOUND RATING CHARACTERISTICS

The first and most important characteristic of any rating system is accuracy. Since the majority of rating techniques rely on the judgment of the time study observer, however, perfect consistency in rating is impossible. Yet, rating procedures that readily permit time study analysts to study different operators employing the same method to arrive at standards that do not deviate more than 5 percent from the standard average established by the group are considered adequate. The rating plan with variations greater than ±5 percent should either be improved or replaced.

Other factors being similar, the rating plan that gives the most consistent results is the most useful. Inconsistency, more than anything else, destroys the operator's confidence in the time study procedure. For example, assume that through stopwatch study, an analyst developed a standard that allowed 7.88 hours to mill 100 castings. At some later date, the analyst studied a similar casting entailing a slightly longer cut and established a standard of 7.22 hours. Considerable complaint would be registered by the operator, even though both studies were within the ±5 percent accuracy criterion. If these inequities are within the 5 percent tolerance range, few grievances develop. However, if an operator can earn as much as a 50 percent premium on one job, and not even make standard on another, employees will be dissatisfied over the whole rating program.

Inequities in different rates are not necessarily due entirely to poor performance rating. Some may be due to methods improvements inaugurated without conducting a new time study.

A rating plan that is consistent for all study analysts in a given plant, yet is outside the accepted definition of normal accuracy, can be corrected. A rating plan that is inconsistent for the different analysts will fail. Time study analysts, who find it difficult to rate consistently, after being trained, should probably not continue to conduct such studies. It is not difficult to correct the rating habits of an analyst who consistently rates high or consistently low, but it is very difficult to correct the rating ability of an analyst who is inconsistent, rating too high today and too low tomorrow. Finally, a rating system that is simple, concise, easily explained, and keyed to well-established benchmarks is more successful than a complex rating system requiring involved adjustment factors and computational techniques that may confuse the average shop employee.

Workstation Rating

Performance rating should only be done during the observation of elemental times. As the operator progresses from one element to the next, using the prescribed method, analysts should carefully evaluate speed, dexterity, false moves, rhythm, coordination, effectiveness, and the other factors influencing output. At this time, the operator's performance compared to normal performance is most clearly evident. Once the performance has been judged and recorded, it should not be changed. However, this does not imply that the observer always has perfect judgment. If the rating is questioned, the job or operation should be restudied to prove or disprove the recorded evaluation.

Immediately after completing the study and recording the final performance factor (if overall rating was used), the observer should advise the operator of the performance rating. Even if elemental rating is used, the analyst can approximate the operator's performance. This gives the operator an opportunity to express his or her opinion about the fairness of the performance factor, and to give this opinion directly to the person responsible for its development. Hopefully, the operator will understand the rating and its implications prior to computation of the standard. Where this procedure is practiced, time study analysts find that they receive greater respect from the operators and that they tend to be more conscientious in their rating of operator performance. Also, considerably fewer rate grievances are submitted, because most rates have been agreed upon, or at least satisfactorily explained, before they are issued.

Rating Elements Versus Overall Study

The question of how frequently to evaluate the performance during the course of a study comes up often. Although no set rule can be established regarding the interval limit that permits concise rating, the more frequently the study is rated, the more accurate is the evaluation of demonstrated operator performance.

On short-cycle repetitive operations, little deviation in operator performance is realized during the course of the average-length study (15 to 30 minutes). In such cases, it is perfectly satisfactory to evaluate the performance of the entire study and record the rating factor for each element. Remember, however, that power-fed or machine-controlled elements are rated normal, or 100, as their speed cannot be changed or modified at will by the operators. In short-cycle studies, an observer who endeavors to performance rate each element in the study will be so busy recording values that he or she will be unable to effectively observe, analyze, and evaluate the operator's performance.

When the study is relatively long (over 30 minutes), or is made up of several long elements, operator performance may vary during the course of the study. In such studies, analysts should periodically evaluate and rate said performance. They can consistently and accurately rate elements longer than 0.10 minute as they occur. However, if a study is comprised of a series of elements shorter than 0.10 minute, then no effort should be made to evaluate each element of each cycle of the study, as time does not permit such action. It is satisfactory to rate the overall time of each cycle or a group of cycles. As examples, the time studies shown in Figures 8–8, 8–9, and 8–11 rate each element, while the time study shown in Figure 8–12 rates the overall study.

Operator Selection

To eliminate the performance rating step when calculating the standard, some companies select the operators and then consider the average observed time to be the normal time. Using this method, an analyst studies more than one operator and observes enough cycles to calculate a reliable average time (within ±5 percent of the population average). The success of this method depends on the selection of the employees and their performance during the study. If the performances of the operators are slower than normal, then too liberal a standard results; conversely, if the observed operators produce at a pace more rapid than normal, then the standard will be unduly tight. There is always the possibility that only one or two operators will be available and they may differ from the norm. If, in an effort to avoid delay in establishing a standard, the observer goes ahead with the study, the result will be a poor time standard.

RATING METHODS

The Westinghouse System

One of the oldest and most widely used rating systems was developed by the Westinghouse Electric Corporation. It is outlined in detail in Lowry, Maynard, and Stegemerten (1940). This method considers four factors in evaluating the performance of the operator: skill, effort, conditions, and consistency.

Lowry, et al., defines skill as "proficiency at following a given method," and further relates it to expertise, as demonstrated by the proper coordination of mind and hands. The skill of an operator results from experience and inherent aptitudes, such as natural coordination and rhythm. Practice develops and contributes to skill, but it cannot entirely compensate for deficiencies in natural aptitude.

A person's skill in a given operation increases over time, because increased familiarity with the work brings speed, smoothness of motions, and freedom from hesitations and false moves. A decrease in skill is usually caused by some impairment of ability, brought about by physical or psychological factors, such as failing eyesight, failing reflexes, and the loss of muscular strength or coordination. Therefore, a person's skill can vary from job to job, and even from operation to operation on a given job.

The *Westinghouse rating system* lists these six skill degrees or classes that represent an acceptable proficiency for evaluation: poor, fair, average, good, excellent, and super. The observer evaluates the skill displayed by the operator and places it in one of these six classes. Table 9–1 illustrates the characteristics of the various degrees of skill, with their equivalent percentage values. The skill rating is then translated into its equivalent percentage value, which ranges from plus 15 percent for superskill to minus 22 percent for poor skill. This percentage is then combined algebraically with the ratings for effort, conditions, and consistency, to arrive at the final rating, or performance rating factor.

TABLE 9–1
Westinghouse System Skill Ratings

+0.15	A1	Superskill
+0.13	A2	Superskill
+0.11	B1	Excellent
+0.08	B2	Excellent
+0.06	C1	Good
+0.03	C2	Good
0.00	D	Average
−0.05	E1	Fair
−0.10	E2	Fair
−0.16	F1	Poor
−0.22	F2	Poor

Source: S. M. Lowry, H. B. Maynard, and G. J. Stegemerten, *Time and Motion Study and Formulas for Wage Incentives,* 3rd ed. (New York: McGraw-Hill, 1940), p. 233.

TABLE 9–2
Westinghouse System Effort Ratings

+0.13	A1	Excessive
+0.12	A2	Excessive
+0.10	B1	Excellent
+0.08	B2	Excellent
+0.05	C1	Good
+0.02	C2	Good
0.00	D	Average
−0.04	E1	Fair
−0.08	E2	Fair
−0.12	F1	Poor
−0.17	F2	Poor

Source: S. M. Lowry, H. B. Maynard, and G. J. Stegemerten, *Time and Motion Study and Formulas for Wage Incentives,* 3rd ed. (New York: McGraw-Hill, 1940), p. 233.

This rating method defines effort as a "demonstration of the will to work effectively." Effort is representative of the speed with which skill is applied, and can be controlled to a high degree by the operator. When evaluating the operator's effort, the observer must rate only the "effective" effort. To explain, on occasion, an operator will apply misdirected high-speed effort to increase the cycle time of the study and yet retain a liberal rating factor.

The six effort classes for rating purposes are: poor, fair, average, good, excellent, and excessive. Excessive effort is assigned a value of plus 13 percent, and poor effort, a value of minus 17 percent. Table 9–2 gives the numerical values for the different degrees of effort and outlines the characteristics of the various categories.

The conditions referred to in this performance rating procedure affect the operator and not the operation. Time study analysts rate conditions as normal or average in more than a majority of instances, as conditions are evaluated in comparison with the way they are customarily found at the workstation. Elements affecting working conditions include temperature, ventilation, light, and noise. Thus, if the temperature at a given workstation is 60° F, yet it is customarily maintained at 68°F to 74°F, the conditions would be rated as lower than normal. Factors that affect the operation, such as poor tools or materials, would not be considered when applying the performance factor for working conditions.

The six general classes of conditions, with values ranging from +6 percent to −7 percent, are: ideal, excellent, good, average, fair, and poor. Table 9–3 gives the respective values for these conditions.

The last of the four factors that influence the performance rating is operator consistency. Unless the analyst uses the snapback method, or makes and records successive subtractions as the study progresses, the consistency of the operator must be evaluated as the study is worked up. Elemental time values that constantly repeat would have perfect consistency. This situation occurs very infrequently, as there always tends to be dispersion due to the many variables, such as material hardness, tool cutting edge, lubricant, operator skill and effort, erroneous watch readings, and

TABLE 9–3
Westinghouse System Condition Ratings

+0.06	A	Ideal
+0.04	B	Excellent
+0.02	C	Good
0.00	D	Average
−0.03	E	Fair
−0.07	F	Poor

Source: S. M. Lowry, H. B. Maynard, and G. J. Stegemerten, *Time and Motion Study and Formulas for Wage Incentives*, 3rd ed. (New York: McGraw-Hill, 1940), p. 233.

TABLE 9–4
Westinghouse System Consistency Ratings

+0.04	A	Perfect
+0.03	B	Excellent
+0.01	C	Good
0.00	D	Average
−0.02	E	Fair
−0.04	F	Poor

Source: S. M. Lowry, H. B. Maynard, and G. J. Stegemerten, *Time and Motion Study and Formulas for Wage Incentives*, 3rd ed. (New York: McGraw-Hill, 1940), p. 233.

foreign elements. Elements that are mechanically controlled would also have near-perfect consistency, but such elements are not rated.

The six classes of consistency are: perfect, excellent, good, average, fair, and poor. Perfect consistency is rated +4 percent, and poor consistency is rated −4 percent, with the other classes falling in between these values. Table 9–4 summarizes these values.

No fixed rule for rating consistency can be cited. Some short-duration operations are free of delicate positioning manipulations and give relatively consistent results from one cycle to the next. Such operations would demand average consistency more than a long-duration job involving high skill in its positioning, engaging, and aligning elements. The time study analyst's knowledge of the work to a large extent determines the justified range of variation for a particular operation.

Some operators consistently perform poorly, in an effort to deceive observers. This is easily accomplished by counting to themselves, setting a pace that can be accurately followed. Operators familiar with this performance rating procedure sometimes function at a consistent pace that is below the effort rating curve. In other words, they may be performing at a pace that is poorer than poor. In such cases, the operator should be rated. The study should be stopped and the situation brought to the attention of the operator or the supervisor or both.

Once the skill, effort, conditions, and consistency of the operation have been assigned, and their equivalent numerical values established, analysts can determine the overall performance factor by algebraically combining the four values and adding their sum to unity. For example, if a given job is rated C2 on skill, C1 on effort, D on conditions, and E on consistency, the performance factor would be as follows:

Skill	C2	+0.03
Effort	C1	+0.05
Conditions	D	+0.00
Consistency	E	−0.02
Algebraic sum		+0.06
Performance factor		1.06

The performance factor is only applied to the effort, or the manually performed elements; all machine-controlled elements are rated 100.

Many companies have modified the Westinghouse system to include only the skill and effort factors of the overall rating. They contend that consistency is very closely allied to skill, and that conditions are rated average in most instances. If conditions deviate substantially from normal, the study could be postponed, or the effect of the unusual conditions could be taken into consideration in the application of the allowance (see Chapter 10).

In 1949, Westinghouse Electric Corporation developed a new rating method called the *performance rating plan.* In addition to using the operator-related physical attributes, the company attempted to evaluate the relationship between those physical attributes and the basic divisions of work.

The characteristics and attributes that the Westinghouse performance rating plan considers are: (1) dexterity, (2) effectiveness, and (3) physical application. The classifications themselves do not carry numerical weight; rather, they have been assigned attributes that do carry numerical weight. Table 9–5 gives the numerical values of the nine attributes evaluated under this system.

The first major classification, dexterity, has three attributes, the first of which is "Displayed ability in use of equipment and tools, and in assembly of parts." For this attribute, the primary concern is with the "do" portion of the work cycle after the "get" operations (reach, grasp, move).

The second attribute under dexterity is "Certainty of movement." In this attribute, the number and degree of hesitations, pauses, or roundabout moves is important. The basic divisions of accomplishment that give the operator a low rating for this attribute are: direction change, plan, and avoidable delay.

The last attribute considered under dexterity is "Coordination and rhythm." This attribute is evidenced by the degree of the displayed performance, the smoothness of the motions, and freedom from spasmodic spurts and lags.

The second major classification, effectiveness, relates to efficient, orderly procedure. The classification has four individual attributes. The first is "Displayed ability to continually replace and retrieve tools and parts with automaticity and accuracy." Here, concern centers on the worker's ability repeatedly to place tools, materials, and parts in specified locations and positions, and to retrieve them automatically and accurately, eliminating such ineffective work efforts as searching and selecting.

TABLE 9–5
Westinghouse Performance Rating Plan

	+ Above		0 Expected	− Below	
DEXTERITY:					
1. Displayed ability in use of equipment and tools, and in assembly of parts.	6	3	0	2	4
2. Certainty of movement.	6	3	0	2	4
3. Coordination and rhythm.		2	0	2	
EFFECTIVENESS:					
1. Displayed ability to continually replace and retrieve tools and parts with automaticity and accuracy.	6	3	0	2	4
2. Displayed ability to facilitate, eliminate, combine, or shorten motions.	6	3	0	4	8
3. Displayed ability to use both hands with equal ease.	6	3	0	4	8
4. Displayed ability to confine efforts to necessary work.			0	4	8
PHYSICAL APPLICATION:					
1. Work pace.	6	3	0	4	8
2. Attentiveness.			0	2	4

The second attribute in effectiveness is "Displayed ability to facilitate, eliminate, combine, or shorten motions." Scrutiny falls on the proficiency of the basic divisions of position: pre-position, release, and inspect. The transport therbligs are usually predetermined by the established method. However, a skilled worker can eliminate or shorten the pre-position, position, and inspect elements through manipulative ability.

The third attribute under effectiveness is "Displayed ability to use both hands with equal ease." Here, the analyst rates the effective utilization of both hands.

The fourth and last attribute under effectiveness is "Displayed ability to confine efforts to necessary work." This attribute rates the presence of unnecessary work that could not be removed when the study was conducted. It carries only a negative weight, because no percentage is added when the work is confined to the necessary work since this condition is expected.

The third major classification, physical application, is the demonstrated rate of performance, and it has two attributes. The first is "Work pace," which compares the actual speed of movement to preconceived standards for the work under consideration. The second attribute for physical application is "Attentiveness," which rates the degree of concentration displayed.

Both of the Westinghouse rating techniques demand considerable training to differentiate the levels of each attribute. The training entails a 30-hour course, during which approximately 25 hours are spent rating videotapes or films, and discussing the attributes and the degree to which each is displayed. The procedure generally followed is:

1. A film is shown and the operation explained.
2. The film or tape is reshown and rated.
3. The individual ratings are compared and discussed.
4. The film or tape is reshown, and the attributes are pointed out and explained.
5. Step 4 is repeated as often as necessary to reach understanding and agreement.

Westinghouse rating procedures are appropriate for either cycle rating or overall study rating. Elemental rating is not practical using either of the Westinghouse systems. Except for very long elements, analysts would not have time to evaluate the dexterity, effectiveness, and physical application of each element.

Synthetic Rating

In an effort to develop a rating method that would not rely on the judgment of a time study observer and would give consistent results, Morrow (1946) established a procedure known as *synthetic rating*. The synthetic rating procedure determines a performance factor for representative effort elements of the work cycle by comparing actual elemental observed times to times developed through fundamental motion data (see Chapter 13). Thus, the performance factor may be expressed algebraically as:

$$P = \frac{F_t}{O}$$

where: P = Performance or rating factor.
F_t = Fundamental motion time.
O = Observed mean elemental time for the elements used in F_t.

This factor would then be applied to the remainder of the manually controlled elements comprising the study. Again, machine-controlled elements are not rated. A typical illustration of synthetic rating appears in Table 9–6.
For element 1,

$$P = 0.096/0.08 = 120 \text{ percent}$$

and for element 4,

$$P = 0.278/0.22 = 126 \text{ percent}$$

The mean of these is 123 percent, which is the rating factor used for all effort elements.

More than one element should be used to establish a synthetic rating factor, because research has proved that operator performance varies significantly from element to element, especially in complex work. Unfortunately, a major objection to

TABLE 9–6
Examples of Synthetic Ratings

Element no.	Observed average time in minutes	Element type	Fundamental motion time in minutes	Performance factor
1	0.08	Manual	0.096	123
2	0.15	Manual	—	123
3	0.05	Manual	—	123
4	0.22	Manual	0.278	123
5	1.41	Power fed	—	100
6	0.07	Manual	—	123
7	0.11	Manual	—	123
8	0.38	Power fed	—	100
9	0.14	Manual	—	123
10	0.06	Manual	—	123
11	0.20	Manual	—	123
12	0.06	Manual	—	123

the synthetic rating procedure is the time required to construct a left- and right-hand chart of the elements selected for the establishment of basic motion times.

Speed Rating

Speed rating is a performance evaluation method that only considers the rate of accomplishment of the work per unit time. In this method, the observer measures the effectiveness of the operator against the concept of a normal operator doing the same work, and then assigns a percentage to indicate the ratio of the observed performance to normal or standard performance.

This method particularly emphasizes the observer having complete knowledge of the job before doing the study. To illustrate, the pace of machine workers in a plant producing aircraft engine parts would appear considerably slower than the pace of machine workers producing farm machinery components. The greater precision of aircraft work requires such care that the movements of the various operators might appear unduly slow to one not completely familiar with the work.

In the speed rating method, analysts first appraise the performance to determine whether it is above or below normal. They then try to place the performance in the precise position on the rating scale that correctly evaluates the numerical difference between the standard and the performance demonstrated. Thus, 100 percent is usually considered normal. A rating of 110 percent indicates that the operator was performing at a speed 10 percent greater than normal, and a rating of 90 percent would mean that the operator was performing at a speed 90 percent of normal.

All analysts, through years of experience, will eventually develop a mental model of standard performance. However, it may be useful for the novice analyst to consider the performance of several common tasks in order to develop an initial mental model. Two such tasks were suggested by Presgrave (1957): (1) walking at

TABLE 9–7
Speed Rating Guide

Rating	Verbal Anchor Points	Walking Speed (mph)	Cards Dealt per 1/2 Minute
0	No activity	0	0
67	Very slow, clumsy	2	35
100	Steady, deliberate	3	52
133	Brisk, businesslike	4	69
167	Very fast, high degree of dexterity	5	87
200	Upper limit for short period	6	104

3 miles per hour (4.83 km/hr), that is, 100 feet (30.5 m) in 0.38 minutes, and (2) dealing a deck of 52 cards into four equal piles closely spaced in one half minute (note that the opposite thumb feeds the cards to the dealing hand). A guide with specific anchor points for various levels is presented in Table 9–7.

It may also be useful for the novice analyst to begin by rating only by 10's, that is, 80, 90, 100, etc. and then eventually moving onto 5's, etc. Also, it is very important for any analyst to record the rating in the R column on the time study form before snapping and looking at the watch readout. Otherwise, the analyst may be accused of *rating by the watch.*

Some companies use a speed rating technique normalized to 60 percent standard. This is based on the standard hour approach, that is, producing 60 minutes of work every hour. On this basis, a rating of 80 would mean that the operator was working at a speed of 80/60, or 133 percent, which is 33 percent above normal. A rating of 50 would indicate a speed of 50/60, or 83.3 percent of normal.

A speed rating form referred to as "pace rating" has received considerable attention from the basic steel industry. In an effort to identify completely a normal pace on different jobs, the industry provided benchmarks for a broad range of work, including such effort operations as sand shoveling, coremaking, brick handling, and walking. All have been clearly identified as to method, and quantified as to normal rate of production. Once time study analysts become familiar with a series of benchmarks closely allied to the work under study, they are much better equipped to evaluate the performance speed.

Time study analysts use speed rating for elemental, cycle, or overall rating. For example, all of the time studies mentioned in Chapter 8 utilized speed rating.

Objective Rating

The *objective rating* method, developed by Mundel and Danner (1994), eliminates the difficulty of establishing a normal speed criterion for every type of work. This procedure establishes a single work assignment to which the pace of all other jobs is compared. After the judgment of pace, a secondary factor assigned to the job indicates its relative difficulty. Factors influencing the difficulty adjustment are:

(1) amount of body used, (2) foot pedals, (3) bimanualness, (4) eye–hand coordination, (5) handling or sensory requirements, and (6) weight handled or resistance encountered.

Numerical values, resulting from experiments, have been assigned for a range of each factor. The sum of the numerical values for each of the six factors comprises the secondary adjustment. The rating (R) can thus be expressed as follows:

$$R = P \times D$$

where: P = Pace rating factor.
D = Job difficulty adjustment factor.

This performance rating procedure gives consistent results. Comparing the pace of the operation under study to an operation completely familiar to the observer is easier than simultaneously judging all the attributes of an operation and comparing them to a concept of normal for that specific job. The secondary factor does not affect inconsistency, since this factor merely adjusts the rated time by some percentage. Tables of percentage values for the effects of various difficulties in the operation performed are given in Mundel and Danner (1994).

RATING APPLICATION

The value of a rating is written in the R column of the time study form. Typically, the decimal point is omitted and a whole number (i.e., percent) value is written, to save time. After the stopwatch phase is complete, the analyst multiplies the observed time (OT) by the rating (R), scaled by 100, to yield the normal time (NT):

$$NT = OT \times R/100$$

In effect, this rates the operator's performance compared to that of a qualified operator working at the standard performance pace without overexertion, while adhering to the correct method (see Figure 9–2).

RATING ANALYSIS

As is true of all procedures requiring the exercise of judgment, the simpler and more concise the plan, the easier it is to use and, in general, the more valid the results. The performance rating plan that is easiest to apply, easiest to explain, and gives the most valid results is straight speed rating, augmented by synthetic benchmarks. As has been explained, 100 is considered normal in this procedure, and performance greater than normal is indicated by values directly proportional to 100. The speed rating scale usually covers a range of from 50 to 150. Operators performing outside this 3-to-1 productivity range may be studied, but this is not recommended. The closer the performance is to normal, the better the chance of achieving a fair normal time.

FIGURE 9–2
Relationships of observed time, ratings, and normal time.

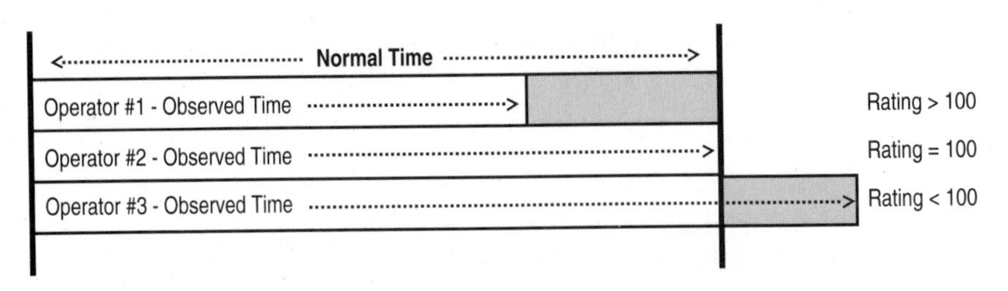

Four criteria determine whether or not time study analysts using speed rating can consistently establish values within 5 percent of the rating average calculated by a group of trained analysts. These are:

1. Experience in the class of work performed.
2. Use of synthetic benchmarks on at least two of the elements performed.
3. Selection of an operator who gives performances somewhere between 85 and 115 percent of normal.
4. Use of the mean value of three or more independent studies.

The most important of these criteria is experience in the class of work performed. From past personal experience, either by observation or operation, analysts should be sufficiently familiar with the work to understand every detail of the method being used. For example, on an assembly job taking place in a fixture, the observer should be familiar with the difficulty in positioning the components in the fixture, should know the class of fit between all mating parts, and should have a clear understanding of the relationship between time and class of fit. The observer should also know the proper sequence of events and the weights of all the parts handled. This does not necessarily mean that the analyst must have been an actual operator in the work being studied, although this would be desirable. On the other hand, an analyst with 10 years experience in the metal trades would have considerable difficulty establishing standards in a women's shoe factory.

Whenever more than one operator is available to be studied, the analyst should select the one who is thoroughly experienced on the job, has a reputation of being receptive to time study practice, and consistently performs at a pace near or slightly better than standard. The closer the operator performs to a normal pace, the easier it will be to rate him or her. For example, if 0.50 minute is normal for dealing a deck of cards into four bridge hands, performance within ± 15 percent of this standard would be fairly easy to identify. However, a performance 50 percent faster or slower than normal would cause considerable difficulty in establishing an accurate rating factor for the performance demonstrated.

The accuracy of the analyst's rating can be checked using synthetic standards. In the time study of Figure 8–12, the average of observed time for element #1 is

2.07/15 = 0.138 minutes. When 0.138 minutes is divided by the MTM conversion factor of 0.0006, an observed time of 230 TMUs (see Chapter 13) is obtained. The fundamental motion time for element #1 is 255 TMUs, which yields a synthetic rating of 255/230 = 111%. In this particular study, the analyst was on target with a speed rating of 110 percent, and the synthetic rating was used as a validation of that judgment.

For rating the overall study, the analyst should take three or more independent samples before arriving at the standard. These can be made on the same operator at different times of the day, or on different operators. The point is that as the number of samples increases, compensating errors diminish the overall error. A recent example clarifies this point. One author, along with two other trained industrial engineers, reviewed performance rating training films involving 15 different operations. The results are shown in Table 9–8. The standard ratings were not disclosed until all 15 operations had been rated. The average deviation of all three engineers for the 15 different operations was only 0.9 on a speed rating scale with a range of 50 points to 150 points. Yet, Engineer C was 30.0 high on operation 7 and Engineer B was 20.0 low on operation 2. When the known ratings were within the 70 to 130 range, the average rating of the three analysts exceeded ±5 points of the known rating in only one case (operation 3).

TABLE 9–8

Performance Ratings of Three Different Engineers Observing 15 Different Operations

		Engineer A		Engineer B		Engineer C		Average of Engineers A, B, C	
Operation	Standard rating	Rating	Deviation	Rating	Deviation	Rating	Deviation	Rating	Deviation
1	110	110	0	115	+5	100	−10	108	−2
2	150	140	−10	130	−20	125	−25	132	−18
3	90	110	+20	100	+10	105	+15	105	+15
4	100	100	0	100	0	100	0	100	0
5	130	120	−10	130	0	115	−15	122	−8
6	120	140	+20	120	0	105	−15	122	+2
7	65	70	+5	70	+5	95	+30	78	+13
8	105	100	−5	110	+5	100	−5	103	−2
9	140	160	+20	145	+5	145	+5	150	+10
10	115	125	+10	125	+10	110	−5	120	+5
11	115	110	−5	120	+10	115	0	115	0
12	125	125	0	125	0	115	−10	122	−3
13	100	100	0	85	−15	110	+10	98	−2
14	65	55	−10	70	+5	90	+25	72	+7
15	150	160	+10	140	−10	140	−10	147	−3
Average of 15 operations	112	115.0		112.3		111.3		112.9	
Average deviation	0	+3.0		+0.3		−0.7		+0.9	

RATING TRAINING

To be successful, analysts must develop track records for setting accurate standards that are accepted by both labor and management. To maintain the respect of all parties, the rates must be consistent.

In general, when studying operators performing somewhere in the range of 0.70 to 1.30 of normal, good analysts should regularly establish standards within ±5 percent of the true rate. Thus, if several operators are performing the same job, and different analysts, each studying a different operator, establish time standards on the job, then the resulting standard from each study should be within ±5 percent of the mean of the group of studies.

To assure rating consistency, both with their own rates and with the rates established by the others, analysts should continually participate in organized training programs. Such training should be more intense for the neophyte time study analyst. One of the most widely used training methods is the observation of videotapes or motion-picture films illustrating diverse operations performed at different productivity levels. Each film has a known level of performance. (Selected training videotapes are listed at the end of this chapter.) After the film is shown, the correct rating is compared with the values established independently by the trainees. If the analysts' values deviate substantially from the correct value, then specific information should be put forward to justify the rating. For example, the analyst may have underrated the operator due to an apparently effortless sequence of motions, whereas the operator's smooth, rhythmic blending of movements may really have been an indication of high dexterity and manipulative ability.

As successive operations are reviewed, analysts should plot their ratings against the known values (see Figure 9–3). A straight line indicates perfection, whereas high irregularities on both sides of the line indicate inconsistency, as well as an inability to evaluate performance. In Figure 9–3, the analyst rated the first film 75, but the correct rating was 55. The second was rated 80, while the proper rating was 70. In all but the first case, the analyst was within the company's established area of correct rating. Note that, due to the nature of confidence intervals, the ±5 percent accuracy criterion is valid only around 100 percent, or the normal level of performance. When performance is below 70 percent of normal or above 130 percent of normal, an experienced time study analyst would expect an error much larger than 5 percent.

It is also helpful to plot successive ratings on the x-axis and indicate the positive or negative magnitude of deviation from the known normal on the y-axis (see Figure 9–4). The closer the time study analyst's rating comes to the x-axis, the more nearly correct he or she is.

To determine quantitatively an analyst's ability to rate performance, compute the percentage of the analyst's rating contained within specified limits of the known ratings. This can be done as follows:

1. Compute the mean difference ($\overline{x}_d$) between the analyst's rating and the actual rating for n tests (n should be at least 15 observations).
2. Compute the standard deviation s_d of the differences in rating.

FIGURE 9–3
Chart showing a record of seven studies, with the analyst tending to rate a little high on studies 1, 2, 4, and 6, and a little low on studies 3 and 7. Only study 1 was rated outside the range of desired accuracy.

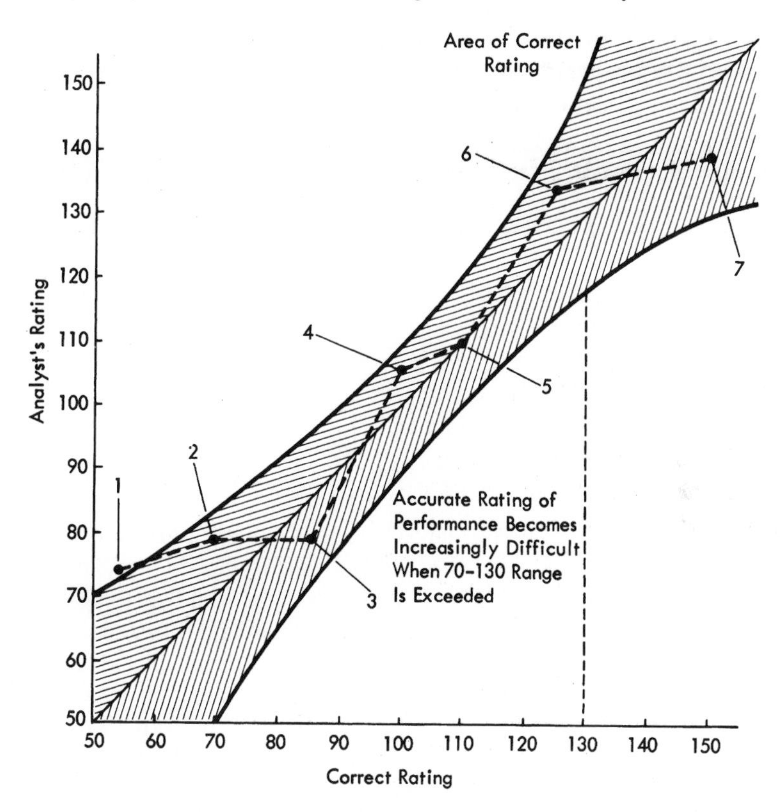

3. Compute the normal deviate z_1, where:

$$z_1 = \frac{+5 \text{ (or some other figure of accuracy)} - \bar{x}_d}{s_d}$$

4. Compute the normal deviate z_2, where:

$$z_2 = \frac{-5 \text{ (or some other figure of accuracy)} - \bar{x}_d}{s_d}$$

5. Compute the area under the normal distribution curve between ± 5 (or some other figure of accuracy) centered at $\bar{x}_d$, which is assumed to be equal to μ_d, and s_d, which is assumed to be equal to σ_d.

A recent statistical study involving the performance rating of 6,720 individual operations by a group of 19 analysts over a period of approximately two years confirmed several facts that had been accepted by most industrial engineers. This study

FIGURE 9–4
Record of an analyst's rating factors on 15 studies.

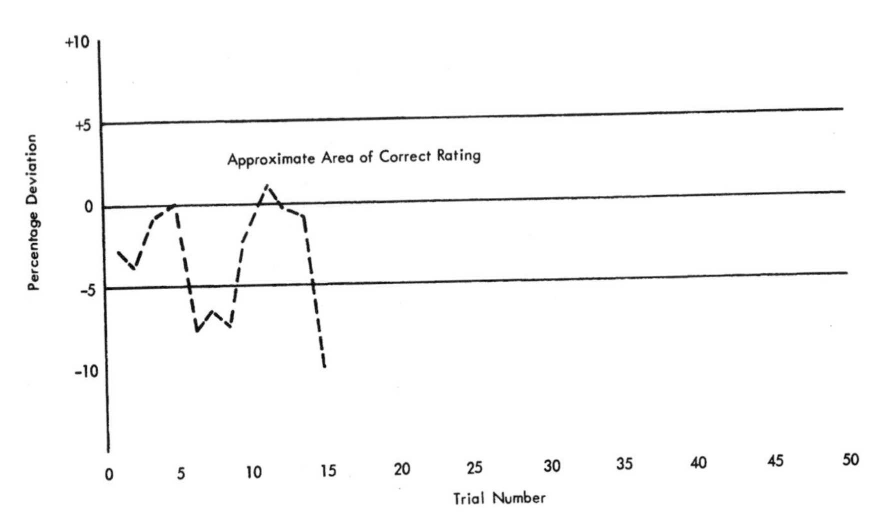

concluded that the "level of performance" is a factor that significantly affects errors in performance rating. Analysts overrated low performance levels and underrated high performance levels. This is typical of novice raters who tend to be conservative raters and are afraid to deviate too far from standard performance. In statistical applications, this tendency is termed *regression toward the mean* and results in a relatively flat line compared to the expected line with a slope of one (see Figure 9–6). The novice rater who rates higher than the true value for performances below standard performance is called a *loose* rater. The result is standard times that are too easy for operators to achieve, which means the company would lose money on that operation. For performances above the standard, a novice rater who rates lower than the true value is called a *tight* rater, which results in times that are difficult for operators to achieve. Some analysts, even after years of experience, consistently tend to rate either tight or loose (see Figure 9–6).

The statistical study also concluded that the operation being examined has an effect on errors in rating performance. Complex operations tend to be more difficult to performance rate than simpler operations, even for experienced analysts. At low performance levels, overrating is greater for difficult operations than for simple operations, while at high performance levels, underrating is greater for the easy-to-perform operations.

Today, no known tests can accurately evaluate the ability of a person to rate performance. Experience has shown, however, that only analysts with a tendency toward consistency and proficiency after a brief training period can do an acceptable job of rating.

A survey by one large manufacturer disclosed that an experienced industrial engineering employee did not rate any more accurately than a less experienced analyst. This survey also indicated that those in the higher work classifications did not

EXAMPLE 9–1
Calculation of Rating Ability.

Suppose that an analyst rated 15 training tapes as shown in Table 9–9. The following computations would be made:

$$\bar{x}_d \text{ (mean difference)} = \frac{\Sigma d}{n} = \frac{50}{15} = 3.33$$

$$s_d \text{ (standard deviation)} = \sqrt{\frac{\Sigma d^2 - \frac{(\Sigma d)^2}{n}}{n-1}} = 7.7$$

$$z_1 = \frac{5.00 - 3.33}{7.7} = 0.217$$

and

$$P(z_1) = \int_0^{z_1} \frac{1}{\sqrt{2\pi}} e^{-\frac{z^2}{2}} dz = 0.0859$$

$$z_2 = \frac{-5 - 3.33}{7.7} = -1.08$$

and

$$P(z_2) = \int_0^{z_2} \frac{1}{\sqrt{2\pi}} e^{-\frac{z^2}{2}} dz = 0.3597$$

$$P(z_1) + P(z_2) = 0.4456$$

In this example, the analyst would receive a rating of 0.4456. This represents the portion of ratings that lies within ±5 rating points of the ideal ratings. This area of correct rating, together with the distribution of the difference between the analyst's rating and the correct rating, is shown in Figure 9–5. If the number of films observed is less than 30, use the t-distribution.

rate any better than those in the lower classifications, and that those in machining areas did no better than those in assembly areas.

A group of undergraduate engineers was given 15 minutes in rating instruction and then requested to performance rate a film illustrating a series of heavy-labor operations. The students were given no orientation in the class of work, and the operations observed differed substantially from the operations used to give the students a concept of proper performance. These students were extremely successful in applying the concepts learned on light operations to heavy-labor operations.

EXAMPLE 9–1 ***(concluded)***

TABLE 9–9
Comparison of Analyst's and Correct Ratings

Tape no.	Correct rating	Analyst's rating	Difference (d)	Difference squared
1	115	105	−10	100
2	125	120	−5	25
3	85	95	+10	100
4	105	105	0	0
5	100	105	+5	25
6	95	110	+15	225
7	120	125	+5	25
8	140	150	+10	100
9	100	105	+5	25
10	60	75	+15	225
11	100	100	0	0
12	110	105	−5	25
13	90	90	0	0
14	130	125	−5	25
15	80	90	+10	100

FIGURE 9–5
Graphic representation of the analyst's correct ratings (±5) and the distribution of all his ratings.

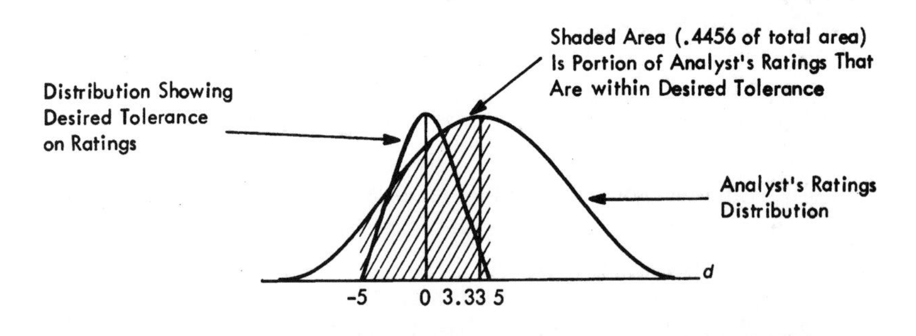

In another instance, 34 undergraduate engineering students performance rated five industrial films involving operations ranging from mold making to bench assembly. The students found that four films were representative of normal performance, and one was 1.15, or 15 percent above normal. The results are shown in Table 9–10. Most students did not deviate greatly from the known values. However, for the film with faster performance, the median rating value for the students was in the 101–105 bracket, an underrating of the true 115 value and an example of conservative rating.

FIGURE 9–6
Examples of loose, tight, and conservative rater.

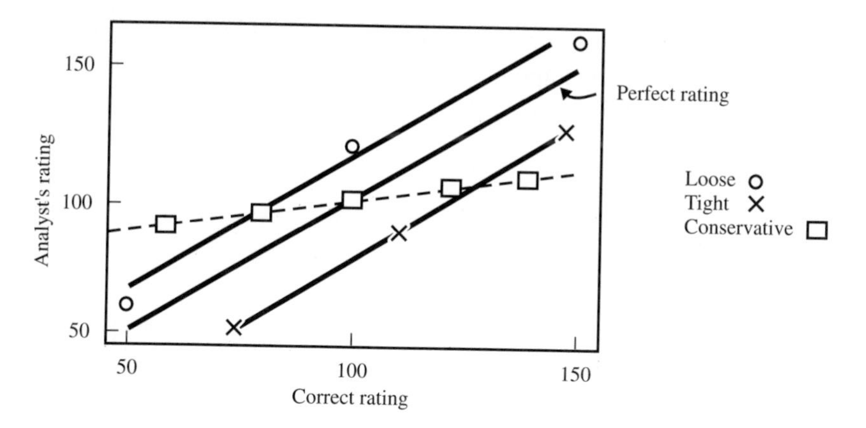

TABLE 9–10
Distribution of Student Ratings

	Number of students in each rating bracket					
Known rating	**85–90**	**91–95**	**96–100**	**101–105**	**106–110**	**111–125**
1.15	0	3	4	12	6	9
1.00	4	5	19	2	3	1
1.00	2	3	7	12	5	5
1.00	7	5	14	8	0	0
1.00	1	2	26	4	1	0

The foregoing studies brought out that: (1) the concept of normal performance can be taught quickly; and (2) the concept is transferable to some degree to dissimilar operations. By actual observation of different work assignments throughout a plant, under the guidance of a time study supervisor, an unskilled employee can achieve excellent training in rating. The supervisor explains in detail the "why" of values after the trainees have independently written their ratings. The independent values should be recorded to determine the consistency of the group and the necessity for additional, and perhaps more intensive, training. Companies with extensive, well-designed training programs in rating have been successful in eliminating the tendency to over- and underrate.

SUMMARY

Performance rating is a means of adjusting the average observed time on a job, to derive the time required for a qualified operator to do the job when working at a normal pace. Only in this manner can a true time standard be established. Since the standard is based entirely on the experience, training, and judgment of the time study analyst, the standard can be subject to criticism. Consequently, many different rating systems have been developed in an attempt to obtain an "objective" system. However, each rating system still ultimately depends on the subjectivity and honesty of the rater. Thus, it is important to train the time study analyst thoroughly in rating properly and consistently. That this can be done successfully has been borne out by many studies.

QUESTIONS

1. Why has industry been unable to develop a universal conception of "normal performance"?
2. Which factors enter into large variances in operator performance?
3. What are the characteristics of a sound rating system?
4. During a time study, when should a rating be given? Why is this important?
5. What governs the frequency of performance rating during a given study?
6. Explain the Westinghouse system of rating.
7. Under the Westinghouse rating system, why are "conditions" evaluated?
8. What is synthetic rating? What is its principal weakness?
9. What is the basis of speed rating, and how does this method differ from the Westinghouse system?
10. Which four criteria are fundamental for doing a good job in speed rating?
11. Why is training in performance rating a continuous process?
12. Why should more than one element be used in the establishment of a synthetic rating factor?
13. Would there be any objection to studying an operator who is performing at an excessive pace? Why or why not?
14. In what ways can an operator give the impression of high effort and yet produce at a mediocre or poor level of performance?

PROBLEMS

1. The work measurement analyst in charge of training time study analysts decided to have all trainees review 20 film loops, and the rate of each loop was known. Each trainee then computed his or her own record, which was based on the proportion of ratings that fell within ±5 points of the known ratings. One analyst computed her average rating difference as −4.08 points on 20 films. The standard deviation was 6.4. What percentage of this analyst's ratings was contained within the desired rating? (Note: Assume that the sample values are the population values.)

2. Using the data in Figure 9–3, determine the average percentage of correct rating within ±5 rating points.

REFERENCES

Barnes, Ralph M. *Motion and Time Study: Design and Measurement of Work.* 7th ed. New York: John Wiley & Sons, 1980.

Hancock, Walton M. "The Learning Curve." In *Handbook of Industrial Engineering,* 2nd ed. Ed. Gavriel Salvendy. New York: John Wiley & Sons, 1992.

Lowry, S. M., H. B. Maynard, and G. J. Stegemerten. *Time and Motion Study and Formulas for Wage Incentives.* 3rd ed. New York: McGraw-Hill, 1940.

Morrow, R. L. *Time Study and Motion Economy.* New York: Ronald Press, 1946.

Mundel, Marvin E. and David L. Danner. *Motion and Time Study: Improving Productivity.* 7th ed. Englewood Cliffs, NJ: Prentice-Hall, 1994.

Nadler, Gerald. *Work Design: A Systems Concept.* Rev. ed. Homewood, IL: Richard D. Irwin, 1970.

Presgrave, R. W. *The Dynamics of Time Study.* 4th ed. Toronto, Canada: The Ryerson Press, 1957.

SELECTED VIDEOTAPES

Workplace Fundamentals. 1/2″ VHS, Tampa Manufacturing Institute, Bradenton, FL.

Workplace Rating Exercises. 1/2″ VHS, Tampa Manufacturing Institute, Bradenton, FL.

Fair Day's Work Concepts. 1/2″ VHS, Tampa Manufacturing Institute, Bradenton, FL.

CHAPTER 10

Allowances

KEY POINTS:

- Use allowances to compensate for fatigue and delays at work.
- Determine allowances either through direct observation or work sampling.
- Provide a minimum of 9–10 percent constant allowance for personal needs and basic fatigue.
- Add allowances to normal time as a percentage of normal time.

After calculation of the normal time, one additional step must be performed to arrive at a fair standard. This last step is the addition of an allowance to account for the many interruptions, delays, and slowdowns caused by fatigue in every work assignment. For example, when planning a road trip of 1,000 miles, we know that the trip cannot be made in 20 hours driving at a speed of 50 miles per hour. An allowance must be added for periodic stops for personal needs, driving fatigue, unavoidable stops due to traffic congestion and stoplights, possible detours and the resulting rough roads, car trouble, and so forth. Thus, we may actually estimate that the trip will take 25 hours, allowing 5 additional hours for all delays. Similarly, analysts must provide an allowance if the resulting standard is to be fair and readily maintainable by an average worker performing at a steady, normal pace.

ALLOWANCE USES

The watch readings of any time study are taken over a relatively short period of time. Therefore, normal time does not include unavoidable delays, which may not even have been observed, and other legitimate lost time. Consequently, analysts must make some adjustment to compensate for such losses. The application of these adjustments, or *allowances,* may be considerably broader in some companies than

TABLE 10–1
Typical Industrial Allowances

Allowance factor	No. of firms	Percent of firms
1. Fatigue	39	93
A. General	19	45
B. Rest periods	13	31
Did not specify A or B	7	17
2. Time required to learn	3	7
3. Unavoidable delay	35	83
A. Man	1	2
B. Machine	7	17
C. Both, man and machine	21	50
Did not specify A, B, or C	6	14
4. Personal needs	32	76
5. Setup or preparation operations	24	57
6. Irregular or unusual operations	16	38

Source: J. O. P. Hummel, "Motion and Time Study in Leading American Industrial Establishments" (Master's thesis, Pennsylvania State University).

in others. For example, Table 10–1 reveals the items that 42 firms included in their allowances.

Allowances are frequently applied carelessly because they have not been established on sound time study data. This is especially true of allowances for fatigue, for which it is difficult, if not impossible, to establish values based on rational theory. Many unions, well aware of this situation, have bargained for additional fatigue allowance as if it were a "fringe" issue. (Fringe benefits are company expenses, such as insurance and pensions, unrelated to employee output.) Allowances must be as accurate and correct as possible; otherwise, the care and precision put into the study could be completely nullified.

Allowances are applied to three parts of the study: (1) the total cycle time; (2) machine time only; and (3) manual effort time only. Allowances applicable to the total cycle time are expressed as a percentage of the cycle time, and compensate for such delays as personal needs, cleaning the workstation, and oiling the machine. Machine time allowances include time for tool maintenance and power variance, while representative delays covered by effort allowances are fatigue and certain unavoidable delays.

Two methods are frequently used for developing standard allowance data. One is the production study, which requires observers to study two, or perhaps three, operations over a long period of time. Observers record the duration of and reason for each idle interval (see Figure 10–1). After establishing a reasonably representative sample, observers summarize their findings to determine the percent allowance for each applicable characteristic. Data obtained in this fashion, like that for any time study, must be adjusted to the normal performance level. Since observers must spend a long time directly observing one or more operations, this method is exceptionally tedious, not only to analysts, but also to the operators. Another disadvantage is the tendency to take too small a sample, which may result in biased results.

FIGURE 10–1
Lost time analysis chart.

LOST TIME ANALYSIS OF TIME STUDY

Dwg.________ Part ________ Date ________
Operation ________
Symbol ________
A. Personal ________
B. Start work late ________
C. Stop work early ________
D. Talk with foreman or instructor ________
E. Talk with other persons ________
F. Search for tools ________
G. Search for drawings ________
H. Rework fault of operator ________
I. Rework fault of another operator ________
J. Rework fault of machine or fixtures ________
K. Idle-wait for crane (excess over allowed) ________
L. Idle-wait for inspector (excess over allowed) ________
M. Wait in line at tool crib (excess over allowed) ________
N. Wait in line at dispatch office (excess over allowed) ________
O. Wait in line at B/P station ________
P. Tool maintenance ________
Q. Oil machine ________
R. Clean work station ________
S. Circled readings (circled reading minus ave. for ele.) ________
T. Miscellaneous minor delays ________
U. Lost time developing methods during study ________
V. ________
W. ________
X. ________
Y. ________
Z. ________
Total ________

1. Gross over-all ________ Mins. ________ Hrs.
2. Total lost ________ " ________ "
3. % lost time compared with net actual (2 ÷ 4) ________
4. Net actual or productive ________ Mins. ________ Hrs.
5. Allowed time ________ " ________ "

Note--Place lost time symbol alongside description of lost time on study and staple this card to study.

The second technique involves work sampling studies (see Chapter 14). This method requires taking a large number of random observations, thus requiring only part-time, or at least intermittent, services of the observer. In this method, no stopwatch is used, as observers merely walk through the area under study at random times and note briefly what each operator is doing. The number of delays recorded, divided by the total number of observations during which the operator is engaged in productive work, approximate the allowance required by the operator to accommodate the normal delays encountered.

In using work sampling studies for the determination of allowances, observers must practice several precautionary measures. First, they must not anticipate observations, but record only the actual happenings. Second, a given study should not cover dissimilar work, but should be confined to similar operations on the same general type of equipment. Third, the larger the number of observations and the

FIGURE 10–2
Allowances by function.

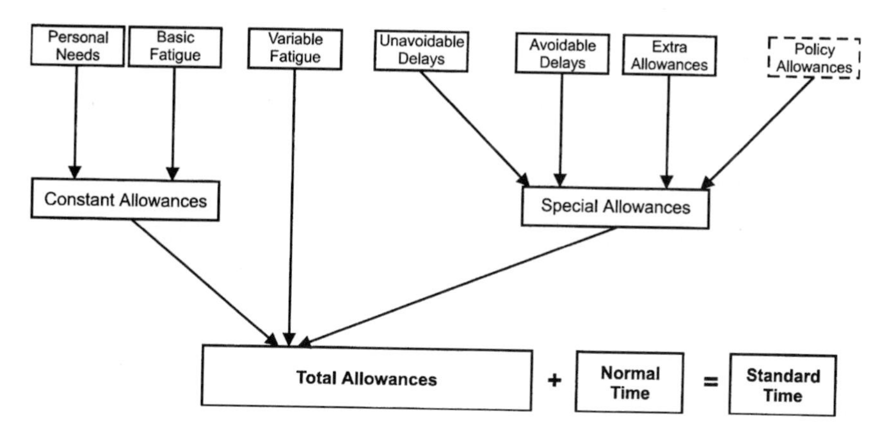

longer the period of time over which the data is taken, the more valid are the results. Analysts should take daily observations over a span of at least two weeks.

Figure 10–2 attempts to provide a scheme for ordering the various types of allowances according to function. The main division is *fatigue* versus *special* allowances. Fatigue allowances, as the name implies, provide time for the worker to recover from fatigue incurred as a result of the job or work environment, and these are subdivided into *constant* and *variable fatigue* allowances. Special allowances include many different factors related to the process, equipment, materials, etc. and are termed unavoidable delays, avoidable delays, *extra,* and *policy* allowances.

CONSTANT ALLOWANCES

Personal Needs

Personal needs include those cessations in work necessary for maintaining the general well-being of the employee; examples are trips to the drinking fountain and the restroom. The general working conditions and class of work influence the time necessary for personal delays. For example, working conditions involving heavy work performed at high temperatures, such as that done in the pressroom of a rubber-molding department or in a hot-forge shop, would require greater allowance for personal needs than light work performed in comfortable temperature areas.

There is no scientific basis for a numerical percent to give; they are intended for the individual. Detailed production checks have demonstrated that a 5 percent allowance for personal time, or approximately 24 minutes in 8 hours, is appropriate for typical shop working conditions. Lazarus (1968) reported that of 235 plants in 23 industries, the personal allowance ranged from 4.6 to 6.5 percent. Thus, the 5 percent figure appears to be adequate for the majority of workers.

Basic Fatigue

The *basic fatigue allowance* is a constant to account for the energy expended to carry out the work and to alleviate monotony. A value of 4 percent of normal time is considered adequate for an operator who is doing light work, while seated, under good working conditions, with no special demands on the sensory or motor systems (ILO, 1957).

Between the 5 percent personal needs allowance and the 4 percent basic fatigue allowance, most operators are given an initial 9 percent basic allowance, to which other allowances may be added, if necessary.

VARIABLE FATIGUE ALLOWANCES

Basic Principles

Closely associated with the allowance for personal needs is the allowance for fatigue, although this allowance is usually applied only to the effort portions of the study. Fatigue allowances have not reached the state where their qualifications are completely based on sound, rational theories. Consequently, next to performance rating, the fatigue allowance is the least defensible and the most open to argument of all the factors making up a time standard. For example, many proponents of the MTM system (see Chapter 13) feel that no fatigue allowance should be used in the development of the majority of standards, since MTM values are based on a work rate that can be sustained for an 8-hour working day by the average healthy employee. However, fair fatigue allowances for different classes of work can be approximated by empirical means.

Fatigue is not homogeneous in any respect. It ranges from strictly physical to purely psychological and includes combinations of the two. Also, it has a marked influence on some people, yet has apparently little or no effect on others.

Whether the fatigue is physical or mental, the results are similar; there is a lessening in the will to work. The major factors that affect fatigue are well known and clearly established. Some of these include: working conditions, especially noise, heat, and humidity; the nature of the work, such as posture, muscular exertion, and tediousness, and the general health of the worker. Although heavy manual work, and thus muscular fatigue, is diminishing in industry, due to mechanization, other fatigue components, such as mental stress and monotony, may be increasing. Because all fatigue cannot be eliminated, proper allowance must be made for the working conditions and repetitiveness of the work.

One method of determining the fatigue allowance is to measure the decline in production throughout the working period. The production rate for every quarter of an hour during the course of the working day may be measured. Any decline in production that cannot be attributed to methods changes or personal or unavoidable delays may be attributed to fatigue and expressed as a percentage. Brey (1928) expressed the coefficient of fatigue as follows:

$$F = (T - t) \times 100/T$$

where: F = Coefficient of fatigue.
T = Time required to perform the operation at the end of continuous work.
t = Time required to perform the operation at the beginning of continuous work.

Many attempts have been made to measure this fatigue through various physical, chemical, and physiological means, none of which have so far been completely successful. Therefore, the International Labour Office (ILO, 1957) has tabulated the effect of various working conditions, to arrive at appropriate allowance factors (see Table 10–2) with a more detailed point system introduced in a later edition (ILO, 1979). These factors include: standing versus sitting; abnormal positions; use of force; illumination; atmospheric conditions; required job attention; noise level; mental strain; monotony; and tediousness. To use this table, the analyst would determine the allowance factors for each element of the study and then sum them for a total variable fatigue allowance, which is then added to the constant fatigue allowance.

These ILO recommendations were developed through consensus agreements between management and workers across many industries and have not been directly substantiated. On the other hand, since the 1960s, much work has been done in developing specific standards for the health and safety of the U.S. worker. Here, we examine how well these standards compare with the ILO fatigue allowances.

Abnormal Posture

Allowances for posture are based on metabolic considerations and can be supported by metabolic models that have been developed for various activities (Garg, et al., 1978). Three basic equations for sitting, standing, and bending can be used to predict and compare the energy expended for various postures. Using an average adult (both male and female) body weight of 152 pounds (69 kg) and adding an additional energy expenditure of 2.2 kcal/minute for manual hand work (Garg, et al., 1978), we obtain energy expenditures of 3.8 kcal/min, 3.86 kcal/min, and 4.16 kcal/min, for sitting, standing, and bending, respectively. Since sitting is a basic comfortable posture that can be maintained for extended time periods, the other postures are compared to sitting. The ratio of standing to sitting energy expenditures is 1.02, or an allowance of 2 percent, while the ratio of bending to sitting energy expenditures

TABLE 10–2
ILO Recommended Allowances

Allowance	
A. Constant allowances:	
1. Personal allowance	5
2. Basic fatigue allowance	4
B. Variable allowances:	
1. Standing allowance	2
2. Abnormal position allowance:	
a. Slightly awkward	0
b. Awkward (bending)	2
c. Very awkward (lying, stretching)	7
3. Use of force, or muscular energy (lifting, pulling, or pushing): Weight lifted, pounds:	
5	0
10	1
15	2
20	3
25	4
30	5
35	7
40	9
45	11
50	13
60	17
70	22
4. Bad light:	
a. Slightly below recommended	0
b. Well below	2
c. Quite inadequate	5
5. Atmospheric conditions (heat and humidity)—variable	0–100
6. Close attention:	
a. Fairly fine work	0
b. Fine or exacting	2
c. Very fine or very exacting	5
7. Noise level:	
a. Continuous	0
b. Intermittent—loud	2
c. Intermittent—very loud	5
d. High-pitched—loud	5
8. Mental strain:	
a. Fairly complex process	1
b. Complex or wide span of attention	4
c. Very complex	8
9. Monotony:	
a. Low	0
b. Medium	1
c. High	4
10. Tediousness:	
a. Rather tedious	0
b. Tedious	2
c. Very tedious	5

is 1.10, or an allowance of 10 percent. The first is identical to the ILO recommendations. The second is slightly larger than the ILO value of 7 percent, but may represent an extreme case of posture which cannot be maintained for an extended period of time.

Muscular Force

Fatigue, or more properly termed *relaxation allowances,* can be formulated on two important physiological principles: muscle fatigue and muscle recovery after fatigue. The most immediate result of muscle fatigue is the significant reduction in muscle strength. Rohmert (1960) quantified these principles as follows:

1. Reduction in maximum strength occurs if the static holding force exceeds 15 percent of maximum strength.
2. The longer the static muscular contraction, the greater the reduction in muscle strength.
3. Individual or specific muscle variations are minimized if forces as normalized to the individual's maximum strength for that muscle.
4. Recovery is a function of the degree of fatigue; that is, a given percent decrease in maximal strength will require a given amount of recovery.

These concepts of fatigue and recovery were further quantified by Rohmert (1973) into a series of curves for relaxation allowances (RA) as a function of force and holding time (see Figure 10–3):

$$RA = 1800 \times (t/T)^{1.4} \times (f/F - 0.15)^{0.5}$$

where: RA = Relaxation allowance in percent of time t.
t = Duration of holding time (minutes).
f = Holding force (pounds).
F = Maximum holding force (pounds).
T = Maximum holding time for holding force f (min), defined as:

$$T = 1.2/(f/F - 0.15)^{0.618} - 1.21$$

The maximum holding force (F) can be approximated based on data collected on 1522 industrial male and female workers (Chaffin, et al., 1987). The average of the three basic standardized lifting strengths (arm, leg, torso) is approximately 100 pounds (45.5 kg.). Using this value of maximum holding force for infrequent lifts (less than one lift every five minutes) of short duration yields the allowances tabulated in Table 10–3. For more frequent lifts (more than one lift every five minutes), metabolic considerations predominate, and the NIOSH lifting guidelines (see Chapter 4) should be utilized in determining the limitation for lifting. Also, loads above 51 pounds (23.2 kg.) are not allowed by the NIOSH lifting guidelines.

FIGURE 10–3
Percentage rest allowances for various combinations of holding forces and time. (*From:* Rohmert, 1973)

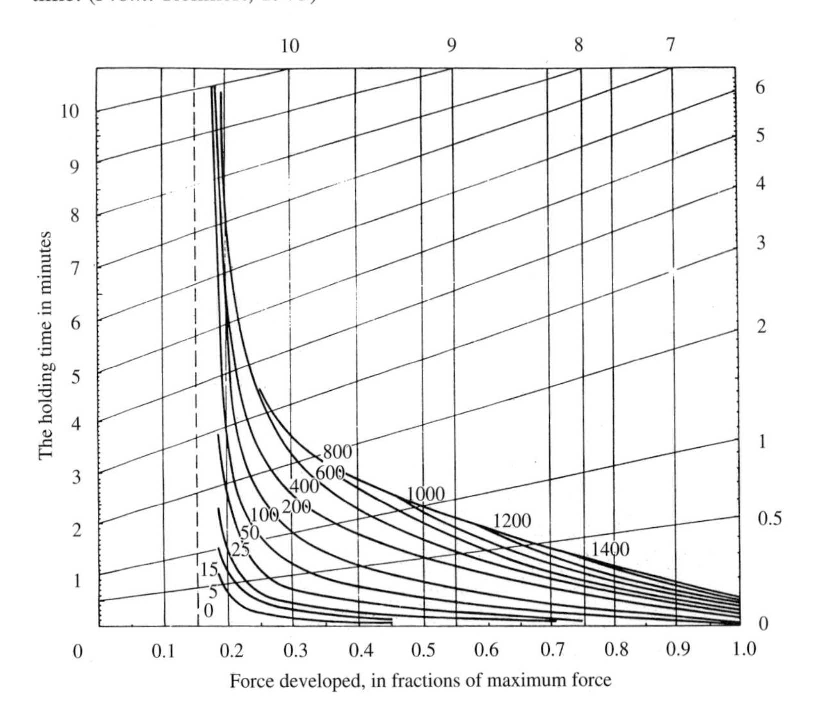

TABLE 10–3
Comparison of ILO and Calculated Allowances for Use of Muscular Force

Load (lbs)	ILO	Calculated
5	0	0
10	1	0
15	2	0
20	3	0.5
25	4	1.3
30	5	2.7
35	7	4.5
40	9	7.0
45	11	10.2
50	13	14.4
60	17	NA
70	22	NA

For muscular energy, we consider the formula for determining the amount of rest required for heavy work presented in Chapter 4:

$$R = (W - 5.33)/(W - 1.33)$$

EXAMPLE 10–1
Calculation of Relaxation Allowance for Infrequent Use of Muscular Force.

Consider a worker lifting a 40 pound (18.2 kg) load less than once every five minutes. First, calculate the maximum holding time for a load that is 40/100 = 40% of an average person's maximum strength capability:

$$T = 1.2/(0.4 - 0.15)^{0.618} - 1.21 = 1.62 \text{ min}$$

Then, substitute a short-duration exertion of 0.05 min and a maximum holding time of 1.62 minutes into the relaxation equation:

$$RA = 1800 \times (0.05/1.62)^{1.4} \times (0.4 - 0.15)^{0.5} = 1800 \times (0.0768) \times (0.5) = 6.96 \approx 7\%$$

The resulting variable relaxation allowance of 7% is added to the typical 9% constant allowance for a total allowance of 16%.

where: R = Time required for rest, as percent of total time.
W = Average energy expenditure during work (kcal/min).

This expression can be reformulated into a relaxation allowance:

$$RA = (\Delta W/4 - 1) \times 100$$

where: RA = Rest allowance as percentage added to normal time.
ΔW = Energy expenditure increment for work = W − 1.33 kcal/min).

Measuring the heart rate is easier than measuring energy expenditure. The relaxation allowance can therefore be reformulated with the heart rate as follows:

$$RA = (\Delta HR/40 - 1) \times 100$$

where: RA = Rest allowance, as percentage added to work time.
ΔHR = Difference between working heart rate and resting heart rate.

Atmospheric Conditions

Modeling the human body and its responses to atmospheric conditions is a very difficult task. Many attempts have been made to combine the physiological manifestations and the changes of several environmental conditions into one simple index (Freivalds, 1987). However, no such index can suffice, and considerable variability in allowances can result. The ILO allowances are based on an outdated concept of cooling power, and they greatly under-predict required relaxation allowances. Thus, there is considerable deviation in the ILO allowances from true stress levels. This is explained in detail in Freivalds and Goldberg (1988).

A better approach is to consider the more recent guidelines developed by NIOSH (1986), utilizing the wet-bulb globe temperature (WBGT, described in

EXAMPLE 10–2
Calculation of Relaxation Allowance for Overall Fatigue.

Consider the strenuous task presented in Chapter 4 for shoveling coal into a hopper, with an energy expenditure of 9.33 kcal/min. The required rest allowance RA is:

$$RA = [(9.33 - 1.33)/4 - 1] \times 100 = 100\%$$

Therefore, to provide adequate time for recovery from fatigue, the worker would need to spend 4 hours out of an 8-hour shift resting. Note that the acceptable ΔW for males is 5.33 − 1.33 = 4 kcal/min. (For females, substitute the values 4.0 for 5.33 and 1.0 for 1.33.)

Chapter 6) and working energy expenditures. The resulting relaxation allowances (from Figure 6-11) can be quantified through a least-squares regression, resulting in:

$$\text{RA} = e^{(-41.5 + 0.0161\text{W} + 0.497\ \text{WBGT})}$$

where: W = Working energy expenditure.
WBGT = Wet-bulb globe temperature (°F).

Noise Level

The Occupational Safety and Health Administration (OSHA, 1983) has established permissible noise exposures for workers in industry. The permissible levels depend on the duration of the exposure, as shown in Table 10–4.

Should the total daily exposure consist of exposures to several different noise levels, then the combined exposure is calculated using the equation:

TABLE 10–4
OSHA Permissible Noise Levels

Noise Level (dBA)	Permissible Time (hr)
80	32
85	16
90	8
95	4
100	2
105	1
110	0.5
115	0.25
120	0.125
125	0.063
130	0.031

EXAMPLE 10–3
Calculation of a Relaxation Allowance for Atmospheric Conditions.

A worker is performing manual assembly at a seated workstation and is expending roughly 200 kcal/min. If the $WBGT_{IN}$ is 88.5° F, then:

$$RA = e^{\{-41.5 + 0.0161(200) + 0.497(88.5)\}} = e^{\{5.7045\}} \approx 300\%$$

Based on a relaxation allowance of 300 percent, the worker would need 45 minutes of rest for each 15 minutes of work.

$$D = C_1/T_1 + C_2/T_2 + .. \leq 1$$

where: D = Noise dose (decimal value).
C = Time spent at specified noise level (hours).
T = Time permitted (Table 10–4) at specified noise level (hours).

and the required relaxation allowance (in %) is simply:

$$RA = 100 \times (D - 1)$$

Thus, the total exposure to various noise levels cannot exceed a 100-percent dose. For example, a worker may be exposed to 95 dBA noise for 3 hours and 90 dBA noise for 5 hours. Although each exposure is separately permissible, the combined dose is not:

$$D = 3/4 + 5/8 = 1.375 > 1$$

Therefore, the required relaxation allowance for OSHA compliance is:

$$RA = 100 \times (1.375 - 1) = 37.5\%$$

Thus, 90 dBA is the maximum permissible level for an 8-hour day, and any sound level above 90 dBA will require a relaxation allowance.

For computation of the noise dose, all sound levels between 80–130 dBA must be included in the computations (although continuous levels above 115 dBA are not allowed at all). Since Table 10–4 provides only certain key times, the following computational formula can be used for intermediate noise levels:

$$T = \frac{32}{2^{\left(\frac{L-80}{5}\right)}}$$

where: L = Noise level (dBA).

Illumination Levels

Reconciliation of the ILO (1957) allowances and the IES (1981) recommended illumination levels (see Chapter 6) can be approached as follows. For relaxation allowances, a task that is slightly below recommended guidelines can be considered

EXAMPLE 10–4
Calculation of a Relaxation Allowance for Noise.

A worker is exposed to the noise levels shown in Table 10–5 during an 8-hour work day.
The last entry is obtained as follows:

$$T = \frac{32}{2^{\left(\frac{96-80}{5}\right)}} = 3.48$$

The noise dose is:

$$D = 1/32 + 4/8 + 3/3.48 = 1.393$$

and the rest allowance is:

$$RA = 100 \times (1.393 - 1) = 39.3\%$$

Since the 8-hour noise exposure dosage exceeds OSHA requirements, the worker must be provided with a 39.3-percent rest allowance. Note that, again, the ILO recommended values greatly under-predict required allowances.

TABLE 10–5
Noise Levels Over an 8-Hour Day

Noise Level (L) dBA	Time Spent (C) hr	Time Permitted (T) hr
80	1	32
90	4	8
96	3	3.48

to be within the same illumination sub-category, perhaps slightly substandard, at the low end of the range, and is assigned a 0 percent allowance. A task that is well below adequate illumination may be defined as being one sub-category beneath its recommended illumination and is assigned a 2 percent allowance. A task with quite inadequate illumination may be defined as being two or more sub-categories below its recommended level and receives a 5 percent allowance. These definitions are fairly realistic, as human perceptions of illumination are logarithmic; that is, as illuminance increases, we require a greater intensity difference before a change is noted (IES, 1981).

The literature contains some evidence that increased task illumination results in better-skilled performance. The most germane performance measure is task completion time under varying illumination conditions. However, it should be kept in mind that performance accuracy is also important. For example, Bennett, Chitlnagio, and Pangrekar (1977) reported the time to read a 450-word, pencil-written article as a cubic function of task illumination:

TABLE 10–6
Modeled Times

Illumination (fc)	Modeled Time (sec)	% Change from 75 fc	ILO Category	Allowance (%)
75	207.3	—	(Recommended)	0
50	210.0	1.3	slightly below	0
30	213.9	3.2	well below	2
20	217.2	4.8	well below	2
15	219.8	6.0	inadequate	5
10	223.6	7.9	inadequate	5

(From Bennett, et al., 1977)

$$\text{Time} = 251.8 - 33.96 \log FC + 6.15 (\log FC)^2 - 0.37 (\log FC)^3$$

where: Time = Mean reading time (seconds).
FC = Task illumination (foot candles).

The recommended IES (1981) illumination for reading this pencil-written material is category E (from Table 6.2), 50–100 foot-candles (500–1,000 lux). The weighting factors total zero (from Table 6.3), so the recommended illumination is 75 foot-candles (750 lux), the middle value in category E. Table 10–6 compares modeled reading times as a function of decreasing illumination and allowances. Decreasing the illumination to the next lower sub-category increased performance times by 3–5 percent, which is not too far from the 2 percent ILO allowance for well below recommended lighting. The next lower illumination sub-category produced times that were 6–8 percent greater than that at the recommended levels, which is somewhat greater than the 5 percent allowance for inadequate illumination. Overall, this study (see Table 10–6) came reasonably close to supporting the ILO allowances.

Visual Strain

The ILO visual strain relaxation allowance provides no allowance for fairly fine work, a 2 percent allowance for fine or exacting work, and a 5 percent allowance for very fine or very exacting work. These allowances only refer to the precision of the visual task requirements, without mentioning the other task conditions that have a very large effect on visual requirements: illumination (or luminance), glare, flicker, color, viewing time, contrast, etc. Therefore, the ILO allowances are only rough approximations. More specific values can be determined by target detectability, as first quantified by Blackwell (1952) in his visibility curves (see Chapter 6 and Figure 6-2). Four factors have the greatest effect in determining how visible a target in a task will be:

1. *Background luminance of the task.* This is the magnitude of the light reflected from the target's background into the eyes of an observer, measured in foot-Lamberts.
2. *Contrast.* This is the difference between the luminance levels of the target and the background. The contrast also needs to be adjusted (divided) by the following factors: real-world conditions (2.5), movement of the target (2.78), and uncertainty in the location (1.5).
3. *Time available for observation.* This ranges from a few milliseconds to several seconds, and can affect the speed and accuracy of performance.
4. *Size of target, measured as a visual angle in arc minutes.*

Blackwell's visibility curves can be modeled by the following equation (with allowable ranges):

$$\%\text{Det} = 81 \times C^{.2} \times L^{.045} \times T^{-0.003} \times A^{0.199}$$

where: %Det = % targets detected (0-100%).
C = Contrast (0.001–1.8).
L = Background luminance (1–100 foot-Lamberts).
T = Viewing time (0.01–1 sec).
A = Visual angle (1–64 arc min).

The percentage of targets detected may be used to check visual strain allowances by specifying a percentile range for population description abilities. Since highly used percentile ranges are the 50th and 95th percentiles, these may also be applied to target detection, to define relaxation allowance categories. At least a 95-percent target detection defines a visual task without significant problems and can thus define the ILO fairly fine work category with its associated 0 percent allowance. At least a 50-percent target detection defines the fine or exacting work category with its 2-percent allowance. Finally, less than half of targets detected can define very fine or very exacting work with its associated 5 percent allowance.

It must be stressed that Blackwell's model does not directly define the relaxation allowance. Instead, it defines absolute target detection ability, which in turn can be used to define a relaxation allowance. Thus, the relaxation allowance should be inversely proportional to the expected percentage of detected targets.

In general, small visual angles usually produce the lowest performance levels, whereas viewing time only affects performance for the higher contrast levels.

Mental Strain

Mental strain is very difficult to measure clearly across many types of tasks. For mental workload, standardized measures of performance have not yet been clearly defined, and variability between individuals on the same task is high. Also underlying any mental strain definition is an understanding of the factors that make a task complex, on which models are lacking. Investigation of the basis and adequacy of these relaxation allowances thus necessarily requires: (1) an independent indicator

EXAMPLE 10–5
Calculation of Relaxation Allowance for Visual Strain.

The inspection of resistors on electronic circuit boards may be considered exacting work which requires a 2 percent allowance based on ILO guidelines. To confirm this, we use the following calculations. The board is viewed at a distance of 12 inches, without magnification, and the stripes on each resistor are 0.02 inches wide. The required visual angle is $3438 \times 0.02/12 = 5.73$ arc minutes. The resistor body (task background) has a luminance of 10 foot-Lamberts, and the contrast between the stripe and background is 0.5. The contrast is divided by a factor of $1.5 \times 2.5 = 3.75$ to adjust for real-world detection and uncertain location (Freivalds and Goldberg, 1988). The mean time for eye fixation is 0.2 seconds. Plugging these values into the detectability equation yields:

$$\begin{aligned}\%\text{Det} &= 81 \times (0.5/3.75)^{0.2} \times 10^{0.045} \times 0.2^{-0.003} \times 5.73^{0.199} \\ &= 81 \times 0.668 \times 1.109 \times 1.005 \times 1.414 = 85.3\%\end{aligned}$$

A detectability of 85.3% is under 95% and would require a 2 percent allowance.

of task complexity, and (2) objective evidence of changing work output with fatigue or time on the job. Even given this information, experimental differences in motivation can greatly affect observed results, rendering comparisons between studies useless. The vagueness of the ILO relaxation allowance complicates matters even further: 1 percent for a fairly complex process; 4 percent for a process requiring a complex or wide span of attention; and 8 percent for a very complex process. At best, a controlled study with timed reading or mental arithmetic tasks, such as those by Okogbaa and Shell (1986), can serve as a crude check of these allowances. Both of these tasks could be considered complex and requiring wide attention span and thus deserving of a 4 percent relaxation allowance. However, reading performance decreased at a rate of 3.5 percent per hour, while arithmetic performance decreased at a rate of approximately 2 percent per hour. Thus, the ILO (1957) guidelines support performance decrements due to mental strain for one hour, but are inadequate for longer time periods and may need to be modified.

Monotony

Assignment of a monotony relaxation allowance, as defined by ILO (1957), is most appropriate as "the result of repeated use of certain mental faculties, as in mental arithmetic." Tasks with low monotony receive no additional allowance; tasks with medium monotony receive 1 percent, and tasks with high monotony receive a 4 percent allowance. Since the cognitive tasks of Okogbaa and Shell (1986) were performed over four hours, perhaps they should also receive the monotony allowance. However, even the addition of the maximum 4 percent allowance would only extend the period of adequacy to two hours. Vigilance tasks present another example of monotonous work. Baker, Ware, and Sipowicz (1962) noted that subjects detected 90 percent of short light interruptions in a lamp after one hour of continuous testing. By the end of 10 hours, the subjects were only detecting about 70 percent of signals, or a drop in performance of 2 percent per hour. Again, the ILO allowance is not sufficient to compensate performance decrements that occur over an entire shift and better allowances need to be developed.

Tediousness

Allowances for task tediousness (or task repetition) are 0 percent for a rather tedious task, 2 percent for a tedious task and 5 percent for a very tedious task. As defined by ILO (1957), this allowance is applied to elements in which there is "repeated use of certain members of the body, such as fingers, hands, arms or legs." In other words, a tedious task repeatedly utilizes the same physical movements, whereas a monotonous task repeatedly uses the same mental faculties. A methods study used to simplify work and make it more efficient also tends to make it more tedious or repetitive for skilled workers, making it more likely that the workers will be prone to work-related musculoskeletal disorders (see Chapter 5).

Developmental work on risk assessment models for CTD (ANSI, 1995; Seth, Weston, and Freivalds, 1998) has found that the frequency of motions, postures of the hand and wrist, and the forces exerted by the hand are key factors in increasing the risk for CTD. However, these relatively crude models are far from reliable and are not validated over a wide range of jobs and industries. Still, epidemiological data from NIOSH (1989) have indicated that 10,000 damaging wrist motions per shift is a threshold point at which CTD cases increase noticeably, and that at 20,000 motions, the number of cases increases significantly. This would seem to imply that 10,000 motions is a limit for unimpaired performance and for relaxation allowances of up to 100 percent, which is much greater than recommended by ILO (1957). Obviously, most of these models are very much in the developmental stages, and considerable validation must be performed before specific values for allowances can be set.

SPECIAL ALLOWANCES

Unavoidable Delays

This class of delays applies to effort elements and includes: interruptions from the supervisor, dispatcher, time study analyst, and others; material irregularities; difficulty in maintaining tolerances and specifications; and interference delays where multiple machine assignments are made.

As can be expected, every operator experiences numerous interruptions during the course of the work day. The supervisor or group leader may interrupt the operator to give instructions or to clarify certain written information. The inspector may interrupt to point out the reasons for some defective work that passed through the operator's workstation. Interruptions also come from planners, expediters, fellow workers, production personnel, and others.

Unavoidable delays are frequently a result of material irregularities. For example, the material may be in the wrong location; or it may be running too soft or too hard, or too short or too long; or it may have excessive stock on it, as in forgings when the dies begin to wash out, or on castings due to incomplete removal of risers. When material deviates substantially from standard specifications, the customary unavoidable delay allowance may prove inadequate. The analyst must then restudy the job and allow time for the extra elements introduced by the irregular material.

If more than one facility is assigned to an operator during the work day, one facility or more must wait until the operator completes work on another facility. As more facilities are assigned to the operator, the interference time delay increases. In practice, machine interference predominantly occurs from 10 to 30 percent of the total working time, with extremes of 0 to 50 percent (Maynard, 1956). The amount of machine interference depends on the number of machines assigned, the randomness of the required servicing time, the proportion of service time to the running time, the length of the running time, and the mean length of the service time.

When from two to six machines are assigned, the use of empirical curves, as illustrated in Figure 10–4, is recommended. For seven or more machines, the following expression can be used:

$$I = 50[\sqrt{[(1 + X - N)^2 + 2N]} - (1 + X - N)]$$

where: I = Interference, expressed as a percentage of the mean attention time.
X = Ratio of mean machine running time to mean machine attention time.
N = Number of machine units assigned to one operator.

The amount of interference is related to the performance of the operator. Thus, the operator demonstrating a low level of effort experiences more machine interference than the operator who, through higher effort, reduces the time spent in attending the stopped machine. The analyst determines the normal interference time that, when added to the machine running time required to produce one unit and the normal time spent by the operator in servicing the stopped machine, equals the cycle time. The cycle time divided into the running time of each machine, multiplied by the number of machines assigned to the operator, yields the average machine running hours per hour. Thus, we have the expression:

$$O = \frac{NT_1}{C}$$

where: O = Machine running hours per hour.
N = Number of machines assigned to the operator.

FIGURE 10–4
Interference in the percentage of attention time when the number of facilities assigned to one operator is six or less.

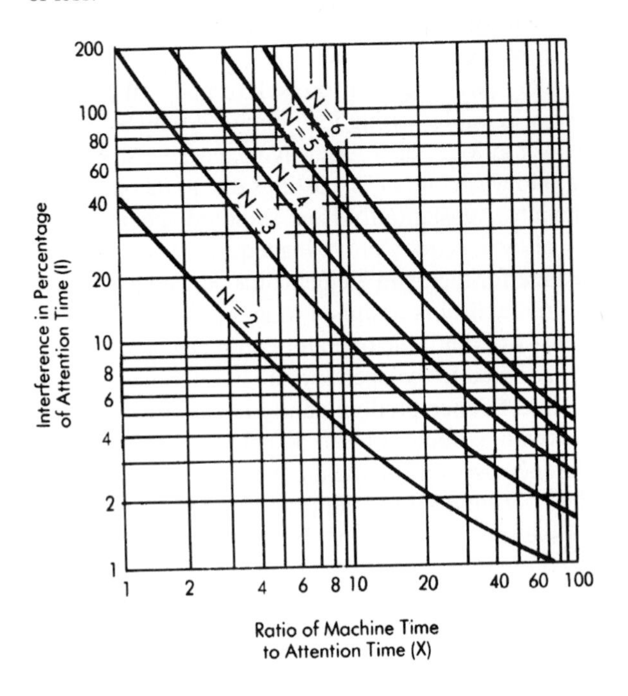

T_1 = Running time (hours) to produce one piece.
C = Cycle time to produce one piece.

and

$$C = T_1 + T_2 + T_3$$

where: T_2 = Time (hours) spent by normal operator attending the stopped facility.
T_3 = Time lost by normal operator working at normal pace because of interference.

Using queuing theory, analysts have developed tables in which the interval between service times is exponential and the service time is either constant or exponential. Table A3-13, Appendix 3, gives these values for various values of *k,* which is the ratio of service time to facility running time, $k = T_2/T_1$.

Avoidable Delays

It is not customary to provide any allowance for avoidable delays, such as: visits with other operators for social reasons, uncalled-for work stoppages, and idleness

EXAMPLE 10–6
Relaxation Allowance for Machine Interference.

In quilling production, an operator is assigned 60 spindles. The mean running time per package (unit of output), determined by stopwatch study, is 150 minutes. The standard mean attention time per package, developed by time study, is 3 minutes. The computation of the machine interference, expressed as a percentage of the mean operator attention time, is:

$$I = 50[\sqrt{[(1 + X - N)^2 + 2N]} - (1 + X - N)]$$

$$= 50\left[\sqrt{\left(1 + \frac{150}{3.00} - 60\right)^2 + 120} - \left(1 + \frac{150}{3.00} - 60\right)\right]$$

$$I = 50[\sqrt{(1 + 50 - 60)^2 + 120} - (1 + 50 - 60)]$$

$$I = 1{,}160\%$$

Thus, we would have:

Machine running time	150.00 min
Attention time, including the personal, fatigue, and unavoidable delay allowances	3.00 min
Interference time (11.6)(3.00)	34.80 min
Standard time for 60 packages	187.80 min
Standard time per package	$\frac{187.80}{60} = 3.13$ min

other than rest to overcome fatigue. While operators may take these delays at the expense of output, no allowance for these cessations of work is provided in the development of the standard.

Extra Allowances

In the typical metal trade and related operations, the allowance for personal, unavoidable, and fatigue delays usually approximates 15 percent. However, in certain cases, an extra allowance may be needed for a fair standard. For example, for a substandard lot of raw material, analysts may need to add an extra allowance to account for an unduly high number of rejects. Or, a situation may arise in which, because of the breakage of a jib crane, the operator is obliged to place a 50-pound casting in the chuck of the machine. An extra allowance would be needed for the additional fatigue in manually handling the work.

Whenever practical, the analyst should establish allowed time for additional work by breaking the operation down into elements and then including the

EXAMPLE 10–7
Relaxation Allowance for Machine Interference, Using Queuing Theory.

With reference to Example 10–6:

$$k = \frac{3.00}{150.00} = 0.02$$

$$N = 60$$

From Table A3-13, Appendix 3, with exponential service time and $k = 0.02$ and $N = 60$, we have a waiting time (interference delay) of 16.8 percent of the cycle time. Denoting the interference time by T_3, we have $T_3 = 0.168C$, where C is the cycle time to produce one unit per spindle. Then:

$$150 + 3.00 + 0.168C = C$$
$$0.832C = 153$$
$$C = 184 \text{ minutes}$$

and

$$T_3 = 0.168C = 30.9 \text{ minutes}$$

The interference time computed by the equation closely agrees with that developed by the queuing model. However, as N (the number of machines assigned) becomes smaller, the proportional difference between the two techniques increases.

allowance times in the specific elements. If this is not practical, the analyst would provide an extra allowance for the operation as a whole.

One extra allowance frequently used, especially in the steel industry, is a percentage added to a portion or all of the cycle time to account for the operator observing the process to maintain efficient progress of the operation. This allowance is frequently referred to as "attention time" allowance and may cover such situations as: an inspector observing tin plate coming off the line; a first helper observing the conditions of a molten bath or receiving instructions from the melter; or a crane operator receiving directions from the crane hooker. For example, the U.S.X. Company adds a 35-percent attention allowance to the actual required time.

This extra allowance provides the incentive for the operator to keep the facility productively employed during the entire working period. Without this extra allowance, such operators would find it impossible to make the same earnings as fellow employees. For example, if the machine-controlled portion of a cycle is 2 minutes and the operator-controlled portion is 1 normal minute, operators would have to work at a pace 25 percent above normal to realize a 7 percent increase in productivity, as follows:

$$\frac{3 \text{ min (normal cycle time)}}{2 \text{ min (mach. control)} + \dfrac{1 \text{ min (normal effort time)}}{1.25 \text{ (operator performance)}}} = 1.07$$

$$1.07 - 1.00 = 0.07$$

which is the increase in productivity when the operator is working at 25 percent above normal during the effort part of the cycle. If management adds an extra allowance of 25 percent to the machine-controlled portion of the cycle, operators would be able to achieve 25 percent incentive earnings if they worked at a pace 25 percent above normal and did not utilize more allowance than was provided for personal delays, unavoidable delays, and fatigue.

Workstation cleaning and machine oiling time.

The time required to clean the workstation and lubricate the operator's machine may be classified as an unavoidable delay. When these elements are the responsibilities of the operator, management must provide an applicable allowance. Analysts often include this time as a total cycle time allowance when these functions are performed by the operator. The type and size of equipment and the material being fabricated have a considerable effect on the time required to do these tasks. One company has established a table of allowances to cover these items (see Tables 10–7,10–8 and 10–9). Supervisors frequently give operators 10 or 15 minutes at the end of the day to perform these elements. When this is done, the established standards would not include any allowance for cleaning and oiling the machine.

Power feed machine time.

The allowance required for power feed elements usually differs from that required for effort elements. Consider the following two factors in power feed: shutdowns and tool maintenance. Allowances are made for shutdowns due to minor repairs. If a major facility repair is needed, an extra allowance would be provided. This extra allowance would not be applied within the standard, but would be an independent standard covering machine repair.

TABLE 10–7
Clean Machine Allowance Chart

Item	Percent per machine		
	Large	Medium	Small
1. Clean machine when lubricant is used	1	3/4	1/2
2. Clean machine when lubricant is not used	3/4	1/2	1/4
3. Clean and put away large amounts of tools or equipment	1/2	1/2	1/2
4. Clean and put away small amounts of tools or equipment	1/4	1/4	1/4
5. Shut machine down for cleaning (this percentage is for machines equipped with chip pans, which are stopped at intervals to permit sweeper to clean away large chips)	1	3/4	1/2

TABLE 10–8
Machine Classification

Large machine	Medium machine
1. Turret lathe (20-in chuck or over)	1. Turret lathe (10-in to 20-in chuck)
2. Boring mill (60 in and over)	2. Boring mill (under 60 in)
3. Punch press (100T and over)	3. Punch press (40T to 100T)
4. Planer (over 48 in)	

TABLE 10–9
Oil Machine Allowance Chart

	Percent per machine		
Item	Large	Medium	Small
1. Machine oiled or greased by hand	1½	1	½
2. Machine oiled automatically	½	½	½

A tool maintenance allowance provides time for the operator to maintain tools after the original setup. In the setup time, the operator is expected to provide first-class tools properly ground. Generally, little tool maintenance takes place during the course of the average production run. In long runs, tools have to be sharpened periodically. The percentage allowance for tool maintenance varies directly with the number of perishable tools in the setup. As an example, one manufacturer's tool maintenance allowance is shown in Table 10–10.

Policy Allowances

A policy allowance is used to provide a satisfactory level of earnings for a specified level of performance under exceptional circumstances. Such allowances could cover new employees, the differently abled, workers on light duty, etc. These are typically decided by management, perhaps with union negotiations.

ALLOWANCE APPLICATIONS

The fundamental purpose of all allowances is to add enough time to normal production time to enable the average worker to meet the standard when performing at standard performance. There are two ways of applying allowances. The most common is to add a percentage to the normal time, so that the allowance is based on a percentage of the productive time only. It is also customary to express the allowance as a multiplier, so that the normal time (*NT*) can be readily adjusted to the standard time (*ST*):

TABLE 10–10
Allowance for Tool Maintenance

	Percent
1. One or more tools ground in tool crib	1
2. One tool sharpened by the operator	3
3. Two or more tools cutting at one time sharpened by the operator	6

$$ST = NT + NT \times \text{allowance} = NT \times (1 + \text{allowance})$$

where: ST = Standard time
NT = Normal time

Thus, if a 10 percent allowance were provided on a given operation, the multiplier would be $1 + 0.1 = 1.1$.

For example, the computation of a total allowance might be:

Personal needs	5.0 percent
Basic fatigue	4.0
Unavoidable delay	1.0
Total	10.0 percent

Normal time would then be multiplied by 1.1 to determine the standard time. Using the time study example in Figure 8-12, the average normal time of 0.177 minutes for element #1 is multiplied by 1.1, corresponding to a 10 percent allowance, to yield a standard time of 0.195 minutes. Out of a 480-minute work day, the operator would work 480/1.1 = 436 minutes and would be allowed 44 minutes rest. (For the proper allocation of this rest, see Chapter 4.)

Some companies apply the percent allowance to the total working day, since the actual production time might not be known. For the previous example, the multiplier of normal times becomes $100/(100 - 10) = 1.11$ (instead of 1.1), and the standard time for element 1 becomes 0.196. Of the 480-minute work day, $480 \times 0.1 = 48$ minutes (instead of 44 minutes) would be allocated to rest. Although the difference between the two approaches is not large, it could add up to over a year for several hundred workers. This would then become a policy decision for the company.

SUMMARY

In establishing time study allowance values, the analyst should exercise the same care as for individual studies. It would be useless to divide a job carefully into elements, precisely measure the duration of each element in hundredths of a minute, accurately evaluate the performance of the operator, and then arbitrarily assign a random allowance. If the allowances are too high, manufacturing costs are unduly inflated; if the allowances are too low, tight standards result, causing poor labor relations and eventual failure of the system.

Typical allowances used in industry are found in the industrial survey of 42 different plants mentioned in the beginning of the chapter. The smallest average total allowance was 10 percent, found in a plant producing household electrical appliances. The greatest average allowance was 35 percent, in effect in two different steel plants. The average allowance of all the plants was 17.7 percent.

Some guidelines are provided in Table 10–11 for allocating allowances in a more quantitative manner than has typically been done. These guidelines are especially appropriate for abnormal position, use of force, atmospheric, and other work environment conditions. The allowances for visual strain, mental strain, monotony and tediousness are currently less reliable and need to be developed in greater detail.

QUESTIONS

1. What main areas are allowances intended to cover?
2. What are the two methods used in developing standard allowance data? Briefly explain the application of each technique.
3. Give several examples of personal delays. Which percentage allowance seems adequate for personal delays, under typical shop conditions?
4. What are some of the major factors that affect fatigue?
5. Why do many proponents of the MTM fundamental motion data system advocate that no allowance be provided for fatigue?
6. Which operator interruptions would be covered by the unavoidable delays allowance?
7. What percentage allowance is usually provided for avoidable delays?
8. When are extra allowances provided?
9. Why is fatigue allowance frequently applied only to the effort areas of the work cycle?
10. Why are allowances based on a percentage of the productive time?
11. What are the advantages of having operators oil and clean their own machines?
12. When would it be desirable to add to the allowances shown in Table 10–2?
13. Give several reasons for not applying an extra allowance to operations, if the major part of the cycle is machine controlled and the internal time is small compared to the cycle time.

TABLE 10–11
Revised Table of Allowances

CONSTANT ALLOWANCES	
Personal Needs	5
Basic Fatigue	4
VARIABLE RELAXATION ALLOWANCES	
Posture Allowances	
Standing	2
Awkward (bending, lying, crouching)	10
Illumination Levels	
One level (one IES subcategory) below recommended	1
Two levels below recommended	3
Three levels (full IES category) below recommended	5
Visual Strain (Close Attention)	
Fine or exacting work	2
Very fine or exacting work	5
Mental Strain	
First hour	2
Second hour	4
Each succeeding hour	+2
Monotony	
First hour	2
Second hour	4
Each succeeding hour	+2
Use of Muscular Force or Energy	
Repeated grasping or infrequent lifting (<1 lift per 5 minutes)	$RA = 1800*(t/T)^{1.4}*(f/F-0.15)^{0.5}$, where $T = 1.2/(f/F-0.15)^{0.618} - 1.21$
Frequent Lifting (>1 lift per 5 minutes)	Use NIOSH Lifting Guidelines with LI < 1.0
Overall whole body activities	$RA = (\Delta W/4 - 1)*100$
Atmospheric Conditions	$RA = \exp(-41.5 + 0.0161W + 0.497\,WBGT)$
Noise Level	$RA = 100*(D-1)$, where $D = C_1/T_1 + C_2/T_2 + ..$
Repetitiveness (Tediousness)	
No established standards yet	Use CTD risk analysis and keep risk index < 1.0

PROBLEMS

1. Develop an allowance factor for an assembly element for which the operator stands in a slightly awkward position, regularly lifts a weight of 15 pounds, and has good light and atmospheric conditions. The attention required is fine, the noise level is continuous at 70 dBA, and the mental strain is low, as is the monotony and the tediousness of the work.

2. Calculate the fatigue allowance for an operation for which the operator loads and unloads a 25-pound gray iron casting once every five minutes at a height of 30 inches.

3. What would be the allowance for problem #2 if the frequency increased to 5 per minute?

4. In the XYZ Co., an all-day study revealed the following noise sources: 0.5 hrs, 100 dBA; 1 hr, less than 80 dBA; 3.5 hrs, 90 dBA; 3 hrs, 92 dBA. Calculate the rest allowance.

5. What fatigue allowance should be given to a job if it took 1.542 minutes to perform the operation at the end of continuous work, but only 1.480 minutes at the beginning of continuous work?

6. Based on the ILO tabulation, what would the allowance factor be on a work element involving a 42-pound pulling force in inadequate light, in which exacting work was required?

7. Calculate the fatigue allowance for a worker shoveling scrap metal into a bin. The operator's working heart rate is approximately 130 beats/min and resting heart rate is 70 beats/min.

8. Calculate the fatigue allowance for a 200-pound worker monitoring a steel furnace while standing next to it. The WBGT index indicates 92 °F.

9. Compare the interference delay allowance using the queuing model and the machine interference equation, where N = 20, mean facility running time is 120 minutes, and attention time is 3 minutes.

REFERENCES

Åberg, V., K. Elgstrand, P. Magnus, and A. Lindholm. "Analysis of Components and Prediction of Energy Expenditure in Manual Tasks." *The International Journal of Production Research,* 6, no. 3 (1968), pp. 189–196.

ANSI. *Control of Work-Related Cumulative Trauma Disorders -Part I: Upper Extremities.* ANSI Z-365 Working Draft. Itasca, NY: American National Standards Institute, 1995.

Baker, R. A., J. R. Ware, and R. R. Sipowicz. "Signal Detection by Multiple Monitors." *Psychological Record,* 12, no. 2 (April 1962), pp. 133–137.

Bennett, C. A., A. Chitlangia, and A. Pangrekar. "Illumination Levels and Performance of Practical Visual Tasks." *Proceedings of the 21st Annual Meeting of the Human Factors Society* (1977), pp. 322–325.

Blackwell, H. R. "Brightness Discrimination Data for the Specification of the Quantity and Quality of Illumination." *Illuminating Engineering* (1952), p. 602.

Brey, E. E. "Fatigue Research in Its Relation to Time Study Practice." *Proceedings, Time Study Conference.* Chicago, IL: Society of Industrial Engineers, February 14, 1928.

Chaffin, D. B., A. Freivalds, and S. R. Evans. "On the Validity of an Isometric Biomechanical Model of Worker Strengths." *IIE Transactions,* 19, no. 3 (September 1987), pp. 280–288.

Davis, H. L., T. W. Faulkner, and C. L. Miller. "Work Physiology." *Human Factors,* 11, no. 2 (April 1969), pp. 157–165.

Freivalds, A. "Development of an Intelligent Knowledge Base for Heat Stress Evaluation." *International Journal of Industrial Engineering*, 2, no. 1 (November 1987), pp. 27–35.

Freivalds, A., and J. Goldberg. "Specification of Bases for Variable Relaxation Allowances." *The Journal of Methods-Time Measurement,* 14 (1988), pp. 2–29.

Garg, A., D. B. Chaffin, and G. D. Herrin. "Prediction of Metabolic Rates for Manual Materials Handling Jobs." *American Industrial Hygiene Association Journal*, 39, no. 12 (December 1978), pp. 661–674.

IES. *IES Lighting Handbook Reference Volume*. New York: Illuminating Engineering Society of North America, 1981.

ILO. *Introduction to Work Study.* 1st ed. Geneva, Switzerland: International Labour Office, 1957.

ILO. *Introduction to Work Study*. 3rd ed. Geneva, Switzerland: International Labour Office, 1979.

Konz, S. *Work Design*. 4th ed. Scottsdale, AZ: Publishing Horizons, Inc., 1995.

Lazarus, I. "Inaccurate Allowances Are Crippling Work Measurements." *Factory* (April 1968), pp. 77–79.

Maynard, H. B. *Industrial Engineering Handbook*. New York: McGraw-Hill, 1956.

Moodie, Colin L. "Assembly Line Balancing." In *Handbook of Industrial Engineering*, 2nd ed. Ed. Gavriel Salvendy. New York: John Wiley & Sons, 1992.

MTM Association. *Work Measurement Allowance and Survey*. Fair Lawn, NJ: MTM Association, 1976.

Murrell, K. F. H. *Human Performance in Industry*. New York: Reinhold Publishing, 1965.

NIOSH. *Criteria for a Recommended Standard for Occupational Exposure to Hot Environments*. Washington, DC: National Institute for Occupational Safety and Health, Superintendent of Documents, 1986.

NIOSH. *Health Hazard Evaluation - Eagle Convex Glass Co.* HETA 89-137-2005. Cincinnati, OH: National Institute for Occupational Safety and Health, 1989.

Okogbaa, O. G., and R. L. Shell. "The Measurement of Knowledge Worker Fatigue." *IIE Transactions*, 12, no. 4 (December 1986), pp. 335–342.

Rohmert, W. "Problems in Determining Rest Allowances, Part I: Use of Modern Methods to Evaluate Stress and Strain in Static Muscular Work." *Applied Ergonomics*, 4, no. 2 (June 1973), pp. 91–95.

Rohmert, W. "Ermittlung von Erholungspausen für statische Arbeit des Mensche." *Internationale Zeitschrift für Angewandte Physiologie einschließlich Arbeitsphysiologie*, 18 (1960), pp. 123–140.

Seth, V., R. Weston, and A. Freivalds. "Development of a Cumulative Trauma Disorder Risk Assessment Model." To appear in *International Journal of Industrial Ergonomics*, 1998.

Silverstein, B. A., L. J. Fine, and T. J. Armstrong. "Occupational Factors and Carpal Tunnel Syndrome." *American Journal of Industrial Medicine*, 11, no. 3 (1987), pp. 343–358.

Stecke, Kathryn E. "Machine Interference: Assignment of Machines to Operators." In *Handbook of Industrial Engineering*, 2nd ed. Ed. Gavriel Salvendy. New York: John Wiley & Sons, 1992.

CHAPTER 11

Standard Data

KEY POINTS:

- Use standard data to comprise tabular or graphical collections of normal times for motions or work elements.
- Keep set-up and cyclical elements separate.
- Keep constant and variable elements separate.
- Add allowances after summation of element times for a new standard time.

Standard time data are element time standards, from time studies, that have been proven to be accurate and reliable. Analysts classify and file element standards so they can readily be abstracted when needed. For example, using such standards, an analyst might determine how long it should take the normal operator to pick up a small casting, place it in a jig, close the jig, lock the part with a quick-acting clamp, advance the spindle of the drill press, and perform the remainder of the elements required to produce the part.

The application of standard time data is fundamentally an extension of the same process used to arrive at allowed times through a stopwatch time study. The principle of applying standard data is not new; many years ago, Frederick W. Taylor proposed that each element time established be properly indexed so that it could be used to establish time standards for future work. When we speak of standard data today, we refer to all the tabulated element standards, curves, alignment charts, and tables that allow the measurement of a specific job without the use of a timing device, such as a stopwatch.

Standard data can have several levels of refinement: motion, element, and task. The more refined the standard data element, the broader its range of usage. Thus, motion standard data has the greatest application, but it takes longer to develop such a standard than either element or task standard data. Element standard data is widely applicable and allows the faster development of a standard than motion data. This chapter is devoted to element standard data, and Chapter 13 addresses motion standard data and its use to predetermine standard time.

Work standards calculated from standard data are relatively consistent in that the tabulated elements result from many proven stopwatch time studies. Since the values are tabulated, it is only necessary to accumulate the required elements to establish a standard, and the various time study personnel within a given company should all arrive at identical performance standards for a given method. Consistency therefore is assured for standards established by different analysts in a plant, as well as for the various standards computed by a given time study observer.

Standards on new work can usually be computed more quickly through standard data than by a stopwatch time study. This allows the establishment of standards for indirect labor operations which is usually impractical if stopwatch studies are required. Typically, one work measurement analyst can establish five rates per day using stopwatch methods. This compares to 25 rates per day using standard data. Also, standard data permit the establishment of time standards over a wide range of work. Table 11–1 illustrates the coverage possible when element standard data are determined.

STANDARD TIME DATA DEVELOPMENT

General Approach

To develop standard time data, analysts must distinguish constant elements from variable elements. A constant element is one for which the allowed time remains approximately the same for any part, within a specific range. A variable element is one for which the allowed time varies within a specified range of work. Thus, the element "start machine" would be a constant, while the element "drill $^{3}/_{8}$-inch diameter hole" would vary with the depth of the hole, the feed, and the speed of the drill.

Standard data are indexed and filed as they are developed. Also, setup elements are kept separate from elements incorporated into each piece time, and constant

TABLE 11–1
Efficiency of Standard Data

Operation	Number of time studies taken for development of standard data	Number of standards set in one year from standard data developed	Percent of standards set that would be covered by stopwatch studies
Coremaking	60	7,500	0.8
Snag grinding	40	656	6.1
Visual inspection	53	422	12.6
Turret lathe operation	100	600	16.7

Source: Phil Carroll, Jr., "Notes on Standard Elemental Data," *Modern Machine Shop,* April 1950, p. 176.

elements are separated from variable elements. Typical standard data for machine operation would be tabulated as follows:

1. Setup.
   a. Constants.
   b. Variables.
2. Each piece.
   a. Constants.
   b. Variables.

Standard data are compiled from different elements in time studies of a given process over a period of time. Only those studies proved valid through use are included in the data. In tabulating standard data, the analyst must be careful to define the endpoints clearly. Otherwise, there may be a time overlap in the recorded data. For example, in the element "out stock to stop" on a bar feed No. 3 Warner & Swasey turret lathe, the element could include: reaching for the feed lever, grasping the lever, feeding the bar stock through the collet to a stock stop located in the hex turret, closing the collet, and reaching for the turret handle. Then again, this element may involve only the feeding of bar stock through the collet to a stock stop. Since standard data elements are compiled from a great number of studies taken by different time study observers, the limits or end points of each element should be carefully defined. Figure 11–1 illustrates a form for summarizing data taken from an individual time study to develop standard data on die-casting machines.

To fill a specific need in a standard data tabulation, analysts may resort to work measurements of the element in question. This can be handled accurately by using an electronic stopwatch (see Chapter 8) accurate to 0.001 minute. In this analysis, the snapback method records the elapsed element time, as analysts are usually interested in determining the allowed time for only a few of the elements comprising the study. Upon completion of the observations, the analyst summarizes the element elapsed times and determines the mean, as in the case of a typical time study. Next, the analyst performance rates the average values and adds an allowance, to arrive at fair allowed times.

Sometimes, because of the brevity of individual elements, measuring their durations separately is impossible. However, the analyst can determine their individual values by timing groups collectively and using simultaneous equations to solve for the individual elements, as shown in Example 11–1.

In determining standard data elements by simultaneous equations, the analyst must be consistent when reading the watch at the terminal points of the established elements. As previously stated, inconsistency in establishing the terminal points results in erroneous standard data elements.

Calculation of Cutting Times

By learning the feeds and speeds for different types of material, analysts can calculate and tabulate the cutting times for different machining operations. Table 11–2 gives recommended speeds and feeds for high-speed and tungsten carbide drills

FIGURE 11–1
Standard data development form.

DIE-CASTING MACHINE

Part No. ____ Mach. No. & Type __________ Operator ______________ Date __________

Of __________ No. of Parts in Tote Pan ________ Method of Placing Parts in Tote Pan ______ Total Wt. of Flsh, Parts, Gate & Sprue ____

__________ No. of Parts per Shot __________ Liquid Metal ______ Plastic Metal ______ Chill ____ Skim ____ Drain ____

Capacity in Lbs. Holding Pot __________ Describe Greasing ______________________________

Describe Loosening of Part ______________________________

Describe Location ______________________________

ELEMENTS	TIME	END POINTS
Get metal in holding pot	______	All waiting time while metal is being poured in pot.
Chill metal	______	From time operator starts adding cold metal to liquid metal in pot until operator stops adding cold metal to liquid metal in pot.
Skim metal	______	From time operator starts skimming until all scum has been removed.
Get ladleful of metal	______	From time ladle starts to dip down into metal until ladleful of metal reaches edge of machine or until ladle starts to tip for draining.
Drain metal	______	From time ladle starts to tip for draining until ladleful reaches edge of machine.
Pour ladle of metal in machine	______	From time ladleful of metal reaches edge of machine until foot starts to trip press.
Trip press	______	From time foot starts moving toward pedal until press starts downward.
Press time	______	Complete turnover of press.
Hold plunger down	______	From time plunger stops downward motion until plunger starts moving upward.
Press button and raise slug	______	From time plunger stops moving until slug is raised out of cavity.
Remove slug--drop slug--lift	______	From time slug is raised out of cavity until slug is pushed into tote pan or pot.
Trip pedal to open dies	______	From time foot starts moving to pedal until dies start to open.
Wait for dies to open	______	From time dies start to open until die stops moving.
Remove part from die	______	From time die stops moving until part is free of die cavity.
Place part in tote pan	______	From time part is free of die cavity until part is placed in tote pan.

used on various kinds of material. A number of technical handbooks provided by cutting tool manufacturers also include this information. The use of formula construction tools (see Chapter 12) may also be helpful here.

EXAMPLE 11–1
Calculation of Brief Element Times.

Element a is "pick up small casting," element b is "place in leaf jig," c is "close cover of jig," d "position jig," e "advance spindle," and so on. These elements are timed in groups, as follows:

$$a + b + c = \text{element } 1 = 0.070 \text{ min} = A \quad (1)$$

$$b + c + d = \text{element } 3 = 0.067 \text{ min} = B \quad (2)$$

$$c + d + e = \text{element } 5 = 0.073 \text{ min} = C \quad (3)$$

$$d + e + a = \text{element } 2 = 0.061 \text{ min} = D \quad (4)$$

$$e + a + b = \text{element } 4 = 0.068 \text{ min} = E \quad (5)$$

First, we add these five equations:

$$3a + 3b + 3c + 3d + 3e = A + B + C + D + E$$

Then, let

$$A + B + C + D + E = T$$
$$3a + 3b + 3c + 3d + 3e = T = 0.339 \text{ min}$$

and

$$a + b + c + d + e = \frac{0.339}{3} = 0.113 \text{ min}$$

Therefore:

$$A + d + e = 0.113 \text{ min}$$

Then:

$$d + e = 0.113 \text{ min} - 0.07 \text{ min} = 0.043 \text{ min}$$

since

$$c + d + e = 0.073 \text{ min}$$
$$c = 0.073 \text{ min} - 0.043 = 0.03 \text{ min}$$

Likewise:

$$d + e + a = 0.061$$

and

$$a = 0.061 - 0.043 = 0.018 \text{ min}$$

Substituting in Equation 1, we get:

$$b = 0.070 - (0.03 + 0.018) = 0.022$$

Substituting in Equation 2, we see that:

$$d = 0.067 - (0.022 + 0.03) = 0.015 \text{ min}$$

Substituting in Equation 3, we arrive at:

$$e = 0.073 - (0.015 + 0.03) = 0.028 \text{ min}$$

Drill press work

A drill is a fluted end-cutting tool used to originate or enlarge a hole in solid material. In drilling operations on a flat surface, the axis of the drill is at 90 degrees to the surface being drilled. When a hole is drilled completely through a part, the analyst must add the lead of the drill to the length of the hole to determine the entire distance the drill must travel to make the hole. When a blind hole is drilled, the dis-

TABLE 11–2
Typical Operating Conditions for Producing Holes in Various Materials with Twist Drills

Material	Tool Material	Speed (sfm)	Feed Rate 1/8″D	1/4″D	1/2″D	3/4″D	1″D
Aluminum and its alloys	High speed steel	350	0.003	0.006	0.010	0.0155	0.0190
Bakelite	Tungsten carbide	60–70	0.0015	0.003	0.0035	0.0045	0.005
Copper and its alloys							
(high machinability)	High speed steel	200	0.003	0.006	0.010	0.0155	0.0190
(low machinability)	High speed steel	70	0.003	0.006	0.010	0.0155	0.0190
Fiberglass-epoxy	Tungsten carbide	650	0.0025	0.0030	0.0035	0.0042	0.0050
Glass	Tungsten carbide	15–25	light	light	light	light	light
High temperature alloys							
cobalt based	High speed steel with cobalt	20	0.0015	0.003	0.0035	0.0045	0.005
Iron based	High speed steel with cobalt	25	0.002	0.0035	0.006	0.0085	0.0105
Iron							
cast (soft)	High speed steel	140–150	0.003	0.006	0.010	0.0155	0.019
	Tungsten carbide	90–165	0.003	0.005	0.008	0.0105	0.0125
cast (medium hard)	High speed steel	80–110	0.003	0.005	0.008	0.0105	0.0125
hard chilled	Tungsten carbide	30	0.002	0.0035	0.006	0.0085	0.0105
malleable	High speed steel	90–120	0.003	0.005	0.008	0.0105	0.0125
	Tungsten carbide	100–150	0.002	0.0035	0.006	0.0085	0.0105
ductile	High speed steel	60	0.003	0.005	0.008	0.0105	0.0125
	Tungsten carbide	80–100	0.002	0.0035	0.006	0.0085	0.0105
Magnesium and its alloys	High speed steel	150–400	0.003	0.006	0.010	0.0155	0.0190
Plastics	High speed steel	100	0.003	0.005	0.008	0.0105	0.0125
	Tungsten carbide	100–200	0.002	0.0035	0.006	0.0085	0.0105
Rubber (hard)	High speed steel	100–300	0.002	0.0035	0.006	0.0085	0.0105
Steel							
plain to 0.25 C	High speed steel	80	0.003	0.005	0.008	0.0105	0.0125
plain 0.25 to 0.50 C	High speed steel	65	0.003	0.005	0.008	0.0105	0.0125
plain 0.5 to 0.9 C	High speed steel	55	0.003	0.005	0.008	0.0105	0.0125
alloy, low C	High speed steel	70	0.003	0.006	0.010	0.0155	0.019
alloy, med C	High speed steel	50–60	0.002	0.0035	0.006	0.0085	0.0105
maraging	High speed steel	55	0.003	0.005	0.008	0.0105	0.0125
stainless (austenitic)	High speed steel with cobalt	55	0.002	0.0035	0.006	0.0085	0.0105
stainless (ferritic)	High speed steel	65	0.002	0.0035	0.006	0.0085	0.0105
stainless (martensitic)	High speed steel with cobalt	65	0.003	0.006	0.010	0.0155	0.019
Zinc alloys	High speed steel	250	0.003	0.006	0.010	0.0155	0.0190

Source: Tool and Manufacturing Engineers Handbook, 4th ed. (Dearborn, Mich.: Society of Manufacturing Engineers, 1983), Vol. 1, *Machining,* Table 9–20, pp. 9-90–9-91.

tance from the surface being drilled to the deepest penetration of the drill is the distance that the drill must travel (see Figure 11–2).

Since the commercial standard for the included angle of drill points is 118 degrees, the lead of the drill may be readily found through the following expression:

$$l = \frac{r}{\tan A}$$

FIGURE 11–2
Drill travel distance.
Distance L indicates the distance the drill must travel when drilling through (illustration at left) and when drilling blind holes (illustration at right) (lead of drill is shown by distance l).

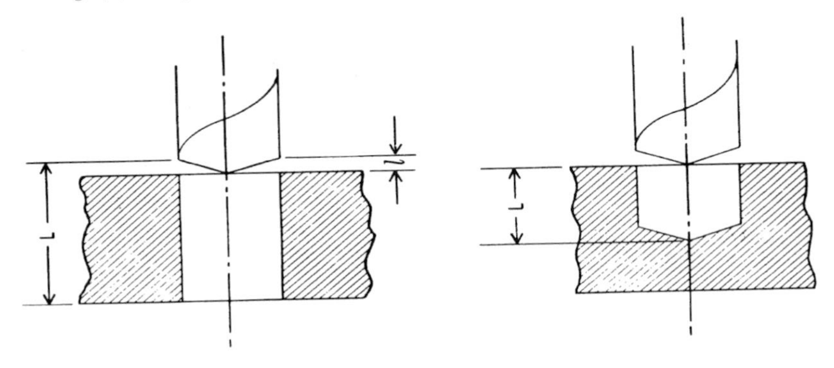

where: l = Lead of drill.
r = Radius of drill.
tan A = Tangent of one half the included angle of the drill.

To illustrate, calculate the lead of a general purpose drill 1 inch in diameter:

$$l = \frac{0.5}{\tan 59°}$$

$$l = \frac{0.5}{1.6643}$$

$$l = 0.3 \text{ inch lead}$$

After determining the total length that the drill must move, we divide this distance by the feed of the drill in inches per minute, to find the drill cutting time in minutes.

Drill speed is expressed in feet per minute (fpm), and feed in thousandths of an inch per revolution. To change the feed into inches per minute when the feed per revolution and the speed in feet per minute are known, the following equation can be used:

$$F_m = \frac{3.82\,(f)(Sf)}{d}$$

where: F_m = Feed in inches per minute.
f = Feed in inches per revolution.
Sf = Surface feet per minute.
d = Diameter of drill in inches.

For example, to determine the feed in inches per minute of a 1-inch drill running at a surface speed of 100 feet per minute and a feed of 0.013 inch per revolution, we have:

$$F_m = \frac{(3.82)(0.013)(100)}{1} = 4.97 \text{ inches per minute}$$

To determine how long it would take for this 1-inch drill running at the same speed and feed to drill through 2 inches of a malleable iron casting, we use the equation:

$$T = \frac{L}{F_m}$$

where: T = Cutting time in minutes.
L = Total length drill must move.
F_m = Feed in inches per minute.

which should yield:

$$T = \frac{2 \text{ (thickness of casting)} + 0.3 \text{ (lead of drill)}}{4.97}$$

$$= 0.464 \text{ minutes cutting time}$$

The cutting time thus calculated does not include an allowance, which must be added to determine the total allowed time. The allowance should include time for variations in material thickness and for tolerance in setting the stops, both of which affect the cycle cutting time. Personal and unavoidable delay allowances should also be added to arrive at an equitable total allowed element time.

Not all speeds may be available on the machine being used. For example, the recommended spindle speed for a given job might be 1,550 rpm, but the machine may be capable of running only 1,200 rpm. In that case, 1,200 rpm should be used as the basis for computing allowed times.

Lathe Work

Many variations of machine tools are classified as lathes. These include the engine lathe, turret lathe, and automatic lathe (automatic screw machine). All of these lathes are used primarily with stationary tools or with tools that translate over the surface to remove material from the revolving work, which includes forgings, castings, or bar stock. In some cases, the tool revolves while the work is stationary, as on certain stations of automatic screw machine work. For example, a slot in a screw head can be machined in the slotting attachment on the automatic lathe.

Many factors alter speeds and feeds, such as the condition and design of the machine tool, the material being cut, the condition and design of the cutting tool, the coolant used for cutting, the method of holding the work, and the method of mounting the cutting tool. Table 11–3 outlines the approximate cuts, feeds, and speeds for certain metallic and nonmetallic turning.

TABLE 11–3
Suggested Operating Parameters for Machining Various Materials with Carbide Tools

Work Material	Cutting Speed, sfm (m/min)		Feed Rate, ipr (mm/rev)		Depth of Cut, in. (mm)	
	Roughing	Finishing	Roughing	Finishing	Roughing	Finishing
Free-machining carbon steels: AISI 1100 and 1200 series, 140–190 Bhn	250–1,100 (76–335)	1,000–2,000 (305–610)	0.010–0.085 (0.25–2.16)	0.005–0.015 (0.13–0.38)	0.125–0.675 & up (3.18–17.15)	Up to 0.180 (4.57)
Plain carbon steels: AISI 1000 series, 185–240 Bhn	200–800 (61–244)	700–1,600 (213–488)	0.010–0.085 (0.25–2.16)	0.005–0.015 (0.13–0.38)	0.125–0.675 & up (3.18–17.15)	up to 0.180 (4.57)
Alloy steels: AISI 1300, 4000, 5000, 8000, and 9000 series, 190–240 Bhn	175–600 (53–183)	550–1,200 (168–366)	0.010–0.085 (0.25–2.16)	0.005–0.015 (0.13–0.38)	0.125–0.675 & up (3.18–17.15)	up to 0.180 (4.57)
Cast irons: gray, nodular, and malleable, 150–210 Bhn	200–1,200 (61–366)	200–750 (61–229)	0.010–0.055 (0.25–1.40)	0.005–0.015 (0.13–0.38)	0.125–0.675 & up (3.18–17.15)	Up to 0.180 (4.57)
Martensitic stainless steels: wrought 400 and 500 series, and PH types, 175–210 Bhn	175–450 (53–137)	450–850 (137–259)	0.010–0.040 (0.25–1.02)	0.005–0.015 (0.13–0.38)	0.125–0.500 (3.18–12.70)	Up to 0.180 (4.57)
Austenitic stainless steels: wrought 200 and 300 series, 140–190 Bhn	125–425 (38–130)	425–650 (130–198)	0.010–0.040 (0.25–1.02)	0.005–0.015 (0.13–0.38)	0.125–0.500 (3.18–12.70)	Up to 0.180 (4.57)
Superalloys: iron, nickel, titanium, and cobalt-based alloys, 240–300 Bhn	30–150 (9–46)	150–400 (46–122)	0.010–0.025 (0.25–1.02)	0.005–0.015 (0.13–0.38)	0.100–0.300 (2.54–7.62)	Up to 0.180 (4.57)
Tool steels: wrought high-speed, shock resistant, and hot and cold work, 210–240 Bhn	100–300 (30–91)	275–750 (84–229)	0.010–0.065 (0.25–1.65)	0.005–0.015 (0.13–0.38)	0.125–0.675 & up (3.18–17.15)	Up to 0.180 (4.57)
Nonferrous free-matching alloys: aluminum, copper, zinc, and brass alloys, 80–120 Bhn	400–1,200 (122–366)	1,000–2,000 (305–610)	0.010–0.085 (0.25–2.16)	0.005–0.015 (0.13–0.38)	0.125–0.675 & up (3.18–17.15)	Up to 0.180 (4.57)
Nonmetallics: nylons, acrylics, and phenolic resins	350–800 (107–244)	800–1,500 (244–457)	0.010–0.040 (0.25–1.02)	0.005–0.015 (0.13–0.38)	0.125–0.500 (3.18–12.70)	Up to 0.180 (4.57)

Source: Tool and Manufacturing Engineers Handbook, 4th ed. (Dearborn, Mich.: Society of Manufacturing Engineers, 1983), Vol. 1, *Machining,* Table 3–12, p. 3–25; Kennametal, Inc.

As in drill press work, feeds are expressed in thousandths of an inch per revolution, and speeds in surface feet per minute. To determine the cutting time for L inches of cut, the length of cut in inches is divided by the feed in inches per minute, or

$$T = \frac{L}{F_m}$$

where: T = Cutting time in minutes.
L = Total length of cut.
F_m = Feed in inches per minute.

and

$$F_m = \frac{3.82\,(Sf)(f)}{d}$$

where: f = Feed in inches per revolution.
Sf = Speed in surface feet per minute.
d = Diameter of work in inches.

Milling machine work

Milling refers to the removal of material with a rotating multiple-toothed cutter. While the cutter rotates, the work is fed past the cutter. This differs from a drill press, for which the work is usually stationary. In addition to machining plane and irregular surfaces, operators use a milling machine for cutting threads, slotting, and cutting gears.

In milling work, as in drill press and lathe work, the speed of the cutter is expressed in surface feet per minute. Feeds or table travel are usually expressed in thousandths of an inch per tooth. To determine the cutter speed in revolutions per minute from the surface feet per minute and the diameter of the cutter, use the following expression:

$$N_r = \frac{3.82S_f}{d}$$

where: N_r = Cutter speed in revolutions per minute.
S_f = Cutter speed in feet per minute.
d = Outside diameter of cutter in inches.

To determine the feed of the work in inches per minute into the cutter, use the expression:

$$F_m = fn_tN_r$$

where: F_m = Feed of the work into the cutter in inches per minute.
f = Feed of cutter in inches per tooth.
n_t = Number of cutter teeth.
N_r = Cutter speed in revolutions per minute.

The number of cutter teeth suitable for a particular application may be expressed as:

$$n_t = \frac{F_m}{F_t \times N_r}$$

where: F_t = Chip thickness.

Table 11–4 gives suggested feeds and speeds for milling under average conditions.

To compute the cutting time on milling operations, the analyst must take into consideration the lead of the milling cutter when figuring the total length of cut

TABLE 11–4

Typical Speeds and Feeds for Milling Various Materials with Representative Tool Materials (feed per tooth—in; cutting speed—sfm)

Material milled	Tool Material	Face Mills	Slab Mills	Form Mills
Aluminum alloys	High speed steel	0.010–0.025 300–1,200	0.015–0.025 300–1,200	0.010–0.020 300–1,200
	Uncoated carbide	0.010–0.020 2,000–4,000	0.010–0.020 2,000–4,000	0.008–0.015 2,000–4,000
	Coated carbide	0.010–0.020 4,000–6,000	0.010–0.020 4,000–6,000	0.008–0.015 4,000–6,000
Brass	High speed steel	0.010–0.025 150–300	0.008–0.020 100–300	0.008–0.015 100–300
	Uncoated carbide	0.010–0.020 500–1,500	0.010–0.020 500–1,500	0.008–0.015 500–1,500
	Coated carbide	0.010–0.020 1,500–3,000	0.008–0.020 1,500–3,000	0.008–0.015 1,500–3,000
Bronze	High speed steel	0.010–0.025 50–225	0.008–0.020 50–200	0.008–0.015 50–200
	Uncoated carbide	0.010–0.020 300–1,500	0.010–0.020 300–1,400	0.008–0.015 200–1,400
	Coated carbide	0.010–0.020 1,500–2,700	0.010–0.020 1,400–2,500	0.008–0.015 1,400–2,500
Cast irons 150–180 Brinell	High speed steel	0.010–0.025 80–120	0.010–0.025 70–110	0.010–0.015 60–80
	Uncoated carbide	0.010–0.016 275–800	0.010–0.020 275–900	0.008–0.015 250–800
	Coated carbide	0.010–0.016 800–1,100	0.010–0.020 900–1,200	0.008–0.015 800–1,100
Cast irons 180–225 Brinell	High speed steel	0.010–0.020 60–80	0.008–0.015 50–70	0.008–0.012 50–60
	Uncoated carbide	0.008–0.015 250–500	0.008–0.015 225–500	0.006–0.012 225–500
	Coated carbide	0.008–0.015 500–800	0.008–0.015 500–750	0.006–0.012 500–750
Cast irons	High speed steel	0.005–0.012 40–60	0.005–0.01 35–50	0.005–0.01 35–50
	Uncoated carbide	0.005–0.010 200–400	0.005–0.010 200–400	0.005–0.010 200–400
	Coated carbide	0.005–0.010 400–600	0.005–0.010 400–600	0.005–0.010 400–600
Steels 100–150 Brinell	High speed steel	0.015–0.020 80–130	0.008–0.015 80–130	0.008–0.010 70–100
	Uncoated carbide	0.010–0.018 400–900	0.008–0.015 350–800	0.004–0.010 350–800
	Coated carbide	0.010–0.018 900–1,500	0.008–0.015 800–1,300	0.004–0.010 800–1,300
Steels 150–250 Brinell	High speed steel	0.010–0.020 50–70	0.008–0.015 50–70	0.006–0.010 50–70
	Uncoated carbide	0.010–0.015 300–700	0.008–0.015 300–700	0.004–0.010 300–700
	Coated carbide	0.010–0.015 700–1,200	0.008–0.015 700–1,200	0.004–0.010 700–1,200
Steels 250–350 Brinell	High speed steel	0.005–0.010 35–60	0.005–0.010 35–50	0.005–0.010 35–50
	Uncoated carbide	0.008–0.015 225–600	0.007–0.012 200–600	0.003–0.008 200–600
	Coated carbide	0.008–0.015 600–1,000	0.007–0.015 600–1,000	0.003–0.008 600–1,000
Steels 350–450 Brinell	High speed steel	0.003–0.008 20–35	0.005–0.008 20–35	0.003–0.008 20–35
	Uncoated carbide	0.005–0.012 180–400	0.007–0.012 150–400	0.003–0.008 150–400
	Coated carbide	0.005–0.012 400–600	0.007–0.012 400–600	0.003–0.008 400–600

Source: Tool and Manufacturing Engineers Handbook, 4th ed. (Dearborn, Mich.: Society of Manufacturing Engineers, 1983), Vol. 1, *Machining,* Table 10–5, p. 10–52, Valenite Div., Valeron Corp.

under power feed. This can be determined by triangulation, as illustrated in Figure 11–3, which shows the slab-milling of a pad.

In this case, to arrive at the total length that must be fed past the cutter, the lead *BC* is added to the length of the work (8 inches). Clearance for removal of the work after the machining cut is handled as a separate element, because greater feed under rapid table traverse is used. By knowing the diameter of the cutter, you can determine *AC* as being the cutter radius, and you can then calculate the height of the right triangle *ABC* by subtracting the depth of cut *BE* from the cutter radius *AE,* as follows:

$$BC = \sqrt{AC^2 - AB^2}$$

In the preceding example, suppose we assume that the cutter diameter is 4 inches and that it has 22 teeth. The feed per tooth is 0.008 inch, and the cutting speed is 60 feet per minute . We can compute the cutting time by using the equation:

$$T = \frac{L}{F_m}$$

where: T = Cutting time in minutes.
L = Total length of cut under power feed.
F_m = Feed in inches per minute.

Then, L would be equal to (8 inches + BC) and

$$BC = \sqrt{4 - 3.06} = 0.975$$

Therefore:

$$L = 8.975$$
$$F_m = fn_tN_r$$
$$F_m = (0.008)(22)N_r$$

FIGURE 11–3
Slab-milling a casting 8 inches in length.

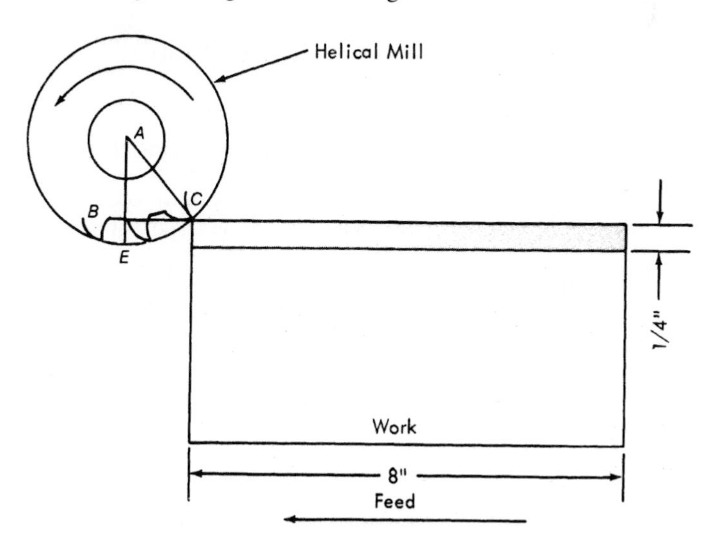

or $$N_r = \frac{3.82\ S_f}{d} = \frac{(3.82)(60)}{4} = 57.3 \text{ rpm}$$

Then $$F_m = (0.008)(22)(57.3) = 10.1 \text{ inches per minute}$$

and $$T = \frac{8.975}{10.1} = 0.888 \text{ minutes cutting time}$$

Through a knowledge of feeds and speeds, analysts can determine the required cutting or processing time for various work performed in their plants. The illustrations cited in drill press, lathe, and milling work are representative of the techniques used to establish raw cutting times. The necessary applicable allowances must be added to these values to create fair total element allowed values.

Determining Horsepower Requirements

When developing standard data times for machine elements, the analyst must tabulate horsepower requirements for various materials in relation to depth of cut, cutting speeds, and feeds. Frequently, standard data can be used for planning new work. To avoid overloading existing equipment, the analyst should have information on the workload being assigned to each machine for the conditions under which the material is being removed. For example, in the machining of high-alloy steel forgings on a lathe capable of a developed horsepower of 10, it would not be feasible to take a 3/8-inch depth of cut while operating at a feed of 0.011 inch per revolution and a speed of 200 surface feet per minute. Table 11–5 indicates a horsepower requirement of 10.6 for these conditions. Consequently, the work would need to be planned for a feed of 0.009 inch at a speed of 200 surface feet; this would only require a horsepower rating of 8.7. (See Table A3–12, Appendix 3, for horsepower requirements.)

Plotting Curves

Because of space limitations, tabularizing values for variable elements is not always convenient. By plotting a curve or a system of curves in the form of an

TABLE 11–5
Horsepower Requirements for Turning High-Alloy Steel Forgings for Cuts 3/8 Inch and 1/2 Inch Deep at Varying Speeds and Feeds

Surface feet	3/8-in. depth cut (feeds, in./rev.)						1/2-in. depth cut (feeds, in./rev.)					
	0.009	0.011	0.015	0.018	0.020	0.022	0.009	0.011	0.015	0.018	0.020	0.022
150	6.5	8.0	10.9	13.0	14.5	16.0	8.7	10.6	14.5	17.3	19.3	21.3
175	8.0	9.3	12.7	15.2	16.9	18.6	10.1	12.4	16.9	20.2	22.5	24.8
200	8.7	10.6	14.5	17.4	19.3	21.3	11.6	14.1	19.3	23.1	25.7	28.4
225	9.8	11.9	16.3	19.6	21.7	23.9	13.0	15.9	21.7	26.1	28.9	31.8
250	10.9	13.2	18.1	21.8	24.1	26.6	14.5	17.7	24.1	29.0	32.1	35.4
275	12.0	14.6	19.9	23.9	26.5	29.3	15.9	19.4	26.5	31.8	35.3	39.0
300	13.0	16.0	21.8	26.1	29.0	31.9	17.4	21.2	29.0	34.7	38.6	42.5
400	17.4	21.4	29.1	34.8	38.7	42.5	23.2	28.2	38.7	46.3	51.5	56.7

alignment chart, the analyst can express considerable standard data graphically on one page.

Figure 11–4 illustrates a nomogram for determining turning and facing time. For example, if the problem is to determine the production in pieces per hour to turn 5 linear inches of a 4-inch diameter shaft of medium carbon steel on a machine utilizing 0.015-inch feed per revolution and having a cutting time of 55 percent of the

FIGURE 11–4
Nomogram for determining facing and turning time. (Crobalt, Inc.)

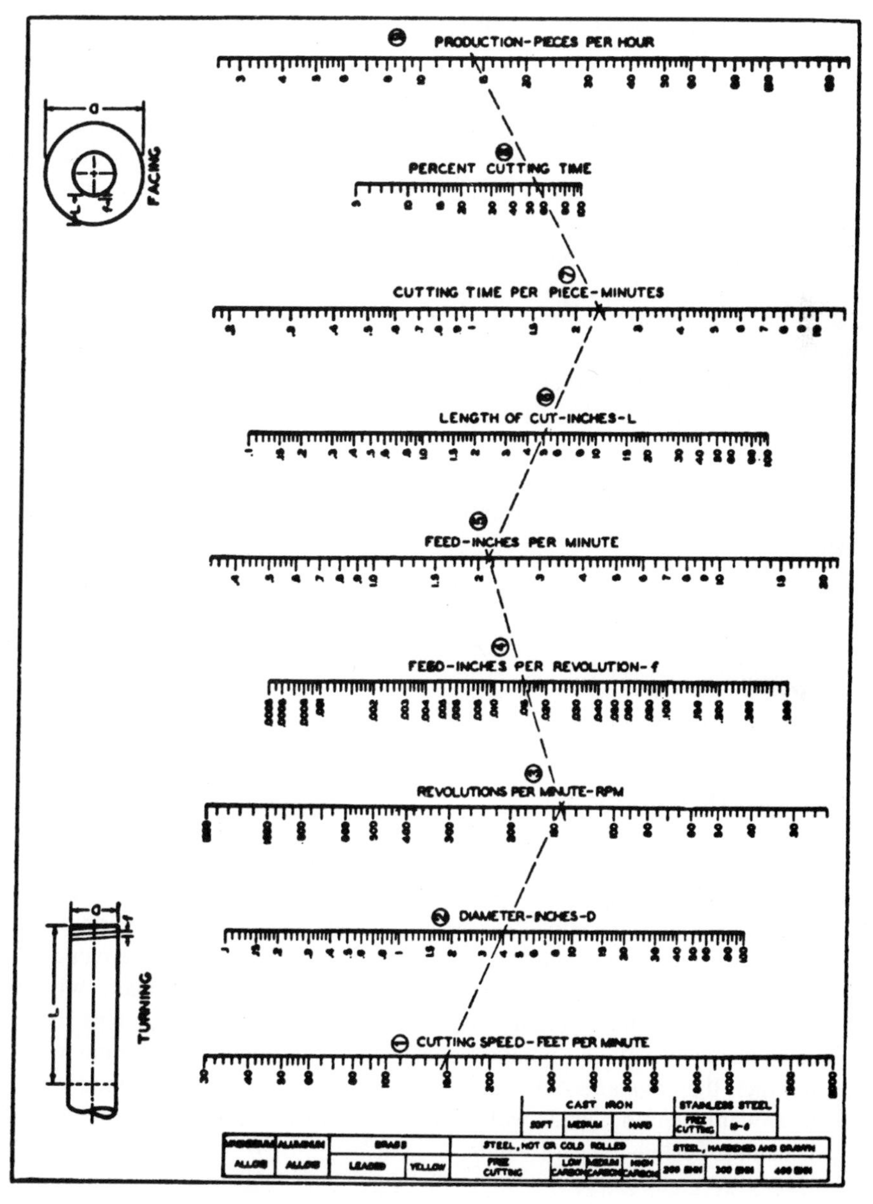

cycle time, the answer could be readily determined graphically. Connecting a recommended cutting speed of 150 feet per minute for medium carbon steel, shown on scale 1, to the 4-inch diameter of the work, shown on scale 2, results in a speed of 143 rpm, shown on scale 3. The 143-rpm point is connected with the 0.015-inch feed per revolution, shown on scale 4. This line extended to scale 5 shows a feed of 2.15 inches per minute. This feed point connected with the length of cut, shown on scale 6 (5 inches), gives the required cutting time, on scale 7. Finally, this cutting time of 2.35 minutes, connected with the percentage of cutting time, shown on scale 8 (in this case, 55 percent), gives the production in pieces per hour, on scale 9 (in this case, 16).

Using curves has some distinct disadvantages. First, it is easy to introduce an error in reading from the curve, because of the amount of interpolation usually required. Second, there is the chance of outright error through incorrect reading or misalignment of the intersections on the various scales.

Curves that show the relationship between time and the variables that affect time may take the form of a single straight line, a curved line, a system of straight lines, as in the ray chart, or a special arrangement of lines characteristic of an alignment chart or nomogram. In plotting simple, one-line curves, certain standard procedures are observed. First, time is plotted on the ordinate, and the independent variables are plotted on the abscissa. Second, if practical, all scales begin at zero, to show their true proportions. Finally, the range of the scale selected for the independent variable should be sufficient to fill the sheet on which the curve is plotted.

Figure 11–5 illustrates a chart that expresses forming time in hours per hundred pieces for a certain gage of stock over a range of sizes expressed in square inches. Each of the 12 points in this chart represents a separate time study. The plotted points indicate a straight-line relationship between the various studies. This relationship can be expressed as:

$$\text{Allowed time} = 0.088 + 0.00038\text{Area}$$

For details on solving for the slope and intercept of the line, using the method of least squares, see Chapter 12.

STANDARD DATA USAGE

For easy reference, constant standard data elements should be tabularized and filed under the machine or the process. Variable data can either be tabularized or expressed as a curve or equation, and then filed under the facility or operation class.

In some instances, for which standard data are broken down to cover a given machine and class of operation, it may be desirable to combine constants with variables and tabularize the summary. These quick-reference data express the time allowed to perform a given operation completely. Figure 11–6 illustrates welding data in which the constants "change electrode" and "arc" are combined with the variables "weld cleaning" and "welding." The result is the work-hours required to weld 1 inch for various sizes of welds.

FIGURE 11–5
Forming time for different stock sizes.

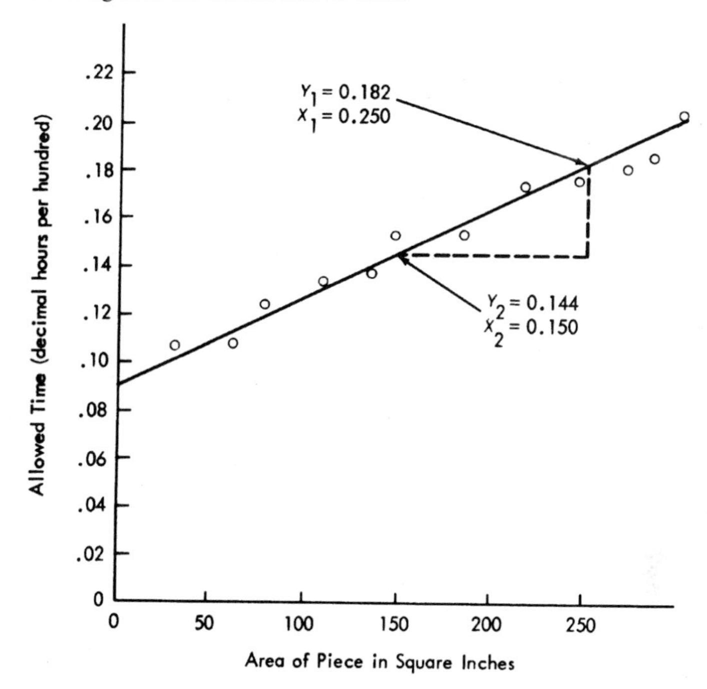

Table 11–6 illustrates standard data for a given facility and operation class for which elements have been combined. By identifying the job according to the distance that the strip of sheet stock is moved per piece, the analyst can find the allowed time for the complete operation.

Frequently, setup elements are combined, or tabularized in combinations, to diminish the time required to summarize a series of such elements. Table 11–7 illustrates standard setup data for No. 5 Warner & Swasey turret lathes at a specific plant. To determine the setup time with these data, the analyst would visualize the tooling in the square and hex turret and would then refer to the table. For example, if a certain job required a chamfering tool, turning tool, and facing tool in the square turret, and needed two boring tools, one reamer, and a collapsible tap in the hex turret, the setup standard time would be 69.70 minutes plus 25.89 minutes, or 95.59 minutes. To arrive at this solution, the analyst would find the value of the relevant tooling under the square turret column (line 8) and the most time-consuming applicable tooling in the hex turret section, which in this case is tapping. This would give a value of 69.7 minutes. Since three additional tools are in the hex turret (first bore, second bore, and ream), the analyst would then multiply 8.63 by 3 to get 25.89 minutes. Finally, adding 25.89 minutes to 69.70 minutes yields the total required setup time.

More frequently, standard data are not combined, but are left in their elemental form, allowing greater flexibility in the development of time standards. Representative standard data applicable to a given plant would appear as shown in Table 11–8. The data include the applicable personal delay and fatigue allowance. Such

FIGURE 11–6
Welding standard data.

CLASSIFICATION

KIND Fillet
TYPE Flat Position
ELECTRODE E-6020 D.H.

WELDING PROCEDURE GENERAL
PROCESS OF WELDING Shielded Metallic Arc POSITION Flat
MATERIAL Mild Steel to Mild Steel P.D.S. 1550, 1555 S.A.E. 1010
ELECTRODE E-6020 (A.W.S. CLASS) D.H. (TYPE) Convex (SHAPE OF WELD) Heavy (COATING)
POWER SOURCE (A.C. OR D.C. – AND POLARITY IF D.C.) D.C. Straight
BACKING None PEENING None CHIPPING None
PREHEAT None
STRESS RELIEVING None

WELDING PROCEDURE–DETAILS

SIZE OF WELD	SIZE OF ELECTRODE	THICKNESS OF PLATE	NUMBER OF PASSES	WELDING CURRENT (AMPERES)	WELDING VOLTAGE (@ ARC)	*MAN HOURS PER INCH WELD	*WELDING SPEED FT./HR.
1/8	1/8	1/8	1	160-190	26-28	.0025	33.3
3/16	5/32	3/16	1	160-190	26-28	.0028	29.8
1/4	3/16	1/4	1	180-230	32-36	.0033	25.3
3/8	1/4	3/4	1	280-330	32-36	.0050	16.7
1/2	1/4	3/4	2	280-330	32-36	.0078	10.7
5/8	1/4	1"	2	280-330	32-36	.0123	6.8
3/4	1/4	1 1/2	4	280-330	32-36	.0196	4.3
1	1/4	1 1/2	6	280-330	32-36	.0318	2.6

* NOTE: INCLUDES CHANGE ELECTRODE TIME, ARC TIME, WELD CLEANING TIME AND WELDING TIME.

standard time data could be used to establish the allowed setup and piece times to drill two hold-down bolt holes in a cast-iron housing casting, for example, as illustrated in Figure 11–7.

The allowed setup time would equal

$$B + C + D + F + G + H + I = 8.70 \text{ minutes}$$

Element A is not included in the setup time because of the simplicity of the job. The drawing is not issued to the operator, because the operation card shows the drill jig number, drill size, plug gage number, and feed and spindle speed. Element E is not included in the setup since, in this case, first-piece inspection is not performed by the inspector. According to the operation card issued, the operator periodically inspects every 10th piece with the go/no-go gage provided. Since the casting would weigh less than 4 pounds, and two parts can easily be handled in one hand, the data outlined would be applicable.

TABLE 11–6

Standard Data for Blanking and Piercing Strip Stock Hand Feed with Piece Automatically Removed on Toledo 76 Punch Press

L (distance in inches)	T (time in hours per hundred hits)
1	0.075
2	0.082
3	0.088
4	0.095
5	0.103
6	0.110
7	0.117
8	0.123
9	0.130
10	0.137

After investigating the design of the drill jig to be used and making an analytical motion study of the job to determine the elements required to perform the operation, the analyst compiles an element summary, as shown in Table 11–9.

The actual drilling time to drill the two holes must be added to this 0.278-minute time. As previously outlined, this can be readily determined. For a $^1/_2$-inch diameter drill used for drilling cast iron, a surface speed of 100 feet per minute and a feed of 0.008 inch per revolution would be used. One hundred surface feet per minute would equal 764 revolutions per minute.

$$\text{rpm} = \frac{12S_f}{\pi d}$$

where: S_f = Surface feet per minute.
$\pi = 3.14$
d = Diameter of drill in inches.

However, investigation of the drill press for which this work has been routed reveals that 600 rpm or 900 rpm are the closest speeds available to the recommended 764 rpm. The analyst proposes to use the slower speed, due to the condition of the machine, and determines the drilling time as follows:

$$T = \frac{L}{F_m} \times 2$$

$$L = 0.437 + \frac{0.25}{\tan 59^\circ} = 0.588 \text{ inches}$$

$$F_m = (600)(0.008)$$
$$= 0.48 \text{ inches per minute}$$

$$T = \left(\frac{0.588}{4.8}\right) 2 = 0.244 \text{ minute}$$

TABLE 11–7
Standard Setup Data for No.5 Turret Lathes

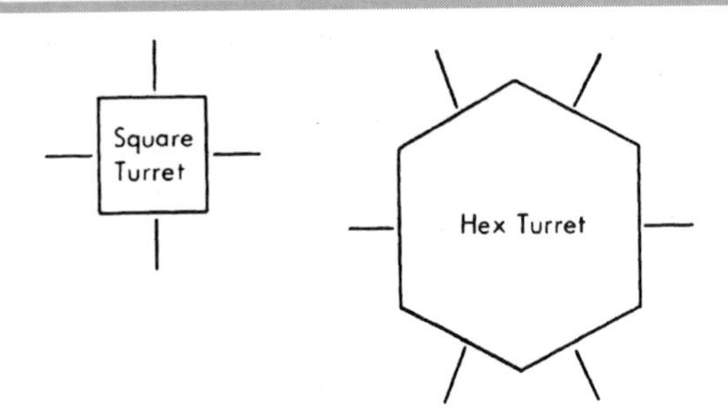

Basic tooling

		Hex turret						
No.	Square turret	Partial	Cham-fer	Bore or turn	Drill	S. tap or ream	C. tap	C. die
1.	Partial	31.5	39.6	44.5	48.0	47.6	50.5	58.5
2.	Chamfer...............	38.2	39.6	46.8	49.5	50.5	53.0	61.2
3.	Face or cut off	36.0	44.2	48.6	51.3	52.2	55.0	63.0
4.	Tn bo grv rad...........	40.5	49.5	50.5	53.0	54.0	55.8	63.9
5.	Face and chf............	37.8	45.9	51.3	54.0	54.5	56.6	64.8
6.	Fa and cut off...........	39.6	48.6	53.0	55.0	56.0	58.5	66.6
7.	Fa and tn or tn and cut off	45.0	53.1	55.0	56.7	57.6	60.5	68.4
8.	Fa, tn, and chf	47.7	55.7	57.6	59.5	60.5	69.7	78.4
9.	Fa, tn, and cut off........	48.6	57.6	57.5	60.0	62.2	71.5	80.1
10.	Fa, tn, and grv	49.5	58.0	59.5	61.5	64.0	73.5	81.6

11. Circled basic tooling from above .... ________
12. Each additional tool in square .... $4.20x$________________ = ________
13. Each additional tool in hex .... $8.63x$________________ = ________
14. Remove and set-up three jaws .... 5.9 ________
15. Set up subassembly or fixture .... 18.7 ________
16. Set up between centers .... 11.0 ________
17. Change lead screw .... 6.6 ________

Total setup ____________ min.

An appropriate allowance must be added to this 0.244-minute cutting time. If a 10-percent allowance on the actual machining time is used, the cutting time would be 0.268 minute and the handling time would be 0.278 minute, for a total time of 0.546 minute to drill one casting on the single-spindle press available. This standard, supplemented with the setup time of 8.70 minutes, would be released as:

Setup: 0.145 hours
Each piece: 0.91 hours per hundred

Thus, an operator who set up this job and ran 1,000 pieces in an 8-hour working day would be performing at an efficiency of 116 percent:

$$\frac{(0.91)(10) + 0.145}{8} = 116\%$$

TABLE 11–8
Standard Data

Setup elements:	**Minutes**
A. Study drawing	1.25
B. Get material and tools and return and place ready for work	3.75
C. Adjust height of table	1.31
D. Start and stop machine	0.09
E. First-piece inspection (includes normal wait time for inspector)	5.25
F. Tally production and post on voucher	1.50
G. Clean off table and jig	1.75
H. Insert drill in spindle	0.16
I. Remove drill from spindle	0.14
Each piece element:	
1. Grind drill (prorate)	0.78
2. Insert drill in spindle	0.16
3. Insert drill in spindle (quick-change chuck)	0.05
4. Set spindle	0.42
5. Change spindle speed	0.72
6. Remove tool from spindle	0.14
7. Remove tool from spindle (quick-change chuck)	0.035
8. Pick up part and place in jig	
a. Quick-acting clamp	0.070
b. Thumbscrew	0.080
9. Remove part from jig	
a. Quick-acting clamp	0.050
b. Thumbscrew	0.060
10. Position part and advance drill	0.042
11. Advance drill	0.035
12. Clear drill	0.023
13. Clear drill, reposition part, and advance drill (same spindle)	0.048
14. Clear drill, reposition part, and advance drill (adjacent spindle)	0.090
15. Insert drill bushing	0.046
16. Remove drill bushing	0.035
17. Lay part aside	0.022
18. Blow out jig and lay part aside	0.081
19. Plug gage part	0.12 per hole

Application: Allen 17-inch vertical single-spindle drill.
Work size: Small work—up to four pounds in weight and such that two or more parts can be handled in each hand.

COMPUTERIZED STANDARD DATA

Today, computers store standard data and retrieve, accumulate, and develop standards for new work in advance of production. Using a computer, standard data, whether stored in the motion, element, or task form, are easily retrieved, accumulated, and adjusted for applicable allowances. Several software systems have a fundamental motion database. For example, MOST, WOCOM, 4 and Univation all use motion data as a base for standards development. Other systems, such as CSD, do not have a built-in database, but allow any company to introduce its own.

FIGURE 11–7
Cast-iron housing.

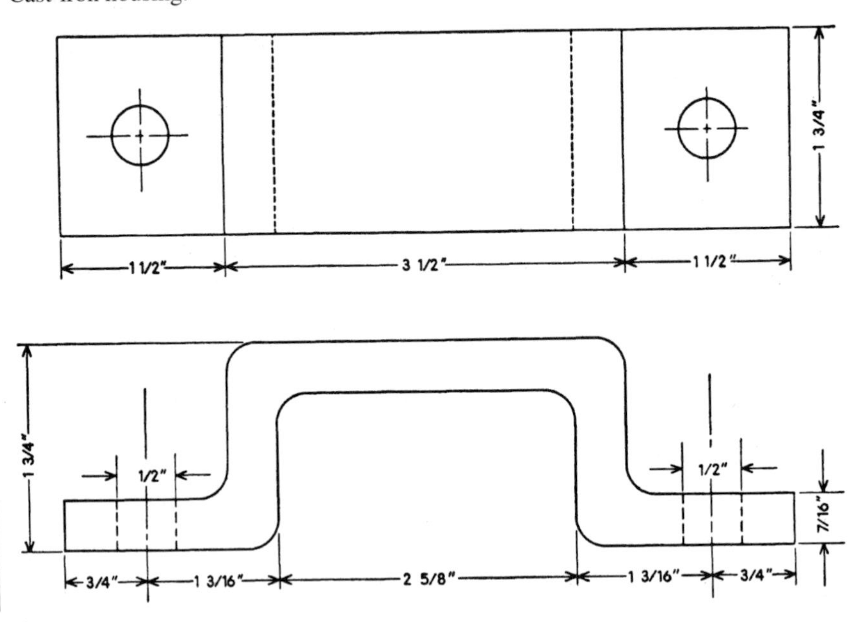

TABLE 11–9
Summary of Jig Standard Data

Element number	Element	Allowed time (minutes)
8	Pk. pt. pl. jig (quick-acting clamp)	0.070
9	Rem. pt. jig (quick-acting clamp)	0.050
10	Pos. pt. & adv. dr.	0.042
13	Cl. dr. rep. pt. & adv. dr.	0.048
12	Cl. dr.	0.023
17	Lay pt. aside	0.022
19	Plug gage pt. (10%)	0.012
1	Gr. dr. (once per 100 pcs.)	0.008
2	Insert dr. in sp. (once per 100 pcs.)	0.002
6	Rem. tool from sp. (once per 100 pcs.)	0.001
		0.278 min

For maximum efficiency and for the development of the best standards, data are stored in both the motion and element forms. Software can select, retrieve, and modify the appropriate motions and/or elements to generate a work standard.

For the processing elements, such as metal cutting or metal joining, logic modules can serve as simple process time calculators, or they can optimize the process parameters, with various degrees of sophistication. For example, analysts can enter the speed, feed, number of cuts, and length of cuts into the computer, which then performs the necessary arithmetic operations to calculate the time for the cut. In the advanced machining logic modules, the input includes the machine scheduled to perform the operation, the workpiece material, and the characteristics of the cut. The computer selects the optimum speed, feed, and number of cuts, consistent with the horsepower available and the condition of the machine. The system may also select the tooling and the proper sequence of cuts.

Assembly logic modules may be used in conjunction with standard data stored on the computer. In such cases, the system can analyze several parameters relating to an assembly operation and can determine which work elements should be applied. The applicable work elements are summed, allowances are applied, and the final job standard and operator instructions are created. The work measurement analyst interacts with the computer by responding to its queries. The computer then selects the appropriate standard data based on the input responses of the analyst.

SUMMARY

When properly applied, standard data permit the establishment of accurate time standards before the job is performed. This feature makes their use especially attractive for estimating the cost of new work, for cost quotes, and for subcontracting purposes.

Time standards can be calculated much more rapidly using standard data, and the consistency of the standards thus established can be assured. Consequently, this technique facilitates the economical development of indirect labor standards. Standards developed from standard data tend to be fair to both the worker and management in that they are the result of already proven standards. The element values used in arriving at the standards have already proved satisfactory as components of established and acceptable standards in use throughout the plant.

The use of standard data simplifies many managerial and administrative problems in plants where unions operate as bargaining agents. Union contracts contain many clauses pertaining to such matters as the type of study to be taken (continuous or snapback), the number of cycles to be studied, the operator(s) to be studied, and the observer to conduct the study. These restrictions frequently make it difficult for analysts to arrive at a standard that is equitable to both the company and the operator. By using the standard data technique, analysts may avoid restrictive details. Not only is determination of a standard simplified, but also some sources of tension between labor and management are alleviated.

In general, the more refined the element times, the greater the coverage of the data. Consequently, in job shops, it is practicable to have individual element values, as well as grouped or combined values, so that the data for a given facility will have enough flexibility to allow rates to be set for all types of work scheduled for a machine.

The use of fundamental motion data for establishing standards, particularly for short-cycle jobs, is becoming more widespread. These data are so basic that standards on practically any class of manual elements can be predetermined. The "objective" or "do" basic divisions, such as "use," must be handled as variables, and tabularized data, curves, or algebraic expressions must be developed.

As the examples indicate, the application of standard data is an exacting technique. Careful and thorough training in methods and shop practice are fundamental before analysts can accurately establish shop standards using standard data. Analysts must know and recognize the need for each element in the class of work. Supplementing this background, the analysts should be analytic, accurate, thorough, conscientious, and completely dependable.

QUESTIONS

1. What do we mean by "standard data"?
2. What is the approximate ratio of the time required to set standards by stopwatch methods and the time required using standard data methods?
3. What are the advantages to establishing time standards by using standard data rather than taking individual studies?
4. What would the time for the element "mill slot" depend on?
5. How are feeds usually expressed in lathe work?
6. What are some of the disadvantages of using curves to tabulate standard data?
7. Which standard procedures should be followed in the plotting of simple curves?

PROBLEMS

1. What would be the horsepower requirements of turning a mild steel shaft 3 inches in diameter if a cut of 1/4 inch with a feed of 0.022 inch per revolution at a spindle speed of 250 rpm were established?

2. How long would it take to turn 6 inches of 1-inch bar stock on a No. 3 W. & S. turret lathe running at 300 feet per minute and feeding at the rate of 0.005 inch per revolution?

3. A plain milling cutter 3 inches in diameter with a face width of 2 inches is being used to mill a piece of cold-rolled steel 1.5 inches wide and 4 inches long. The depth of cut is 3/16 inch. How long would it take to make the cut if the feed per tooth is 0.010 inch and a 16-tooth cutter running at a surface speed of 120 feet per minute is used?

4. Compute the times for elements a, b, c, d, and e when: a + b + c is timed at 0.057 minute; and b + c + d is timed at 0.078 minute; c + d + e equals 0.097 minute; d + e + a is 0.095 minute; and e + a + b is 0.069 minute.

5. What would be the lead of a 3/4-inch diameter drill with an included angle of 118 degrees?

6. What would be the feed in inches per minute of a 3/4-inch drill running at a surface speed of 80 feet per minute and a feed of 0.008 inch per revolution?

7. How long would it take the preceding drill to drill through a casting 2.25 inches thick?

8. The analyst in the Dorben Company made 10 independent time studies in the hand paintspraying section of the finishing department. The product line under study revealed a direct relation between spraying time and product surface area. The following data were collected:

Study no.	Leveling factor	Product surface area	Standard time
1	0.95	170	0.32
2	1.00	12	0.11
3	1.05	150	0.31
4	0.80	41	0.14
5	1.20	130	0.27
6	1.00	50	0.18
7	0.85	120	0.24
8	0.90	70	0.23
9	1.00	105	0.25
10	1.10	95	0.22

Compute the slope and intercept constant, using regression line equations. How much spray time would you allow for a new part with a surface area of 250 square inches?

9. The work measurement analyst in the Dorben Company wants to develop an accurate equation for estimating the cutting of various configurations in sheet metal with a band saw. The data from eight time studies for the actual cutting element provided the following information:

No.	Lineal inches	Standard time
1	10	0.40
2	42	0.80
3	13	0.54
4	35	0.71
5	20	0.55
6	32	0.66
7	22	0.60
8	27	0.61

What would be the relation between the length of cut and the standard time, using the least squares technique?

10. The work measurement analyst in the XYZ Company wishes to develop standard data involving fast, repetitive manual motions, for use in a light assembly department. Because of the shortness of the desired standard data elements, the analyst is obliged to measure them in groups as they are performed on the factory floor. On a certain study, this analyst is endeavoring to develop standard data for five elements, denoted A, B, C, D, and E. Using a fast (0.001) decimal minute watch, the analyst studied a variety of assembly operations and arrived at the following data:

$$A + B + C = 0.131 \text{ min}$$
$$B + C + D = 0.114 \text{ min}$$
$$C + D + E = 0.074 \text{ min}$$
$$D + E + A = 0.085 \text{ min}$$
$$E + A + B = 0.118 \text{ min}$$

Compute the standard data values for each of the elements A, B, C, D, and E.

11. The work measurement analyst in the Dorben Company is developing standard data for prepricing work in the drill press department. Based on the following recommended speeds and feeds, compute the power feed cutting time of 1/2-inch high-speed drills with a 118 degree included angle to drill through material that is 1-inch thick. Include a 10 percent allowance for personal needs and fatigue.

Material	Recommended speed (ft/min)	Feed (in/rev)
Al (copper alloy)	300	0.006
Cast iron	125	0.005
Monel (R)	50	0.004
Steel (1112)	150	0.005

REFERENCES

Brisley, C. L., and R. J. Dossett. "Computer Use and Non-Direct Labor Measurement Will Transform Profession in the Next Decade." *Industrial Engineering,* 12, no. 8 (August 1980), pp. 34–43.

Cywar, Adam W. "Development and Use of Standard Data." In *Handbook of Industrial Engineering.* ed. Gavriel Salvendy. New York: John Wiley & Sons, 1982, pp. 4.8.1–4.8.19.

Fein, Mitchell. "Establishing Time Standards by Parameters." *Proceedings of the Spring Conference of the American Institute of Industrial Engineers.* Norcross, GA: American Institute of Industrial Engineers, 1978.

Metcut Research Associates. *Machining Data Handbook.* Cincinnati, OH: Metcut Research Associates, 1966.

Pappas, Frank G., and Robert A. Dimberg. *Practical Work Standards.* New York: McGraw-Hill, 1962.

Rotroff, Virgil H. *Work Measurement.* New York: Reinhold Publishing, 1959.

CHAPTER 12

Formula Construction

KEY POINTS:

- Use formulas to provide quick and consistent estimates of time standards.
- Use care in defining dependent and independent variables (typically, time is dependent).
- Plot the data to identify initial trends.
- Use log-log or semi-log paper to establish power or exponential relationships.
- Use statistical software packages for calculations.
- Keep formulas as simple as possible.

A time study formula represents a simplification of standard data, as presented in Chapter 11. Such formulas have particular application in nonrepetitive work for which it is impractical to establish standards for each job using an individual time study. Formula construction involves the design of an algebraic expression or a system of curves that establishes a time standard in advance of production by substituting known values peculiar to the job for the variable elements.

FORMULA APPLICATION

Time formulas are applicable to practically all work. They have been successfully used in office operations, foundry work, maintenance work, painting, machine work, forging, coil winding, grass cutting, window washing, floor sweeping, welding, and many other areas. If the collection of time studies involving standardized elements is sufficient to give a reliable sampling of data, it is possible to design a formula for a given range of work in any type of job. However, analysts should only apply the formula to jobs that fall within the limits of the data used in developing it. If they exceed the boundaries of the formula without the supporting proof of individual time studies, erroneous standards, with all the dangers brought about by inequitable rates, may result.

Formula Advantages and Disadvantages

The advantages of using formulas for setting standards rather than individual time studies parallel those of using standard data (see Chapter 11). They may be summarized as follows:

1. More consistent time standards result.
2. Duplicate time studies on similar operations are eliminated.
3. Standards are established much more rapidly.
4. Less experienced, less trained persons can calculate time standards.
5. Accurate, rapid estimates for labor costs may be made before production is begun.

Probably the most significant advantage of the formula over the time study method is that a less costly person can work with formulas than with the standard data resulting from time studies. Any high school graduate proficient in algebra can work out a time study formula and solve for the allowed time required to perform a task. In addition, standards can be established more rapidly using formulas than by accumulating standard data elements. Since columns of figures must be added in the standard data method, there is a greater chance for omission or arithmetic error in setting a standard than in using a formula.

However, caution must be exercised in the treatment of constants in the development of formulas. There is a natural tendency to treat more elements as constants than is valid. Errors can thus creep into the formula design. It is true that a formula will give consistent results; it is therefore also true that if the formula is not accurate, it will yield standards that are consistently wrong.

Another disadvantage of the formula lies in its application. Sometimes, to arrive at a standard at the earliest possible time, work measurement analysts use formulas in instances where the variables are beyond the range of the data used in developing the formula. Thus, the formula may be used where it is not applicable, and the resulting value will not be valid.

Usable Formula Characteristics

A time study formula must be both completely reliable and practical. For the expression to be reliable, it must always give accurate results. If the formula developed is accurate, it should verify the individual standards used in its development, within ±5 percent. The greater the number of studies that can be used in developing the formula, the better the chances for ending up with a reliable formula. In addition, all mathematical calculations must be free from error before the formula can be used with complete confidence.

A practical formula is as clear, concise, and simple as possible. The simpler the formula, the better it can be understood and applied. Cumbersome expressions involving terms to powers should be avoided. Symbols of unknowns should not be repeated throughout the formula, but should appear in only one place, together with their applicable suffixes, prefixes, and coefficients. The area of work that each symbol represents should be specifically identified. Liberal substantiating data should be included in the formula report, so that any qualified, interested party can clearly identify the derivation of the formula. The limitations of the formula must be noted by describing its applicable range in detail. Finally, the analyst should take note of the maxim "garbage in - garbage out." The formula will only be as accurate as the data used to construct it. Analysts using formulas thus constructed can apply them rapidly and accurately, with little difficulty in obtaining the required information.

FORMULA CONSTRUCTION STEPS

The first, and most basic, step in formula construction is the determination of the class of work involved and the range of work to be measured. For example, a formula might be developed for curing bonded rubber parts between 2 and 8 ounces in weight. The class of work would be "curing molded parts," and the range of work would be 2 to 8 ounces.

After overall analysis is finished and analyzed, the next step is collecting the formula data. This step involves gathering former studies with standardized work elements and standard data elements that prove satisfactory, as well as taking new studies, to obtain a sufficiently large sample to cover the range of work for the formula. It is important that like elements (standardized elements) in the different studies have consistent endpoints. This is essential in determining the variables that influence time, as well as in arriving at an accurate value for constant elements.

Next, the element time study data are posted on a worksheet for analysis of the constants and variables. The constants are combined and the variables analyzed so that the factors influencing time can be expressed either algebraically or graphically. Once the constant values have been selected and the variable elements equated, the expression is simplified by combining constants and unknowns where possible.

The next procedure is to develop the synthesis in which the derivation of the formula is fully explained, so that persons using it, and any other interested party, will be able to understand its application and development.

Before the formula is used, it is thoroughly checked for accuracy, consistency, and ease of application. After the formula report is written, describing the method used, working conditions, and application limitations, the formula is ready for installation.

In developing the formula, it is perfectly acceptable for the observer to use existing time studies if: they have been satisfactory; the elements are standardized; the constant and variable elements in the studies have been properly separated; and the studies were taken under prevailing conditions and methods. For any new studies, they should first be broken down into like elements, with endpoints terminating at identical places in the work cycle, thus standardizing the elements. Also, different operators should be studied, to get as large a cross section as possible, and jobs should be selected from the entire range of the proposed formula.

The number of studies needed to construct a formula is influenced by the range of work for which the formula is to be used, the relative consistency of like constant elements in the various studies, and the number of factors that influence the time required to perform the variable elements. At least 10 studies should be available before a formula is constructed. If fewer than 10 are used, the accuracy of the formula may be impaired through incorrect curve construction and data not representative of typical performance. The more studies used, the more data will be available, and the more normal will be the conditions reflected.

Analyze Elements

After a sufficient number of time studies have been gathered, the data are summarized on one worksheet for analysis purposes. Figure 12–1 illustrates a "Master Table of Detail Time Studies" form designed for this purpose. In addition to the information called for on the form, the analyst should post any specific information affecting the variable elements, such as surface area, volume, length, diameter, hardness, radius, and weight, under its corresponding study number in the "Job Characteristics" section of the form.

The analyst enters the name of the operator studied, the rating factor, and the part number of the job in the applicable columns for each time study. In the left-hand column headed "Symbol," is placed an identifying term for each element. Analysts frequently use letters followed by a suffix number for this purpose: A-1, B-1, or C-1. When more than 26 elements are involved, the alphabet is repeated and the suffix 2 is used. These symbols link elements from the Master Table to elements grouped in the synthesis.

Under "Operation Description," the analyst records every element that has occurred in the individual time studies. The element description should be clear so that any interested person can visualize exactly the work content of the element. A "C" or a "V" in the column headed "Operation Class" indicates whether an element is a constant or a variable.

Next, the analyst enters the normal element times from the individual studies in the appropriate spaces. After these data have been posted, the analyst compares the element time values for each element and determines the reasons for variance. As can be expected, even in the constant elements, a certain amount of variation occurs, because of inconsistency in operator method and in performance rating. In general, however, the constant elements should not deviate substantially. The allowed time for each constant is determined by averaging the values of the different studies. This average time is then posted in the "Normal Time" column, and the word "average" is entered in the "Reference" column.

Variable elements tend to vary in proportion to some characteristic(s) of the work, such as size, shape, or hardness. These elements should be carefully studied to determine which factors influence the time, and to what extent. By plotting a

FIGURE 12-1
Master Table of Detail Time Studies.

SHEET NO. 1
OF 1 SHEETS

MASTER TABLE OF

FORMULA 73
DATE June 10,
PART Cylindrical Core
OPERATION Made core from oil sand mix
PERFORMED ON Bench
COMPILED BY J. Bodesky

JOB CHARACTERISTICS

STUDY	S-1	S-2	S-3	S-4
OPERATOR	Petrecca	Winters	Ekey	Kumpf
Performance Factor	110	110	90	110
Diameter of Core	1 7/8	2 1/4	1 7/8	2 3/4
Length of Core	7 7/8	6 1/2	9 1/4	4 1/4
Number of Clamps	1	1	2	1
L/D Ratio	4.2	2.89	4.93	1.55
Area	2.76	3.97	2.78	5.93
Volume	21.7	25.8	25.6	25.2
Cl + Dl + Fl	1.242	1.244	1.499	.890

SYMBOL	OPERATION DESCRIPTION	NORMAL TIME MINUTES	REFERENCE	Operation Class	S-1	S-2	S-3	S-4
A-1	Close core box	.046	Average	C	.049	.041	.050	.053
B-1	Clamp core box (C- clamps)	.112 N	Time vs.No.Cl.	V	.110	.111	.243	.120
C-1	Fill partly full of sand	Y x CT	Time v. L/D v.Vol.	V	.085	.094	.093	.119
D-1	Ram	Y x CT	Time v. L/D v.Vol.	V	.225	.255	.242	.248
E-1	Place rod and wire	.0153L+.03	Time vs.Length	V	.150	.103	.168	.088
F-1	Fill and ram	Y x CT	Time v. L/D v.Vol.	V	.35	.668	.900	.280
G-1	Strike off with slick	.0157A+.07	Time vs. Area	V	.115	.135	.129	.152
H-1	Remove vent wire	.047	Average	C	.048	.050	.048	.056
J-1	Rap box	.043	Average	C	.039	.042	.045	.050
K-1	Remove clamps	.061N	Time vs.No.Cl.	V	.057	.062	.116	.048
M-1	Open box	.046	Average	C	.046	.040	.038	.052
N-1	Roll out core	.0057L+.045	Time vs.Length	V	.098	.082	.102	.075
P-1	Clean box	$\sqrt{.0067+.000016V^2}$	Time vs.Volume	V	.107	.112	.120	.110

curve of time versus the independent variable, the analyst may deduce an algebraic expression of the time representing the element. This procedure is explained in Chapter 11. If an equation can be computed, it is posted in the "Normal Time" column, with notes about the curve(s) used in its deduction placed in the adjacent "Reference" column.

Compute Variable Expressions

If the analysis of element data reveals that one variable characteristic governs the element time, the analyst should construct a graphic relationship, following the procedure outlined in Chapter 11. Time would be the dependent variable, since it is the value to be predicted. For those cases in which time is estimated in terms of a single independent variable, y is the random variable, and time, whose distribution depends on the independent variable, is x. In most relations, x is not random; for all practical purposes, it is fixed and the analyst is concerned with the mean of the corresponding distribution of time y (the successive elements observed through stopwatch analysis) for given x (see Figure 12–9).

Plotted data may take a number of forms: straight line, parabola, hyperbola, ellipse, exponential forms, or no regular geometric form. Graphic procedures may be very useful for predicting equations. If the data plots as a straight line on uniform graph paper, then time (T) is equal to a function of the slope times the variable (v) and the y intercept.

DETAIL TIME STUDIES

S-5	S-6	S-7	S-8	S-9	S-10	S-11	S-12	S-13	S-14	S-15	S-16	S-17
Plasan	Markley	Noyes	Kinachan	Winters	Kumpf	Petrecca	Judd	Judd	Geiger	Plasan	Ekey	Noyes
105	105	105	98	120	100	108	95	102	117	85	.95	100
1 3/4	2	1 7/8	1 1/8	1	1 3/4	1 1/4	1 1/4	1 1/4	7/8	1 1/8	2	1 7/8
9 1/2	8	4 1/4	10	13 1/8	5 1/8	6 3/4	7 5/8	4 1/8	8 1/4	6 1/4	10	12 5/8
2	1	1	2	1	1	1	1	1	1	1	2	2
5.43	4	2.27	8.89	13.13	2.73	5.40	6.10	3.30	9.43	5.56	5.0	6.31
2.40	3.14	2.78	.99	.78	2.40	1.23	1.23	1.23	.60	.99	3.14	2.78
22.8	25.1	11.8	9.90	10.3	12.3	8.27	9.33	5.05	4.96	6.19	31.4	35.1
1.557	1.316	.733	1.489	2.138	.804	1.159	1.181	.797	1.277	1.032	1.740	2.299
.045	.042	.047	.050	.041	.043	.042	.047	.048	.049	.046	.043	.048
.250	109	.087	.232	.218	.098	.114	.097	.128	.091	.099	.229	.240
.121	.141	.082	.129	206	.073	.102	.079	.062	.133	.100	.107	.195
.272	.245	.160	.344	.436	.194	.281	.252	.090	.342	.162	.285	.480
.171	.142	.093	.180	.222	.110	.125	.140	.090	.142	.120	.182	.210
.932	.725	.258	.889	1.290	.392	.665	.850	.432	.802	.626	1.089	1.390
.109	120	104	.078	.077	.198	.089	.084	.086	.074	.085	.119	.106
.047	.042	.041	.052	.051	.043	.042	.042	.044	.040	.051	.052	.045
.035	.041	.045	.057	.048	.040	.039	.042	.041	.050	.038	.047	.041
.120	.039	.062	.116	.120	.071	.059	.056	.070	.058	.058	.140	.145
.038	.041	.045	.054	.046	.066	.038	.065	.045	.040	.035	.038	.052
.109	.091	.068	.108	.120	.080	.082	.090	.065	.090	.071	.110	.118
.115	.118	.090	.089	.088	.090	.087	.085	.085	.085	.088	.138	.150

Another frequent linear relationship is the reciprocal function expressed by the equation:

$$y = \frac{1}{b + mx}$$

The reciprocal function involves a linear relationship for values of $\frac{1}{y}$ and x, since, by taking the reciprocal of both sides of the equation, we get:

$$\frac{1}{y} = b + mx$$

The hyperbolic curve takes the form illustrated in Figure 12–2 and is expressed by the equation:

$$\frac{y^2}{a^2} - \frac{x^2}{k^2} = 1$$

If the plotted data take the form of a segment of a hyperbola, k is computed graphically by drawing a tangent to the curve through the origin and by using the resulting line as the diagonal of the rectangle whose height is a, which is the distance from the origin to the y intercept. Substituting time and the variable characteristic for y and x, respectively, would give:

$$T = \sqrt{a^2 + \frac{(v^2)(a^2)}{k^2}}$$

FIGURE 12-2
Segment of hyperbolic curve.

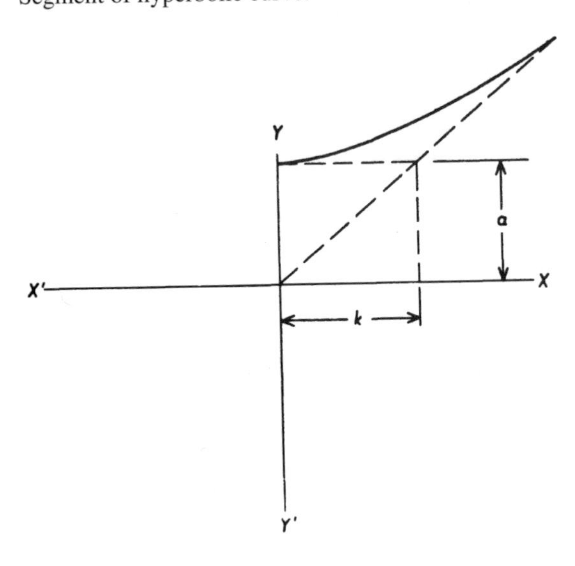

The ellipse segment shown in Figure 12–3 is expressed by the equation:

$$\frac{x^2}{a^2} + \frac{y^2}{b^2} = 1$$

When the plotted time values take the form of an ellipse, a and b can be computed graphically and their values substituted in the expression:

$$T = \sqrt{b^2 - \frac{b^2 v^2}{a^2}}$$

In other instances, the data may take the form of a parabola, as illustrated in Figure 12–4. This curve is expressed by the equation:

$$x^2 = \frac{a^2 y}{b}$$

After substituting graphic solutions for a and b, we would have the equation:

$$T = \frac{b v^2}{a^2}$$

As in previous illustrations, T is equal to time and v is equal to the variable characteristic governing the element time.

One additional relationship that is applicable to formula development is the power function:

$$y = bm^x$$

FIGURE 12–3
Segment of ellipse.

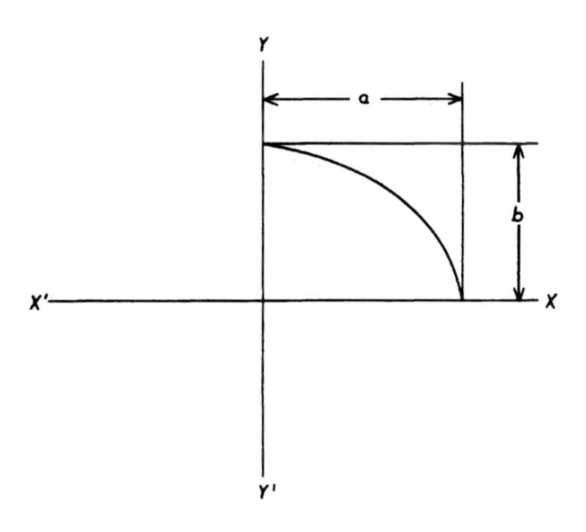

FIGURE 12-4
Segment of parabola.

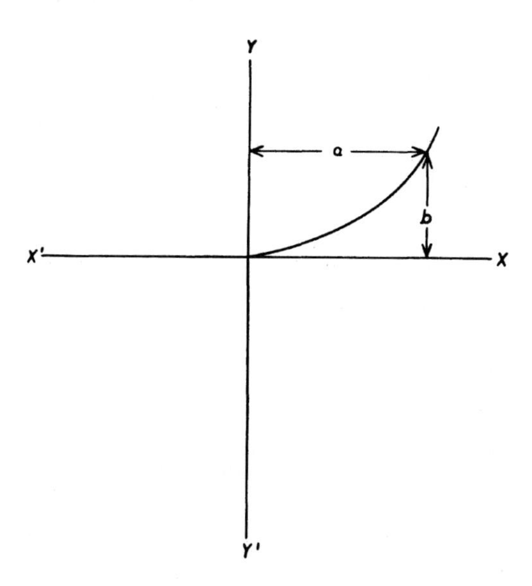

To explain, when the data plotted on uniform graph paper do not follow a straight line, parabola, hyperbola, or circle, semilogarithmic paper may be useful for determining whether the points fall close to a straight line on a transformed scale. If the paired data yield a straight line when plotted on semilog paper, then the curve of y on x is exponential. Therefore, for any given x, the mean of the distribution of y (time in our work) is given by bm^x.

As an example, in the element "strike arc and weld," analysts obtained the following data from 10 detailed studies:

Study no.	**Size of weld**	**Minutes per inch of weld**
1	1/8	0.12
2	3/16	0.13
3	1/4	0.15
4	3/8	0.24
5	1/2	0.37
6	5/8	0.59
7	11/16	0.80
8	3/4	0.93
9	7/8	1.14
10	1	1.52

Plotting the data on rectangular coordinate paper resulted in a smooth curve (see Figure 12–5). Plotting the same data on semilogarithmic paper created the straight line, as shown in Figure 12–6.

The derivation of the equation for this plot was as follows. Select any two points on the straight line:

Point 1: $X = 1 \qquad Y = \log 1.52$
Point 2: $X = 3/8 \qquad Y = \log 0.24$

Solve for the slope m:

$$\begin{aligned} m &= \frac{Y_1 - Y_2}{X_1 - X_2} \\ &= \frac{\log 1.52 - \log 0.24}{1 - 3/8} \\ &= 1.28 \end{aligned}$$

Determine the equation of the line from the equation:

$$Y - Y_1 = m(X - X^1)$$

We get:

$$\begin{aligned} \log Y - \log 1.52 &= 1.28(X - 1) \\ \log Y - 0.18184 &= 1.28X - 1.28 \\ \log Y &= 1.28X - 1.10 \end{aligned}$$

From the exponential form, we get:

$$Y = AB^x$$

FIGURE 12-5
Curve plotted on regular coordinate paper takes exponential form.

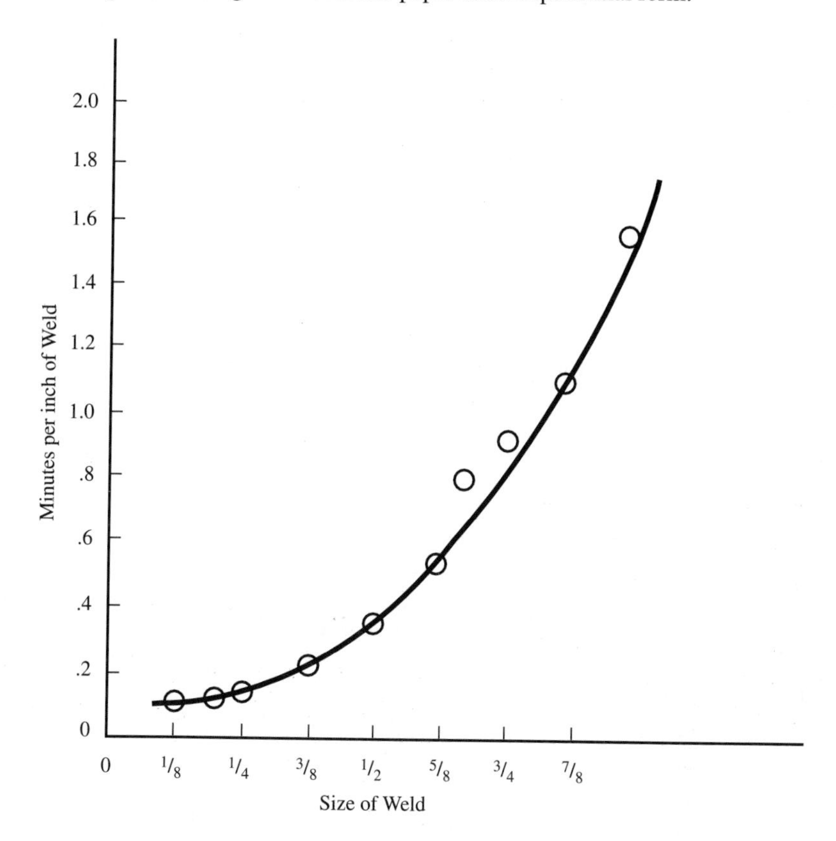

$$\log Y = \log A + X \log B$$

and

$$\log A = -1.10$$
$$A = 0.07944$$
$$\log B = 1.28$$
$$B = 19.05$$

Therefore: $$Y = (0.07944)(19.05)^x$$

Rounding off: $$Y = (0.08)(19)^x$$

This equation would become a component of the formula and would be less cumbersome to use than having to refer to the curve in Figure 12–6.

The equation can be checked as follows. With a $^1/_2$-inch weld:

$$\text{Time} = (0.08)(19)^{0.5} = 0.35 \text{ minutes}$$

This checks quite closely with the time study value of 0.37 minutes.

FIGURE 12-6
Data plotted on semilogarithmic paper take the form of a straight line with the equation $Y = AB^{X}$.

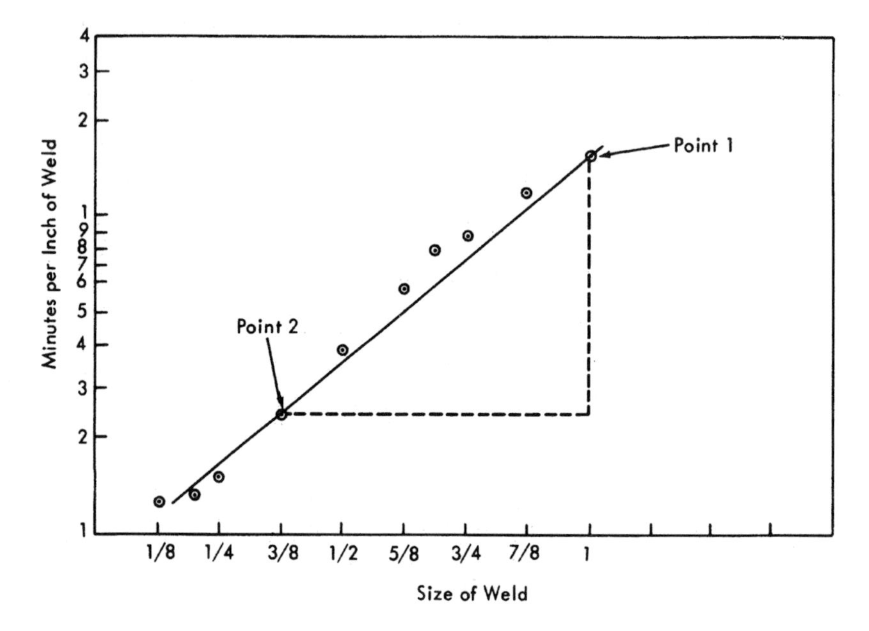

Sometimes, plotting the data on logarithmic paper results in a straight line. For example, the equation $y = ax^m$ written in logarithmic form becomes:

$$\log y = \log a + m\log x$$

Here, log y is linear with x; a straight line results from plotting the data on logarithmic paper.

Plot Graphic Solutions for More Than One Variable

When more than one variable affects time, graphic techniques should be used, even if the relationships are nonlinear. For example, when one or more variable is nonlinear, you could solve for time with two or more variables by constructing a chart for each variable. The first chart would plot the relationship between time and one of the variables in selected studies in which the values of the other variables tend to remain constant. The second chart would show the relationship between the second variable and time, adjusted to remove the influence of the first variable. The procedure is extended with charts until all variables have thus been handled.

The following example illustrates the application of this procedure. Analysts want to construct a formula for rolling various widths and lengths of 0.072-inch-thick cold-rolled sheetmetal on a bench roll. Element 1, shown on the Master Table of Detail Time Studies, is "pk. up pc. & pos.," and element 3 is "lay aside pc. &

pos." Since both elements involve handling the same amount of stock, their time values can be combined to simplify the final algebraic expression. These values are as follows:

Study no.	Width	Length	Element 1 + element 3
1	2	16	0.18
2	3	18	0.13
3	8	1	0.09
4	10 1/2	32	0.14
5	7	84	0.55
6	8 1/2	69	0.38
7	20	30	0.22
8	11	55	0.26
9	10	75	0.68
10	20 1/2	41	0.66
11	10 1/2	90	1.80
12	16	64	1.44

First inspection seems to suggest that the total time for element 1 plus element 3 varies with the area of the part handled. However, plotting indicates that something else is influencing the time required to perform these elements. Further analysis reveals that the long and narrow pieces require considerably more time than the nearly square parts, even though their respective areas are about the same. Analysts decide that two variables are affecting time: the area of the part, and the increased difficulty of handling the longer pieces. The latter may be expressed as a ratio of length to width.

Four of the 12 studies (i.e., studies 5, 6, 7, and 8) on the Master Worksheet involve stock having areas of about 600 square inches. These four studies have a relatively wide spread of "length divided by width" ratios (L/W) which is apparently responsible for the range in allowed element times. Since these four studies have about the same areas, a simple curve shows the effect of L/W (see Figure 12–7).

Now, to see the effect of area on the element time, it is necessary to adjust all the tabularized time study values to take into account the influence of the variable L/W. To get a factor that can be used to divide the time study values, to arrive at an adjusted time that varies with area, the analysts construct a scale parallel to the y axis and call it "Factor A" (adjustment factor). By taking the lowest point on the L/W curve and extending it horizontally to the adjustment axis, the analysts get a distance from the origin that can be evaluated as unity on the adjustment axis. Proportional values can then be constructed to get a scale on the adjustment axis.

By taking the L/W value of all the studies appearing on the master table, the analysts can graphically determine adjustment factors for each study, moving horizontally to the Factor A axis from the specific point on the L/W versus time curve. Once the adjustment values for each study are determined, the analysts can compute the adjusted time by dividing the allowed element time values by the adjustment factor. The resulting adjusted times are then plotted against area (see Figure 12–8).

The system of curves thus developed represents the graphic solution, which can be used for establishing standards within their range, in the following manner: First,

FIGURE 12-7
Time plotted against the ratio of length to width for constant areas of about 600 square inches.

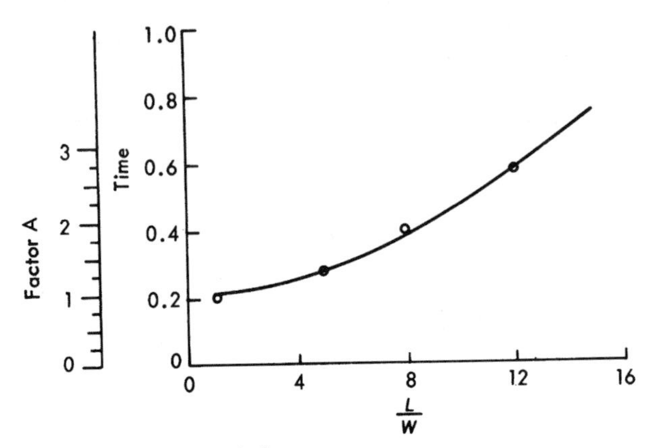

Area = C = 600 square inches.

FIGURE 12-8
Adjusted time plotted against area.

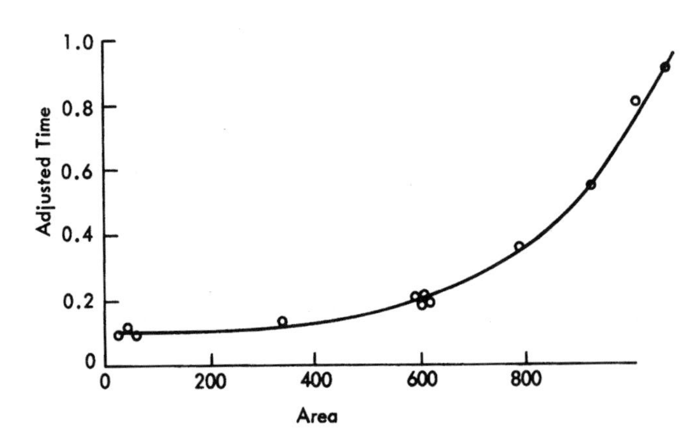

the area and the *L/W* ratio of the sheet are calculated. Using the *L/W* ratio, the analysts can refer to the first curve to obtain an adjustment factor. Referring to the second curve yields an adjusted time. The product of the adjusted time and the adjustment factor yields the allowed time to perform elements 1 and 3. By extending this procedure, the analysts can graphically solve element time values when more than two variables govern the element time.

Sometimes, variables remain relatively constant within a specific group, but show a pronounced change in value once the limits of the specific group are extended. For example, 1-inch white pine boards sent to a planer might be classified as follows:

Group	Allowed time
Small (up to 300 square inches)	0.070 min
Medium (300 to 750 square inches)	0.095 min
Large (750 to 1,800 square inches)	0.144 min

However, this method of grouping gives erroneous values at the extremities of each group. For example, a board of 295 square inches would be allowed 0.070 minute handling time, while one of 305 square inches would be given 0.095 minute. The chances are that in the former case, 0.070 minute would represent a tight standard, and in the latter case, 0.095 minute would be somewhat loose. The rule therefore is, when elements are grouped, the grouping must be clearly and specifically defined so that there are no questions as to the category in which an element belongs.

Use Least Squares and Regression Techniques

Although graphical procedures may be quite useful for establishing a predicting equation for one dependent variable as a function of one independent variable, the analyst may want to use more sophisticated methods of curve fitting, such as least squares and regression techniques. This is especially true when one dependent variable y is viewed as a function of more than one independent variable. However, least squares and regression techniques are often mistakenly or erroneously employed by the novice, with the expected poor results.

As an example, we are interested in determining the mean of the distribution of time values for a given variable x. There is a distribution of the random variable time for a given value of the independent variable x. Figure 12–9 illustrates such a relationship, with the true straight line of the equation $y = B_0 + B_1x$ passing through the mean of each of the four distributions of y. This relationship is referred to as the regression curve of y on x.

While data are being gathered, the analyst may only be able to approximate the mean of the time distributions. Any observed time may differ from the true mean by some finite amount. To determine the value of the estimates b_0 and b_1 of the parameters B_0 and B_1 for a given set of data (n paired observations x_iy_i), where the regression of y on x is assumed to be linear, we must determine the equation of the line that best fits the data. By best fit, we mean the equation of the line $y' = b_0 + b_1x$, for which the sum of the squared vertical distances $(y - y')$ for the points comprising the data is a minimum.

For the parameters b_1 and b_0, the normal equations whose values are the line of best fit of the data, are:

$$\sum_{i=1}^{n} y_i = b_0n + b_1 \sum_{i=1}^{n} x_i$$

$$\sum_{i=1}^{n} x_iy_i = b_0 \sum_{i=1}^{n} x_i + b_1 \sum_{i=1}^{n} x_i^2$$

These equations can readily be solved simultaneously for the parameters b_1 and b_0:

FIGURE 12-9
Four distributions of time for four values of x and the resulting true curve passing through the means of these distributions.

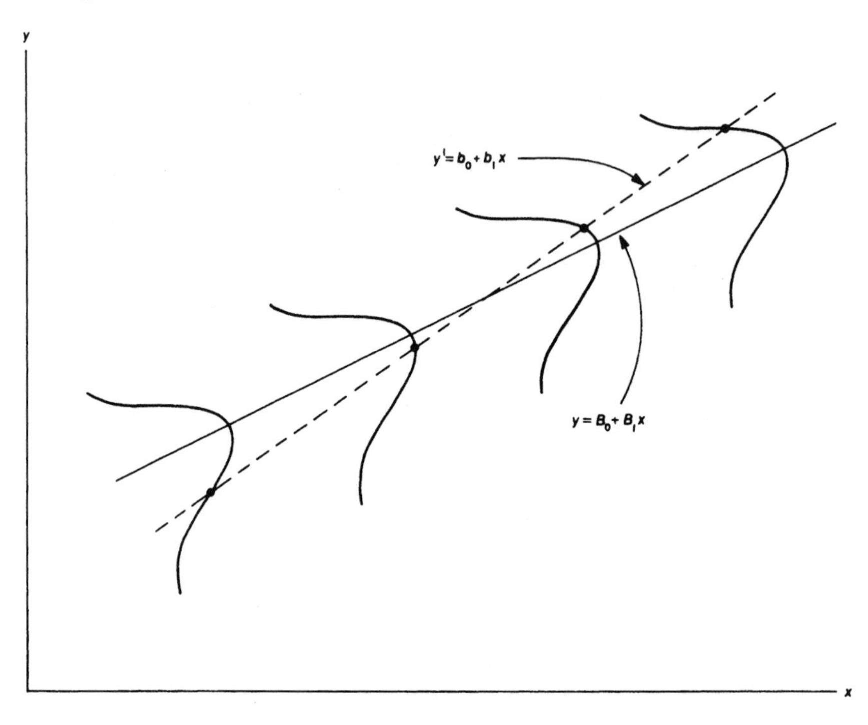

$$b_0 = \frac{(\Sigma x^2)(\Sigma y) - (\Sigma x)(\Sigma xy)}{(n)(\Sigma x^2) - (\Sigma x)^2}$$

$$b_1 = \frac{(n)(\Sigma xy) - (\Sigma x)(\Sigma y)}{(n)(\Sigma x^2) - (\Sigma x)^2}$$

The least squares technique is also used for plotting curves and exponential relationships as straight lines on semilog and log-log paper. An example is the exponential relationship involving semilog paper where:

$$\log y = \log b_0 + x\log b_1$$

Solve for the y intercept b_0 from the equation:

$$\log b_0 = \frac{(\Sigma x^2)(\Sigma \log y) - (\Sigma x)(\Sigma x \log y)}{(n)(\Sigma x^2) - (\Sigma x)^2}$$

and for the slope b_1 from the equation:

$$\log b_1 = \frac{(n)(\Sigma x \log y) - (\Sigma x)(\Sigma \log y)}{(n)(\Sigma x^2) - (\Sigma x)^2}$$

EXAMPLE 12–1.
Sample Linear Regression Calculation

Use the normal equations and the following data:

Study	x or area	y or time	xy	x^2
1	25	0.104	2.60	625
2	65	0.109	7.09	4,225
3	77	0.126	9.70	5,929
4	112	0.134	15.01	12,544
5	135	0.138	18.63	18,225
6	147	0.150	22.05	21,609
7	185	0.153	28.31	34,225
8	220	0.174	38.28	48,400
9	245	0.176	43.12	60,025
10	275	0.182	50.05	75,625
11	287	0.186	53.38	82,369
12	300	0.202	60.60	90,000
	2,073	1.834	348.82	453,801

we obtain:

$$1{,}834 = 22b_0 + 2{,}073b_1$$
$$348.82 = 2{,}073b_0 + 453{,}801b_1$$

Solving for b_0: $b_0 = (1.834 - 2{,}073b_1)/12$

Substituting into the second equation yields:

$$b_1 = 0.000334$$
$$b_0 = 0.095$$

At times analysts recognize that more than one independent variable is influencing the dependent variable (which, in this chapter, is time). If that is the case, multiple regression techniques must be applied. These calculations are quite tedious and are best left to commercially available statistical software packages.

In summary, regression methods are very useful tools for deriving formulas for the development of standards. However, these techniques may present difficulties with the statistical and engineering analyses used in the selection of the independent variables and with the significance of each variable in the predicting equation. The form of the equation selected (that is, linear or nonlinear) and the inclusion of cross products of the variables in the equation when interaction effects are present may further complicate the analysis. Under these circumstances, the analyst should consult a statistician competent in regression techniques.

In the interest of ease of application, it is generally good practice to make predicting equations as simple as possible (i.e., Occam's razor). From a statistical viewpoint, complicated predicting equations may result in time estimates that are little more reliable than simpler equations.

Much of the drudgery of the preceding calculations is easily avoided by using the statistical routines provided in commercially available spreadsheet programs, such as Microsoft Excel. More detailed statistical procedures (especially for multivariable approaches) are available in specialized software, such as MINITAB or SAS.

Develop Synthesis

The purpose of the synthesis in the formula report is to explain the derivation of the various components in the formula to facilitate its use. A clearly developed synthesis helps explain and verify the suitability of the formula.

For the final expression to be in its simplest form, all constants and symbols should be combined wherever possible, without compromising the accuracy and flexibility of the formula. Element operations should be classified under specific headings, such as preparations, handling by hand, handling by jib crane, and operation. Constant values should be entered under each element class. These values should then be combined to simplify the final formulas. Details of the combinations should be explained in the synthesis. A typical illustration of a synthesis for a constant element would be:

$$\begin{aligned} A1 + C1 + \text{Constant from equation 4} + H1 &= 0.11 + 0.17 + 0.09 + 0.05 \\ &= 0.42 \text{ minute} \end{aligned}$$

The treatment of variable elements would require a more detailed explanation, due to the added complexities associated with this class of elements. For example, to combine elements $B1$ and $D1$, which have been influenced by the ratio of L/D, the two equations may be:

$$B1 = 0.1\frac{L}{D} + 0.08$$

$$D1 = 0.05\frac{L}{D}$$

and the combination of these would give

$$B1 + D1 = 0.15\frac{L}{D} + 0.08$$

The synthesis would thus clearly show where the $0.15\ L/D$ came from.

Compute Expression

The final expression may not be entirely in algebraic form. It may be more convenient to express some of the variables as systems of curves, nomographs, or single curves. Then, too, some of the variable data may be entered into tables referred to by a single symbol in the formula. A typical example of such a formula would be:

$$\text{Normal time} = 0.07 + \text{Chart 1} + \text{Curve 1} + \frac{0.555N}{C}$$

where: N = Number of cavities per mold.
C = Cure time in minutes.

To apply this formula, analysts would refer to Chart 1 for the applicable tabularized value. They would also refer to Curve 1 to obtain the numerical value of another variable in the formula. Knowing the number of cavities in the mold and the cure time in minutes, analysts would substitute these values in the latter part of the expression and would then summarize the values of the three variables with the constant time of 0.07 minute, to get the normal time.

Check for Accuracy

Upon completion of the formula, analysts should verify it before releasing it for use. The easiest and fastest way to check the formula is to use it to check existing time studies. This can best be done by tabularizing the results under the following headings: "Part Number," "Time Study Value," "New Formula Value," "Difference," and "Percentage Difference."

The analyst should investigate any marked differences between the formula value and the time study value and determine the cause. The formula should show an average difference of less than 5 percent from the values of the time studies used in its derivation. If the formula does not have the expected validity, the analyst should accumulate additional data by taking more stopwatch and/or standard data studies. After testing the formula over a range of work, the analyst should explain it to those who are directly and indirectly affected by it, so they become familiar with it. The analyst should especially go over the formula with the supervisor of the department where it will be used. By understanding its derivation and application and verifying its reliability, the supervisor should feel comfortable backing the standards resulting from the formula's use.

Write Formula Report

The final step in the formula development process is to write the formula report. The analyst should consolidate all data, calculations, derivations, and applications of the formula and present this information in a complete report prior to putting the formula into use. This will make available all the facts regarding the process employed, the operating conditions, and the scope of formula.

As an example, the Westinghouse Electric Corporation includes 14 separate sections in its formula reports. These are:

1. Formula number, division, data, and sheet number.
2. Part.
3. Operation.
4. Workstation.
5. Normal time.
6. Application.

7. Analysis.
8. Procedure.
9. Time studies.
10. Table of detail elements.
11. Synthesis.
12. Inspection.
13. Wage payment.
14. Signatures of constructor and approver.

Formula number

For ease of reference, all paperwork used to develop the formula is identified by an assigned number. This formula number should have a prefix that identifies the department or division in which the formula will be used, and a number that gives the chronological arrangement of the formula with respect to other formulas being used in the department. Thus, the formula 11N-56 would be the 56th formula designed for use in department 11N. Also included is the date that the formula was put into use to connect working conditions in effect at the time it was designed with standards established through its application.

Part

Here, the analyst would list the part number or numbers, together with drawing numbers, with a concise description of the work, clearly defining the products for which the formula has application. An example would be "Parts J-1101, J-1146, J-1172, J-1496, side plates ranging from 12 × 24 inches to 48 × 96 inches."

Operation and workstation

Next, the analyst should clearly describe the operation covered by the formula, such as "roll radius," or "fabricate inner and outer members," or "broach keyways," or "assemble farm tank." In addition, the analyst would describe the workstation, including such equipment information as jigs, fixtures, and gages, as well as their size, condition, and serial numbers. A photograph often helps to define the method being used at the time the formula is compiled.

Normal time

When expressing the formula for the normal time, the analyst should use separate equations for the "setup" time and the "each piece" time. Immediately following the formula, the analyst adds a key that outlines the meaning of the symbols used in each equation. Also included are all tables, nomograms, and systems of curves, as well as a sample solution, so that the person using the formula can understand it clearly.

Application

After concisely stating the expressions for the allowed time, the analyst gives a clear explanation of the application of the formula. This should describe in detail the nature of the work on which the formula may be used, and should specifically state the limits within which the formula may be applied. An example application statement for a formula for curing bonded rubber parts might be: "This formula applies to all curing operations done in 24 × 28-inch platen presses when the number of cavities ranges between 8 and 100 and the cubic inches of rubber per cavity range between 0.25 and 3."

Analysis

Under the analysis, the analyst gives a detailed account of the entire method employed, including: tools, fixtures, jigs, and gages, and their application; workplace layout; methods of material handling; method of obtaining supply materials; and nature of setup, including how the operator is assigned the job, the distance to the tool crib, and other pertinent data.

Also included in the analysis section of the report is a breakdown of the allowances used in the formula, including the reasons for any special or extra allowance. Personal time, fatigue, and unavoidable delay allowances are given independently, so that any question about the inclusion of an allowance in the formula can be answered more easily.

Procedure

After completing the analysis of the job, the analyst writes up the procedure used by the operator in performing the work. The best way of doing this is to include all the elements appearing on the master worksheet, in their correct chronological sequence. While abbreviations may be used liberally on the time studies and on the master table, their use in the written description of the operator's procedure should be avoided. Also, individual elements should be covered in exact detail, so that anyone can understand which work elements fall within the scope of the formula.

Time studies

The formula report need not include the actual time studies used in the compilation of the formula, as they are usually available in their own independent files. However, they should referred to, and this can be done in tabular form, as follows:

EXAMPLE 12–2
Representative Formula Report.

FORMULA NO.: M–11–No. 15
DATE: September 15, 19–
SHEET: 1 of 15

PART: Cylindrical oil sand mix cores $7/8''$ diameter to $2\,3/4''$ diameter and 4″ long to 13″ long.
OPERATION: Make core complete in wood core box.
WORKSTATION: 30″ × 52″ bench 36″ high.
NORMAL TIME: Piece time in decimal minutes equals:

(continued)

EXAMPLE 12–2 ***(continued)***

$$0.173N + 0.0210L + \sqrt{0.0067 + 0.000016V^2} + Y \times CT + 0.0157A + 0.327$$

where: N = Number of "C" clamps
L = Length of core in inches
V = Volume of core in cubic inches
Y = Adjustment factor (Curve 1)
CT = Adjusted time (Curve 2)
A = Cross-sectional area of core

EXAMPLE:

Calculate normal time to make a green sand core 2" in diameter and 8" long. Volume would be 25.1 cubic inches, and with 8" length, only one clamp would be required. L/D would be equal to 8/2 or 4.
Then:

$$\text{Normal time} = (0.173)(1) + (0.0210)(8) + \sqrt{0.0067 + (0.000016)(25.1^2)} + (1.5)(.76) + (0.0157)(3.14) + 0.327 = 1.987 \text{ minutes.}$$

APPLICATION

This formula applies to all cylindrical cores made of oil sand mix with diameters ranging from $^7/_8$ inch to $2^3/_4$ inches and lengths ranging from 4 inches to 13 inches. This work is performed manually in wooden split core boxes.

ANALYSIS

The following hand tools are provided to the operator for making cores covered by this formula: C-clamps, slick, and rammer. The work is done at a bench, with the operator alternately sitting and standing. Oil sand mix for producing the cores is piled on the floor by the move worker, approximately 4 feet from the operator. Periodically, the operator replenishes a supply of sand on the bench from the inventory on the floor, through the use of a short-handled shovel. Rods and reinforcing wires are requisitioned from the storeroom by the operator, who usually acquires a full day's supply with each requisition.

The work voucher placed above the operator's workstation by the departmental supervisor indicates the sequence of jobs to be performed and is the authority for the operator to begin a specific job. Operation cards and drawings must be obtained by the operator from the tool crib.

The standard allowance that must be added when applying this formula is 17 percent. This involves 5 percent for personal delays, 6 percent for unavoidable delays, and 6 percent for fatigue.

PROCEDURE

The working procedure followed in making oil sand mix cores covered by this formula includes 13 elements, exclusive of setup and put-away elements. These are:

1. Pick up two sections of core box and close together.
2. Clamp core box shut, using one clamp for cores 9 inches in length or less and two clamps for cores greater than 9 inches in length.
3. Fill core box partly full of sand (about one-third, depending on core).
4. Ram the sand solid in the core box.
5. Place the vent rod and reinforcing wire.
6. Fill the remainder of the core box with sand and ram solid.
7. With slick, strike off sand on both ends of core box so sand core is flush with box.
8. Remove the vent wire from the core and lay aside.
9. Rap box lightly.
10. Remove clamps and lay aside.
11. Open core box.
12. Roll out core on rack beside workstation.
13. Clean core box with kerosene rag.

TIME STUDIES

The following summary includes the time studies used in developing this formula:

Time study no.	Part number	Plant	Date taken	Taken by
S-13	P-1472 PB-1472	A	6-15	Black
S-111	P-1106 PB-1106	A	12-11	Black
S-45	P-1901 PB-1901	A	7-13	Hirsch
S-46	P-1907 PB-1907	B	7-14	Black
S-47	P-1908 PA-1908	B	7-14	Black
S-32	P-1219 PA-1219	A	8-10	Black
S-76	P-1711 PA-1711	A	11-12	Obe
S-70	P-1701 PB-1701	B	11-9	Obe
S-17	P-1311 PB-1311	B	6-16	Black
S-18	P-1312 PB-1312	B	6-16	Black
S-59	P-1506 PB-1506	A	7-26	Hirsch

(continued)

EXAMPLE 12–2 ***(continued)***

Time study no.	Part number	Plant	Date taken	Taken by
S-60	P-1507 PB-1507	A	7-26	Hirsch
S-50	P-1497 PB-1497	B	7-19	Obe
S-51	P-1498 PA-1498	B	7-20	Obe
S-52	P-1499 PA-1499	B	7-21	Obe
S-53	P-1500 PA-1500	B	7-21	Obe
S-54	P-1501 PA-1501	B	7-22	Obe

Table of Detail Elements

Symbol	Element description	Decimal minute normal time	Reference
A-1	Close core box	0.046	Average
B-1	Clamp core box	$0.112N$	Time vs. no clamps
C-1	Fill partly full of sand	$Y \times CT$	Time vs. *L*/*D* & vol. Time vs. vol.
D-1	Ram	$Y \times CT$	Time vs. *L*/*D* & vol.
E-1	Place rod and wire	$0.0153L + 0.03$	Time vs. length
F-1	Fill and ram	$Y \times CT$	Time vs. *L*/*D* & vol.
G-1	Strike off	$0.0157A + 0.07$	Time vs. area
H-1	Remove vent wire	0.047	Average
J-1	Rap box	0.043	Average
K-1	Remove clamps	$0.061N$	Time vs. no clamps
M-1	Open box	0.046	Average
N-1	Roll out core	$0.057L + 0.045$	Time vs. length
P-1	Clean box	$\sqrt{0.0067 + 0.000016V^2}$	Time vs. volume

SYNTHESIS

$$\begin{aligned}\text{Standard time} &= \text{Sum of allowed elemental time}\\ &= \text{A-1} + \text{B-1} + \text{C-1} + \text{D-1} + \text{E-1} + \text{F-1} + \text{G-1}\\ &\quad + \text{H-1} + \text{J-1} + \text{K-1} + \text{M-1} + \text{N-1} + \text{P-1}\end{aligned}$$

Since elements C-1, D-1, and F-1 are all dependent on both volume and length/diameter, their values may be combined before plotting on coordinate paper. The combined values for these three elements for the 17 time studies are:

Study no.	C-1 + D-1 + F-1
S-13	1.045
S-111	1.017
S-45	1.235

Study no.	C-1 + D-1 + F-1
S-46	0.647
S-47	1.325
S-32	1.110
S-76	0.500
S-70	1.362
S-17	1.932
S-18	0.659
S-59	0.988
S-60	1.181
S-50	0.584
S-51	1.277
S-52	0.888
S-53	1.481
S-54	2.065

For reference purposes, the combination of C-1 + D-1 + F-1 is designated R-1 (see Figures 12–10 and 12–11). Curves 1 and 2 are used to solve for the allowed time for these three elements by first getting the adjustment fac-

FIGURE 12-10
Curve 1 Time versus L/D

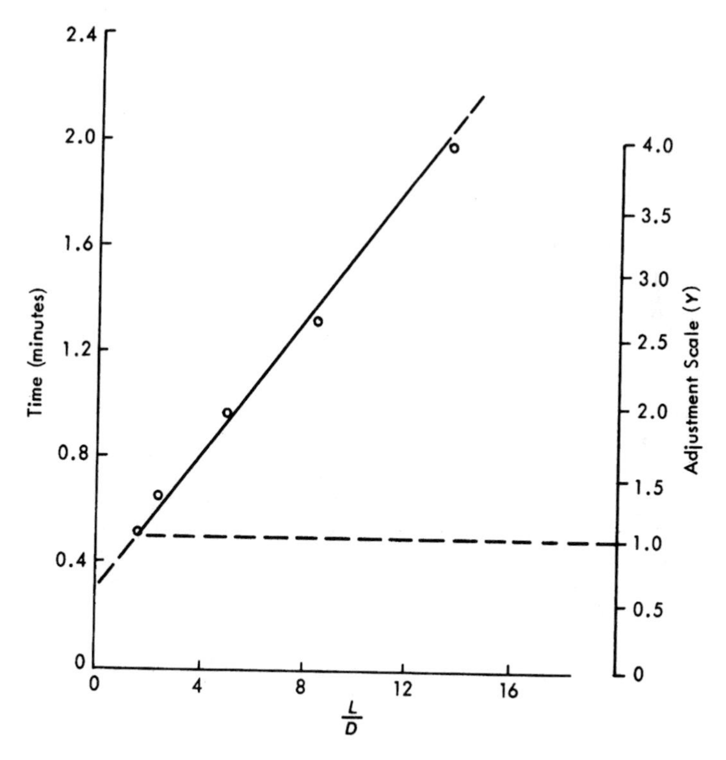

(continued)

EXAMPLE 12–2 ***(continued)***

FIGURE 12-11
Curve 2

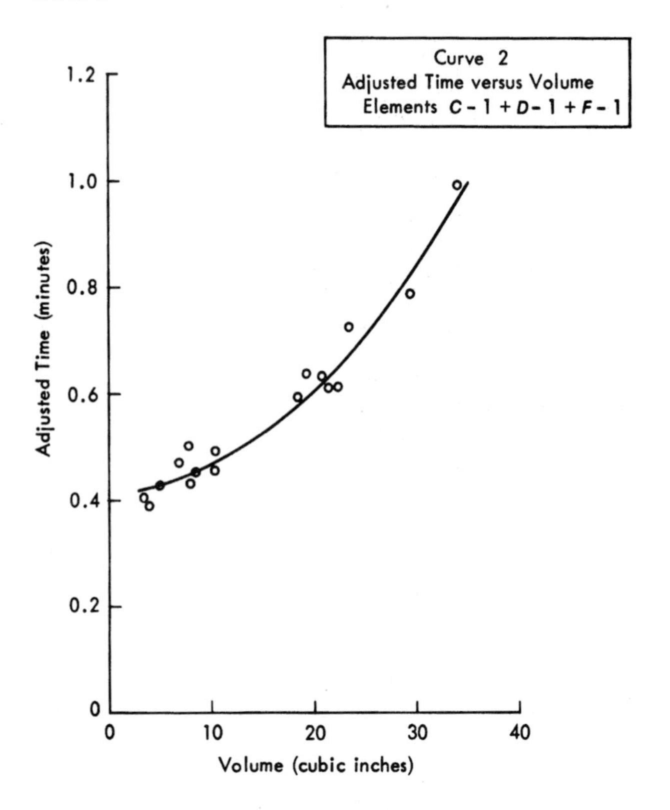

tor from curve 1 and then getting the adjusted time from curve 2. The product of these two values gives the allowed time for C-1 + D-1 + F-1.

The time for element E-1 (place rod and wire) has been plotted on curve 3 (see Figure 12–12), where its relationship to core length is shown.

This curve can be expressed algebraically by the equation:

$$T = fv + C$$

or $$\text{Time} = (\text{slope})(\text{length}) + y \text{ intercept}$$

Solving graphically:

$$f = \frac{0.210 - 0.100}{12 - 4.8} = \frac{0.110}{7.20}$$
$$= 0.0153$$
$$C = 0.03 \text{ (from graph)}$$

Thus: $$\text{Time} = 0.0153L + 0.03$$

FIGURE 12-12
Curve 3.

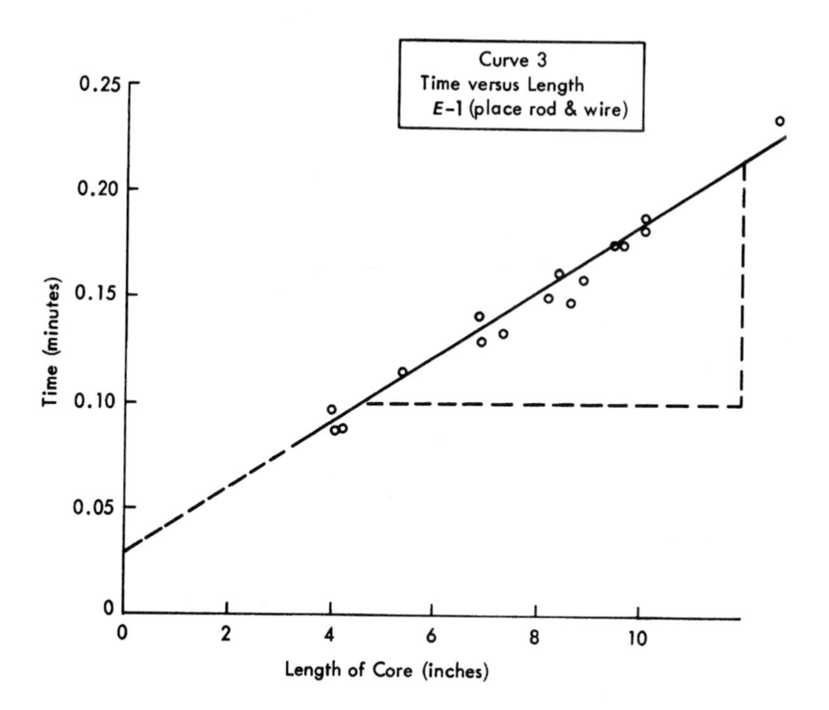

Element G-1 (strike off) has also been solved as a straight-line relationship. Here a cross-sectional area of the core has been plotted against time (see curve 4 in Figure 12–13). In this case

$$\text{Time} = (f)(\text{area}) + 0.07$$

and

$$f = \frac{0.132 - 0.08}{4 - 0.7} = \frac{0.052}{3.3}$$

$$f = 0.0157$$

Then:

$$\text{Time} = 0.0157A + 0.07$$

Element N-1 (roll out core) is shown on curve 5 (see Figure 12–14). Here,

$$\text{Time} = f(\text{length}) + 0.045$$

and

$$f = \frac{0.125 - 0.068}{14 - 4}$$

$$= 0.0057$$

Then:

$$\text{Time} = 0.0057L + 0.045$$

(continued)

EXAMPLE 12–2 ***(continued)***

FIGURE 12-13
Curve 4.

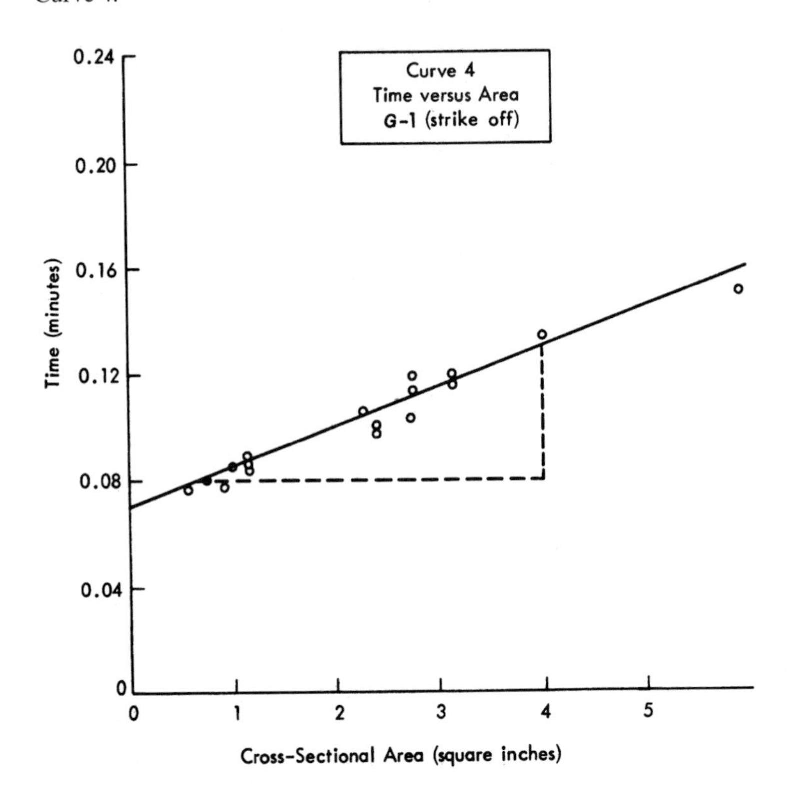

The time for element P-1 (clean box) when plotted against the volume of the core, gives a hyperbolic relationship as shown in curve 6 (Figure 12–15). This relationship with time can be expressed algebraically as follows:

$$\text{Time} = \sqrt{a^2 + \frac{v^2a^2}{k^2}}$$

$$= \sqrt{0.082^2 + \frac{(v^2)(0.082)^2}{400}}$$

$$= \sqrt{0.0067 + 0.000016v^2}$$

Elements A-1, H-1, J-1, M-1 were classified as constants with the following respective allowed times determined by taking their average values: 0.046,

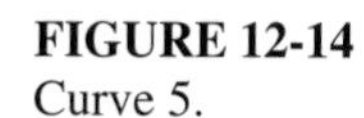

FIGURE 12-14
Curve 5.

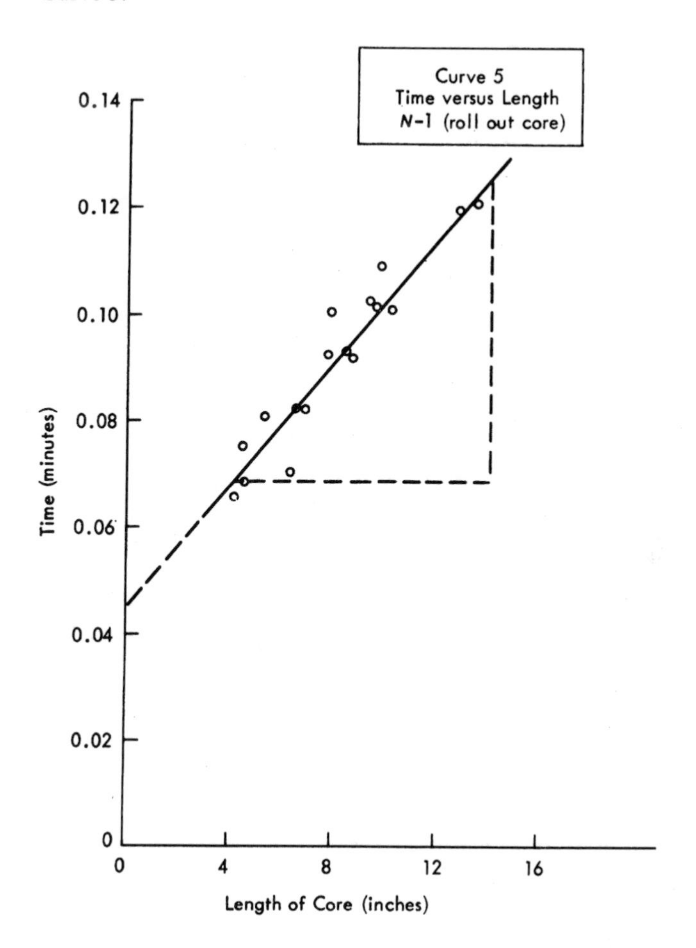

0.047, 0.043, 0.046. These constants are added to the sum of the y intercept values of Curves 3, 4 and 5, to determine a total constant time of 0.327 minute:

A-1	0.046
Curve 3 intercept	0.030
H-1	0.047
Curve 4 intercept	0.070
J-1	0.043
M-1	0.046
Curve 5 intercept	0.045
	C_T = 0.327 minute

(continued)

EXAMPLE 12–2 ***(continued)***

FIGURE 12-15
Curve 6.

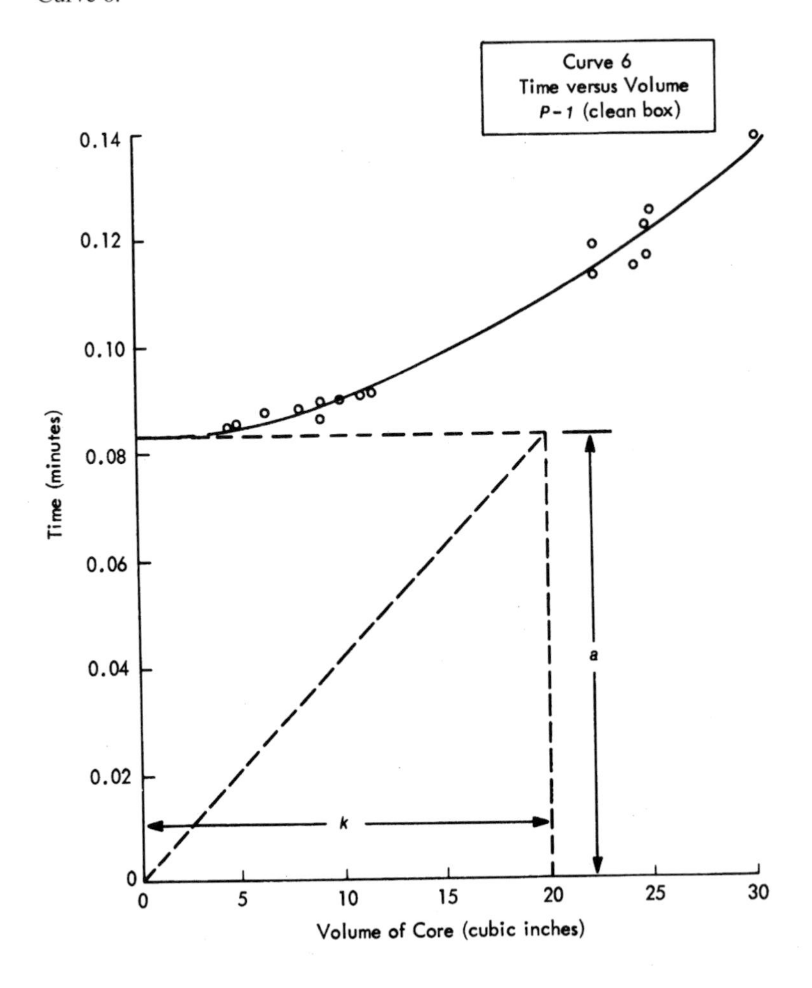

Both elements B-1 and K-1 are proportional to the number of clamps N. Thus, these two elements may be combined for simplicity as follows:

$$B\text{-}1 = 0.112N$$
$$K\text{-}1 = 0.061N$$
$$S\text{-}1 = 0.173N$$

Since elements E-1 (place rod and wire) and N-1 (roll out core) are proportional to length of core, we may combine their equated relationships with length and designate the combined total by the symbol T-1.

Thus:

$$E\text{-}1 = 0.0153L$$
$$N\text{-}1 = 0.0057L$$
$$T\text{-}1 = 0.0210L$$

Simplifying our initial equation:

$$\text{Normal time} = \text{S-1} + \text{T-1} + \text{P-1} + \text{R-1} + \text{G-1} + C_T$$

$$\text{Normal time} = 0.173N + 0.0210L + \sqrt{0.0067 + 0.000016V^2} + Y \times C_T + 0.0157A + 0.327$$

INSPECTION

The only inspection requirement on the cores covered by this formula is visual inspection done by the operator at the time the core is rolled free from the core box.

Time study no.	Part	Drawing	Plant	Date taken	Taken by
S-112	J-1102	JB-1102	A	9–15-	J. B. Smith
S-147	J-1476	JA-1476	B	10–24-	A. B. Jones
S-92	J-1105	JB-1105	B	6–11-	J. B. Smith

Table of detail elements

This table is a reference source for information about the normal element time and its derivation. The information for this table comes from the Master Table of Detailed Elements, as follows:

Symbol	Element description	Normal time	Reference
A1	Close core box	0.046 min.	Average
C1	Fill partly full of sand	A × C.T.	All studies
J1	Rap box	0.043 min.	Average

Synthesis

After recording the table of detailed elements, the analyst should include a synthesis of the report. The synthesis should explain the manner in which the allowed time was derived.

Inspection, payment, and signatures

Upon completion of the synthesis, the analyst should record the following information, in order: the inspection requirements appearing on the drawings of the jobs covered by the formula; the type of wage payment plans for which the formula will be used, such as daywork, piecework, and group incentive; and finally, the analyst's signature and that of the analyst's supervisor.

SUMMARY

A time study formula can establish standards in a fraction of the time required for individual studies. However, before a formula is released for use, the mathematics used in its development should be carefully checked to assure that the expression is correct. In addition, several test cases should be tried to ensure that the formula establishes true, consistent standards. The formula's range of application should be clearly identified, and the standards established should never be based on data beyond the scope of the formula.

The following steps represent the chronological procedure for designing time study formula.

1. Collect the data.
   a. Use time studies with standardized elements that are already available.
   b. Use standard data based on standardized elements that are already available.
   c. Establish element values from basic motion data for standardized elements.
   d. Take new time studies. Be careful to break new studies down into standardized elements.
2. Compile a master work sheet and identify the formula.
3. Analyze and classify all elements.
   a. Constants.
   b. Variables.
4. Develop the synthesis.
5. Compute the final expression.
6. Check the mathematics of the developed formula.
7. Test the formula.
8. Write the formula report.
9. Use the formula.

By systematically following these nine steps, analysts should have little difficulty in designing reliable time study formulas. Once a formula has been designed, it can be programmed on a digital computer and solved for all possible variations within its scope. Thus, tabulated standards can be made available for broad variations in the work assignment.

QUESTIONS

1. What advantages does the formula offer over standard data in establishing time standards?

2. Is the use of time study formulas restricted to machine shop operations where feeds and speeds influence allowed times? Explain.

3. What are the characteristics of a sound time study formula?

4. What is the danger of using too few studies in the derivation of a formula?

5. What is the function of the synthesis in the formula report?

6. List the 14 sections that make up the formula report.

7. Which nine steps represent the chronological procedure in the design of time study formulas?

8. Explain in detail how it is possible to solve graphically for time when two variables are influential and they cannot be combined.

PROBLEMS

1. In the Dorben Company, the work measurement analyst planned to develop standard data on a new milling machine that was recently installed. In one group of studies, the material being cut involved a 1.5-inch width of cut, with lengths varying from 4 inches to 30 inches. For this work, a plain carbide-tipped milling cutter 3 inches in diameter with a face width of 2 inches was used. Depths of cut ranged from $3/16$ inch to $7/16$ inch. Give the equation that can be programmed to provide cutting time in terms of d (depth of cut) and l (length of casting being milled). In all cases, the feed per tooth is 0.010 inch and the 16-tooth cutter is running at a surface speed of 80 feet per minute.

2. The work measurement analyst in the Dorben Company was studying the hand-filing and polishing of external radii. Six studies provided the following information:

Study	Size of radii	Minutes per inch
1	$3/8$	0.24
2	$1/2$	0.37
3	$5/8$	0.59
4	$11/16$	0.80
5	$3/4$	0.93
6	1	1.52

These data plotted as a straight line on semilog paper, with time (the dependent variable) plotted on the logarithmic scale. Develop an algebraic equation for estimating the time for filing and polishing various radii.

3. In the assembly department of the Dorben Company, various hardware is bagged for shipment. The time study analyst wishes to design a formula or a system of curves to establish standards for this work. The master table of detailed time studies contains the following information for the second element, "place hardware components in bag":

Study no.	Time (min)	Weight of components (lb)	Bag size	No. of components
1	0.264	6.62	4	11
2	0.130	1.15	2	6
3	0.186	5.61	3	8
4	0.169	2.91	2	6
5	0.126	4.01	2	4

(continued)

Study no.	Time (min)	Weight of components (lb)	Bag size	No. of components
6	0.220	6.14	3	9
7	0.200	5.50	3	6
8	0.332	7.25	4	14
9	0.222	7.02	4	6
10	0.155	1.75	2	8
11	0.345	5.45	4	17
12	0.256	6.91	4	10

From these data, design a formula or a system of curves for establishing the standard for this element.

4. In the blanking of various leather components from animal skins, the analyst noted a relationship between standard time and the area of the component. After five independent time studies, the analyst observed the following:

Study no.	Area of leather component (sq. in.)	Standard time (min)
1	5.0	0.07
2	7.5	0.10
3	15.5	0.13
4	25.0	0.20
5	34.0	0.24

Derive an algebraic expression to preprice the blanking of the various leather components.

5. The analyst in the Dorben Company decides to develop a formula for prepricing a particular assembly operation involving different sizes of work. The assembly operation involves three constant elements and one variable element. The constant elements are determined from MTM data. The classifications of the fundamental motions on the constant elements are as follows:

Element 1: One R10C, one G1B, one P2SSD, one AF (max), one M20C with 2# weight, and one RL (case 1).
Element 2: One eye travel with T = 20 and D = 10, one R12B, one G1A, one M10C, one P1SSE, and ten T 30° 2#.
Element 3: One R10A, one G1B, one M20B, and one RL2.

The variable element is based on the following data:

Study number	Standard time (min)	Surface area (sq. in.)
1	0.282	7
2	0.163	5
3	0.022	2
4	0.120	4.5
5	0.227	6

Develop the algebraic expression for establishing standards for this operation, for parts with surface areas of up to 7 square inches.

6. The industrial engineer in the Dorben Company is in the process of building a formula to preprice the manufacture of a line of specialty forgings. The data indicate a nonlinear relationship between forging volume and standard time, for the variable elements related to positioning in the die and the actual forging. The data are as follows:

Study number	**Time (min)**	**Forging volume (cu. in.)**
1	0.130	30
2	0.110	24
3	0.103	20
4	0.088	10
5	0.083	5
6	0.120	27

Develop an algebraic expression for the computation of standard time for any forging having a volume of between 5 and 30 cubic inches.

7. Write the equation of the ellipse with its center at the origin and with axes along the coordinate axes and passing through (2, 3) and $(-1, 4)$.

8. Find the equation for the hyperbola with: center at (0, 0), foci $a = 4$, $b = 5$, and on the y axis.

9. Develop an algebraic expression for the relationship between time and area from the following data:

Study no.	1	2	3	4	5
Time	4	7	11	15	21
Area	28.6	79.4	182	318	589

REFERENCES

Denbow, Carl H., and Victor Goedicke. *Foundations of Mathematics.* New York: Harper & Row, 1959.

Lowry, Stewart M., Harold B. Maynard, and B. J. Stegemerten. *Motion and Time Study.* 3rd ed. New York: McGraw-Hill, 1940.

Neter, J., M. Wasserman, and M. H. Kutner. *Applied Linear Statistical Models.* 3rd ed. Homewood, IL: Richard D. Irwin, 1990.

Rotroff, Virgil H. *Work Measurement.* New York: Reinhold Publishing, 1959.

SELECTED SOFTWARE

MINITAB. v. 11, State College, PA, 1997.

SAS. Cary, NC: SAS Institute, 1998.

CHAPTER 13

Predetermined Time Systems

KEY POINTS:

- Use predetermined time systems to predict standard times for new or existing jobs.
- Predetermined time systems are a collection of basic motion times.
- Accurate systems require more time to complete.
- Quick, simple systems are usually less accurate.
- Consider not only the main motion, but also complexities or interactions with other motions.
- Use predetermined time systems to improve methods analysis.

Since the time of Frederick W. Taylor, management has realized the desirability of assigning standard times to the basic elements of work. These times are referred to as *basic motion times, synthetic times,* or *predetermined times*. They are assigned to fundamental motions and groups of motions that cannot be precisely evaluated with ordinary stopwatch time study procedures. They are also the result of studying a large sample of diversified operations with a timing device, such as a motion-picture camera or videotape machine, capable of measuring very short elements. The time values are synthetic in that they are often the result of logical combinations of therbligs; they are basic in that further refinement is both difficult and impractical; they are predetermined because they are used to predict standard times for new work resulting from methods changes.

Since 1945, there has been a growing interest in the use of basic motion times (refer to Chapter 4) as a method of establishing rates quickly and accurately without using the stopwatch or other time recording devices. One byproduct of predetermined time standards has been the development of *methods consciousness* associated with the principles of motion economy and work design.

Today, practicing methods analysts can obtain information from approximately 50 different systems of established synthetic values. Essentially, these predetermined time systems are sets of motion–time tables with explanatory rules and instructions on the use of the motion–time values.

Considerable specialized training is essential to the practical application of these techniques. In fact, most companies require certification before analysts are permitted to establish standards using the Work-Factor, MTM, or MOST systems.

A trained analyst establishing a standard on a given method by using two different predetermined time systems will probably arrive at two different answers. The reason is that different concepts of "normal performance" may have been utilized in the development of the standard data. For example, the work of Maynard, Stegemerten, and Schwab (1948), who developed Methods Time Measurement (MTM), was done on sensitive drill presses. Anyone familiar with metal trade operations recognizes that normal performance here is not at all difficult, and performances of 125 percent or more can readily be attained if the operator demonstrates high effort, has good skill qualifications, and has been thoroughly trained.

On the other hand, during the Great Depression (early 1930s) when practically everyone felt it was not only desirable but absolutely necessary to work very hard to hold a job, concepts of normal performance were tighter than what was considered normal 15 years later. This was especially true in such industries as the garment industry and those involving light assembly work, such as the fabrication of radios, washing machines, refrigerators, and similar white ware. Studies there (Quick, Shea, and Koehler, 1962) resulted in the Work-Factor predetermined time system, which was based on a different concept of normal performance.

All predetermined time systems fall into one of three groups (Sellie, 1992):

1. *Acceleration–deceleration systems.* These systems recognize that different body motions move at different velocities. Values determined using this approach suggest that 40 percent of the total time is used during the acceleration period, 20 percent for constant velocity, and 40 percent for deceleration. Today, acceleration–deceleration systems are not widely used for setting standards.
2. *Average-motion systems.* Here, recognition is given to representative or average motion difficulties that are usually encountered in industrial operations.
3. *Additive systems.* Under these systems, basic time values are used. Time percentages for motion difficulties encountered are added to these basic values. These additives range from 10 to 50 percent.

To give the reader a good overview of the field of basic motion times, we will review MTM in some detail, since it is the pioneer in the average-motion system classification, as well as a quicker subsystem of MTM, called MTM-2. In addition, the Maynard Operation Sequence Technique (MOST), and the MACROMotion Analyses systems will be briefly discussed. The reader should understand that MOST is derived from MTM, so it would be representative of average-motion systems. MACROMotion Analyses use input from both Work-Factor, an additive system, and MTM.

MTM-2 and MOST are the most commonly used predetermined time systems, according to 141 industrial engineers surveyed. Figure 13–1 illustrates the derivations of all of these predetermined time systems.

FIGURE 13–1
Family tree of predetermined times. (*Courtesy* of Standards International, Chicago, Illinois)

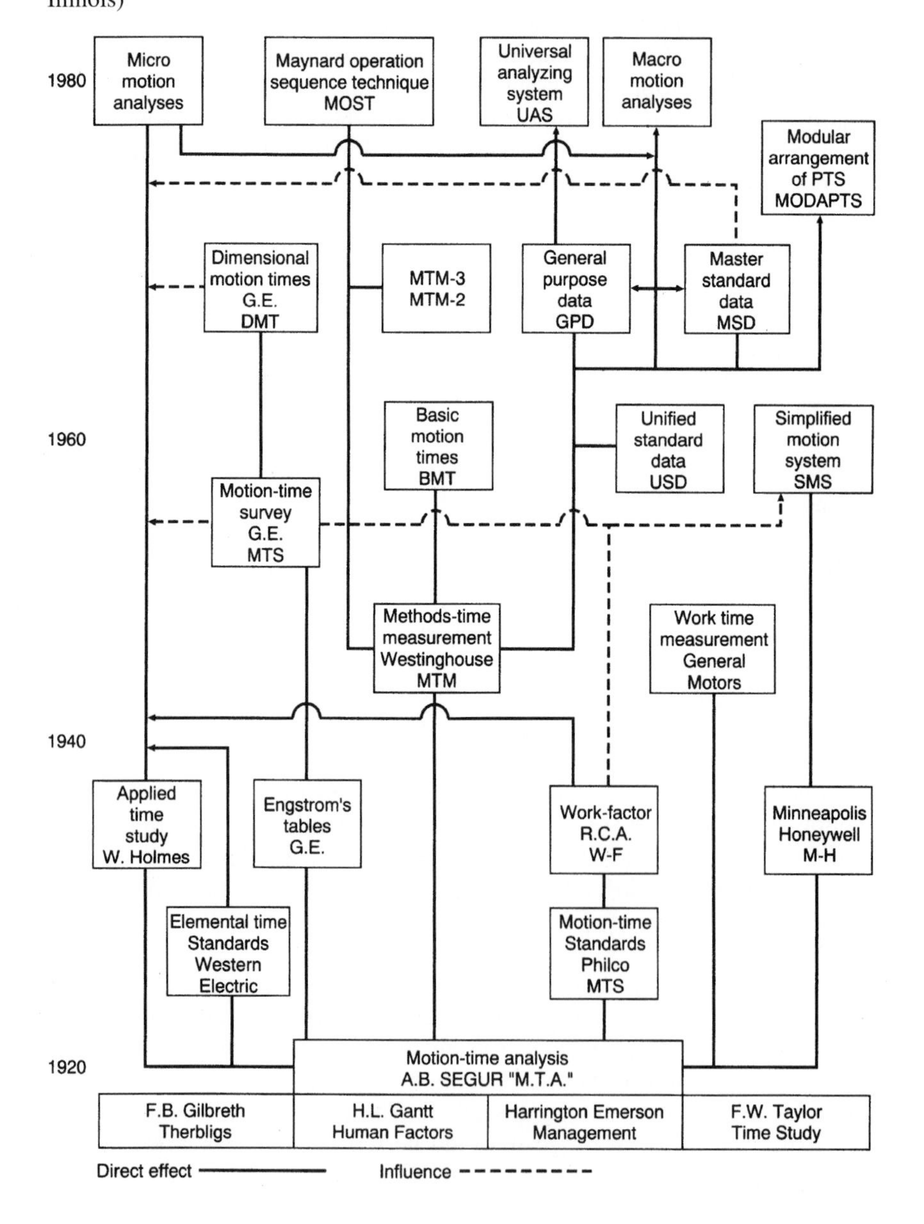

METHODS–TIME MEASUREMENT

MTM-1

Methods-Time Measurement (Maynard, Stegemarten, and Schwab, 1948) gives time values for the fundamental motions of: reach, move, turn, grasp, position, disengage, and release. The authors defined MTM as "a procedure which analyzes any manual operation or method into the basic motions required to perform it, and assigns to each motion a pre-determined time standard which is determined by the nature of the motion and the conditions under which it is made."

MTM-1 data are the result of frame-by-frame analyses of motion-picture films of diversified areas of work. The data taken from the various films were rated by the Westinghouse technique (see Chapter 9), tabulated, and analyzed to determine the degree of difficulty caused by variable characteristics. For example, both the distance and the type of reach affect reach time. Further analysis categorized five distinct cases of reach, each requiring a different time allotment for a given distance. These are:

1. Reach to object in fixed location, or to object in other hand, or to object on which other hand rests.
2. Reach to single object in location that may vary slightly from cycle to cycle.
3. Reach to object jumbled with other objects, requiring search and select as well.
4. Reach to a very small object, or an object for which accurate grasp is required.
5. Reach to indefinite location to get hand in position for body balance, or for next motion, or out of the way.

In addition, they found that move time was influenced by both the distance and the weight of the object being moved, as well as by the specific type of move. The three cases of move are:

1. Move object to other hand, or against stop.
2. Move object to approximate or indefinite location.
3. Move object to exact location.

Finally, two cases of release and 18 cases of position also influence time.

Table 13–1 summarizes the MTM-1 values. The time values of the therblig grasp vary from 2.0 TMU to 12.9 TMU (1 TMU [Time-Measurement Unit] equals 0.00001 hour), depending on the classification of the grasp.

First, the analyst summarizes all left-hand and right-hand motions required to perform the job properly. Then, the rated times in TMU for each motion are determined from the methods–time data tables. To determine the time required for a normal performance of the task, the nonlimiting motion values are either circled or deleted, as only the limiting motions will be summarized, provided that it is "easy" to perform the two motions simultaneously (see Table X of Table 13–1). For example, if the right hand must reach 20 inches (50 cm) to pick up a nut, the classification would be R20C and the time value would be 19.8 TMU. If, at the same time, the left hand must reach 10 inches (25 cm) to pick up a cap screw, a

TABLE 13–1
Summary of MTM-1 Data

TABLE I—REACH—R

Distance Moved Inches	Time TMU A	Time TMU B	Time TMU C or D	Time TMU E	Hand in Motion A	Hand in Motion B
¾ or less	2.0	2.0	2.0	2.0	1.6	1.6
1	2.5	2.5	3.6	2.4	2.3	2.3
2	4.0	4.0	5.9	3.8	3.5	2.7
3	5.3	5.3	7.3	5.3	4.5	3.6
4	6.1	6.4	8.4	6.8	4.9	4.3
5	6.5	7.8	9.4	7.4	5.3	5.0
6	7.0	8.6	10.1	8.0	5.7	5.7
7	7.4	9.3	10.8	8.7	6.1	6.5
8	7.9	10.1	11.5	9.3	6.5	7.2
9	8.3	10.8	12.2	9.9	6.9	7.9
10	8.7	11.5	12.9	10.5	7.3	8.6
12	9.6	12.9	14.2	11.8	8.1	10.1
14	10.5	14.4	15.6	13.0	8.9	11.5
16	11.4	15.8	17.0	14.2	9.7	12.9
18	12.3	17.2	18.4	15.5	10.5	14.4
20	13.1	18.6	19.8	16.7	11.3	15.8
22	14.0	20.1	21.2	18.0	12.1	17.3
24	14.9	21.5	22.5	19.2	12.9	18.8
26	15.8	22.9	23.9	20.4	13.7	20.2
28	16.7	24.4	25.3	21.7	14.5	21.7
30	17.5	25.8	26.7	22.9	15.3	23.2

CASE AND DESCRIPTION
A Reach to object in fixed location, or to object in other hand or on which other hand rests.
B Reach to single object in location which may vary slightly from cycle to cycle.
C Reach to object jumbled with other objects in a group so that search and select occur.
D Reach to a very small object or where accurate grasp is required.
E Reach to indefinite location to get hand in position for body balance or next motion or out of way.

TABLE II—MOVE—M

Distance Moved Inches	Time TMU A	Time TMU B	Time TMU C	Hand in Motion B
¾ or less	2.0	2.0	2.0	1.7
1	2.5	2.9	3.4	2.3
2	3.6	4.6	5.2	2.9
3	4.9	5.7	6.7	3.6
4	6.1	6.9	8.0	4.3
5	7.3	8.0	9.2	5.0
6	8.1	8.9	10.3	5.7
7	8.9	9.7	11.1	6.5
8	9.7	10.6	11.8	7.2
9	10.5	11.5	12.7	7.9
10	11.3	12.2	13.5	8.6
12	12.9	13.4	15.2	10.0
14	14.4	14.6	16.9	11.4
16	16.0	15.8	18.7	12.8
18	17.6	17.0	20.4	14.2
20	19.2	18.2	22.1	15.6
22	20.8	19.4	23.8	17.0
24	22.4	20.6	25.5	18.4
26	24.0	21.8	27.3	19.8
28	25.5	23.1	29.0	21.2
30	27.1	24.3	30.7	22.7

Wt. Allowance: Wt. (lb.) Up to	Factor	Constant TMU
2.5	0	0
7.5	1.06	2.2
12.5	1.11	3.9
17.5	1.17	5.6
22.5	1.22	7.4
27.5	1.28	9.1
32.5	1.33	10.8
37.5	1.39	12.5
42.5	1.44	14.3
47.5	1.50	16.0

CASE AND DESCRIPTION
A Move object to other hand or against stop.
B Move object to approximate or indefinite location.
C Move object to exact location.

TABLE III—TURN AND APPLY PRESSURE—T AND AP

Weight	Time TMU for Degrees Turned 30°	45°	60°	75°	90°	105°	120°	135°	150°	165°	180°
Small— 0 to 2 Pounds	2.8	3.5	4.1	4.8	5.4	6.1	6.8	7.4	8.1	8.7	9.4
Medium—2.1 to 10 Pounds	4.4	5.5	6.5	7.5	8.5	9.6	10.6	11.6	12.7	13.7	14.8
Large— 10.1 to 35 Pounds	8.4	10.5	12.3	14.4	16.2	18.3	20.4	22.2	24.3	26.1	28.2

APPLY PRESSURE CASE A—10.6 TMU. APPLY PRESSURE CASE B—16.2 TMU

TABLE 13–1 *(continued)*

Summary of MTM-1 Data

TABLE IV—GRASP—G

Case	Time TMU	DESCRIPTION
1A	2.0	**Pick Up Grasp**—Small, medium or large object by itself, easily grasped.
1B	3.5	Very small object or object lying close against a flat surface.
1C1	7.3	Interference with grasp on bottom and one side of nearly cylindrical object. Diameter larger than ½".
1C2	8.7	Interference with grasp on bottom and one side of nearly cylindrical object. Diameter ¼" to ½".
1C3	10.8	Interference with grasp on bottom and one side of nearly cylindrical object. Diameter less than ¼".
2	5.6	**Regrasp.**
3	5.6	**Transfer Grasp.**
4A	7.3	Object jumbled with other objects so search and select occur. Larger than 1" x 1" x 1".
4B	9.1	Object jumbled with other objects so search and select occur. ¼" x ¼" x ⅛" to 1" x 1" x 1".
4C	12.9	Object jumbled with other objects so search and select occur. Smaller than ¼" x ¼" x ⅛".
5	0	Contact, sliding or hook grasp.

TABLE V—POSITION*—P

CLASS OF FIT		Symmetry	Easy To Handle	Difficult To Handle
1—Loose	No pressure required	S	5.6	11.2
		SS	9.1	14.7
		NS	10.4	16.0
2—Close	Light pressure required	S	16.2	21.8
		SS	19.7	25.3
		NS	21.0	26.6
3—Exact	Heavy pressure required.	S	43.0	48.6
		SS	46.5	52.1
		NS	47.8	53.4

*Distance moved to engage—1" or less.

TABLE VI—RELEASE—RL

Case	Time TMU	DESCRIPTION
1	2.0	Normal release performed by opening fingers as independent motion.
2	0	Contact Release.

TABLE VII—DISENGAGE—D

CLASS OF FIT	Easy to Handle	Difficult to Handle
1—Loose—Very slight effort, blends with subsequent move.	4.0	5.7
2—Close — Normal effort, slight recoil.	7.5	11.8
3—Tight — Considerable effort, hand recoils markedly.	22.9	34.7

TABLE VIII—EYE TRAVEL TIME AND EYE FOCUS—ET AND EF

Eye Travel Time $= 15.2 \times \frac{T}{D}$ TMU, with a maximum value of 20 TMU.

where T = the distance between points from and to which the eye travels.
D = the perpendicular distance from the eye to the line of travel T.

Eye Focus Time = 7.3 TMU.

TABLE 13–1 *(concluded)*

Summary of MTM-1 Data

TABLE IX—BODY, LEG, AND FOOT MOTIONS

DESCRIPTION	SYMBOL	DISTANCE	TIME TMU
Foot Motion—Hinged at Ankle.	FM	Up to 4"	8.5
With heavy pressure.	FMP		19.1
Leg or Foreleg Motion.	LM —	Up to 6"	7.1
		Each add'l. inch	1.2
Sidestep—Case 1—Complete when leading leg contacts floor.	SS-C1	Less than 12"	Use REACH or MOVE Time
		12"	17.0
		Each add'l. inch	.6
Case 2—Lagging leg must contact floor before next motion can be made.	SS-C2	12"	34.1
		Each add'l. inch	1.1
Bend, Stoop, or Kneel on One Knee.	B,S,KOK		29.0
Arise.	AB,AS,AKOK		31.9
Kneel on Floor—Both Knees.	KBK		69.4
Arise.	AKBK		76.7
Sit.	SIT		34.7
Stand from Sitting Position.	STD		43.4
Turn Body 45 to 90 degrees—Case 1—Complete when leading leg contacts floor.	TBC1		18.6
Case 2—Lagging leg must contact floor before next motion can be made.	TBC2		37.2
Walk.	W-FT.	Per Foot	5.3
Walk.	W-P	Per Pace	15.0

TABLE X—SIMULTANEOUS MOTIONS

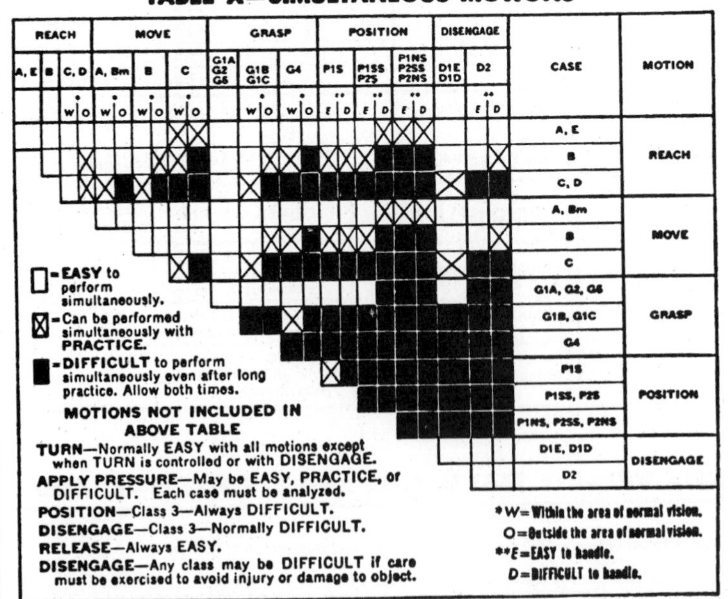

MTM Association for Standards and Research, Fair Lawn, New Jersey

designation of R10C with a TMU value of 12.9 would be in effect. The right hand value would be the limiting value, and the 12.9 value of the left hand would not be used in calculating the normal time.

The tabulated values do not carry any allowance for personal delays, fatigue, or unavoidable delays. When analysts use these values to establish time standards, they must add appropriate allowances to the summary of the synthetic basic motion times. Proponents of MTM-1 state that no fatigue allowance is needed in the vast majority of applications, because the MTM-1 values are based on a work rate that can be sustained at steady-state for the working life of a healthy employee.

An example of an MTM-1 analysis of a clerical operation (replacing a page in a three-ring binder) is shown in Table 13–2.

Today, MTM has received worldwide recognition. In the United States, it is administered, advanced, and controlled by the MTM Association for Standards and Research. This nonprofit association is one of 12 associations comprised by the International MTM Directorate. Much of MTM systems' success is the result of an active committee structure made up of members of the association.

The MTM family of systems continues to grow. In addition to MTM-1, the association has introduced MTM-2, MTM-3, MTM-V, MTM-C, MTM-M, MTM-Link, MTM-MEK, and MTM-UAS.

MTM-2

In an effort to further the application of MTM to work areas where the detail of MTM-1 would economically preclude its use, the International MTM Directorate initiated a research project to develop less refined data suitable for the majority of motion sequences. The result of this effort was MTM-2. Defined by the MTM Association of the United Kingdom, it is a system of synthesized MTM data and is the second general level of MTM data. It is based exclusively on MTM and consists of:

1. Single basic MTM motions.
2. Combinations of basic MTM motions.

The data are adapted to the operator and are independent of the workplace or equipment used. In general, MTM-2 should find application in work assignments where:

1. The effort portion of the work cycle is more than one minute.
2. The cycle is not highly repetitive.
3. The manual portion of the work cycle does not involve a large number of either complex or simultaneous hand motions.

The variability between MTM-1 and MTM-2 depends to a large extent on the length of the cycle. This is reflected in Figure 13–2, which shows the range of percentage deviation of MTM-2 from MTM-1. This range of error is considered to be the expected range for 95 percent of the time.

TABLE 13–2
MTM-1 A1

MTM-1 ANALYSIS OF MTM-C

VALIDATION

MTM ASSOCIATION FOR STANDARDS AND RESEARCH

ELEMENT TITLE: Replace page in 3-ring binder
STARTS: Get binder from shelf at left
INCLUDES: Get binder, open cover, locate correct page, open rings, replace old sheet, close rings, Aside binder to shelf
ENDS:

ANALYST:

DATE:

LEFT HAND DESCRIPTION	F	LH MOTION	TMU	RH MOTION	F	RIGHT HAND DESCRIPTION
1. GET BINDER—OPEN COVER						
Reach to binder		R30B	25.8			
Grasp binder		G1A	2.0			
Move to desk		M30B	24.3			
Release		RL1	2.0			
Reach to cover		R7B	9.3			
Grasp edge		G1A	2.0			
Open cover		M16B	15.8			
Release		RL1	2.0			
			83.2			
2. LOCATE CORRECT PAGE			14.6	EF	2	Read first page data
Reach to edge	3	R3D	21.9			
Grasp	3	G1B	10.5			
Move up	3	M4B	20.7			
Regrasp		G2	—			
			43.8	EF	2×3	Identify pages
Move pages back		M8B	10.6			
Release		RL1	2.0			
Reach to hold		R8B	10.1	R4B (circled)		To edge of page
Grasp		G5	0.0	G5		Contact
			8.0	MfB	4	Slide back up
Contact	3	G5	0.0	RL2	4	Release
Move	3	MfB (circled)	7.5	R1B	3	To corner
			0.0	G5	3	Contact
			5.6			
Regrasp pages		G2	87.6	EF	4×3	Identify pages
Move pages back		M8B	10.6			
Release		RL1	2.0			
			255.5			

TABLE 13–2 *(concluded)*

MTM-1 A1

MTM-1 ANALYSIS OF MTM-C

VALIDATION

MTM ASSOCIATION FOR STANDARDS AND RESEARCH

ELEMENT TITLE: Replace page in 3-ring binder
STARTS:
INCLUDES: Continues
ENDS:

ANALYST:

DATE:

LEFT HAND DESCRIPTION	F	LH MOTION	TMU	RH MOTION	F	RIGHT HAND DESCRIPTION
3. REPLACE PAGE						
To ring		R7A	7.4	R7A		To ring
Grasp		G1A	2.0	G1A		Grasp
Pull open		APB	16.2	APB		Pull open
Open		MfA	2.0	MfA		Open
Release		RL1	2.0	RL1		Release
To edge of paper		R6D	10.1			
Grasp		G1B	3.5			
To basket		M30B	24.3	R-E		
Release		RL1	2.0			
			10.1	R6D		To new sheet
			3.5	G1B		Grasp
			15.2	M12C		To rings
			16.2	P2SE		Align to ring
			2.0	MfC		To ring
			16.2	P2SE		Align
			2.0	MfA		Down on rings
			2.0	RL1		Release
To center ring		R4B	8.6	R6B		To center ring
Grasp		G1A	2.0	G1A		Grasp
Press to close		APB	16.2	APB		Press to close
Close		MfA	2.0	MfA		Close
Release		RL1	2.0	RL1		Release
			167.5			
4. CLOSE COVER AND ASIDE BINDER						
Reach to cover		R7B	9.3			
Grasp edge		G1A	2.0			
Close cover		M16B	15.8			
Release		RL1	2.0			
Reach to binder		R6B	8.6			
Grasp		G1A	2.0			
Regrasp		G2	5.6			
Move to shelf		M30B	24.3			
Release		RL1	2.0			
			71.6			
ELEMENT SUMMARY						
1. Get Binder-Open Cover			83.2			
2. Locate Correct Page			255.5			
3. Replace Page			167.5			
4. Close Cover and Aside Binder			71.6			
		TOTAL	577.8			

FIGURE 13–2
Percentage variation of MTM-1 when compared with MTM-2 at increasing cycle lengths. (Courtesy: MTM Association)

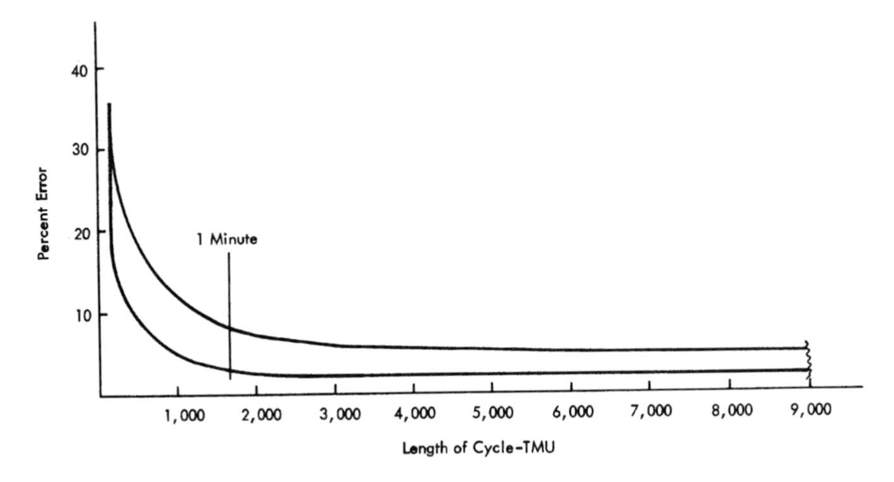

MTM-2 recognizes 11 classes of actions, which are referred to as categories. These 11 categories and their symbols are:

GET	G
PUT	P
GET WEIGHT	GW
PUT WEIGHT	PW
REGRASP	R
APPLY PRESSURE	A
EYE ACTION	E
FOOT ACTION	F
STEP	S
BEND & ARISE	B
CRANK	C

In using MTM-2, analysts estimate distances by classes; these distances affect the times of the GET and PUT categories. As in MTM-1, the analysts base the distance moved on the path traveled by the knuckle at the base of the index finger for hand motions, and the path traveled by the fingertips, if only the fingers move. The codes for the five tabulated distance classes correspond to the five levels of classification of motions, as discussed in Chapter 4 (see Table 13–3).

Three variables affect the time required to perform a GET: the case involved, the distance traveled, and the weight handled. GET can be considered a composite of the therbligs reach, grasp, and release, while PUT is a combination of the therbligs move and position.

The three cases of GET are A, B, and C. Case A implies a simple contact grasp, such as the fingers pushing an object across a desk. If an object, such as a pencil, is picked up by simply enclosing the fingers with a single movement, this is a B grasp. If the type of grasp is neither an A nor a B, then it is a case C GET. Analysts can

TABLE 13–3
Summary of MTM-2 Data

MTM - 2

RANGE	Code	GA	GB	GC	PA	PB	PC
Up to 2"	-2	3	7	14	3	10	21
Over 2" - 6"	-6	6	10	19	6	15	26
Over 6" - 12"	-12	9	14	23	11	19	30
Over 12" - 18"	-18	13	18	27	15	24	36
Over 18"	-32	17	23	32	20	30	41
MTM ASSOCIATION ®	GW 1 - per 2 lb				PW 1 - per 10 lb		
	A	R	E	C	S	F	B
	14	6	7	15	18	9	61

(Courtesy: MTM Association)

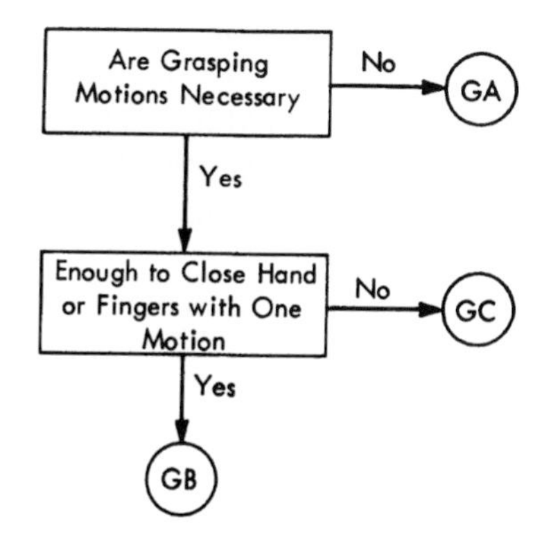

FIGURE 13–3
Algorithm for determining case of GET. (Courtesy: MTM Association)

resort to the decision diagram (see Figure 13–3) to assist in determining the correct case of GET. Tabular values in TMU of the three cases of GET, as applied to each of the five coded distances, appear in Table 13–3.

PUT involves moving an object to a destination with the hand or fingers. It starts with an object grasped and under control at the initial place and includes all transporting and correcting motions necessary to place the object at the destination. PUT ends with the object still under control at the intended place. The same case, distance, and weight variables affect PUT as for GET.

Just as there are three cases of GET, there are three cases of PUT. The PUT case depends on the number of correcting motions required. A correction is an unintentional stop, hesitation, or change in direction at the terminal point.

1. *PA: No correction.* This is a smooth motion from start to finish and is the action employed in laying an object aside, or placing it against a stop or in an approximate location. This is the most common PUT.

2. *PB: One correction.* This PUT occurs most often in positioning easy-to-handle objects where a loose fit occurs. It is difficult to recognize. The decision diagram shown in Figure 13–4 is designed to identify this PUT by exception.
3. *PC: More than one correction.* Multiple corrections, or several very short, unintentional motions, are usually obvious. These unintentional motions are generally caused by handling difficulty, close fits, nonsymmetry of engaging parts, or uncomfortable working positions.

Analysts identify cases of PUT by the decision model shown in Figure 13–4. When in doubt, analysts assign the higher class. If an engagement of parts follows a correction, and if the engagement distance exceeds 1 inch (2.5 cm), an additional PUT is used. The tabular values for the three cases of PUT as applied to the five coded distances (the same as for GET) are given in Table 13–3.

A final technicality involving PUTs is that a PUT can be accomplished in one of two ways: insertion, or alignment. An insertion involves placing one object into another, such as a shaft into a sleeve. With an insertion, the terminal point for a correction is the point of insertion. An alignment involves orienting a part on a surface, such as bringing a rule up to a line. Table 13–4 may be helpful in assisting the analyst in better identifying the appropriate case.

Weight in MTM-2 is determined similarly to MTM-1. The time value addition for GET WEIGHT is 1 TMU per two pounds (1 kg). Thus, if a load of 12 pounds (6 kg) is handled by both hands, the time addition due to weight would be 3 TMU, since the effective weight per hand would be 6 pounds (3 kg). For PUT WEIGHT, additions are 1 TMU per 10 pounds (5 kg) of effective weight, up to a maximum of 40 pounds (20 kg).

The category REGRASP is also similar to MTM-1. Here, however, a time of 6 TMU is assigned. The authors of MTM-2 point out that for a REGRASP to be in effect, the hand must retain control.

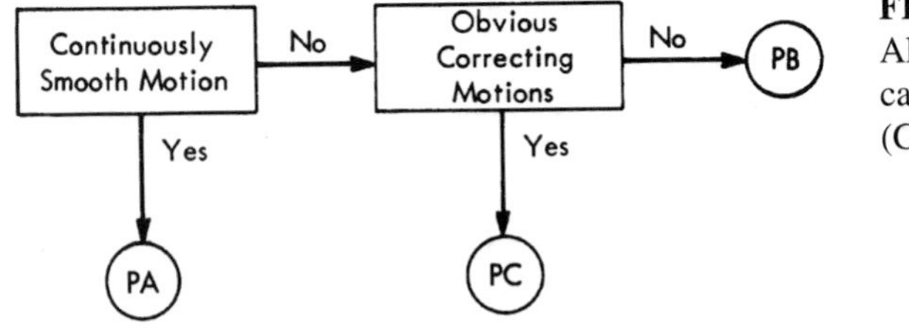

FIGURE 13–4
Algorithm for determining case of PUT.
(Courtesy: MTM Association)

TABLE 13–4
Comparison of Insertion and Alignment PUTs [Inches (mm)]

	PA	PB	PC
Insertion	Clearance > 0.4 in (10.2 mm)	Clearance < 0.4 in (10.2 mm)	Tight fit
Alignment	Tolerance > 0.25 in (6.3 mm)	0.0625 in (1.6 mm) < Tolerance < 0.25 in (6.3 mm)	Tolerance < 0.0625 in (1.6 mm)

APPLY PRESSURE has a time of 14 TMU. The authors point out that this category can be applied by any member of the body and that the maximum permissible movement for an apply pressure is 1/4 inch (6.4 mm).

EYE ACTION is allowed under either of the following cases:

1. The eyes must move to see various aspects of the operation involving more than one specific section of the work area. This eye movement is defined as moving beyond a 4-inch (10-cm) diameter circle at a typical viewing distance of 16 inches (40 cm). Note that this is equivalent to the primary visual field defined in Chapter 4.
2. The eyes must concentrate on an object to recognize a distinguishable characteristic.

The estimated value of EYE ACTION is 7 TMU. The value is only allowed when EYE ACTION is independent of hand or body motions.

CRANK occurs when the hands or fingers move an object in a circular path of more than half an evolution. For less than half a revolution, a PUT is used instead. Under MTM-2, the category CRANK has only two variables: the number of revolutions, and the weight or resistance. A time of 15 TMU is allotted for each complete revolution. Where weight or resistance is significant, PUT WEIGHT is applied to each revolution.

FOOT movements are 9 TMU, and STEP movements are 18 TMU. The time for STEP movement is based on a 34-inch pace (85 cm). The decision diagram (see Figure 13–5) can be helpful in ascertaining whether a given movement is a STEP or a FOOT movement.

The category BEND & ARISE occurs when the body changes its vertical position. Typical movements characteristic of BEND & ARISE include: sitting down, standing up, and kneeling. A time value of 61 TMU is assigned to this category. The authors indicate that when an operator kneels on both knees, the movement should be classed as 2 B. Table 13–3 summarizes the applicable MTM-2 values.

There are several special situations of which the analyst should be aware in performing a correct MTM-2 analysis. Motions performed with both hands simultaneously cannot always be performed in the same time as motions performed by one hand only. Figure 13–6 reflects motion patterns for which the time required for simultaneous motions is the same as that required for easy

FIGURE 13–5
Algorithm for differentiating between STEP (S) and FOOT motion (F). (Courtesy: MTM Association)

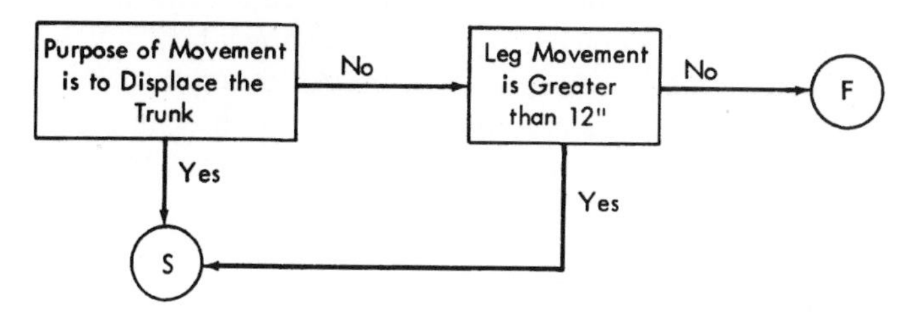

MOTION		GET			PUT			
	CASE	GA	GB	GC	PA	PB O*	PB W	PC
GET	GA					PRACTICE	PRACTICE	PRACTICE
GET	GB					PRACTICE	PRACTICE	DIFFICULT
GET	GC			DIFFICULT	PRACTICE	DIFFICULT	DIFFICULT	DIFFICULT
PUT	PA					PRACTICE	PRACTICE	PRACTICE
PUT	PB	PRACTICE	PRACTICE	DIFFICULT	PRACTICE	DIFFICULT	PRACTICE	DIFFICULT
PUT	PC	PRACTICE	DIFFICULT	DIFFICULT	PRACTICE	DIFFICULT	DIFFICULT	DIFFICULT

EASY (empty box) PRACTICE (box with X) DIFFICULT (shaded box)

*O = OUTSIDE; W = WITHIN NORMAL VISION

FIGURE 13–6
Difficulty of simultaneous motions.
(Courtesy: MTM Association)

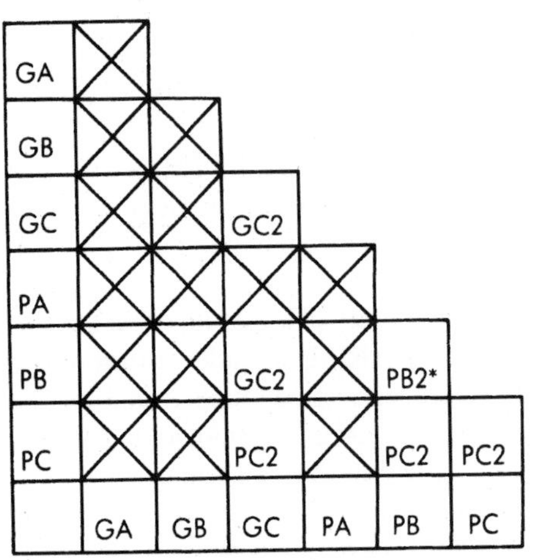

* If PB ______ is performed simultaneously with PB ______, an addition of PB2 is made only if the actions are outside the normal area of vision.

FIGURE 13–7
MTM-2 simultaneous hand motion allowances.

motions performed by one hand. In these instances, an open rectangle appears. An X in the rectangle indicates that, with practice, simultaneous motions can be made. A darkened rectangle indicates that it is difficult, even with practice, to perform the motions simultaneously. Figure 13–7 shows how much additional time these difficult simultaneous motions require. An example of the application of this *principle of simultaneous motions* is marked with ① in Figure 13–8. A PC2 is added to the overall time because of two simultaneous PC motions in the right and left hands.

A second situation involves the *principle of limiting motions:* for two different motions performed simultaneously by the left and right hands, the longer time pre-

FIGURE 13–8
Examples of MTM-2 conventions.

MTM Methods Analysis Page of

Operation:	Remarks:
Study No.:	
Date:	
Analyst:	

Description	No.	LH	TMU	RH	No.	Description
①		PC6	26	PC6		
		PC2	21	PC2		
②		(GC12)	23	GB18		
③		GA2	18	GB18		
				F		
④		GC12	18	GB18		
		R1	6	R		

Summary	Total TMU:	Conversion:	% Allowance:	Standard time:

dominates. This is shown as a ② in Figure 13–8, with the longer element circled (GC12). A GC12 is 23 TMUs, as compared to 18 TMUs for a GB18.

If one hand performs two motions simultaneously, the longer time again predominates, because of the *principle of combined motions.* This would apply for case ③; a GB18 is longer than an F. (A curve connects the two lines to indicate a combined motion.)

A similar situation appears for case ④, but it is more complicated because of the combined motion in the left hand. In this case, R is already assumed to be part of any C-type motion and is thus crossed out. Therefore, the overall time for the first two lines is 24 TMUs, because of the right hands' GB18 (18 TMUs) and R (6 TMUs) rather than GC12 (23 TMUs). As an example, Figure 13–9 shows a complete MTM-2 analysis for a flashlight assembly.

FIGURE 13–9
MTM-2 analysis of flashlight assembly.

MTM Methods Analysis Page of

Operation: BLUE FLASHLIGHT ASSEMBLY
Study No.: 1-2
Date: 1-27-98
Analyst: AF
Remarks:

Description	No.	LH	TMU	RH	No.	Description
BODY TO WORK AREA		GB6	10	GB6		1ST BATTERY TO BODY
		PA6	15	(PB6)		
			~~6~~	R		
			10	GB6		2ND BATTER TO BODY
			15	PB6		
			~~6~~	R		
			10	GB6		CAP TO WORK AREA/BODY
			21	PC6		
				~~R~~		
			3	PA2		PRESS AGAINTS SPRING
			3	PA2		BACK TURN
			12	PA2	4	SCREW DOWN
			28	GB2	4	
			14	A		TIGHTEN
PUT DOWN		PA6	6			

Summary	Total TMU: 147	Conversion: .0006	% Allowance: 10	Standard time: .097

MTM-3

The third level of Methods–Time Measurement is MTM-3. This level was developed to supplement MTM-1 and MTM-2. MTM-3 is helpful in work situations where an interest in saving time at the expense of some accuracy makes it the best alternative. The accuracy of MTM-3 is within ±5 percent, with a 95 percent confidence level, when compared to MTM-1 analysis for cycles of approximately 4 minutes. It has been estimated that MTM-3 can be applied in about one-seventh the time of MTM-1. However, MTM-3 cannot be used for operations that require either eye focus or eye travel time, since the data do not consider those motions.

The MTM-3 system consists of only four categories of manual motions:

1. *Handle.* A motion sequence with the purpose of getting control of an object with the hand or fingers and placing the object in a new location.
2. *Transport.* A motion with the purpose of moving an object to a new location with the hand or fingers.
3. *Step and Foot Motions.* These are the same as defined in MTM-2.
4. *Bend and Arise.* These, too, are the same as defined in MTM-2.

Table 13–5 presents MTM-3 data. Ten time standards, ranging from 7 to 61 TMU, form the basis for the development of any standard, subject to the limitations noted earlier.

MTM-V

MTM-V was developed by Svenska MTM Gruppen, the Swedish MTM Association, for use in metal cutting operations. It is of particular use with short runs in machine shops. MTM-V provides for work elements involved in: (1) bringing the work to the jig, fixture, or chuck, removing the work from the machine, and placing it aside; (2) operating the machine; (3) checking the work to assure quality of output; and (4) cleaning the nip point area of the machine, to maintain facility output and product quality. MTM-V does not cover process time involving feeds and speeds.

Analysts use this system to establish setup times for all typical machine tools. Therefore, standard times for such elements as setting up and dismounting fixtures,

TABLE 13–5
Summary of MTM-3 Data

		Handle		Transport	
Inches	**Code**	**HA**	**HB**	**TA**	**TB**
6	6	18	34	7	21
6	32	34	48	16	29
		SF 18		B 61	

jigs, stops, cutting tools, and indicators can be calculated. All manual cycle times of 24 minutes (40,000 TMU) or more established by MTM-V are within ±5 percent of that produced by MTM-1 at the 95 percent confidence level. MTM-V is about 23 times faster than MTM-1.

MTM-C

MTM-C, which is widely used in the banking and insurance industries, is a two-level standard data system used to establish time standards for clerical-related work tasks, such as keypunching, filing, data entry, and typing. Both levels of MTM-C are traceable to MTM-1 data.

The system provides three distinct ranges for reach and move (Get Place). A six-place numeric coding system (similar to MTM-V) provides a detailed description of the operation being studied.

MTM-C develops standards in the same way as the other MTM systems. Analysts can combine it with existing proven standard data or standard data developed from other sources or techniques. MTM-C is available in manual or automated forms. For the latter, an MTM-C data set can be incorporated into MTM-Link.

The nine level 1 categories used in MTM-C, shown in Table 13–6, are:

1. *Get Place*. This category includes the basic work divisions required to get an object, move it aside without relinquishing control, and release it. For example, the coding and description of an element in this category might be: 112210—Get small stack with medium move.
2. *Open Close.* Characteristic of this category are such operations as opening or closing books, doors, drawers, binder rings, zippered objects, envelopes, and files. An example of the coding for a representative operation would be: 212100—Open hinge cover, medium.
3. *Fasten Unfasten.* This category includes attaching and removing clips, clamps, bands, and staples, all used to join materials. A representative coding for this work element would be: 312130—Fasten with a large paper clip.

TABLE 13–6
MTM-C Level-1 Elements

Level 1 Elements	Symbol
Get Place	11 X X X X
Open Close	21 X X X X
Fasten Unfasten	31 X X X X
Organize File	4 X X X X X
Read Write	5 X X X X X
Typing	6 X X X X X
Handling	7 X X X X X
Walk Body Motions	8 X X X X X
Machines	9 X X X X X

4. *Organize File.* This category includes the basic elements involved with filing activities and some of the organizational handling of work directly or indirectly related to filing. An example of the coding and description of this category is: 410400—Arrange a pile into a stack.
5. *Read Write.* This category includes a prose reading speed of 330 words per minute. Writing time values have been developed for letters, numerals, and symbols. Values are a weighted average based on the frequency of occurrence of each type of character in normal prose. An example of coding and a representative description would be: 510600-Read average prose, per word.
6. *Typing.* This category includes all of the actions related to the preparation for typing, the manual typing functions, and the related process times of the machine. An example of the coding and the description in this category follows: 613530-Insert single object in typewriter, long distance.
7. *Handling.* This category includes all of the clerical activities not covered in the other categories. An example of the coding and the description of an element in this category might be: 760600-Adhere envelope flap.
8. *Walk Body Motions.* This category includes walking values based on pace. Body motions include sitting, standing, and horizontal and vertical body movements while in a chair. An example of the coding and description of an element in this category follows: 860002-Move while in swivel chair.
9. *Machines.* The machine data are representative of a group of similar types of equipment. Data for keyboard calculators and keypunch machines are typical examples of this category.

Level 2 data are directly traceable to Level 1 and to MTM-1. Level 2 elements and their symbols are shown in Table 13–7. The Level 2 elements are:

1. *Get/Place/Aside.* These elements are applied collectively or separately. The coding and element example of this category, with collective basic divisions, would be: G5PA2-Get a pencil for use and set it aside for later.

TABLE 13–7
MTM-C Level-2 Elements

Level 2 Elements	Symbol
Aside	A
Body Motions	B
Close	C
Fasten	F
Get	G
Handling	H
Identify	I
Locate File	L
Open	O
Place	P
Read	R
Typing	T
Unfasten	U
Write	W

2. *Open/Close.* Included in these data is getting the object being opened or closed. These data are applied individually or in combination, as follows: C65-Close string, tie envelope; or OC4-Open and close binder rings.
3. *Fasten/Unfasten.* For fasten (F), the element is made up of getting the objects and fastening them. For unfasten (U), the element includes getting the objects and unfastening them.
4. *Identify.* The data for this element include eye travel time values, along with eye focuses required to identify (I) single or multiple words and sets of numbers.
5. *Locate File.* The data for this element cover typical filing activities. The first position of the coding is L, for locate. The second position is also a letter which corresponds to the filing activity, such as LI (insert), LR (remove), LT (tilt and replace).
6. *Read/Write.* The reading data include the reading of words and single numbers and/or characters. Read also contains detailed Read-and-Compare, as well as Read-and-Transcribe data. The writing data include common clerical items, such as address, date, initials, and names. The coding and element description for two representative elements would be: RW20-Read 20 words; RCN25-Read and compare 25 numbers.
7. *Handling.* This element includes the actual paper handling activities from Level 1: organize data and handling data. In the majority of elements, objects have been obtained with a "Get," as well as the action handling elements. In the coding for handling elements, H is the first coding position. The second position is the initial letter of the element activity. An example of coding for folding a sheet with two folds would be: HF12.
8. *Body Motions.* These elements include walking, sitting and standing, bending, and arising, and the horizontal body motions alone or in a chair.
9. *Typing.* These elements include three major sections of data: Handling, Keystroke, and Correction. Examples of the coding and descriptions are: THI32-Put three sheets and two carbons in machine and remove; TKE17E-Type one 7-inch (17.5 cm) line with an electric machine, elite; TCL41-Correct single error on four sheets, with correction fluid.

MTM-C Level 1 can be calculated faster than MTM-2. Also, the speed of MTM-C Level 2 is faster than MTM-3. Compare the standards for replacing a page in a three-ring binder, developed first using MTM-1 (see Table 13–2), then using MTM-C Level 1 (see Table 13–8), and finally using MTM-C Level 2 (see Table 13–9). Note how closely the three standards agree (see Table 13–10).

MTM-M

MTM-M is a predetermined time system for evaluating operator work using a stereoscopic microscope. In the development of MTM-M, basic times from MTM-1 were not used, although the beginning and endpoint definitions of motion elements were compatible with MTM-1. The data used were original data developed through the efforts of the United States/Canada MTM Association. In general, MTM-M is a higher level system, similar to MTM-2.

TABLE 13–8

MTM-C Operation Analysis (Level 1)

MTM-C OPERATION ANALYSIS

VALIDATION

Sheet of

MTM ASSOCIATION FOR STANDARDS AND RESEARCH

MTM-C LEVEL 1

Replace page in 3-ring binder

DEPARTMENT: Clerical	ANALYST: CNR	DATE: 11/77

No.	Description	Reference	Element TMU	Occurrence per Cycle	TMU per Cycle
1.	OPEN BINDER				
	Get binder from shelf	113 520	21	1	21
	Aside to desk	123 002	22	1	22
	Get cover	112 520	14	1	14
	Open cover	212 100	15	1	15
2.	LOCATE CORRECT PAGE				
	Read on first page	510 000	7	2	14
	Locate approximate	451 120	16	3	48
	Identify page number	440 630	22	3	66
	Locate correct page	450 130	18	4	72
	Identify pages	440 630	22	3	66
3.	REPLACE PAGES				
	Get binder rings	112 520	14	1	14
	Open rings	210 400	21	1	21
	Get old sheet	111 100	10	1	10
	Aside sheet to basket	123 002	22	1	22
	Get new sheet	111 100	10	1	10
	Insert sheet in binder	462 104	64	1	64
	Get rings	112 520	14	1	14
	Close rings	222 400	21	1	21
4.	CLOSE COVER AND ASIDE BINDER				
	Get cover	111 520	8	1	8
	Close cover	222 100	13	1	13
	Get binder	112 520	14	1	14
	Aside binder to shelf	123 002	22	1	22
		TOTAL TMU PER CYCLE			571
		ALLOWANCES ____ %			
		STANDARD HOURS PER ____ UNIT			
		UNITS PER HOUR			

TABLE 13–9

MTM-C Operation Analysis (Level 2)

MTM-C OPERATION ANALYSIS

VALIDATION

Sheet of

MTM-C LEVEL 2

MTM ASSOCIATION FOR STANDARDS AND RESEARCH

Replace page in 3-ring binder

DEPARTMENT: Clerical | ANALYST: CNR | DATE: 2/77

No.	Description	Reference	Element TMU	Occurrence per Cycle	TMU per Cycle
	Get and aside binder	G5A2	29	1	29
	Open cover	O1	29	1	29
	Read first page	RN2	14	1	14
	Locate pages	LC12	129	1	129
	Identify pages	130	22	6	132
	Open rings	O4	35	1	35
	Remove sheet	G1A2	32	1	32
	New sheet on rings	HI14	84	1	84
	Close rings	C4	35	1	35
	Close cover	C1	27	1	27
	Aside binder	G5A2	29	1	29
			TOTAL TMU PER CYCLE		575

TABLE 13–10

Comparison of MTM-1, MTM-C (1) and MTM-C (2)

Techniques	Number of elements	Standard
MTM-1	57	577.8
MTM-C Level 1	21	577
MTM-C Level 2	11	575

This system has four major tables and one subtable. Analysts must consider four variables when selecting the appropriate data: (1) type of tool; (2) condition of tool; (3) terminating characteristic of the motion; and (4) distance/tolerance ratio. Additional factors, other than the direction of movement and these four variables, that have an impact on motion performance time include:

1. Tool load state, empty or loaded.
2. Microscope power.
3. Distance moved.
4. Positioning tolerance.
5. Purpose of the motion, as determined by the manipulations involved at the motion termination. (For example, workers may use tweezers to contact grasp an object, or to pick up an object.)
6. Simultaneous motions.

With the growing amount of microminiature manufacturing, the application of fundamental data similar to MTM-M will expand. Such data allow analysts to establish equitable standards that would be difficult to establish by stopwatch procedures. Establishing reliable elemental standards by direct observation is impossible. It is only possible to establish sound elemental and operation standards for microscope work by using standard data similar to MTM-M or by micromotion procedures.

Other Specialized MTM Systems

Three other specialized MTM systems are: MTM-TE, MTM-MEK, and MTM-UAS. The first of these, MTM-TE, was developed for electronic tests. This system has two levels of data that were developed from MTM-1 for basic test application. Level 1 includes the elements: get, move, body motions, identify, adjust, and miscellaneous data. Level 2 includes: get and place, read and identify, adjust, body motions, and writing. A third level of data is also available in the form of synthesized Level 1 data. MTM-TE data do not cover "trouble shoot" relative to electronic test operations. They do, however, provide guidelines for investigations and recommendations for work measurement for this activity.

The second specialized system, MTM-MEK, was designed to measure one-of-a-kind and small-lot production. This two-level system developed from MTM-1 can analyze all manual activities, as long as the following requirements are met:

1. The operation is not highly repetitive or organized, although it may contain similar elements that require different methods. The method used to perform a given operation typically varies from cycle to cycle.
2. The workplace, tools, and equipment used are universal in character.
3. The task is complex and necessitates employee training; yet the lack of a specific method requires a high degree of versatility by the operator.

The objectives of MTM-MEK are to:

1. Provide accurate measurement of an activity connected with one-of-a-kind or small-lot production.
2. Provide an easily definable description of unorganized work, thus generally identifying a procedure.
3. Provide fast application.
4. Provide accuracy relative to MTM-1.
5. Require minimum training and application practice.

The data in MTM-MEK consist of 51 time values in the following eight categories: get and place, handle tool, place, operate, motion cycles, fasten or loosen, body motions, and visual control. In addition, there are standard data for a wide range of assembly tasks in one-of-a-kind and small-lot production. These data consist of 290 time values in the following categories: fasten, clamp and unclamp, clean and/or apply lubricant/ adhesive, assemble standard parts, inspect and measure, mark and transport.

The third specialized system, MTM-UAS, is a third level system. The authors developed it to provide a process description, as well as to determine the allowed times in any activity related to batch production. MTM-UAS is applicable to various activities, as long as the following characteristics of batch production are present:

1. Similar tasks.
2. Workplace specifically designed for the task.
3. Good levels of work organization.
4. Detailed instructions.
5. Well-trained operators.

The MTM-UAS analyzing system consists of 77 time values in seven of the eight categories used in MTM-MEK. These are: get and place, place, handle tool, operate, motion cycles, body motions, and visual control. MTM-UAS is about eight times faster than MTM-1. At cycle times of 4.6 minutes or more, the standard produced by MTM-UAS is within ±5 percent of that produced by MTM-1, with a 95 percent confidence level.

MTM Systems Comparison

Figure 13–10 illustrates the total absolute accuracy, at a 90 percent confidence level, of all the MTM systems. Table 13–11 compares the level of detail, such as, the number of therbligs utilized, the time to analyze a job (expressed as a multiple of the job cycle time), and the accuracy, for the three basic MTM systems. Overall, MTM-2 may be a good compromise between the excessive time required for MTM-1 and the poor accuracy of MTM-3. A 6-minute job will take approximately 600 minutes to analyze with MTM-2 and will be off by no more than 0.24 minutes.

MAYNARD OPERATION SEQUENCE TECHNIQUE (MOST)

An outgrowth of MTM, called Maynard Operation Sequence Technique (MOST), is a simplified system developed by Zandin (1980) and originally applied at Saab-Scania in Sweden in 1967. With MOST, analysts can establish standards at least five times faster than with MTM-1, with little if any sacrifice in accuracy.

MOST utilizes larger blocks of fundamental motions than MTM-2; consequently, the analysis of the work content of an operation can be made faster. In contrast to MTM-2, which is built around 37 time values for describing manual work, MOST utilizes only 16 time fragments. MOST identifies three basic sequence models: general move, controlled move, and tool use.

The general move sequence identifies the spatial free movement of an object through the air, while the controlled move sequence describes the movement of an object when it either remains in contact with a surface or remains attached to another object during the movement. The tool use sequence is aimed at the use of common hand tools.

FIGURE 13–10

Total absolute accuracy at 90 percent confidence level of the various MTM systems.

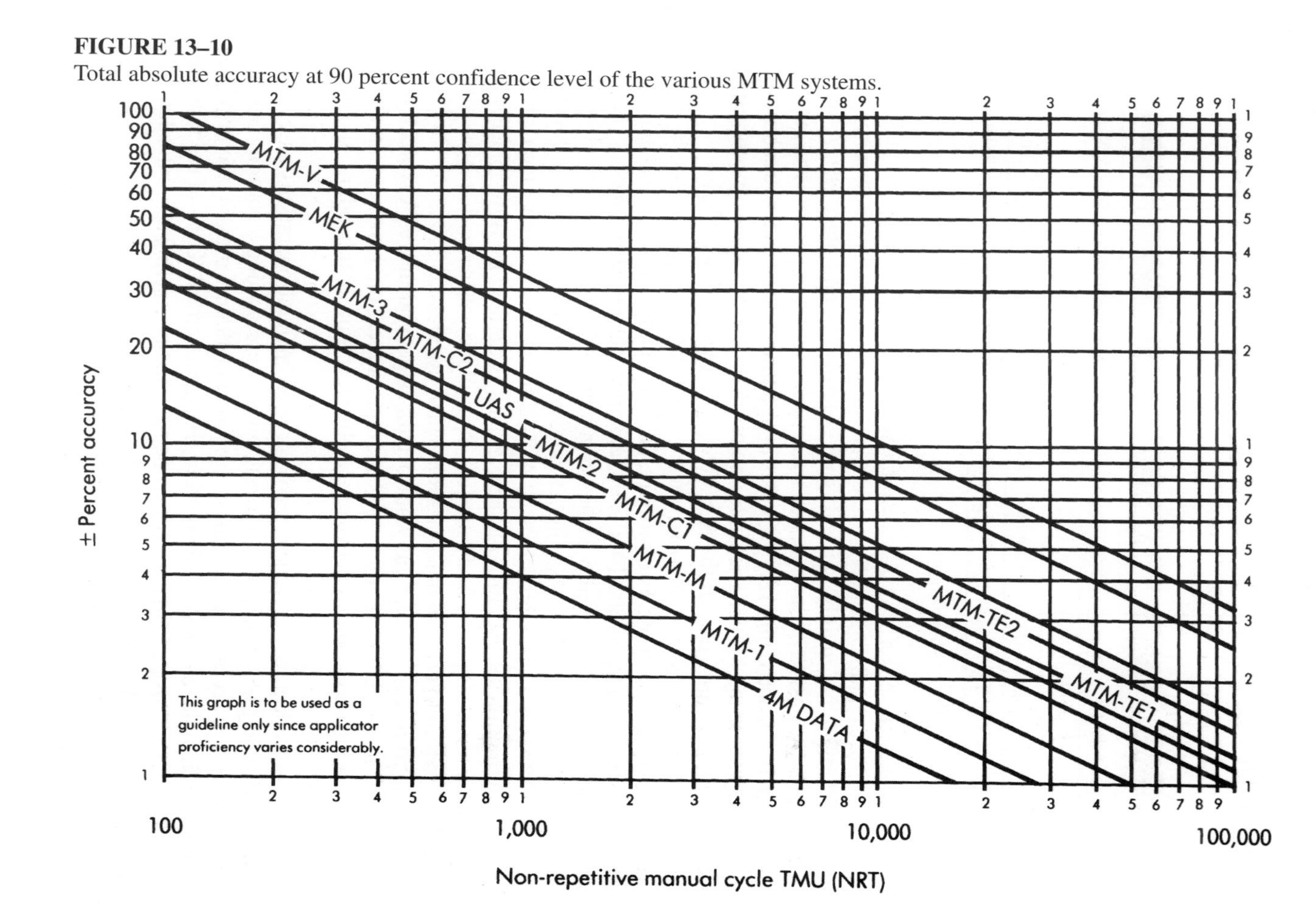

TABLE 13–11
Comparison of MTM-1, MTM-2, and MTM-3

	MTM-1	MTM-2	MTM-3
Therbligs utilized	Release	Get	Handle
	Reach		
	Grasp		
	Move	Put	
	Position		
Time to analyze job	250 × cycle time	100 × cycle time	35 × cycle time
Relative speed	1	2.5	7
Time/accuracy - 100 TMUs	15 min/± 21%	6 min/±40%	2 min/±70%
Time/accuracy - 10,000 TMUs	1500 min/±2.1%	600 min/±4%	200 min/±7%

To identify the exact way a general move is performed, analysts consider four parameters: action distance (A), which is primarily horizontal distance; body motion (B), which is mainly vertical; gain control (G); and place (P). A specific move sequence consists of three phases, each with a subset of parameters: get (A, B, G), put (A, B, P), and return (A). Analysts assign time-related index numbers to the applicable parameter. MOST uses index numbers 0, 1, 3, 6, 10, and 16, corresponding to the relative difficulty of the parameter, as shown in Table 13–12.

It is relatively easy to memorize these values and their application to the four parameters of general move. Furthermore, the index numbers, when scaled by a factor of 10, yield the appropriate TMUs. For example, get a washer 5 inches (12.5 cm) away, place on a bolt located 5 inches away, and return to the original position, would yield: $A_1B_0G_1A_1B_0P_1A_1$ with a total time of $(1 + 0 + 1 + 1 + 0 + 1 + 1) \times 10 = 50$ TMUs. The get is defined by A_1 = reach to washer with 5 inch travel, B_0 = no body motion, G_1 = grasp washer; the put is defined by A_1 = place washer with 5 inch travel, B_0 = no body motion, P_1 = place washer with a loose fit. The final A_1 = return to original position with 5 inch travel.

About 50 percent of manual work occurs as general move. A typical general move may include the parameters of walking to a location, bending to pick up an object, reaching and gaining control of the object, arising after bending, and placing the object.

The controlled move sequence covers such manual operations as cranking, pulling a starting lever, turning a steering wheel, or engaging a starting switch. In controlled move sequences, the following parameters may prevail: the previously defined action distance (A), body motion (B), gain control (G); and the new parameters, move controlled (M), process time (X), and align (I). The controlled move sequence parameters are further defined in Table 13–12.

The final sequence in MOST is tool use/equipment use. Cutting, gaging, fastening, and writing with tools are all covered by this sequence. The tool use/equipment use model embraces a combination of general move and controlled

TABLE 13–12
Basic Most Data Card

ATKFVLVPTA — **Manual Crane**

Index x 10	A Action Distance Steps	T Transportation Up to 2 Ton Feet (m.) Empty	L Loaded	K Hook-up and Unhook	F Free Object	V Vertical Move Inches (cm.)	P Placement	Index x 10
3	2				Without direction change	9 (20)	Without direction change	3
6	4				With single direction change	15 (40)	Align with one hand	6
10	7	5 (1.5)	5 (1.5)		With double direction change	30 (75)	Align with two hands	10
16	10	13 (4)	12 (3.5)		With one or more direction changes, care in handling or apply pressure	45 (115)	Align and place with one adjustment	16
24	15	20 (6)	18 (5.5)	Single or double hook		60 (150)	Align and place with several adjustments	24
32	20	30 (9)	26 (8)	Sling			Align and place with several adjustments and apply pressure	32
42	26	40 (12)	35 (10)					42
54	33	50 (15)	45 (13)					54

Index	Interval Mean TMU	MOST Interval Limits TMU
0	0	0
1	10	1 - 17
3	30	18 - 42
6	60	43 - 77
10	100	78 - 126
16	160	127 - 196
24	240	197 - 277
32	320	278 - 366
42	420	367 - 476
54	540	477 - 601
67	670	602 - 736
81	810	737 - 881
96	960	882 - 1041
113	1130	1042 - 1216
131	1310	1217 - 1411
152	1520	1412 - 1621
173	1730	1622 - 1841
196	1960	1842 - 2076
220	2200	2077 - 2321
245	2450	2322 - 2571
270	2700	2572 - 2846
300	3000	2847 - 3146
330	3300	3147 - 3446

MOST®
Work Measurement System
BasicMOST® DATA CARD

WARNING

Do not attempt to apply the data contained in these tables unless trained by a certified instructor

MAYNARD

H. B. MAYNARD & COMPANY, INC.
Eight Parkway Center, Pittsburgh, PA 15220
Phone: 412.921.2400 Fax: 412.921.4575
www.hbmaynard.com

1 TMU = .00001 hour	**1 hour = 100,000 TMU**
= .0006 minute	**1 minute = 1,667 TMU**
= .036 second	**1 second = 27.8 TMU**

Source: Reprinted with permission of H.B. Maynard and Company.

TABLE 13–12 *(continued)*

Basic Most Data Card

General Move

ABG (Get) ABP (Put) A (Return)

Index x 10	A Action Distance	B Body Motion	G Gain Control	P Placement	Index x 10
0	≤ 2 in. (5 cm.)	No Body Motion	No Gain Control Hold	No Placement Hold Toss	0
1	Within Reach		Grasp Light Object Grasp Light Objects Simo	Lay Aside Loose Fit	1
3	1 - 2 Steps	Sit without adjustments Stand without adjustments Bend and Arise 50% occ.	Get Non-simo Get Heavy/Bulky Get Blind Get Obstructed Free Interlocked Disengage Collect	Loose Fit Blind Place with Adjustments Place with Light Pressure Place with Double Placement	3
6	3 - 4 Steps	Bend and Arise		Position with Care Position with Precision Position Blind Position Obstructed Position with Heavy Pressure Position with Intermediate Moves	6
10	5 - 7 Steps	Sit Stand			10
16	8 - 10 Steps	Bend and Sit Climb on Climb off Stand and Bend Through Door			16

A Action Distance Extended Values

Index	Steps	Distance (ft.)	Distance (m.)
24	11-15	38	12
32	16-20	50	15
42	21-26	65	20
54	27-33	83	25
67	34-40	100	30
81	41-49	123	38
96	50-57	143	44
113	58-67	168	51
131	68-78	195	59
152	79-90	225	69
173	91-102	255	78
196	103-115	288	88
220	116-128	320	98
245	129-142	355	108
270	143-158	395	120
300	159-174	435	133
330	175-191	478	146

Controlled Move

ABG (Get) MXI (Move/Actuate) A (Return)

Index x 10	M Move Controlled: Push/Pull/Pivot	M Move Controlled: Crank	X Process Time: Seconds	X Process Time: Minutes	X Process Time: Hours	I Alignment	Index x 10
0	No Action	No Action	No Process Time			No Alignment	0
1	Push/Pull/Pivot ≤ 12 in. (30 cm.) Push Button Push or Pull Switch Rotate Knob		.5 sec.	.01 min.	.0001hr.	Align to 1 Point	1
3	Push/Pull/Pivot > 12 in. (30 cm.) Push/Pull with Resistance Seat Unseat Push/Pull with High Control Push/Pull 2 Stages ≤ 12 in.(30 cm.) Push/Pull 2 Stages ≤ 24 in. Total	1 Rev.	1.5 sec.	.02 min.	.0004 hr.	Align to 2 Points ≤ 4 in. (10 cm.)	3
6	Push/Pull 2 Stages > 12 in. (30 cm.) Push/Pull 2 Stages > 24 in. Total Push with 1 - 2 Steps	2 - 3 Revs.	2.5 sec.	.04 min.	.0007 hr.	Align to 2 Points > 4 in. (10 cm.)	6
10	Push/Pull 3 - 4 Stages Push with 3 - 5 Steps	4 - 6 Revs.	4.5 sec.	.07 min.	.0012 hr.		10
16	Push with 6 - 9 Steps	7 - 11 Revs.	7.0 sec.	.11 min.	.0019 hr.	Align with Precision	16

M Push or Pull Extended Values

Index	Steps
24	10-13
32	14-17
42	18-22
54	23-28
67	29-34

Crank Extended Values

Index	Revs.
24	12-16
32	17-21
42	22-28
54	29-36

X Process Time Extended Values

Index	Seconds	Minutes	Hours
24	9.5	.16	.0027
32	13.0	.21	.0036
42	17.0	.28	.0047
54	21.5	.36	.0060
67	26.0	.44	.0073
81	31.5	.52	.0088
96	37.0	.62	.0104
113	43.5	.72	.0121
131	50.5	.84	.0141
152	58.0	.97	.0162
173	66.0	1.10	.0184
196	74.5	1.24	.0207
220	83.5	1.39	.0232
245	92.5	1.54	.0257
270	102.0	1.70	.0284
300	113.0	1.88	.0314
330	124.0	2.06	.0344

TABLE 13–12 *(concluded)*
Basic Most Data Card

ABG Get Tool, ABP Put Tool, * Use Tool, ABP Aside Tool, A Return

Tool Use

F L — Fasten or Loosen

Index x 10	Finger Action: Spins (Fingers, Screw-driver)	Wrist Action: Turns (Hand, Screw-driver, Ratchet, T-Wrench)	Wrist Action: Strokes (Wrench, Allen key)	Wrist Action: Cranks (Wrench, Allen key, Ratchet)	Wrist Action: Taps (Hand, Hammer)	Arm Action: Turns (Ratchet)	Arm Action: Turns (T-Wrench 2-Hands)	Arm Action: Strokes (Wrench, Allen key)	Arm Action: Cranks (Wrench, Allen key, Ratchet)	Arm Action: Strikes (Hand, Hammer)	Tool Action: Screw Dia. (Power Wrench)	Index x 10
1	1	-	-	-	1	-	-	-	-	-	-	1
3	2	1	1	1	3	1	-	1	-	1	1/4" (6mm)	3
6	3	3	2	3	6	2	1	-	1	3	1" (25mm)	6
10	8	5	3	5	10	4	-	2	2	5		10
16	16	9	5	8	16	6	3	3	3	8		16
24	25	13	8	11	23	9	6	4	5	12		24
32	35	17	10		30	12	8	6		16		32
42	47	23	13		39	15	11	8		21		42
54	61	29	17		50	20	15	10		27		54

P Tool Placement

Tool	Index
Hammer	0 (1)
Fingers or Hand	1 (3 or 6)
Knife	1 (3)
Scissors	1 (3)
Pliers	1 (3)
Writing Instrument	1
Measuring Device	1
Surface Treating Device	1
Screwdriver	3
Ratchet	3
T-Wrench	3
Fixed End Wrench	3
Allen Wrench	3
Power Wrench	3
Adjustable Wrench	6

I Alignment of Machining Tools

Index	Align to
3	Workpiece
6	Scale Mark
10	Indicator Dial

Alignment of Nontypical Objects

Index	Positioning Method
0	Against stop(s)
3	1 adjustment to stop
6	2 adjustments to stop(s) 1 adjustment to 2 stops
10	3 adjustments to stop(s) 2-3 adjustments to linemark

Nontypical Object Characteristics
Flat, Large, Flimsy, Sharp, Difficult to Handle

ABG Get Tool, ABP Put Tool, * Use Tool, ABP Aside Tool, A Return

Tool Use

Index x 10	C Cut: Twist/Bend (Pliers)	C Cut: Cutoff (Pliers) Wire	C Cut: Cut (Scissors) Cut(s)	C Cut: Slice (Knife) Slice(s)	S Surface Treat: Air-Clean (Nozzle) sq. ft. (0,1 m.²)	S Surface Treat: Brush-Clean (Brush) sq. ft. (0,1 m.²)	S Surface Treat: Wipe (Cloth) sq. ft. (0,1 m.²)	M Measure: Measure (Measuring Device) in.(cm.) ft.(m.)	R Record: Write (Pencil) Digits	R Record: Write (Pencil) Words	R Record: Mark (Marker) Digits	T Think: Inspect (Eyes, Fingers) Points	T Think: Read (Eyes) Digits, Single Words	T Think: Read (Eyes) Text of Words	Index x 10
1	Grip		1	-	-	-	-		1	-	Check Mark	1	1	3	1
3		Soft	2	1	-	-	1/2		2	-	1 Scribe Line	3	3 Gauge	8	3
6	Twist Bend-Loop	Medium	4	-	1 Spot Point Cavity	1 Small Object	-		4	1	2	5 Touch for Heat	6 Scale Value Date or Time	15	6
10		Hard	7	3	-	-	1	Profile-Gauge	6	-	3	9 Feel for Defect	12 Vernier-Scale	24	10
16	Bend Cotter Pin		11	4	3	2	2	Fixed Scale Caliper 12 in. (30 cm.)	9 Signature or Date	2	5		Table Value	38	16
24			15	6	4	3	-	Feeler-Gauge	13	3	7			54	24
32			20	9	7	5	5	Steel-Tape 6 ft. (2 m.) Depth Micrometer	18	4	10			72	32
42			27	11	10	7	7	OD-Micrometer 4 in. (10 cm.)	23	5	13			94	42
54			33					ID-Micrometer 4 in. (10 cm.)	29	7	16			119	54

move activities. Other parameters unique to this activity include: fasten (F), loosen (L), cut (C), surface treat (S), record (R), think (T), and measure (M), which are further defined in Table 13–12.

MOST work measurement systems have two adaptations: Mini and Maxi MOST. Mini MOST measures identical, short-cycle operations, while Maxi MOST measures long-cycle operations with significant variation in actual method from cycle to cycle.

MOST is available in both manual and computerized versions. The computerized version permits the retrieval of suboperation data and the arithmetic operations involved in developing a standard of performance for the input characteristics of the method under study.

Another relatively new feature of the computerized version of MOST is ErgoMOST, which allows the engineer to analyze ergonomic problems in the workplace. ErgoMOST uses a biomechanical model to calculate stresses for push/pull and lifts, to highlight awkward postures and repetitive body movements and to quantify the relative risk on the job, using the Ergonomic Stress Indices. In addition, ergonomic corrections are identified and reports are generated.

It has been estimated that the use of a computerized system should result in application speeds five to ten times faster than manual application speeds. Computer-developed standards are also more error free, since the system does not accept an input that is not logical, and it incorporates tables of simultaneous motions, handles frequencies of occurrence, and calculates elemental times, as well as cumulative times of successive elements.

MACROMOTION ANALYSES

Standards International, Inc., a work management consulting firm headquartered in Chicago, Illinois, has developed two specialized predetermined time systems: MICRO Motion Analyses, for precise methods specifications and time standards, and MACRO Motion Analyses, for general purpose data. These systems were developed to provide improvements over MTM and Work-Factor, which the Standards International firm had been using for several years in conjunction with their consulting work. Specifically, the existing systems did not appear to be adequate for some special types of motions, and describing these motions and assigning appropriate time values for these motions were subject to the individual user's judgment. Also, some individual analysts found some of the tables difficult to use, since interpolation was often required. The resulting MACRO Motion tables were developed with much input from several of their clients (see Tables 13–13A through 13–13D). As in other systems, these values have been proven to be valid in thousands of applications.

BASIC MOTION TIMES APPLICATION

Standard Data Development

One of the most important uses of basic motion times is the development of standard data elements. With standard data, standard times for operations can be set

TABLE 13–13A

MACRO™
MOTION ANALYSES

Do not attempt to use these tables to determine standards unless you understand the proper application of the data.

This note of caution is presented to prevent the difficulties that may result from misapplication of the time values. Values shown are in decimal minutes (to 4 places) and at required time.

Compatible selections: MTS, W-F, DMT, ETS or MICRO. For interchange with MTM, UAS, MODAPTS, or MSD, add 25% allowance and convert to decimal hours.

NOTE: These are condensed tables; more detail and fuller tables are provided in the manuals.

STANDARDS, INTERNATIONAL INC.
RESEARCH ENGINEERS & MANAGEMENT CONSULTANTS
Chicago, Illinois

TABLE 13–13B

MACRO™ MOTION ANALYSES

Frequently used General Purpose Data Elements recommended for:

1. Evaluation of manual motions (and costs) by methods personnel and production supervisors.
2. Development of standards for short run, long cycle operations by methods/standards personnel.

OBTAIN AND PLACE			DIST. RANGE IN INCHES			
WT.	**CONDITIONS OF OBTAIN**	**PLACE ACCURACY**	**CODE**	**6″**	**18″**	**30″**
2 LBS. OR LESS	EASY GRASP	Approximate	OEA	90	150	190
		Close	OEC	110	170	210
		Tight	OET	130	190	230
	DIFFICULT GRASP	Approximate	ODA	100	160	200
		Close	ODC	120	180	220
		Tight	ODT	140	200	240
	GRASP HANDFUL	Approximate	OHA	150	190	230
OVER 2 LBS. THRU 18 LBS.		Approximate	OWHA	160	230	270
		Close	OWHC	170	250	290
		Tight	OWHT	180	270	310
OVER 18 LBS. THRU 48 LBS.		Approximate	OW2HA	190	290	320
		Close	OW2HC	200	310	340
		Tight	OW2HT	210	330	360

PLACE ONLY	CODE	6″	18″	30″
Approximate	PA	40	60	80
Close	PC	60	80	100
Tight	PT	70	100	130
Add for: Weights: OVER 2 LBS THRU 24 LBS. 40; OVER 24 LBS. 60				

TABLE 13–13C

ASSEMBLE NON-ROUND OBJECT	CODE	UP TO & INCL. 18 LBS.	OVER 18 LBS. THRU 48 LBS.
UP TO & INCL. 1/2" CLEARANCE	ANT	90	120
OVER 1/2" CLEARANCE	ANC	50	70

ASSEMBLE	CODE	PLUG		
		UP TO & INCL. 1/4"	OVER 1/4" THRU 1/2"	OVER 1/2" THRU 1"
Loose Fit	ASYL	60	50	40
Normal Fit	ASYN	80	70	60
Add for: Simo (S) 15; Temporary Blind (TB) 15; Apply Pressure (AP) 40; Regrasp & Apply Pressure (RAP) 60				

CIRCULAR MOTIONS	CODE	DIAMETER		
		3" & UNDER	OVER 3" THRU 12"	OVER 12"
Revolution without Deceleration	CR	50	80	90
Revolution to General Location	CRG	60	110	120
Revolution to Exact Location	CRE	80	140	150

MOTION-PATTERNS

WALKING						(Measured at Toe or Heel)
NUMBER OF STEPS	OPEN/BASIC (WO)		CONFINED (WC)		RESTRICTED (WR)	
	DIST. FT.	TIME	DIST. FT.	TIME	DIST. FT.	TIME
1 (1 Leg)	2.5′	120	2.5′	130	2′	140
1 (Both Legs)	2.5′	200	2.5′	220	2′	240
2	5′	260	5′	280	4′	300
5	13.5′	500	13.5′	550	11′	600
10	26′	900	26′	1000	21′	1100
Each Add'l Pace — Add	2.5′	80	2.5′	90	2′	100
Add for Turn Over 120°	—	100	—	100	—	100

Stairs/Up (SU) 130	Bend (B) 160	Sit (SI) 230
/Down (SD) 100	Arise (AB) 130	Stand (ST) 280

TABLE 13–13D

VISUAL & MENTAL PROCESSES	CODE	TIME
READ — Per Word	ER	50
INSPECT — Per Criteria — Per Occurrence	EI	50
RECALL, DECIDE, REACT or CALCULATE — Per Criteria — Per Occurrence	MA	50
MOVE HEAD TO MICROSCOPE & SEE	HM	130
WRITE — Per Character	WC	90

MOTION-PATTERNS

TOOL HANDLING					
TYPE OF TOOL	MOTION PATTERN	DISTANCE 9″	18″	30″	36″
Screw Driver, Mallet Spin-Tites, Knives File (w/Rd. Handle)	OX-RH-E OX-RH-G	210 180	270 230	340 290	360 310
Pliers, Wire Strippers Scissors, Cutters	OX-DH-E OX-DH-G	240 200	300 260	360 310	380 330
Pencils, Brushes	OX-TH-E OX-TH-G	210 170	270 220	330 280	350 300
Soldering Iron: — in Holster	OX-SI-E OX-SI-G	310 280	380 340	440 390	470 410
Air Tool: — on Bench	OX-ATB-E OX-ATB-G	260 240	340 300	400 360	430 380
Air Tool: — in Holster	OX-ATH-E OX-ATH-G	370 340	450 410	520 470	540 490
Air Tool: — Suspended	OX-ATS-E OX-ATS-G	170 140	220 170	260 210	270 220

Motion Pattern E: Pick up tool; move directly to work; place tool aside.
Motion Pattern G: Pick up tool; move to general area (move to assemble, etc., must be added); place tool aside.

RELATED ELEMENTS	CODE	TIME
TWEEZERS GRASP	GT	140
PALM & UNPALM TOOL	PXT	30
START THREADS	STT	90
THREAD ON OR OFF — Per Twist	TO	30
USE: SCREWDRIVER — Per Twist NUT RUNNER — Per Twist BOX OR END WRENCH — Per Turn RATCHET — Per Turn HAMMER — Per Hit	SDT NRT BEW RW HB	70 70 180 60 90
TIGHTEN OR LOOSEN w/HAND TOOL	ASYX	100

much faster than by the laborious procedure of summarizing long columns of fundamental motion times. In addition, standard data usually reduces clerical errors, since less arithmetic is involved.

With sound standard data, it is economically feasible to establish standards on indirect work, such as maintenance, material handling, clerical and office, inspection, and similar expense operations. Thus, with standard data, analysts can economically calculate operation times involving long cycles and consisting of many short-duration elements. For example, one company developed standard data applicable to radial drill operations in its tool room. Time study analysts developed standard data for the elements required to move the tool from one hole to the next and to present and back off the drill. They then combined these standard data elements into one multivariable chart, so that the data could be rapidly summarized.

An example illustrating the flexibility of basic motion times is the development of a standard time for a clerical operation. This formula for sorting time slips includes the following elements:

1. Pick up pack of departmental time slips and remove rubber band.
2. Sort time slips into direct labor (incentive), indirect labor, and daywork.
3. Record the total number of time slips.
4. Get pile of time slips, put rubber band around pack, and put aside.
5. Get pile of time slips and bunch.
6. Sort incentive time slips into "parts" time slips.
7. Count piles of incentive "parts" time slips.
8. Record the number of "parts" time slips and the number of incentive time slips.
9. Sort "parts" time slips into numerical sequence.
10. Bunch piles of numerical time slips and place in one pile on desk.

Methods analysts break each element down into the fundamental motions. Once the basic values are assigned and the variables are determined, the resulting algebraic equation allows the rapid calculation of time for the clerical operation. A stopwatch can frequently be helpful in developing standard data elements. Some portions of an element may be more readily determined by basic motion times, and other portions may be better adapted to stopwatch measurement. Also, standard data elements can be verified with a stopwatch. Predetermined basic motion times are converted to the verified standard data elements, to establish fair standards that are more consistent than those established through stopwatch procedures alone.

Untrained, inexperienced analysts should not establish time standards to be used for rate purposes. Analysts must know whether the distance moved is the linear distance taken by the hand or the circumferential distance taken by the hand as it makes an arc. They must know whether the distance is measured from the center of the hand, the knuckles, or the fingertips. They must know when the application of pressure prevails and when it does not. They must clearly understand how the elements of alignment and orientation affect positioning time. These and many other considerations must be mastered before analysts can expect to establish consistent and accurate time standards.

Methods Analysis

An equally important use for any synthetic basic motion-time technique is methods analysis. Analysts who appreciate these techniques look more critically at each and every workstation, thinking about how improvements may be made. Using a predetermined time system is simply developing a motion or methods analysis in greater numerical detail, identifying better ways of eliminating ineffective therbligs, and reducing the times on the remaining effective therbligs. Another checklist has therefore been developed (see Figure 13–11) to aid the analyst in better methods analysis. Key opportunities for simplifying the method (using the MTM-2 system as an example) include:

1. Elimination of body motions, such as Bend and Arise, with a large time value of 61 TMUs.

FIGURE 13-11
MTM-2 methods analysis checklist (Adapted from A. D. Brown 1976, Apply Pressure, *Journal of the Methods-Time-Measurement Association-UK,* 14)

GETs (G)	Yes	No
1. Can GETs be performed simultaneously with other GETs or PUTs without penalty?	❑	❑
2. Can GETs be performed during a machine cycle?	❑	❑
3. Can jigs/fixtures, gravity-feed devices, or bins be used to simplify GETs (i.e., from GC to GB or to GA)?	❑	❑
4. Can GAs be used and objects slid into position?	❑	❑
5. Can the transfer of objects from one hand to another be avoided?	❑	❑
6. Can tools be prepositioned to simplify GETs?	❑	❑
7. Can tools be palmed while performing other work (instead of being set down and later retrieved)?	❑	❑
8. Can more than one object be grasped at the same time?	❑	❑
9. Can travel distances be reduced (i.e., to lower motion classification levels)?	❑	❑
10. Are hand motions balanced in terms of case and distance?	❑	❑

PUTs (P)	Yes	No
1. Can PUTs be performed simultaneously with other GETs or PUTs, without penalty?	❑	❑
2. Can tight tolerances or the accurate location of an object be avoided?	❑	❑
3. Can the delivery point of an object be chamfered or funneled?	❑	❑
4. Can fixed guides or stops be utilized?	❑	❑
5. Can the object be made symmetrical?	❑	❑
6. Can the depth of insertion be reduced?	❑	❑
7. Can the other hand assist in complex PUTs?	❑	❑
8. Can objects be PUT together mechanically?	❑	❑
9. Can drop deliveries be utilized to simplify PUTs (i.e., from PC to PB or to PA)?	❑	❑
10. Can objects be slid to a location (i.e., use a PA)?	❑	❑
11. Are destination points in the normal area of vision?	❑	❑

Apply Pressure (A)	Yes	No
1. Can As be avoided by improved design or better processing (e.g., eliminate burrs or tight spots)?	❑	❑
2. Can unnecessary tightening from previous operations be avoided?	❑	❑
3. Can tight tolerances be avoided?	❑	❑
4. Can the contamination of parts due to filings, dust, dirt, etc., causing As be avoided?	❑	❑
5. Can momentum be used to eliminate As?	❑	❑
6. Are the largest muscle groups used to best advantage in applying pressure?	❑	❑
7. Can clamping devices or mechanical actions be used to eliminate As?	❑	❑

FIGURE 13-11 ***(concluded)***

MTM-2 methods analysis checklist (Adapted from A. D. Brown 1976, Apply Pressure, *Journal of the Methods-Time-Measurement Association-UK,* 14)

Regrasp (R)	**Yes**	**No**
1. Can Rs be avoided during PUTs?	❑	❑
2. Can tools be prepositioned in the desired orientation?	❑	❑
3. Can magazine feeds, stacking devices, vibratory feeders, etc., be used to present the part properly?	❑	❑
4. Can parts be made symmetrical to avoid the need for Rs?	❑	❑
5. Can parts be pre-positioned during a machine cycle?	❑	❑

Eye Action (E)	**Yes**	**No**
1. Can objects and displays be placed in the normal area of vision to avoid Es?	❑	❑
2. Is there sufficient illumination to avoid Es?	❑	❑
3. Are bins and parts correctly identified, perhaps by use of color?	❑	❑
4. Can parts be made symmetrical and positioned properly to avoid Es?	❑	❑
5. Can visual checks of assembly parts be avoided (i.e., use detents and tactile feel)?	❑	❑
6. Can visual interpretation of dial settings be avoided (i.e., use ON/OFF or status indicators)?	❑	❑
7. Can Es be performed during preceding manual motions without penalty?	❑	❑

Crank (C)	**Yes**	**No**
1. Can the wheel or crank be spun?	❑	❑
2. Can the number of revolutions be reduced (i.e., larger thread size used)?	❑	❑
3. Can resistance during cranking be eliminated?	❑	❑
4. Can the crank be power driven?	❑	❑

Step (S)	**Yes**	**No**
1. Is the shortest route or best layout being utilized?	❑	❑
2. Are floor surfaces even and clear of obstructions?	❑	❑
3. Are the most commonly used parts located close by?	❑	❑
4. Is any necessary information and tooling located at the workstation (i.e., avoid unnecessary Ss)?	❑	❑
5. Can materials and parts be brought mechanically (via conveyors) to and from the workstation?	❑	❑
6. Can vehicular transport (carts) be used?	❑	❑

Foot Motion (F)	**Yes**	**No**
1. Can Fs be performed simultaneously with other motions?	❑	❑
2. Can the foot rest comfortably on the switch or pedal during the operation?	❑	❑
3. Is the body weight supported by a stool (weight off the load-bearing leg)?	❑	❑
4. Can either foot operate the pedal alternately?	❑	❑

Bend and Arise (B)	**Yes**	**No**
1. Can drop deliveries be utilized to avoid Bs?	❑	❑
2. Are materials and products located between elbow and knuckle height to minimize Bs?	❑	❑
3. Are proper lifting procedures (squat lifting, etc.) being utilized?	❑	❑
4. Can the too frequent entry and exit of a seated workstation be avoided?	❑	❑

2. Reduction of case levels, especially case C motions, resulting in a 39 percent decrease in basic motion times.
3. Minimizing reach distances, with a 5 TMU decrease for each shorter distance code.
4. Avoiding the lifting of heavy parts, where each 2 pounds (1 kg) drops 1 TMU.

EXAMPLE 13–1
Methods Improvement in T-Shirt Turning.

This example considers both the productivity and health/safety aspects obtained through an MTM-2 analysis of T-shirt turning (Freivalds and Yun, 1994). Garments are sewn "inside out" so that the seams can be stitched. Once the garment is completed, it must be "turned" or inverted.

Workers on this job were highly susceptible to various cumulative trauma disorders. An MTM-2 analysis of the present method, shown in Figure 13–12a, indicated that the results were a total of 141 TMUs. An obvious characteristic of this job was the high usage of case C motions. Was it possible to reduce the GETs and PUTs (questions #3 and #4 under GET in Figure 13–11)?

The proposed solution was to build a vacuum-powered device to draw the t-shirt into a pipe. Once the vacuum was shut off, the t-shirt could be removed in an inverted state. The MTM-2 analysis of the improved method (see Figure 13–12b) yielded a total of only 108 TMUs. For the complete operation of turning, inspection, and folding (total of 360 TMUs), this resulted in a (141 - 108)/360 = 9.2% decrease in time. Overall, the difficult and injurious case C hand motions were eliminated, with a simultaneous improvement in productivity.

FIGURE 13–12
MTM-2 analysis of T-shirt turning.

a) Present method

MTM Methods Analysis Page of

Operation: T-SHIRT TURNING
Study No.: (MANUAL)
Date: 2-12-93
Analyst: AF
Remarks: MANUAL HANDLING TOTAL OF 141 TMUs

Description	No.	LH	TMU	RH	No.	Description
GET T-SHIRT		GB18	18	GB18		GET T-SHIRT
REACH INSIDE, PINCH CLOTH		GC12	23	GC12		REACH INSIDE, PINCH CLOTH
SIMULTANEOUS MOTION		GC2	14	GC2		SIMULTANEOUS MOTION ALLOW.
PULL SLEEVE UP AND OUT		PC32	41	PC32		PULL SLEEVE UP AND OUT
SIMULTANEOUS MOTION ALLOW.		PC2	21	PC2		SIMULTANEOUS MOTION
SET T-SHIRT DOWN		PB18	24	PB18		SET T-SHIRT DOWN
			141			

(continued)

EXAMPLE 13–1 ***Concluded***

b) Proposed method

MTM Methods Analysis Page of

Operation: T-SHIRT TURNING
Study No.: (AUTOMATED)
Date: 2-12-93
Analyst: AF
Remarks: USING VACUUM DEVICE TOTAL OF 108 TMUs

Description	No.	LH	TMU	RH	No.	Description
GET T-SHIRT		GB18	18	GB18		GET T-SHIRT
PULL T-SHIRT UP PIPE		PA32	20	PA32		PULL T-SHIRT UP PIPE
			9	F		ACTIVATE PEDAL
PULL T-SHIRT DOWN		PB32	30	PB32		PULL T-SHIRT DOWN
AGAINST RESISTANCE		PW10	1	PW10		AGAINST RESISTANCE
SET T-SHIRT DOWN		PB32	30	PB32		SET T-SHIRT DOWN
			108			

5. Eliminating operations that require eye travel and eye focus, which eliminates 7 TMUs each.
6. Pre-positioning tools, parts, and materials.

In one company, $40,000 was allocated for advanced tooling to increase the rate of production on a brazing operation. Prior to retooling, analysts conducted a work measurement study of the existing method. Using fundamental motion data, they discovered that by providing a simple fixture and rearranging the loading and unloading area, the company production could increase from 750 to 1,000 pieces per hour. The total cost of the synthetic basic motion-time study was $40. As a result of the study, the company avoided the costly $40,000 retooling program.

SUMMARY

This chapter discusses several of the more popular predetermined or basic motion time systems. There are many others, including several proprietary systems developed by industry. Many years ago, Frederick W. Taylor visualized the development of standards for basic divisions of work similar to those still in use. In his paper on "Scientific Management," he predicted that the time would come when a sufficient volume of basic standards would be developed to make further time studies unnecessary. We have just about reached this state in connection with effort elements. Today, the vast majority of standards are developed by using standard data and/or basic motion times.

Basic motion-time values are becoming more accurate as additional studies are made. However, there is still a need for further research, testing, and refinement. For example, there is a question as to the validity of adding basic motion times to determine elemental times, since therblig times may vary once the sequence is changed. Thus, the time for the basic element "reach 20 inches," may be affected by the preceding and succeeding elements and may not be entirely dependent on the class of reach and distance. In general, all successful fundamental motion techniques have made an effort to address the additivity of the motion elements. MTM, for instance, has provided three motions for the movement of the hand, as well as several cases, according to the nature of a move or reach. Work-Factor recognizes five elements of difficulty. Similar observations might be made about other fundamental motion techniques.

When analyzing motion patterns with existing data, the analyst should consider the main purpose of the motion pattern, as well as its complexity and characteristics. For example, if the hand is empty while moving toward an object, the motion is classified as a reach. If the hand holds an object while moving toward another object, the analyst must consider not only the main purpose of the motion, but also what the hand does with the object during the motion. If the hand is palming an object while reaching for something else, the motion cannot be classified as a basic reach. The time necessary to perform the combined motion depends on factors other than distance. One such is the physical characteristics of the motion. When an object is palmed while the hand is moving, a simultaneous operation takes place in addition to the move. The result might be a reduction in the average speed. This would allow the hand to establish control of the object over the distance moved. The longer the distance, the more time the hand has to palm the object. Thus, the longer the combined motion, the more the motion approaches the time required for a simple reach of the same distance.

Additional study is necessary for operations that involve simultaneous motions. Intuitively, there should be a relationship between the degree of bimanualness of the normal operator and the motion pattern employed.

Predetermined time systems have a most important place in the field of methods and work measurement. There are several compelling reasons to use them. They can be used to define a standard time before production begins and to estimate production costs ahead of time, when no job exists to time study. However, these systems are only as good as the people using them. The analyst must be very careful to understand the assumptions behind the systems and use them in the proper manner. They should not be installed without professional help or a complete understanding of their application.

QUESTIONS

1. Who was originally responsible for thinking in terms of developing standards for basic work divisions? What was his contribution?
2. What are the advantages of using basic motion times?
3. What other two terms are frequently used to identify basic motion times?
4. Who pioneered the MTM system?
5. What is the time value of one TMU?
6. Would it be easy or difficult to perform a GB get with the left hand while simultaneously performing a PC place with the right hand? Explain.
7. Why was MTM-2 developed? Where does MTM-2 have special application?
8. Are MTM-1 and MTM-2 consistent in their handling of simultaneous motions?
9. If MTM-3 were used to study an operation of approximately 3 minutes' duration, what could you say about the accuracy of the standard?
10. How is MTM related to the method analysis?
11. Explain the relationship of basic motion times to standard data.
12. Explain why most companies today require certification before they utilize basic motion times for the establishment of standards.

PROBLEMS

1. Determine the time for the dynamic component of M20 B20.
2. A 30-pound bucket of sand having a coefficient of friction of 0.40 is pushed 15 inches away from the operator, with both hands. What would be the normal time for the move?
3. A 3/4-inch-diameter coin is placed within a 1-inch-diameter circle. What would be the normal time for the position element?
4. Calculate the equivalent in TMUs of 0.0075 hours per piece, 0.248 minutes per piece, 0.0622 hours per hundred, 0.421 seconds per piece, and 10 pieces per minute.
5. If you have finished drilling a hole 3 inches deep using a Western radial drill, how long would it take to present the drill and drill a second hole in a steel forging1/2 inch in diameter and 3 inches deep? Traverse, 6 inches; swing of head, 8 inches; feed 0.007 inch, surface speed 50 feet per minute.

6. The following MTM-2 analysis describes a simple operation in which each hand gets a part, the right hand regrasps it, and then the right hand puts it into a fixture. Pressure is applied to seat it. Next, a pin is grasped, regrasped, and inserted into the assembly. A handwheel is cranked six revolutions under resistance until a pointer is aligned exactly. Identify any errors by circling them. Rewrite the analysis correctly and explain each correction.

FIGURE 13–13
MTM-2 analysis of simple assembly for problem 6.

MTM Methods Analysis Page of

Operation: ASSEMBLY
Study No.: PROBLEM #6
Date: 1-27-98
Analyst: AF
Remarks:

Description	No.	LH	TMU	RH	No.	Description
		GC12	18	GB18		
PART TO FIXTURE		R	6	R		PART TO FIXTURE
		PC12	30	PC6		
SEAT PART		A	14	A		SEAT PART
		GC12	23	GC12		
GET & ASSEMBLE PIN		PC12	30	PC12		GET & ASSEMBLE PIN
		R	6	R		
			10	GB6		
			5	GW10		CRANK AGAINST
			90	C	6	RESISTANCE
			5	PW5		
			21	PC2		ALIGN POINTER

Summary	Total TMU:	Conversion:	% Allowance:	Standard time:
	259	.0006	10	.171

REFERENCES

Antis, William, John M. Honeycutt, Jr., and Edward N. Koch. *The Basic Motions of MTM*. 3rd ed. Pittsburgh, PA: The Maynard Foundation, 1971.

Baily, Gerald B., and Ralph Presgrave. *Basic Motion Time-Study*. New York: McGraw-Hill, 1958.

Birn, Serge A., Richard M. Crossan, and Ralph W. Eastwood. *Measurement and Control of Office Costs*. New York: McGraw-Hill, 1961.

Brown, A. D. "Apply Pressure." *Journal of the Methods-Time Measurement Association*, 14, (1976).

Freivalds, A., and M. H. Yun. "Productivity and Health Issues in the Automation of T-Shirt Turning." *International Journal of Industrial Engineering*, 1, no. 2 (June 1994), pp. 103–108.

Geppinger, H. C. *Dimensional Motion Times*. New York: John Wiley & Sons, 1955.

Karger, Delmar W., and Franklin H. Bayha. *Engineered Work Measurement*. New York: Industrial Press, 1957.

Karger, Delmar W., and Walton M. Hancock. *Advanced Work Measurement*. New York: Industrial Press, 1982.

Maynard, Harold B., G. J. Stegemerten, and John L. Schwab. *Methods Time Measurement*. New York: McGraw-Hill, 1948.

Quick, Joseph H., James H. Duncan, and James A. Malcolm. *Work-Factor Time Standards*. New York: McGraw-Hill, 1962.

Sellie, Clifford N. "Predetermined Motion-Time Systems and the Development and Use of Standard Data." In *Handbook of Industrial Engineering*. 2nd ed. Ed. Gavriel Salvendy. New York: John Wiley & Sons, 1992.

Zandin, Kjell B. *MOST Work Measurement Systems*. New York: Marcel Dekker, 1980.

SELECTED SOFTWARE

ErgoMOST, Pittsburgh, PA: H. B. Maynard and Co., 1997.

MOST, Pittsburgh, PA: H. B. Maynard and Co., 1997.

MTM Link. Des Plaines, IL: MTM Association, 1998.

CHAPTER 14

Work Sampling

KEY POINTS:

- Work sampling is a method for analyzing work by taking a large number of observations at random times.
- Use work sampling to:
  - Determine machine utilization.
  - Determine allowances.
  - Establish time standards.
- Use as many observations as practical.
- Take observations at random times over two or more weeks.

Work sampling is a technique used to investigate the proportions of total time devoted to the various activities that constitute a job or work situation. The results of work sampling are effective for determining: machine and personnel utilization; allowances applicable to the job; and production standards. Although the same information can be obtained by time study procedures, work sampling frequently provides the same information faster and at considerably less cost.

In conducting work sampling studies, analysts take a comparatively large number of observations at random intervals. The ratio of observations of a given activity to the total observations approximates the percentage of time that the process is in that state of activity. For example, if 1,000 observations taken at random intervals over several weeks showed that an automatic screw machine was turning out work in 700 instances, but was idle for miscellaneous reasons in 300 instances, then the downtime of the machine would be 30 percent of the working day.

Work sampling was first applied in the British textile industry. Later, under the name *ratio-delay study*, the technique was brought to the United States (Morrow, 1946). The accuracy of the data determined by work sampling depends on the number of observations and the period over which the random observations are taken.

Unless the sample size is sufficiently high, and the sampling period represents typical conditions, inaccurate results may occur.

The work sampling method has several advantages over the conventional time study procedure:

1. It does not require continuous observation by an analyst over a long period of time.
2. Clerical time is diminished.
3. The total work-hours expended by the analyst are usually much fewer.
4. The operator is not subjected to long-period stopwatch observations.
5. Crew operations can be readily studied by a single analyst.

THE THEORY OF WORK SAMPLING

The theory of work sampling is based on the fundamental law of probability: at a given instant, an event can either be present or absent. Statisticians have derived the following expression to show the probability of x occurrences of an event in n observations:

$$(p + q)^n = 1$$

where: p = Probability of a single occurrence.
$q = (1 - p)$ = The probability of an absence of occurrence.
n = Number of observations.

If this expression, $(p + q)^n = 1$, is expanded according to the *binomial theorem*, the first term of the expansion gives the probability $x = 0$, the second term $x = 1$, and so on. The distribution of these probabilities is known as the *binomial distribution*. Statisticians have shown that the mean of this distribution is equal to np, and the variance is equal to npq. The standard deviation is equal to the square root of the variance.

According to elementary statistics, as n becomes large, the binomial distribution approaches the normal distribution. Since work sampling studies involve large sample sizes, the normal distribution is a satisfactory approximation of the binomial distribution. Rather than use the binomial distribution, it is more convenient to use the distribution of a proportion, with a mean of p (i.e., np/n) and a standard deviation of

$$\sqrt{\frac{pq}{n}} \left(\text{i.e., } \frac{\sqrt{npq}}{n}\right)$$

as the approximately normally distributed random variable.

In work sampling studies, we take a sample of size n in an attempt to estimate p. We know from elementary sampling theory that we cannot expect $\hat{p}$ ($\hat{p}$ = the proportion based on a sample) for each sample to be the true value of p. We do, however, expect the $\hat{p}$ of any sample to fall within the range of $p \pm 2$ standard deviations approximately 95 percent of the time. In other words, if p is the true percentage of a given condition, we can expect the $\hat{p}$ of any sample to fall outside the limits $p \pm 2$ standard deviations only about 5 times in 100 due to chance alone.

This theory can be used to estimate the total sample size needed to achieve a certain degree of accuracy. The expression for the standard deviation σ_p of a sample proportion is:

$$\sigma_p = \sqrt{\frac{pq}{n}} = \sqrt{\frac{p(1-p)}{n}}$$

where: σ_p = Standard deviation of a percentage.
p = True percentage occurrence of the element being sought, expressed as a decimal.
n = Total number of random observations upon which p is based.

Sometimes, an alternate, simpler expression is used. Consider the term 1.96σ as the acceptable limit of error ℓ at a 95 percent confidence error, where:

$$\ell = 1.96\sigma = 1.96\sqrt{pq/n} \qquad (1)$$

Rounding 1.96 to 2, squaring both sides, and solving for n yields:

$$n = 4pq/\ell^2 = 4p(1-p)/\ell^2 \qquad (2)$$

EXAMPLE 14–1
Normal Approximation of the Binomial Distribution.

To clarify the fundamental theory of work sampling, it would be helpful to interpret the results of an experiment. Assume the following circumstances: one machine with random breakdowns was observed for a 100-day period. During this period, eight random observations were taken per day.

Let: n = Number of observations per day.
k = Total number of days that observations were taken.
x_i = Number of breakdown observations observed in n random observations on day i ($i = 1, 2, \ldots, k$).
N = Total number of random observations.
N_x = Number of days that the experiment showed the number of breakdowns equal to x ($x = 0, 1, 2, \ldots, n$).

The probability, $P(x)$, that the machine is down x times in n observations is given by the binomial distribution:

$$P(x) = \frac{n!}{x!\,(n-x)!}\, p^x q^{n-x}$$

where: p = Probability of machine being down.
q = Probability of machine running.

and: $$p + q = 1$$

For our example, $n = 8$ observations per day, $k = 100$ days of observations, and $N = 800$ total observations. An all-day time study for several days revealed that $p = 0.5$. The following table shows the number of days in which x breakdowns were observed in the work sampling study ($x = 0, 1, 2, 3, \ldots, n$), and the expected number of breakdowns given by our binomial model, using $p = 0.5$ from the all-day time study.

X	N_x	$P(x)$	**100** $P(x)$
0	0	0.0039	0.39
1	4	0.0312	3.12
2	11	0.1050	10.5
3	23	0.2190	21.9
4	27	0.2730	27.3
5	22	0.2190	21.9
6	10	0.1050	10.5
7	3	0.0312	3.12
8	0	0.0039	0.39
	100	1.00*	100*

*Approximately

There is close agreement between the observed days that a specified number of breakdowns occurred (N_x) and the expected number computed theoretically as $kP(x)$.

$$\bar{P}_i = \frac{x_i}{n} = \text{Observed proportion of downtime on day } i \text{, where}$$

$$i = 1,2,3, \ldots, k$$

$$\hat{P} = \frac{\sum_{i=1}^{k} \bar{P}_i}{k} = \frac{\sum_{i=1}^{k} x_i}{n \cdot k}$$

$$= \frac{\sum_{i=1}^{k} x_i}{N} = \text{Estimated proportion of machine downtime, based on a work sampling experiment.}$$

The hypothesis is that the theoretical information shows close enough agreement to the observed information for the theoretical binomial to be accepted. This may be tested using the chi-square (χ^2) distribution. The (χ^2) distribution tests whether the observed distribution frequencies differ significantly from the expected frequencies.

In the example, the observed frequency is N_x and the expected frequency is $kP(x)$, and we have:

(continued)

EXAMPLE 14–1 ***(continued)***

$$\chi^2 = \sum_{k=0}^{k} \frac{[N_x - 100\, P(x)]^2}{100\, P(x)}$$

The quantity under the summation is distributed approximately as χ^2 for k degrees of freedom. In this example, $\chi^2 = 0.206$.

Analysts must determine whether the calculated value of χ^2 is sufficiently large to refute a null hypothesis, that is, the difference between the observed frequencies and the computed frequencies is due to chance alone. This experimental value of χ^2 is so small that it could easily have occurred through chance. Therefore, we accept the hypothesis that the experimental data "fits" the theoretical binomial distribution.

In typical industrial situations, p (which was known to have a value of 0.5) is unknown to analysts. The best estimate of p is $\hat{p}$, which may be computed as $\frac{\sum_{i=1}^{k} x_i}{N}$. As the number of random observations per day (n) increases and/or the number of days increases, $\hat{p}$ will approach p. However, with limited observations, analysts are concerned with the accuracy of $\hat{p}$.

If a plot of $P(x)$ versus x were made from our example, it would appear as shown in Figure 14–1.

FIGURE 14–1
Probability distribution of breakdown observations.

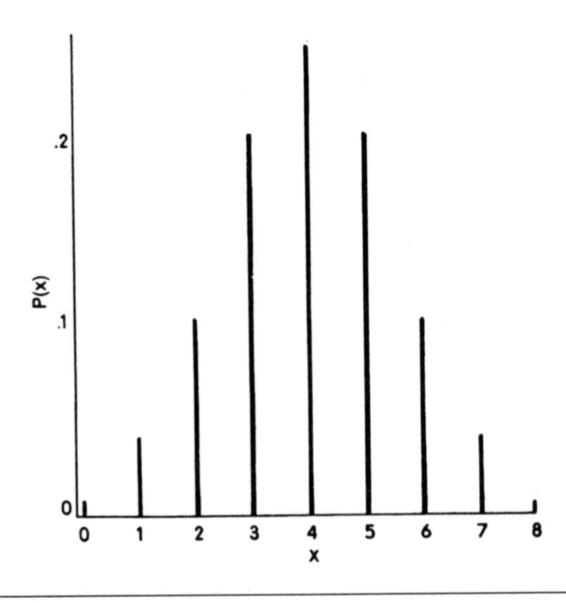

When n is sufficiently large, regardless of the actual value of p, the binomial distribution very closely approximates the normal distribution. This tendency can be seen in the example when p is approximately 0.5. When p is near 0.5, n may be small and the normal can be a good approximation to the binomial.

When using the normal approximation, set

$$\mu = p$$

and

$$\sigma_p = \sqrt{\frac{pq}{n}}$$

To approximate the binomial distribution, the variable z is used for entry in the normal distribution (see Table A3–2, Appendix 3) and it takes the following form:

$$z = \frac{\hat{p} - p}{\sqrt{\frac{pq}{n}}}$$

Although p is unknown in the practical case, we can estimate p from $\hat{p}$, and determine the interval within which p lies, using confidence limits. For example, imagine that the interval defined by

$$\hat{p} - 2\sqrt{\frac{\hat{p}\hat{q}}{n}}$$

and

$$\hat{p} + 2\sqrt{\frac{\hat{p}\hat{q}}{n}}$$

contains p 95 percent of the time.

Graphically, this may be represented as:

$$\hat{p} - 2\sqrt{\frac{\hat{p}\hat{q}}{n}} \qquad \hat{p} \qquad \hat{p} + 2\sqrt{\frac{\hat{p}\hat{q}}{n}}$$

We derive the expression for finding a confidence interval for p as follows: Suppose that we want an interval that contains p 95 percent of the time; that is, a 95 percent confidence interval. For n sufficiently large, the expression

$$z = \frac{\hat{p} - p}{\sqrt{\frac{\hat{p}\hat{q}}{n}}}$$

(continued)

EXAMPLE 14–1 ***(continued)***

is approximately a standard normal variable. Therefore, we set the probability

$$P\left(z_{0.025} < \frac{\hat{p} - p}{\sqrt{\frac{\hat{p}\hat{q}}{n}}} < z_{0.975}\right) = 0.95$$

Rearranging the inequalities and remembering that $-z_{0.025} = z_{0.975} = 1.96$, or approximately 2, we have:

$$P\left(\hat{p} - z_{0.975}\sqrt{\frac{\hat{p}\hat{q}}{n}} < p < \hat{p} + z_{0.975}\sqrt{\frac{\hat{p}\hat{q}}{n}}\right) = 0.95$$

The interval with approximately a 95 percent chance of containing p is then:

$$\hat{p} - 2\sqrt{\frac{\hat{p}\hat{q}}{n}} < p < \hat{p} + 2\sqrt{\frac{\hat{p}\hat{q}}{n}}$$

These limits imply that the interval defined contains p with 95 percent confidence, since z has been selected as having a value of 2.

The underlying assumptions of the binomial are that p, the probability of a success (the occurrence of downtime), is constant for each random instant that we observe the process. Therefore, it is always necessary to take random observations when doing a work sampling study. This reduces any bias introduced by worker anticipation of observation times.

WORK SAMPLING ACCEPTANCE

Before beginning a work sampling program, the analyst must "sell" its use and reliability to all members of the organization who will be affected by the results. If the program will establish allowances, it should be sold to the union and the supervisor, as well as to company management. This can be done by conducting several short sessions with representatives of the various interested parties and explaining examples of the law of probability, thus illustrating why ratio-delay procedures work. Both unions and workers favor work sampling techniques, once the procedure is fully explained, since work sampling is completely impersonal, does not utilize a stopwatch, and is based on accepted mathematical and statistical methods.

In the initial session, the analyst should create a simple study by tossing unbiased coins. All participants should readily recognize that a single coin toss stands a 50-50 chance of being heads. When asked how they would determine the probability of heads versus tails, they will undoubtedly propose tossing a coin a few times to find out. When asked whether two times is adequate, they will say no. Ten times may be suggested, but they may still think that is not adequate. When larger num-

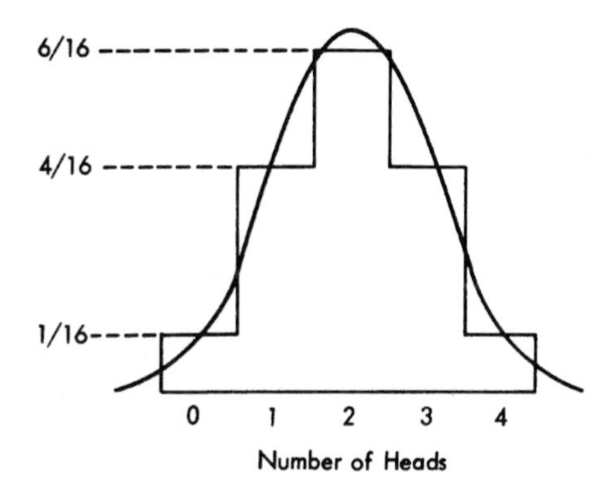

FIGURE 14–2
Distribution of number of heads with infinite number of tosses, using four unbiased coins.

bers are suggested, they will probably agree that 100 or more times is sufficient to achieve the desired result with some degree of assurance. This example firmly implants the principle of work sampling: adequate sample size to ensure statistical significance.

Next, the analyst should discuss the probable results of tossing four unbiased coins. Here, there is only one arrangement in which the coins can fall to show no heads, and only one arrangement that permits all heads. However, three heads or one head can result from four possible arrangements. Six possible arrangements can give two heads. With all 16 possibilities thus named, four unbiased coins tossed continually will distribute themselves as shown in Figure 14–2.

After this explanation and a demonstration of this distribution, that is, by making several tosses and recording the results, the audience should agree that 100 tosses could demonstrate a normal distribution. A thousand tosses would probably approach a normal distribution more closely, and 100,000 would give a nearly perfect distribution. However, such a distribution is not sufficiently more accurate than the 1,000-toss distribution, and it is not economically worth the extra effort. This establishes the idea that significant accuracy is approached rapidly at first, and then at a diminishing rate.

Next the analyst should point out that a machine or operator could figuratively be in a heads or tails state. For example, a machine could be running (heads) or idle (tails). A cumulative plot of "running" would eventually level off, giving an indication of when it would be safe to stop taking readings (see Figure 14–3). Also, "idle" machine time could be broken down into the various interruptions and delays, for a more detailed understanding of such time.

WORK SAMPLING STUDY PLANS

Detailed planning must be done before actual work sampling study observations are made. The plans start with a preliminary estimate of the activities on which information is sought. This estimate may involve one or more activities, and the estimate can frequently be made from historical data. If the analyst cannot make a reasonable estimate, he or she should work sample the area for two or three days and use that information as the basis for these estimates.

FIGURE 14–3
Cumulative percentage of running time.

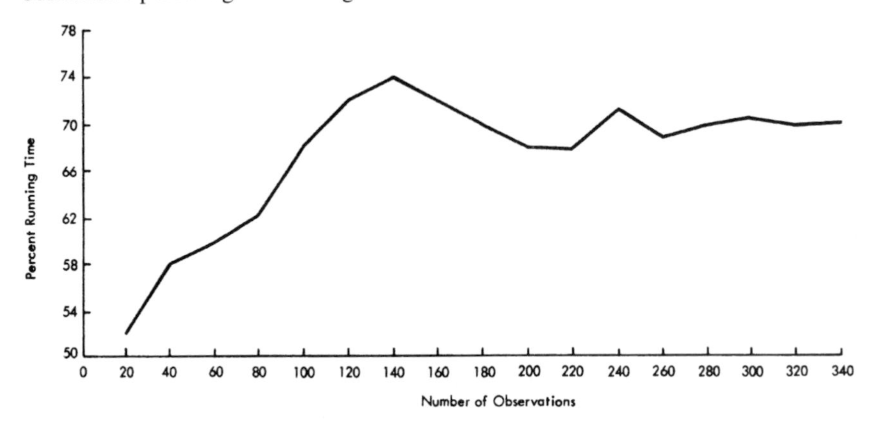

Once the preliminary estimates have been made, the analyst can determine the desired accuracy of the results. This can best be expressed as a tolerance, or limit of error, within a stated confidence level. Next, the analyst must estimate the number of observations to be made and determine the frequency of observations. Finally, the analyst designs the work sampling form on which to tabulate the data, as well as the control charts used in conjunction with the study.

Determining the Observations Needed

To determine the number of observations needed, the analyst must know the desired accuracy of the results. The more observations, the more valid the final answer. Three thousand observations give considerably more reliable results than 300. However, because of the cost of obtaining so many observations and the marginal improvement in accuracy, 300 observations may be ample.

For example, suppose you want to determine the number of observations required, with 95 percent confidence, such that the true proportion of personal and unavoidable delay time is within the interval 6–10 percent. The unavoidable and personal delay time is expected to be 8 percent. These assumptions are expressed graphically in Figure 14–4.

In this case, $\hat{p}$ would equal 0.08, and ℓ would equal 2 percent, or 0.02. Using these values, we can solve for n as follows:

$$n = \frac{4 \times 0.08 \times (1 - 0.08)}{0.02^2} = 736 \text{ observations}$$

If the analyst does not have the time or capability to collect 736 observations, but can only collect 500 data points, equation 2 can be inverted to solve for the resulting error limit:

FIGURE 14–4
Tolerance range of the percentage of unavoidable delay allowance required within a given section of a plant.

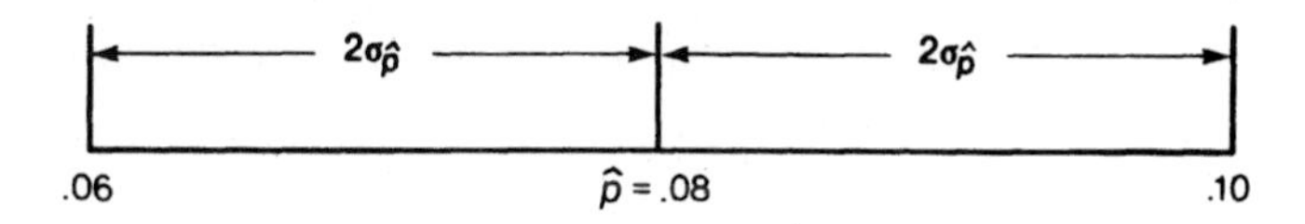

$$\ell = \frac{\sqrt{4p(1-p)}}{n} = \frac{\sqrt{4(0.92)(0.08)}}{500} = 0.024$$

With 500 observations, then, the accuracy of the study would be ±2.4 percent. Thus, there is a direct trade-off between the error or accuracy of the study and the number of observations collected.

Software for determining the observations required for a work sampling study is readily available today. These programs perform all the statistical calculations required to determine sample sizes and confidence intervals. For example, they can calculate the 90 percent, 95 percent, and 99 percent confidence intervals for a sample. They can also provide the number of samples necessary to achieve 90 percent, 95 percent, and 99 percent confidence for a specified degree of accuracy.

Determining Observation Frequency

The frequency of the observations depends, for the most part, on the number of observations required and the time available to develop the data. For example, for

EXAMPLE 14-2
Determination of the Required Number of Observations.

An analyst wishes to determine the amount of downtime due to tool problems in an area involving 10 CNC machining centers where very fine drilling is taking place. An initial pilot study indicated that out of 25 observations, only one CNC machine was down, for a $\hat{p}$ of 0.04. The analyst wishes a more accurate study with an estimate within ±1 percent of the true value, with a 95 percent confidence. The number of observations needed is then:

$$n = \frac{4 \times 0.04 \times (1 - 0.04)}{0.01^2} = 1536$$

Note that the analyst would take 154 trips to the plant floor and 10 simultaneous observations on each trip.

3,600 observations to be completed in 20 calendar days, the analyst would need to obtain approximately 3,600/20 = 180 observations per day.

Of course, the number of analysts available and the nature of the work being studied also influence the frequency of the observations. For example, if only one analyst is available to accumulate the data in the previous example it may be impractical for that person to take 180 observations during one day.

After determining the number of observations per day, the analyst must select the actual time needed to record the observations. To obtain a representative sample, observations are taken at all times of the day. For our example, we assume that one analyst is available to make the 180 observations on a battery of 20 turret lathes which are completely independent of each other, to determine the personal and unavoidable delay allowance. Since there are 20 machines to observe, the observer must make nine random trips to the machine floor each day for 20 days. The time of day selected for these nine observations is also chosen at random daily. Thus, the analyst establishes no set day-to-day observation pattern on the production floor.

There are many ways of randomizing the occurrence of the observations. In one approach, the analyst may select nine numbers daily from a statistical table of random numbers, ranging from 1 to 48 (see Appendix 3, Table A3–4). If each number carries a value, in minutes, equivalent to 10 times its size, the numbers selected can then set the time, in minutes, from the beginning of the day to the time for taking the observations. For example, the random number 20 would mean that the analyst should make a series of observations 200 minutes after the beginning of the shift. If the day begins at 8 AM, then at 20 minutes after 11 AM, an inspection of the 20 turret lathe operators would begin.

Another approach considers four adjacent digits in the random number table. Digit #1 is the day identifier, with numbers 1 to 5 identifying the workday Monday through Friday. Digit #2 is the hour identifier, with numbers 0 to 8 added to the starting time of work (e.g., 7:00 AM). Digits # 3 and #4 are the minutes identifiers, with numbers between 0 and 60 acceptable.

Computers can also be used to determine the schedule of daily observations. For example, work sampling programs for the DataMyte collector described in Chapter 8 can print random time schedules. Based on 20 random observations during an 8 AM to 5 PM working day, the generated observation schedule could be similar to the following:

Observation 1 at 8:01	Observation 11 at 11:25
Observation 2 at 8:29	Observation 12 at 11:41
Observation 3 at 8:54	Observation 13 at 13:14
Observation 4 at 9:11	Observation 14 at 13:41
Observation 5 at 9:30	Observation 15 at 14:15
Observation 6 at 9:41	Observation 16 at 14:54
Observation 7 at 9:58	Observation 17 at 15:15
Observation 8 at 10:22	Observation 18 at 15:32
Observation 9 at 10:47	Observation 19 at 16:16
Observation 10 at 11:00	Observation 20 at 16:57

The study should be long enough to include normal fluctuations in production. The longer the overall study, the better the chance of observing average conditions.

Usually, work sampling studies are made over a block of time ranging from two to four weeks.

Another alternative to help analysts decide when to make daily observations is a random reminder. This pocket-sized instrument beeps at random times, letting analysts know when to make the next observation. The user preselects an average sampling rate (observations per hour, or observations per day) and responds with a trip to the data collection area upon hearing the beep. Typically, the instrument can be preset at any of the following average beeps per hour: 0.64, 0.80, 1.0, 1.3, 1.6, 2.0, 2.5, 3.2, 4.0, 5.0, 6.4, and 8.0. This instrument is especially useful for self-observation, discussed later in this chapter. A table with times prepared in advance can require too much of the analyst's time when attempting to record data conscientiously at the listed times.

Using an Electronic Work Measurement Recorder

Electronic work measurement machines with optional work sampling software can be extremely helpful. For example, the OS-3 Plus recorder (see Chapter 8) is available with a work sampling program that permits the scheduling of random observations and the performance rating of individual readings, and it produces summary statistics and formatted printed reports. Figure 14–5 illustrates a printout in which the random schedule of tours for the current shift is shown at the top and a portion of a work sampling summary report is shown at the bottom.

Designing the Work Sampling Form

The analyst should design an observation form to record the data to be gathered during the work sampling study. A standard form is usually not acceptable, since each work sampling study is unique from the standpoint of the total observations needed, the random times that observations are made, and the information being sought. The best form is tailored to the study objectives.

Figure 14–6 is an example of a work sampling study form. An analyst designed this form to determine the time utilized for various productive and nonproductive states in a maintenance repair shop. The form accommodates 20 random observations during the workday. Some analysts prefer to use a specially designed card that allows observations to be made without the attention caused by a clipboard. The card can be sized so that it can be carried conveniently in the shirt or coat pocket. For instance, the form shown in Figure 14–6 could easily be split into two sections, with one section on each side of a 3-inch by 5-inch card that could be carried in the shirt pocket.

Using Control Charts

The control chart techniques used in statistical quality control work can readily be applied to work sampling studies. Since these studies deal exclusively with percentages or proportions, analysts use the p chart most frequently.

FIGURE 14–5

Sample schedule printout. *(The schedule printout lists the random schedule of tours for the current shift when the OS-3's scheduler is used.)*

```
Your Company Name Here              OS-3 WORK SAMPLING  03/22/92  18:24
---------------------------------------------------------------------------

---------------------------------------------------------------------------
SHIFT#  START     END         BREAK   START     END
  1     08:00:00  17:00:00
                                1     10:00:00  10:15:00
                                2     12:00:00  12:30:00
                                3     14:00:00  14:15:00

TOUR TIME    MIN START TO START     MAX START TO START   SEED
00:30:00      00:30:00               00:50:00             28348
---------------------------------------------------------------------------
sample   08:17:43
sample   08:55:29
 break            10:00:00
sample   10:15:45
sample   10:56:55
sample   00:00:00
```

FIGURE 14–5 *(continued)*

Sample schedule printout. *(The schedule printout lists the random schedule of tours for the current shift when the OS-3's scheduler is used.)* (*Courtesy* of Gage Talker Corporation formerly Observational Systems)

```
Your Company Name Here        OS-3 WORK SAMPLING  03/22/83 18:57   Page 1
---------------------------------------------------------------------------

---------------------------------------------------------------------------
OBSERVER         PLANT #              MO/DY/YR      HR:MN
John Anderson    1255                 03/22/92      18:24
---------------------------------------------------------------------------

---------------------------------------------------------------------------
FRED
---------------------------------------------------------------------------

     MACHINE           #=  5  15%
     WELD              #= 10  30%
     FIT PIPE          #=  6  18%
     CLEAN UP          #=  4  12%
productive TOTAL                         #= 25   76%

     GET TOOLS         #=  4  12%
     WAIT JOB          #=  0   0%
     CONFIR FOREMAN    #=  2   6%
     IDLE              #=  1   3%
     PERSONAL          #=  1   3%
nonproductive TOTAL                      #=  8   24%

     ABSENT            #=  0   0%
not coded   TOTAL                        #=  0    0%
```

FIGURE 14–6
Work sampling study form.

WORK SAMPLING STUDY

MAIN REPAIR SHOP

Number Working This Study ________ Date ________ By ________

Remarks ________

Obs. Nos.	Random Time	Productive Occurrences							Nonproductive Occurrences							Total Observations	Percentage Productive	Percentage Nonproductive
		Mch	Weld	Pipe Fit	Gen. Labor	Elect.	Carpen.	Janitor	Get Tools	Grind Tools	Wait Job	Wait Crane	Confer Foreman	Personal	Idle			
1																		
2																		
3																		
4																		
5																		
6																		
7																		
8																		
9																		
10																		
11																		
12																		
13																		
14																		
15																		
16																		
17																		
18																		
19																		
20																		
	TOTAL																	

FIGURE 14–7
Sample control chart.

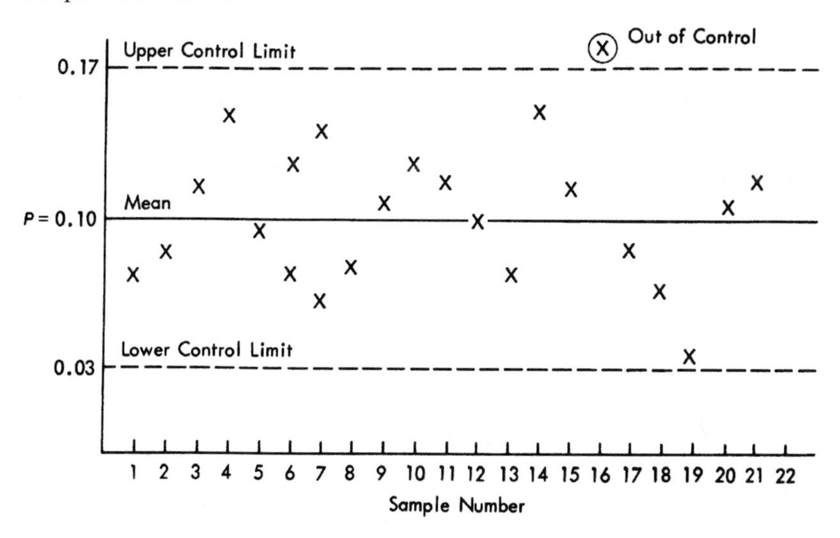

The first problem in setting up a control chart is the choice of limits. In general, a balance must be sought between the cost of looking for an assignable cause when none is present and the cost of not looking for an assignable cause when one is present. As an arbitrary choice, the analyst should use the $\pm 3\sigma$ limits as control limits on the p chart. Substituting 3σ for 1.96σ in equation 1 yields:

$$\ell = 3\sigma = 3\sqrt{p(1-p)/n}$$

Suppose that p for a given condition is 0.10 and that 180 observations are taken each day. Solving for ℓ yields:

$$\ell = 3 \times [0.1 \times 0.9/180]^{1/2} = 0.067 \approx 0.07$$

A control chart similar to Figure 14–7 could then be constructed, and the p' values for each day would be plotted on that chart.

In quality control work, the control chart indicates whether or not the process is in control. In a similar manner, in work sampling, the analyst considers points beyond $\pm 3\sigma$ limits of p as being out of control. Thus, a sample that yields a value of p' is assumed to have been drawn from a population with an expected value of p if p' falls within the $\pm 3\sigma$ limits of p. Expressed another way, if a sample has a value p' that falls outside the $\pm 3\sigma$ limits, the sample is assumed to be from some different population, or the original population is assumed to be changed.

As in quality control work, points other than those out of control may be of some statistical significance. For example, it is more likely that a point will fall outside the $\pm 3\sigma$ limits than that two successive points will fall between the $\pm 2\sigma$ and $\pm 3\sigma$ limits. Hence, two successive points between these limits would indicate that the population has changed. A series of significant sets of points has been derived. This idea is discussed in most statistical quality control texts under the heading "runs."

EXAMPLE 14–3
Use of Control Charts in Work Sampling.

The Dorben Company wishes to measure the percentage of machine downtime in the lathe department. An original estimate showed the downtime to be approximately 0.20. The desired results were to be within ±5 percent of p, with a confidence level of 0.95. Analysts took 6,400 readings over 16 days at the rate of 400 readings per day. They computed a p' value for each daily sample of 400 and set up a p chart for $p = 0.20$ and subsample size $N = 400$ (see Figure 14–8).

Each day, they took readings and plotted p'. On the third day, the point for p' went above the upper control limit. An investigation revealed that following an accident in the plant, several workers left their machines to assist the injured employee. Since an assignable cause of error was discovered, they discarded this point from the study. If they had not used a control chart, these observations would have been included in the final estimate of p.

On the fourth day, the point for p' fell below the lower control limit. No assignable cause could be found for this occurrence. The industrial engineer in charge of the project also noted that the p' values for the first two days were below the mean p and decided to compute a new value for p, using the values from days 1, 2, and 4. The new estimate of p turned out to be 0.15. To obtain the desired accuracy, n was then 8,830 observations. The control limits also changed (see Figure 14–9).

The analysts took observations for 12 more days and plotted the individual p' values on the new chart. As can be seen, all the points fell within the

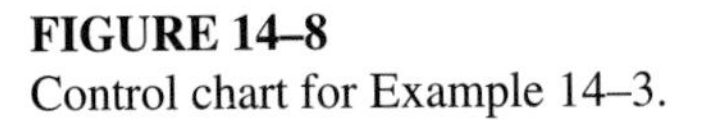

FIGURE 14–8
Control chart for Example 14–3.

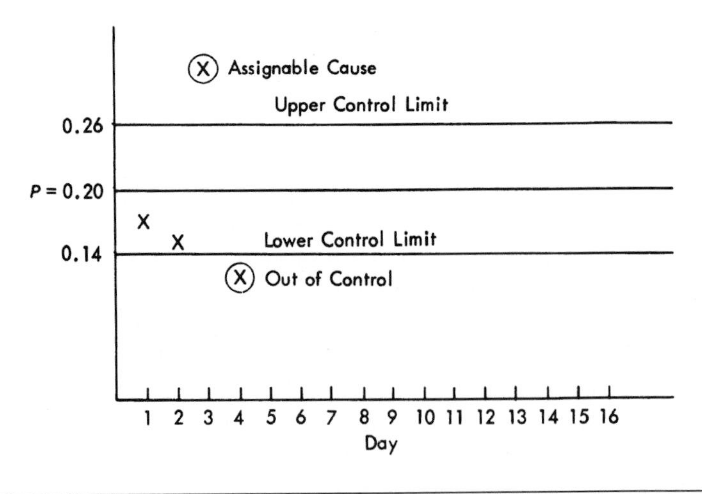

control limits. They then calculated a more accurate value of p, using all 6,000 observations, and determined that the new estimate of p was 0.14. A recalculation of achieved accuracy showed it to be slightly better than the desired accuracy. As a final check, the analysts computed new control limits, using p equal to 0.14. The dashed lines superimposed on Figure 14–9 show that all points were still in control using the new limits. If a point had fallen out of control, the analysts would have eliminated it and computed a new value of p. They would have repeated this process until the desired accuracy was achieved and all p' values were in control.

FIGURE 14–9
Revised control chart for Example 14–3.

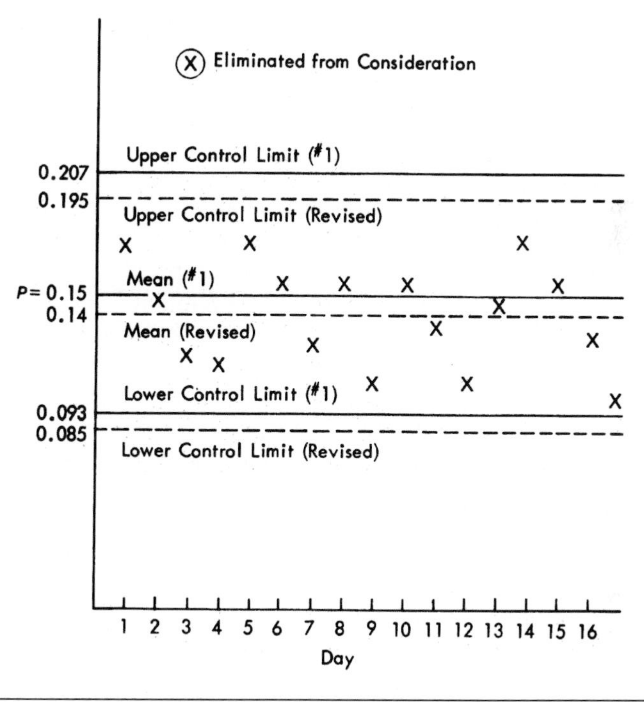

Do not assume that the percentage downtime will remain the same forever. Improvement should be a continuing process, and percentage downtime should diminish. One purpose of work sampling is to determine work areas that might be improved. After discovering such areas, the analyst should make an attempt to improve the situation. Control charts can show the progressive improvement of work areas. This idea is especially important if work sampling studies are used to establish standard times, since such standards must change whenever conditions change if they are to remain realistic.

OBSERVATIONS AND DATA RECORDING

A representative sample form for a shift study appears in Figure 14–10. Here, analysts made six random observations of each facility per shift. A digit in the space provided for the state of each facility designated the particular observation. Since 14 facilities were studied, the analysts made a total of 84 observations per shift.

In approaching the work area, the analyst must not anticipate the expected recording. The analyst should walk to a point a given distance from the facility, make the observation, and record the facts. It might be helpful to make an actual mark on the floor to show where to stand before making the observation. If the operator or machine being studied is idle, the analyst must determine the reason for the idleness, confirming the reason with the line supervisor before entering the data on the form. The analyst must learn to take the visual observations, and then make the written entries after leaving the work area. This minimizes the shop workers' feelings of being watched, and allows them to perform in their accustomed manner.

Even if the analyst observes the requisites of work sampling, the data tend to be biased when the technique is only used for studying people. The arrival of an observer at the work center immediately influences the activity of the operator. The operator becomes productively engaged as soon as he or she sees the analyst approaching the work center. Then, too, there is a natural tendency for the observer to record what has just happened or what will be happening, rather than what is actually happening at the exact moment of the observation.

A video camera can be very useful in performing unbiased work sampling studies involving only people. A work sampling study was made by the authors over a 10-day period on a data processing installation where the workers keypunched and verified various quantities of data processing cards. The study was concerned only with the elements "working" and "not working." Working time included the elements adjusting cards, removing cards, punching cards, and so on, and not working time involved absence from the workstation and idleness. The 2,520 observations collected indicated a statistically significant ($p < 0.001$) difference of a 12.3 percent greater "not working" average with the random activity analysis camera than with the personal observation method.

MACHINE UTILIZATION

Observers use the work sampling technique to determine machine utilization in the same manner as that used in establishing allowances. The following case history details the steps involved.

Analysts were told to gather information on machine utilization in a specific section of a heavy machine shop. Management estimated that the actual cutting time in this section should be 60 percent of the workday, to comply with the quotations

FIGURE 14–10
Work sampling summary sheet.

DATE 7-15
OBSERVER R Guild

MACHINE	DWG.	CUTTING	SETUP	MACHINE IDLE	CRANE WAIT	WAIT-INSPECTION	AID INSPECTION	WAIT-TOOLS NOT AVAILABLE	WAIT-TOOL TROUBLE	CONFER WITH OTHER SHIFT	TOOL HANDLING	GET OR GRIND TOOLS	CONFER WITH FOREMAN, INSP.	WAIT FOR JOB	REMOVE CHIPS CLEAN TABLE	MISCELLANEOUS	NO OPERATOR	
20′ VBM		101	7	14	2	3		1		2	37	5	3			6	35	216
16′ VBM		102	34	14	15	3	1	1		1	28	5	1	7	4			216
28′ VBM		119	34	10	5	5	2				18	2	1	2			18	216
12′ VBM		109	24	12	13	6	1			3	26	6	2	3	3	2	6	216
16′ PLANER		127	17	6	9	2					22		2	15		4	12	216
8′ IMM		64	18	17	16	3				2	30	7	3			28	28	216
16′ VBM		147	19	10	14	3	1				15	2			1	1	3	216
14′ PLANER		140	8	5	7	2				2	17	3		3		11	18	216
72″ E. LATHE		99	13	12	7	3				1	32	8	2			3	36	216
96″ E. LATHE		89	9	29	18	11	1			2	29	8	3	4		3	10	216
96″ E. LATHE		109	14	12	8	10		3	3		32	9	8	2		1	5	216
160″ E. LATHE		72	34	13	14	6	2	1		4	21	3	3	1	1	4	37	216
11-1/2′ PLNR		106	35	11	10	4				1	11	4	5	3	2	8	16	216
32′ VBM		151	23	8	7	1				1	10	2	1	5	2	5		216
		1535	289	173	145	62	8	6	3	19	328	64	34	45	13	76	224 =	3024
	%	50.7	9.6	5.9	4.8	2.1	.3	.2	.1	.6	10.8	2.1	1.1	1.5	.4	2.5	7.4 =	100%

FIGURE 14–11
Cumulative percentage machine cutting.

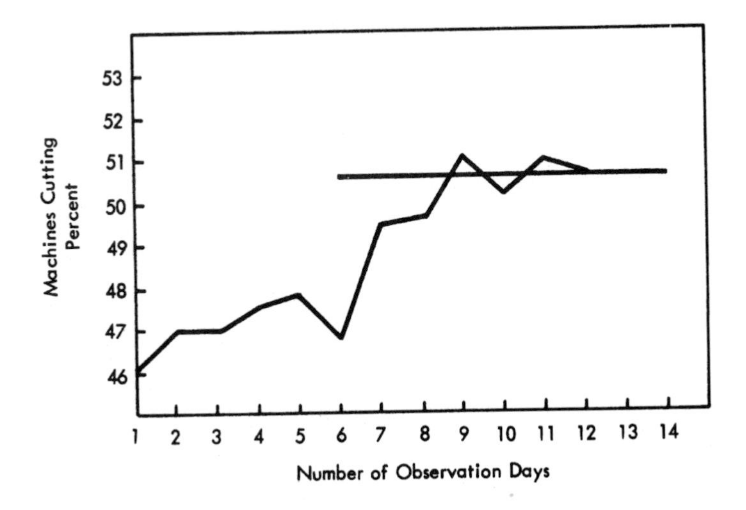

being submitted. There were 14 facilities involved, and the analysts had to take approximately 3,000 observations to get the accuracy desired.

Observers designed a work sampling form (see Figure 14-10) to accommodate the 16 possible states that each of the 14 facilities might be in at the time of an observation. To assure random observations, they set up a random pattern of visitation to the shop area. They made 6 observations of the 14 facilities during each shift. To get the required total number of observations, they observed 36 separate shifts, with 84 observations on each shift. Each trip took one analyst about 15 to 20 minutes. This work occupied only about 2 hours per shift, leaving that analyst free to perform other work during the remaining 6 hours.

Since the principal purpose of the study was to learn the status of the actual cutting time in this section, an analyst kept a cumulative percentage machine cutting chart (see Figure 14-11). At the beginning of each day's study, the analysts took a ratio of all previous cutting observations to the total observations to date. By the end of the 10th day of the study, the percentage of machine cutting time began to level off at 50.5 percent.

After studying 36 shifts, the analysts divided the sum of all observations in each category by the total number of observations. This resulted in percentages that represented the distribution of cutting time, setup time, and the various delay times listed. Figure 14–10 illustrates the summary sheet for this study. Cutting time amounted to 50.7 percent. The percentage of time required by the various delays indicated areas for methods improvement that would help increase the cutting time.

ALLOWANCES DETERMINATION

The determination of time allowances must be correct, if fair standards are to be developed. Prior to the introduction of work sampling, analysts frequently determined allowances for personal reasons and unavoidable delays by taking a series of all-day studies on several operations and then averaging the results. Thus, they recorded, timed, and analyzed trips to the restroom, trips to the drinking fountain, interruptions, and so forth, and determined a fair allowance. Although this method provided an answer, it was costly and time-consuming, and was fatiguing to both the analyst and the operator.

Through a work sampling study, analysts take a great number of observations (usually, over 2,000) at different times of the day and of different operators. They can then divide the total number of legitimate nonwork occurrences that involve normal operators by the total number of working observations. The result equals the percentage allowance that should be given to the operator for the class of work being studied. The different elements that enter into personal and unavoidable delays can be kept separate, and an equitable allowance can be determined for each class or category.

Figure 14–12 illustrates a summary of a work sampling study for determining unavoidable delay allowances on bench, bench machine, machine, and spray operations. There were interferences in 26 cases out of 2,895 observations made on the bench operations. This indicated an unavoidable delay allowance of 0.95 percent on this class of work.

STANDARD TIME DETERMINATION

Work sampling can be very useful for establishing time standards on both direct and indirect labor operations. The technique is the same as that used for determining allowances. The analyst must take a large number of random observations. The percentage of the total observations that the facility or operation is working approximates the percentage of total time that it is truly in that state.

More specifically, the observed time (*OT*) (see Chapter 8) for a given element is calculated from the working time divided by the number of units produced during that time:

$$OT = \frac{T \times n_i}{P \times n}$$

where: T = Total time.
n_i = Number of occurrences for element i.
n = Total number of observations.
P = Total production for period studied.

The normal time (*NT*) is found by multiplying the observed time by the average rating:

FIGURE 14–12
Summary of interruptions on various classes of work taken from a work sampling study for determining unavoidable delay allowance.

XYZ ELECTRIC PRODUCTS, INC.

PLANT 26
DEPARTMENT 4
SUMMARY SHEET WORK SAMPLING
DATE 4/22/

OPERATION	Engineer	Supply	Quality	Mechanic	Supervision	Turning on Light	Miscellaneous	Working		Number of Interferences	TOTAL OBSER.	PERCENT ALLOWANCE
	NUMBER OF OBSERVATIONS											
Bench	1	11	1	0	12	1	119	2,750		26	2,895	.95
Bench Machine	0	2	0	0	5	0	69	984		7	1,060	.71
Machine	0	1	6	11	9	0	29	1,172		27	1,228	2.30
Spray	0	0	0	7	36	0	262	1,407		43	1,712	3.06

REMARKS: SUMMARY FOR INTERFERENCE ALLOWANCES. SEE OBSERVATION SHEET FOR DETAILS OF MISCELLANEOUS. MISCELLANEOUS WAS INCLUDED IN WORKING CATEGORY BECAUSE IN ALL CASES DOWNTIME WAS CREDITED TO THE OPERATORS.

NOTE: ALL OBSERVATIONS ARE TO BE TAKEN AT RANDOM.

$$NT = OT \times \overline{R}/100$$

where: $\overline{R}$ = average rating = $\Sigma R/n_i$

Finally, the standard time is found by adding allowances to the normal time.

SELF-OBSERVATION

Conscientious administrators periodically take work samples of their own work to evaluate the effectiveness of their time usage. In the majority of cases, administrators spend less time on the important aspects than they think they are spending. They also spend more time on unimportant aspects, such as personal and avoidable delays, than they believe they are spending. Once administrators learn how much time is being taken by functions that could be readily be handled by subordinates and clerical personnel, they can take positive action.

For example, a university professor may decide to conduct a personal work sample to learn just how he or she is spending time. The professor decides to take

EXAMPLE 14–4
Calculation of Standard Time for a Complete Operation.

Table 14–1 lists the information necessary for the calculations, the sources for the information, and the specific data used in this example of a drill press operator.

$$OT = \frac{T \times n_i}{P \times n} = \frac{480}{420} \times 0.85 = 0.971 \text{ min}$$

The normal time (*NT*) is found by scaling the observed time by the average rating ($\bar{R}$):

$$NT = OT \times \bar{R}/100 = 0.971 \times 110/100 = 1.069 \text{ min}$$

Finally, the standard time is found by adding allowances (using the multiplier approach) to the normal time:

$$ST = NT \times (1 + \text{Allowance}) = 1.069 \times (1.15) = 1.229 \text{ min}$$

TABLE 14–1
Information on Drill Press Operator

Information	Source	Data
Total working day (working + idle)	Time card	480 minutes
Number of units drilled	Inspection department	420 units
Working fraction	Work sampling	85 percent
Average rating	Work sampling	110 percent
Allowances	Work sampling	15 percent

random samples over an 8-week period during the academic year. This period should supply typical data not subject to seasonal variation. The professor sets a random reminder to provide 2.0 samples per hour on average. Thus, over the 8-week study period, the professor would have 640 observations (8 weeks × 40 hours per week × 2 observations per hour). The professor could have chosen to take samples at a higher rate on randomly selected days within the study interval. For example, he or she may have chosen to take 4 samples an hour over 20 days randomly selected from the 8-week period. This would also provide 640 observations.

To record the data, the professor designed a form similar to that shown in Figure 14–13. The form was designed to contain one week of daily random observations. Each time the random reminder beeped, the professor recorded the code letter for the applicable category and the time.

At the end of the 8-week study, the professor summarized the weekly data sheets. The professor found that 80 of the total 640 observations were code I (committee participation), which meant that about 12.5 percent of work time was spent in committee participation. The 95 percent confidence interval would be:

FIGURE 14–13
A specially designed work sampling form.

Name A. B. Jones   Week of 3/25   Study No. B-47

Code	Activity	Code	Activity	Code	Activity
T	Teaching	C	Continuing Ed.	D	Personal
R	Research	A	Student Advising	I	Committee
P	Preparation	S	Professional Development		

Mon.		Tues.		Wed.		Thur.		Fri.		Sat.		Sun.		Notes
Time	CO	Time	CO	Time	CO	Time	CO	Time	CO	Time	CO	Time	CO	
8:17	T	8:06	C	8:58	P	8:02	I	8:49	T					I - Exec. Com.
8:52	T	8:32	C	9:08	R	8:31	A	9:07	C					
9:04	D	8:58	A	9:25	R	8:45	S	9:17	C					
9:27	R	9:32	P	10:01	I	9:32	T	9:51	I					I - Res. Com. + Personnel Policy
9:50	R	10:11	T	10:50	S	10:17	T	10:11	R					
10:11	I	11:00	S	10:57	S	10:40	S	10:32	R					I - Personnel Policy
10:18	A	11:05	S	11:26	A	11:35	P	10:53	A					
11:01	P	11:55	I	11:40	P	11:59	D	11:17	P					I - Dept. Curriculum
11:25	P	1:42	P	1:17	D	1:04	R	11:42	P					
1:05	P	1:59	P	2:05	T	1:27	R	1:11	I					I - Dept Curriculum
2:01	T	2:11	R	2:35	T	1:47	I	1:47	I					I - Dept. Curriculum
2:35	T	2:37	R	3:00	I	2:17	R	2:15	T					I - Entertainment Committee
2:55	S	3:25	R	3:24	S	2:46	A	2:45	T					
3:45	S	3:40	S	4:14	S	3:40	P	3:00	T					
4:11	P	3:57	R	4:38	P	4:11	S	4:02	S					
4:42	R	4:15	A	5:00	P	4:37	S	4:25	D					

Week Summary

T	13	P	15	A	7	D	4		
R	14	C	4	S	14	I	9	Total	80

$$\pm 1.96 \sqrt{\frac{0.125(1 - 0.125)}{640}} = \pm 0.013$$

The professor is 95 percent confident that committee work is occupying 12.5 ± 1.3 percent of the time. The professor followed the same procedure for all the coded activities. Using the results of this study, the professor altered the calendar to utilize time and energy in a more positive manner.

EXAMPLE 14–5
Calculation of Standard Time for Multiple Elements.

An analyst made 30 observations over 15 minutes on a work assignment involving three elements, during which time 12 units were produced. The resulting data appear in Table 14–2. The observed times were respectively:

$$OT_1 = \frac{15 \times 9}{12 \times 30} = 0.375 \text{ min}$$

$$OT_2 = \frac{15 \times 7}{12 \times 30} = 0.292 \text{ min}$$

TABLE 14–2
Tabular Data of a Three-Element Work Sampling Study

Observation number	Performance rating observed			
	Element 1	Element 2	Element 3	Idle
1	90			
2				100
3		110		
4	95			
5	100			
6		100		
7			105	
8	90			
9			110	
10	85			
11			95	
12		90		
13			100	
14			95	
15	80			
16			110	
17		105		
18			90	
19	100			
20			85	
21			90	
22			90	
23	110			
24			100	
25		95		
26				100
27		105		
28		100		
29			110	
30	110			
Σ Rating	860	705	1180	100

(continued)

EXAMPLE 14–5 ***(continued)***

$$OT_3 = \frac{15 \times 12}{12 \times 30} = 0.500 \text{ min}$$

and the respective normal times were:

$$\text{NT}_1 = 0.375 \times \frac{860}{9 \times 100} = 0.358 \text{ min}$$

$$\text{NT}_2 = 0.292 \times \frac{705}{7 \times 100} = 0.294 \text{ min}$$

$$\text{NT}_3 = 0.500 \times \frac{1180}{11 \times 100} = 0.492 \text{ min}$$

Assuming a constant 10 percent allowance for all elements, the final standard time was:

$$ST = (0.358 + 0.294 + 0.492)(1 + 0.10) = 1.258 \text{ min}$$

COMPUTERIZED WORK SAMPLING

Using a computer can save an estimated 35 percent of the total work sampling study cost, because of the high percentage of clerical effort relative to actual observation time. The majority of the effort involved in summarizing work sampling data is clerical: calculating percentages and accuracies, plotting data on control charts, determining the number of observations required, determining the daily observations required, determining the number of trips to the area being studied per day, determining the time of day for each trip, and so on.

By mechanizing the repetitive calculations, computers can calculate not only daily results, but also cumulative results, such as the maintenance of control charts. For example, one company developed the Mechanized Activity Sampling Technique (MAST), which automates the clerical work, including the mathematical calculations associated with recording the observations, computing the element percentages, developing the performance ratings, checking the statistical accuracies, preparing and maintaining control charts, and extrapolating the data into equivalent staffing needs and/or machines and annual costs.

Users of MAST claim the following benefits:

1. The amount of industrial engineering time is increased through the reduction of clerical routines.

2. Results of the study are achieved more rapidly, and the data are presented in a professional manner.
3. The cost of conducting work sampling studies is significantly reduced.
4. Accuracy of the computations is improved.
5. Fewer errors are committed by analysts.
6. The automated system provides an incentive to make greater use of the work sampling technique.

SUMMARY

Work sampling is another tool that allows time and methods analysts to obtain the facts easier and more quickly. Performance rated work sampling is especially useful in determining the amount of time that should be allocated for unavoidable delays, work stoppages, and the like. The extent of these interruptions is a suitable area for study to improve productivity. Typically, in the machining industry, these interruptions amount to between 24 and 45 percent of the total time. Work sampling is also used more heavily for establishing standards on production support labor, maintenance, and service labor.

Everyone in the field of methods-time study and wage payment should become familiar with the advantages, limitations, uses, and applications of this technique. In summary, the following considerations should be kept in mind:

1. Explain and "sell" the work sampling method before using it.
2. Confine individual studies to similar groups of machines or operations.
3. Use as large a sample size as is practicable.
4. Take individual observations at random times, so that observations are recorded for all hours of the day.
5. Take the observations over two weeks or more.

QUESTIONS

1. Where was work sampling first used?

2. What advantages are claimed for the work sampling procedure?

3. In what areas is work sampling applicable?

4. How can you determine the time of day to make the various observations, such that biased results do not occur?

5. What considerations should be kept in mind when doing work sampling studies?

6. What is meant by stratifying the data collected? Explain when it would be desirable to stratify data.

7. What are the principal advantages of using a random reminder in connection with gathering data for a work sampling study?

8. Over how long a period is it desirable to continue to acquire sampling data?

9. How biased can we expect work sampling data to be? Will this bias vary with the work situation? Explain.

10. How can the validity of work sampling be "sold" to the employee not familiar with probability and statistical procedures?

11. What are the pros and cons for using work sampling to establish performance standards?

PROBLEMS

1. The analyst in the Dorben Reference Library decides to use the work sampling technique to establish standards. Twenty employees are involved. The operations include: cataloging, charging books out, returning books to their proper location, cleaning books, recordkeeping, packing books for shipment, and handling correspondence. A preliminary investigation resulted in the estimate that 30 percent of the group's time was spent in cataloging. How many work sampling observations would be made if it were desirable to be 95 percent confident that the observed data were within a tolerance of ±10 percent of the population data? Describe how the random observations should be made.

   The following table illustrates some of the data gathered from 6 of the 20 employees. The number of volumes cataloged equals 14,612. From these data, determine a standard, in hours per hundred, for cataloging. Then, design a control chart based on $\pm 3\sigma$ limits for the daily observations.

	Operators					
Item	**Smith**	**Apple**	**Brown**	**Green**	**Baird**	**Thomas**
Total hours worked	78	80	80	65	72	75
Total observations (all elements)	152	170	181	114	143	158
Observations involving cataloging	50	55	48	29	40	45
Average rating	90	95	105	85	90	100

2. The work measurement analyst in the Dorben Company is planning to establish standards for indirect labor, using the work sampling technique. This study will provide the following information:

T = Total operator time represented by the study.
N = Total number of observations involved in the study.
n = Total observations of the element under study.
P = Production for the period under study.
R = Average performance rating factor during the study.

Derive the equation for estimating the normal elemental time for an operation.

3. The analyst in the Dorben Company wishes to measure the percentage of downtime in the drop hammer section of the forge shop. The superintendent estimates the downtime to be about 30 percent. The desired results, using a work sampling study, are to be within ±5 percent of p, with a confidence level of 0.95.

The analyst decides to take 300 random readings a day for three weeks. Develop a p chart for $p = 0.30$ and subsample size $N = 300$. Explain the use of this p chart.

4. The Dorben Company is using the work sampling technique to establish standards for its typing pool. This pool has varied responsibilities, including typing from tape recordings, filing, Kardex posting, and copying. The pool has six typists who each work a 40-hour week. Over a 4-week period, 1700 random observations were made. During the period, the typists produced 1,852 pages of routine typing. Of the random observations, 1,225 showed that typing was taking place. Assuming a 20 percent allowance and an adjusted performance rating factor of 0.85, calculate the hourly standard per page of typing.

5. How many observations should be recorded in determining the allowance for personal delays in a forge shop, if it is expected that a 5 percent personal allowance will suffice, and if this value is to remain between 4 and 6 percent across 95 percent of the time?

6. To get ±5 percent precision on work that is estimated to take 80 percent of the workers' time, how many random observations are required at the 95 percent confidence level?

7. If the average handling activity during a 10-day study is 82 percent, and the number of daily observations is 48, how much tolerance can be allowed on each day's percentage activity?

8. If 4 seconds of time would be desirable for each random observation, and if a sample size of 2,000 is necessary, how often would the analyst need to service the random activity analysis camera?

REFERENCES

Barnes, R. M. *Work Sampling.* 2nd ed. New York: John Wiley, & Sons, 1957.
Morrow, Robert Lee. *Time Study and Motion Economy,* New York: Ronald Press, 1946.
Pape, Elinor S. "Work Sampling." In *Handbook of Industrial Engineering.* 2nd ed. Ed. Gavriel Salvendy. New York: John Wiley & Sons, 1992.

Richardson, W. J. *Cost Improvement, Work Sampling and Short Interval Scheduling*. Reston, VA: Reston Publishing, 1976.

SELECTED SOFTWARE

MAST. Rochester, NY: Applied Research Laboratories Division of Bausch & Lomb Inc., 1988.

CHAPTER 15

Indirect and Expense Labor Standards

KEY POINTS:

- Use both time studies and predetermined time systems to develop standards for relatively predictable indirect labor.
- Use slots or similar job categories to establish standards for relatively unpredictable indirect and expense labor.
- Utilize work sampling and historical records to establish standards for professional expense labor.
- Use queuing theory to calculate waiting times for jobs.
- Use Monte Carlo simulation to predict delays or downtimes on jobs.

Since 1900, the percentage increase of indirect and expense workers has more than doubled that of direct labor workers. Groups usually classified as *indirect labor* include: shipping and receiving, trucking, stores, inspection, material handling, toolroom, janitorial, and maintenance. *Expense labor* includes all positions not coming under direct or indirect: office clerical, accounting, sales, management, engineering, and so on.

The rapid growth in office workers, maintenance workers, and other indirect and expense employees is due to several factors. First, the increased mechanization of industry and the complete automation of many processes, including the use of robots, have decreased the need for craftspeople and operators. This trend toward increased mechanization has resulted in a huge demand for electronics specialists, electricians, instrument makers, fluid mechanics technicians, and other service people. Also, the design of complicated machines and controls has resulted in a greater demand for engineers, designers, and draftspersons.

Second, the tremendous increase in paperwork brought on by federal, state, and local legislation is responsible, to a large extent, for an ever increasing number of clerical employees. Third, office and maintenance work has not been subjected to the methods study and technical advances that have been applied so effectively to direct labor in industrial processes. With a large share of most payrolls being earmarked for indirect and expense labor, progressive management is beginning to realize the opportunities for the application of methods and standards in this area.

INDIRECT AND EXPENSE WORK STANDARDS

The systematic approach to methods standards and wage payment is just as applicable to the indirect and expense areas as it is to direct labor. Careful fact-finding, analysis, proposed method development, presentation, installation, and job analysis development should precede a program for establishing standards on indirect and expense labor. The methods analysis procedure in itself can introduce economies.

Work sampling is a good technique for determining the severity of a problem and the savings potential in the indirect and expense areas. It is not unusual to find that the workforce is productively engaged only 40 to 50 percent of the time, or even less. For example, in maintenance work, which represents a large share of the total indirect cost, analysts may find the following reasons for much of the time lost during the workday:

1. *Inadequate communication.* It is quite common to find incomplete and even incorrect job instructions on work orders. This necessitates additional trips to the toolroom and supply room to obtain parts and tools that should have been available when the work was started. The work order that merely states "Repair leak in oil system" is indicative of poor planning and inadequate communication. The worker needs to know whether a new valve or pipe is needed, whether a new gasket can do the job, or whether the valve needs to be repacked.
2. *Unavailability of parts, tools, or equipment.* If the craftspeople do not have the facilities and parts to do the job, they are obliged to improvise, which usually wastes time and frequently results in inferior work. Proper planning ensures that the correct tools and equipment are available and that suitable spares are supplied to the job site.
3. *Interference of production employees.* With improper scheduling, maintenance employees may find that they are unable to begin a repair, service, or overhaul operation, because the facility is still being used by production employees. This can result in the craftspeople idly waiting until the production department is ready to turn over the equipment.
4. *Overstaffing of the maintenance job.* This has been one of the principal causes of lost time in maintenance work. Too often, a crew of three or four is supplied, when only two are needed.
5. *Unsatisfactory work that must be redone.* Poor planning frequently results in an attitude of "this will get by" on the part of the mechanic. This results in the repair work having to be redone.

6. *Instructions for the next job not ready*. With improper planning, "wait for work" idle time can be significant. Good planning assures that there is ample work ahead of the craftsperson, who seldom has to wait for the next job.

After good methods are developed, they should be standardized and taught to the workers who will be using them. This obvious step is too often neglected by management. Giving as much attention to operation improvement for indirect work as for direct work can decrease costs.

Indirect Labor Standards

Standards for indirect labor departments, such as clerical, maintenance, and toolmaking should be developed from standard data or formulas. Time would not permit using stopwatch techniques for each and every standard developed. However, both time studies and fundamental motion data can be used.

Analysts can calculate time standards on any operation or group of operations that can be quantified and measured. If the work elements performed by the electrician, blacksmith, boilermaker, sheet metal worker, painter, carpenter, millwright, welder, pipe fitter, material handler, toolmaker, and others doing indirect work are broken down and studied, it is usually possible to evolve equitable standards by summarizing the direct, transportation, and indirect elements. The tools used for establishing standards for indirect and expense work are identical to those used in establishing standards for direct work: time study, predetermined motion time standards, standard data, time formulas, and work sampling.

Analysts can therefore establish standards for such tasks as hanging a door, rewinding a 1-horsepower motor, painting a centerless grinder, sweeping the chips from a department, or delivering a skid of 200 forgings. For each of these operations, analysts establish standard times by measuring the time required for the operator to perform the job. They then performance rate the study and apply an appropriate allowance.

Careful study and analysis often reveal that crew balance and interference cause more unavoidable delays in indirect work than in direct work. Crew balance is the delay time encountered by one member of a crew while waiting for other crew members to perform other job elements. Interference time is the time that a worker waits for others to do necessary work. Both crew balance and interference delays are unavoidable delays; however, they are usually characteristic of indirect labor operations only, such as those performed by maintenance workers. Queuing theory, explained later in this chapter, is a useful tool for estimating the magnitude of waiting time.

Because of the high degree of variability characteristic of most maintenance and material handling operations, it is necessary to conduct sufficient independent time studies of each operation to ensure that the resulting standard is representative of the time needed for the normal operator to do the job under average conditions. For example, if a study indicates that it takes 47 minutes to sweep a machine floor 60 feet wide by 80 feet long, observers must assure that average conditions prevail during the study. The work of a sweeper will be considerably more time consuming if the shop is machining cast iron rather than an alloy steel, because alloy steel is much cleaner, and alloy steel chips are easier to handle. Also, the use of slower speeds and feeds results in fewer chips. Therefore, a 47-minute standard established when the shop is working with alloy steel, will be inadequate for producing cast-iron parts. Additional time studies would assure that average conditions prevailed and that the resulting standards represented those conditions.

Similarly, analysts can establish standards for painting. A standard for overhead painting is based on square footage. Vertical and floor painting can be measured, and time standards can also be developed based on square footage.

Just as the automotive industry has established time standards for common repair jobs, such as grinding valves, replacing piston rings, and adjusting brakes, observers can establish standards for typical maintenance and repair operations performed on the machine tools within a plant. They can determine time standards for rewinding a 1/8-HP motor, a 1/4-HP motor, a 1/2-HP motor, a 3/4-HP motor, and so on.

Toolroom work is similar to the work done in job shops. Analysts can therefore predetermine the method by which tools such as a drill jig, milling fixture, form tool, or die can be made. Analysts use time study and/or predetermined elemental times to establish a sequence of elements and to measure the normal time required for each element. Work sampling provides an adequate tool for determining the allowances that must be added for fatigue, personal, unavoidable, and special delays. The standard elemental times thus developed can be tabularized in the form of standard data and then used to design time formulas for pricing future work.

Factors Affecting Indirect and Expense Standards

All indirect and expense work is a combination of four divisions: (1) direct work, (2) transportation, (3) indirect work, (4) and unnecessary work and delays.

Direct work is that segment of the operation that discernibly advances the progress of the work. For example, in the installation of a door, the direct work elements may include: cut door to rough size, plane to finish size, locate and mark hinge areas, chisel out hinge areas, mark for screws, install screws, mark for lock, drill out for lock, and install lock. Such direct work can be easily measured using conventional techniques, such as a stopwatch time study, standard data, or fundamental motion data.

Transportation is the work performed in movements during the course of the job or from job to job. Transportation may be horizontal or vertical, or both. Typical transportation elements include: walk up and down stairs, ride elevator, walk, carry load, push truck, and ride on motor truck. Transportation elements are also easy to measure, and to establish as standard data. For example, one company uses 0.50 minutes per 100-foot zone as its standard for horizontal travel time and 0.30 minutes as its standard time for 10 feet of vertical travel.

As a general rule, analysts cannot evaluate the indirect portion of indirect or expense labor by physical evidence in the completed job, or at any stage during the work, except by deductive inferences from certain features of the job. Indirect work elements may be separated into three divisions: (a) tooling, (b) material, and (c) planning.

Tooling work elements include the acquisition, disposition, and maintenance of all tools needed to perform an operation. Typical elements under this category would include: getting and checking tools and equipment; returning tools and equipment at the completion of the job; cleaning tools; and repairing, adjusting, and sharpening tools. The tooling work elements are easy to measure by conventional means; statistical records provide data on the frequency of their occurrence. Waiting line or queuing theory provides information on the expected waiting time at supply centers.

Material work elements involve acquiring and checking the material used in an operation and disposing of scrap. Making minor repairs to materials, picking up and disposing of scrap, and getting and checking materials, are characteristic of material work elements. As with tooling elements, material elements are readily measured, and their frequency can be accurately determined through historical records. Queuing theory provides the best estimate of waiting time for acquiring material from storerooms.

The planning elements represent the most difficult area in which to establish standards. These elements include: consulting with the supervisor, planning work procedures, inspecting, checking, and testing. Work sampling techniques provide a basis for determining the time required to perform the planning elements. Analysts also measure planning elements by stopwatch procedures, but their frequency of occurrence is difficult to determine; consequently, work sampling is the more practical tool for this class of elements.

Planning and methods improvement can eliminate unnecessary work and delays, which may represent as much as 40 percent of the indirect and expense payroll. Much of this wasted time is management oriented; for this reason, analysts should follow the systematic procedure recommended in Chapter 1 prior to establishing standards. In this work, analysts should get and analyze the facts, and develop and install the method, before establishing the standard.

Much of the delay time encountered in indirect and expense work is due to queues. Workers are obliged to stand in line at the tool crib, the storeroom, or the stockroom, waiting for a forklift truck, a desk calculator, or some other piece of equipment. Through the application of queuing theory, analysts can frequently determine the optimum number of servicing units required in the given circumstances.

Basic Queuing Theory

Queuing system problems may occur when the flow of arriving traffic (people, facilities, and so on) establishes a random demand for service at facilities with a limited service capacity. The time interval between arrival and service varies inversely with the level of the service capacity. The greater the number of service stations and the faster the rate of servicing, the smaller the time interval between arrival and service.

Methods and work measurement analysts should select an operating procedure that minimizes the total cost of operation. There must be an economic balance between waiting times and service capacity. The following four characteristics define queuing problems:

1. *The pattern of arrival rates.* The arrival rate (for example, a machine breaking down for repair) may be either constant or random. If random, the pattern is a probability distribution of the values of the intervals between successive arrivals. Also, probability distribution of the random pattern may be definable or undefinable.
2. *The pattern of the servicing rate.* The servicing time may also be either constant or random. If random, analysts should define the probability distribution that fits the random pattern.
3. *The number of servicing units.* In general, multiple-service queuing problems are more complex than those of single-service systems. However, most problems are the multiple-service type, such as the number of mechanics required to keep a battery of machines in operation.
4. *The pattern of selection for service.* Service is usually on a first-come, first-served basis; however, in some cases, the selection may be completely random, or may follow some set of priorities.

The solutions to queuing problems fall into two broad categories: analytic and simulation. The analytic category covers a wide range of problems for which mathematical probability and analytic techniques have provided equations representing systems with various assumptions about queuing characteristics. One of the most common assumptions about the arrival pattern, or arrivals per unit time interval, is that they follow Poisson's probability distribution:

$$p(k) = \frac{a^k e^{-a}}{k!}$$

where: $a =$ The mean arrival rate.
$k =$ The number of arrivals per time interval.

A helpful graphic presentation of the cumulative Poisson probabilities appears in Figure 15–1.

A further consequence of a Poisson-type arrival pattern is that the random variable time between arrivals obeys an exponential distribution with the same parameter a. Being a continuous distribution, the exponential distribution has a density function:

$$f(x) = ae^{-ax}$$

The exponential distribution has a mean $\mu = \frac{1}{a}$, and a variance $= \frac{1}{a^2}$, and μ can be recognized as the mean interval between arrivals. In some queuing systems, the number of services per unit time may follow the Poisson pattern; the service times subsequently follow the exponential distribution. Figure 15–2 illustrates the exponential curve $F(x) = e^{-x}$, which shows the probabilities of exceeding various multiples of any specified service time.

The basic equations governing the queue applicable to Poisson arrivals and service, in arrival order, fall into these five categories:

1. Any service time distribution and a single server.
2. Exponential service time and a single server.
3. Exponential service time and finite servers.
4. Constant service time and a single server.
5. Constant service time and finite servers.

Equations have been developed for each of these categories. These equations provide quantitative answers to such problems as mean delay time in the waiting line and mean number of arrivals in the waiting line.

Two examples demonstrate the application of queuing theory. The first (Example 15–1) is the determination of a standard for an inspection process. This fits into the first of the five categories, a single server with any service time distribution. The second example (Example 15–2) applies to toolroom delays and falls into the second category, a single server with exponential service time. Many industrial problems fall into this category.

Monte Carlo Simulation

The simulation method used to solve queuing problems does not employ formal mathematical models. Such models may be too complex or the systems too unique in light of the mathematical models in existence. Instead, a series of numerical statements creates a model of the queuing system. The analyst introduces a sample set of input values drawn from specified arrival and service time distributions. These input data generate a sample output distribution of waiting line results for the period.

A Monte Carlo simulation estimates the expected waiting time and service times and develops an optimum solution through a proper balance of service stations, servicing rates, and arrival rates. This technique is most helpful for analyzing the waiting line problem involved in the centralized–decentralized storage location of tools, supplies, and service facilities.

The Monte Carlo simulation technique is also applicable to problems with a Poisson arrival pattern. However, Monte Carlo is generally used for problems from which neither standard nor empirical formulas are obtainable. Example 15–3 demonstrates the use of the technique to determine the minimum number of operators needed to set up machines in an automatic screw machine section.

Expense Standards

More and more, management is recognizing its responsibility to determine accurately the appropriate office force for a given volume of work. To control office

FIGURE 15–1
Poisson distribution of arrivals.

$$P(c, a) = \sum_{x=c}^{\infty} \frac{e^{-a}a^{x}}{x!}$$

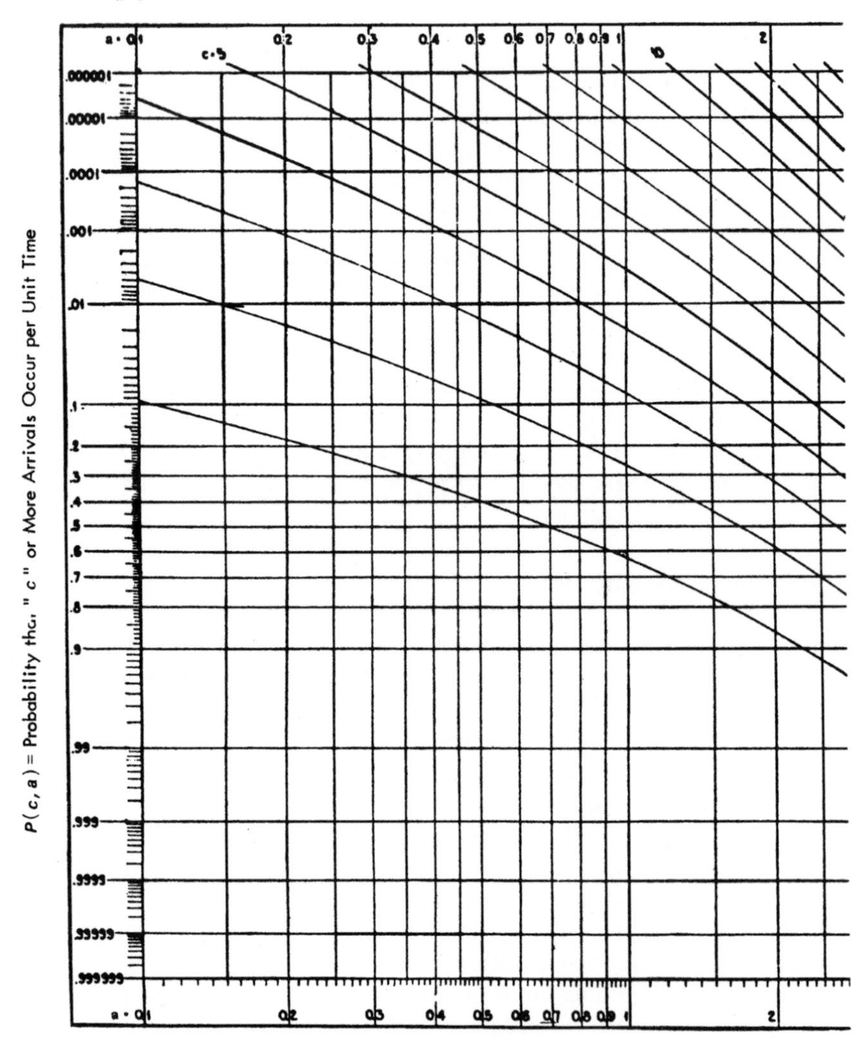

payroll, management must develop time standards, since they are the only reliable yardsticks for evaluating the size of any task.

As for other work, methods analysis should precede work measurement in all expense operations. The flow process chart is the ideal tool for displaying the facts of the present method, allowing that method to be critically reviewed. Using the primary approaches to operation analysis, the analyst then considers such factors as the

FIGURE 15–1 *(concluded)*

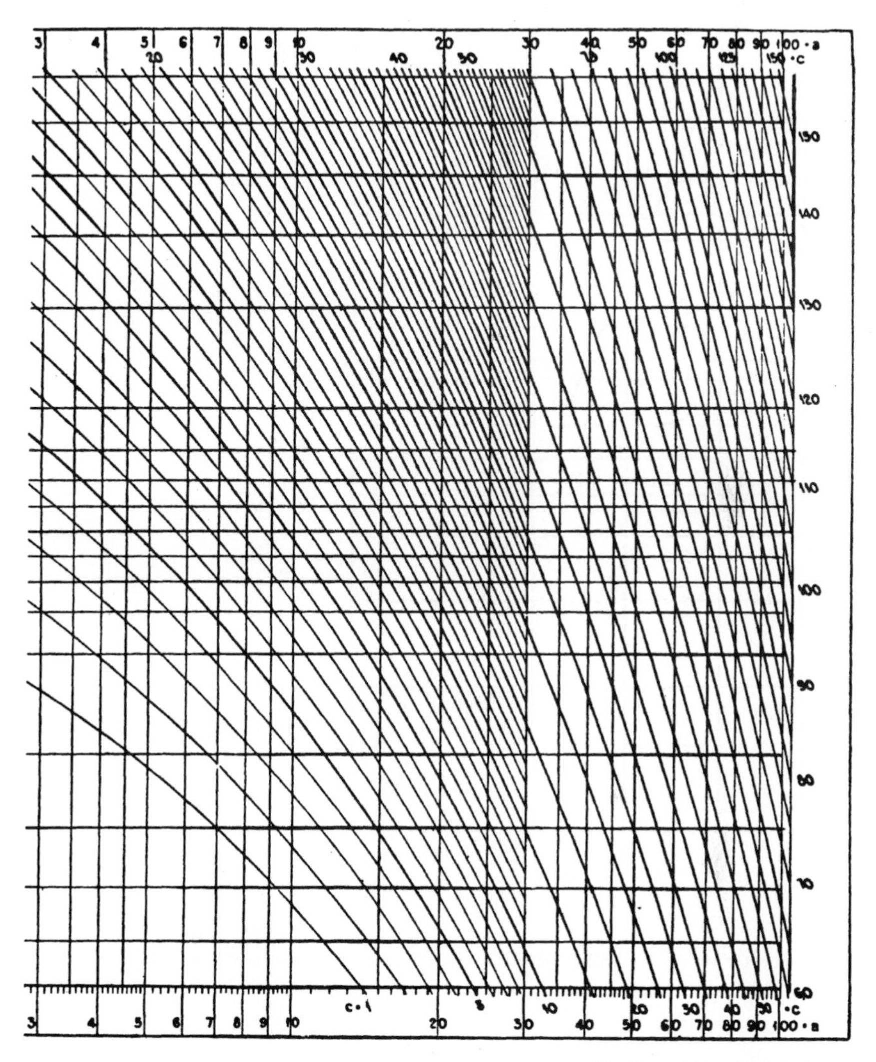

The Port of New York Authority

purpose of the operation, design of forms, office layout, elimination of delays resulting from poor planning and scheduling, and the adequacy of existing equipment.

After the completion of a thorough methods program, standards development can commence. Many office jobs are repetitive; consequently, it is not particularly difficult to set fair standards. Word processing centers, billing groups, file clerks, and duplicating machine operators are representative groups that readily lend

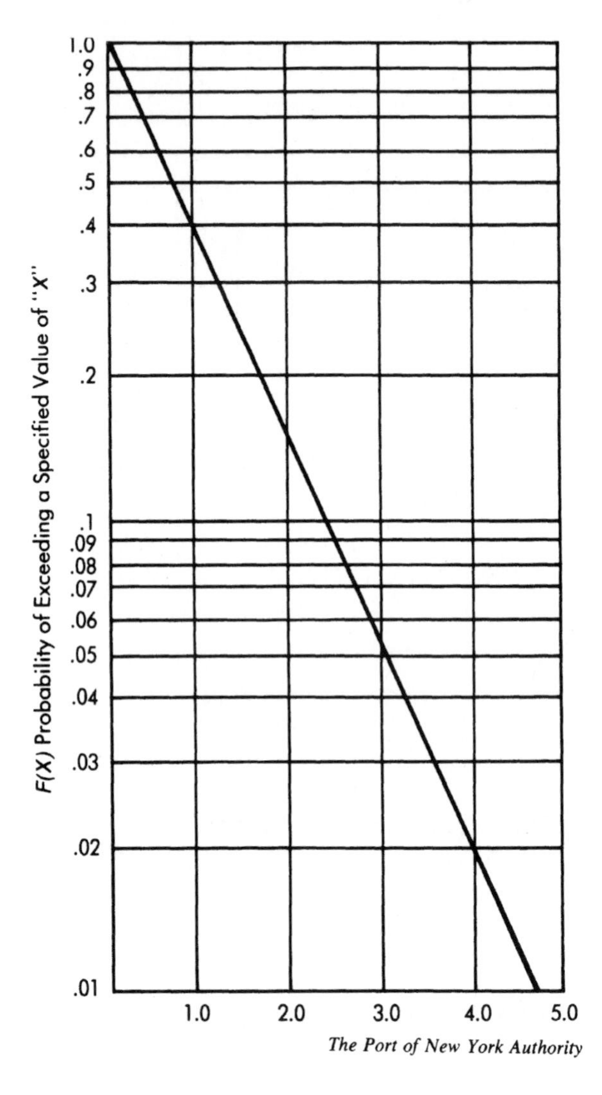

FIGURE 15–2
The exponential distribution.

themselves to work measurement by stopwatch, standard data, and basic motion data techniques.

In studying office work, analysts should carefully identify element endpoints, so that standard data may be established for pricing future work. For example, in typing production orders, the following elements of work normally occur for each page of each order typed:

1. Pick up production order from pile and position in typewriter.
2. Pick up sheet to be copied from, and place in Copy Right.
3. Read product order instructions.
4. Type heading on order:
   a. Date.
   b. Number of pieces.

EXAMPLE 15–1

Use of Queuing Theory to Establish a Standard Time for Inspection.

An analyst wishes to determine a standard for inspecting the hardness of large motor armatures. The time is composed of two distinct quantities: the time the inspector takes to make his Rockwell observations, and the time the operator must wait until the next armature shaft is made available for inspection. The following assumptions apply: (1) single server; (2) Poisson arrivals; (3) arbitrary servicing time; and (4) first-come, first-served discipline. This is a situation that fits into the first of the five categories, and the following equations apply:

$$(a)\ P > 0 = u = \frac{ah}{s}$$

$$(b)\ w = \left[\frac{uh}{2(1-u)}\right]\left[1 + \left(\frac{\sigma}{h}\right)^2\right]$$

$$(c)\ m = \frac{w}{P > 0} = \frac{w}{u}$$

where:

a = Average number of arrivals per unit of time.
h = Mean servicing time.
w = Mean waiting time of all arrivals.
m = Mean waiting time of delayed arrivals.
n = Number of arrivals present (both waiting and being served) at any given time.
s = Number of servers.
u = Servers' occupancy ratio $= \frac{ah}{s}$.
σ = Standard deviation of the servicing time.
$P(n)$ = Probability of n arrivals being present at any random time.
$P(\geqq n)$ = Probability of at least n arrivals being present at any random time.
t = Unit of time.
$P > t/h$ = Probability of a delay greater than t/h multiples of the mean holding time.
$P(d > 0)$ = Probability of any delay (delay greater than 0).
L = Mean number of waiting individuals among all individuals.

A stopwatch time study establishes a normal time of 4.58 minutes per piece for the actual hardness testing. The standard deviation of the service time is 0.82 minutes, and 75 tests are made in every 8-hour workday. From these data, we have:

(continued)

EXAMPLE 15–1 ***(concluded)***

$$s = 1$$

$$a = \frac{75}{480} = 0.156$$

$$h = 4.58$$

$$\sigma = 0.82$$

$$u = (0.156)(4.58) = 0.714$$

$$w = \left[\frac{(0.714)(4.58)}{2(1 - 0.714)}\right]\left[1 + \left(\frac{0.82}{4.58}\right)^2\right]$$

$$= 5.95 \text{ minutes mean waiting time of an arrival}$$

Thus, the analyst is able to determine a total time of 4.58 + 5.95 = 10.53 minutes per shaft.

c. Material.
d. Department.

Once analysts have developed standard data for most of the common elements used in an office, they can calculate time standards rapidly and economically. Of course, many clerical positions comprise a series of diversified activities that do not readily lend themselves to measurement. Such work is not made up of a series of standard cycles that continually repeat themselves; consequently, the work is more difficult to measure than direct labor operations. Because of this characteristic of some office routines, it is necessary to take many time studies, each of which may only be one cycle in duration. Then, by calculating all the studies taken, analysts can develop a standard for typical or average conditions. For example, they can calculate a time standard based on a page of copy typing. Granted, some pages of technical typing require symbols, radicals, fractions, formulas, and other special characters or spacing and therefore take considerably longer than routine pages. Yet, if the technical typing is not representative of average conditions, it does not result in an unfair influence on the operator's performance over a period of time; simple typing and shorter than standard pages balance out the extra time needed to type complex documents.

It is usually not practical to establish standards on office positions that require creative thinking. Such jobs as tool or product designing should be carefully considered before trying to establish time standards for those jobs. If such standards are established, they should be used for scheduling, control, or labor budgeting, not for incentive wage payment. Putting pressure on these employees retards creative thinking. The result may be inferior designs that can ultimately be more costly to the business than the amount saved through the greater productivity of the designer.

EXAMPLE 15–2
Use of Queuing Theory to Establish a Standard Time for Tool Room Service.

Toolroom service time can be modeled as a single server with Poisson arrivals with exponential service. The equations that apply are:

$$(a)\ P > 0 = u$$
$$(b)\ P > (t/h) = ue^{(u-1)(t/h)}$$
$$(c)\ P(n) = (1 - u(u)^n$$
$$(d)\ P(\geqq n) = u^n$$
$$(e)\ w = \frac{h(P > 0)}{(1-u)} = \frac{uh}{(1-u)}$$
$$(f)\ m = \frac{w}{(P > 0)} = \frac{h}{(1-u)}$$
$$(g)\ L = \frac{m}{h} = \frac{1}{(1-u)}$$

The arrivals at a tool crib are considered to be Poisson, with an average time of 7 minutes between one arrival and the next. The length of the service time at the crib window is distributed exponentially, with a mean of 2.52 minutes determined through a stopwatch time study. The analyst wishes to determine the probability that a person arriving at the crib will have to wait, and the average length of the queues that form from time to time. Using this information, the analyst can evaluate the practicality of opening a second tool-dispensing window.

$a = 0.14$ average number of arrivals per minute
$h = 2.52$ minutes mean service time
$s = 1$ server
$P > 0 = u$
$u = \frac{ah}{s} = 0.35$ probability of a person arriving at crib having to wait
$L = \frac{1}{(1-u)} = 1.52$ average length of the queue formed

Supervisory Standards

It is possible to establish standards for supervisory work (see Example 15–4). The work sampling technique is an especially good tool for determining equitable supervisor loads and for maintaining a proper balance between supervisors, facilities, clerical employees, and direct labor. Observers could obtain the same information through all-day time studies, but the cost of reliable data would usually be prohibitive. Supervisory standards can be expressed in effective machine running hours or some other benchmark.

EXAMPLE 15-3
Use of Monte Carlo Simulation to Determine the Optimum Number of Operators.

Fifteen machines are presently being serviced by three operators. The labor rate is $12.00 per hour, while the machine rate is $48.00 per hour. An analysis of past records reveals the following probability distributions of work stoppages per hour, and the time required to service a machine:

Work stoppages per hour	Probability
0	0.108
1	0.193
2	0.361
3	0.186
4	0.082
5	0.040
6	0.018
7	0.009
8	0.003
	1.000

Hours to get machine into operation	Probability
0.100	0.111
0.200	0.254
0.300	0.009
0.400	0.007
0.500	0.005
0.600	0.008
0.700	0.105
0.800	0.122
0.900	0.170
1.000	0.131
1.100	0.075
1.200	0.003
	1.000

The time to get a machine into operation results in a bimodal distribution and does not conform to any standard distribution. The analyst assigns random blocks of three-digit numbers (000 to 999), in direct proportion to the probabilities associated with the data for both the arrival and the service rates, to simulate the expected behavior in the screw machine section over a period of time. The analyst takes eight random observations, to simulate the work stoppages occurring during one day of activity (8 working hours) on the floor. These eight random numbers yield the following:

Hour	Random number	Work stoppages
1	221	1
2	193	1
3	167	1
4	784	3
5	032	0
6	932	5
7	787	3
8	236	1
9	153	1
10	587	2
11	573	2

To estimate the time required to get a machine into operation after it has stopped, the analyst selects a different set of random numbers as input for each work stoppage.

Hour	Random number	Hours to get machine into operation
1	341	0.200
2	112	0.200
3	273	0.200
4	106	0.100
5	597	0.800
6	337	0.200
7	871	1.000
8	728	0.900
9	739	0.900
10	799	1.000
11	202	0.200
12	854	1.000
13	599	0.800
14	726	0.900
15	880	1.000

Table 15–1 shows the predicted machine downtime because of insufficient operators, based on the indicated outcomes for 8 hours of operation in the automatic screw machine department.

The simulated model indicates 2.8 hours of machine downtime every day because of the lack of an operator. At a machine rate of \$48 per hour, this amounts to a daily cost of $2.8 \times 48 = \$134.40$. Since the cost of a fourth operator only results in an added daily direct labor cost of $8 \times 12 = \$96$, it appears that three workers are not economically optimum to service the operation.

(continued)

EXAMPLE 15–3 ***(concluded)***

TABLE 15–1
Results of Monte Carlo Simulation of Machine Downtime

Hour	Random number	Work stoppages	Random number	Hours to get machine into operation	Operators available for next work stoppage	Downtime hours because of lack of operators for servicing
1	221	1	341	0.200	2	
2	193	1	112	0.200	2	
3	167	1	273	0.200	2	
4	784	3	106	0.100	2	
			597	0.800	1	
			337	0.200	0	
5	032	0	—	—	3	
6	932	5	871	1.000	2	
			728	0.900	1	
			739	0.900	0	
			799	1.000	0	0.9
			202	0.200	0	0.9
7	787	3	854	1.000	0	
			599	0.800	0	0.9
			726	0.900	0	0.1
8	236	1	880	1.000	1	
9	153	1	495	0.700	2	
10	587	2	128	0.200	2	
			794	1.000	1	
					Total	2.8

Note, however, that eight random numbers is a very small simulation and may lead to incorrect results. With larger numbers, there is a greater chance of converging to an optimum solution. In fact, with 80 random numbers, the present setup of three workers to service the 15 automatic screw machines does appear to be the optimum solution, as only 19.6 hours (instead of the extrapolated 28) are lost over the course of 80 hours of operation.

STANDARD INDIRECT AND EXPENSE LABOR DATA

Fundamentals

Developing standard data to establish standards on indirect and expense labor operations is quite feasible. In view of the diversification of indirect labor operations, standard data are probably more appropriate for office, maintenance, and other indirect work than for standardized production operations.

EXAMPLE 15–4
The Setting of Supervisory Standards.

A study of a vacuum tube manufacturer revealed that 0.223 supervisory hours were required per machine running hour in a given department (see Figure 15-3). The work sampling study showed that out of 616 observations, the supervisor was working with the grid machines, inspecting grids, doing desk work, supplying material, walking, or engaging in activities classified as miscellaneous allowances for a total of 519 times. Converted to prorated hours, this figure revealed that 518 indirect hours were required for 2,461 machine running hours. Adding a 6 percent personal allowance, analysts computed a standard of 0.223 supervisory hours per machine running hour.

$$\frac{518}{(2{,}461)(1.00 - 0.06)} = 0.223$$

Thus, in a department operating 192 machine running hours a week, a supervisor's efficiency would be:

$$\frac{192 \times 0.223}{40 \text{ (hrs/week)}} = 107 \text{ percent}$$

FIGURE 15–3
Summary of supervisory work.

INDIRECT LABOR STANDARD

JOB- SUPERVISION DEPT.- GRID DATE- 4-16

Cost Center	Number of Observations	Percent of Observations	Prorated Hours	Base Indirect Hours	Effective Machine Running Hours (EMRH)	Direct Labor Hours		Base Indirect Hours per Machine Running Hour (includes 6% personal)
Grid Machines	129	21	130	130	2,461			0.223
Inspect Grids	161	26	160	160				
Desk Work	54	9	56	56				
Supply Material	18	3	18	18				
Misc. Allow.	150	24	148	148				
Walking	7	1	6	6				
Out of Dept.	11	2	12					
Idle	86	14	86					
Total	616	100	616	518	2,461			0.223

As individual time standards are calculated, the analyst should tabularize the elements and their respective allowed times, for future reference. As the inventory of standard data is built up, the cost of developing new time standards declines proportionally. For example, tabularizing standard data for forklift truck operations can be based on six different elements: travel, brake, raise fork, lower fork, tilt fork, and the manual elements required to operate the truck. Once standard data have been accumulated for each of these elements, through the required range, analysts can determine the standard time required for any fork truck operation by summarizing the applicable elements. In a similar manner, standard data can readily be established on janitorial work elements, such as sweep floor, wax and buff floor, dry mop, wet mop, vacuum rugs, or clean, dust, and mop lounge.

The maintenance job of "inspecting seven fire doors in a plant and making minor adjustments" can be estimated from standard data. For example, the Department of the Navy developed the following standard:

Operation	Unit time (hours)	No. units	Total time (hours)
Inspection of fire door, roll-up type (manual chain, crankshaft, or electrically operated) fusible link. Includes minor adjustment	0.170	7	1.190
Walk 100 feet between each door, obstructed walking	0.00009	600	0.054
			1.24

The standard data time of 0.00009 hours per foot of obstructed walking was established from fundamental motion-time data, and the inspection time of 0.170 hours per fire door was established from a stopwatch time study.

Several manufacturers of material handling equipment have taken detailed studies of their product, and they provide standard data applicable to their equipment when it is purchased. This information saves analysts many hours in developing material handling time standards.

Universal Indirect Labor Standards (UILS)

Where maintenance and other indirect operations are numerous and diversified, efforts have been made to reduce the number of different time standards for indirect operations, through universal indirect labor standards. The principle behind universal standards is the assignment of the major proportion of indirect operations (perhaps as much as 90 percent) to appropriate groups. Each group has its own standard, which is the average time for all indirect operations assigned to the group. For example, group A may include the following indirect operations: replacing defective union, repairing door (replace two hinges), replacing limit switch, and replacing two sections (14 feet of 1-inch pipe). The standard time for any indirect operation performed in group A may be 48 minutes. This time represents the mean (x) of all jobs within the group, and the dispersion of the jobs within the group for $\pm 2\sigma$ is some predetermined percentage of $\bar{x}$ (perhaps ± 10 percent).

The three principal steps in introducing a universal indirect labor system, expressed as *slotting*, are:

1. Determining the number of standards (groups or slots) to do a satisfactory job. (Twenty slots should be used when the range is up to 40 hours.)
2. Determining the numerical standard representative of each group of operations contained in each slot.
3. Assigning the standard to the appropriate slot of indirect labor work as it occurs.

The initial step is to determine good benchmark standards, based on measurements of an adequate sample of the indirect labor for which the UILS system is being developed. This is the most time-consuming and costly step in the process. The analyst must establish a relatively large number of standards (200 or more) that are representative of the entire population of indirect work. Competent analysts can develop these measured benchmark standards by using proven industrial engineering tools, including stopwatch time study, standard data, formulas, fundamental motion data, and work sampling.

Once established, the benchmark standards are arranged in numerical sequence. Thus, assuming 200 benchmark standards, the shortest would be listed first, the next shortest second, and so on, ending with the longest. If there are 20 slots, and if a uniform distribution is used, the time standard for the first slot (UILS One) is computed by calculating the mean of the first 10 benchmark standards. Similarly, the value of UILS Two is calculated by taking the mean of benchmark standards 11 through 20. The last UILS (20) would be equal to the average of the benchmark standards 191 through 200. Engineers have used this procedure extensively in the development of UILS.

More reliable UILS result from using the normal, rather than the uniform, distribution. For 20 slots, the 200 standards would not be assigned as 10 per slot. Instead, the standard normal variable would be divided into 20 equal intervals (truncation of the two tails allows this). For example, the standard normal variable may have a truncated range of

$$-3.0 \leq z \leq +3.0$$

which accounts for 99.87 percent of the area under the curve. The range of each interval would be 0.3. The benckmark standards used in the compilation of the mean of each of the 20 slots (intervals) would equal:

$$P(z \epsilon \text{ interval})(200)/0.9987$$

Slot numbers 1 and 20 (because of symmetry) would have

$$\frac{\text{P}(-3.0 \geqq z \leq -2.7)(200)}{0.9987} = \frac{\text{P}(2.7 \leq z \leq 3.0)(200)}{0.9987}$$

$$= \frac{(0.9987 - 0.9965)(200)}{0.9987} = 0.4406 \text{ standards}$$

and slot numbers 10 and 11 would have

$$\frac{P(-0.3 \geqq z\ 0.0)(200)}{0.9987} = \frac{P(0.0 \leq z\ 0.3)(200)}{0.9987}$$
$$= \frac{(0.6179 - 0.5000)(200)}{0.9987} = 23.61 \text{ standards}$$

Rounding off the fractions is dictated by the fact that all 200 jobs must be assigned to a slot. The universal standard time for each slot is the average of the benchmark standards assigned to the slot.

When studying a new part to be made, analysts can fit jobs to categories where similar jobs have been studied and standards established. Although the normal distribution is superior to the uniform distribution, a skewed distribution may outperform the normal in view of the plotting of the data (see Figure 15–4).

The gamma distribution is a positively-skewed distribution with a probability density function given by:

$$f(x) = \begin{cases} \dfrac{1}{\beta^2\, \Gamma(\alpha)} x^{\alpha-1}\, e^{-x\beta} & \text{for } x, \alpha, \beta > 0 \\ 0 \end{cases}$$

where $\Gamma(\alpha)$ is a value of the gamma function given by

$$\Gamma(\alpha) = \int_0^\infty x^{\alpha-1}\, e^{-x}\, dx$$
$$= (\alpha - 1)\, \Gamma(\alpha - 1)$$

FIGURE 15–4
Distribution of 20 Universal Indirect Labor Standards ranging in time from 0.625 hours to 24.38 hours.

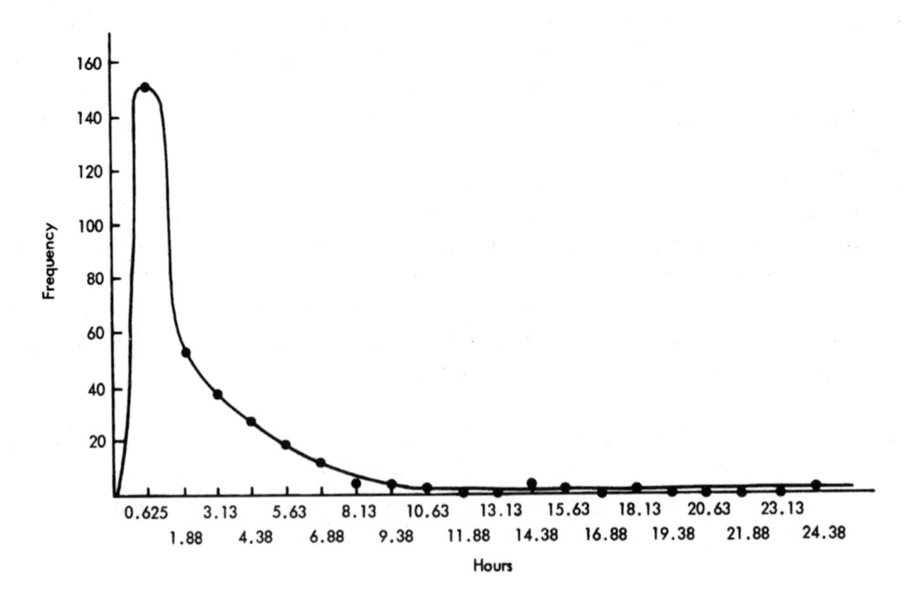

The skewness of the gamma distribution decreases as α increases for a fixed β.

The mean and variance of this distribution equal:

$$\mu = \alpha\beta$$
$$\sigma^2 = \alpha\beta^2$$

For a given set of data estimates of α and β, $\hat{\alpha}$ and $\hat{\beta}$ are determined by first obtaining the mean and variance of the data, then calculating the estimates:

$$\hat{\alpha} = \mu^2/\sigma^2$$
$$\hat{\beta} = \mu/\hat{\alpha}$$

To determine the relative accuracy of the uniform, normal, and gamma techniques, analysts used 270 maintenance standards developed by the Department of the Navy as benchmark standards. They divided these benchmark standards into 20 slots, using the three approaches. The gamma distribution and expected frequencies for the 20 slots are shown in Table 15–2.

To compare the results of the uniform, normal, and gamma techniques, a simulation was done. For each of 25 weeks, analysts selected jobs at random until the sum of the actual standard times exceeded or equaled 40 hours. They then determined the universal maintenance standard for each job and calculated the weekly sum. They assumed that each job was properly slotted.

TABLE 15–2

The Gamma Distribution Probabilities and Expected Frequencies for the 20 Slots

Slot (hours)	Cumulative probabilities	Probabilities	Actual frequency	Expected frequency
0.0 — 1.25	0.5404	0.5405	150	145.94
1.26— 2.50	0.7146	0.1741	50	47.00
2.51— 3.75	0.8108	0.0962	29	25.97
3.76— 5.00	0.8698	0.0590	13	15.93
5.01— 6.25	0.9077	0.0379	11	10.23
6.26— 7.50	0.9327	0.0250	6	6.75
7.51— 8.75	0.9495	0.0168	2	4.54
8.76—10.00	0.9610	0.0115	2	3.11
10.01—11.25	0.9688	0.0078	2	2.11
11.26—12.50	0.9743	0.0055	0	1.49
12.51—13.75	0.9780	0.0037	0	1.00
13.76—15.00	0.9807	0.0027	2	0.73
15.01—16.25	0.9825	0.0018	1	0.49
16.26—17.50	0.9838	0.0013	0	0.35
17.51—18.75	0.9847	0.0009	1	0.24
18.76—20.00	0.9854	0.0007	0	0.19
20.01—21.25	0.9859	0.0005	0	0.13
21.26—22.50	0.9862	0.0003	0	0.08
22.51—23.75	0.9864	0.0002	0	0.05
23.76—25.00	0.9866	0.0002	1	0.05
25.01—x	1.000	0.0134	0	3.62
Total			270	270.00

For each week, an error was calculated as:

$$\left| \frac{\text{Actual standard time} - \text{Universal standard time}}{\text{Actual standard time}} \right| \times 100 \text{ percent}$$

The results of the simulations for the uniform, normal, and gamma distributions are given in Table 15–3. This study confirmed that the gamma distribution offers some improvement over the normal, and that the normal gives better results than the uniform.

Increasing the pay period from one week (40 hours) to two weeks (80 hours) would markedly reduce the cumulative error per pay period. The magnitude of the error would also decrease as the number of groups (slots) increases.

UILS offers an opportunity to introduce standards for a majority of indirect operations at a moderate cost, and it minimizes the cost of maintaining the indirect standards system.

TABLE 15–3
Results of 25-Week Simulation

	Absolute percent error		
Week number	**Uniform**	**Normal**	**Gamma**
1	5.97	7.18	2.57
2	16.01	6.93	0.27
3	8.49	6.42	5.74
4	10.94	4.03	3.32
5	25.78	1.67	1.85
6	2.61	0.47	5.25
7	4.79	6.08	3.90
8	0.88	3.37	3.21
9	4.51	5.34	5.36
10	0.05	6.45	3.07
11	30.78	0.32	1.79
12	21.93	1.75	0.64
13	8.23	4.24	1.62
14	6.67	7.55	5.59
15	2.37	2.37	1.53
16	0.06	0.87	1.24
17	12.53	2.88	1.79
18	3.73	5.21	5.86
19	6.85	1.52	5.35
20	11.50	2.29	0.58
21	20.18	2.48	0.05
22	6.44	8.31	0.92
23	3.46	6.72	5.11
24	2.96	0.45	3.09
25	11.74	1.01	1.13
Mean	9.18	3.84	2.83
Variance	151.78	21.62	11.34
Std. Dev.	12.32	4.65	3.37

PROFESSIONAL PERFORMANCE STANDARDS

The cost of professionals is a sizable proportion of the total expense budget. In most manufacturing and business operations, the professional salaries of employees in engineering, accounting, purchasing, sales, and general management represent a significant proportion of total cost. If the productivity of these employees can be improved by as little as a few percent, the overall impact on the firm's business is consequential. Establishing standards for professional employees and utilizing these standards as achievable goals inevitably enhances productivity.

The difficulties in developing professional standards are: first, determining what to count, and second, determining the method for counting these outputs. In determining what to count, the analyst can begin by stating the objectives of the professional employees' positions. For example, buyers in the purchasing department might have the following objective: "to procure quality components and raw materials at the lowest price, in time to meet company production and delivery schedules." To be effective, a count of buyers' outputs must consider five things: (1) proportion of deliveries made on schedule; (2) proportion of deliveries that meet or exceed quality requirements; (3) proportion of shipments that represent the lowest available price; (4) number of orders placed during some interval of time, such as one month; and (5) total dollar value of the purchases made during a period of time.

The next problem is establishing achievable goals. In such instances, using historical records, supplemented with work sampling analyses, to determine how time is utilized can serve as the basis for the development of professional standards.

Returning to the example of establishing buyers' standards, it would not be difficult to identify the purchases made by the various buyers, and to review what proportion of these orders was delivered on schedule over a six-month period. The historical data study may reveal something analogous to the following:

Buyer	**Proportion of orders delivered on or before schedule (precent)**
A	70
B	82
C	75
D	50
E	80

Based on this record, skilled buyers should be able to procure at least 72 percent of their purchases on schedule (the mean of their performance).

Similarly, a quality review of the purchases made by the five buyers may disclose the following:

Buyer	**Proportion of orders delivered with less than 5 percent rejects (percent)**
A	85
B	90
C	80
D	95
E	80

Here, the quality standard could be that 86 percent of orders received have less than 5 percent rejects (the mean of the past performance of the five buyers).

For the proportion of procurements purchased at the lowest available price, historical records can once again provide a comparison of the performance of the five buyers. Assume the following performance record applied:

Buyer	Proportion of orders procured at the lowest available price (percent)
A	45
B	50
C	60
D	47
E	40

The average of these values, 48.4 percent of the orders placed at the lowest available price, could be used as the normal performance.

Historical records might also indicate that, on average, a buyer placed 120 orders per month, with a total monetary worth of $120,840.

These five criteria could then be used to develop an overall performance standard: delivery, quality, price, number of orders, and order value. For example, one method would add the means of the first three criteria (0.72 + 0.86 + 0.484) plus 0.002 times the mean of the order placed plus 0.000001 times the average monetary worth of purchases. The buyers standard in this fictitious operation would then be:

$$0.72 + 0.86 + 0.484 + 0.24 + 0.12 = 2.424$$

Another example may help clarify how performance standards may be developed for managerial personnel. Consider the position of director of personnel administration. An analysis may suggest four specific objectives of this position:

1. Establish a methodology for identifying both the quantity and quality of the company's human resources.
2. Establish a procedure for attracting, employing, and retaining the kinds and numbers of employees required for the successful operation of the company.
3. Establish policies, programs, and practices that facilitate the achievement of departmental objectives and maintain employee morale.
4. Administer and maintain the company's benefit program.

Now that the objectives have been stated, it is relatively easy to develop a performance standard in terms of time. For example, the standard for objective 1 may be: "Within the next three months, train staff representatives to conduct an audit of the company's personnel, to determine projected needs from the standpoint of both numbers and type."

The performance standard for objective 2 might be: "Within the next 12 months, employ (a) 2 Ph.D. chemists; (b) 7 M.S. degrees in industrial and/or mechanical engineering; and (c) 35 B.S. degrees, with a distribution of 10 in business, 20 in engineering, and 5 in liberal arts. Employ (based on anticipated turnover and expansion) 75 hourly employees. Investigate the turnover rate of professional employees in the past year, and prepare a report showing how turnover may be reduced."

For objective 3, the performance standard might be as follows: "Within the next three months, update the current management handbook, bringing the salary administration program up to date. Within the next six months, develop and distribute a booklet for all hourly employees, describing the new grievance procedure established in the new labor contract. The booklet should not only explain the importance of reducing the number of grievances, but also how this can be done."

The performance standard for objective 4 could be: "Within the next 12 months, review the company's entire fringe benefit program and compare our benefits to those of similar-sized companies in this area. Make any appropriate recommendations to management."

These objectives identify performance standards for finite time periods. The standards may change as time passes, since each standard is result based. The standards established for succeeding periods may include different work assignments to meet the stated objectives.

In the establishment of professional performance standards, professionals should assist in identifying the objectives of each position, gathering the historical performance records, and developing the standards. Performance standards developed without the complete involvement of the professionals are seldom realistic.

When gathering historical data to facilitate the development of professional standards, the analyst should take a work sampling study during the period that serves as the basis of the historical record data. This work sampling study can reveal how much working time was spent on the various necessary work routines, or on work assignments that could better be handled by clerical or semiprofessional employees. It could also reveal how much time was literally wasted. After reviewing the work sampling study, the analyst can performance rate the average data gathered over the historical period, to obtain a standard that is more representative of normal professional experience.

In the development of professional standards, the following guidelines should be observed:

1. Each manager should be involved with setting the standards for his or her professional subordinates. Professional standards should be developed jointly by employees and their supervisors.
2. Standards should be result based and should be worded to include measurement references.
3. Standards must be realistically attainable by at least one-half of the group concerned.
4. Standards should be periodically audited and revised if necessary.
5. It is helpful to work sample managers to assure that they have adequate clerical and administrative aid support and are using their time judiciously.

INDIRECT WORK STANDARDS ADVANTAGES

Standards on indirect work offer distinct advantages to both the employer and the employee. Some of these advantages are:

1. The installation of standards leads to many operating improvements.
2. The mere fact that standards are established results in better performance.
3. Indirect labor costs are related to the workload, regardless of fluctuations in the overall workload.
4. Labor loads can be budgeted.
5. The efficiency of various indirect labor departments can be determined.
6. The costs of such items as specific repairs, reports, and documents are allocated. This frequently results in the elimination of needless reports and procedures.
7. System improvements can be evaluated prior to installation. It is therefore possible to avoid costly mistakes by choosing the right procedure.
8. It is possible to install incentive wage payment plans on indirect work, thus allowing employees to increase their earnings.
9. Accurate planning and scheduling of all indirect labor work gets the jobs done on time.
10. Employees require less supervision, as a program of work standards tends to enforce itself. Employees who know what is required do not arbitrarily waste time.

SUMMARY

It is more difficult to study and determine representative standard times for nonrepetitive tasks, which are characteristic of most indirect labor operations, than for repetitive tasks. Since indirect labor operations are difficult to standardize and study, they are only infrequently subjected to methods analysis. Consequently, this area usually offers a greater potential for reducing costs and increasing profits through methods and time study than does any other.

After good methods have been introduced and operator training has taken place, it is both readily possible and practical to establish standards on indirect labor operations. The usual procedure is to take a sufficiently large sample of stopwatch time studies to ensure the representation of average conditions, and then to tabularize the allowed elemental times in the form of standard data. Fundamental motion data also have wide application in establishing standards on indirect work. This is especially true of those systems that utilize larger blocks of fundamental motions, such as Work-Factor, MTM-2, and MOST.

Once standards are developed, they should be stored in a computer as standard data. Computers make calculations faster than electronic hand-held calculators. They establish allowed operation times consistently, accurately, and economically.

Companies representing the fundamental motion data systems, such as the Work-Factor Company, H. B. Maynard and Company, Inc., and Standards International, have developed software to accompany their systems for standards development using the computer. In such cases, analysts only need to dictate workplace and method data, which may be transcribed by a typist for input into the computer. The results may be displayed almost immediately on the monitor. Software programs also produce hard-copy printouts of the developed standard, which can include both allowances and process time, along with a sketch of the workplace, which serves as a record of the method used to develop the standard.

Analysts can accurately estimate indirect elements involving waiting time by using waiting line or queuing theory. Analysts must understand elementary waiting line theory to establish the mathematical model that fits the parameters of the problem. Where the problem does not fit established waiting line equations, analysts can use Monte Carlo simulation as a tool for determining the extent of the waiting line problem in the work area.

There are no automatic means for determining a standard job time for a maintenance or other type of indirect or expense job. Analysts must review each job request and determine both the labor and material requirements. They usually go through the following steps:

1. Examine the work request in detail and consult with the initiator of the work request, or even visit the jobsite, to determine the exact requirements of the job.
2. Prepare a material estimate for the job.
3. Study the job from the standpoint of the work elements performed.
4. Select the appropriate direct work or task times and the applicable tooling, material, and planning times from standard data tables or universal time standards.
5. Assign an estimated time from similar work shown on the master table of universal standard times, or on a spread sheet if no specifically applicable times exist.
6. Add standard times to cover the transportation times for the work order being analyzed, as well as the necessary allowance to cover unavoidable delays, personal delays, and fatigue.

To establish standards on indirect and expense work, Table 15-4 can serve as a guide for choosing the appropriate method.

QUESTIONS

1. Differentiate between indirect labor and expense labor.

2. Explain queuing theory.

3. Which four divisions constitute indirect and expense work?

4. How are standards established on the "unnecessary and delays" portion of indirect and expense work?

5. Why has there been a marked increase in indirect workers?

6. Why do more unavoidable delays occur in maintenance operations than on production work?

7. What is meant by crew balance? By interference time?

TABLE 15–4
Guide for Establishing Indirect Labor and Expense Standard

Indirect and expense type work	Recommended method of establishing standards
Routine maintenance. Work standards 0.5 to 3 hrs	Standard data, MTM-2, MTM-3, Work-Factor, MOST, Macro Motion Analyses
Complicated maintenance, standards 3 hrs to 40 hrs	Slotting based on universal indirect labor standards
Shipping and receiving	Standard data MTM-2, MTM-3, Work-Factor, MOST, Macro Motion analyses
Toolroom	Slotting based on universal indirect labor standards
Inspection	Standard data, MTM-2, Work-Factor, MOST, Macro Motion Analyses
Tool design	Slotting based on universal indirect labor standards
Buying	Standards basedon historical records, analysis, and work sampling
Accounting	Standards based on historical records, analysis, and work sampling
Plant engineering	Standards based on historical records, analysis, and work sampling
Clerical	Standards based on standard data, Work-Factor, MTM-2, Work-Factor, MOST, Micro and Macro Motion Analyses
Janitorial	Standard data, slotting based on universal indirect labor standards
General management	Standards based on historical records, analysis, and work sampling

8. Explain how time standards would be established on janitorial operations.
9. Which office operations are readily time studied?
10. Why are standard data especially applicable to indirect labor operations?
11. Summarize the advantages of standards established on indirect work.
12. Why is work sampling the best technique for establishing supervisory standards?
13. Explain the application of slotting for indirect or expense labor.
14. Why will a universal standards system involving as few as 20 benchmark standards work in a large maintenance department where thousands of different jobs are performed each year?

PROBLEMS

1. Work measurement procedures establish an average time of 6.24 minutes per piece on the inspection of a complex forging. The standard deviation of the inspection time is 0.71 minutes. Usually, 60 forgings are delivered to the inspection station on the line every 8-hour turn. One operator performs this inspection. Assuming that the castings arrive in Poisson fashion and that the service time is exponential, what would be the mean waiting time of a casting at the inspection station? What would be the average length of the casting queue?

2. In the tool and die room of the Dorben Company, the work measurement analyst wishes to determine a standard for the jig boring of holes on a variety of molds. The standard will be used to estimate mold costs only. It will be based on operator wait time for molds coming from a surface grinding section and on operator machining time. The wait time is based on: a single server, Poisson arrivals, exponential service time, and first-come, first-served discipline. A study revealed the average time between arrivals was 58 minutes. The average jig-boring time was 46 minutes. What is the possibility of a delay of a mold at the jig borer? What is the average number of molds in back of the jig borer?

3. What would be the expected waiting time per shipment if a stopwatch time study established that the normal time to prepare a shipment was 15.6 minutes? Twenty-one shipments are made every shift (8 hours). The standard deviation of the service time has been estimated as 1.75 minutes. It is assumed that the arrivals are Poisson distributed and that the servicing time is arbitrary.

4. Using Monte Carlo methods, what would be the expected downtime hours due to the lack of an operator for servicing if four operators were assigned to the work situation described in Example 15–3?

5. The arrivals at the company cafeteria are Poisson, with an average time of 1.75 minutes between arrivals during the lunch period. The average time for a customer to obtain lunch is 2.81 minutes, and this service time is distributed exponentially. What is the probability that a person arriving at the cafeteria will have to wait? How long?

REFERENCES

Crossman, Richard M., and Harold W. Nance. *Master Standard Data: The Economic Approach to Work Measurement*. Rev. ed. New York: McGraw-Hill, 1972.

Knott, Kenneth. "Indirect Operations: Measurement and Control." In *Handbook of Industrial Engineering*. 2nd ed. Ed. Gavriel Salvendy. New York: John Wiley & Sons, 1992.

Lewis, Bernard T. *Developing Maintenance Time Standards*. Boston, MA: Cahners, 1967.

Nance, Harold W., and Robert E. Nolan. *Office Work Measurement*. New York: McGraw-Hill, 1971.

Newbrough, E. T. *Effective Maintenance Management*. New York: McGraw-Hill, 1967.

Pappas, Frank G., and Robert A. Dimberg. *Practical Work Standards*. New York: McGraw-Hill, 1962.

Raghavachari, M. "Queuing Theory." In *Handbook of Industrial Engineering.* 2nd ed. Ed. Gavriel Salvendy. New York: John Wiley & Sons, 1992

SELECTED VIDEOTAPES

Manufacturing Insights Videotape Series. Simulation 1/29 VHS VT 253-1368 and 3/49 U-Matic VT 253U-1368. Dearborn, MI: Society of Manufacturing Engineers, 1987.

CHAPTER 16

Standards Follow-Up and Uses

KEY POINTS:

- Follow-up of methods and standards is necessary for both fairness to workers and profitability.
- Use the appropriate approach for setting and revising standards.
- Use standards to:
  - Set wage incentives.
  - Compare methods.
  - Determine plant capacity.
  - Determine labor costs and budgets.
  - Enforce quality standards.
  - Improve customer service.

Follow-up is the eighth and last of the systematic steps in installing a methods improvement program. Although follow-up is as important as any of the other steps, it is frequently the most neglected. Analysts have a natural tendency to consider the methods improvement program complete after developing the time standards. However, a methods installation and resultant standard should never be considered complete.

Follow-up is necessary to assure that the proposed method is being followed, that the established standards are being realized, and that the new method is being supported by labor, supervision, the union, and management. Follow-up usually results in additional benefits accruing from new ideas and new approaches, which eventually stimulate the desire to improve a methods engineering program for an existing design or process. The procedure is to repeat the methods improvement cycle shortly after it is completed, so that each process and each design are continually being scrutinized for possible further improvement. This is a necessary component of any sound continuous improvement program.

Without follow-up, it is easy for the proposed methods to revert to the original procedures. We have made innumerable methods studies where follow-up revealed that the method under study was slowly reverting to, or had reverted to, the original method. Humans are creatures of habit, and a workforce must develop the habit of the proposed method, if it is to be preserved. Continual follow-up is the only way to assure that the new method is maintained long enough for all those associated with its details to become completely familiar with its routines.

STANDARD TIMES FOLLOW-UP AND MAINTENANCE

Both labor and management have emphasized the necessity of establishing fair standard times. Once fair standards are introduced, it is equally important that they be maintained. Although it is the normal function of production supervision to spot check and monitor standards, the extensiveness of this job seldom permits adequate time for completely effective follow-up. Consequently, the methods and standards department should schedule regular follow-ups.

The initial follow-up or audit for production jobs should take place approximately one month after the development of time standards. A second audit should be made two months later, and a third three to nine months after that. The frequency of the audits should be based on the expected hours of application per year. As an example, one large company uses the data shown in Table 16–1 to determine the frequency of the audit of methods and standards.

On each follow-up, analysts should review the original method report as well as the development of the standard, to be certain that all aspects of the new method are being followed. At times, they may find that portions of the new method are being neglected and that workers have reverted to the old ways. Workers sometimes conceal methods changes for which they are responsible, so they can increase their earnings or diminish their effort while achieving the same production. Of course, methods changes that increase the time required to perform the task may develop. These changes may be initiated by the supervisor or inspector and, in their opinion, may be of insufficient consequence to adjust the standard.

TABLE 16–1
Frequency of Audits

Hours of application per standard per year	Frequency of audit
0–10	Once per three years
10–50	Once per two years
50–600	Once per year
Over 600	Twice per year

Courtesy Industrial Engineers Division, Procter & Gamble, Co.

FIGURE 16–1
Distribution of performance in plant where standards are loose compared to expected distribution of average performance of 115 percent of normal.

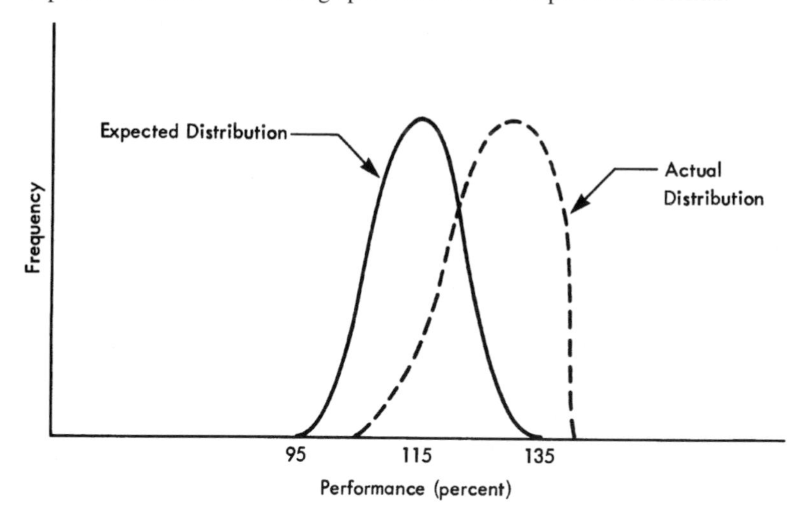

When this happens, the supervisor should be contacted immediately and the analyst should attempt to determine why the unauthorized change has taken place. If no satisfactory reasons can be given for going back to the old method, the analyst should insist that the correct procedure be followed. Tact, coupled with firmness, is essential; analysts need to display both sales ability and technical competence.

Both the method and the performance of the operator should be followed up. The worker's daily efficiency should be checked. The worker's performance should be equal to or greater than standard. Performance should be evaluated using typical learning curves for the class of work. If the operator is not making the progress anticipated, a careful study, including a conference with the operator, should be conducted to determine if any unforeseen difficulties have been encountered.

Usually, workers' performance approximates the normal curve, as described in Chapter 9. However, several common variations from the normal curve are symptomatic of restrictions in output, and they indicate the desirability of conducting an audit. Figure 16–1 illustrates that the standards are loose and that workers are holding back so that they do not earn at a rate above 140 percent, feeling that if they perform beyond this point, the time standards will be adjusted downward.

Figure 16–2 illustrates the output in an environment where the method has not been standardized. Variation in material is another cause of this flat distribution. In both of these instances, an audit can ensure that the best method is being used. Thus, the developed standard reflects the time required for average experienced operators, working with good skill and effort, to perform an operation at a pace that can be maintained for 8 hours, allowing for personal and unavoidable delays and fatigue. A standard does not get out of line if the method that was time studied is maintained by the operators. If the methods study has developed the ideal method, and if this

FIGURE 16–2
Distribution of performance in plant where methods and/or material have not been standardized.

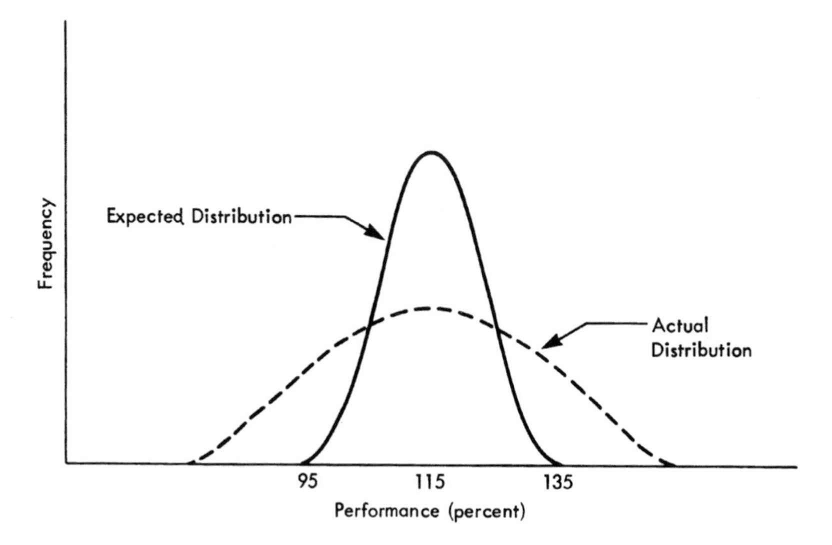

method is standardized and followed by the operators, then there is less need to maintain time standards.

Frequently, however, both favorable and unfavorable methods changes are introduced. If these changes are extensive, they are brought to the attention of management. Tight standards are brought to management's attention by operators. Standards that become very loose are brought to management's attention through the payroll department, where excessive earnings by workers are reported. However, minor accruing methods changes frequently take place unnoticed, and they weaken the entire standards structure.

To maintain standards properly, the time study department should periodically verify that the method being used is the method that was studied when the standard was established. This can readily be done by referring to the original time study, which includes a complete description of the method. The analyst can then observe the operation as performed to determine the correctness of the description, sequences, frequencies, conditions, and standard time allowances. The accuracy of the time standard can be checked by measuring several cycles of the overall time, performance rating the data, and adding an appropriate allowance. A time study analyst should also enlist the cooperation of the supervisor, who is closer to the operators, to verify standards that may be out of line.

If the overall cycle time and the existing time standards vary by more than ±5 percent, a detailed time study should be conducted to identify the cause of the discrepancy. In the majority of cases where the audit reveals a discrepancy of more than ±5 percent, a change in method is the cause. If the operator is using an inferior method, then he or she should immediately be instructed in the ideal method. If the operator has developed a better method than the one that existed when the standard was developed, the new method should become the accepted standard. If the operator has developed the improvement, he or she should be rewarded accordingly.

The analyst should review all factory layouts to ensure that the ideal flow of materials and product is taking place. The industrial engineering department should maintain detailed flow process charts for the major products being produced. If new equipment has been acquired in conjunction with the method, its capability should be audited regularly, to ensure that the anticipated productivity and performance are being realized.

Also, the analyst should audit the job evaluation after the worker has performed the new method for six months. This review should assure that the compensation of all employees associated with the developed method is competitive with equivalent jobs in the area. Absentee rates should also be audited, to obtain an additional measure of operator acceptance. Although auditing methods and standards require time and expense, a thorough and regular follow-up system will assure the success of the program.

STANDARDS USAGE

Standards Review

Time standards are fundamental to the operation of any manufacturing enterprise or business. Time is the one common denominator from which all elements of cost evolve. In fact, everyone uses time standards for practically everything they do or want anyone else to do. Examples in everyday life include: a worker who allows one hour to wash, shave, dress, eat breakfast, and get to work; the student who schedules so many minutes to cover an assignment; and the bus driver who follows a specific schedule of arrivals and departures.

We are particularly interested in time standards used in the effective operation of a manufacturing enterprise or business, and in the results of time study. Such time standards may be determined in one or more of the following ways:

1. By estimate (see Chapter 8).
2. By performance records (see Chapter 7).
3. By stopwatch time study (see Chapter 8).
4. By standard data (see Chapter 11).
5. By time study formulas (see Chapter 12).
6. By predetermined time systems (see Chapter 13).
7. By work sampling studies (see Chapter 14).
8. By queuing theory (see Chapter 15).

Methods 3, 4, 5, 6, and 7 give considerably more reliable results than methods 1, 2 or 8. If standards are used for wage payment, they should be determined as accurately as possible. Consequently, standards determined by estimate or by performance records do not suffice. Of course, standards developed by performance records and estimates are better than no standards at all, and can frequently be used to exercise controls throughout an organization.

All of these methods are applicable under certain conditions, and all have limitations on accuracy and installation cost. For the more reliable methods, the summaries presented in Tables 16–2 and 16–3 may prove helpful in selecting the appropriate approach.

Wage Incentive Plans Basis

Time standards are usually thought of in relation to wage payment (see Chapter 17). While standards have many other uses in the operation of an enterprise, the need for reliable and consistent standards is most pronounced in the wage payment area. Without equitable standards, no incentive plan that compensates in proportion to output can possibly succeed. Without a yardstick, it is not possible to measure individual performance. Standardized methods and times comprise the yardstick for wage incentive applications.

Similarly, any type of supervisory bonus keyed to productivity depends directly on equitable time standards. Since workers receive more and better supervisory attention under a plan where the supervisory bonus is related to output, the majority of supervisory plans consider worker productivity as the principal criterion for the supervisory bonus. Other factors usually are: indirect labor costs, scrap cost, product quality, and methods improvements.

Methods Comparisons

Since time is a common measure for all jobs, time standards are a basis for comparing various methods of doing the same work. For example, suppose an operator thought it might be advantageous to install broaching on a close-tolerance inside diameter, rather than ream the part to size, as is currently being done. To make a sound decision on the practicality of the change, analysts would develop time standards for each procedure and then compare the results.

Effective Space Utilization

Time is the basis for determining how much of each kind of equipment is needed. Management can only make the best possible utilization of space by knowing the exact requirements for facilities. If a company requires 10 milling machines, 20 drill presses, 30 turret lathes, and 6 grinders in one machining department, the manager can plan for the best layout of this equipment. Without time standards, the company could over-provide for one facility and under-provide for another, inefficiently utilizing the space available.

As another example, to determine the size of storage and inventory areas, managers consider the length of time a part will be in storage, as well as the demand for

TABLE 16–2

Comparison of Different Methods for Establishing a Time Standard

Advantages	Disadvantages
Stopwatch Time Study	
1. Only method directly measuring operator times	1. Requires rating worker performance
2. Allows detailed observation of complete cycle and method	2. Does not force a detailed record of the method, motions, tools, etc. used
3. May cover relatively infrequent elements	3. May not properly evaluate noncyclic elements
4. Provides quick and accurate values for machine-controlled elements	4. Bases standard on the bias of one analyst studying one worker using one method
5. Is relatively simple to learn and explain	5. Requires on-going production work
Predetermined Time Systems	
1. Forces a detailed record of the method, motions, tools, etc. used	1. Depends on complete description of method, motions, tools, etc. for accurate standard
2. Encourages work simplification	2. Requires more training of analysts
3. Eliminates performance rating	3. Is more difficult to explain to workers
4. Permits establishment of standard before actual production	4. Requires more time to establish standards
5. Permits easy adjustment of standard due to methods changes	5. Requires other data sources for process- or machine-controlled elements
6. Establishes consistent standards	
Standard Data, Formulas and Queuing Methods	
1. Eliminates performance rating	1. May require more training of analysts
2. Establishes consistent standards	2. Are more difficult to explain to workers
3. Permits establishment of standard before actual production	3. May not accommodate small variations in the method
4. Permits easy adjustment of standard due to methods changes	4. May be inaccurate if extended beyond the scope of the data used in their development
Work Sampling	
1. Reduces tension caused by constant observation of the worker	1. Impairs accuracy of performance rating
2. Establishes an average standard over varying conditions	2. Requires relatively large number of random observations for accuracy
3. Permits simultaneous development of standards for a variety of operations	3. Requires accurate records of working hours and units produced
4. Best for analysis of machine utilization, work activities and delays	4. Assumes that worker uses a standard method

TABLE 16–3
Guide for Selecting Appropriate Time Standard Method

Best method:	For work:
Stopwatch time study	1. With repetitive cycles of any duration 2. With wide variety of dissimilar elements 3. With process- or machine-controlled elements
Predetermined time systems	1. With operator-controlled elements 2. With repetitive cycles of short to medium duration 3. Not yet in production 4. With controversy over rating and consistency of standards
Standard data, formulas and queuing methods	1. With similar elements of any duration 2. With controversy over rating and consistency of standards
Work sampling	1. With large variations from cycle to cycle 2. With controversy over stopwatch use 3. With controversy over constant observation of worker 4. Where machine utilization, activity levels and delay allowances are needed

the part. Here again, time standards are the basis for determining the size of such areas.

Plant Capacity Determination

Through time standards, machine, department, and plant capacity can all be determined. Once the available facility hours and the time required to produce a unit of product are known, it is a matter of simple arithmetic to estimate product potential. For example, if the bottleneck operation requires 15 minutes per piece, and if 10 facilities for this operation exist, then based on a 40-hour per week operation, the plant capacity for this product would be:

$$\frac{40 \text{ hours} \times 10}{0.25 \text{ hours}} = 1{,}600 \text{ pieces per week}$$

Figure 16–3 illustrates a weekly graphic analysis of the direct labor requirements for a specific industrial plant. This chart clearly indicates when the plant is in a position to produce new orders.

New Equipment Purchase Basis

Since time standards allow analysts to determine machine, department, and plant capacity, they also provide the information necessary to determine how many of which facilities are needed for a given production volume. Accurate comparative time standards also highlight the advantages of one facility over its competitors. For

FIGURE 16–3
Time standards allow the determination of projected direct labor requirements.

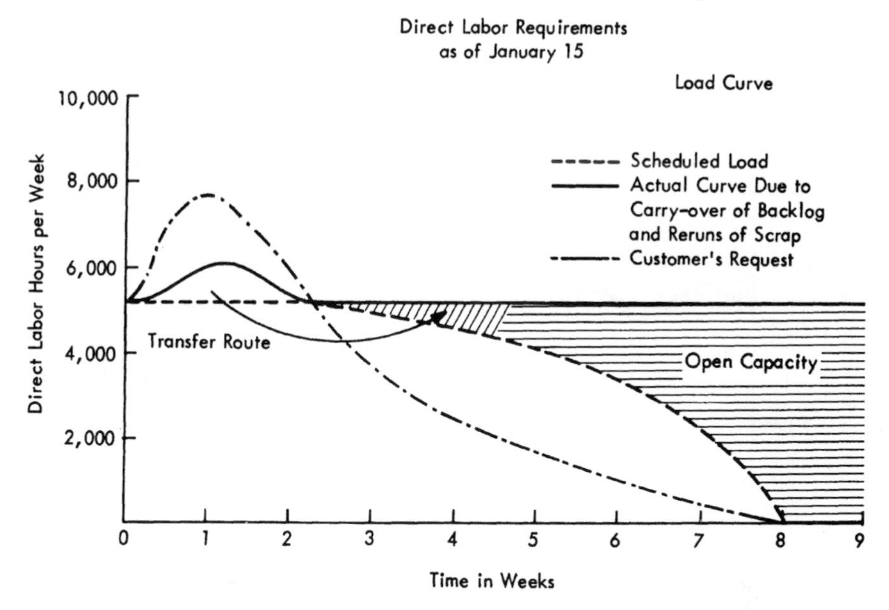

example, a plant may find it necessary to purchase three additional single-spindle, bench-type drill presses. By reviewing available standards, plant managers can procure the style and design of drill press that produces the most favorable output per unit of time.

Workforce versus Available Work

Having concrete information on the required production volume, as well as the time needed to produce a unit, enables analysts to determine the required labor force. For example, if the production load for a given week is 4,420 hours, the company needs 4,420/40, or 111 operators. This use of standards is especially important in a retrenching market, in which production volume is going down. Without a yardstick to determine the actual number of people needed to perform the reduced load, when overall volume diminishes, the entire workforce may slow down to make the available work last. Unless the workforce balances with the available work volume, unit costs progressively rise. Under these circumstances, it is only a matter of time until production operations are performed at a substantial loss, necessitating increased selling prices and further reductions in volume. The cycle repeats until the plant closes due to the losses.

In an expanding market, it is equally important to be able to budget labor. Rising customer demands necessitate a greater volume of personnel. Companies must determine the exact number and type of personnel to add to the payroll, so that

FIGURE 16–4
Chart illustrating actual projected work-hour load and budgeted work-hour load.

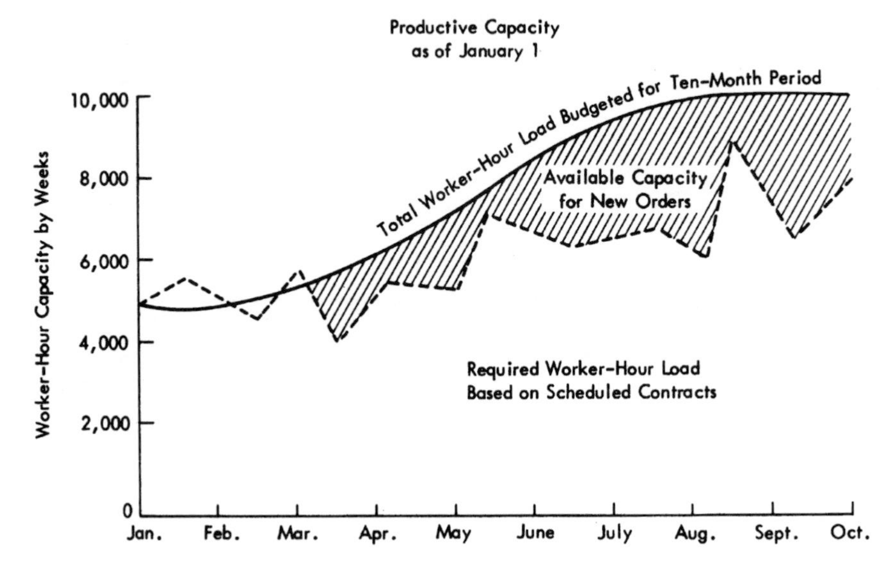

these workers can be recruited in sufficient time to meet customer schedules. If accurate time standards exist, it is a matter of simple arithmetic to convert product requirements to departmental work-hours.

Figure 16–4 illustrates how overall plant capacity may be increased in an expanding market. Here, the plant anticipates doubling its work-hour capacity between January and November. This budget projects the scheduled contracts in terms of work-hours and allows a reasonable cushion (crosshatched section) for receiving additional orders.

Besides allocating plant labor requirements, time standards help in budgeting the labor needs of specific departments. Valid time standards keep the workforce in proportion to the volume of production required, thus controlling costs and maintaining operation in a competitive market. For example, Figure 16–5 illustrates using a Gantt chart (see Chapter 2) to budget spindle hours in a multiple winding department. Four product areas are served: streamline, serving, straight edge, and fishing rod. For present customer requirements, 75 spindles are allotted for the multiple winding of streamline products, 42 spindles for serving, 14 for straight edge, and 15 for fishing rod. Based on this budgeting of spindles and work-hours, the load for streamline extends to the end of the first week in November, serving to the middle of October, straight edge to the middle of November, and fishing rod to the last week in October. Although customer requirements fluctuate, time standards make it possible to adjust the labor budget of this department to meet all customer requirements in the best manner feasible. The solid bar on this chart shows the material that has been allocated to the various product classes, as of September 1.

FIGURE 16–5
Chart illustrating the budgeting of the spindle hours of a specific department to best meet customer needs.

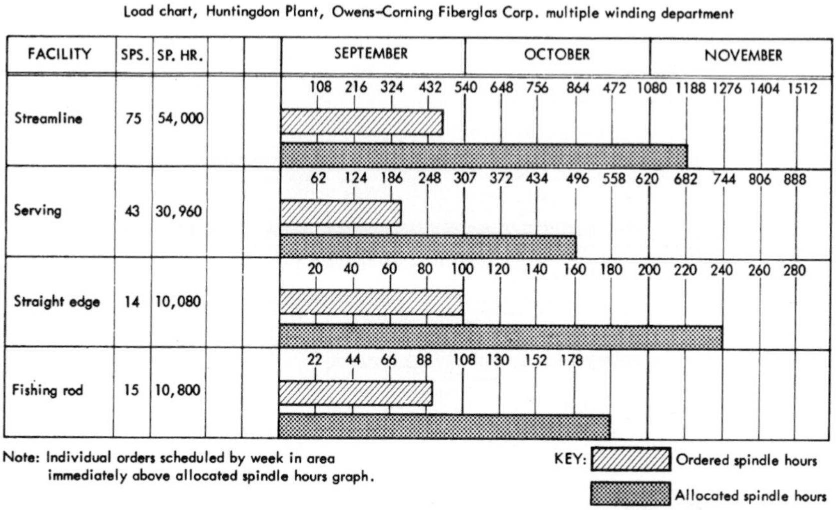

Load chart, Huntingdon Plant, Owens-Corning Fiberglas Corp. multiple winding department

FACILITY	SPS.	SP. HR.	SEPTEMBER / OCTOBER / NOVEMBER
Streamline	75	54,000	108 216 324 432 540 648 756 864 472 1080 1188 1276 1404 1512
Serving	43	30,960	62 124 186 248 307 372 434 496 558 620 682 744 806 888
Straight edge	14	10,080	20 40 60 80 100 120 140 160 180 200 220 240 260 280
Fishing rod	15	10,800	22 44 66 88 108 130 152 178

Note: Individual orders scheduled by week in area immediately above allocated spindle hours graph.

KEY: Ordered spindle hours; Allocated spindle hours

Improving Production Control

Production control is the phase of an operation that schedules, routes, expedites, and follows up production orders, in an effort to achieve operating economies and satisfy customer requirements. The whole function of production control is based on determining where and when the work can be done. This cannot be achieved without a concrete idea of "how long."

Scheduling, one of the major functions of production control, is usually handled in three degrees of refinement: (1) long-range, or master, scheduling; (2) firm order scheduling; and (3) detailed operation scheduling, or machine loading.

Long-range scheduling is based on the existing and anticipated production volume. In this case, specific orders are not given any particular sequence, but are merely lumped together and scheduled in appropriate time periods.

Firm order scheduling involves scheduling existing orders to meet customer demands, while still operating in an economical fashion. Here, workers assign degrees of priority to specific orders, and anticipated shipping promises evolve from this schedule.

Detailed operation scheduling, or machine loading, involves assigning specific operations to individual machines day-by-day. The scheduling is planned to minimize setup time and machine downtime, while meeting firm order schedules. Figure 16–6 illustrates the machine loading of a specific department for one week. It shows that considerable capacity exists on milling machines, drill presses, and internal thread grinders.

FIGURE 16–6
Machine loading of a machining department for one week. (Notice that several schedules depend on receiving additional raw material.)

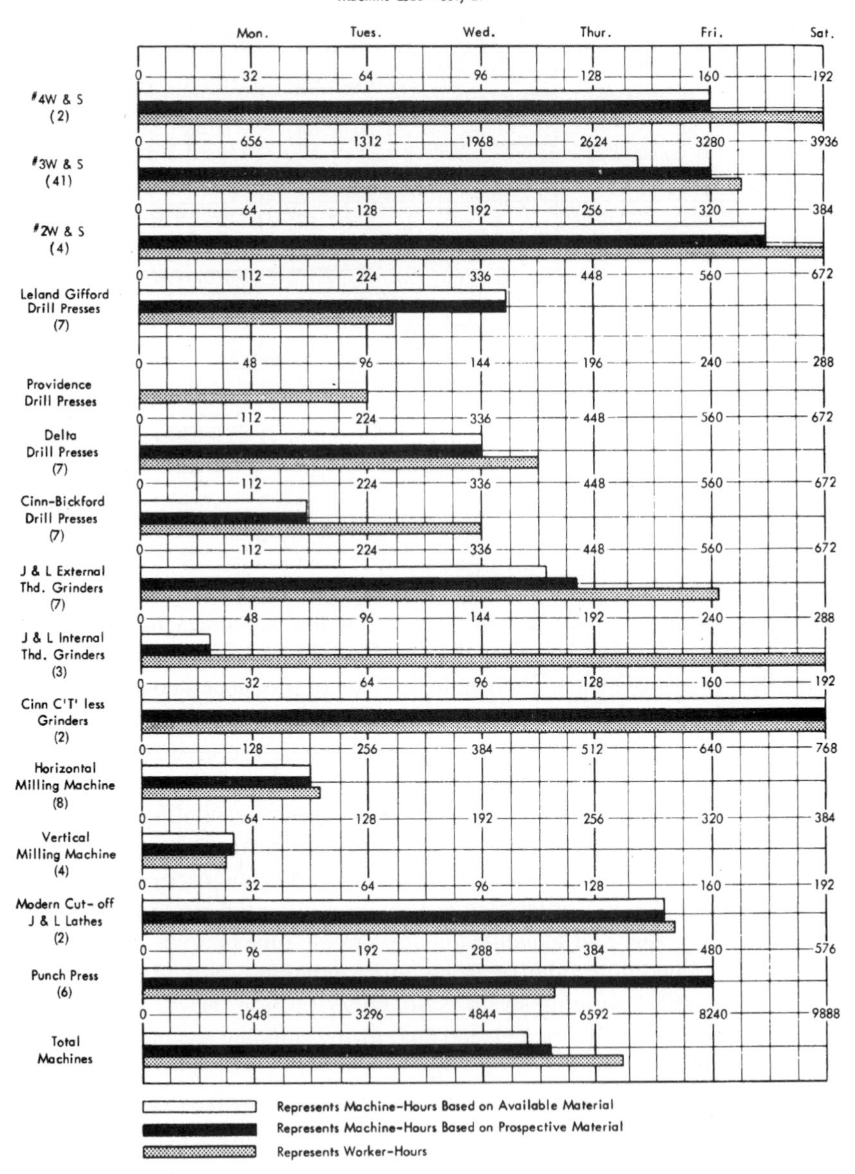

Regardless of the degree of refinement of the scheduling procedure, scheduling would be impossible without time standards. Time standards help predetermine the flow of materials and work in progress, thus forming a basis for accurate scheduling. The success of any schedule is directly related to the accuracy of the time values used in determining the schedule. If time standards do not exist, schedules formulated on judgment only cannot be expected to be reliable.

Work center routings provide process information to the shop floor, and convey time data to the shop floor control system. They are the preferred means of disseminating job standards to the employees. Expediting and follow-up involve performance reporting. Modern production control systems utilize time standards from a variety of sources to generate performance reports. Today, in many plants, shop floor time data collection devices are computers that allow line supervisors or staff to review the status of any job based on the most recent data submitted. These modern follow-up devices help ensure quality production control with improved edit checks and less paperwork.

Determining Labor Efficiency

With reliable time standards, a plant does not have to have an incentive wage payment system to determine and control its labor costs. The ratio of departmental earned production hours to departmental clock hours reveals the efficiency of that department. The reciprocal of the efficiency multiplied by the average hourly rate gives the hourly cost of standard production. For example, the finishing department in a plant using straight daywork may have 812 clock hours (H_c) of labor time for 876 earned hours (H_e) of production. The departmental efficiency would then be:

$$E = \frac{H_e}{H_c} = \frac{876}{812} = 108 \text{ percent}$$

If the average daywork (see Chapter 17) hourly rate in the department is \$16.80, then the hourly direct labor cost, based on standard production, would be:

$$\frac{1}{1.08} \times \$16.80 = \$15.56$$

In a second example, assume that in another department, the clock hours are 2,840 and the earned hours of production for the period are only 2,760. In this case, the efficiency would be:

$$\frac{2{,}760}{2{,}840} = 97 \text{ percent}$$

and the hourly direct labor cost, based on standard production, with an average daywork rate of \$16.80, would equal:

$$\frac{1}{0.97} \times \$16.80 = \$17.32$$

In the latter case, management would realize that its labor costs were running \$0.52 per hour more than base rates, and it could increase supervision to bring total labor costs into line. In the first example, labor costs were running less than standard, which would allow a downward price revision, increasing the production volume, or making some other adjustment suitable to both management and labor. Figure 16–7 illustrates a direct labor variance report, indicating departmental performance above and below standard.

Costing Basis

Costing refers to the procedure of accurately determining costs in advance of production. The advantage of being able to predetermine cost is obvious. Most contracts today are let on a "firm cost" basis, which means that the producer must predetermine production costs in order to set the firm price high enough to make a margin of profit. By having time standards on direct labor operations, producers can pre-price those elements entering into the *prime cost* of the product. Prime cost is usually thought of as the sum of the direct material and direct labor costs.

Costs are the basis of actions within an organization. When the costs of processing a part become too high compared to competitive production methods, consideration will be given to making a change. Invariably, several alternatives exist for producing a given functional design, alternatives that compete on a cost basis. For example, casting will compete with forging, reaming with broaching, die casting with plastic molding, powdered metal with automatic screw machine, etc.

Manufacturing costs can be classified into four groups: direct material costs, direct labor costs, factory expense, and general expense. The first two are directly involved with production, while the latter two are expenses beyond production costs, sometimes termed *overhead. Direct material* costs include the costs of: raw materials, purchased subcomponents, standard commercial items (fasteners, wires, connectors, etc.), and subcontracted items. The industrial engineer begins by calculating the basic quantity required for the design. To this value are added: losses for scrap, either from manufacturing or process errors; waste, from design errors; and shrinkage due to theft or environmental effects. The resulting increased quantity, multiplied by unit price, yields final material costs, with a factor subtracted for any anticipated salvage:

$$Cost_{materials} = Q \times (1 + L_{sc} + L_w + L_{sh}) \times C - S$$

where: Q = Base quantity in weight, volume, area, length, etc.
L_{sc} = Loss factor due to scrap (same units).
L_w = Loss factor due to waste (same units).
L_{sh} = Loss factor due to shrinkage (same units).
C = Unit cost of materials.
S = Value of salvaged materials.

Direct labor refers to workers that are involved in direct production of the product. Direct labor costs are calculated from the time required to produce the product (the standard time, as discussed in previous chapters) multiplied by the wage rate.

FIGURE 16–7

Weekly report illustrating department performance in a specific manufacturing plant. (Regular print indicates hours and percentages earned over standard; italics indicate to what degree standard has not been achieved.)

DIRECT LABOR VARIANCE HOURS

Week Ending June 3

			Efficiency variance					**Percent total variance over-under standards**				
			Week ending			**Weekly average**		**Week ending**				
No.	**Name**	**Allowed direct labor**	**6/3**	**5/26**	**5/19**	**First qtr.**	**Apr.**	**6/5**	**5/26**	**5/19**	**4 weeks, April**	**First qtr.***
11	Machine shop	892	204	*29*	*110*	*33*	3	22.9	*2.5*	*9.2*	0.2	*2.2*
12	Wire brush	178	—	—	—	—	—	—	—	—	—	—
19	Punch press	41	18	*8*	—	*6*	*3*	43.9	*9.5*	—	*4.5*	*9.1*
20	Rubber milling	21	*101*	43	*124*	*21*	*51*	*481.0*	18.1	*172.2*	*37.4*	*13.8*
31	Rubber fabricating	1,183	36	29	*12*	116	59	3.0	1.5	*0.7*	3.2	5.2
35	Pilot plant	53	—	—	—	—	—	—	—	—	—	—
39F	Finishing	339	60	107	27	42	50	17.7	18.5	5.8	10.0	6.6
39P	Paint	23	*1*	*9*	12	*8*	*3*	*4.3*	*12.7*	23.1	*11.0*	*26.4*
40	Assembly	13	1	15	15	14	4	7.7	28.3	25.9	6.0	19.9
50	Reclaim	20	—	—	—	—	—	—	—	—	—	—
65	Toolroom	—	—	—	—	—	—	—	—	—	—	—
	Total—This week	2,763	217	148	*192*	104	59	7.9%	3.7%	*4.5%*	1.2%	1.9%
	Last week	4,462										

Note: The latest planning and method changes are reflected in all groups.
*First quarter includes the 13 weeks from January 1 through March 31.

EXAMPLE 16–1

Estimating the Prime Cost of a Component.

In estimating the prime cost to produce an ABS component that is injection molded, the analyst first multiplies the cost per pound of the ABS resin by the weight in pounds of the product, with due allowance for the sprue, runners, and normal shrinkage (typically, 3 to 7 percent on complex thermoplastic parts). To this figure, the analyst adds the direct labor cost. For example, if the injection press operator services five machines and has an hourly rate (including the cost of fringe benefits) of $18/hour, the direct labor cost per injection press would be $3.60/hour, or $0.06/minute. If the cycle time for molding the part is 0.5 minutes, the direct labor cost per piece would be $0.03.

Let us assume the cost of the resin to be $1.20 per pound, and the weight of each piece to be 1 ounce. Let us also assume that the weight of the runners and sprue is 0.1 ounce, and that 5 percent shrinkage is characteristic of this particular molding room. Then, the estimated material cost would be:

$$\$1.20 \times 1.1 \times 1.05/16 = \$0.087$$

In this example, the prime cost per piece would be:

$$\$0.03 + 0.087 = \$0.117$$

Factory expense includes such costs as indirect labor, tooling, machine, and power costs. *Indirect labor* typically includes shipping and receiving, trucking, warehousing, maintenance, and janitorial services. Indirect labor, machine, and tool costs may have more influence on the selection of a particular process than material and direct labor costs. For example, in the previous illustration, the single cavity mold may cost $30,000, and the machine rate (cost of operating the injection press, exclusive of the press operator) may be $20.00/hour. (In a complex machining center, machine rates are often as low as a few dollars per hour and range as high as $50 or more per hour.)

The allocation of tool costs also has a significant relationship with the quantity to be produced. Going back to our example, let us assume that 10,000 pieces are to be produced. This would give a unit tool cost of $30,000/10,000, or $3.00 per piece. This is much greater than the combined material and direct labor cost (more than 10 times). If 1,000,000 pieces are to be produced, the unit cost for the mold will only be $0.03 (about one third the cost of direct labor and material). Let us assume that the machine rate of this equipment (not including the mold cost) is $20.00/hr, or $0.333/min, and that a million pieces are desired. Then, the total factory cost (direct material + direct labor + factory expense) would be:

$$\$0.087 + \$0.03 + \$0.333/2 = \$0.2835$$

General expense includes such costs as *expense labor* (accounting, administration, clerical, engineering, sales, etc.), rent, insurance, utilities, etc. The industrial

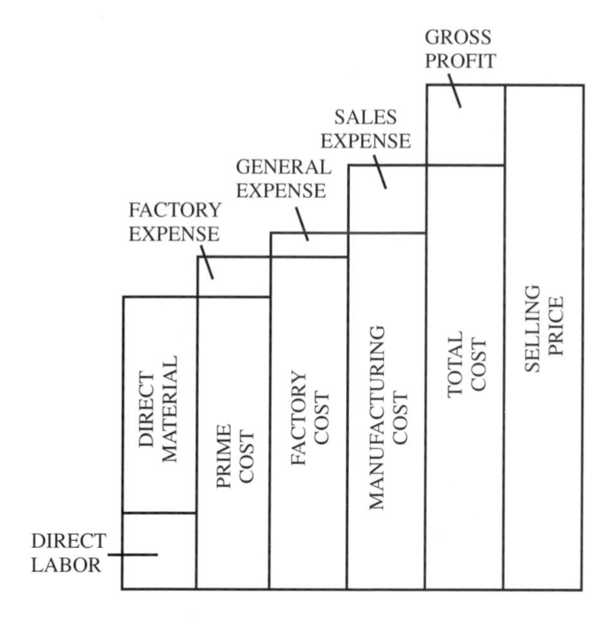

FIGURE 16–8
Elements of cost and gross profit (profit before taxes) entering into the development of selling price.
Note that, in this particular product, material cost is approximately 53 percent of total cost, and 17.5 percent of total cost is anticipated as gross profit.

engineer is primarily concerned with factory cost, since this is the cost that impacts the choice of alternative ways of producing a given design.

Figure 16–8 illustrates the various cost and profit elements in the development of selling price. An understanding of the basis of cost will enable the engineer to select the materials, processes, and functions to create the best product. An increase in perfection from 90 to 95 percent may result in a 50 percent increase in development and product cost and would destroy the sales value of the product. Cost, quality, and degree of perfection should be carefully considered to obtain the greatest profit over a given period. Cost is usually the deciding factor. The relationship between cost, sales, profit or loss, and volume is best revealed in a break-even or crossover chart (from Chapter 7). Figure 16–9 illustrates a typical break-even chart.

The distribution of cost factors will vary dramatically with the number of units to be produced. This has been demonstrated in connection with fixed expense, as shown in Figure 16–8, and the distribution of mold cost referred to in the previous example. When quantities are low, the proportion of development costs is high compared to the cost of manufacturing expense, direct labor, raw material, and purchased parts. The development cost includes design, drawings preparation, manufacturing information compilation, tool design and construction, testing, inspection, and many other items incidental to placing the first parts into production. As the number of units increases, the emphasis centers on a reduction in factory overhead, direct labor, and material costs, through advanced process engineering and manufacturing methods. When the quantities to be produced are low, expenditures on tooling, automation, robotics, and engineering refinements will net less of a return in cost reduction. When quantities are high, an expenditure of engineering effort will result in reduced labor, material, and overhead costs per

FIGURE 16–9
Break-even chart indicating relationship between cost, sales, profit or loss, and volume.
Note that for each category the previous category is added in to yield a cumulative sum.

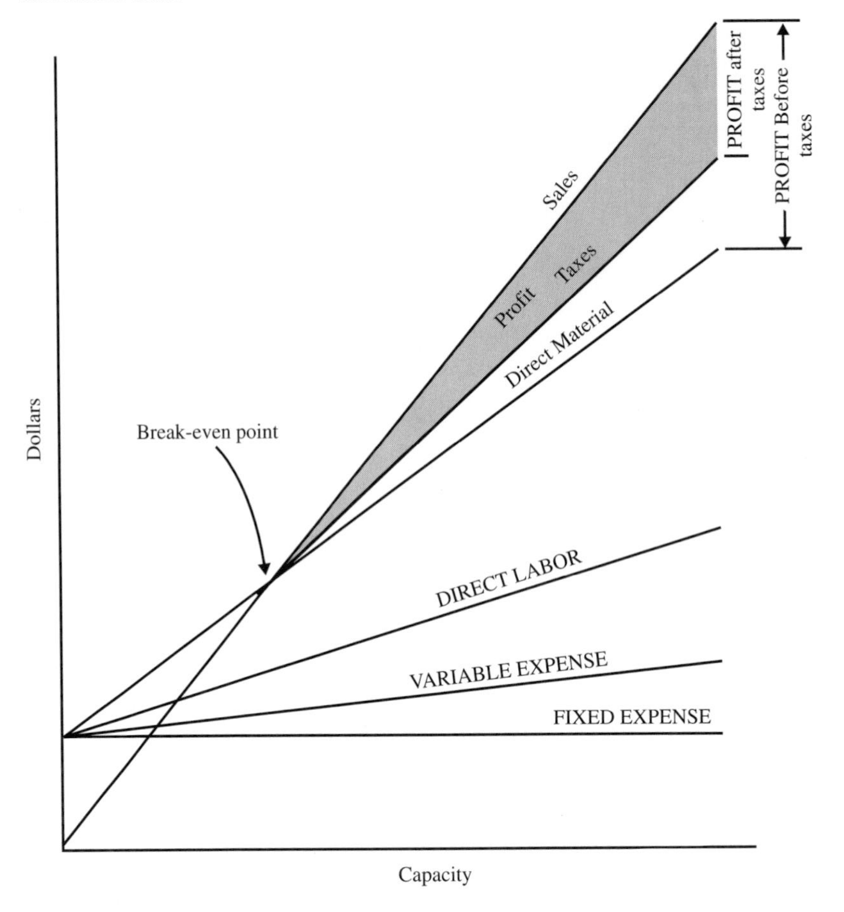

unit of output, and will result in large returns for even small savings per unit. For example, if an automobile manufacturer produces two million cars per year, of which one million have four cylinders per engine and the other one million have six cylinders per engine, and if there are four piston rings per cylinder, then 40 million piston rings must be produced per year. A savings of $0.01 on each ring would equal $400,000 per year. Therefore, to obtain the minimum cost, considerable engineering effort can be profitably applied to the production effort, beginning with raw material and continuing up to the installation of the finished product.

There is constant competition between materials and processes, based on costs that are influenced by the number of pieces made during a period of time. The parts activity affects the amount of time the activity is operated, compared with the hours available. The ratio of hours operated to hours available has considerable effect upon cost. Consider the following representative example.

TABLE 16–4

Reduction in Direct Labor on a Part May Cause Hourly Overhead to Exceed Budgeted Hourly Overhead

		New method: Coated carbide tools	
	Old method	Overhead based on established hourly rate	Overhead based on true costs
Allowed time for turning sheave in NC vertical boring mill	5.00 h	2.5 h	2.5 h
Direct labor rate per hour	$ 20.00	$ 20.00	$ 20.00
Direct labor costs	$100.00	$ 50.00	$ 50.00
Overhead costs for inspection, repairs, material handling, depreciation, etc. per hour	$ 40.00	$ 40.00	$ 80.00*
Cost of overhead	$200.00	$100.00	$200.00
Cost of direct labor overhead	$300.00	$150.00	$250.00

Apparent savings per piece = $300 – $150.00 = $150.00
Actual savings per piece = $300.00 – $250.00 = $50.00

*Since overhead costs per piece will tend to remain constant, the actual overhead costs for the new method will be about the same as for the old method.

A large hydraulic extrusion press, including the hydraulic pumps and the building to house the press, costs $3,000,000. Depreciation, maintenance, and interest on investment amount to 20 percent ($600,000) per year. Normally there are 2,000 working hours a year in one shift (8 hours/day × 5 days/week = 40 hours/week × 50 weeks/year = 2000 hours/year). Three shifts would represent 6000 hours/year available. The minimum cost of the facility during a 24-hour day would therefore be 600,000/6000 = $100/hour. Actually, the sales department can sell only enough to keep the equipment busy 8 hours a day. Therefore, the machine costs 600,000/2000 = $300/hour. If sales decrease, the machine cost per hour will increase, making it difficult to operate the business at a profit.

It is important that the engineer engaged in cost reduction activities understands the company's accounting system. For example (see Table 16–4), when budgets and cost ratios are based on direct labor hours, a line supervisor who is progressive and introduces new or improved methods may be penalized because of the system. He may have a part that is bored and turned on a vertical turret lathe in 5 hours. By using coated carbides, he can reduce the machining time to 2.5 hours. This is 50 percent less labor, over which he must spread the additional cost of maintenance (since the machine is working harder) and the additional material handling, inspection, supplies, production clerks, etc., which vary according to the actual number of parts handled per unit of time, rather than direct labor hours. When the line supervisor's budgets go into the red, the supervisor can only point out that the department is turning parts out 50 percent faster than before and that the departmental expenses are therefore higher in proportion to direct labor.

Suppose that by introducing coated carbides and other improvements throughout the department, the line supervisor reduces all his direct labor costs by 50 percent and still produces the same number of parts. Since his budgets are based on direct labor, budget adjustments should be made as improvements are made. When the pressure is to increase budgets and cost rates (which has been the characteristic situation during the past 20 years), methods engineering can often be effective by decreasing indirect costs. As an example, better material handling facilities and improved layout can reduce overhead expense. The efficient shop usually has a low direct labor cost and a high overhead rate. A low cost rate may not indicate an efficient shop. When making a cost reduction, the manufacturing engineer should study the efficiency of the overhead departments (maintenance, shipping, receiving, material handling, stock, and stores), as well as the direct labor areas, and should thus obtain overall reductions in cost.

EXAMPLE 16–2

Costing a Picture Window.

As an example, consider new-window construction. Hess Manufacturing Co. produces a large variety of vinyl replacement windows (see Figure 16–10). A standard data system and costing are a necessity for Hess to consider any changes in the styles, shapes, and features of windows and doors. Consider the simplest element in window construction, the picture window, or the fixed pane of glass in an aluminum frame in a 3-lite casement, angle bay, or bow window. One of the smaller picture windows is nominally 2 × 3 feet in size, but is actually 24 × 35.5 inches in size.

Direct materials for a picture window include the vinyl-clad 6069 aluminum extrusions for the head, sill, two jambs, and four glazing stops; two glazing blocks and two weep-hole covers; a double pane of glass separated by a butyl spacer for insulation purposes; and packaging materials. By separate count (see Table 16–5), these total \$18.42 in direct material costs. Note that several different scrap factors are utilized in calculating the material costs. For the aluminum extrusions, an 8 percent scrap factor increases the cut length and cost by 8 percent:

$$24.25 \text{ in} \times (1 + .08) \times \$.08275/\text{in} = \$2.167$$

A flow process chart of the manufacturing operation indicates that the following basic operations are needed: cutting the aluminum extrusions and glazing stops to size, punching weep holes, welding the frame, cleaning the corners, assembling the final product, and packaging. From the standard data of Table 16–6 and required components, a total standard assembly time of 19.193 minutes for a 2 × 3 foot picture window can be calculated (see Table 16–7). Of these 19.193 minutes, 14.522 minutes are due to manual elements

FIGURE 16–10
A variety of window styles.
(*Courtesy:* Hess Manufacturing Co.)

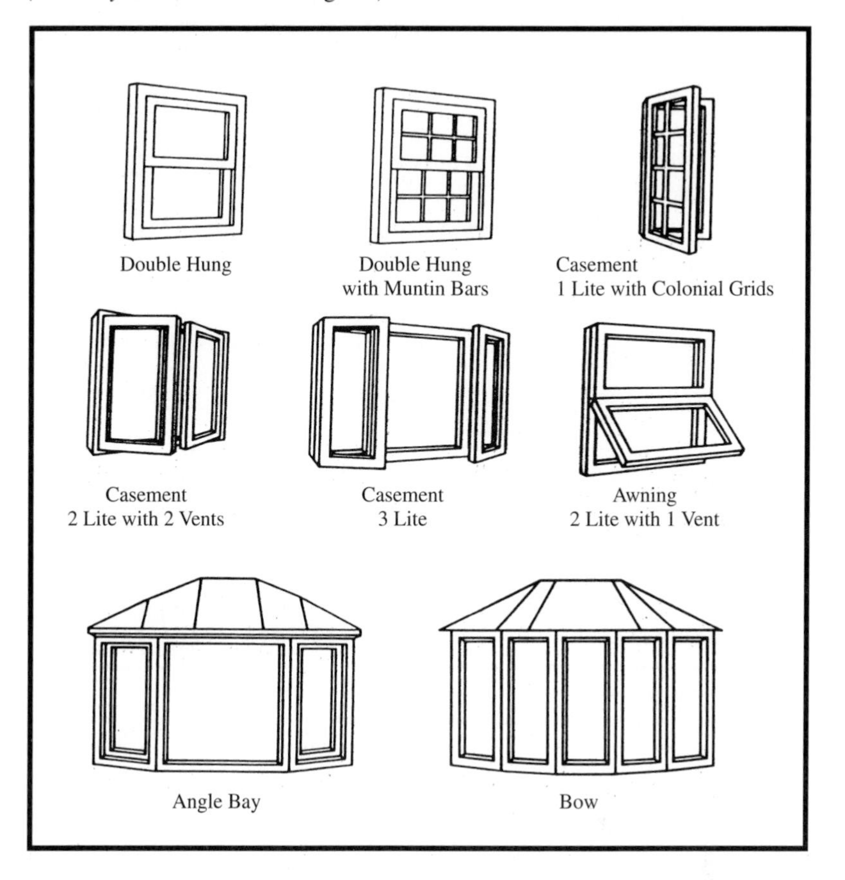

of the operator (with a 20 percent allowance, $12.102 \times 1.2 = 14.522$ minutes), and 4.671 minutes are due to machine elements (with a 5 percent allowance for machine malfunction and maintenance, $4.449 \times 1.05 = 4.671$ minutes).

These 19.193 minutes correspond to 0.32 hours (19.193/60) of direct labor, which, at an average rate of $7.21/hr, yields $2.31 in direct labor costs. An overhead rate of 136 percent of direct labor costs yields $3.14 in factory expense: $(0.32 \times 1.36 \times 7.21)$. Adding direct material costs of $18.42 to the two previous costs yields a total factory cost of $23.87 (see Table 16–8). Based on this cost, Hess can determine a suggested retail cost that will maintain their budget and desired profitability.

(continued)

EXAMPLE 16–2 ***(concluded)***

TABLE 16–5
Bill of Materials—New-Construction Picture (2′ × 3′) Window

Part-Materials	Units	Length (inches)	Scrap Factor	Total Length (in)	Unit cost ($/in)	Cost ($)
Extrusions						
Head, 6069	1	24.250	0.08	26.190	0.08275	2.167
Sill, 6069	1	24.250	0.08	26.190	0.08275	2.167
Jamb, 6069	2	35.697	0.08	38.552	0.08275	6.380
Glazing stop, top	2	15.290	0.08	16.513	0.00733	0.242
Glazing stop, sides	2	19.750	0.08	21.330	0.00733	0.313
Hardware						
Glazing block	2	—	0.01	—	0.019	0.0388
Weep hole cover	2	—	0.01	—	0.085	0.1717
Glass						
Clear glass[1]	2	5.92	0.10	6.51	0.258	3.360
Swiggle[2]	1	119.00	0.10	130.90	0.0246	3.220
Packaging						
Corner boots	4	—	0.03	—	0.056	0.231
Stretch wrap[3]	1	—	—	—	0.131	0.131
				Total Direct Material Costs		18.42

[1]Glass size is given as an area in feet and unit cost as $/foot.

[2]A swiggle is the butyl spacer between the double panes of glass.

[3]Not specifically measured; an average value is used.

TABLE 16–6

Standard Data (Minutes) for New-Construction (2′ x 3′) Windows

Operation	Oper. Code	Window Type				Observed Time		Operator Rating	Normal Time	
		sh	sci	dh	pic	Open	Mach		Open	Mach
Cut	CT	•	•	•		0.125	—	115	0.144	—
	CT	•	•	•	•	0.232	—	102	0.236	—
	CT	•	•	•	•	0.432	—	122.5	0.529	—
Mill	ML	•	•			0.305	—	125	0.381	—
Drill	DL	•	•	•		0.275	—	115	0.316	—
	DL	•		•		0.242	—	117	0.283	—
Punch	PC	•	•	•		0.145	—	115	0.167	—
	PC	•	•	•	•	0.208	—	122.5	0.255	—
Balance assembly	BA	•		•		0.757	—	120	0.908	—
Reinforcing bar	RB	•	•	•		1.233	—	115	1.418	—
Wool pile	WP	•	•	•		0.163	—	115	0.187	—
Weld	WD	•	•	•	•	0.767	0.717	107.5	0.825	0.717
Corner clean	CC	•	•	•		1.133	—	122.5	1.388	—
	CC	•	•	•	•	0.220	2.942	100	0.220	2.942
Hardware	HW	•	•	•		1.673	—	112.5	1.882	—
Drop-in glazing	DG	•	•	•	•	3.210	—	107.5	3.451	—
Final assembly	FA	•	•	•	•	3.390	—	115	3.899	—
Packaging	PK	•	•	•	•	0.373	0.790	105	0.392	0.790

sh = single-hung window
sci = slider single-hung window
dh = double-hung window
pic = picture window

(continued)

EXAMPLE 16–2 ***(concluded)***

TABLE 16–7
Assembly Time (Minutes) Analysis—New-Construction Picture (2′ × 3′) Window

	Manual Elements				
Process	**Head**	**Sill**	**Jamb**	**Glazing Stop**	**Machine Elements**
Cut	0.529	0.529	0.529	0.236	—
Punch weep holes	—	0.255	—	—	—
# Parts/Frame	1	1	2	4	—
Subtotal time		3.315			—
Weld		0.825			0.717
Corner clean		0.220			2.942
Drop-in glazing		3.451			—
Final assembly		3.899			—
Packaging		0.392			0.790
Total Assembly Time		12.102			4.449
Allowances		20%[1]			5%[2]
Standard assembly time		14.522			4.671
		Total Standard Assembly Time			19.193

[1]Includes 5 percent personal allowance, 5 percent basic fatigue allowance, 5 percent delay allowance, 5 percent material handling allowance
[2]Considers malfunctioning and maintenance of machine

TABLE 16–8
Costing of New Construction Picture (2′ × 3′) Window

Type of cost	Minutes	Hours	Rate ($/hr)	Cost ($)
Direct materials	—	—	—	18.42
Direct labor	19.193	0.320	7.21	2.31
Factory expense	19.193	0.320	9.81	3.14
			Total Factory Cost	23.87

Note that these *standard costs* are carefully predetermined target costs that should be attained. They are used to determine product costs, evaluate performance, and, in general, form budgets. As work is done, however, actual costs are incurred, which may reveal *variances* when compared to the standard costs. These variances are considered favorable when actual costs are less than the budgeted or standard costs; or unfavorable when actual costs exceed standard costs. The variances provide feedback on what should be modified to run a productive line.

Quality Standards Enforcement

Time standards force quality requirements to be maintained. Since production standards are based on the quantity of acceptable pieces produced in a unit of time, and since no credit is given for defective work, there is a constant intense effort by all workers to produce only good parts. If an incentive wage payment plan is in effect, operators are only compensated for good parts; to keep their earnings up, they keep their scrap down. Sampling inspection is invariably more effective under incentive conditions. The operator has already assured that the quality of each piece turned out is satisfactory before the piece is released. When some of the pieces are defective, either the operator who produced the parts is held responsible for the salvage, or the worker's earnings are adjusted so that compensation is only received for satisfactory parts.

Management Problems

Time standards are accompanied by many control measures, such as scheduling, routing, material control, budgeting, forecasting, planning, and standard costs. Having controls on practically every phase of an enterprise, including production, engineering, sales, and cost, minimizes management problems. By exercising the "exception principle," in which attention is given only to those items deviating from the planned course of events, management can confine its efforts to only a small segment of the total activity of the enterprise.

For example, Figure 16–11 illustrates a Weekly Lost Time Analysis developed so that management can take positive action when scheduled hours are not attained as planned. Note that a goal of 9.50 percent hours lost of scheduled time was established for 1999, as compared to 10.6 percent hours lost in 1998.

Figure 16–12 illustrates helpful management information in connection with failure analysis for seven major subsystems of a large compressor system. The data obtained from a work sampling study illustrate the Pareto distribution (a distribution that reflects that the major part of an activity is accounted for by a minority).

Government operations have also found time standards to be extremely helpful. This added emphasis on standards was started in 1949 by a Congressional act outlining specific provisions for an efficiency awards system in each government agency. Later, in 1986, MIL-STD 1567A required the application of a disciplined work measurement program as a management tool to improve productivity and to evaluate individuals or groups submitting proposals.

Customer Service

Experience has proven that companies that have developed sound standards based on measurement are more likely to meet scheduled delivery dates for their products.

FIGURE 16–11
ETF weekly lost time analysis. July 14–20, 1991. (*Courtesy* of Ramesh C. Gulati, Sverdrup Technology, Inc. AEDC Group.)

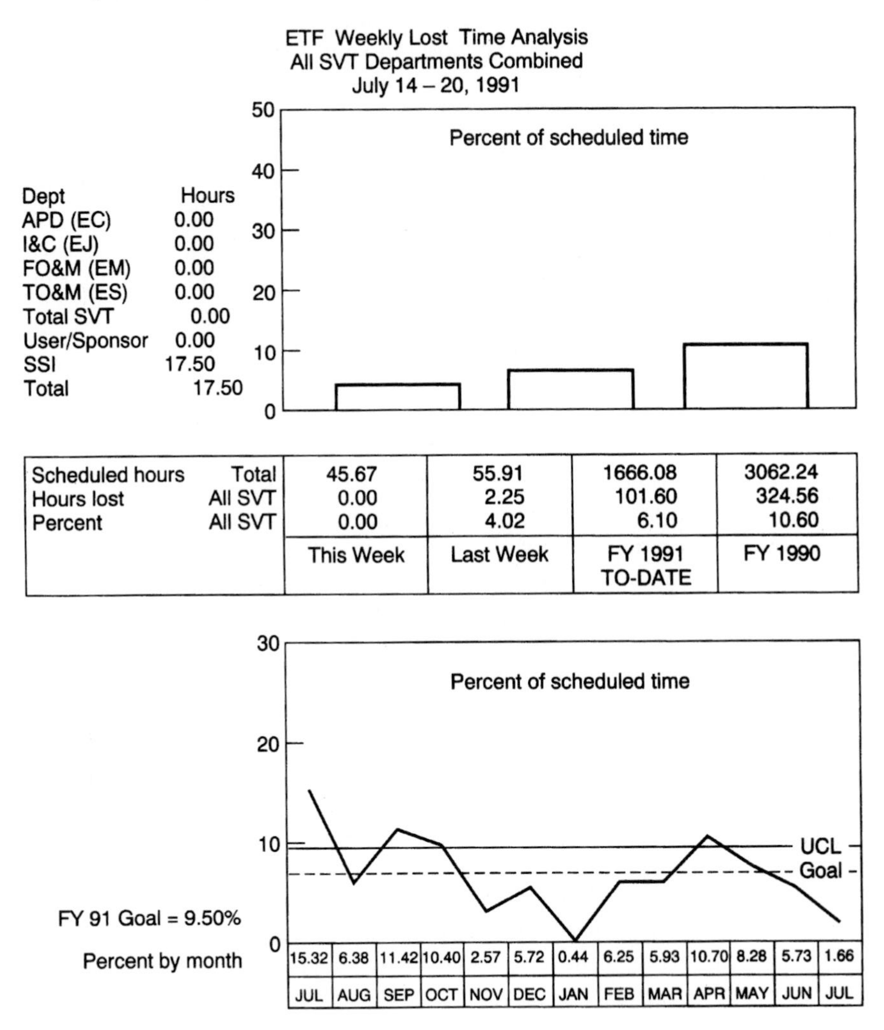

The use of time standards allows the introduction of up-to-date production control procedures, with the resulting advantage to customers who get their merchandise when they want and need it. Also, time standards tend to make any company more time and cost conscious; this usually results in lower selling prices. As has been explained, quality is maintained under a work standards plan, thus assuring the customers of more parts made to required specifications.

FIGURE 16–12
Failure analysis for seven major subsystems of a large compressor system and human error. (*Courtesy* of Ramesh C. Gulati, Sverdrup Technology, Inc. AEDC Group.)

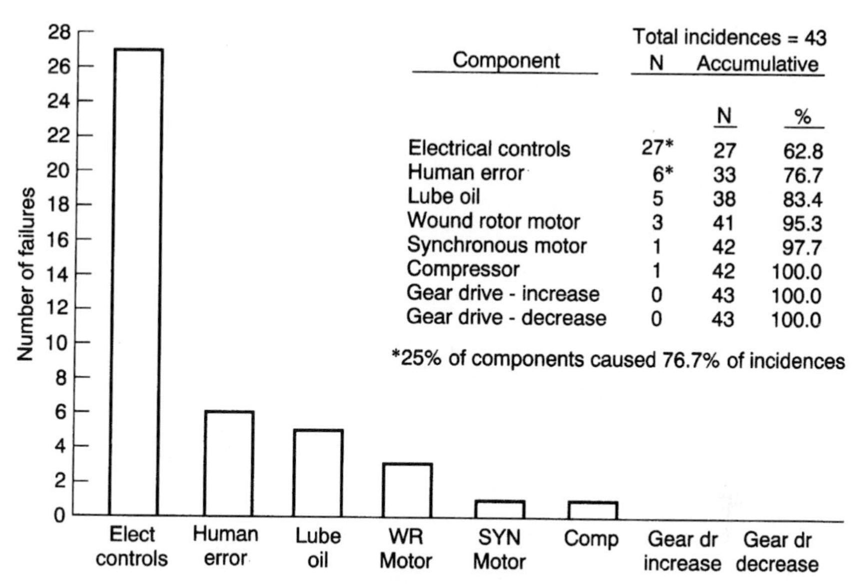

SUMMARY

Thorough and regular follow-up assures the expected benefits from the new method. This calls for maintaining time standards to ensure a satisfactory rate structure. Periodically, all standards should be checked to verify that the methods being employed are identical to those in use at the time the standards were established. A continuing methods analysis is a must.

There are many uses of time standards in all areas of any enterprise. Probably the most significant result of time standards is the maintenance of overall plant efficiency. If efficiency cannot be measured, it cannot be controlled, and without control, it will markedly diminish. Once efficiency goes down, labor costs rapidly rise, and the result is eventual loss of competitive position in the market. Figure 16–13 illustrates the relationship between labor costs and efficiency in one leading manufacturer's automobile accessories business. By establishing and maintaining effective standards, a business can standardize direct labor costs and control overall costs.

QUESTIONS

1. Compare and contrast the different ways of determining a standard time.
2. How can valid time standards help develop an ideal plant layout?

FIGURE 16–13
Relationship of labor costs to efficiency. (Developed from data furnished by AC Spark Plug Division, General Motors Corp.)

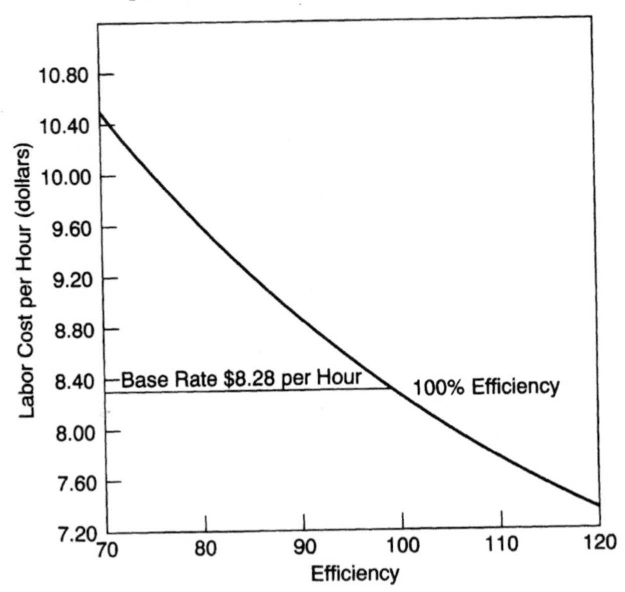

3. Explain the relationship between time standards and plant capacity.

4. In what way are time standards used for effective production control?

5. How do time standards allow the accurate determination of labor costs?

6. How does developing time standards help maintain the product quality?

7. In what way is customer service improved through valid time standards?

8. What is the relationship between labor cost and efficiency?

9. How are management problems simplified through the application of time standards?

10. Explain how inventory and storage areas can be accurately predicted.

11. Explain what is meant by the Pareto distribution. Give some examples of this distribution with which you are familiar.

12. If an audit revealed that a standard as originally established, was 20 percent loose, explain the methodology for rectifying the rate.

13. What is the relationship between the accuracy of time standards and production control? Does the law of diminishing returns apply?

14. How does work measurement improve the selection and placement of personnel?

15. When is it no longer necessary to follow up the installed method?

PROBLEMS

1. In the XYZ Manufacturing Company, direct labor cost is based on the efficiency relationship illustrated in Figure 16–13. For a given product line, the selling price is based on the company running at 95 percent efficiency. How much additional profit does the company realize if the actual efficiency turns out to be 110 percent? Profit was originally estimated at 10 percent of the total cost. On this product line, the total overhead was estimated to be 100 percent of prime cost (direct labor plus direct material). Material cost averaged $5 per unit of output, and direct labor averaged 0.50 hours per unit of output.

2. In the XYZ Company, management is considering going from two 8-hour shifts per day to three 8-hour shifts per day, or two 10-hour shifts per day, in order to increase capacity. Management realizes that shift start-up results in a loss of productivity that averages 0.5 hour per employee. The premium for the third shift is 15 percent per hour. Time over 8 hours worked per day gives the operator 50 percent more pay. To meet projected demands, it is necessary to increase the work-hours of production by 25 percent. In view of insufficient space and capital equipment, this increase cannot be accommodated by increasing the employees on either the first or the second shift. How should management proceed?

3. A time standard is established allowing the operator 11.28 minutes per piece. The sales department expects to sell at least 2,000 of these parts in the next year. How many audits of this standard would you recommend be scheduled during the next 12 months?

4. If a daywork shop was paying an average rate of $12.75 per hour and had 250 direct labor employees working, what would be the true direct labor cost per hour if, during a normal month, 40,000 hours of work were produced?

REFERENCES

Buffa, Elwood S. *Modern Production Operations Management*. 6th ed. New York: John Wiley & Sons, 1980.

Graham, C. F. *Work Measurement and Cost Control*. Oxford, England: Pergamon Press, 1965.

Greene, J. H. *Production and Inventory Control: Systems and Decisions*. Rev. ed. Homewood, IL: Richard D. Irwin, 1967.

Magee, J. F., and D. M. Boodman. *Production Planning and Inventory Control*. 2nd ed. New York: McGraw-Hill, 1967.

Moore, F. G., and R. Jablonski. *Production Control*. 3rd ed. New York: McGraw-Hill, 1969.

Mundel, Marvin E. *A Conceptual Framework for the Management Sciences*. New York: McGraw-Hill, 1967.

Panico, Joseph A. "Work Standards: Establishment, Documentation, Usage and Maintenance." In *Handbook of Industrial Engineering*. 2nd ed. Ed. Gavriel Salvendy. New York: John Wiley & Sons, 1992.

SELECTED SOFTWARE

Nicks, J. E. *Cost Estimating, Basic Programming Solutions for Manufacturing*. Dearborn, MI: Society of Manufacturing Engineers, 1982.

CHAPTER 17

Wage Payment

KEY POINTS:

- Set simple but fair incentives based on proven standards.
- Guarantee basic hourly rates.
- Provide individual incentives above the base rates.
- Tie incentives directly to increased production.
- Remember to tie in product quality to the incentive scheme.

The principal factors in creating highly productive and satisfied workers are: reward and recognition for effective performance. The reward must be meaningful to the employees, whether it is financial, psychological, or both. Experience has proven that workers do not give extra or sustained effort unless some incentive, either direct or indirect, is in the offing. Incentives, in one form or another, have been used for many years. Today, with the increasing need for American business and industry to improve productivity to retard inflation and maintain or improve their position in the world market, management should not overlook the advantages of wage incentives. Only about 25 percent of manufacturing employees are now on incentives. If this figure were doubled in the next decade, the improvement in our national productivity could be phenomenal.

With fringe benefits becoming increasingly significant (today, they average about 40 percent of direct labor), these costs must be spread over more units of output. At present, fringe benefits include not only basic features such as pensions, vacation time, and medical benefits (the costs of which have been skyrocketing), but also added features such as disability insurance and educational benefits. Typical fringe benefits are shown in Table 17–1.

In a broad sense, all incentive plans that increase the employee's production may be referred to as *flexible compensation plans*. Four types of flexible plans will be discussed briefly in this chapter: (1) piecework and standard labor hour plans;

TABLE 17–1
Typical Fringe Benefits Provided by a Company

Benefit	Approximate percentage*
Health insurance	13–18
Vision insurance	½–1
Dental insurance	2–4
Vacations (up to 4 weeks per year)	20–25
Personal leave (up to 5 days per year)	2–5
Holidays (up to 10 per year)	10–12
Term life insurance	2–5
Long term disability	1–3
Pension	25–30
Educational expense reimbursement	1–2

*as percentage of total fringe benefits

(2) gainsharing plans (Scanlon, Rucker, Improshare); (3) employee stock ownership plans (ESOPs); and (4) profit-sharing plans. Before analysts design wage payment plans for specific plants, they should review the strengths and weaknesses of past plans, including day work and all nonfinancial plans.

DAY WORK PLANS

Day work plans compensate the employee on the basis of number of hours worked times an established hourly base rate. Company policies that stimulate employee morale and result in good productivity, without directly relating compensation to production, fall into the sound day rate classification. Such overall company policies are fair, and their relatively high base rates, based upon job evaluation, merit rating, sound suggestion systems, a guaranteed annual wage, and relatively high fringe benefits, build healthy employee attitudes, which tend to stimulate and increase productivity.

From the company's perspective, it would seem that a day work plan is ideal. Unit labor costs (employee wages divided by productivity or worker performance) decrease as worker productivity increases (see Figure 17–1). Mathematically, the unit labor costs (y_c) can be expressed as:

$$y_c = y_w/x$$

where: y_w = The normalized hourly base rate. (= 1)
x = Normalized productivity or performance.

Unfortunately, all day work plans have one weakness: they allow too broad a gap between employee benefits and productivity. After a period of time, employees

FIGURE 17–1
Relationship between costs, wages, and productivity in day work plans. Adapted from Fein, (1982) (Reprinted by permission of John Wiley & Sons, Inc.)

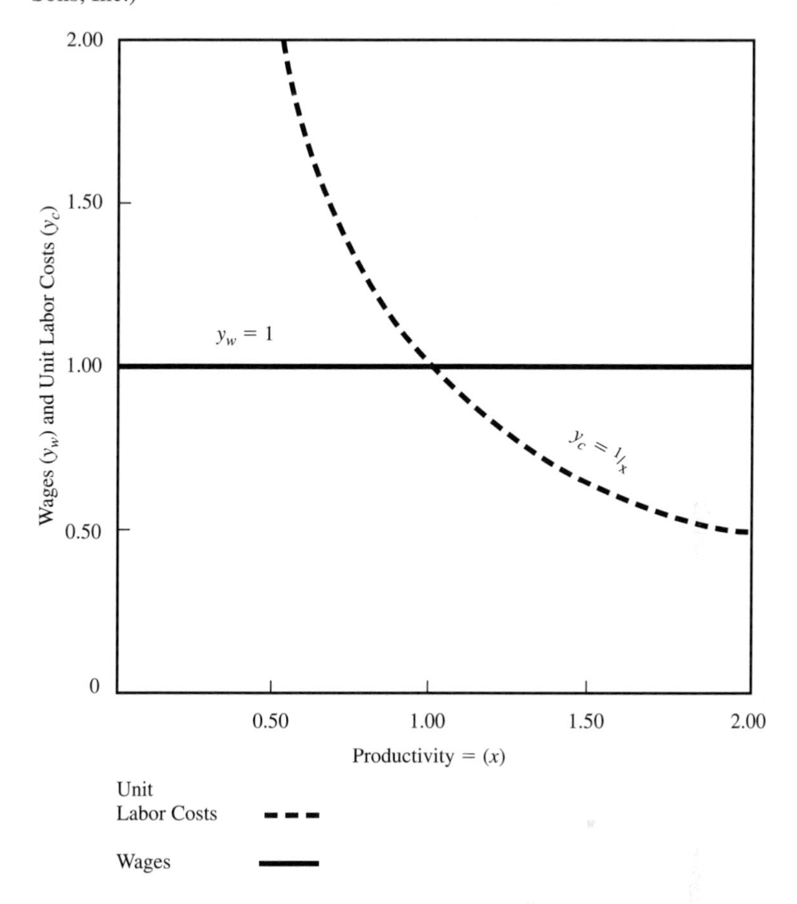

take the benefits for granted, and the company never realizes the expected lower unit labor costs. The theories, philosophies, and techniques of day rate plans are beyond the scope of this text; for more information in this area, refer to books on personnel administration.

FLEXIBLE COMPENSATION PLANS

Flexible compensation plans include all plans in which the worker's compensation is related to output. This category includes both simple, individual incentive plans and group incentive plans. In simple individual plans, each employee's performance for the period governs that employee's compensation. Group plans are applicable to two or more persons that are dependent on one another working as a team. In these plans, each employee's compensation within the group is based on his or her base rate and on the performance of the entire group for the period.

The incentive for high or prolonged individual effort is not nearly as great in group plans as it is in individual plans. Hence, industry favors individual incentive methods. In addition to lower overall productivity, group plans have other drawbacks: (1) personnel problems brought about by nonuniformity of production, coupled with uniformity of pay; and (2) difficulties in justifying base rate differentials for the various opportunities within the group.

However, group plans do offer some decided advantages over individual incentives: (1) ease of installation, due to ease of measuring group, rather than individual, output; and (2) reduction of administration cost due to reduced paperwork, less verification of inventory in process, and less in-process inspection.

In general, individual incentive plans foster higher production rates and lower product unit costs. If it is practical to install, the individual incentive plan should be given preference over group systems. On the other hand, the group approach works well where it is difficult to measure individual output and where individual work is variable and frequently performed in cooperation with another employee. For example, if four operators are working together in the operation of an extrusion press, it would be virtually impossible to install an individual incentive system; rather, a group plan would be applicable.

Piecework Plan

Under the *piecework* plan all standards are expressed in money and the operators are rewarded in direct proportion to output. Under piecework, the day rate is not guaranteed. Since federal law requires a minimum guaranteed hourly rate, piecework is no longer used in the United States. Prior to World War II, piecework was used more extensively than any other incentive plan. The reasons for the popularity of piecework are that it is easily understood by the worker, easily applied, and one of the oldest wage incentive plans. Figure 17–2 illustrates graphically the relationship between an operator's wages and unit direct labor costs under a piecework plan.

Since the unit labor costs remain constant regardless of worker productivity, the company does not appear to benefit from a piecework plan. However, that is not true, if the reader remembers the different costs that enter into factory expenses (see Chapter 16). Relatively constant overhead costs would decrease if considered on a unit cost basis.

Standard Hour Plan

The *standard hour* plan with a guaranteed base rate, established by job evaluation, is by far the most popular incentive plan in use today. The fundamental difference between the standard hour plan and the piecework plan is that under the former, standards are expressed in time rather than money, and operators are rewarded in direct proportion to output.

FIGURE 17–2
Relationship between costs, wages, and productivity in piecework.

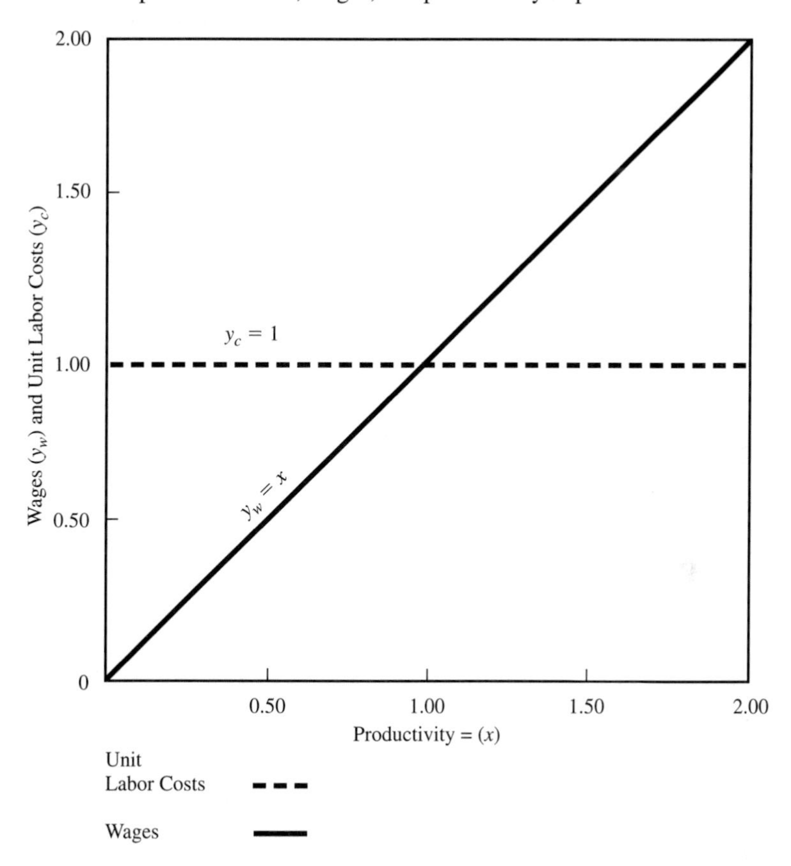

Graphically, the relationship between operator wages and unit direct labor cost, plotted against productivity, is a combination of Figures 17–1 and 17–2 (see Figure 17–3). The worker operates under a day work plan up to 100 percent productivity, and under piecework, beyond 100 percent productivity. For example, a standard may be expressed as 0.02142 hours per piece or 373 pieces per 8-hour shift. Once the base rate is known, it is easy to calculate either the money rate or the operator's wages. If the operator has a base rate of \$12, then the money rate of this job would be: $12.00 \times 0.02142 = \$0.257$ per piece. If the operator produced 412 pieces in an 8-hour workday, wages for the day would be: $412 \times 0.257 = \$105.88$, and hourly wages would be: $\$105.88/8 = \$13.24$. The operator's efficiency for the day, in this case, would then be: 412/373, or 110 percent.

The standard hour plan offers all the advantages of piecework and eliminates the major disadvantages. However, under this plan, it is somewhat more difficult for workers to compute earnings than if standards were expressed in money. The principal advantage is that the standards are not changed when base rates are altered. Thus, over time, this plan reduces clerical work when compared to the piecework plan. Moreover, the term "standard hour" is more palatable to workers than the term

FIGURE 17–3
Relationship between costs, wages, and productivity in the standard hour plan. (Adapted from Fein, 1982) (Reprinted by permission of John Wiley & Sons, Inc.)

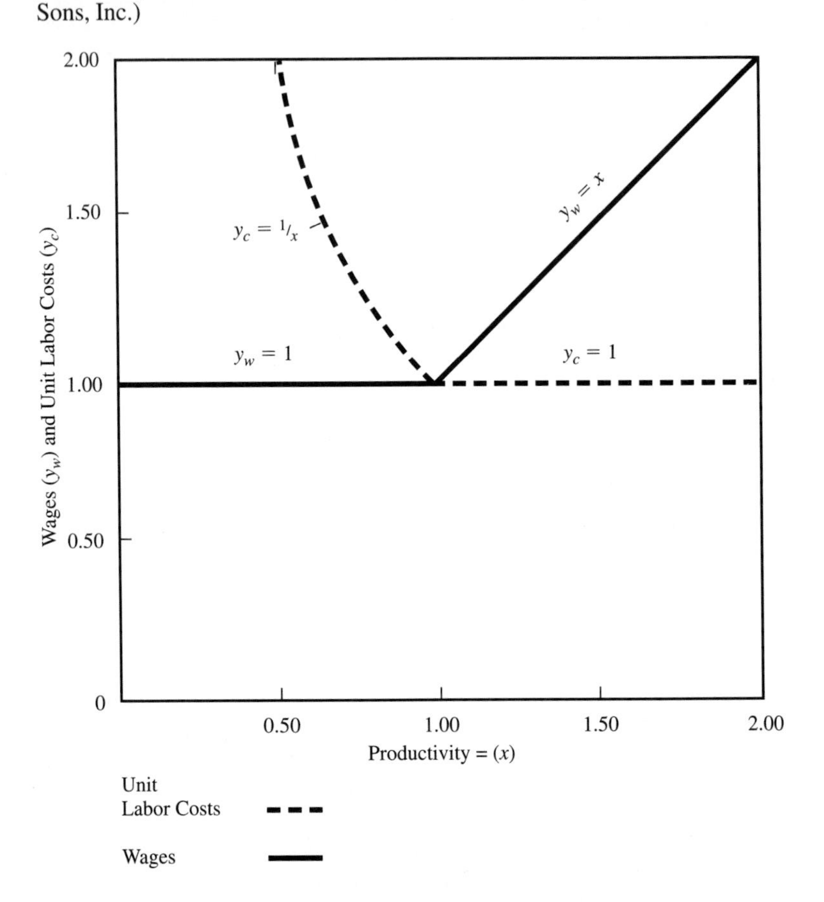

"piecework," and with standards expressed in time, the money earned by workers is not so closely linked with time study practices. For these reasons, there has been a marked increase in the popularity of standard hour plans.

A variation of the standard hour plan is a scheme whereby the incentives are applied to each worker based on group output, creating a group incentive scheme. This is especially useful for work cells (as part of job enlargement, discussed in Chapter 18) or in situations where individual performance cannot easily be measured (e.g., shipbuilding, aircraft manufacturing, etc.). These schemes have some advantages in allowing greater flexibility for workers, reducing competition, and encouraging group spirit and teamwork. On the other hand, individual incentive is reduced, and better workers may become discouraged.

Measured Day Work

During the early 1930s, shortly after the era of efficiency experts, organized labor tried to get away from time study practice and, in particular, piecework. At that time, *measured day work* became popular as an incentive system that broadened the gap between the establishment of a standard and the worker's earnings. Many modifications of measured day work installations are in operation today, and the majority of them follow a specific pattern. First, job evaluations establish base rates for all opportunities falling under the plan. Second, some form of work measurement determines standards for all operations. Third, analysts keep a progressive record of each employee's efficiency for usually one to three months. This efficiency, multiplied by the base rate, forms the basis of a guaranteed base rate for the next period. For example, the base rate of a given operator may be $12.00 per hour. Assume that the governing performance period is one month, or 173 working hours. If, during the month, the operator earned 190 standard hours, his or her efficiency for the period would be 190/173, or 110 percent. Then, in view of the performance, the operator would receive a base rate of $1.10 \times 12.00 = \$13.20$, for every hour worked during the next period, regardless of performance. However, achievement during this period would govern the base rate for the succeeding period.

In all measured day work plans today, the base rate is guaranteed; thus, an operator falling below standard (100 percent) for any given period would receive the base rate for the following period. The length of time used in determining performance usually runs three months, to diminish the clerical work of calculating and installing new guaranteed base rates. Of course, the longer the period, the less incentive effort can be expected. When the spread between performance and realization is too great, the effect of incentive performance diminishes.

The principal advantage of measured day work is that it takes the immediate pressure off workers. They know what their base rates are, and they realize that, regardless of performance, they will receive that amount for the period.

The limitations of measured day work are apparent. First, because of the length of the performance period, the incentive feature is not particularly strong. Second, to be effective, the plan places a heavy responsibility on supervisors for maintaining production above standard. Otherwise, the employee's performance drops, thus lowering the base rate for the following period and causing employee dissatisfaction. Third, keeping detailed rate records and making periodic adjustments is costly in all base rates. In fact, as much clerical work is involved under measured day work as under any straight incentive plan in which employees are rewarded according to output.

Gain Sharing Plans

Gain sharing plans, also known as productivity sharing plans, are characterized by sharing the benefits of improved productivity, cost reduction, and/or quality improvement. Many firms throughout the United States today have some form of gain sharing plan. In many of these, plants add sharing supplements, rather than replacing existing compensation systems. Most progressive management accepts the principle of rewarding employees for improvements in productivity and/or cost, whether or not the improvements are due to performance above normal or improvements in work methods.

Under plans of this type, management computes incentives on a monthly basis. Customarily, only two-thirds of the incentive earned in a given pay period is distributed. The remaining third is placed in a reserve fund to be used for any month in which performance falls below standard. The three productivity sharing plans discussed here are: Scanlon, Rucker, and IMPROSHARE. They differ in the formula used to compute productivity savings and in the implementation method. The Scanlon and Rucker plans measure the payroll of the firm against total dollar sales and compare the result to the average of the past several years. The IMPROSHARE plan measures output against total hours worked. Thus, the Scanlon and Rucker plans use dollars as the measurement unit, while IMPROSHARE uses hours. All three of these productivity plans are flexible regarding the personnel included in the plan. Direct and indirect workers, as well as all levels of management, may be included.

Scanlon Plan

During the Great Depression, Joseph Scanlon developed the *Scanlon plan* to save a failing company. Three fundamental principles form the basis of this plan: bonus payment, identity with the company or firm, and employee involvement. Scanlon plans recognize the value and contribution of each member of the firm, encourage decentralized decisionmaking, and seek to get each employee to identify with the organization's objectives through financial participation.

Before the bonus is calculated, a base ratio must be computed. This traditionally is:

$$\text{Base ratio} = \frac{\text{Payroll costs to be included}}{\text{Value of production}}$$

Analysts make a historical study of approximately one year to gather data prior to calculating a proper base ratio. For example, if the base ratio is 15 percent and if during the past month, the value of production (sales plus or minus inventory) equals $2 million, then allowed labor equals $300,000 ($0.15 \times 2{,}000{,}000$). An actual labor cost of $270,000 generates a bonus pool of $30,000. Typically, the company keeps a portion of this pool to provide for capital expenditures. The remainder is distributed to the employees as a monthly bonus, based on a percentage of their wages.

To stimulate identity with the company, the Scanlon plan recommends a continuing program of management development in which all employees, through effective communication, learn about the goals, objectives, opportunities, and problem areas characteristic of the plan. The Scanlon plan incorporates most "quality of work life" variables, including job enlargement, job enrichment, feeling of achievement, and recognition.

Employee involvement is typically accomplished through a formalized suggestion system and two overlapping committee systems. Elected employee representatives meet at least monthly with their departmental supervisors to review productivity, cost reductions, and quality improvement suggestions. These committees frequently have certain decisionmaking authority for less costly suggestions. More costly suggestions, or those affecting another department, are referred to a higher-level committee.

Rucker Plan

This plan came into being during the early 1940s. It was conceived by Allen W. Rucker, who noted the relationship between payroll costs and actual net sales plus or minus inventory changes minus purchased materials and services.

Like the Scanlon plan, the Rucker plan emphasizes identity with the company and employee involvement, through the establishment of a suggestion system, Rucker committees, and good labor–management communications. The Rucker plan provides a bonus in which everyone, excluding top administration, shares a percentage of the gains. In the evaluation of the bonus, a historical relationship between labor and value added must be established. For example:

Net sales (for period of one year)	$1,500,000
Inventory change (decrease)	200,000
	$1,300,000
Less materials and supplies used	700,000
Production value added	$ 600,000

$$\text{Rucker standard} = \frac{\text{Payroll costs included in group}}{\text{Production value}}$$

Assuming that the labor costs in the base one-year period are $350,000, the Rucker standard becomes:

$$\frac{\$350{,}000}{\$600{,}000} = 0.583$$

Thus, in any future period (usually one month) that the actual labor costs are less than 0.583 of production value, employees earn bonuses. Typically, 30 percent of this bonus is reserved for deficit months, a portion is kept by the company for future improvements, and the remainder (often 50 percent) is distributed to employees. With 50 percent of the bonus distributed to employees and 30 percent retained for deficit months, gainsharing additives for rework production and delivery performance can often utilize the remaining 20 percent of the bonus, rather than that amount being kept by the company to provide improvements.

Since materials and supplies used are deducted from net sales, the Rucker plan calculation partially accounts for variables, such as product mix. This plan also encourages employees to conserve supplies and materials, since the employees would benefit from these savings.

IMPROSHARE

The IMproved PROductivity through SHARing plan was developed by Mitchell Fein in 1974. Its goal is to produce more products in fewer hours of direct and indirect labor. Unlike the Scanlon and Rucker plans, *IMPROSHARE* does not emphasize employee involvement, but rather measures performance and encourages workers to improve productivity.

IMPROSHARE compares the work hours saved for a given number of units produced to the hours required to produce the same number of units during a base period. The savings are shared by the company and the direct and indirect employees

EXAMPLE 17–1
IMPROSHARE Incentive Plan.

Assume that in a single-product plant, 122 employees produced 65,500 units over a 50-week period. If the total hours worked were 244,000, the work hour standard would be:

$$\frac{244{,}000}{65{,}500} = 3.725 \text{ hrs./unit}$$

If, in a week, 125 employees worked a total of 4,908 hours and produced 1,650 units, the value of the output would be 1,650 × 3.725 = 6,146.25 hours. The gain would be 6,146.25 − 4,908 = 1,238.25 hours. Typically, one-half of this amount, or 619.125 hours, goes to the employees. This would be a 12.6 percent (619.125/4,908) bonus or additional pay to each employee.

The company also benefits, since labor costs are reduced. The unit labor cost of 3.725 hours established for the base period is reduced to (4,908 + 619.25)/1,650 = 3.350 hours per unit.

involved with the production of the product. Base productivity is measured by comparing the labor hour value of completed production to the total labor input for this production. Only acceptable products are counted. Thus:

$$\text{Work hour standard} = \frac{\text{Total production work hours}}{\text{Units produced}}$$

Graphically, the IMPROSHARE plan can be considered a variation of the standard hour plan of Figure 17–3, except that the slope of the piecework segment is not 1, but a fraction less than 1 (see Figure 17–4). This fraction, or slope p, is the participation fraction and could vary between companies. If the split between the company and employees is 50/50 (as discussed previously), then p is equal to 0.5. In the standard hour plan, the participation is 100 percent and $p = 1$.

Employee Stock-Ownership Plans (ESOP)

Employee stock-ownership plans have become increasingly popular in the past decade. The Bureau of National Affairs' 1984 survey of 195 employers to determine type of productivity improvement programs administered indicated that 37, or 19 percent, had installed employee stock-ownership plans. These plans involved the creation of a trust that holds company stock for its employees. Although 100-percent worker ownership plants are rare, an ESOP can be used to develop such an organization.

Profit Sharing

Profit sharing can be defined as a procedure under which, in addition to regular pay, an employer pays all employees special current or deferred sums based on the prosperity of the company. No one specific type of profit sharing has received gen-

FIGURE 17–4
Relationship between costs, wages, and productivity in incentive plans with variable participation (p = participation fraction). (Adapted from Fein, 1982) (Reprinted by permission of John Wiley & Sons, Inc.)

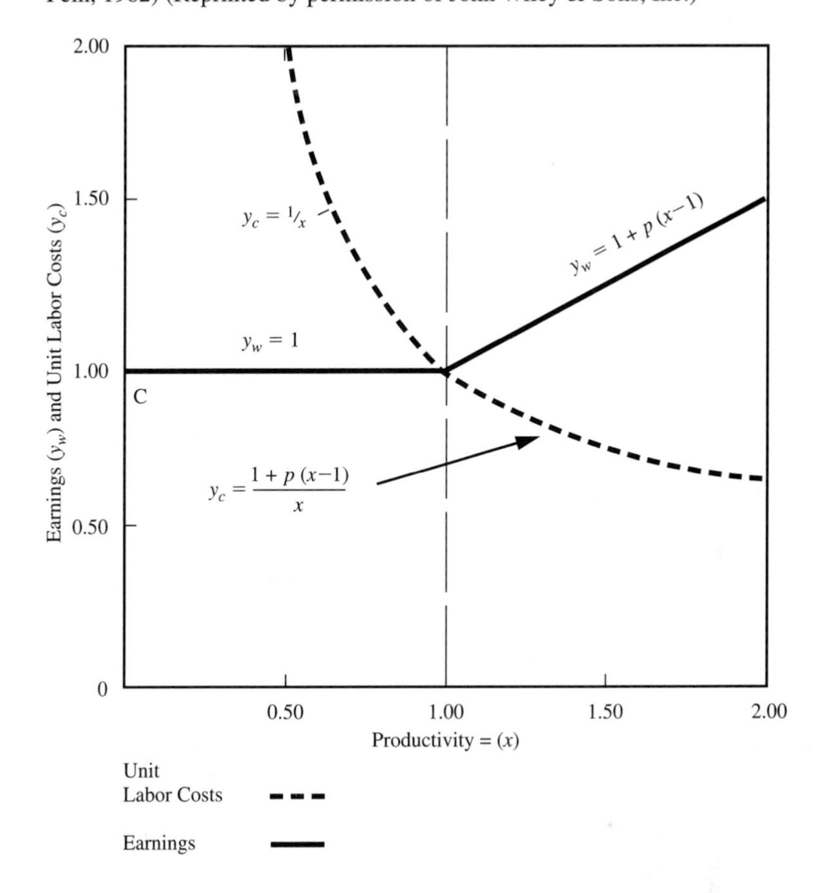

eral industrial acceptance. In fact, just about every installation has certain "tailor-made" features that distinguish it from others. However, the majority of profit sharing systems fall into one of the following broad categories: (1) cash plans, (2) deferred plans, and (3) combined plans.

As the name implies, the straight cash plan involves the periodic distribution of money from the profits of the business to the employees. The payment is not included with the regular pay envelope, but is made separately, to identify it as an extra reward, brought about by the individual and combined efforts of the entire operating force. The amount of the cash distribution is based on the degree of financial success of the enterprise for the bonus period. The shorter the period, the closer the connection between effort and financial reward to the employees. Longer periods are selected because they average out the vagaries of business cycles.

Deferred profit sharing plans feature the periodic investment of portions of the profits for employees. Upon retirement or separation from the company, employees have a source of income other than wages. Deferred profit sharing plans obviously do not provide the incentive stimulus that cash plans do. However, deferred profit sharing plans do offer the advantage of being easier to install and administer. Also, plans of this type offer more security than the cash reward plans. This makes them especially appealing to stable workers.

Combined plans arrange to have some of the profits invested for retirement and similar benefits, and some distributed as cash rewards. This class of plans can realize the advantages of both the deferred plans and those employing the straight cash system. A representative installation might provide for sharing half the profits with the employees. Of this amount, one-third may be distributed to employees as extra bonus checks, one-third may be held in reserve to be given out during a less successful financial period, and the remaining third may be placed with a trustee for deferred distribution.

There are three methods for determining the amount of money to be given to individual employees from the company's profits. The first and least used is the "share and share alike" plan. Here, each employee, regardless of job class, receives an equal amount of the profits, after attaining the prescribed period of company service. Proponents of this method believe that individual base rates already take care of the relative importance of the different workers to the company. The "share and share alike" plan supplies a feeling of teamwork and importance to each employee, no matter what his or her position in the plant.

The most commonly used method of distribution under profit sharing is based on the regular compensation paid to workers. The theory is that the employee who was paid the most during the period contributed the greatest to the company's profits and consequently should share in them to the greatest extent. For example, a toolmaker earning $15,000 during a six-month period would receive a greater share of the company's profits than a chip-hauler paid $7,000 during the same period.

Another popular means of profit distribution involves the allocation of points. Points are given for each year of seniority, each $100 of pay, and other factors such as attendance, cooperation, etc. The number of points accumulated for the period determines each employee's share of the profits. Perhaps the principal disadvantage of the point method is the difficulty of maintaining and administering the complex and detailed records.

For a profit sharing plan to succeed, worker representation and union cooperation are essential. The emphasis should be placed on partnership and not on management benevolence. Management should recognize that the plan should be dynamic and is not a panacea for all problems. Profit sharing should not be used as an excuse for paying lower than prevailing wages.

For the most part, union officials have not supported profit sharing. There can be no doubt that profit sharing, when practiced with perfect harmony between labor and management, minimizes the necessity of a union in the eyes of the employees and diminishes the union's prestige, power, and income.

A successful profit sharing program depends on the profits of the company, which are frequently not under the control of the direct labor force. In periods of low profits, or of losses, the plan may actually weaken rather than strengthen employee morale. As a result of the length of time between performance and reward, the incentive effect can also be weakened.

EXAMPLE 17–2
Comparison of Two Incentive Schemes.

A company would like to evaluate two incentive schemes. The first is similar to the IMPROSHARE gain sharing plan, with a 50/50 split above 100 percent productivity. The second is constant day work up to 100 percent productivity, a *kicker* (a step increase in wages to induce workers to reach a certain level of productivity), and then a participative gain sharing plan but with the workers only receiving 20 percent and the company 80 percent. The plans and unit labor costs are shown in Figure 17–5). The company would like to know the point at which the two plans break even above 100 percent productivity.

The first plan's unit labor costs can be expressed as:

$$y_{c_1} = 0.5 + 0.5/x$$

while the second plan's are:

$$y_{c_2} = 0.2 + 1/x$$

Equating the two equations and solving for x yields $x = 1.67$. Therefore, with plan #1, the company benefits up to 167 percent productivity, and with plan #2, with higher productivity levels. The company must decide whether it is reasonable to expect workers to reach such high levels of productivity.

FIGURE 17–5
Comparison of two incentive schemes.

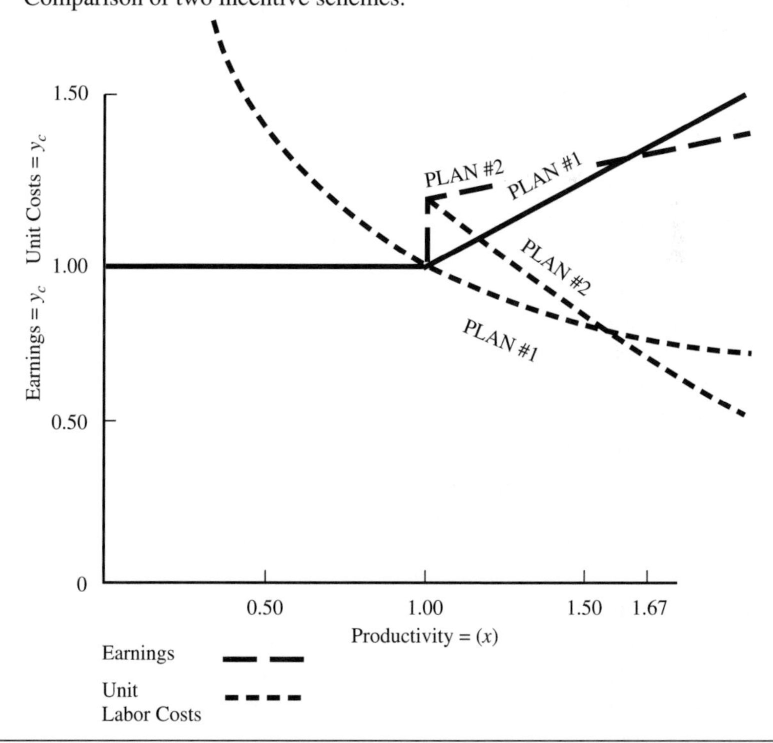

Perhaps the greatest objection to profit sharing is the employee taking for granted that an extra check will be received at the end of the year. Employees will come to expect these checks, and they may even feel cheated if the company has a lean year and cannot pay them. For these reasons, any employer should be very cautious before embarking on a profit sharing program. On the other hand, many companies today are experiencing high worker efficiency, decreased costs, scrap reduction, and better worker morale as a result of profit sharing.

INDIRECT FINANCIAL PLANS

Company policies that stimulate employee morale and increase productivity, without directly relating compensation to production, fall in the indirect financial plans classification. Overall company policies such as fair and relatively high base rates, equitable promotion practices, sound suggestion systems, a guaranteed annual wage, and relatively high fringe benefits build healthy employee attitudes, which stimulate and increase productivity. Thus, they are classified as indirect financial plans.

The weakness of all indirect incentive methods is that they allow too broad a gap between employee benefits and productivity. After a period of time, employees take the benefits for granted and fail to realize that continuance of those benefits must result entirely from employee productivity. The theories, philosophies, and techniques of indirect incentives are beyond the scope of this text; for more information in this area, refer to books on personnel administration.

UNION ATTITUDES

The subject of wage incentives has always been controversial to employees, unions, and management. Industries that have assured a good living wage and then applied easily-calculated incentive earnings for extra or prolonged effort find that their employees are receptive to wage incentives and insist on their continuance. On the other hand, in industries or businesses where workers find it necessary to work at an incentive pace to earn the necessities of life, they are not enthusiastic about any form of wage incentive payment. The majority of union officials with whom the writers have been in contact oppose the installation of incentive wage payment in plants where incentives do not exist.

In an AFL-CIO Collective Bargaining Report that is still being used as that union's position paper about incentives, past president George Meany stated: "Wage incentive plans, that is, plans which offer more wages for more production, present a host of special problems which normally far outweigh any possible benefits. With few exceptions, unions are opposed to wage incentive systems, both because of the damaging past experience with the abuses under such plans, and because of the difficulties and ill effects inherent in incentive plans. Unions which have accepted them or permitted them to continue have usually done so only with reluctance and misgivings. It simply has not always been practical or expedient actively to oppose or to eliminate such plans."

The unions' principal objections to incentive plans arise from the fear that given a fixed production of goods and services, a reduction in personnel will result from high effort. Their leaders also believe that incentive installations de-emphasize the necessity of unions since the unions' major role is to achieve higher wages, which the incentive plan does automatically. Union officials have stated that they object to incentives because they "pit worker against worker." They state that when one worker makes high earnings and another makes low wages, a feeling of distrust and suspicion permeates the working group and disrupts the partnership relations among workers.

There are some unions that approve of incentives. In fact, the late Philip Murray, a past president of the CIO, favored incentive wage payment and believed that practically any sound system of wage payment could be made to work when a harmonious relationship prevailed between labor and management. United Steelworkers of America arbitration summary of 1969 recommended that all jobs be covered by incentives.

In summary, most unions fight to keep incentives where they already exist. Where incentives do not exist, unions resist attempts to install them.

WAGE INCENTIVE PLAN PREREQUISITES

The majority of companies that have incentive plans installed favor their continuance and believe that their plans are: (1) increasing the production rate, (2) lowering overall unit costs, (3) reducing supervision costs, and (4) promoting increased employee earnings. However, in a National Metal Trades Association survey of 160 companies practicing incentive wage payment, 84 replies from managers implied that they felt their plans to be only fair, and that additional improvement could be made. Five plant managers felt that their incentive systems were poor, and that some changes must be made to warrant their continuance.

Before installing a wage incentive program, management should survey its plant to be sure that the plant is ready for an incentive plan. Initially, a policy of methods standardization must be introduced so that valid work measurement can be accomplished. If different operators follow different patterns while performing their work, and if the sequence of elements has not been standardized, the organization is not ready for the installation of wage incentives.

Work schedules should create a backlog of orders for each operator, so that the chances of running out of work are minimized. This implies that adequate inventories of material are available, and that machines and tools are properly maintained. Also, established base rates should be fair, and should provide for a sufficient spread between job classes, to recognize the positions that demand more skill, effort, and responsibility. Preferably, management will have established base rates through a sound job evaluation program.

Finally, fair performance standards must be developed before wage incentives are installed. These rates should never be set by judgment or past performance records. To be sure that the rates are correct, some form of work measurement, based on time study, fundamental motion data, standard data, formula, or work sampling procedure, should be used.

Once these prerequisites are complete, and management is fully sold on incentive wage payment, the company is in a position to design the system.

WAGE INCENTIVE PLAN DESIGN

To be successful, an incentive plan must be fair to both the company and its operators. The plan should give operators the opportunity to earn approximately 20 to 35 percent above base rate, if they are normally skilled and continuously execute high effort. Management benefits from the added productivity by being able to prorate fixed costs over a larger number of pieces, thus reducing total cost.

Next to fairness, the most important qualification of a good incentive plan is simplicity. To be successful, the plan must be completely sold to the employee, the union, and management itself. The simpler the plan, the more easily it is understood by all parties, and understanding enhances the chances for approval. Individual incentive plans are more easily understood and they work the best, if individual output can be measured.

The plan should guarantee the basic hourly rate set by job evaluation, and the rate should be a good living wage comparable to the prevailing wage rate of the area for each job in question. There should be a range of pay rates for each job, and these should be related to total performance. Total performance encompasses quality, reliability, safety, and attendance, as well as output. At periodic intervals, such as every six months or every year, management should review employees' rate steps in relation to their total performance. For performance greater than standard, operators should be compensated in direct proportion to output, thus discouraging any restriction of production.

To help employees associate effort with compensation, paycheck stubs should clearly show both the regular and the incentive earnings. It is also advisable to indicate, on a separate form placed in the pay envelope, the efficiency of the operator for the past pay period. This is calculated as the ratio of the standard hours produced during the period to the hours worked during the period.

Once the plan has been installed, management must accept responsibility for maintaining it. Administration of the plan calls for keen judgment in making decisions, and for close analysis of all grievances submitted. Management must exercise its right to change the standards when the methods and/or equipment are changed. Employees must be guaranteed an opportunity to present their suggestions, and the advisability of their requests must be proved before any changes are made. Compromising on standards must be avoided, or it will lead to a failure of the plan.

A checklist for fundamental principles to be applied in a sound wage incentive plan is given in Table 17–2.

TABLE 17–2
Checklist for Sound Wage Incentive Plans.

	Yes	No
1. Is there agreement between management and labor on general principles?	❑	❑
2. Is there a sound foundation of job evaluations and wage rate structures?	❑	❑
3. Are there individual, group, or plant-wide incentives?	❑	❑
a. Is the most weight applied to individual incentives?	❑	❑
4. Are incentives in direct proportion to increased production?	❑	❑
5. Is the plan as simple as possible?	❑	❑
6. Is quality tied to incentives?	❑	❑
7. Is the establishment of incentives preceded by methods improvements?	❑	❑
8. Are the incentives based on proven techniques:		
a. From detailed time studies?	❑	❑
b. From basic motion data or predetermined time systems?	❑	❑
c. From standard data or formulas?	❑	❑
9. Are the standards based on standard performance under normal conditions?	❑	❑
10. Are standards changed when methods change?	❑	❑
a. By mutual agreement between management and labor representatives?	❑	❑
11. Are temporary standards kept to a minimum?	❑	❑
12. Are basic hourly rates guaranteed?	❑	❑
13. Are incentives established for indirect workers?	❑	❑
14. Are accurate records kept for piece counts, unmeasured work, setup, and downtime?	❑	❑
15. Are good human relations maintained?	❑	❑

Incentive Effort Motivation

Methods analysts should recognize that unless operators perform at good effort, the most favorably designed workstation cannot result in productivity levels synonymous with company objectives. To achieve high productivity levels, the conditions surrounding the work should encourage all employees to do their best to realize company objectives.

Basically, most people want to work and achieve, and they expect to be rewarded for their contributions. They want to be involved in the achievement of the goals established by their organization, and they perform better if given both independence and control in their work situation. A motivational climate must accompany any formal incentive plan.

Perhaps the first requirement in establishing the proper motivational climate is the development of a management style that emphasizes a supporting role, rather than a directive role. The goal should be to have all the workers feel that it is their responsibility to meet the objectives of the business and that it is the supervisors' responsibility to assist the workers as best they can.

Second, the goals of the enterprise should be clearly established, and should be broken down into division, department, work center, and individual goals. It is important that established goals be realistic, that they emphasize both quality and quantity, as well as reliability, and any other characteristic essential to the success of the business. All workers should understand the objectives of the company and the goals related to their work. These goals should be quantified in such a manner that workers are cognizant of their achievement in relation to the established goals.

Third, there should be regular feedback to all employees. Timely reporting should inform workers about the results of their efforts and the impact of these efforts on the established goals. Fourth, every work situation should be designed so that operators are in a position to control, to a large extent, the assignments they are given. A sense of responsibility is an important source of motivation, as is recognition for achievement. More details on motivational theories and approaches are presented in Chapter 18.

Incentive Plan Failure

An incentive plan may be classified as a failure when it costs more for its maintenance than the plan actually saves; it thus must be discontinued. Usually, it is not possible to put a finger on the precise cause of failure for a given incentive installation. If the facts were completely known, numerous reasons would be found for a plan's lack of success. One survey (Britton, 1953; see Table 17–3) listed the principal causes of plan failure as poor fundamentals, inept human relations, and poor administration, resulting in too costly a program.

Companies should discontinue any plan when its cost of maintenance exceeds the benefits derived through its use. For the most part, the reasons given in the sur-

TABLE 17–3
Most Common Reasons for Incentive Plan Failures

	Percent
Fundamental deficiencies	**41.5**
Poor standards	11.0
Low incentive coverage of direct productive work	8.6
Ceiling on earnings	7.0
No indirect incentives	6.8
No supervisory incentives	6.1
Complicated pay formula	2.0
Inept human relations	**32.5**
Insufficient supervisor training	6.9
No guarantee of standards	5.7
A fair day's work not required	5.0
Standards negotiated with the union	4.8
Plan not understood	4.1
Lack of top-management support	3.6
Poorly trained operators	2.4
Poor technical administration	**26.0**
Method changes not coordinated with standards	7.8
Faulty base rates	5.1
Poor administration, i.e., poor grievance procedure	4.9
Poor production planning	3.2
Large group on incentive	2.8
Poor quality control	2.2

vey are symptoms of a sick plan, a plan doomed to failure, they are not really the causes. The actual cause of the failure of any plan is incompetent management—management that permits the installation of a plan with poor scheduling, unsatisfactory methods, a lack of standardization or loose standards, and the compromising of standards.

Unfortunately, all the requisites of a sound incentive system may be met, but the plan may still be unsatisfactory, because of a failure to promote good industrial relations with respect to the program. The complete cooperation of the employees, the union, and management must be won to foster the team spirit so necessary to attain ultimate success with an incentive plan installation.

WAGE INCENTIVE SYSTEM ADMINISTRATION

To be successful, an incentive system must be adequately maintained; it cannot maintain itself. To maintain a plan effectively, management must keep all employees aware of how the plan works and of any changes to the plan. One technique frequently used is to distribute to all employees an "Operating Instructions" manual detailing both company policy relative to the plan, and all working details, with examples. The manual should thoroughly explain the basis of job classifications, time standards, performance rating procedure, allowances, and grievance procedures. It should also describe the technique of handling any unusual situation. Finally, it should present the objectives of the organization, and the role of each employee in the fulfillment of those objectives.

Administrators of the plan should make a daily check of low and excessively high performance, in an effort to determine their causes. Low performance is not only costly to management in view of the guaranteed hourly rate, but it leads to employee unrest and dissatisfaction. Unduly high performance is a symptom of loose standards, or the introduction of a methods change for which no standard revision has been made. In addition, a loose rate would lead to the dissatisfaction of any employees in the immediate vicinity of an operator carrying the low standard. A sufficient number of such poor standards can cause the whole incentive plan to fail. Frequently, operators who have the loose rate restrict their daily production, in fear that management will adjust the standard. This restriction of output is costly to the operators and to the company, and it results in dissatisfaction among neighboring workers, who see fellow employees on a "soft job."

Management should make a continuous effort to include a greater share of the employees in the incentive plan. When only a portion of the plant is on standard, there will be a lack of harmony among operating personnel, because of significant differentials in take-home pay. However, work generally should not be put on incentive unless:

1. It can readily be measured.
2. The volume of available work is sufficient to justify economically an incentive installation.
3. The cost of measuring the output is not excessive.

Periodically, management should review old standards to assure their validity. On standards that have proved to be satisfactory, elemental values should be recapped for standard data purposes, so that even greater utilization of the time values may be made. Analysts can thus achieve greater coverage of the plant relative to the use of standards.

Fundamental to the administration of any wage incentive plan keyed to production is the constant adjustment of standards in response to changes in the work. No matter how insignificant a methods change may be, the standard should be reviewed for possible adjustment. Several minor methods improvements, in aggregate, can amount to a sufficient time differential, causing a loose rate, if the standard is not changed. When revising time standards due to methods changes, it is only necessary to study those elements affected by the changes.

To keep the incentive plan healthy, the company should arrange periodic meetings with operating supervisors to discuss fundamental weaknesses of the plan and possible improvements in the installation. At these meetings, departmental performance should be compared, and specific standards that appear unsatisfactory should be brought to light and discussed.

The company should keep progress reports showing such pertinent information as departmental efficiency, overall plant efficiency, the number of workers not achieving standard performance, and the highest individual performance. These reports highlight areas that need attention, as well as areas where the plan is working satisfactorily.

Effective administration of the plan requires a continuing effort to minimize the nonproductive hours of direct labor. This nonproductive time, for which allowance must be given, represents lost time due to machine breakdowns, material shortages, tool difficulties, and long interruptions of any sort not covered in the allowances applied to the individual time standards. Managers must watch this time—frequently referred to as "blue ticket time" or "extra allowance time"—carefully, or it will destroy the purpose of the entire plan.

Under incentive effort, production performance is considerably higher than under day work operation. With the shorter accompanying in-process time of materials, very careful inventory control is needed to prevent material shortages. Likewise, management should introduce a program of preventive maintenance to assure the continuous operation of all machine tools. Equally important to material control is the control of all nondurable tools, so that shortages, with resulting operator delays, do not develop.

An effective technique often employed to control the "extra allowance" is to key the supervisor's bonus to the amount of this nonproductive time credited to the operator. The more of this time turned in for the pay period, the less would be the supervisor's compensation. Since supervisors are in an ideal position to watch schedules and material inventories and to maintain facilities, they can control nonproductive downtime better than anyone else in the plant.

In addition to the control of the "extra allowance" or day work time, it is essential that exact piece counts be recorded at each workstation. The piece count that determines the operator's earnings is usually done by the operator. However, controls must be established to prevent operators from falsifying their production output.

EXAMPLE 17–3
Administration of a Wage Incentive Plan.

Assume that on a certain job, the production rate is 10 pieces per hour, and that an hourly rate of $12 is in effect under a straight day work operation. Thus, the unit direct labor cost is $1.20. Now, this shop changes over to incentive wage payment for which the day rate of $12 per hour is guaranteed, and, above task, the operator is compensated in direct proportion to his or her output. Let us assume that the standard developed through time study is 12 pieces per hour, and that for the first 5 hours of the workday, a certain operator averages 14 pieces per hour. His earnings for this period would then be:

$$(\$12,00)(5)\left(\frac{14}{12}\right) = \$70.00$$

Now, assume that for the remainder of the workday, the operator could not be productively engaged in work, due to a material shortage. The worker would then expect at least the base rate, or:

$$(3)(\$12.00) = \$36.00$$

which would give earnings for the day of:

$$\$70.00 + \$36.00 = \$106.00$$

This would result in a unit direct labor cost of:

$$\frac{\$106.00}{70} = \$1.514$$

Under day work, even with the low performance, the operator would have produced the 70 pieces in less than the working day. Here, the earnings would be: $8 \times \$12.00$, or $96.00, and the unit direct labor cost would be: $96.00/70, or $1.371. Therefore, any nonproductive time should be carefully controlled.

Where the work is small (several pieces can be held in one hand), operators can make a "weigh" count of their production at the end of the day or the end of the production run, whichever is shorter. This weigh count is verified by the immediate supervisor, who initials the production report.

On larger work, one technique frequently employed is to have a tray or box with built-in compartments to hold the work. The box holds round numbers of the work, such as 10, 20, or 50. Thus, at the end of the shift, it is a simple matter for the operators' supervisors to authenticate production reports by counting the boxes and multiplying by the number each box holds.

Basically, management establishes wage incentive plans to increase productivity. In a sound and properly maintained installation, the percentage of incentive earnings of those workers on incentive would remain relatively constant over time. If an analysis shows that incentive earnings continue to rise over a period of years, the installation probably has problems that will ultimately erode the effectiveness of the plan. If, for example, the average incentive earnings increases from 17 percent to 40 percent over a period of 10 years, the 23 percent rise is most likely not due to a proportionate increase in productivity, but to a creeping looseness in standards.

NONFINANCIAL PERFORMANCE MOTIVATION PLANS

Nonfinancial incentives include any rewards that have no relation to pay, and yet they improve employees' spirits to such an extent that added effort is evident. Elements or company policies that fall under this category include: periodic shop conferences, quality control circles, frequent talks between supervisors and employees, proper employee placement, job enrichment, job enlargement (see Chapter 18), nonfinancial suggestion plans, ideal working conditions, and the posting of individual production records. Effective supervisors and capable, conscientious managers also use many other techniques, such as treating the employee and spouse to dinner, providing tickets to sporting events or the theater, or arranging special trips to trade shows or other companies for exposure to state-of-the-art technology. All of these approaches seek to motivate by improving the work environment. They are frequently referred to as "quality of work life" plans.

The management team also needs to set an example of high performance and the pursuit of excellence. Thus, employees will understand that the culture of their company is top performance in the manufacture of products of the best quality. The results of this philosophy by all workers will be a feeling of great pride in their work. In concert with this philosophy should be individual and group programs that recognize teamwork and team results.

SUMMARY

The only acceptable wage incentive plan applied to individual workers today is the standard hour plan with a guaranteed day rate. Group plans must guarantee their respective day rates to all members of the group, and should reward group members in direct proportion to their productivity once standard performance is achieved.

Profit sharing, employee stock ownership, and other related cost improvement savings plans have met with success in many cases. In general, they tend to be more effective when they are installed in addition to, rather than instead of, a simple incentive plan. (See Table 17–4.)

Incentive principles have been applied in job shops and production shops; in the manufacture of both hard goods and soft goods; in manufacturing and service industries; and in direct and indirect labor operations. Incentives have been used to increase productivity, improve product quality and reliability, reduce waste, improve safety, and stimulate good working habits, such as punctuality and regular attendance.

Table 17–5 illustrates the thinking of 508 personnel/industrial relations managers as to the characterization of flexible compensation plans. The majority feel that simple incentives—piece work, standard hour plans, and measured day work—are the best from the standpoint of raising productivity and being easy to explain.

Soundly administered incentive systems possess important advantages, both for workers and management. The chief benefit to employees is that these plans make it possible for the employees to increase their total wages, not at some indefinite time in the future, but immediately—in their next paycheck. Management obtains a greater output and, assuming that some profit is being made on each unit produced, a greater volume of profits. Normally, profits

TABLE 17–4
Survey of Profit Sharing Plans

	Profit sharing		ESOP	
Respondent type	**Have plan (%)**	**Do not have plan (%)**	**Have plan (%)**	**Do not have plan (%)**
All respondents	52	48	25	75
With ESOP	54	46	—	—
With profit sharing	—	—	25	75
With gain sharing	48	52	16	84
With simple incentive	58	42	28	72

Source: Based on 508 surveys returned 10/15/86. Adapted from data by R. Broderick and D. J. B. Mitchell, "Who Has Flexible Wage Plans and Why Aren't There More of Them?" *IRRA 29th Annual Proceedings*, pp. 163–64.

TABLE 17–5
Characterization of Flexible Compensation Plans by Survey Respondents

	Profit sharing (%)	Employee stock ownership plans (%)	Gain sharing Scanlon, Rucker, IMPROSHARE (%)	Simple std. hr. pc. rate (%)
Best for:				
Raising productivity	28	5	26	41
Increasing loyalty	48	17	19	14
Providing for retirement	80	13	n.a.	n.a.
Linking labor cost to performance	53	n.a.	28	19
Easiest to:				
Explain to employees	32	9	4	49
Administer	40	7	4	38

*Refers to tax-deferred profit sharing plans; n. a. not asked.

Source: Adapted from data by R. Broderick and D. J. B. Mitchell, "Who Has Flexible Wage Plans and Why Aren't There More of Them?" *IRRA 29th Annual Proceedings*, pp. 163–64.

increase not in proportion to production, but when a higher rate of production occurs, so that overhead costs per unit decrease. Also, the higher wages that result from incentive plans improve employee morale and tend to reduce labor turnover, absenteeism, and tardiness.

Since the proper functioning of incentive systems implies the existence of many prerequisites, such as good methods, standards, scheduling, and management practices, the installation of incentives normally results in important improvements in production and supervisory methods. The activities that bring about these improvements should be performed even when incentives are not introduced; the improvements, therefore, are not necessarily attributable to the employment of the incentive plan.

In general, the harder the work is to measure, the more difficult it will be to install a successful wage incentive plan. Usually, it is not advantageous to install incentives unless the work can reasonably be accurately measured. Furthermore, it is usually not advantageous to introduce incentives if the availability of work is limited to less than 120 percent of normal.

Due to recent research that has produced more reliable fundamental motion data, the reliability of standards is assured. Furthermore, we now are able to apply good standards to the vast majority of work opportunities in industry and business. Poor standards and insufficient coverage were probably the principal reasons for the failure of incentive systems in the past. Now that we can avoid these pitfalls, we are more conscious of the necessity for good human relations and technical administration, recognizing that sound wage incentives can do much to stimulate productivity and reduce inflation. Wages incentives permit workers to increase their standard of living, in spite of inflation.

Well-planned and skillfully administered incentives increase production and decrease total unit cost. Usually, they more than compensate for the increased costs of industrial engineering, quality control, and timekeeping that may result from their use. Table 17–2 gives the fundamental principles that apply in the development of sound wage incentive plans.

One important caveat is that there is a definite tradeoff between increasing the work pace with an incentive plan and increasing the risk of injury from repetitive motion, especially if the job or workplace has not been ergonomically designed. The authors have seen many instances, especially in the garment industry, in which jobs with low base rates but high incentives (so that to achieve a decent wage, the sewers must perform at very high rates—well above 150 percent) have experienced high injury rates. Undoubtedly a better designed job may decrease the injury rate. However, even with the best of conditions, high rates (over 20,000 hand motions per 8-hour shift) may still lead to some injuries. Therefore, even neglecting worker health and safety issues, the standards engineer must decide whether the added medical costs, under today's escalating conditions, offset the gains obtained with a given incentive plan.

QUESTIONS

1. What are the three general categories under which the majority of wage incentive plans may be classified?
2. Differentiate between individual wage payment plans and group-type plans.
3. What is meant by fringe benefits?
4. Which company policies are included under nonfinancial incentives?
5. What are the characteristics of piecework? Plot the unit cost curve and operator earning curve for day work and piecework on the same set of coordinates.
6. Why did measured day work become popular in the 1930s?
7. What advantages does the Rucker plan have over the Scanlon plan?
8. How does IMPROSHARE differ from the Rucker and the Scanlon plans?
9. Define profit sharing.
10. Which specific type of profit sharing plan has received general acceptance?
11. Which three broad categories cover the majority of profit sharing installations?
12. What does the amount of money distributed depend on, under the cash plan?
13. What determines the length of the period between bonus payments under the cash plan? Why is it poor practice to have the period too long? What disadvantages are there to the short period?
14. What are the characteristic features of the deferred profit sharing plan?
15. Why is the "share and share alike" method of distribution not particularly common? On what basis is this technique advocated by its proponents?
16. Why have many unions been antagonistic toward any form of incentive wage payment?
17. Why is it usually advisable to have a range of pay rates that are applicable to each job?
18. What are the four important hypotheses related to motivation?
19. What are the fundamental prerequisites of a successful wage incentive plan?
20. Why is it fundamental to keep time standards up to date if a wage incentive plan is to succeed?
21. What does unduly high performance indicate?
22. How would you go about establishing a climate to increase worker motivation?

PROBLEMS

1. In a single-product plant where IMPROSHARE was installed, 411 employees produced 14,762 product units over a one-year period, and recorded 802,000 clock hours. In a given week, 425 employees worked a total of 16,150 hours and produced 348 units. What would be the hourly value of this output? What percentage bonus would each of these 425 workers receive? What would be the unit labor cost in hours for this week's production?

2. Analysts established an allowed time of 0.0125 hours/piece for machining a small component. A setup time of 0.32 hour was also established, as the operator performed the necessary setup work on incentive. Compute the following:
a. Total time allowed to complete an order of 860 pieces.
b. Operator efficiency, if job is completed in an 8-hour day.
c. Efficiency of the operator who requires 12 hours to complete the job.

3. A "one-for-one" or 100-percent time premium plan for incentive payment is in operation. The operator base rate for this class of work is $10.40. The base rate is guaranteed. Compute:
a. Total earnings for the job at the efficiency determined in problem 2(b).
b. Hourly earnings.
c. Total earnings for the job, at the efficiency determined in problem 2(c).
d. Direct labor cost per piece from (a), excluding setup.
e. Direct labor cost per piece from (c), excluding setup.

4. A rate of 0.42 minute per piece is set for a forging operation. The operator works on the job for a full 8-hour day and produces 1,500 pieces.
a. How many standard hours does the operator earn?
b. What is the operator's efficiency for the day?
c. If the base rate is $9.80 per hour, compute the earnings for the day. (Use a 100 percent time premium plan.)
d. What is the direct labor cost per piece at this efficiency?
e. What would be the proper piece rate (expressed in dollars) for this job, assuming that the time standard is correct?

5. A 60–40 gain sharing plan is used in a plant (base rate is guaranteed and operator receives 60 percent of proportional gain after exceeding 100 percent). The established time value on a certain job is 0.75 minute, and the base rate is $8.80. What is the direct labor cost per piece when operator efficiency is:
a. 50 percent of standard?
b. 80 percent of standard?
c. 100 percent of standard?
d. 120 percent of standard?
e. 160 percent of standard?

6. In a plant where all the rates are set on a money basis (piece rates), a worker is regularly employed at a job for which the guaranteed base rate is $8.80. This worker's regular earnings are in excess of $88 per day. Due to the pressure of work, the operator is asked to help out on another job, classified to pay $10 per hour. This employee works three days on this job and earns $80 each day.

a. How much should the operator be paid for each day's work on this new job? Why?
b. Would it make any difference if the operator had worked on a new job for which the base rate was $8 per hour and had earned $72? Explain.

7. An incentive plan employing a "low-rate high-rate" differential is in use. A certain class of work has the guaranteed "low-rate" of $6 per hour, and the "high-rate" for work on standard is $9.20 per hour. A job is studied and a rate of 0.036 hour per piece is set. What is the direct labor cost per piece at the following efficiencies:
a. 50 percent?
b. 80 percent?
c. 98 percent?
d. 105 percent?
e. 150 percent?

8. A company would like to evaluate two incentive schemes that take effect once the worker exceeds normal performance. In the first case, the benefits are split 50/50 between the worker and the company. In the second case, the worker receives a kicker up to 120 percent earnings and then maintains level performance up to 150 percent, after which all of the earnings go to the worker.
a. Plot the unit labor costs for each scheme.
b. Derive the equations for worker earnings and unit labor costs for each scheme.
c. Find the point at which the two plans break even.
d. Which do you think the company would prefer?

9. A company would like to evaluate two incentive schemes that take effect once the worker exceeds normal performance. In the first case, the benefits are split 30 percent to the worker and 70 percent to the company up to 120 percent performance. If the worker exceeds 120 percent performance, all of the earnings go to the worker. In the second case, all earnings beyond normal performance are split 50/50 between the worker and the company.
a. Plot the unit labor costs for each scheme.
b. Derive the equations for worker earnings and unit labor costs for each scheme.
c. Find the point at which the two plans break even.
d. Which do you think the company would prefer?

REFERENCES

Britton, Charles E. *Incentives in Industry*. New York: Esso Standard Oil Co., 1953.

Campion, Michael A., and Gina J. Medsker. "Job Design." In *Handbook of Industrial Engineering* 2nd ed. Ed. Gavriel Salvendy. New York: John Wiley & Sons, 1992.

Dingus, Victor R., and Russell E. Justice. "Celebrating Quality." *Quality Progress*, November 1989, p. 74.

Fay, Charles H., and Richard W. Beatty. *The Compensation Source Book*. Amherst, MA: Human Resource Development Press, 1988.

Fein, M. "Financial Motivation." In *Handbook of Industrial Engineering*. Ed. Gavriel Salvendy. New York: John Wiley & Sons, 1982, pp. 2.3.1–2.3.40.

Lokiec, Mitchell. *Productivity and Incentives*. Columbia, SC: Bobbin Publications, 1977.

U.S. General Accounting Office. *Productivity Sharing Programs: Can They Contribute to Productivity Improvement?* Gaithersburg, MD: U.S. Printing Office, 1981.

Von Kaas, H. K. *Making Wage Incentives Work.* New York: American Management Associations, 1971.

Zollitsch, Herbert G., and Adolph Langsner. *Wage and Salary Administration.* 2nd ed. Cincinnati, OH: South-Western Publishing, 1970.

CHAPTER 18

Training and Other Management Practices

KEY POINTS:

- Train workers to minimize injuries and to reach the standard time more quickly.
- Use learning curves to adjust standards for new workers and small batches.
- Recognize and understand worker needs.
- Use job enlargement and job rotation to minimize repetitive injuries and increase worker self-esteem.

In a 1954 survey on subjects found in industrial engineering curriculums, educators ranked motion and time study first in a list of 41 subject areas (Balyeat, 1954). A somewhat similar survey 10 years later of more than 8,700 nonclerical employees in industrial engineering in 250 large United States manufacturing companies found that industrial engineers still spent most of their time on work measurement (Anonymous, 1964). A more recent survey (Freivalds, et al., 1997) of 139 practicing industrial engineers indicated that traditional work measurement topics (time study, standard data, work sampling) were no longer at the top of the list (although methods tools were). On the other hand, several nontraditional work organizational items (teamwork, job evaluations and training) jumped to the top ten. Because of this demand, a greater emphasis has been placed on these topics in this book.

Another trend is the spread of industrial engineering techniques to all areas of modern business, including marketing, finance, sales, and top management. Also, the importance of work measurement in such indirect areas as office activities, maintenance, shipping and receiving, sales, inspection, and the tool room continues to grow. To meet the demand, and to reap the benefits of training in this field more quickly, many industries have embarked on education programs conducted in their own plants on company time. For example, an extensive survey of over 5,300 United States companies revealed that 80 percent were providing formal training programs for first-line supervisors, and 42 percent of these training programs dealt with work simplification and methods. Therefore, these topics are also emphasized in this chapter.

OPERATOR TRAINING

Training Approaches

A company's labor force is one of its main resources. Without skilled workers, production rates would be slower, product quality would be poorer, and overall productivity would be lower. Therefore, once the new method is installed and the proper standard is set, the operators must be trained appropriately to follow the prescribed method and attain the desired standard. If this is done, the operators will have little difficulty in meeting or exceeding the standard. Many excellent sources of training material, programs, and consultants are readily available and are not discussed in detail here. However, it is important to be aware of some of the major options in training programs, such as those that follow.

On-the-Job Learning

Putting the operators directly on a new job without any training is a sink-or-swim approach. Although the company may think it is saving money, it definitely is not. Some of the operators will muddle along and eventually adapt to the new technique, theoretically "learning." However, they may learn the incorrect method and may not attain the desired standard. Or they may take a longer time to reach the proper standard. This means a longer learning curve. Other operators may watch or ask questions of their co-workers and eventually learn the new method. However, during this period, they will have slowed down both the other operators and overall production. Worse yet, the co-workers may be using an incorrect method, which would then be passed on to the new operator. In addition, the new operator may experience considerable anxiety during the whole learning process, which may hinder that process.

Written Instructions

Simple written descriptions of the correct method are an improvement over on-the-job learning, but only for relatively simple operations, or in situations where the operator is relatively knowledgeable of the process and only needs to adjust for minor variations. This assumes that the operator understands the language in which the instructions are written or has sufficient education to read well. In these days, with greater diversity in the workplace, neither can be assumed.

Pictorial Instructions

Still pictures or photographs used with written instructions have proven to be very effective in training operators. This also allows less educated workers and those speaking another language to acquire the new method more easily. Line drawings generally have an advantage over photographs in emphasizing specific details, omitting extraneous detail and allowing exploded views. On the other hand, photographs are easier to produce and store and are more true to life (Konz, 1995), if properly exposed and focused.

Videotapes

Moving pictures can show the dynamics of the process, such as the interrelationships of motions, parts, tools, etc., much better than still pictures. Videotapes are inexpensive and easy to produce and show. Furthermore, videotapes allow the operator the freedom to control the time and rate of viewing, backing up if necessary, and reviewing procedures. Also, videotapes can be stored, erased, and rerecorded.

Physical Training

Training involving physical models, simulators, or real equipment is best for complex tasks. This allows the trainee to perform the job activities under valid real-life conditions, experience emergency conditions under safe controls, and have performance monitored for feedback. Such physical training is best exemplified by the high fidelity flight simulators for pilot training used by several airlines, and the simulated coal mine for roof bolting or continuous miner operator training at the Bureau of Mines Bruceton research facility near Pittsburgh, Pennsylvania.

One advantage of physical training is that the operators are also *work hardened* in the process, that is, they perform muscle exertions or wrist motions under controlled conditions and reduced frequencies, so that the body can gradually build up to the more extreme conditions found on the job. This procedure has been quite successful, for example, in reducing work-related musculoskeletal disorders for meat packers, mentioned by OSHA in their meatpacking guidelines (OSHA, 1990) and recommended by the American Meat Institute in its Ergonomics and Safety Guidelines.

The Learning Curve

Industrial engineers, human behavior engineers, and other professionals interested in the study of human behavior recognize that learning is time dependent. Even the simplest operation may take hours to master. Complicated work may take days and even weeks before the operator achieves the coordinated mental and physical qualities enabling him or her to proceed from one element to another without hesitation or delay. This time period and the related level of learning form the learning curve, a typical one of which appears in Figure 18–1.

Once the operator reaches the flatter section of the learning curve, the problem of performance rating is simplified. However, it is not always convenient to wait this long to develop a standard. Analysts may be obliged to establish the standard at the point where the slope of the curve is greatest. In such cases, analysts must have acute powers of observation and must be able to execute mature judgment based on thorough training so that an equitable normal time is computed.

FIGURE 18–1
Typical productivity increase graph.

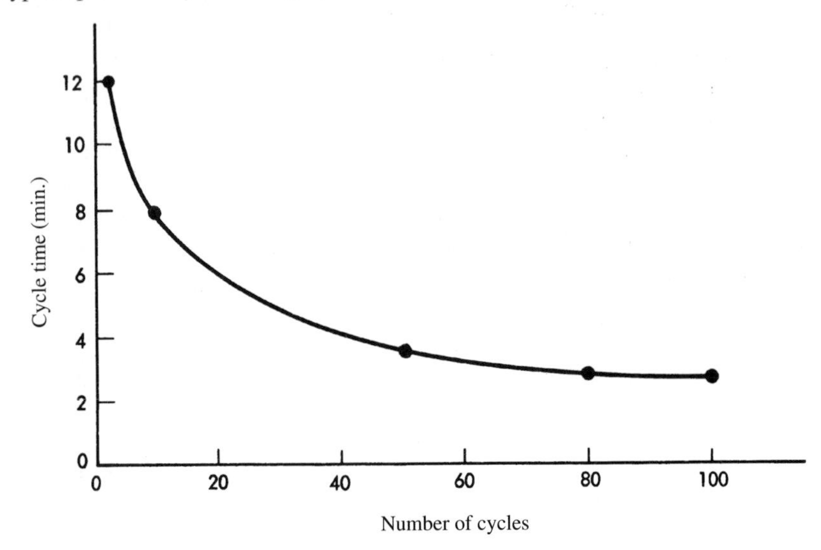

It is helpful to have available learning curves representative of the various classes of work being performed in the company. This information can be useful, both for determining the production stage at which it would be desirable to establish the standard, and for providing a guide as to the expected productivity level of the average operator with a known degree of familiarity with the operation, after having produced a fixed number of parts.

By plotting learning curve data on logarithmic paper, analysts may linearize that data, making it easier to use. For example, plotting both the dependent variable (cycle time) and the independent variable (number of cycles) shown in Figure 18–1 on log-log paper results in a straight line, as seen in Figure 18–2.

A new learning curve situation does not necessarily result every time a new design is put into production. Former designs similar to the new designs have an effect on the point at which the learning curve begins to flatten. Thus, if a company introduces a completely new design of a complex electronic panel, the assembly of this panel would involve a much different learning curve than the introduction of a panel similar to one that had been in production for the past five years.

The theory of the learning curve proposes that as the total quantity of units produced doubles, the time per unit declines at some constant percentage. For example, if analysts expect a 90 percent rate of learning, then as production doubles, the average time per unit declines 10 percent. Table 18–1 illustrates the decline in the cycle time as the number of cycles increases; with successive doubling of the cycles, a 90 percent rate of improvement is realized.

The smaller the percentage rate of improvement, the greater the progressive improvement with production output. Typical rates of learning are as follows: large or fine assembly work (such as aircraft), 70–80 percent; welding, 80–90 percent; machining, 90–95 percent.

FIGURE 18–2
Estimated cycle times based on a 20 percent reduction each time the quantity doubles.

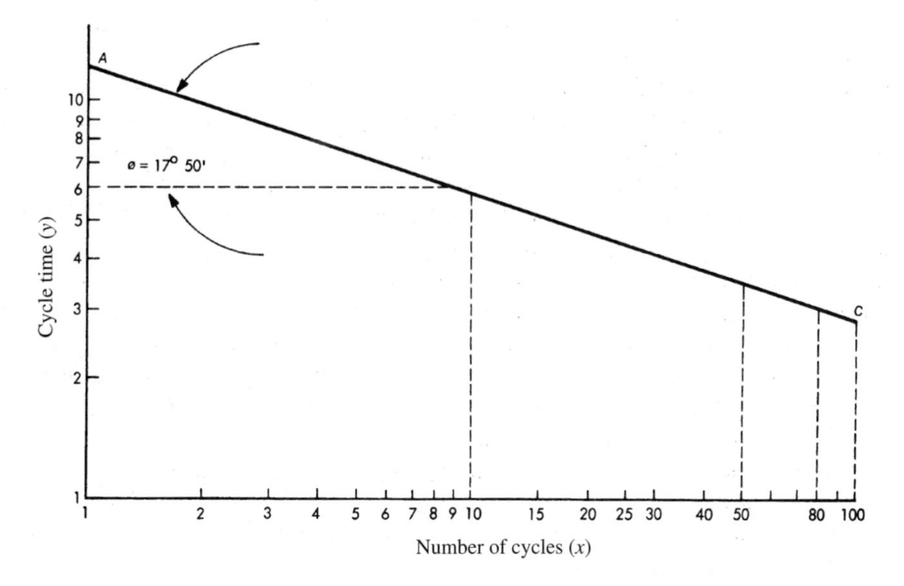

TABLE 18–1
Learning Data in a Tabular Form

Number of cycles	Cycle time (hrs)	Ratio to previous time
1	100.0	—
2	90.0	90
4	81.0	90
8	72.9	90
16	65.6	90
32	59.0	90
64	53.1	90
128	47.8	90

When linear graph paper is used, the learning curve is a power curve of the form $y = kx^n$. On log-log paper, the curve is represented by:

$$\log_{10} y = \log_{10} k + n \times \log_{10} x$$

where: y = Cycle time.
x = Number of cycles or of units produced.
n = Exponent representing the slope.
k = Value of the first cycle time.

By definition, the learning in percent is then equal to:

$$\frac{k(2x)^n}{kx^n} = 2^n$$

Taking the log of both sides:

$$n = \frac{\log_{10}(\text{learning percent})}{\log_{10} 2}$$

For 80 percent learning:

$$n = \frac{\log_{10}(0.80)}{\log_{10} 2} = \frac{-0.0969}{0.301} = -0.322$$

Also, n can be found directly from the slope:

$$n = \frac{\Delta y}{\Delta x} = \frac{(\log_{10} y_1 - \log_{10} y_2)}{(\log_{10} x_1 - \log_{10} x_2)}$$

Table 18–2 presents the slopes of the common learning curves as a function of the learning percentages. Example 18–1 should help clarify these relationships.

The analyst in Example 18–1 may wish to determine how many cycles are needed to reach a specific time, for example, a standard time of 10 minutes. Substitute $y = 10$ minutes into the learning equation, take the logs of both sides, and solve for x, we get:

$$10 = 101.5\, x^{-0.4152}$$
$$\log_{10}(10/101.5) = -0.4152 \log_{10} x$$
$$\log_{10} x = -1.006/-0.4152 = 2.423$$
$$x = 10^{2.423} = 264.8 \approx 265 \text{ cycles (always round up)}$$

Thus, it would take the worker 265 cycles to reach the standard time.

Next, the analyst may desire to know how long it takes in actual time to reach a standard time of 10 minutes. This is the area under the learning curve, which may be found by integrating under the curve:

$$\text{Total time} = \int_{x_1-1/2}^{x_2+1/2} kx^n\, dx = k\{(x_2 + 1/2)^{n+1} - (x_1 - 1/2)^{n+1}\}/(n + 1)$$
$$= 101.5\{265.5^{0.5848} - 0.5^{0.5848}\}/0.5848 = 4{,}424 \text{ min}$$

Thus, for Example 18–1, it takes a total of 4,424 minutes, or approximately 73.7 hours, to reach a cycle time of 10 minutes. The average cycle time would be $4424/265 = 16.7$ minutes.

TABLE 18–2
Relationship of the Slope of the Learning Curve to the Learning Curve Percentage

Learning curve percentage	Slope
70	–0.514
75	–0.415
80	–0.322
85	–0.234
90	–0.152
95	–0.074

EXAMPLE 18–1
Calculation of Learning Curve.

Assume that it takes 20 hours to produce the 50th unit and 15 hours to produce the 100th unit. What is the learning curve?

$$n = \frac{\Delta y}{\Delta x} = \frac{(\log_{10} 20 - \log_{10} 15)}{(\log_{10} 50 - \log_{10} 100)} = \frac{1.301 - 1.176}{1.699 - 2.000} = -0.4152$$

The learning curve percentage is:

$$2^{-0.4152} = 75\%$$

To complete the learning curve equation, we substitute one of the data points, such as (20,50), into the equation and solve for k:

$$k = y/x^{-n} = 20/50^{-.4152} = 101.5$$

Thus, the analyst's costs for the first units produced would be based upon 101.5 minutes of time to produce one assembly, not the 10 minutes developed from standard data.

An interesting question involves what happens if the operator takes a vacation. Does the operator forget some of the learning? This does in fact happen and is known as *remission* (Hancock and Bayha, 1982). The amount of remission is a function of the operator's position on the learning curve when the break occurs. The amount of remission is approximated by extrapolating a straight line from the time of the first cycle to the standard time (see Figure 18–3). The equation for this remission line is:

$$y = k + \frac{(k - s)(x - 1)}{(1 - x_s)}$$

where: s = Standard time.
x_s = Number of cycles to standard time.

Being able to estimate the time for the first unit produced and the time for successive units can be extremely helpful in estimating relatively low quantities if the analyst has the standard data and learning curve information. Since standard data are usually based on worker performance when learning has leveled off or reached the flat portion of the learning curve, the data need to be adjusted upward to ensure that adequate time is allowed per unit under low-quantity conditions. For example, let us assume that the analyst wants to know the time needed to produce the first unit of a complex assembly. The standard data analysis suggests a time of 1.47 hours, which is the cycle time for the nth unit, or the point where the learning curve begins to flatten. The nth unit, in this case, is estimated to be 300 assemblies. Based on other similar jobs, the analyst expects a 95 percent learning rate. From Table 18–2, the exponent n, representing the slope, is – 0.074. Then k, the value for the first cycle time, is:

$$k = 1.47/300^{-0.074} = 2.24 \text{ hours}$$

FIGURE 18–3
The effect of breaks on operator learning (*From:* Hancock and Bayha, 1982) (Reprinted by permission of John Wiley & Sons, Inc.)

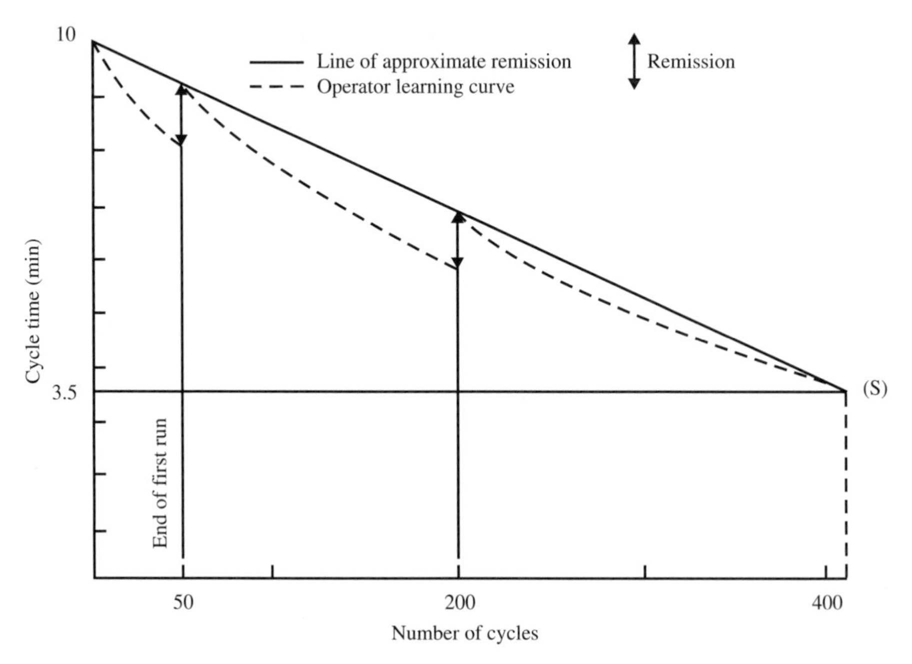

EXAMPLE 18–2
Calculation of Learning Curve with Remission.

In Example 18–1, the operator stops after 50 cycles for a two-week vacation. His cycle time for the 51st cycle will be determined from the remission function:

$$y = 101.5 + \frac{(101.5 - 10)(51 - 1)}{(1 - 265)} = 84.17$$

The operator's cycle time without a break would have been:

$$y = 101.5x^{-0.4152} = 101.5 \times 51^{-0.4152} = 19.84$$

Therefore, there has been a remission of 84.17 – 19.84 = 64.33 minutes, and a new learning curve with a new $k = 84.17$ starts. The 51st cycle now becomes the first cycle of the new learning curve of $y = 84.17x^{-0.4152}$.

Thus, the analyst's costs would be based on 2.24 hours of time to produce one assembly, not the 1.47 hours developed from the standard data.

Many factors affect human learning. Job complexity is very important. The longer the cycle length, the greater the uncertainty in movements, the more C-type or simultaneous motions (see Chapter 13)—the more training will be required. Similarly, individual capabilities, such as age (rate of learning declines with age), prior training, and physical capabilities all affect the ability to learn.

PLANT TRAINING PROGRAMS

When properly undertaken, any methods training program is self-supporting. Several companies that have introduced such training have almost immediately realized substantial savings in all plant areas. One farm machinery manufacturer gave a 64-hour course in methods analysis over 21 weeks. In attendance were 42 supervisors, assistant supervisors, time study analysts, and other key operating personnel. At the termination of the course, 28 methods projects were turned in, representing methods improvement solutions that could be introduced in the plant. These ideas included: machine coupling; redesigning tools and products; simplifying paperwork; improving layout; handling material better; improving dies, jigs, and fixtures; eliminating operations; adjusting tolerances and specifications; and many others that resulted in a savings of thousands of dollars annually.

In addition to the immediate gains mentioned, the training in operations analysis and work simplification developed an analytical awareness and ability on the part of the operating personnel, so that in the future, they were continually on the alert to find a better way. They developed an appreciation for the cost of manufacture, and, at the completion of the course, they were more cognizant of the relationship between output and selling price. By providing the means for on-the-job training in the field of methods and related areas, this company went a long way toward assuring its place in an extremely competitive market.

Methods and Time Study Training

The lack of success of some time and methods study programs is due in part to a lack of understanding of the techniques by both management and operating personnel. One of the easiest ways to ensure the success of any practical innovation is to inform all affected parties as to how and why it will operate. When the theories, techniques, and economic necessity of methods, work measurement, and employee motivation are understood by all parties, little difficulty is usually encountered in their application. In unionized shops, it is especially important that the local union officers at the local level understand the need for and the steps involved in establishing performance standards. Training in the areas of performance rating, allowances application, standard data methods, and job evaluation are especially essential. Companies that have provided training in the elements of time study for union officials and stewards, as well as for representatives of management, have experienced harmonious relationships with respect to methods, standards, and wage payments.

Management should sponsor training to acquaint various operating and supervisory personnel with the philosophies and techniques of time and motion study. In addition, industry must provide training for those who plan to make time and motion study their life's work. Also, experienced analysts should be checked continually to make certain that their conception of normal is not deviating from the standard. Periodic verification of the rating ability of the time study staff is fundamental.

New developments are constantly being made. As they are recognized, personnel in the methods, time study, and wage payment sections should be trained accordingly.

Any company that has, or plans to have, a program of work simplification or methods analysis, time study, work measurement, and incentive wage payment should include a continuing training program as part of its installation. A 2-hour training period once a week for supervisors, union stewards, direct labor, and management, can be well worth the time and money spent.

Creativity Development

Creative work is not confined to a particular field or a few individuals, but is carried on in varying degrees by people in many occupations: the artist sketches, the newspaper writer promotes an idea, the teacher encourages student development, the scientist experiments with a theory, and the methods and time study analyst develops improved methods of doing work.

Creativity implies newness, but it is often just as concerned with the improvement of old products. A "how to produce something better" attitude, tempered with good judgment, is an important characteristic of effective methods and time study analysts.

Developing creativity in practicing methods and time study analysts is a continuing challenge. Knowledge of the fundamental principles of physics, chemistry, mathematics, and engineering is a good foundation for creative thinking. If practicing analysts do not have this basic background, they should acquire it either by education or through self study. Of course, knowledge is only a basis for creative thinking; it does not necessarily stimulate it.

The personal characteristics of curiosity, intuition, perception, ingenuity, initiative, and persistence contribute to creative thinking. Curiosity seems to stimulate more ideas than any other personal characteristic. One aid in the development or restoration of curiosity is careful observation. Methods and time study analysts should get into the habit of asking such questions as how a particular object was made, what materials were used in the construction, why it was designed for a particular size and shape, why and how it was finished as it was, and how much it cost. If they cannot answer these questions themselves, they should seek the answers, either through analysis or by references to source materials and other experts. These observations lead creative thinkers to see ways in which products or processes can be improved through cost reduction, quality improvement, ease of maintenance, or improved aesthetic appeal.

One significant creative idea usually opens up fields of activities that lead to many new ideas. Frequently, one idea that is applicable to one product or process is equally applicable to other products and similar processes.

EMPLOYEES AND MOTIVATION

Employee Reactions

In addition to having an understanding of the unions' objectives and its attitudes toward the methods, standards, and wage payment approach, analysts must have a clear understanding of the psychological and sociological reactions of operators. Three points should always be recognized:

1. Most people do not respond favorably to change.
2. Job security is uppermost in most workers' minds.
3. People have a need for affiliation and are consequently influenced by the group to which they belong.

Most people, regardless of their positions, have an inherent resistance to changing anything associated with their work patterns or work centers. This is due to several psychological factors. First, change indicates dissatisfaction with the present situation. But the natural tendency is to defend the present way, since it is intimately associated with the individual. No one likes others to be dissatisfied with their work; if a change is even suggested, the immediate reaction is to expound on why the proposed change will not work.

Second, people tend to be creatures of habit. Once a habit is acquired, it is difficult to give up, and there is resentment if someone endeavors to alter the habit. For example, anyone in the habit of eating at a certain place is reluctant to change to another restaurant, even though the food may be better and less expensive.

Third, people naturally desire security in their position, which is just as basic as the instinct for self-preservation. In fact, security and self-preservation are related. Most workers prefer job security over high wages when choosing a place to work.

Fourth, to the worker, all methods and standards changes appear to be an effort to increase productivity. The immediate and understandable reaction is to believe that if production goes up, the demand will be filled in a shorter period; and without demand, there will be fewer jobs.

The solution to the need for job security lies principally in the sincerity of management. When methods improvement results in job displacement, management is responsible for making an honest effort to relocate those who have been displaced. This may include providing for retraining. Some companies have gone as far as to guarantee that no one will lose employment as a result of methods improvement. Since the labor turnover rate is usually greater than the improvement rate, the natural attrition through resignation and retirement can usually absorb any people displaced as a result of improvement.

Fifth, the sociological need for affiliation and the resulting impact of "behaving as the group wants everyone to behave" also influence change. Frequently, the worker, as a union member, feels that he or she is expected to resist any change that has been instituted by management; consequently, the worker is reluctant to cooperate with any contemplated changes resulting from methods and standards work. Another factor is the resistance toward anybody who is not part of one's own group. A company represents a "group" that has several groups within its major boundaries. These individual groups respond to basic sociological laws. Change proposed by someone outside one's own group is often received with open hostility. The worker is associated with a different group than that of the methods and standards practitioner, and tends to resist any effort from analysts that might interfere with the usual performance within the group.

Maslow's Hierarchy of Human Needs

Psychosocial factors such as stress, needs, or rewards can be very important aspects of worker productivity. Workers naturally want to work with the least amount of stress and the greatest amount of rewards. Maslow (1954) quantified these wants into a hierarchy comparable to a set of steps leading to the top of a pyramid, or the ultimate goal (see Figure 18–4). Each lower want or need must be satisfied before a worker will seek rewards at the next higher level. The lowest level includes the *physiological* needs corresponding to survival, food, water, and health. Job related factors at this level could be sufficient pay or other monetary rewards.

Once these physiological wants are satisfied, the second level, *safety* needs, becomes important. Safety needs include the need for security, both in the physical and psychological sense. These could be as simple as trying to avoid physical injury on the job, or as complex as seeking a "nice" supervisor who doesn't threaten or demean the worker. With the prevalence of company downsizing in the late 1990s, safety needs could include job security and seniority rights.

The third level, *social* needs, includes the need for attention, friendship, social belonging, and meaningful relationships with co-workers. In the fourth level, *self-esteem* needs, workers strive for competence and achievement, express a desire for self respect, or seek to satisfy their egos.

At the top of the pyramid, the final or fifth level is *self-fulfillment.* The workers have finally achieved all of their needs, they are personally fulfilled, and their egos are satisfied. This level can vary considerably from individual to individual. Where some people may be satisfied making widgets day in and day out, others may only be satisfied running their own businesses.

The industrial engineer may wonder what purpose Maslow's hierarchy serves on the plant floor, or how these wants can be satisfied for the production worker. Consider the first level of physiological needs. One tactic, though very negative in terms of labor–management relationships, is the threat of termination for failure to meet production quotas, or for violation of safety rules. Other scare procedures or hard-sell approaches fall in the same category. A more positive approach is the implementation of wage incentives (Chapter 17). This is classic condition-

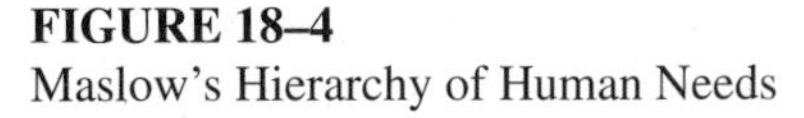

FIGURE 18–4
Maslow's Hierarchy of Human Needs

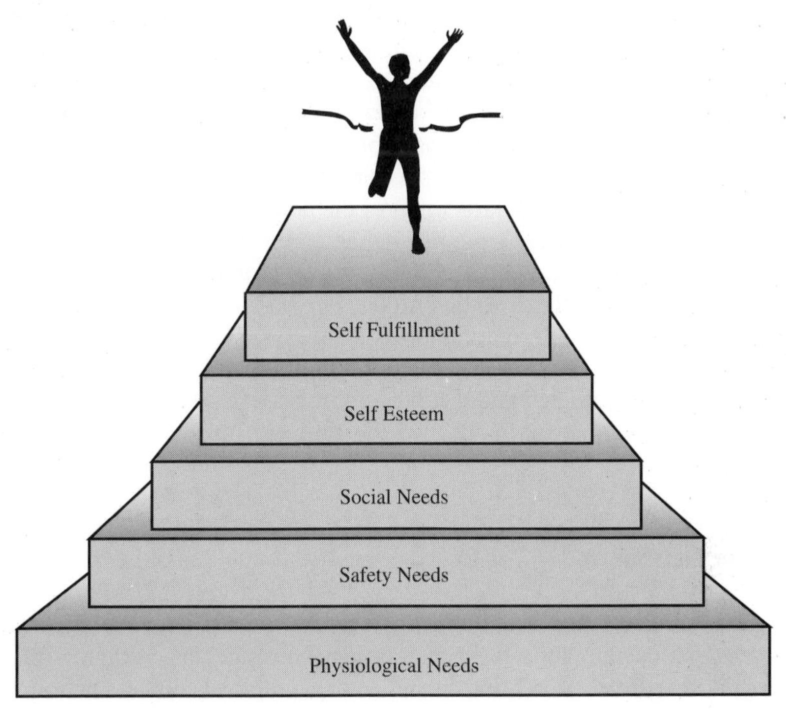

ing, or positive reinforcement, at its simplest. Many workers are willing to work at relatively tedious jobs, or at higher production rates, given sufficient monetary incentive. Thus, workers trade increased satisfaction with the job, provided by the extra pay, for decreased satisfaction on the job. Unfortunately, with increasing wealth and progressive income taxes, additional income becomes less meaningful, and the industrial engineer may need to proceed to higher levels of Maslow's hierarchy.

At the second level, safety or security needs, overall job security is the concern, especially with the increasing trend toward downsizing or rightsizing. Traditionally, in other cultures, especially in Japan, a job was a lifetime guarantee with that company. In the United States, it is not unusual for a worker to change jobs every five to six years and work for a half dozen employers during his or her lifetime. Employment can perhaps be guaranteed for a fixed number of years. At the worksite level, specific regulations regarding work practices, physical guarding of unsafe machines, or safety contests can improve the overall safety or the working climate.

At the third level, social needs, workers seek "belonging" within a social system. In terms of work, this could imply having friendly co-workers, comfortable interaction with management, participation on ergonomics or safety committees, etc. Such formal organizations are much more common in Japan with their Quality Circles, in Germany, where workers elect a work council (Betriebsrat) to handle grievances and negotiate with management, and in Swedish auto plants with their work groups (arbetsgrupper).

At the fourth level, the workers seek an increase in self-esteem. This could be provided by making the work more challenging, adding more responsibility, and providing greater variety. The latter can be done through *job enlargement,* a horizontal expansion of work. Instead of just tightening one set of bolts all day, the worker could perform the complete assembly. This not only increases the worker's sense of responsibility, it also utilizes a variety of his or her muscles and joints, dividing the work stress over a larger part of the body, and thus reducing the risk of cumulative trauma disorders. Tied into job enlargement is also *job enrichment,* a vertical expansion of work, which allows workers both to start and to complete a given task, diversifying duties so that no one person has all of a boring task, delegating decision making, and rotating job assignments. *Job rotation* is similar to job enlargement in that any worker gets the opportunity to do a variety of tasks, while adhering to a more rigid schedule. Job rotation has effects similar to job enrichment in varying job stressors and allowing fatigued muscles and body parts to recover.

Volvo Approach

All of these concepts (job enlargement, job enrichment, job rotation, and work groups) were pioneered in the 1960s in Sweden. The impetus was increasing absenteeism, wildcat strikes, worker unrest, and general employee dissatisfaction. Drastic changes had to be made. Therefore, under the direction of its president Pehr Gyllenhammer, Volvo devised a revolutionary plan and built a completely new auto assembly plant at Kalmar in 1974. The traditional conveyor line was replaced by an automated guided vehicle (AGV) system on which the car assembly took place. The AGV was guided by an electronic system of cables imbedded in the floor. A central computer controlled the movement of the AGVs throughout the plant, but could be overridden by the employees at any time. In addition, there was a drastic change in work organization: employees were fully involved and formed work groups that received and examined factory orders, decided exactly which group member would do which task for the given day, inspected their own work, completed paperwork after assembly, and, at the end of the day, had a brief discussion of the day's happenings and problems. Job enlargement was carried out to the highest degree in that one group of workers assembled over 25 percent of one car.

The Kalmar design was successful from the beginning, as the work became more meaningful and workers assumed more responsibility. Absenteeism and employee turnover were greatly reduced, while cost and production targets were met. Because of the success at Kalmar, similar new plants were opened at Uddevalla and Göteborg (Torslunda). Unfortunately, due to a shifting market and radically lower sales figures, Volvo eventually closed the Uddevalla and Kalmar plants. In 1997, the Uddevalla plant reopened with the production of a new sports car.

Note that all three forms of job reorganization—job enlargement, job enrichment and job rotation—were in place in the Volvo plants. Cycle times increased to many hours, decreasing the repetitiveness of motion for any one limb or set of muscles.

At the fifth and highest level of Maslow's hierarchy, the worker would be expected to devote himself completely to the company. Other than in Japan, this is probably not feasible in any large-scale company. On the other hand, in small, start-up companies, not only the owner, but also some of the closest colleagues, may put in most of the waking hours in keeping the company afloat. Then, the company and work truly becomes one's self actualization.

Motivation

An interesting *motivation-maintenance theory* was developed by Herzberg (1966), based on a survey of factors leading to satisfaction or dissatisfaction for 1500 employees in 12 different organizations. Similar to Maslow's theories, Herzberg found two basic but different needs in individuals. If workers were dissatisfied with their jobs, their main concern was the working environment. However, if they were satisfied with their jobs, their satisfaction dealt with the actual work itself.

Herzberg classified the environmental factors as *extrinsic* and potential dissatisfiers. These included such factors as the administration, supervision, working conditions, salary, and interpersonal relations. The potential satisfiers or motivators, which included achievement, recognition, responsibility and advancement, he termed *intrinsic* factors. The extrinsic factors had little positive effect, but could be strong dissatisfiers, leading to large negative feeling. The intrinsic factors encouraged workers to be more productive and satisfied. Therefore, it is in the manager's interest to maximize the intrinsic factors and minimize the negative effects of the extrinsic factors.

One of the most effective intrinsic motivation techniques is job enrichment, which is the opposite of job simplification. With work methods and the principles of motion economy, the typical goal of an industrial engineer is job simplification. If the job is simple and repetitive, little learning is required and workers can be easily interchanged. This approach was developed for the machine-like consistency required on an assembly line. However, workers are not machines, and when subjected to such conditions, they may become bored and dissatisfied, leading to increased absences and job changes. Even worse, as recent statistics show, are increased stress levels leading to increases in cumulative trauma disorders. It is not worth saving pennies on more repetitive jobs when thousands of dollars are lost in the resulting injuries.

Herzberg also found some interesting deviations in the survey results, depending on the populations examined. These could be used to a company's advantage, depending on the composition of the worker population. For example, younger workers were less concerned about job security than older workers and were generally more satisfied with the whole organizational reward system. More highly educated and more highly paid workers favored the intrinsic rewards. Extrinsic rewards ranked higher overall than intrinsic rewards, but were most prized by less educated, lower paid, and older workers.

HUMAN INTERACTIONS

Interactions between employees at the workplace are an important component of morale and productivity. Several approaches can be used to deal and communicate with people, two of which are discussed here: transactional analysis and the Dale Carnegie approach.

Transactional Analysis

Transactional analysis, as developed by Berne (1964), consists of several components: (1) ego states, (2) transactions, (3) stroking and stamps, and (4) more complex games and lifestyles. There are three ego states, which are found to some degree in all people at all times. The *parent ego state* reflects attitudes and values absorbed from parents as authority figures and produces a statement such as "That's really a dumb mistake." Bill Cosby of television's *Cosby Show* would be a good example of the parent ego state. The *adult ego state* logically analyzes facts, makes rational decisions and conclusions, and operates with phrases such as, "Let's examine that problem carefully." Mr. Spock of the *Star Trek* series would be a perfect example of the adult ego state. The *child ego state* is more complex and may take up to three different forms. A naive state produces responses such as, "Oh, I didn't know that." The adaptive state establishes internal rules based on social conditioning, such as "respect your elders." The manipulative state may fake injuries to get out of something unpleasant, such as pretending to have a cold to get out of school.

Interactions between the ego states occur in the form of transactions. Participants can both send and receive messages from any of the three ego states. If the messages are sent and received at the same ego state level, such as adult to adult, the transactions are termed *complementary* and are considered to result in a positive and successful exchange (Figure 18–5). A parent to child transaction (Figure 18–6), if occurring at a parallel level, is still considered complementary, but may not be as effective as a transaction occurring at the same level.

A crossed transaction occurs when each party assumes a different transaction level, and this often results in anger or hostile feelings (Figure 18–7). Ulterior transactions, although appearing logical on the surface, always have a hidden meaning, and they form the basis for games (Figure 18–8). As an example, a line supervisor conducts an adult-to-adult transaction on the surface, but in reality only goes through the motions and produces a parent-to-child transaction. This may be how an operator wants to be treated; if not, the line supervisor should not be surprised when the employees complain that nobody ever listens to them.

Transactional analysis stresses that all people feel the need to be recognized in some manner. This need (the fourth step in Maslow's hierarchy) probably starts in childhood and continues well into adulthood. Recognition can come as either positive or negative strokes, positive being based on good attributes, such as recognition for intelligence, helpfulness, compassion, etc., while negative is based on bad attributes, such as deceit, selfishness, etc. Only positive strokes (you're OK) keep a per-

FIGURE 18–5
Complementary transaction: adult to adult—the message is sent and received appropriately (Adapted from Berne, 1964).

Production Manager:
"The grinding station needs to get back up to rate."

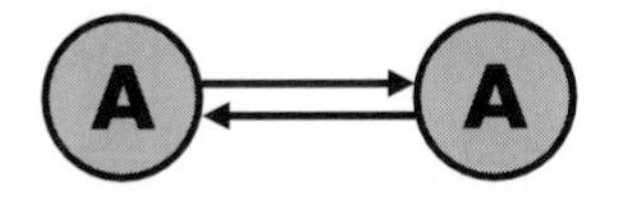

Line Supervisor:
"Yes, I'll go over and take care of it."

FIGURE 18–6
Complementary transaction: parent to child—this is not as effective as the adult to adult complementary transaction, but it is still useful (Adapted from Berne, 1964).

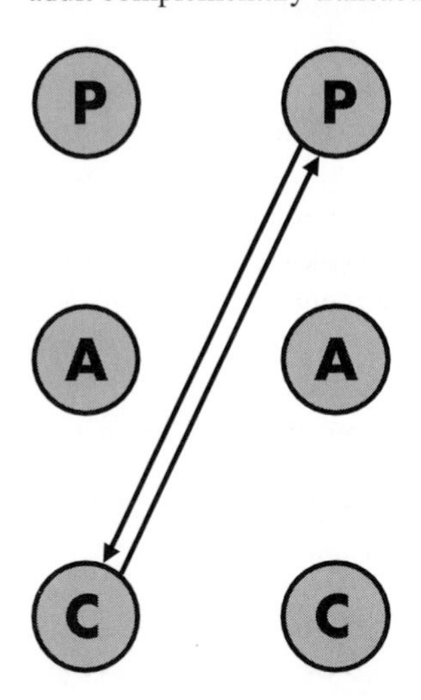

Production Manager (with a troubled facial expression):
"I was wondering if you could get those grinders back up to rate."

Line Supervisor (in a patronizing tone):
"Now don't you worry, I'll take care of it."

son mentally healthy. Negative strokes may leave a person with a chip on the shoulder and a bad view of the world. Excessive negative stroking (criticism) in childhood may carry into adulthood, with the person seeking transactions leading to sympathy or dependency. Some individuals may become extreme positive stroke seekers.

FIGURE 18–7
Crossed transaction—this can result in anger and hostile feelings (Adapted from Berne, 1964).

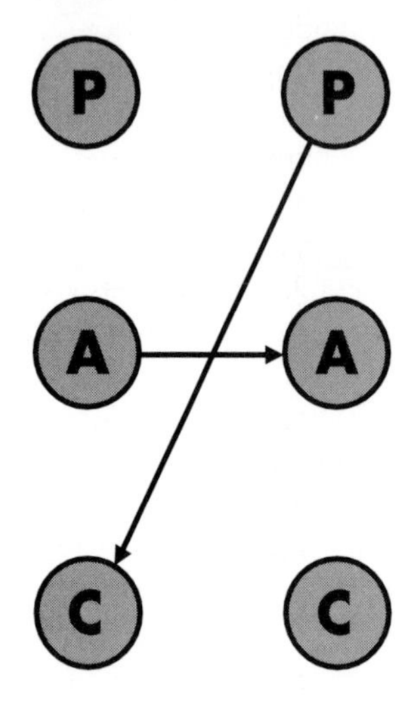

Production Manager (as adult to adult):
"What has been done to correct the material handling problem?"

Line Supervisor (as parent to child):
"Don't you have anything better to do but bug me all the time?"

FIGURE 18–8
Ulterior transaction—although this approach appears logical, it can have a hidden meaning and form the basis for games (Adapted from Berne, 1964).

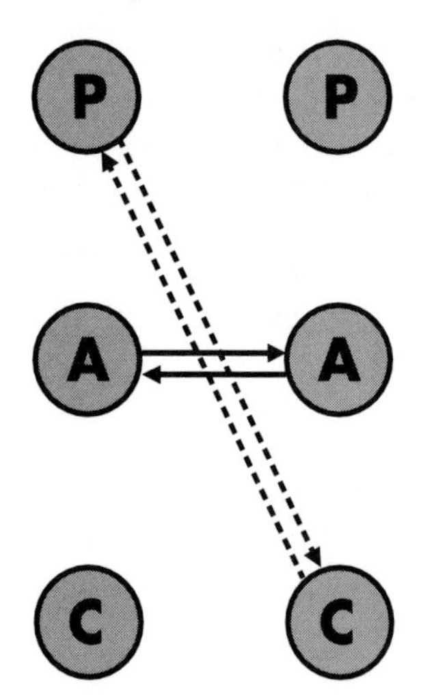

Production Manager (in apparent adult to adult transaction):
"I'll find out what's causing that machine to jam up and get back to you."

But, is really thinking:
"Oh, all right, I'll check on it and get back to you."

Operator hears a condescending tone and sees it as a parent to child transaction.

As the transactions become more complex, they take the form of rituals, pastimes, and games. Rituals are the simplest cultural ties, such as simple morning greetings, "Hi, how's it going?" Pastimes are more complex interactions such as conversations regarding work, sports, or friends in social functions. Games are the most complex transactional interactions, which may replace intimacy in private life, or may produce accident prone behavior (a child seeking forgiveness) at work.

In general, the industrial engineer or manager should try to have a basic understanding of transactional analysis, to interact better with production workers and other personnel. Complex games should be avoided by switching from a parent or child ego to the adult ego. This works in "yes, but" situations in which, because of an ulterior motive for one of the participants, the problem-solving effectiveness of the situation is reduced. The manager should sense that when all suggestions for improving the workstation design are rejected with "yes, but can't" comments, the transaction has proceeded into a parent-to-child mode. Switching to an adult-to-adult transaction with "Yes, that is indeed difficult—what can you do about it?" will short-circuit the game and get directly to the problem. In other words, in crossed transactions, it is better to change egos, even if they are at a parent-to-child level and are thus less effective than adult-to-adult transactions. Finally, it may even be necessary to participate at a lower level of games, such as giving or receiving strokes. In many companies, there is a "Calamity Jane or Joe" who is always involved in one sort of problem or another, whether jamming a machine or damaging a tool. These individuals may have carried a quest for negative strokes and forgiveness through from childhood into adulthood. Switching to an adult ego with, "I accept responsibility for assigning you to that job," would eventually stop the game, but may also create an enemy. Another approach may be to counteract the negative strokes by providing more positive strokes in the form of recognition of the good things done by that operator (e.g., above average performance, high quality, etc.) (Denton, 1982).

Above all, the industrial engineer should take time to talk to the operators and get their ideas and reactions. The work progresses much more smoothly and effectively if operators become part of the team. However, they must be asked, not instructed, to "join the team." The operators are closer to their job situations than anyone else and usually have more specific knowledge of details than anyone else. This knowledge should be realized, respected, and utilized. Accept operators' suggestions gratefully; if they are practical and worthwhile, put them into effect as soon as possible. If they are used, be sure that the operators are appropriately rewarded. If they cannot be used at present, give a complete explanation as to why they cannot be used. At all times, analysts should imagine themselves in the workers' place and then use the approach they would like used toward themselves. Friendliness, courtesy, cheerfulness, and respect, tempered with firmness, are the human characteristics that must be practiced to be successful in this work. In short, the golden rule must be applied.

Dale Carnegie Approach

The human approach to handling people, making people like you, influencing the thinking of people, and changing people, has been developed to a fine art by Dale Carnegie in his series of courses. Carnegie's principles and thoughts are summarized in Table 18–3.

TABLE 18–3
The Dale Carnegie Approach

Fundamental Techniques in Handling People

1. Instead of criticizing people, try to understand them.
2. Remember that all people need to feel important; therefore, try to figure out the other person's good points. Forget flattery; give honest, sincere appreciation.
3. Remember that all people are interested in their own needs; therefore, talk about what they want and show them how to get it.

Six Ways to Make People Like You

1. Become genuinely interested in other people.
2. Smile.
3. Remember that a person's name is to him or her the sweetest and most important sound in the English language.
4. Be a good listener. Encourage others to talk about themselves.
5. Talk in terms of the other person's interest.
6. Make the other person feel important—and do it sincerely.

Twelve Ways to Win People to Your Way of Thinking

1. The only way to get the best of an argument is to avoid it.
2. Show respect for the other person's opinions. Never tell anyone they are wrong.
3. If you are wrong, admit it quickly and emphatically.
4. Begin in a friendly way.
5. Get the other person saying yes, immediately.
6. Let the other person feel that the idea is his or hers.
7. Let the other person do a great deal of the talking.
8. Try honestly to see things from the other person's point of view.
9. Be sympathetic with the other person's ideas and desires.
10. Appeal to the nobler motives.
11. Dramatize your ideas.
12. Throw down a challenge.

Nine Ways to Change People Without Giving Offense or Arousing Resentment

1. Begin with praise and honest appreciation.
2. Call attention to people's mistakes indirectly.
3. Talk about your own mistakes before criticizing the other person.
4. Ask questions instead of giving direct orders.
5. Let the other person save face.
6. Praise the slightest improvement and praise every improvement. Be hearty in your approbation and lavish in your praise.
7. Give the other person a fine reputation to live up to.
8. Use encouragement. Make the fault seem easy to correct.
9. Make the other person happy about doing the thing you suggest.

COMMUNICATIONS

Industrial engineers as middle-level managers spend a considerable amount of their time in interpersonal communications. Therefore, mastery of the ability to communicate effectively goes a long way toward selling an argument or a design, even though it may be worthy on its own merit. Communications can be divided into five major types: verbal, nonverbal, one-to-one, small group, and large audience (Denton, 1982).

Verbal Communications

In verbal communications, words are very powerful and their meanings become very important. Thus, the word "production" is all powerful, while others such as "safety" or "human factors" may carry a negative connotation, because they imply, whether true or not, coddling the workers or slowing down the production. A person's name (and family member names) is very important to that person. Therefore, managers should know the workers' names (and a little about their background), to stimulate the interest of the opposing party and to make the conversation more rewarding.

One problem in any language is the specific meaning of a given word. With greater diversity in the workplace, there is a greater chance that the other individual may assign a slightly different meaning, make a different inference, or perhaps not even understand the meaning of some words.

Managers must also be careful not to dichotomize the world. Categorizing things as either good or bad, safe or unsafe, etc., polarizes events and causes individuals to concentrate on differences rather similarities.

Nonverbal Communications

Some data indicate that more than 50 percent of a message, especially related to feelings, is presented through nonverbal channels, including voice characteristics, facial expressions, body language, etc. In voice characteristics, a rapid speech pattern indicates excitement, while a slower rate with pauses indicates passive emotion. Facial expressions and body language involve such nonverbal behavior as head nodding to indicate attention to the other person's discourse, raising of the eyebrows to indicate surprise, maintaining eye contact to indicate trust, crossing of arms or clenching of fists to indicate a defensive attitude, crossing of the legs to indicate superiority or a lack of involvement, etc.

Other factors, such as the amount of space around the individual, can also affect communications. For example, people try to maintain a certain amount of open space around them; closing this space forces a greater amount of discomfort even though it may increase interaction.

One-to-One

One-to-one or *dyadic* communications occur frequently between a manager and a worker in a face-to-face situation. The purpose of such communication is generally to bring about an understanding of the goals between the two individuals. One of the two may then seek to obtain approval for a proposed idea, and available solutions may need to be presented. To obtain the expected solution, it may be necessary to use motivational techniques, such as guided questioning, which can either be leading questions that guide the answers in a certain direction, or closed-end (yes/no or limited choice) questions to elicit commitment, or open-end questions to elicit discussion.

Unfortunately, conflicts can arise during the conversation. Simple conflicts arise when each side knows the other's goals, but neither can win without the other failing. In such cases, delaying further discussion until both sides can cool down and find a rational solution may be appropriate. Pseudo-conflicts arise because of ineffective communication and can only be diffused when accurate data are provided and distortions eliminated. The worst conflicts are ego conflicts that relate to the previously discussed Berne's (1964) transactional analyses.

Small Groups

Typically, small-group communications are centered around problem solving. Problems can be quite complex, and no one individual may have all the solutions. Therefore, the concept of having a group of individuals working on a problem seems logical. Additional benefits are that extreme individual judgments tend to be moderated, overall judgments tend to improve in accuracy, and a wider range of information or opinions is included in the discussions. There are also tradeoffs. By their very nature, small groups are time consuming. Also, lack of coordination, low motivation, and personality conflicts among the group members can result in the failure of the group to meet is objectives. Therefore, it is important to organize and administer small groups effectively.

The basic problem-solving procedures (Chapter 2) must be followed in small-group communications. The group must identify the problem, analyze the details, develop a variety of ideas, select specific ideas for further development, evaluate the different alternatives, and then specify and sell the solution. To improve this process, the group's facilitator must allow easy access to information, to encourage the building of trust. High standards and proper planning are also very important, as are specific interactional techniques that can increase the effectiveness of the process.

Facilitating agreement

Facilitating agreement among all members, such as obtaining a consensus, can be improved by positively involving all group members, reinforcing their self esteem, using open-ended questions, summarizing each individual's comments before proceeding to the next person, and summarizing the pros and cons of a discussion before proceeding to the next topic.

Role playing

Role playing can help strengthen a group's problem-solving ability by presenting appropriate situations or events. This can be followed by further participation and discussion through *buzz groups* (smaller subgroups). One group member may act as recorder to write down the group's ideas quickly. This can easily lead to *brainstorming* sessions, for which the basic guidelines are: ideas are encouraged, regardless of how wild; the more ideas, the better; no ideas are criticized (sometimes the contributors are not identified); and participants are encouraged to build on or combine previous ideas. Usually, a time limit of 10 minutes is set, after which the ideas are ranked, with possible solutions being included. The pros and cons of each idea are discussed and the potential solutions are voted upon. The top vote getters are further reviewed and voted upon again, until the process of elimination leaves only the best solutions (Denton, 1982).

Quality circles

Quality circles are a small-group format developed in Japan in 1963 to assist in solving quality control problems. The essence is participative problem solving in groups of eight to ten people, including workers, engineers, and managers. It is important to have participants from the different departments involved with this product. These volunteers are given special training in statistical quality control techniques and typically hold meetings once or twice a month. With the help of a facilitator, the group selects a problem that is a cause for product defects and could potentially have a solution. Typically, exploratory operational tools, such as Pareto distributions and fish diagrams (Chapter 2), are used to help in identifying the problem and the factors involved. The group then recommends potential solutions, such as improved procedures or design changes and it then attempts to implement the solution. All of this is done with the cooperation of management (Konz, 1995).

Ergonomics teams

A logical extension of quality circles, to combat the high rates of musculoskeletal disorders in U.S. companies, is the ergonomics team. These are typically interdisciplinary teams consisting of an ergonomist (if there is one on staff), an industrial engineer, a safety specialist, a medical person (typically, the plant nurse), several interested production workers, a labor union member, and perhaps a representative from higher management. This committee typically meets once or twice a month and follows a procedure similar to that used by quality circles in seeking solutions to problem-causing jobs. In the authors' experience, many companies, including such large companies as auto manufacturers and smaller companies of fewer than 500 employees, have experienced considerable success in using such teams.

Large Audiences

Industrial engineers or middle managers rarely present information to large groups, and this topic will not be considered here. Considerable information on producing presentations and using effective delivery techniques for large audiences is available in other sources.

LABOR RELATIONS AND WORK MEASUREMENT

Every business owner recognizes the importance of harmonious labor–management relations. Sound work measurement philosophies and practices do a great deal to promote good relations between labor and management. In contrast, a lack of consideration of the human element in work measurement procedures causes sufficient turmoil to make the profitable operation of a business impossible. Management should identify and implement conditions most likely to enable employees to achieve the organization's objectives.

To understand the relationship between work measurement and labor relations, analysts must understand the objectives of the typical labor union. Briefly, the principal objectives of the typical union are to secure for its members higher wage levels, decreased working hours per workweek, increased social and fringe benefits, improved working conditions, and job security. The philosophy underlying the union movement has, in the past, had much to do with opposition to incentive systems. Unions formerly looked upon themselves primarily as fighting units that united workers by seeking ends common to all members. It was not to the advantage of the early unions to emphasize differences in workers' abilities and interests; to do this would increase rivalries and jealousies among their members and potential members. Consequently, organized labor usually sought percentage wage increases for all members of a group, rather than means by which remuneration could be adjusted to the worth of the individual worker. The work of methods, standards, and wage payment analysts came to be looked upon as means by which management sought to destroy the solidarity of workers by stressing the differences in their capabilities.

The enactment of government legislation, however, has changed the status of labor unions. Management today recognizes the union as a bargaining agent for its employees. Unions, therefore, are less fighting units than bodies concerned with the orderly negotiation of wage contracts for their members. Also, many union members are not content with wage negotiation concerned only with obtaining high minimum wages for the entire group, while leaving the determination of extra rewards for more valuable workers entirely up to management. To satisfy most of their members, unions must obtain equitable wages (recognizing workers' different skills and qualities), as well as high wages, for all. In fact, unions have already done this in many instances. Methods, time standards, job evaluation, merit rating, and incentive systems are tools for assuring equitable wages and good working conditions. They may soon be as important to organized labor as to management.

Labor executives, in bargaining over wage contracts, are in an excellent position to obtain safeguards for the proper development of work methods and standards and for fair wage payment practices. They may, for example, demand that provisions be inserted in contracts: (1) forbidding the reduction of standard times without a methods change; (2) requiring minimum hourly rates and payment for time lost due to no fault of the workers; (3) establishing grievance procedures for handling all workers' complaints arising out of the functioning of job evaluation, merit rating, and wage incentives; and (4) even giving labor the right to participate in job and worker rating activities, the setting of standard times, and the determination of piece rates.

Many unions today train their own work measurement personnel. In most instances, however, these time study people are employed to check standard times and to explain them to workers, rather than to take part in their initial establishment. However, the training that the union time study analysts receive usually treats the concepts, philosophies, and techniques of methods, work measurement, and wage payment from a different point of view than that of management's time study personnel.

In many instances, company training of union time study representatives has been quite successful as a means of promoting a more cooperative atmosphere in the installation and maintenance of methods, standards, and wage payment systems. This procedure provides for the joint training of both company and union personnel. Having received this training, union representatives are much more qualified to evaluate the fairness and accuracy of the technique and to discuss any technical points relative to a specific case.

MODERN MANAGEMENT PRACTICES

Toyota Production System

Because of its use of quality circles, which emphasize respect for the worker, while improving productivity through the methods discussed previously, the Toyota Production System (TPS) deserves special mention. The Toyota Production System was developed by the Toyota Motor Corporation as a means of eliminating waste in the aftermath of the 1973 oil embargo. Its primary purpose is the improvement of productivity and the reduction of costs following the footsteps of the Taylor system of scientific management and the Ford mass-assembly line. Yet, it is much broader in concept, targeting not only manufacturing costs, but also sales, administrative and capital costs. Toyota felt that it would be dangerous to follow the Ford mass-production system blindly, which worked fine in times of high-growth. In times of lower growth, it was important to give more attention to cutting waste, decreasing costs, and increasing efficiency.

TPS highlights seven types of *muda* or waste (Shingo, 1981): (1) overproduction, (2) waiting, (3) transportation, (4) processing, (5) stocks, (6) motion and (7) defective products. These are very similar to the operations analysis techniques and methods study approaches presented in Chapters 2 and 3. For example, waiting and transportation are elements examined directly with flow process charts for potential elimination or improvement. Waste motion summarizes the Gilbreths' lifelong work in motion study, culminating in the principles of work design and motion economy. It also includes gross movements of the operators, which can be minimized through more efficient layout of the workstation or facilities. The wastes of overproduction and stocks are based on common sense in the additional storage requirements and material handling requirements to move items in and out of storage. Finally, the waste of defective products is obvious, requiring rework.

Other key elements of TPS include: (1) elimination of overproduction, including excessive inventory and excessive capital investment; (2) quantity and quality control techniques; (3) just-in-time production (JIT) with its associated autonomous defects control, such as never allowing defective units to disrupt a subsequent process; (4) the *kanban* system, a tag-like card with product information that follows the product completely through the production cycle, to maintain JIT; (5) flexible workforce, such as varying the number of workers in response to demand changes; (6) *kaizen* or continuous improvement activities (Imai, 1986); and (7) respect for the worker and a "creative thinking" worker suggestion system.

A necessary component of JIT is the single-minute exchange of die or SMED. SMED is a series of techniques pioneered by Shingo (1981) for changeovers of production machinery in less than 10 minutes. Obviously, the long-term objective is zero setup, in which changeovers are instantaneous and do not interfere in any way with continuous work flow.

Remarkable successes have been shown with the implementation of TPS, ranging from Toyota itself to tiny suppliers, such as Showa (Womack and Jones, 1996). For more details on the Toyota Production System, consult Shingo (1981), Imai (1986), and Ono (1988) for the original sources, and Monden (1993), and Womack and Jones (1996) for easier reading and understanding.

Total Quality

Quality is a concept that everybody intuitively understands, but it is difficult to define. Everybody can relate to eating out in a restaurant and judging its quality by the taste of the food, the promptness and courtesy of the service, cost and ambience. Two aspects that cross all these factors are: results and customer satisfaction. In other words, does the product or service meet or exceed customer satisfaction? Further, quality is an everchanging state that must be continually maintained through a *continuous improvement* program. *Total quality* is a much broader concept that encompasses not just the results aspect, but also the quality of the process, materials, environment. and people.

The total quality movement, like work measurement, could be considered to have evolved from F. W. Taylor's *Principles of Scientific Management*. Later development came about because of the impact of World War II on the United States and Japanese industries. While as U.S. companies were focused more on meeting delivery dates than on quality, which continued well past the war, Japanese companies were forced to compete with established companies in the rest of the world. This could only be done by emphasizing the quality of its products over the next 20 years.

The Japanese effort on continuous quality improvement and quality circles was initiated primarily by the philosophies and works of three individuals: W. E. Deming, J. M. Duran and A.V. Feigenbaum. Following his work in the United States during World War II, Deming became a consultant to Japanese industries and convinced their top management of the power of statistical methods and the importance of quality as a competitive weapon. He is best known for his 14 points (see Table 18–4) and the Deming prize for quality established by the Japanese Union of Scientists and Engineers.

Juran is one of the founders of statistical quality control and is best known for his *Quality Control Handbook* (Juran, 1951), a standard reference in the area. The Juran philosophy is based on the organization and implementation of improvements through "managerial breakthroughs" highlighted in the 10 steps to quality improvement (see Table 18–5).

Feigenbaum was the first to introduce the concept of a company-wide quality control program, in his book *Total Quality Control* (1991, 3rd ed.), which was widely used in Japan in the 1950s. Only in the late 1980s and early 1990s did the

TABLE 18–4
Deming's Fourteen Points

1. Create constancy of purpose toward the improvement of products and services in order to become competitive, stay in business, and provide jobs.
2. Adopt the new philosophy. Management must learn that it is a new economic age and awaken to the challenge, learn their responsibilities, and take on leadership for change.
3. Stop depending on inspection to achieve quality. Build in quality from the start.
4. Stop awarding contracts on the basis of low bids.
5. Improve continuously and forever the system of production and service, to improve quality and productivity, and thus constantly reduce costs.
6. Institute training on the job.
7. Institute leadership. The purpose of leadership should be able to help people and technology work better.
8. Drive out fear so that everyone may work effectively.
9. Break down barriers between departments so that people can work as a team.
10. Eliminate slogans, exhortations, and targets for the workforce. They create adversarial relationships.
11. Eliminate quotas and management by objectives. Substitute leadership.
12. Remove barriers that rob employees of their pride of workmanship.
13. Institute a vigorous program of education and self-improvement.
14. Make the transformation everyone's job and put everyone to work on it.

TABLE 18–5
Juran's Ten Steps to Quality Improvement

1. Build awareness of both the need for improvement and opportunities for improvement.
2. Set goals for improvement.
3. Organize to meet the goals that have been set.
4. Provide training.
5. Implement projects aimed at solving problems.
6. Report progress.
7. Give recognition.
8. Communicate results.
9. Keep score.
10. Maintain momentum by building improvement into the company's regular systems.

total quality concept start gaining wide acceptance in the United States under the names of total quality management (TQM), total quality assurance (TQA), or more specialized company specific programs, such as Motorola's Six Sigma.

In general, total quality (TQ) is a way of doing business that maximizes the competitiveness of a company through the continuous improvement of its products, service, people, process, and environment. The key elements of TQ include a company-wide strategic focus, even obsession, on quality, with the customer as its driver. TQ utilizes a scientific approach, employee involvement (especially teamwork), education and training, a long-term commitment, and unity of purpose. The process is not always easy to achieve and must be continually worked on to achieve improvements. Also, cost reductions through better awareness of life-cycle costs, improved product/process designs, and better process controls through the whole manufacturing process are important factors in the success of total quality. More details on total quality and specific program components can be found in Goetsch and Davis (1997).

ISO 9000

Related to total quality is ISO 9000 certification. ISO 9000 is a standard for quality control developed by the International Standards Organization (ISO 9000, 1993). (It now actually comprises a set of five standards, ISO 9000–9004.) By definition, ISO 9000 is concerned only with quality management procedures for contract review, design, development, production, installation, and servicing of products and/or services.

Certification for ISO 9000 ensures that a company's products and/or services are consistently up to a certain level of quality. In the United States, such certification is done by a Registration Accreditation Board staffed jointly by the American National Standards Institute and the American Society for Quality Control. However, this is a private volunteer group, and it does not carry the weight of governmental authorization, as in other countries. Where ISO 9000 is primarily limited to the processes used in a company, total quality encompasses every aspect of that company, including the workforce and the environment. Thus, ISO 9000 is compatible with and typically a subset of total quality, serving to ensure that a company is competitive in a global marketplace (Goetsch and Davis, 1998).

Automation and CAD/CAM

The term automation may be defined as "increased mechanization." A completely automated manufacturing process is capable of operating for prolonged periods without human effort. Automation implies the use of robots. Today, we use robots for such tedious operations as screwing light bulbs on dashboards, tightening bolts, spot welding, and spray painting. Few industries, or even processes within an industry, have been completely automated, However, there has been a pronounced tendency toward semi automation in American, Japanese, and European industry. With the increasing demand for greater productivity, more and more industries are automating. Frequently, robots perform routine, boring operations, as well as operations that involve a dangerous environment. Such areas may be contaminated with radiation, be excessively noisy, or be extremely hot or cold.

An automation program begins with the integration of a fully automatic machine, such as an automatic screw machine, with automatic transfer handling devices, so that a series of operations may be performed automatically. To determine the justifiable extent of automation, two factors should be considered: (1) the quantity requirements for the product, and (2) the nature or design of the product itself. If the quantity requirements for the product are large, the design engineer endeavors to design the product such that it lends itself to automation. Frequently, adding something to a part, such as a lug, fin, extension, or hole, creates a means of mechanical handling to and from the workstation. Such a redesign can accommodate indexing at a workstation for successive production operations.

Complete automation is feasible for the continuous processing of chemical products, such as gasoline, oil, and detergents. Partial automation is now widespread for mass-produced items—food products (such as cookies, cereals, and pretzels), light bulbs, transistors, automobile parts, cigarettes, and so forth.

Factors that encourage plant automation include:

1. The increasing cost of labor.
2. Increasing foreign and domestic competition, which reduces selling prices and decreases profits.
3. The prospects for market expansion through cost and price reduction.

Among the principal factors that may discourage automation in a particular plant are:

1. The large capital investment necessary, which must subsequently be absorbed in operating profit.
2. A market potential presently inadequate to absorb the increased output.
3. An existing technology unable to provide automation to produce a particular design.
4. Opposition by production workers, and adverse community relations caused by a reduction in the labor force.

There is no doubt that the present trend toward automation will continue. As automation equipment is developed, the need for effective preventive maintenance becomes apparent. The failure of a single minor component may cause the shutdown of a complete process, or even an entire plant. This fact, together with the complexity of automated equipment that uses pneumatic, hydraulic, and electronic controls, explains the growth of indirect workers, both in numbers and in diversification of occupation. As companies modernize, they must ensure the efficient use of machinery. Workers in today's automated plants not only need problem solving skills, but must also be able to work proficiently under stress as members of dedicated manufacturing teams.

As companies automate, they usually find it advantageous to compress management hierarchy and convert traditional direct labor to involved workers capable of making advanced manufacturing systems a success. The result should be an improvement in productivity from a more motivated and knowledgeable workforce.

With the increasing use of computers in computer-aided design and computer-aided manufacturing (CAD/CAM), the opportunity arises to design for producibility, thus assuring ideal methods, and also to develop work standards as part of the CAD/CAM process. However, more research and development must take place to identify and store the necessary information in the computer, so that the best sequence of the correct operations can be selected. For example, dynamic decision equations need to be developed to evaluate competing processes to determine the most advantageous way to perform a given operation. Also needed are parameters identified and quantified for computer computation, including: quantity to be produced, material being processed, size of part being processed, geometrical configuration desired by the operation, tolerance needed, and so on.

Industry has already proven that for complete, successful CAD/CAM installation, there must be cooperation between the functional designer, manufacturing engineer, industrial engineer, quality assurance personnel, and data processing personnel. As each of these entities learns about the others' problems and responsibilities, the computer system's effectiveness can increase in the complete planning for producibility, which involves good methods and standards determination in the planning stage.

SUMMARY

To a large extent, the work of motion and time study analysts influences labor relations within an enterprise. Therefore, analysts must understand the objectives of the union that represents workers in their plants. Analysts should know the nature of the training that officers in their company's locals are receiving. With this information, they can understand the attitudes and problems of workers. Today and in the immediate future, output quality must be paramount to both labor and management. At all times, analysts must be cognizant of the necessity of using the human approach. They must always ask for and develop methods, procedures, and standards that are fair to both the company and the operator. Improvement in both quality and output needs to be a way of life.

QUESTIONS

1. Why is training necessary for operators?
2. How is learning quantified?
3. What is remission and how does it affect learning?

4. How should the standards analyst utilize learning curves?
5. Outline the objectives of a typical labor union.
6. Why have unions in the past sought flat across-the-board wage increases for their members?
7. Which five states related to the psychological and sociological reactions of the operator should be recognized by the analyst?
8. What do we mean by the human approach?
9. Name 12 ways you can get people to agree with your ideas.
10. Why is plantwide training in the areas of methods and time study a healthy management step?
11. Why should experienced analysts be continually checked on their ability to performance rate?
12. How can a person develop creative ability?
13. What government legislation has changed the status of labor unions?
14. Why do unions often train their own time study analysts?
15. What are the ego states in transactional analysis?
16. What is a crossed transaction?
17. What levels of transactions work best in dealing with workers?
18. What is an ulterior transaction?
19. What is a quality circle?
20. Compare and contrast intrinsic and extrinsic motivators.
21. How does role-playing enter into ergonomic teams?
22. How does total quality management enter into modern management practices?
23. How does job enrichment differ from job enlargement?
24. What are the seven types of waste?
25. What is total quality?
26. What is continuous improvement and why is it important?

PROBLEMS

1. Based on the cost relationships presented in this chapter, what would be the estimated total annual dollar savings of a company with an annual direct labor payroll of $2,500,000 and a fixed overhead rate of 150 percent of direct labor, if the company initiated a methods and standards program?

2. A company employing straight day work as a method of wage payment is compensating its employees an average of $18.00 per hour. In addition, the cost of fringe benefits is running 30 percent of direct labor. Overhead in this company is 125 percent of direct labor. A methods, standards, and incentive plan is being contemplated in which the average incentive earnings have been estimated to equal 20 percent of base wages. What payoff will the proposed plan yield?

3. A new employee at the Dorben Co. took 186 and 140 minutes to assemble the fourth and eight assemblies, respectively. The standard time for assembling this product is 100 minutes.
a. Calculate this worker's learning curve.
b. How many assemblies does it take the worker to achieve the standard time? How long is this?

4. A training expert suggests that one should allocate a minimum of 40 hours of learning. What time would the worker in problem #3 have achieved at this point?

5. Workers new to carburetor assembly take 15 minutes to complete their first assembly. Assuming a 95 percent learning curve, how long would it take them to reach a standard time of 10 minutes?

REFERENCES

Anonymous. "Just What Do You Do? Mr. Industrial Engineer." *Factory*, 122 (January 1964), pp. 83–84.
Balyeat, R. E. "A Survey: Concepts and Practices in Industrial Engineering."*Journal of Industrial Engineering*. 5 (May 1954), pp. 19–21.
Berne, E. *Games People Play*. New York: Grove Press, 1964.
Deming, W. Edwards. *Out of the Crisis*. Cambridge, MA: MIT Center for Advanced Engineering Study, 1986.
Denton, K. *Safety Management, Improving Performance*. New York: McGraw-Hill, 1982.
Feigenbaum, A.V. *Total Quality Control*. 3rd ed. New York: McGraw-Hill, 1991.
Freivalds, A., S. Konz, A. Yurgec, and J. H. Goldberg. "Work Design: Are We Satisfying Customer Needs?" The *Proceedings of the 41st Annual Conference of the Human Factors and Ergonomics Society*, Santa Monica, CA, 1997, pp. 1398.
Goetsch, D. L., and S. B. Davis. *Introduction to Total Quality*. Upper Saddle River, NJ: Prentice Hall, 1997.
Goetsch, D. L. and S. B. Davis. *Understanding and Implementing ISO 9000 and ISO Standards*. Upper Saddle River, NJ: Prentice Hall, 1998.

Hancock, W. M. and F. H. Bayha. "The Learning Curve." In *The Handbook of Industrial Engineering*. Ed. G. Salvendy. New York: John Wiley & Sons, 1982.
Herzberg, F. *Motivation and Personality*. New York: Harper & Row, 1954.
Imai, M. *Kaizen* New York: Random House, 1986.
ISO 9000: International Standards for Quality Management. 3rd ed. Geneva, Switzerland: International Standards Organization, 1993.
Juran, J. M. *Quality Control Handbook*. New York: McGraw-Hill, 1951.
Konz, S. *Work Design*. 4th ed. Scottsdale, AZ: Publishing Horizons, 1995.
Majchrzak, A. "Management of Technological and Organizational Change." In *Handbook of Industrial Engineering*. 2nd ed. Ed. Gavriel Salvendy. New York: John Wiley & Sons, 1992.
Maslow, A. *Motivation and Personality*. 2nd ed. New York: Harper & Row, 1970.
Monden, Y. *Toyota Production System*. Norcross. GA: Industrial Engineering and Management Press, 1993.
Ohno, T. *Toyota Production System: Beyond Large-Scale Production*. Cambridge, MA: Productivity Press, 1988.
Shingo, S. *Study of Toyota Production System from Industrial Engineering Viewpoint*. Tokyo, Japan: Japan Management Association, 1981.
Taylor, F. W. *The Principles of Scientific Management*. New York: Harper, 1911.
Womack, J. P., and D. T. Jones. *Lean Thinking*. New York: Simon & Schuster, 1996.

APPENDIX 1

Glossary

A

abnormal time Elemental time values that are either considerably higher or lower than the mean of the majority of observations taken during a time study.
activity sampling *See Work sampling.*
actual time The average elemental time actually taken by the operator during a time study.
aerobic Muscular work for which the oxygen intake is adequate.
affordance A perceived property that results in the desired action, for example, a door with a handle and the door pulls open.
agonist The primary muscle involved in the desired motion.
algorithm Step-by-step specifications of the solution to a problem, usually represented by a flowchart, which eventually is translated into a program.
alignment chart *See Nomogram.*
allowance The time added to normal time to provide for personal delays, unavoidable delays, and fatigue.
allowed time The time the normal operator takes to perform an operation while working at a standard rate of performance, with due allowance for personal and unavoidable delays and fatigue.
alphanumeric Set of all machine-processable alphabetic letters (a–z), numeric digits (0–9), and special characters (such as those that appear on a typewriter).
anaerobic Muscular work for which the oxygen intake is inadequate.
antagonist The muscle that opposes the agonist and the desired motion.
anthropometry The science that deals with measuring the physical size of the human.
ATP Adenosine triphosphate, the immediate energy unit for muscle contraction.
anaerobic efficiency The efficiency (ratio of work done, in calories, to net energy used, in calories) of the body during heavy work.

assemble The act of bringing two mating parts together.

assignable cause A source of variation that can be isolated in a process or operation.

automation Increased mechanization to produce goods and services.

available machine time That portion of a time cycle during which a machine could be performing useful work.

average cycle time The sum of all average elemental times divided by the number of cycle observations.

average elemental time The mean elemental time taken by the operator to perform the task during a time study.

average hourly earnings The mean dollar-and-cent moneys paid to an operator on an hourly basis, determined by dividing the hours worked per period into the total wages paid for the period.

avoidable delay A cessation of productive work due entirely to the operator and not occurring in the regular work cycle.

B

balanced motion pattern A sequence of motions made simultaneously by both the right and left hands in directions that facilitate rhythm and coordination.

ballistic movement The motion of arms (usually) or legs with smooth, flowing, rapid muscle action from the start to the termination of the action.

base wage rate The hourly money rate paid for a given work assignment performed at a standard pace by a normal operator.

basic fatigue allowance Constant allowance given to account for: the energy expended while carrying out typical work, and the alleviation of monotony.

basic motion A fundamental motion related to primary physiological and/or biomechanical performance capabilities of body members.

benchmark A standard that is identified with characteristics in sufficient detail so that other classifications can be compared as being above, below, or comparable to the identified standard.

binomial distribution A discrete probability distribution with mean = np and variance = np(1 − p) having a probability function

$$C_{n,k}\, p^k\, (1 - p)^{n-k}$$

biomechanics The application of mechanical principles, such as levers, mechanical advantage, and forces, to the analysis of body part structure and movement.

body discomfort chart A method of assessing a worker's health status by checking the level of discomfort for various body parts.

bonus earnings Those moneys paid in addition to the regular wage or salary.

Borg RPE scale Means for assessing the perceived exertion during dynamic whole-body activities, based on a scale from 6 through 20, which corresponds directly to the heart rate (divided by 10).

brainstorming Discussion sessions in which ideas are encouraged, regardless of how wild they are.

breakeven chart *See Crossover chart.*

breakpoint A readily distinguishable point in the work cycle selected as the boundary between the completion of one element and the beginning of another element.

buzz groups Small discussion subgroups.

C

CAD Computer-aided design.

candelas A measure of the luminous intensity of a light source.

carpal tunnel syndrome Median nerve compression due to inflammation within the carpal tunnel of the wrist, causing pain and loss of sensation and motor control.

cause–effect diagrams *See Fish diagrams.*

cervical The part of the vertebral column located in the neck.

change direction A basic motion, characterized by a slight hesitation when the hand alters its directional course while reaching or moving.

changeover time The time required to modify or replace an existing workplace. Includes both the teardown time for the existing condition and the setup of the new condition.

check study A review of a job with either a stopwatch or a regular wristwatch, to determine the appropriateness of a standard.

check time Sum of the time elapsed before the study and the time elapsed after the study.

chronocyclegraph A photographic record of body motion that may be used to determine both the speed and the direction of body motion patterns.

circadian rhythms The roughly 24-hr variation in bodily functions in humans.

classification method A method of job evaluation based on a series of definitions to differentiate between jobs.

clo unit A measure of the thermal insulation provided by clothing. One clo is 0.16 degrees Celsius per watt per square meter of body surface area.

color rendering The closeness with which the perceived colors of an object being observed match the perceived colors of the same object when illuminated by standard light sources.

combined motions Two or more elemental motions performed simultaneously by the same body member.

compatibility The relationship between controls and displays consistent with human expectations; for example, a red light is associated with danger or stopping.

complementary transaction Transaction sent and received at the same level of ego states.

consistency The absence of noticeable or significant variation in behavioral or numerical data.

constant element An element whose performance time does not vary significantly when changes in the process or dimensional changes in the product occur.

constant fatigue allowance The combination of personal needs and basic fatigue allowances which typically are constant for all workers within a company.

continuous improvement An ongoing process assuring total quality in a company.

contrast The ability of a target to stand out from its background; typically measured as the difference in luminances between target and background.

continuous-timing method An operation study method in which the stopwatch is kept running continuously during the course of the study and is not snapped back at elemental termination.

control–response ratio Ratio of the amount of movement in a control to the amount of movement in the response; used to define system responsiveness.

control system A system that has as its primary function the collection and analysis of feedback from a given set of functions, for the purpose of controlling the functions.

controlled time Elapsed elemental time that depends entirely on the facility or process.

construction program A facilities layout program generating the best solution from scratch.

costing Procedure for accurately determining costs in advance of production.

coverage The number of jobs that have been assigned a standard during the reporting period, or the number of personnel whose jobs have been assigned a standard during the reporting period.

CP Creatine phosphate, the immediate precursor to ATP.

CR-10 (category ratio) scale Rating of perceived exertion scale in which the worker rates the level of pain or body discomfort for various parts of the body on a logarithmic scale from 0 (nothing at all) to 10 (almost maximum).

criteria of pessimism A decision-making strategy in which the outcome with minimum negative consequence is selected.

crossed transaction Transaction between ego states that are not parallel.

crossover chart A method for plotting the increase in cost as a function of some variable. The point at which the two lines cross is known as the crossover or breakeven point, and the cost for each method is the same.

CTD Cumulative trauma disorders, a variety of injuries due to the repetitive nature of work.

curve A graphic representation of the relation between two factors, one of which is usually time.

cycle A series of elements that occur in regular order and make an operation possible. These elements repeat themselves as the operation is repeated.

cyclegraph timing The use of small lights on the hands or other body members to indicate their motion patterns. The lights are recorded by a still camera in a darkened room, with an exposure time equal to at least one motion cycle.

cycle timing The measurement of the time for a complete work cycle, rather than for the individual elements of the cycle.

D

database A collection of data items that can be processed by a variety of applications.

day work Any work for which the operator is compensated on the basis of time rather than output.

dBA A measure of sound pressure level; most commonly used to assess the noise exposure of workers.

deadman's control A control requiring the continual application of force. Once released, it returns to the zero (or off) position.

deadspace The amount of control movement resulting in no system response.
decibel Unit for sound intensity.
decimal hour stopwatch A stopwatch used for work measurement, the dial of which is graduated in 0.0001 of an hour.
decimal minute stopwatch A stopwatch used for work measurement, the dial of which is graduated in 0.01 of a minute.
delay Any cessation in the work routine that does not occur in the typical work cycle.
design for adjustability Anthropometric design principle typically used for equipment or facilities that can be adjusted to fit a wider range of individuals.
design for averages "One size fits all" anthropometric design principle.
design for extremes Anthropometric design principle in which a specific feature is a limiting factor in determining either the maximum or the minimum value of a population variable to be accommodated, for example, stature for doorways.
differential timing Timing an element by combining it with preceding and/or succeeding elements and then determining the elemental times by solving the simultaneous collective time equations.
direct labor Labor performed on each piece that advances the piece toward its ultimate specifications.
direct lighting Type of lighting that places more of the light on the work surfaces and the floor.
direct material costs Cost of raw materials and components.
disassemble The basic motion that takes place when two mating parts are separated.
discounted flow method Economic tool computing the ratio of the present worth of cash flow to the original investment.
disk herniation Bulging of the intravertebral disk, causing pressure on spinal nerves, with the resulting pain.
division of labor The separation of jobs or tasks into less complex jobs or tasks, usually to allow the use of workers possessing less skill than that required by the overall job or task, or to make use of special skills.
downtime The time represented by operation cessation due to machine or tool breakdown, lack of material, and so on.
drop delivery The disposal of a part by dropping it on a conveyor or a gravity chute, thus minimizing move and position therbligs.
dry-bulb temperature Basic ambient temperature, with the thermometer shielded from radiation.
dyadic communications One-to-one communications, typically face to face.

E

earned hours The standard hours credited to a worker or a workforce as a result of the completion of a job or group of jobs.
effective time Total of all observed time.
effectiveness The ratio of earned hours to actual hours spent on prescribed tasks.
efficiency The ratio of actual output to standard output. Also, light output per unit energy.
effort The will to perform either mental or manual productive work.

effort time The portion of the cycle time that depends on the skill and effort of the operator.

ego states The three psychic stages an individual can achieve: adult, parent, or child.

elapsed time The actual time that has transpired during the course of a study or an operation.

electromyogram Electrical activity in a muscle.

element A division of work that can be measured with stopwatch equipment and that has readily identified terminal points.

EMG Electromyogram, the electrical activity in a muscle.

engineered work standards Time standards based on measurement of work content (as opposed to historical standards) for work performed in the most productive way.

equivalent wind chill temperature The ambient temperature that in calm conditions would produce the equivalent wind chill index as the actual combination of air temperature and wind velocity.

erector spinae Primary muscles of the back that provide the force for lifting loads.

ergonomics The science of fitting the task or workplace to the abilities and limitations of the human operator. (see also *Human factors*)

expense labor Labor not involved in the manufacture of a product, typically engineering, research, sales, clerical, accounting, and other administrative functions.

exponential distribution A continuous probability distribution with mean $= \frac{1}{a}$ and variance $= \frac{1}{a^2}$, and having a density function $= ae^{-ax}$.

extension Joint motion in which the included angle becomes larger.

external time The time required to perform elements of work when the machine or process is not in operation.

external transport Transport between different plants or companies.

extra allowance An allowance to compensate for required work in addition to that which is specified in the standard method.

extrinsic factors Environmental factors such as administration, supervision and working conditions in Herzberg's motivation-maintenance theory acting as dissatisfiers.

F

facilitating agreement Process for obtaining a consensus by positively involving all group members.

factor comparison A method of job evaluation based on comparing various job factors.

factory cost Direct material costs plus direct labor costs plus factory expense.

factory expense Costs such as indirect labor, tooling, machine, and power costs.

fair day's work The amount of work performed by an operator that is fair to both the company and the operator, considering the wages paid. It is the "amount of work that can be produced by a qualified employee when working at a normal pace and effectively utilizing his time where work is not restricted by process limitations."

fatigue A lessening in the capacity to work.

fatigue allowance An amount of time added to the normal time to compensate for fatigue.

feed The speed at which the cutting tool is moved into the work, as in drilling and turning, or the rate that the work is moved past the cutting tool.

film analysis The frame-by-frame observation and study of a film of an operation or process, with the objective of improving that operation or process.

fish (cause–effect) diagrams A method of defining an occurrence of a typically undesirable event or problem, that is, the effect, as the fishhead, and then identifying contributing factors, that is, the causes, as fish bones attached to a backbone and the fishhead.

fixture A tool that is usually clamped to the workstation and that holds the material being worked on.

flexible compensation plans Any incentive plan that increases employee wages or benefits as a function of increased production.

flexion Joint motion in which the included angle becomes smaller.

flextime Shift system in which the starting and stopping times are established by the workers, within limits set up by management.

float The amount of material not directly employed or worked on in a system or process at a given point in time.

flow analysis The detailed examination of the progressive travel, either of personnel or material, from place to place and/or from operation to operation.

flow diagram A pictorial representation of the layout of a process, showing the location of all activities appearing on the flow process chart and the travel paths of the work.

flow process chart A graphic representation of all operations, transportations, inspections, delays, and storages occurring during a process or procedure. The chart includes information considered desirable for analysis, such as the time required and the distance moved.

foot-candle The measure of light falling on a surface. One footcandle equals 10.8 lumens per square meter.

foot-lambert A unit of luminance (emitted or reflected light). One foot-lambert is equal to 3.43 candelas per square meter.

force–length relationship The inverted-U relationship in which muscle force is greatest at its resting length.

force–velocity relationship The trade-off between slower movements providing greater force and faster movements being weaker.

foreign element An interruption in the regular work cycle.

frame The space occupied by a single picture on a motion-picture film or videotape.

frame counter A device that automatically tabulates how many frames have passed the lens of the projector.

frequency function The complete listing of the values of a random variable, together with their probabilities of occurrence.

frequency of use Principle used in laying out controls or displays based on how often each is used.

fringe benefits The portion of tangible compensation that is not paid in wages, salaries, or bonuses given by the employer to employees. These include insurance, retirement funds, and other employee services. They exclude benefits paid for by employees through pay deductions, such as their participating portions of insurance premiums and retirement funds.

from–to chart *See Travel chart.*

functionality Principle used in laying out controls or displays by similar function.

G

gain sharing Any method of wage payment in which the worker participates in all or a portion of the added earnings resulting from above-standard production.

gang process chart A chart of the simultaneous activities of one or more machines and/or one or more workers.

Gantt chart A series of graphs consisting of horizontal lines or bars in positions and lengths that show schedules or quotas and progress plotted on a common time scale.

general expense Cost for expense labor, rent, insurance, etc.

get The act of picking up and gaining control of an object. It consists of the therbligs reach and grasp, and move; it also sometimes includes search and select.

glare Excessive brightness in the field of vision, impairing visibility.

globe temperature Measure of radiative load, using a thermometer in a 6-in-diameter black copper sphere.

glucose The primary carbohydrates component that enters the biochemical pathways for energy production.

grade description plan See classification method.

grasp The elemental hand motion of closing the fingers around a part.

gravity feed Conveyance of materials either to or away from the workstation by using the force of gravity.

H

hand time That part of the work cycle controlled by manual elements, exclusive of power or mechanized pacing elements.

hazard action table Decision table for specifying certain action for a given hazard.

heart rate creep The slow increase in heart rate during heavy work, indicating fatigue.

heart rate recovery The return of the heart rate to resting levels after work.

hertz The unit of frequency, in cycles per second. One Hz equals one cycle per second.

human factors Those axioms and postulates concerned with the physical, mental, and emotional constraints affecting operators' performance.

I

idle time Time the worker is not working.

illumination The amount of light striking a surface, measured in foot-candles.

importance Principle used in laying out controls or displays based on the importance of each.

IMPROSHARE A gain-sharing plan based on employee productivity as measured in working hours.

improvement program Facilities layout program that improves upon an initial layout.

incentive Reward, financial or other, that compensates the worker for high and/or continued performance above standard.

incentive pace A performance that is above normal or standard.

indirect labor Labor that does not directly enter into transforming the material used in making the product, but is necessary to support the manufacture of the product.

indirect lighting Type of lighting in which the ceiling is illuminated, which in turn reflects the light downward.

ineffective time Sum of all foreign element times.

interference time Idle machine time due to insufficient operator time to service one or more machines that need servicing, because the operator is engaged in other assigned work.

internal transport Transport within a company, plant, etc.
internal work Work performed by the operator during the operation of the machine or equipment.
intrinsic factors Potential satisfiers such as achievement, recognition and advancement in Herzberg's motivation-maintenance theory.
irregular element An element that occurs randomly and can be statistically determined.
isoinertial strength Type of muscle contraction in which the muscle contracts at a constant acceleration.
isokinetic strength Type of muscle contraction in which the muscle contracts at a constant velocity.
isometric strength Type of muscle contraction in which the muscle contracts in a fixed static position and produces the maximum force; also known as static strength.
isotonic strength Type of muscle contraction in which the muscle contracts against a constant force; sometimes termed dynamic strength.

J

jig A tool that may or may not be clamped to the workstation and is used both to hold the work and to guide the tool.
job analysis A procedure for making a careful appraisal of each job and then recording the details of the work so that it can be equitably evaluated.
job enlargement A horizontal expansion or diversification of work, to avoid repetitive work.
job enrichment A vertical expansion of work, allowing workers to both start and complete a given task, providing greater diversification and fulfillment.
job evaluation A procedure for determining the relative worth of various work assignments.
job rotation Similar to job enlargement in providing a worker the opportunity to do a variety of tasks to avoid repetitive work, but on a more rigid schedule.

K

kaizen System of continuous improvement activities.
kanban A tag-like card with product information that follows the product completely through the production cycle to maintain JIT.
keiretsu Interlocking relationship between a Japanese manufacturer and its suppliers.
key job A job representative of similar jobs or classes of work in the same plant or industry.
kicker A step increase in earnings to induce workers to reach a certain level of productivity.

L

lactic acid The byproduct of anaerobic metabolism, causing sensations of fatigue.
learning curve A graphic presentation of the progress of production effectiveness over time.

leveling *See Performance rating.*
lighting efficiency Light output per unit energy, typically lumens per watt.
line balancing The problem of determining the ideal number of workers to be assigned to a production line.
loose rate An established allowed time permitting the normal operator to achieve standard performance with less than average effort.
lumbar The area of the back most prone to injuries; approximately at the belt line.
luminaire A lighting source, such as a lamp.
luminance The amount of light reflected from a surface, measured in foot-Lamberts.
luminous flux The total light output of a source, or the amount of incident light on a surface, expressed in lumens.
luminous intensity Light intensity of a source, measured in candelas.
lux The unit of illuminance equal to one lumen per square meter, or 0.093 foot-candle.

M

machine coupling The practice of having one employee operate more than one machine.
machine cycle time The time required for the machine in process to complete one cycle.
machine downtime That time when the machine or process is inoperative because of some breakdown or because of a material shortage.
machine idle time That time when the machine or process is inoperative.
machine pacing The machine or mechanical control over the rate at which the work progresses.
maximum performance The performance resulting in the highest obtainable production.
maximum working area The area readily reached by the operator when the arms are fully extended, while in a normal working position.
mean of x The expected value of x.
measured day work An incentive system in which hourly rates are periodically adjusted on the basis of operator performance during the previous period.
merit rating A method of evaluating an employee's worth to a company in terms of quantity and quality of work, dependability, and general contribution to the company.
method The technique employed to perform an operation.
methods study Analysis of an operation to increase the production per unit of time and consequently reduce the unit cost.
micromotion study The division of a work assignment into therbligs, accomplished by analyzing motion pictures frame by frame and then improving the operation by eliminating unnecessary movements and simplifying the necessary movements.
minimax regret criterion A decision-making strategy in which a matrix of regret values (differences between actual and projected payoffs) is calculated. The analyst selects the minimum of maximum regrets.
minimum time The least amount of time taken by the operator to perform a given element during a time study.
modal time The elapsed elemental time value that occurs most frequently during a time study. Occasionally used in preference to the average elemental time.
motion study The analysis and study of the motions constituting an operation, to improve the motion pattern by eliminating ineffective motions and shortening the effective motions.

motivation-maintenance theory A motivational theory by F. Herzberg in which extrinsic factors (administration, working conditions) act as potential dissatisfiers and intrinsic factors (achievement, recognition) act as satisfiers.

motor unit The functional unit of muscles comprised of a nerve fiber and all the muscle fibers that it innervates.

move Hand movement with a load.

MTM (Methods-Time Measurement) A procedure for analyzing any manual operation or method, to determine the basic motions required to perform the operation, and to assign a predetermined time standard to each motion, based on the nature of the motion and the conditions under which it is made.

muda In Japanese industry, wastes to be eliminated.

multiple criterion decision making A quantitative decision making procedure in the presence of conflicting information.

musculoskeletal system The system of muscle and bones in the body allowing for movement.

myofibrils Subdivision of muscle fiber containing thick and thin protein filaments.

N

natural frequency The internal frequency of vibration within a system, determined by its mass, spring, and dashpot characteristics.

network analysis A planning technique used to analyze the sequence of activities and their interrelationships within a project.

noise Unwanted sounds that interfere with the detection of desired signals.

noise dose Total daily noise exposure, consisting of exposures to several different noise levels, each resulting in partial doses.

noise reduction rating Measure of ear plug effectiveness, in terms of dBs of noise level attenuation.

nomogram A graph that usually contains three parallel scales graduated for different variables, so that when a straight line connects values of any two, the related value may be read directly from the third at the point intersected by the line.

normal distribution A continuous probability distribution with mean = m and variance = σ^2 and having a density function equal to:

$$\frac{1}{\sigma\sqrt{2\pi}} \exp\left[\frac{-(x-m)^2}{2\sigma^2}\right]$$

normal operator An operator who can achieve the established standard of performance when following the prescribed method and working at an average pace.

normal performance The performance expected from the average trained operator when following the prescribed method and working at an average pace, for example, walking at 3 mph.

normal time The time required for the standard operator to perform the operation when working at a standard pace, without delay for personal reasons or unavoidable circumstances.

normal working area The space at the work area that can be reached by either the left or right hand when both elbows are pivoted on the edge of the workstation.

numerical control A method of controlling a machine or facility whereby either a binary or decimal digit system is programmed to carry out operations through electronic circuits and related activating mechanisms.

O

observation The gathering and recording of the time required to perform an element, or one watch reading.

observation board *See Time study board.*

observation form *See Time study form.*

observed time The elemental time for one cycle, obtained either directly or by subtracting successive watch times.

observer The analyst taking a time study of a given operation.

occupational physiology Scientific study of the worker and environment, utilizing physiological principles.

occurrence An incident or event that happens during a time study.

octave band analysis Noise analysis with a special filter attachment to the sound-level meter that decomposes the noise into component frequencies.

operation The intentional changing of a part toward its ultimate desired shape, size, form, and characteristics.

operation analysis An investigative process dealing with operations in factory or office work. Usually, the process leading to operation standardization, including motion and time study.

operation card A form outlining the sequence of operations, the time allowed, and the special tools required in manufacturing a part.

operation process chart A graphic representation of an operation, showing all methods, inspections, time allowances, and materials used in a manufacturing process.

operator attention time That time during the work cycle in which the operator must devote attention to the machine or process.

operator process chart A graphic representation of all movements and delays made by both the right and left hands, and of the relationship between the relative basic divisions of accomplishment performed by the two hands.

output The total production of a machine, process, or worker for a specified unit of time.

overall study Recording cycle time as a verification of a developed time study standard.

overhead Any costs of a business above prime costs.

oxygen debt The increased metabolic activity after work to repay the oxygen deficit.

oxygen deficit The deficit of oxygen incurred during the initial or heavy stages of work; supplied by anaerobic metabolism.

P

pallet A load carrier, usually with a rectangular standardized load carrier.

Pareto's distribution A distribution that reflects the fact that the major part of an activity (usually 80–85 percent) is accounted for by a minority (usually 15–20 percent). For example, 20 percent of the employees account for 80 percent of the absenteeism.

payback method Economical analysis tool that uses the time to return the cost of the original investment.

performance The ratio of the operator's actual production to the standard production.

performance rating The assignment of a percentage to the operator's average observed time, based on the actual performance of the operator as compared to the observer's conception of normal.

personal needs allowance A percentage added to the normal time to accommodate the personal needs of the operator.

PERT (Program Evaluation and Review Technique) chart A planning and control method that graphically portrays the optimum way to attain some predetermined objective, generally in terms of time.

phototropism Tendency for the eyes to be drawn directly to the brightest light source.

physiological needs First step in Maslow's hierarchy of human needs, basic concerns regarding survival, food, water, and health.

picking rate The rate at which a pallet, or some other transport unit, is completely picked.

piece work A standard of performance expressed in money per unit of production.

plan A basic motion involving the mental process of determining the next action.

plunger criterion A decision making strategy of expecting the best outcome and choosing the maximum positive consequences.

point A unit of output identified as the production of one standard operator in one minute. Used as a basis for establishing standards under the point system.

point system A method of job evaluation in which the relative worth of different jobs is determined by totaling the points assigned to the various factors applicable to the different jobs.

policy allowance Allowance to provide a satisfactory level of earnings for a specified level of performance for exceptional circumstances.

Poisson distribution A discrete probability distribution, with mean = λ and variance = λ, and having a probability function equal to:

$$\frac{\lambda^k e^{-\lambda}}{k!}$$

position An element of work that consists of locating an object such that it will be properly oriented in a specific location.

post-lunch dip Dip in performance and circadian rhythms after midday.

power grip Optimal cylindrical grip for power that utilizes all digits and in which the thumb barely overlaps the index finger.

predetermined time system System based on basic motion times used to calculate a standard time.

pre-position A basic motion that consists of positioning an object in a predetermined place so that it may be grasped in the position in which it is to be held when needed.

prime cost Direct material costs plus direct labor costs.

process A series of operations that advances the product toward its ultimate size, shape, and specifications.

process chart A graphic representation of a manufacturing process.

productive time Any time spent in advancing the progress of a product toward its ultimate specifications.

profit sharing Any procedure in which an employer pays to employees special current or deferred sums, in addition to good rates of regular pay, based not only on individual or group performance, but also on the prosperity of the business as a whole.

progress chart A graphical representation of the status or extent of completion of the work in process.

psychophysical strength Type of strength in which the operator subjectively determines the acceptable load to be lifted.

Q

qualified operator An employee with sufficient training and education and a demonstrated level of skill and effort to perform at an acceptable level with respect to both quantity and quality.
quality circles Small groups for participative problem solving.
queuing theory *See Waiting line theory.*

R

rad Unit of radiation dose equivalent to the absorption of 0.01 joules per kilogram.
radial deviation Bending of the wrist such that the thumb moves toward the arm.
random variable A chance number resulting from a trial from among the set of numbers x_1, x_2, and so on.
random servicing The interaction between the operator and machine that occurs on a random basis.
range effect Tendency of overshooting close targets and undershooting far targets, typically resulting from fatigue.
rapid rotation Type of shiftwork in which the worker changes shifts every two or three days.
rate A standard expressed in dollars and cents.
rate setting The act of establishing money rates or time values on any operation.
rating *See performance rating.*
rating method Method of job evaluation based on arranging jobs in order of importance.
rating of perceived exertion Means for assessing exertion during dynamic whole-body activities.
ratio-delay study See *Work sampling.*
rating by the watch An incorrect rating procedure whereby the analyst uses previously observed times to rate the operator.
Raynaud's syndrome Cold-induced occlusion of blood flow to the hands, reducing dexterity.
reciprocal inhibition Type of reflex in which the agonist muscle is activated and the antagonist is inhibited so as to reduce counterproductive muscle contractions.
reflectance Percentage of light reflected from a surface.
regression to the mean Tendency of a novice analyst to rate closer to normal performance than the true performance.
regret matrix *See minimax regret criterion.*
relationship chart Chart expressing the relative degrees of closeness among different activities, areas, departments, rooms, etc., for facilities layout purposes.
remission Increase in cycle time on the learning curve due to forgetfulness.
resonance Situation in which forced vibrations induce larger-amplitude vibrations in a system.

resting length Length of the muscle while it is in a neutral, uncontracted state.
return on investment Method of economic analysis using the ratio of the yearly profit to the life of the product.
return on sales Method of economic analysis using the ratio of the yearly profit to the yearly sales.
roentgen A unit of radiation exposure that measures the amount of ionization produced in air by X or gamma radiation.
Rucker Plan A gain-sharing plan based on employee productivity as measured by one or more of the following: gross production, net sales and inventory changes.
runout time The time required by machine tools, after cutting is completed, for the tool to be cleared from the work, in preparation for the next sequence of work elements.

S

safety needs Second step in Maslow's hierarchy of human needs, the need for security on the job.
Scanlon Plan A gain-sharing plan based on employee productivity as measured by one or more of the following: gross production, net sales and inventory changes.
selected time An elemental time value chosen as representative of the expected performance of the operator being studied.
self esteem needs Fourth step in Maslow's hierarchy of human needs, a desire for competence, achievement and self respect.
self-fulfillment Final achievement of all of the desired needs in Maslow's hierarchy. The worker is personally fulfilled and the ego is satisfied.
setup The preparation of a workstation or a work center to accomplish an operation or a series of operations.
shiftwork Working at times other than daytime hours.
simo chart A two-hand process chart with times measured with a microchronometer as part of a micromotion study.
simultaneous motions Two or more elemental motions performed simultaneously by different body members.
size principle The orderly recruitment of motor units, from small to large.
skeletal muscle The muscles attached to the bones that provide the driving force for motion.
skill Proficiency at following a prescribed method.
sliding filament theory Theory of muscle contraction in which the component filaments slide over one another.
slipped disk *See Disk herniation.*
slotting Use of similar job categories to establish expense standards.
SMED (Single Minute Exchange of Die) A series of techniques for changing over production machinery in less than 10 minutes.
snapback timing Time study technique in which, after the watch is read at the breakpoint of each element, the time is returned to zero.
social needs Third step in Maslow's hierarchy of human needs: need for attention, friendship and social belonging.
sorting A transport terminal activity by which goods are divided into groups.
speed–accuracy trade-off The situation in which ensuring accuracy increases motion time, while increasing the speed of motion decreases the resulting accuracy.

stamps Strokes collected as a form of debt to be repaid.
standard data A structured collection of normal time values for work elements, codified in tabular or graphic form.
standard costs Prepriced or budget costs on which production runs and sales decisions are made.
standard hour plan A wage incentive plan using day work up to 100% performance and piecework beyond 100% performance.
standard performance *See Normal performance.*
standard time A unit time value for a work task, as determined by the proper application of appropriate work measurement techniques by qualified personnel.
stowage Handling for the purpose of positioning and/or securing goods in the space intended.
strokes Positive recognition of an individual and the individual's accomplishments.
synchronous servicing An ideal case in which both the worker and the machine interact on a fixed, repetitive cycle.
synthetic basic motion times A collection of time standards assigned to fundamental motions and groups of motions.

T

temporary standard A standard established for a limited number of pieces or a limited time, to account for the newness of the work or some unusual job condition.
tendinitis Inflammation of a tendon, caused by repetitive work.
tenosynovitis Inflammation of tendon sheaths, caused by repetitive work.
therblig One of 17 basic work elements defined by Gilbreth.
tight rate A time standard that allows less time than that required by a normal operator to do the work while working at a normal pace.
time study The procedure using stopwatch timing to establish standards.
time study board A convenient board used to support the stopwatch and hold the observation form during a time study.
time study form A form designed to accommodate the elements of a given time study, with spaces for recording their durations.
time value of money Economic concept of a dollar today being worth more than a dollar in the future.
time weighted average The sound level that would produce a given noise dose if a worker were continuously exposed to that sound level over an 8-hr workday.
total quality A Japanese management approach that encompasses quality in all aspects of a business (processes, materials, people, environment) through a continuous improvement process.
transactional analysis An approach to dealing and communicating with people using the concepts of: (1) ego states, (2) transactions, and (3) stroking and stamps.
transactions Interactions between ego states in Berne's transaction analysis.
transmission The travel of nerve impulses across the motor end plate in the muscle fiber.
travel chart A table that provides distances traveled between points in a manufacturing or business facility.
trigger finger Tendinitis in the index finger, caused by repetitive triggering of a power tool.
two-hand process chart A chart showing the motions made by one hand in relation to those made by the other hand, and using standard therblig abbreviations or symbols.

U

ulnar deviation Bending of the wrist such that the little finger moves toward the arm.
ulterior transactions Transactions with a hidden meaning.
unavoidable delay An interruption in the continuity of an operation that is beyond the control of an operator.
unit labor costs Employee wages divided by the productivity or worker performance.
unit load A material in a packed state. Frequently, a standardized-size transport unit.
use A basic motion that occurs when either or both hands have control of an object during that part of the cycle when productive work is being performed.

V

value engineering A method for evaluating alternatives using values and weights for alternatives in a payoff matrix.
variable element An element whose time is affected by one or more characteristics, such as size, shape, hardness, or tolerance, such that as these conditions change, the time required to perform the element changes.
variable fatigue allowance Fatigue allowances which typically are adjusted for individual workers within a company depending on job or working conditions.
variance The difference between actual and standard or budgeted costs.
variance of x A measure of the expected dispersion of the values of x about its mean.
vasoconstriction The occlusion of peripheral blood vessels due to cold conditions.
vasodilation Increased peripheral blood flow due to hot conditions.
vertebrae The bones that form the structural support of the back.
visibility Ability to see fine detail.
visual angle The angle subtended at the eye by the target.

W

wage incentive A financial inducement for effort above normal.
wage rate The money rate expressed in dollars and cents per hour, paid to the employee.
waiting line theory Mathematical analysis of the laws governing arrivals, service times, and the order in which arriving units are taken into service.
waiting time The time when the operator is unable to do useful work because of the nature of the process, or because of the immediate lack of material.
warehouse An installation for storing products during long gaps between production stages, or for storing finished products.
Warrick's principle Principle of display design in which points closest on the display and control move in the same direction, providing the best compatibility.
watch time Time recorded from a watch reading.
WBGT (Wet bulb globe temperature) Heat stress index based on a weighted average of wet-bulb, globe, and dry-bulb temperatures.
wet-bulb temperature Measure of evaporative cooling, using a thermometer with a wet wick and natural air movement.

white finger The occlusion of blood flow to the hand due to the effects of vibration. Results in loss of dexterity and feeling.

wild value *See Abnormal time.*

wind chill index Cold stress index describing the rate of heat loss by radiation and convection as a function of ambient temperature and wind velocity.

work cycle The total sequence of motions and events that comprise a single operation.

work design The design process that uses ergonomics to fit the task and workstation to the human operator.

Work Factor Index of the additional time required over and above the basic time, as established by the Work–Factor system of synthetic basic motion times.

work hardening Physical training on a simulated job to acclimatize the worker to production line conditions.

work-hour The standard amount of work performed by one worker in one hour.

work measurement One of several procedures (timestudy, work sampling and predetermined time systems) for establishing standards.

work pace The rate at which an operation or activity is done.

work physiology The specification of the physiological and psychological factors characteristic of a work environment.

work sampling A method of analyzing work by taking a large number of observations at random intervals, to establish standards and improve methods.

work station The area where the worker performs the elements of work in a specific operation.

worker–machine process chart A chart showing the exact relationship in time between the working cycle of the operator and the operating cycle of the machine or machines.

APPENDIX 2

Helpful Formulas

(1) *Quadratic*

$$Ax^2 + Bx + C = 0$$

$$x = \frac{-B \pm \sqrt{B^2 - 4AC}}{2A}$$

(2) *Logarithms*

$$\log ab = \log a + \log b$$

$$\log \frac{a}{b} = \log a - \log b$$

$$\log a^n = n \log a$$

$$\log \sqrt[n]{a} = \frac{1}{n} \log a$$

$$\log 1 = 0$$

$$\log_a a = 1$$

(3) *Binomial theorem*

$$(a + b)^n = a^n + na^{n-1}b + \frac{n(n-1)}{2!} a^{n-2}b^2 + \frac{n(n-1)(n-2)}{3!} a^{n-3}b^3 + \ldots$$

(4) *Circle*

$$\text{Circumference} = 2\pi r$$

$$\text{Area} = \pi r^2$$

(5) *Prism*

$$\text{Volume} = Ba$$

(6) *Pyramid*

$$\text{Volume} = \tfrac{1}{3} Ba$$

(7) *Right circular cylinder*

$$\text{Volume} = \pi r^2 a$$
$$\text{Lateral surface} = 2\pi ra$$
$$\text{Total surface} = 2\pi r(r + a)$$

(8) *Right circular cone*

$$\text{Volume} = \tfrac{1}{3}\pi r^2 a$$
$$\text{Lateral surface} = \pi rs$$
$$\text{Total surface} = \pi r(r + s)$$

(9) *Sphere*

$$\text{Volume} = \tfrac{4}{3}\pi r^3$$
$$\text{Surface} = 4\pi r^2$$

(10) *Frustum of a right circular cone*

$$\text{Volume} = \tfrac{1}{3}\pi a(R^2 + r^2 + Rr)$$
$$\text{Lateral surface} = \pi s(R + r)$$

(11) *Measurement of angles*

$$1 \text{ degree} = \frac{\pi}{180} = 0.0174 \text{ radians}$$

$$1 \text{ radian} = 57.29 \text{ degrees}$$

(12) *Trigonometric functions*

a. Right triangles:

- The sine of the angle A is the quotient of the opposite side divided by the hypotenuse: $\sin A = \frac{a}{c}$.
- The tangent of the angle A is the quotient of the opposite side divided by the adjacent side: $\tan A = \frac{a}{b}$.
- The secant of the angle A is the quotient of the hypotenuse divided by the adjacent side: $\sec A = \frac{c}{b}$.
- The cosine, cotangent, and cosecant of an angle are, respectively, the sine, tangent, and secant of the complement of that angle.

b. Law of sines:

$$\frac{a}{\sin A} = \frac{b}{\sin B} = \frac{c}{\sin C}$$

c. Law of cosines:

$$a^2 = b^2 + c^2 - 2bc \cos A$$

(13) *Equations of straight lines*

a. Slope—intercept form

$$y = mx + b$$

b. Intercept form

$$\frac{x}{a} + \frac{y}{b} = 1$$

APPENDIX 3

Special Tables

TABLE A3–1
Natural Sines and Tangents

Angle	Sin	Tan	Cot	Cos	
0	0.0000	0.0000	∞	1.0000	**90**
1	0.0175	0.0175	57.2900	0.9998	**89**
2	0.0349	0.0349	28.6363	0.9994	**88**
3	0.0523	0.0524	19.0811	0.9986	**87**
4	0.0698	0.0699	14.3007	0.9976	**86**
5	0.0872	0.0875	11.4301	0.9962	**85**
6	0.1045	0.1051	9.5144	0.9945	**84**
7	0.1219	0.1228	8.1443	0.9925	**83**
8	0.1392	0.1405	7.1154	0.9903	**82**
9	0.1564	0.1584	6.3138	0.9877	**81**
10	0.1736	0.1763	5.6713	0.9848	**80**
11	0.1908	0.1944	5.1446	0.9816	**79**
12	0.2079	0.2126	4.7046	0.9781	**78**
13	0.2250	0.2309	4.3315	0.9744	**77**
14	0.2419	0.2493	4.0108	0.9703	**76**
15	0.2588	0.2679	3.7321	0.9659	**75**
16	0.2756	0.2867	3.4874	0.9613	**74**
17	0.2924	0.3057	3.2709	0.9563	**73**
18	0.3090	0.3249	3.0777	0.9511	**72**
19	0.3256	0.3443	2.9042	0.9455	**71**
20	0.3420	0.3640	2.7475	0.9397	**70**
21	0.3584	0.3839	2.6051	0.9336	**69**
22	0.3746	0.4040	2.4751	0.9272	**68**
23	0.3907	0.4245	2.3559	0.9205	**67**
24	0.4067	0.4452	2.2460	0.9135	**66**
	Cos	**Cot**	**Tan**	**Sin**	**Angle**

(continued)

TABLE A3–1 *(concluded)*

Natural Sines and Tangents

Angle	Sin	Tan	Cot	Cos	
25	0.4226	0.4663	2.1445	0.9063	**65**
26	0.4384	0.4877	2.0503	0.8988	**64**
27	0.4540	0.5095	1.9626	0.8910	**63**
28	0.4695	0.5317	1.8807	0.8829	**62**
29	0.4848	0.5543	1.8040	0.8746	**61**
30	0.5000	0.5774	1.7321	0.8660	**60**
31	0.5150	0.6009	1.6643	0.8572	**59**
32	0.5299	0.6249	1.6003	0.8480	**58**
33	0.5446	0.6494	1.5399	0.8387	**57**
34	0.5592	0.6745	1.4826	0.8290	**56**
35	0.5736	0.7002	1.4281	0.8192	**55**
36	0.5878	0.7265	1.3764	0.8090	**54**
37	0.6018	0.7536	1.3270	0.7986	**53**
38	0.6157	0.7813	1.2799	0.7880	**52**
39	0.6293	0.8098	1.2349	0.7771	**51**
40	0.6428	0.8391	1.1918	0.7660	**50**
41	0.6561	0.8693	1.1504	0.7547	**49**
42	0.6691	0.9004	1.1106	0.7431	**48**
43	0.6820	0.9325	1.0724	0.7314	**47**
44	0.6947	0.9657	1.0355	0.7193	**46**
45	0.7071	1.0000	1.0000	0.7071	**45**
	Cos	**Cot**	**Tan**	**Sin**	**Angle**

TABLE A3–2

Cumulative Probabilities of the Standard Normal Distribution

Entry is area A under the standard normal curve from $-\infty$ to $z(A)$.

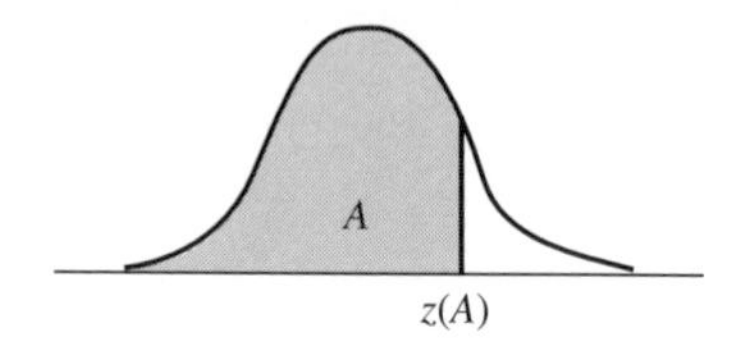

z	0.00	0.01	0.02	0.03	0.04	0.05	0.06	0.07	0.08	0.09
0.0	0.5000	0.5040	0.5080	0.5120	0.5160	0.5199	0.5239	0.5279	0.5319	0.5359
0.1	0.5398	0.5438	0.5478	0.5517	0.5557	0.5596	0.5636	0.5675	0.5714	0.5753
0.2	0.5793	0.5832	0.5871	0.5910	0.5948	0.5987	0.6026	0.6064	0.6103	0.6141
0.3	0.6179	0.6217	0.6255	0.6293	0.6331	0.6368	0.6406	0.6443	0.6480	0.6517
0.4	0.6554	0.6591	0.6628	0.6664	0.6700	0.6736	0.6772	0.6808	0.6844	0.6879
0.5	0.6915	0.6950	0.6985	0.7019	0.7054	0.7088	0.7123	0.7157	0.7190	0.7224
0.6	0.7257	0.7291	0.7324	0.7357	0.7389	0.7422	0.7454	0.7486	0.7517	0.7549
0.7	0.7580	0.7611	0.7642	0.7673	0.7704	0.7734	0.7764	0.7794	0.7823	0.7852
0.8	0.7881	0.7910	0.7939	0.7967	0.7995	0.8023	0.8051	0.8078	0.8106	0.8133
0.9	0.8159	0.8186	0.8212	0.8238	0.8264	0.8289	0.8315	0.8340	0.8365	0.8389
1.0	0.8413	0.8438	0.8461	0.8485	0.8508	0.8531	0.8554	0.8577	0.8599	0.8621
1.1	0.8643	0.8665	0.8686	0.8708	0.8729	0.8749	0.8770	0.8790	0.8810	0.8830
1.2	0.8849	0.8869	0.8888	0.8907	0.8925	0.8944	0.8962	0.8980	0.8997	0.9015
1.3	0.9032	0.9049	0.9066	0.9082	0.9099	0.9115	0.9131	0.9147	0.9162	0.9177
1.4	0.9192	0.9207	0.9222	0.9236	0.9251	0.9265	0.9279	0.9292	0.9306	0.9319
1.5	0.9332	0.9345	0.9357	0.9370	0.9382	0.9394	0.9406	0.9418	0.9429	0.9441
1.6	0.9452	0.9463	0.9474	0.9484	0.9495	0.9505	0.9515	0.9525	0.9535	0.9545
1.7	0.9554	0.9564	0.9573	0.9582	0.9591	0.9599	0.9608	0.9616	0.9625	0.9633
1.8	0.9641	0.9649	0.9656	0.9664	0.9671	0.9678	0.9686	0.9693	0.9699	0.9706
1.9	0.9713	0.9719	0.9726	0.9732	0.9738	0.9744	0.9750	0.9756	0.9761	0.9767
2.0	0.9772	0.9778	0.9783	0.9788	0.9793	0.9798	0.9803	0.9808	0.9812	0.9817
2.1	0.9821	0.9826	0.9830	0.9834	0.9838	0.9842	0.9846	0.9850	0.9854	0.9857
2.2	0.9861	0.9864	0.9868	0.9871	0.9875	0.9878	0.9881	0.9884	0.9887	0.9890
2.3	0.9893	0.9896	0.9898	0.9901	0.9904	0.9906	0.9909	0.9911	0.9913	0.9916
2.4	0.9918	0.9920	0.9922	0.9925	0.9927	0.9929	0.9931	0.9932	0.9934	0.9936
2.5	0.9938	0.9940	0.9941	0.9943	0.9945	0.9946	0.9948	0.9949	0.9951	0.9952
2.6	0.9953	0.9955	0.9956	0.9957	0.9959	0.9960	0.9961	0.9962	0.9963	0.9964
2.7	0.9965	0.9966	0.9967	0.9968	0.9969	0.9970	0.9971	0.9972	0.9973	0.9974
2.8	0.9974	0.9975	0.9976	0.9977	0.9977	0.9978	0.9979	0.9979	0.9980	0.9981
2.9	0.9981	0.9982	0.9982	0.9983	0.9984	0.9984	0.9985	0.9985	0.9986	0.9986
3.0	0.9987	0.9987	0.9987	0.9988	0.9988	0.9989	0.9989	0.9989	0.9990	0.9990
3.1	0.9990	0.9991	0.9991	0.9991	0.9992	0.9992	0.9992	0.9992	0.9993	0.9993
3.2	0.9993	0.9993	0.9994	0.9994	0.9994	0.9994	0.9994	0.9995	0.9995	0.9995
3.3	0.9995	0.9995	0.9995	0.9996	0.9996	0.9996	0.9996	0.9996	0.9996	0.9997
3.4	0.9997	0.9997	0.9997	0.9997	0.9997	0.9997	0.9997	0.9997	0.9997	0.9998

Selected Percentiles

Cumulative probability A:	**0.90**	**0.95**	**0.975**	**0.98**	**0.99**	**0.995**	**0.999**
$z(A)$:	1.282	1.645	1.960	2.054	2.326	2.576	3.090

(*From:* J. Neter, W. Wasserman and M. H. Kutner, Applied Linear Statistical Models, 2nd ed. Homewood, IL: Richard D. Irwin, 1985). (Reproduced with permission of the McGraw-Hill Companies)

TABLE A3–3

Percentage Points of the *t* Distribution (probabilities refer to the sum of the two tail areas; for a single tail, divide the probability by 2)

	Probability (*P*)												
n	0.9	0.8	0.7	0.6	0.5	0.4	0.3	0.2	0.1	0.05	0.02	0.01	0.001
1	0.158	0.325	0.510	0.727	1.000	1.376	1.963	3.078	6.314	12.706	31.821	63.657	636.619
2	0.142	0.289	0.445	0.617	0.816	1.061	1.386	1.886	2.920	4.303	6.965	9.925	31.598
3	0.137	0.277	0.424	0.584	0.765	0.978	1.250	1.638	2.353	3.182	4.541	5.841	12.941
4	0.134	0.271	0.414	0.569	0.741	0.941	1.190	1.533	2.132	2.776	3.747	4.604	8.610
5	0.132	0.267	0.408	0.559	0.727	0.920	1.156	1.476	2.015	2.571	3.365	4.032	6.859
6	0.131	0.265	0.404	0.553	0.718	0.906	1.134	1.440	1.943	2.447	3.143	3.707	5.959
7	0.130	0.263	0.402	0.549	0.711	0.896	1.119	1.415	1.895	2.365	2.998	3.499	5.405
8	0.130	0.262	0.399	0.546	0.706	0.889	1.108	1.397	1.860	2.306	2.896	3.355	5.041
9	0.129	0.261	0.398	0.543	0.703	0.883	1.100	1.383	1.833	2.262	2.821	3.250	4.781
10	0.129	0.260	0.397	0.542	0.700	0.879	1.093	1.372	1.812	2.228	2.764	3.169	4.587
11	0.129	0.260	0.396	0.540	0.697	0.876	1.088	1.363	1.796	2.201	2.718	3.106	4.437
12	0.128	0.259	0.395	0.539	0.695	0.873	1.083	1.356	1.782	2.179	2.681	3.055	4.318
13	0.128	0.259	0.394	0.538	0.694	0.870	1.079	1.350	1.771	2.160	2.650	3.012	4.221
14	0.128	0.258	0.393	0.537	0.692	0.868	1.076	1.345	1.761	2.145	2.624	2.977	4.140
15	0.128	0.258	0.393	0.536	0.691	0.866	1.074	1.341	1.753	2.131	2.602	2.947	4.073
16	0.128	0.258	0.392	0.535	0.690	0.865	1.071	1.337	1.746	2.120	2.583	2.921	4.015
17	0.128	0.257	0.392	0.534	0.689	0.863	1.069	1.333	1.740	2.110	2.567	2.898	3.965
18	0.127	0.257	0.392	0.534	0.688	0.862	1.067	1.330	1.734	2.101	2.552	2.878	3.922
19	0.127	0.257	0.391	0.533	0.688	0.861	1.066	1.328	1.729	2.093	2.539	2.861	3.883
20	0.127	0.257	0.391	0.533	0.687	0.860	1.064	1.325	1.725	2.086	2.528	2.845	3.850
21	0.127	0.257	0.391	0.532	0.686	0.859	1.063	1.323	1.721	2.080	2.518	2.831	3.819
22	0.127	0.256	0.390	0.532	0.686	0.858	1.061	1.321	1.717	2.074	2.508	2.819	3.792
23	0.127	0.256	0.390	0.532	0.685	0.858	1.060	1.319	1.714	2.069	2.500	2.807	3.767
24	0.127	0.256	0.390	0.531	0.685	0.857	1.059	1.318	1.711	2.064	2.492	2.797	3.745
25	0.127	0.256	0.390	0.531	0.684	0.856	1.058	1.316	1.708	2.060	2.485	2.787	3.725
26	0.127	0.256	0.390	0.531	0.684	0.856	1.058	1.315	1.706	2.056	2.479	2.779	3.707
27	0.127	0.256	0.389	0.531	0.684	0.855	1.057	1.314	1.703	2.052	2.473	2.771	3.690
28	0.127	0.256	0.389	0.530	0.683	0.855	1.056	1.313	1.701	2.048	2.467	2.763	3.674
29	0.127	0.256	0.389	0.530	0.683	0.854	1.055	1.311	1.699	2.045	2.462	2.756	3.659
30	0.127	0.256	0.389	0.530	0.683	0.854	1.055	1.310	1.697	2.042	2.457	2.750	3.646
40	0.126	0.255	0.388	0.529	0.681	0.851	1.050	1.303	1.684	2.021	2.423	2.704	3.551
60	0.126	0.254	0.387	0.527	0.679	0.848	1.046	1.296	1.671	2.000	2.390	2.660	3.460
120	0.126	0.254	0.386	0.526	0.677	0.845	1.041	1.289	1.658	1.980	2.358	2.617	3.373
∞	0.126	0.253	0.385	0.524	0.674	0.842	1.036	1.282	1.645	1.960	2.326	2.576	3.291

Reprinted from Table III of R. A. Fisher and F. Yates, *Statistical Tables for Biological, Agricultural, and Medical Research* (Edinburgh: Oliver & Boyd, Ltd.), by permission of the authors and publishers.

TABLE A3–4
Random Numbers III

22 17 68 65 84	68 95 23 92 35	87 02 22 57 51	61 09 43 95 06	58 24 82 03 47
19 36 27 59 46	13 79 93 37 55	39 77 32 77 09	85 52 05 30 62	47 83 51 62 74
16 77 23 02 77	09 61 87 25 21	28 06 24 25 93	16 71 13 59 78	23 05 47 47 25
78 43 76 71 61	20 44 90 32 64	97 67 63 99 61	46 38 03 93 22	69 81 21 99 21
03 28 28 26 08	73 37 32 04 05	69 30 16 09 05	88 69 58 28 99	35 07 44 75 47
93 22 53 64 39	07 10 63 76 35	87 03 04 79 88	08 13 13 85 51	55 34 57 72 69
78 76 58 54 74	92 38 70 96 92	52 06 79 79 45	82 63 18 27 44	69 66 92 19 09
23 68 35 26 00	99 53 93 61 28	52 70 05 48 34	56 65 05 61 86	90 92 10 70 80
15 39 25 70 99	93 86 52 77 65	15 33 59 05 28	22 87 26 07 47	86 96 98 29 06
58 71 96 30 24	18 46 23 34 27	85 13 99 24 44	49 18 09 79 49	74 16 32 23 02
57 35 27 33 72	24 53 63 94 09	41 10 76 47 91	44 04 95 49 66	39 60 04 59 81
48 50 86 54 48	22 06 34 72 52	82 21 15 65 20	33 29 94 71 11	15 91 29 12 03
61 96 48 95 03	07 16 39 33 66	98 56 10 56 79	77 21 30 27 12	90 49 22 23 62
36 93 89 41 26	29 70 83 63 51	99 74 20 52 36	87 09 41 15 09	98 60 16 03 03
18 87 00 42 31	57 90 12 02 07	23 47 37 17 31	54 08 01 88 63	39 41 88 92 10
88 56 53 27 59	33 35 72 67 47	77 34 55 45 70	08 18 27 38 90	16 95 86 70 75
09 72 95 84 29	49 41 31 06 70	42 38 06 45 18	64 84 73 31 65	52 53 37 97 15
12 96 88 17 31	65 19 69 02 83	60 75 86 90 68	24 64 19 35 51	56 61 87 39 12
85 94 57 24 16	92 09 84 38 76	22 00 27 69 85	29 81 94 78 70	21 94 47 90 12
38 64 43 59 98	98 77 87 68 07	91 51 67 62 44	40 98 05 93 78	23 32 65 41 18
53 44 09 42 72	00 41 86 79 79	68 47 22 00 20	35 55 31 51 51	00 83 63 22 55
40 76 66 26 84	57 99 99 90 37	36 63 32 08 58	37 40 13 68 97	87 64 81 07 83
02 17 79 18 05	12 59 52 57 02	22 07 90 47 03	28 14 11 30 79	20 69 22 40 98
95 17 82 06 53	31 51 10 96 46	92 06 88 07 77	56 11 50 81 69	40 23 72 51 39
35 76 22 42 92	96 11 83 44 80	34 68 35 48 77	33 42 40 90 60	73 96 53 97 86
26 29 13 56 41	85 47 04 66 08	34 72 57 59 13	82 43 80 46 15	38 26 61 70 04
77 80 20 75 82	72 82 32 99 90	63 95 73 76 63	89 73 44 99 05	48 67 26 43 18
46 40 66 44 52	91 36 74 43 53	30 82 13 54 00	78 45 63 98 35	55 03 36 67 68
37 56 08 18 09	77 53 84 46 47	31 91 18 95 58	24 16 74 11 53	44 10 13 85 57
61 65 61 68 66	37 27 47 39 19	84 83 70 07 48	53 21 40 06 71	95 06 79 88 54
93 43 69 64 07	34 18 04 52 35	56 27 09 24 86	61 85 53 83 45	19 90 70 99 00
21 96 60 12 99	11 20 99 45 18	48 13 93 55 34	18 37 79 49 90	65 97 38 20 46
95 20 47 97 97	27 37 83 28 71	00 06 41 41 74	45 89 09 39 84	51 67 11 52 49
97 86 21 78 73	10 65 81 92 59	58 76 17 14 97	04 76 62 16 17	17 95 70 45 80
69 92 06 34 13	59 71 74 17 32	27 55 10 24 19	23 71 82 13 74	63 52 52 01 41
04 31 17 21 56	33 73 99 19 87	26 72 39 27 67	53 77 57 68 93	60 61 97 22 61
61 06 98 03 91	87 14 77 43 96	43 00 65 98 50	45 60 33 01 07	98 99 46 50 47
85 93 85 86 88	72 87 08 62 40	16 06 10 89 20	23 21 34 74 97	76 38 03 29 63
21 74 32 47 45	73 96 07 94 52	09 65 90 77 47	25 76 16 19 33	53 05 70 53 30
15 69 53 82 80	79 96 23 53 10	65 39 07 16 29	45 33 02 43 70	02 87 40 41 45
02 89 08 04 49	20 21 14 68 86	87 63 93 95 17	11 29 01 95 80	35 14 97 35 33
87 18 15 89 79	85 43 01 72 73	08 61 74 51 69	89 74 39 82 15	94 51 33 41 67
98 83 71 94 22	59 97 50 99 52	08 52 85 08 40	87 80 61 65 31	91 51 80 32 44
10 08 58 21 66	72 68 49 29 31	89 85 84 46 06	59 73 19 85 23	65 09 29 75 63
47 90 56 10 08	88 02 84 27 83	42 29 72 23 19	66 56 45 65 79	20 71 53 20 25
22 85 61 68 90	49 64 92 85 44	16 40 12 89 88	50 14 49 81 06	01 82 77 45 12
67 80 43 79 33	12 83 11 41 16	25 58 19 68 70	77 02 54 00 52	53 43 37 15 26
27 62 50 96 72	79 44 61 40 15	14 53 40 65 39	27 31 58 50 28	11 39 03 34 25
33 78 80 87 15	38 30 06 38 21	14 47 47 07 26	54 96 87 53 32	40 36 40 96 76
13 13 92 66 99	47 24 49 57 74	32 25 43 62 17	10 97 11 69 84	99 63 22 32 98

Reprinted with permission from Random Numbers III of Table XXXIII of R. A. Fisher and F. Yates, *Statistical Tables for Biological, Agricultural and Medical Research* (Edinburgh: Oliver & Boyd, Ltd.).

TABLE A3–4 *(concluded)*

Random Numbers IV

10 27 53 96 23	71 50 54 36 23	54 31 04 82 98	04 14 12 15 09	26 78 25 47 47
28 41 50 61 88	64 85 27 20 18	83 36 36 05 56	39 71 65 09 62	94 76 62 11 89
34 21 42 57 02	59 19 18 97 48	80 30 03 30 98	05 24 67 70 07	84 97 50 87 46
61 81 77 23 23	82 82 11 54 08	53 28 70 58 96	44 07 39 55 43	42 34 43 39 28
61 15 18 13 54	16 86 20 26 88	90 74 80 55 09	14 53 90 51 17	52 01 63 01 59
91 76 21 64 64	44 91 13 32 97	75 31 62 66 54	84 80 32 75 77	56 08 25 70 29
00 97 79 08 06	37 30 28 59 85	53 56 68 53 40	01 74 39 59 73	30 19 99 85 48
36 46 18 34 94	75 20 80 27 77	78 91 69 16 00	08 43 18 73 68	67 69 61 34 25
88 98 99 60 50	65 95 79 42 94	93 62 40 89 96	43 56 47 71 66	46 76 29 67 02
04 37 59 87 21	05 02 03 24 17	47 97 81 56 51	92 34 86 01 82	55 51 33 12 91
63 62 06 34 41	94 21 78 55 09	72 76 45 16 94	29 95 81 83 83	79 88 01 97 30
78 47 23 53 90	34 41 92 45 71	09 23 70 70 07	12 38 92 79 43	14 85 11 47 23
87 68 62 15 43	53 14 36 59 25	54 47 33 70 15	59 24 48 40 35	50 03 42 99 36
47 60 92 10 77	88 59 53 11 52	66 25 69 07 04	48 68 64 71 06	61 65 70 22 12
56 88 87 59 41	65 28 04 67 53	95 79 88 37 31	50 41 06 94 76	81 83 17 16 33
02 57 45 86 67	73 43 07 34 48	44 26 87 93 29	77 09 61 67 84	06 69 44 77 75
31 54 14 13 17	48 62 11 90 60	68 12 93 64 28	46 24 79 16 76	14 60 25 51 01
28 50 16 43 36	28 97 85 58 99	67 22 52 76 23	24 70 36 54 54	59 28 61 71 96
63 29 62 66 50	02 63 45 52 38	67 63 47 54 75	83 24 78 43 20	92 63 13 47 48
45 65 58 26 51	76 96 59 38 72	86 57 45 71 46	44 67 76 14 55	44 88 01 62 12
39 65 36 63 70	77 45 85 50 51	74 13 39 35 22	30 53 36 02 95	49 34 88 73 61
73 71 98 16 04	29 18 94 51 23	76 51 94 84 86	79 93 96 38 63	08 58 25 58 94
72 20 56 20 11	72 65 71 08 86	79 57 95 13 91	97 48 72 66 48	09 71 17 24 89
75 17 26 99 76	89 37 20 70 01	77 31 61 95 46	26 97 05 73 51	53 33 18 72 87
37 48 60 82 29	81 30 15 39 14	48 38 75 93 29	06 87 37 78 48	45 56 00 84 47
68 08 02 80 72	83 71 46 30 49	89 17 95 88 29	02 39 56 03 46	97 74 06 56 17
14 23 98 61 67	70 52 85 01 50	01 84 02 78 43	10 62 98 19 41	18 83 99 47 99
49 08 96 21 44	25 27 99 41 28	07 41 08 34 66	19 42 74 39 91	41 96 53 78 72
78 37 06 08 43	63 61 62 42 29	39 68 95 10 96	09 24 23 00 62	56 12 80 73 16
37 21 34 17 68	68 96 83 23 56	32 84 60 15 31	44 73 67 34 77	91 15 79 74 58
14 29 09 34 04	87 83 07 55 07	76 58 30 83 64	87 29 25 58 84	86 50 60 00 25
58 43 28 06 36	49 52 83 51 14	47 56 91 29 34	05 87 31 06 95	12 45 57 09 09
10 43 67 29 70	80 62 80 03 42	10 80 21 38 84	90 56 35 03 09	43 12 74 49 14
44 38 88 39 54	86 97 37 44 22	00 95 01 31 76	17 16 29 56 63	38 78 94 49 81
90 69 59 19 51	85 39 52 85 13	07 28 37 07 61	11 16 36 27 03	78 86 72 04 95
41 47 10 25 62	97 05 31 03 61	20 26 36 31 62	68 69 86 95 44	84 95 48 46 45
91 94 14 63 19	75 89 11 47 11	31 56 34 19 09	79 57 92 36 59	14 93 87 81 40
80 06 54 18 66	09 18 94 06 19	98 40 07 17 81	22 45 44 84 11	24 62 20 42 31
67 72 77 63 48	84 08 31 55 58	24 33 45 77 58	80 45 67 93 82	75 70 16 08 24
59 40 24 13 27	79 26 88 86 30	01 31 60 10 39	53 58 47 70 93	85 81 56 39 38
05 90 35 89 95	01 61 16 96 94	50 78 13 69 36	37 68 53 37 31	71 26 35 03 71
44 43 80 69 98	46 68 05 14 82	90 78 50 05 62	77 79 13 57 44	59 60 10 39 66
61 81 31 96 82	00 57 25 60 59	46 72 60 18 77	55 66 12 62 11	08 99 55 64 57
42 88 07 10 05	24 98 65 63 21	47 21 61 88 32	27 80 30 21 60	10 92 35 36 12
77 94 30 05 39	28 10 99 00 27	12 73 73 99 12	49 99 57 94 82	96 88 57 17 91
78 83 19 76 16	94 11 68 84 26	23 54 20 86 85	23 86 66 99 07	36 37 34 92 09
87 76 59 61 81	43 63 64 61 61	65 76 36 95 90	18 48 27 45 68	27 23 65 30 72
91 43 05 96 47	55 78 99 95 24	37 55 85 78 78	01 48 41 19 10	35 19 54 07 73
84 97 77 72 73	09 62 06 65 72	87 12 49 03 60	41 15 20 76 27	50 47 02 29 16
87 91 60 76 83	44 88 96 07 80	83 05 83 38 96	73 70 66 81 90	30 56 10 48 59

TABLE A3–5

Useful Information

To find the circumference of a circle, multiply the diameter by 3.1416.
To find the diameter of a circle, multiply the circumference by 0.31831.
To find the area of a circle, multiply the square of the diameter by 0.7854.
The radius of a circle × 6.283185 = the circumference.
The square of the circumference of a circle × 0.07958 = the area.
Half the circumference of a circle × half its diameter = the area.
The circumference of a circle × 0.159155 = the radius.
The square root of the area of a circle × 0.56419 = the radius.
The square root of the area of a circle × 1.12838 = the diameter.
To find the diameter of a circle equal in area to a given square, multiply a side of the square by 1.12838.
To find the side of a square equal in area to a given circle, multiply the diameter by 0.8862.
To find the side of a square inscribed in a circle, multiply the diameter by 0.7071.
To find the side of a hexagon inscribed in a circle, multiply the diameter of the circle by 0.500.
To find the diameter of a circle inscribed in a hexagon, multiply a side of the hexagon by 1.7321.
To find the side of an equilateral triangle inscribed in a circle, multiply the diameter of the circle by 0.866.
To find the diameter of a circle inscribed in an equilateral triangle, multiply a side of the triangle by 0.57735.
To find the area of the surface of a ball (sphere), multiply the square of the diameter by 3.1416.
To find the volume of a ball (sphere), multiply the cube of the diameter by 0.5236.
Doubling the diameter of a pipe increases its capacity four times.
To find the pressure in pounds per square inch at the base of a column of water, multiply the height of the column in feet by 0.433.
A gallon of water (U.S. standard) weighs 8.336 pounds and contains 231 cubic inches. A cubic foot of water contains 7½ gallons and 1728 cubic inches, and weighs 62.425 pounds at a temperature of about 39° F.
These weights change slightly above and below this temperature.

TABLE A3–6

15% Compound Interest Factors

	Single Payment		Uniform Series			
n	Compound Amount Factor caf′	Present Worth Factor pwf′	Sinking Fund Factor sff	Capital Recovery Factor crf	Compound Amount Factor caf	Present Worth Factor pwf
	Given *P* To find *S* $(1+i)^n$	Given *S* To find *P* $\frac{1}{(1+i)^n}$	Given *S* To find *R* $\frac{i}{(1+i)^n-1}$	Given *P* To find *R* $\frac{i(1+i)^n}{(1+i)^n-1}$	Given *R* To find *S* $\frac{(1+i)^n-1}{i}$	Given *R* To find *P* $\frac{(1+i)^n-1}{i(1+i)^n}$
1	1.150	0.8696	1.00000	1.15000	1.000	0.870
2	1.322	0.7561	0.46512	0.61512	2.150	1.626
3	1.521	0.6575	0.28798	0.43798	3.472	2.283
4	1.749	0.5718	0.20026	0.35027	4.993	2.855
5	2.011	0.4972	0.14832	0.29832	6.742	3.352
6	2.313	0.4323	0.11424	0.26424	8.754	3.784
7	2.660	0.3759	0.09036	0.24036	11.067	4.160
8	3.059	0.3269	0.07285	0.22285	13.727	4.487
9	3.518	0.2843	0.05957	0.20957	16.786	4.772
10	4.046	0.2472	0.04925	0.19925	20.304	5.019
11	4.652	0.2149	0.04107	0.19107	24.349	5.234
12	5.350	0.1869	0.03448	0.18448	29.002	5.421
13	6.153	0.1625	0.02911	0.17911	34.352	5.583
14	7.076	0.1413	0.02469	0.17469	40.505	5.724
15	8.137	0.1229	0.02102	0.17102	47.580	5.847
16	9.358	0.1069	0.01795	0.16795	55.717	5.954
17	10.761	0.0929	0.01537	0.16537	65.075	6.047
18	12.375	0.0808	0.01319	0.16319	75.836	6.128
19	14.232	0.0703	0.01134	0.16134	88.212	6.198
20	16.367	0.0611	0.00976	0.15976	102.443	6.259
21	18.821	0.0531	0.00842	0.15842	118.810	6.312
22	21.645	0.0462	0.00727	0.15727	137.631	6.359
23	24.891	0.0402	0.00628	0.15628	159.276	6.399
24	28.625	0.0349	0.00543	0.15543	184.167	6.434
25	32.919	0.0304	0.00470	0.15470	212.793	6.464
26	37.857	0.0264	0.00407	0.15407	245.711	6.491
27	43.535	0.0230	0.00353	0.15353	283.568	6.514
28	50.065	0.0200	0.00306	0.15306	327.103	6.534
29	57.575	0.0174	0.00265	0.15265	377.169	6.551
30	66.212	0.0151	0.00230	0.15230	434.744	6.566
31	76.143	0.0131	0.00200	0.15200	500.956	6.579
32	87.565	0.0114	0.00173	0.15173	577.099	6.591
33	100.700	0.0099	0.00150	0.15150	664.664	6.600
34	115.805	0.0086	0.00131	0.15131	765.364	6.609
35	133.175	0.0075	0.00113	0.15113	881.168	6.617
40	267.862	0.0037	0.00056	0.15056	1779.1	6.642
45	538.767	0.0019	0.00028	0.15028	3585.1	6.654
50	1083.652	0.0009	0.00014	0.15014	7217.7	6.661
∞				0.15000		6.667

TABLE A3–7

Hourly Production Table

Showing 60% to 80% Efficiency

Sec per Piece	Gross Prod. per Hr	60%	65%	70%	75%	80%	Sec per Piece	Gross Prod. per Hr	60%	65%	70%	75%	80%	Sec per Piece	Gross Prod. per Hr	60%	65%	70%	75%	80%
1/2	7200	4320	4680	5040	5400	5760	12 1/2	288	173	187	202	216	230	50	72	43	47	50	54	58
5/8	5760	3456	3744	4032	4320	4608	13	276	166	179	193	207	221	52	69	41	45	48	52	55
3/4	4800	2880	3120	3360	3600	3840	13 1/2	267	160	174	187	200	214	54	67	40	44	47	50	54
7/8	4114	2468	2674	2880	3086	3291	14	257	154	167	180	193	206	56	64	38	42	45	48	51
1	3600	2160	2340	2520	2700	2880	14 1/2	248	149	161	174	186	198	58	62	37	40	43	47	50
1 1/4	2880	1728	1872	2016	2160	2304	15	240	144	156	168	180	192	60	60	36	39	42	45	48
1 1/2	2400	1440	1560	1680	1800	1920	15 1/2	232	139	151	162	174	186	62	58	35	38	41	44	46
1 3/4	2057	1234	1337	1440	1543	1646	16	225	135	146	158	169	180	64	56	34	36	39	42	45
2	1800	1080	1170	1260	1350	1440	16 1/2	218	131	142	153	164	174	66	54	32	35	38	41	43
2 1/4	1600	960	1040	1120	1200	1280	17	212	127	138	148	159	170	68	53	32	34	37	40	42
2 1/2	1440	864	936	1008	1080	1152	17 1/2	206	124	134	144	155	165	70	51	31	33	36	38	41
2 3/4	1309	785	851	916	982	1047	18	200	120	130	140	150	160	72	50	30	33	35	38	40
3	1200	720	780	840	900	960	18 1/2	195	117	127	137	146	156	74	49	29	32	34	37	39
3 1/4	1107	664	720	775	830	886	19	189	113	123	132	142	151	76	47	28	31	33	35	38
3 1/2	1028	617	668	720	771	822	19 1/2	185	111	120	130	139	148	78	46	28	30	32	35	37
3 3/4	960	576	624	672	720	768	20	180	108	117	126	135	144	80	45	27	29	32	34	36
4	900	540	585	630	675	720	21	171	103	111	120	128	137	82	44	26	29	31	33	35
4 1/4	847	508	551	593	635	678	22	164	98	107	115	123	131	84	43	26	28	30	32	34
4 1/2	800	480	520	560	600	640	23	156	94	101	109	117	125	86	42	25	27	29	32	34
4 3/4	757	454	492	530	568	606	24	150	90	98	105	113	120	88	41	25	27	29	31	33

Sec per Piece	Gross Prod. per Hr	60%	65%	70%	75%	80%	Sec per Piece	Gross Prod. per Hr	60%	65%	70%	75%	80%	Sec per Piece	Gross Prod. per Hr	60%	65%	70%	75%	80%
5	720	432	468	504	540	576	25	144	86	94	101	108	115	90	40	24	26	28	30	32
5 1/4	686	412	446	480	515	549	26	138	83	90	97	104	110	92	39	23	25	27	29	31
5 1/2	654	392	425	458	491	523	27	133	80	86	93	100	106	94	38	23	25	27	29	30
5 3/4	626	376	407	438	470	501	28	128	77	83	90	96	102	96	37	22	24	26	28	30
6	600	360	390	420	450	480	29	124	74	81	87	93	99	98	37	22	24	26	28	30
6 1/4	576	346	374	403	432	461	30	120	72	78	84	90	96	100	36	22	23	25	27	29
6 1/2	553	332	359	387	415	442	31	116	70	75	81	87	93	105	34	20	22	24	26	27
6 3/4	533	320	346	373	400	426	32	112	67	73	78	84	90	110	33	20	21	23	25	26
7	514	308	334	360	386	411	33	109	65	71	76	82	87	115	31	19	20	22	23	25
7 1/4	497	298	323	348	373	398	34	106	64	69	74	80	85	120	30	18	20	21	23	24
7 1/2	480	288	312	336	360	384	35	103	62	67	72	77	82	125	29	17	19	20	22	23
7 3/4	465	279	302	326	349	372	36	100	60	65	70	75	80	130	28	17	18	20	21	22
8	450	270	293	315	338	360	37	97	58	63	68	73	78	135	27	16	18	19	20	22
8 1/4	436	262	283	305	327	349	38	95	57	62	67	71	76	140	26	16	17	18	20	21
8 1/2	423	254	275	296	317	338	39	92	55	60	64	69	74	145	25	15	16	18	19	20
8 3/4	411	247	267	288	308	329	40	90	54	59	63	68	72	150	24	14	16	17	18	19
9	400	240	260	280	300	320	41	88	53	57	62	66	70	155	23	14	15	16	17	18
9 1/4	389	233	253	272	292	311	42	86	52	56	60	65	69	160	22	13	14	15	17	18
9 1/2	379	227	246	265	284	303	43	84	50	55	59	63	67	165	22	13	14	15	17	18
9 3/4	369	221	240	258	277	295	44	82	49	53	57	62	66	170	21	12	14	15	16	17
10	360	216	234	252	270	288	45	80	48	52	56	60	64	175	21	12	14	15	16	17
10 1/2	342	205	222	239	257	274	46	78	47	51	55	59	62	180	20	12	13	14	15	16
11	327	196	213	229	245	262	47	77	46	50	54	58	62	185	20	12	13	14	15	16
11 1/2	313	188	203	219	235	250	48	75	45	49	53	56	60	190	19	11	12	13	14	15
12	300	180	195	210	225	240	49	73	44	47	51	55	58	195	18	11	12	13	14	14

Source: National Twist Drill & Tool Co.

TABLE A3–8

Speed and Feed Calculations for Milling Cutters and Other Rotating Tools

Ft. per Min	30	40	50	60	70	80	90	100	110	120	130	140	150
Diam. In	**Revolutions per Min**												
1/16	1833	2445	3056	3667	4278	4889	....	....	....	....	....	....	....
1/8	917	1222	1528	1833	2139	2445	2750	3056	3361	3667	3973	4278	4584
3/16	611	815	1019	1222	1426	1630	1833	2037	2241	2445	2648	2852	3056
1/4	458	611	764	917	1070	1222	1375	1528	1681	1833	1986	2139	2292
5/16	367	489	611	733	856	978	1100	1222	1345	1467	1589	1711	1833
3/8	306	407	509	611	713	815	917	1019	1120	1222	1324	1426	1528
7/16	262	349	437	524	611	698	786	873	960	1048	1135	1222	1310
1/2	229	306	382	458	535	611	688	764	840	917	993	1070	1146
5/8	183	244	306	367	428	489	550	611	672	733	794	856	917
3/4	153	204	255	306	357	407	458	509	560	611	662	713	764
7/8	131	175	218	262	306	349	393	437	480	524	568	611	655
1	115	153	191	229	267	306	344	382	420	458	497	535	573
1 1/8	102	136	170	204	238	272	306	340	373	407	441	475	509
1 1/4	91.7	122	153	183	214	244	275	306	336	367	397	428	458
1 3/8	83.3	111	139	167	194	222	250	278	306	333	361	389	417
1 1/2	76.4	102	127	153	178	204	229	255	280	306	331	357	382
1 5/8	70.5	94.0	118	141	165	188	212	235	259	282	306	329	353
1 3/4	65.5	87.3	109	131	153	175	196	218	240	262	284	306	327
1 7/8	61.1	81.5	102	122	143	163	183	204	224	244	265	285	306
2	57.3	76.4	95.5	115	134	153	172	191	210	229	248	267	287
2 1/4	50.9	67.9	84.9	102	119	136	153	170	187	204	221	238	255
2 1/2	45.8	61.1	76.4	91.7	107	122	138	153	168	183	199	214	229
23/4	41.7	55.6	69.5	83.3	97.2	111	125	139	153	167	181	194	208
3	38.2	50.9	63.7	76.4	89.1	102	115	127	140	153	166	178	191
3 1/4	35.3	47.0	58.8	70.5	82.3	94.0	106	118	129	141	153	165	176
3 1/2	32.7	43.7	54.6	65.5	76.4	87.3	98.2	109	120	131	142	153	164
3 3/4	30.6	40.7	50.9	61.1	71.3	81.5	91.7	102	112	122	132	143	153
4	28.7	38.2	47.7	57.3	66.8	76.4	85.9	95.5	105	115	124	134	143
4 1/2	25.5	34.0	42.4	50.9	59.4	67.9	76.4	84.9	93.4	102	110	119	127
5	22.9	30.6	38.2	45.8	53.5	61.1	68.8	76.4	84.0	91.7	99.3	107	115
5 1/2	20.8	27.8	34.7	41.7	48.6	55.6	62.5	69.5	76.4	83.3	90.3	97.2	104
6	19.1	25.5	31.8	38.2	44.6	50.9	57.3	63.7	70.0	76.4	82.8	89.1	95.5
6 1/2	17.6	23.5	29.4	35.3	41.1	47.0	52.9	58.8	64.6	70.5	76.4	82.3	88.2
7	16.4	21.8	27.3	32.7	38.2	43.7	49.1	54.6	60.0	65.5	70.9	76.4	81.9
7 1/2	15.3	20.4	25.5	30.6	35.7	40.7	45.8	50.9	56.0	61.1	66.2	71.3	76.4
8	14.3	19.1	23.9	28.7	33.4	38.2	43.0	47.7	52.5	57.3	62.1	66.8	71.6
8 1/2	13.5	18.0	22.5	27.0	31.5	36.0	40.4	44.9	49.4	53.9	58.4	62.9	67.4
9	12.7	17.0	21.2	25.5	29.7	34.0	38.2	42.4	46.7	50.9	55.2	59.4	63.6
9 1/2	12.1	16.1	20.1	24.1	28.2	32.2	36.2	40.2	44.2	48.3	52.3	56.3	60.3
10	11.5	15.3	19.1	22.9	26.7	30.6	34.4	38.2	42.0	45.8	49.7	53.5	57.3
11	10.4	13.9	17.4	20.8	24.3	27.8	31.3	34.7	38.2	41.7	45.1	48.6	52.1
12	9.5	12.7	15.9	19.1	22.3	25.5	28.6	31.8	35.0	38.2	41.4	44.6	47.8

Source: National Twist Drill & Tool Co.

TABLE A3–9

Table of Cutting Speeds for Fractional Sizes

To Find	Having	Formula
Surface (or Periphery) Speed in Feet per Minute = SFM	Diameter of Tool in Inches = D and Revolutions per Minute = RPM	$SFM = \frac{D \times 3.1416 \times RPM}{12}$
Revolutions per Minute = RPM	Surface Speed in Feet per Minutes = SFM and Diameter of Tool in Inches = D	$RPM = \frac{SFM \times 12}{D \times 3.1416}$
Feed per Revolution in Inches = FR	Feed in Inches per Minute = FM and Revolutions per Minute = RPM	$FR = \frac{FM}{RPM}$
Feed in Inches per Minute = FM	Feed per Revolution in Inches = FR and Revolutions per Minute = RPM	$FM = FR \times RPM$
Number of Cutting Teeth per Minute = TM	Number of Teeth in Tool = T and Revolutions per Minute = RPM	$TM = T \times RPM$
Feed per Tooth = FT	Number of Teeth in Tool = T and Feed per Revolution in Inches = FR	$FT = \frac{FR}{T}$
Feed per Tooth = FT	Number of Teeth in Tool = T Feed in Inches per Minute = FM and Speed in Revolutions per Minute = RPM	$FT = \frac{FM}{T \times RPM}$

Source: National Twist Drill & Tool Co.

TABLE A3–10

Comparative Weights of Steel and Brass Bars

Steel—Weights cover hot worked steel about .50% carbon. One cubic inch weighs .2833 lbs. High speed steel 10% heavier.

Brass—One cubic inch weighs .3074 lbs.

Actual weight of stock may be expected to vary somewhat from these figures because of variations in manufacturing processes.

Size, Inches	Weight of Bar One Foot Long, Lbs.					
	Steel			Brass		
	○	□	⬡	○	□	⬡
1/16	.0104	.013	.0115	.0113	.0144	.0125
1/8	.042	.05	.046	.045	.058	.050
3/16	.09	.12	.10	.102	.130	.112
1/4	.17	.21	.19	.18	.23	.20
5/16	.26	.33	.29	.28	.36	.31
3/8	.38	.48	.42	.41	.52	.45
7/16	.51	.65	.56	.55	.71	.61
1/2	.67	.85	.74	.72	.92	.80
9/16	.85	1.08	.94	.92	1.17	1.01
5/8	1.04	1.33	1.15	1.13	1.44	1.25
11/16	1.27	1.61	1.40	1.37	1.74	1.51
3/4	1.50	1.92	1.66	1.63	2.07	1.80
13/16	1.76	2.24	1.94	1.91	2.43	2.11
7/8	2.04	2.60	2.25	2.22	2.82	2.45
15/16	2.35	2.99	2.59	2.55	3.24	2.81
1	2.67	3.40	2.94	2.90	3.69	3.19
1 1/16	3.01	3.84	3.32	3.27	4.16	3.61
1 1/8	3.38	4.30	3.73	3.67	4.67	4.04
1 3/16	3.77	4.80	4.16	4.08	5.20	4.51
1 1/4	4.17	5.31	4.60	4.53	5.76	4.99
1 5/16	4.60	5.86	5.07	4.99	6.35	5.50
1 3/8	5.04	6.43	5.56	5.48	6.97	6.04
1 7/16	5.52	7.03	6.08	5.99	7.62	6.60
1 1/2	6.01	7.65	6.63	6.52	8.30	7.19
1 9/16	6.52	8.30	7.19	7.07	9.01	7.80
1 5/8	7.05	8.98	7.77	7.65	9.74	8.44
1 11/16	7.60	9.68	8.38	8.25	10.51	9.10
1 3/4	8.18	10.41	9.02	8.87	11.30	9.78
1 13/16	8.77	11.17	9.67	9.52	12.12	10.49
1 7/8	9.39	11.95	10.35	10.19	12.97	11.24
1 15/16	10.02	12.76	11.05	10.88	13.85	12.00
2	10.68	13.60	11.78	11.59	14.76	12.78
2 1/16	11.36	14.46	12.53	12.33	15.69	13.60
2 1/8	12.06	15.35	13.30	13.08	16.66	14.42
2 3/16	12.78	16.27	14.09	13.87	17.65	15.29
2 1/4	13.52	17.22	14.91	14.67	18.68	16.17
2 5/16	14.28	18.19	15.75	15.50	19.73	17.09
2 3/8	15.06	19.18	16.62	16.34	20.81	18.02

Source: Brown & Sharpe Manufacturing Co.

TABLE A3–11

S.A.E. Standard Specifications for Steels

S. A. E. Steel Numbering System

A numerical index system is used to identify compositions of S. A. E. steels, which makes it possible to use numerals that are partially descriptive of the composition of materials covered by such numbers. The first digit indicates the type to which the steel belongs. The second digit, in the case of the simple alloy steels, generally indicates the approximate percentage of the predominant alloying element and the last two or three digits indicate the average carbon content in points, or hundredths of 1 percent. Thus, 2340 indicates a nickel steel of approximately 3 percent nickel (3.25 to 3.75) and 0.40 percent carbon (0.35 to 0.45).

In some instances, it is necessary to use the second and third digits of the number to identify the approximate alloy composition of a steel. An instance of such departure is the steel numbers selected for several of the High Speed Steels and corrosion and heat resisting alloys. Thus, 71360 indicates a Tungsten Steel of about 13 percent Tungsten (12 to 15) and 0.60 percent carbon (0.50 to 0.70).

The basic numerals for the various types of S. A. E. steel are listed below:

Type of Steel	Numerals (and Digits)
Carbon Steels	1xxx
Plain Carbon	10xx
Free Cutting, (Screw Stock)	11xx
Free Cutting, Manganese	X 13xx
High Manganese	T 13xx
Nickel Steels	2xxx
0.50 Percent Nickel	20xx
1.50 Percent Nickel	21xx
3.50 Percent Nickel	23xx
5.00 Percent Nickel	25xx
Nickel Chromium Steels	3xxx
1.25 Percent Nickel, 0.60 Percent Chromium	31xx
1.75 Percent Nickel, 1.00 Percent Chromium	32xx
3.50 Percent Nickel, 1.50 Percent Chromium	33xx
3.00 Percent Nickel, 0.80 Percent Chromium	34xx
Corrosion and Heat Resisting Steels	30xxx
Molybdenum Steels	4xxx
Chromium	41xx
Chromium Nickel	43xx
Nickel	46xx and 48xx
Chromium Steels	5xxx
Low Chromium	51xx
Medium Chromium	52xxx
Corrosion and Heat Resisting	51xxx
Chromium Vanadium Steels	6xxx
Tungsten Steels	7xxx and 7xxxx
Silicon Manganese Steels	9xxx

Source: Brown & Sharpe Manufacturing Co.

TABLE A3–12

Horsepower Requirements

For turning

When metal is cut in a lathe, there is a downward pressure on the tool This pressure, called chip pressure, depends on the material cut, shape and sharpness of the tool, and the size and shape of the chip.

For average conditions, a simple formula will give sufficiently accurate results for power estimating purposes, as follows:

$$P = CA$$

where: A = Cross-sectional area of the chip in square inches, which is the product of depth cut and feed per revolution of the work.
C = A constant, depending on material being cut.
P = Chip pressure on the tool in pounds.

Values of C

MATERIAL CUT	CONSTANT C
Low alloy steel	270,000
High alloy steel	350,000
High carbon steel	340,000
Medium carbon steel	300,000
Mild steel	270,000
Cast iron, soft	132,000
Wrought iron	198,000
Malleable iron	170,000
Brass and bronze	110,000

Horsepower may be figured by using the following formula:

$$HP = \frac{P \times S}{33{,}000}$$

where: HP = The horsepower necessary to revolve the work against the cutting pressure.

Example 2: Assuming that SAE 4140 is heat treated so that its strength is 100,000 pounds per square inch, the horsepower necessary to cut it, when other conditions remain the same as in Example 1, is:

$$HP = \frac{3.25 \times 100{,}000 \times \frac{3.14 \times 4}{12} \times 200}{33{,}000} = 7.8 \text{ as before.}$$

Multiplying the result by 1.25, we get 10 horsepower.

For milling

Generally accepted approximate values of power for steel cutting are one horsepower per 3/4 cubic inches of material removed per minute, although 1 cubic inch can be used for rapid estimating purposes. The horsepower is figured using the following formula:

$$HP = KdfNnw$$

where: d = Depth of cut taken in inches.
f = Feed per tooth in inches.
HP = Horsepower necessary to cut.
K = A constant, depending on material cut.
n = Number of teeth in the cutter.
N = Number of revolutions per minutes the cutter makes.
w = The width of cut in inches.

For estimating the horsepower, approximate values of constant K are given below:

MATERIAL CUT	CONSTANT K
Bakelite	0.2
Brass	0.4
Cast iron, soft	0.5
Cast iron, medium hard	0.7

P = Pressure on the tool in pounds.

S = Cutting speed in feet per minute, equal to $\frac{3.14 \times D \times N}{12}$ in which D is the diameter of the work and N is revolutions per minute

Example 1: Determine the horsepower necessary to take a cut $1/4$ inch deep, with feed of $1/64$ inch per revolution, on SAE 4140 steel bar 4 inches in diameter turning 200 times per minute.

Solution: Using C = 325,000

$$P = 325{,}000 \times 1/4 \times 1/64 = 1220 \text{ pounds}$$

$$HP = \frac{1220 \times \frac{3.14 \times 4}{12} \times 200}{33{,}000} = 7.8$$

This should be multiplied by 1.25 to allow for the efficiency of the machine, thus:

$$HP = 7.8 \times 1.25 = 9.7 \text{ or } 10.$$

When the tensile strength of the material cut is known, the following formula may be used for computing the horsepower necessary to cut the material:

$$HP = \frac{3.25\ ATS}{33{,}000}$$

where: A = the cross section area of the chip in square inches, and is equal to the product of depth of cut and feed per revolution.
HP = the horsepower necessary to cut the metal.
S = the cutting speed in fpm.
T = the ultimate strength of the material cut.

Source: Vascoloy-Ramet Corp.

Cast iron, hard	1.0
Steel: 120 Brinell	1.2
150 Brinell	1.4
175 Brinell	1.5
250 Brinell	1.7
300 Brinell	1.9
400 Brinell	2.0
500 Brinell	2.3
600 Brinell	2.5

Note that for a given material cut, fixed width of cut, and fixed number of teeth, the horsepower will vary with the depth of cut, the feed per tooth, and the rpm.

Example: Assuming a width of cut 2 inches, the depth $1/8$ inch, the feed 0.004 in per tooth, what horsepower will be required to mill with a 3-inch 6-tooth cutter running at 600 rpm and cutting steel 250 Brinell hardness.

Solution: K = 1.7 from Table; d = $1/8''$ or $0.125''$; f = $0.004''$; n = 6; N = 600, and w = $2''$.

Substituting these values in the formula, we get:

$$HP = 1.7 \times 0.125 \times 0.004 \times 6 \times 600 \times 2 = 6.14$$

If the machine was powered with a 5 horsepower motor, we could reduce the rpm, and come within the capacity of the machine, using formula: $N = \frac{HP}{Kdfnw}$ in which the symbols have the same meaning as before.

Substituting the known values in this formula, we get:

$$N = \frac{5}{1.7 \times 0.125 \times 0.004 \times 6 \times 2} = 490 \text{ (approx.)}$$

The speed of the machine should not be allowed to drop more than 50 percent from that recommended on page 25, since this would impair the performance of the cutter. 490 rpm for a 3-inch cutter will give us a cutting speed of:

$$S = \frac{3.14 \times 3 \times 490}{12} = 385 \text{ fpm (approx.)}$$

TABLE A3–13

Tables of Waiting Time and Machine Availability for Selected Servicing Constants *†

(Values expressed as percentages of total time, where $T_1 + T_2 + T_3 = 100$ percent)

n	(a) T_3	(a) T_1	(b) T_3	(b) T_1
		$k = 0.01$		
1	0.0	99.0	0.0	99.0
10	0.1	99.0	0.1	98.9
20	0.1	98.9	0.2	98.8
30	0.2	98.8	0.4	98.6
40			0.6	98.4
50			0.9	98.1
60			1.3	97.8
70			1.8	97.2
80			2.7	96.3
85			3.4	95.7
90			4.2	94.9
95			5.2	93.8
100			6.7	92.4
105			8.5	90.6
110			10.7	88.4
115			13.4	85.8
120			16.3	82.9
121			16.9	82.3
122			17.5	81.7
123			18.1	81.1
124			18.8	80.4
125			19.4	79.8
126			20.0	79.2
127			20.6	78.6
128			21.2	78.1
129			21.8	77.5
130			22.4	76.9
131			22.9	76.3
132			23.5	75.7
133			24.1	75.2
134			24.6	74.6
135			25.2	74.1
136			25.7	73.5
137			26.3	73.0
138			26.8	72.5

n	(a) T_1	(a) T_3	(b) T_3	(b) T_1
		$k = 0.01$ (*cont.*)		
139			27.3	71.9
140			27.9	71.4
141			28.4	70.9
142			28.9	70.4
143			29.4	69.9
144			29.9	69.4
		$k = 0.02$		
1	0.0	98.0	0.0	98.0
5	0.1	98.0	0.2	97.9
10	0.2	97.8	0.4	97.6
15	0.4	97.7	0.7	97.4
20	0.6	97.5	1.1	97.0
25	0.8	97.2	1.6	96.5
30	1.2	96.9	2.2	95.9
35			3.1	95.0
40			4.3	93.8
45			6.1	92.0
50			8.7	89.5
51			9.3	88.9
52			10.0	88.3
53			10.7	87.6
54			11.5	86.8
55			12.3	86.0
56			13.1	85.2
57			14.0	84.3
58			14.9	83.4
59			15.9	82.5
60			16.8	81.5
61			17.9	80.5
62			18.9	79.5
63			19.9	78.5
64			21.0	77.5
65			22.0	76.4
66			23.1	75.4

n	(a) T_3	(a) T_1	(b) T_3	(b) T_1
		$k = 0.02$ (*cont.*)		
67			24.2	74.4
68			25.2	73.3
69			26.2	72.3
70			27.2	71.3
71			28.2	70.4
72			29.2	69.4
		$k = 0.03$		
1	0.0	97.1	0.0	97.1
5	0.2	96.9	0.4	96.7
10	0.5	96.6	1.0	96.2
15	1.0	96.2	1.8	95.4
20	1.6	95.5	3.0	94.2
25	2.8	94.4	4.7	92.5
26	3.1	94.1	5.2	92.1
27	3.4	93.7	5.7	91.6
28	3.8	93.4	6.2	91.1
29	4.3	92.9	6.8	90.5
30	4.8	92.4	7.4	89.9
31			8.1	89.2
32			8.9	88.5
33			9.7	87.7
34			10.6	86.8
35			11.6	85.9
36			12.6	84.9
37			13.7	83.8
38			14.9	86.8
39			16.1	81.4
40			17.4	80.2
41			18.8	78.9
42			20.1	77.5
43			21.6	76.2
44			23.0	74.8
45			24.4	73.4
46			25.9	72.0

n	(a) T_3	(a) T_1	(b) T_3	(b) T_1
		$k = 0.03$ (*cont.*)		
47			27.3	70.6
48			28.7	69.2
		$k = 0.04$		
1	0.0	96.2	0.0	96.2
2	0.1	96.1	0.2	96.0
3	0.2	96.0	0.3	95.9
4	0.2	95.9	0.5	95.7
5	0.3	95.8	0.7	95.5
6	0.5	95.7	0.9	95.3
7	0.6	95.6	1.1	95.1
8	0.7	95.5	1.3	94.9
9	0.8	95.4	1.5	94.7
10	1.0	95.2	1.8	94.4
11	1.1	95.1	2.1	94.1
12	1.3	94.9	2.4	93.8
13	1.5	94.7	2.8	93.5
14	1.8	94.5	3.2	93.1
15	2.0	94.2	3.6	92.7
16	2.3	94.0	4.0	92.3
17	2.6	93.6	4.5	91.8
18	3.0	93.3	5.1	91.3
19	3.4	92.9	5.7	90.7
20	3.9	92.4	6.4	90.0
21	4.5	91.8	7.1	89.3
22	5.2	91.2	8.0	88.5
23	6.0	90.4	8.9	87.6
24	6.8	89.6	9.9	86.7
25	7.9	88.6	11.0	85.6
26	9.0	87.5	12.2	84.5
27	10.4	86.2	13.4	83.2
28	11.9	84.7	14.8	81.9
29	13.6	83.0	16.3	80.5
30	15.5	81.3	17.9	79.0
31			19.6	77.4

*All tables assume random calls for service. Column (a) is for constant servicing time and column (b) for an exponential distribution of servicing times. It is hoped that the missing values in column (a) can be secured by approximation in the near future.

† Where no entry appears in column, the figures were not available.

n	(a) T_3	(a) T_1	(b) T_3	(b) T_1
	k = 0.04 (*cont.*)			
32			21.3	75.7
33			23.0	74.0
34			24.8	72.3
35			26.6	70.6
36			28.4	68.9
37			30.1	67.2
	k = 0.05			
1	0.0	95.2	0.0	95.2
2	0.1	95.1	0.2	95.0
3	0.2	95.0	0.5	94.8
4	0.4	94.9	0.7	94.5
5	0.5	94.7	1.0	94.3
6	0.7	94.6	1.4	94.0
7	0.9	94.4	1.7	93.6
8	1.1	94.2	2.1	93.3
9	1.4	93.9	2.5	92.9
10	1.6	93.7	3.0	92.4
11	2.0	93.4	3.5	91.9
12	2.3	93.0	4.1	91.4
13	2.7	92.6	4.7	90.8
14	3.2	92.2	5.4	90.1
15	3.8	91.7	6.2	89.3
16	4.4	91.0	7.1	88.5
17	5.2	90.3	8.1	87.6
18	6.1	89.5	9.1	86.5
19	7.1	88.5	10.4	85.4
20	8.4	87.3	11.7	84.1
21	9.8	85.9	13.1	82.7
22	11.5	84.3	14.7	81.2
23	13.4	82.5	16.5	79.6
24	15.5	80.5	18.3	77.8
25	17.8	78.2	20.2	76.0
26	20.3	75.9	22.2	74.1
27	22.8	73.6	24.3	72.1
28	25.3	71.2	26.4	70.1
29	27.9	68.8	28.5	68.1
	k = 0.06			
1	0.0	94.3	0.0	94.3
2	0.2	94.2	0.3	94.0
3	0.4	94.0	0.7	93.7
4	0.6	93.8	1.1	93.3
5	0.8	93.6	1.5	92.9
6	1.1	93.3	2.0	92.5
7	1.4	93.1	2.5	92.0

n	(a) T_3	(a) T_1	(b) T_3	(b) T_1
	k = 0.06 (*cont.*)			
8	1.7	92.7	3.1	91.4
9	2.1	92.4	3.7	90.8
10	2.6	91.9	4.5	90.1
11	3.1	91.4	5.3	89.4
12	3.8	90.8	6.2	88.5
13	4.5	90.1	7.3	87.5
14	5.4	89.2	8.4	86.4
15	6.5	88.2	9.7	85.2
16	7.8	87.0	11.2	83.8
17	9.3	85.6	12.8	82.3
18	11.1	83.9	14.6	80.6
19	13.2	81.9	16.5	78.8
20	15.6	79.7	18.6	76.8
21			20.8	74.7
22			23.1	72.5
23			25.5	70.3
24			27.9	68.0
25			30.3	65.8
	k = 0.07			
1	0.0	93.5	0.0	93.5
2	0.2	93.2	0.4	93.1
3	0.5	93.0	0.9	92.6
4	0.8	92.7	1.4	92.1
5	1.1	92.4	2.0	91.6
6	1.5	92.1	2.7	91.0
7	1.9	91.7	3.4	90.3
8	2.4	91.2	4.3	89.5
9	3.1	90.6	5.2	88.6
10	3.8	89.9	6.3	87.6
11	4.7	89.1	7.5	86.4
12	5.7	88.1	8.9	85.1
13	7.0	86.9	10.4	83.7
14	8.6	85.4	12.2	82.1
15	10.4	83.7	14.1	80.3
16	12.6	81.6	16.2	78.3
17	15.2	79.3	18.5	76.2
18	18.1	76.6	21.0	73.9
19	21.1	73.7	23.5	71.5
20	24.4	70.7	26.2	69.0
21			28.9	66.5
	k = 0.08			
1	0.0	92.6	0.0	92.6
2	0.3	92.3	0.5	92.1
3	0.6	92.0	1.2	91.5

n	(a) T_3	(a) T_1	(b) T_3	(b) T_1
	k = 0.08 (*cont.*)			
4	1.0	91.7	1.9	90.9
5	1.4	91.2	2.7	90.1
6	2.0	90.8	3.5	89.3
7	2.6	90.2	4.5	88.4
8	3.4	89.5	5.7	87.3
9	4.3	88.6	7.0	86.1
10	5.4	87.6	8.5	84.8
11	6.7	86.4	10.1	83.2
12	8.4	84.8	12.0	81.4
13	10.4	83.0	14.2	79.5
14	12.8	80.8	16.5	77.3
15	15.6	78.2	19.0	75.0
16	18.8	75.2	21.8	72.4
17	22.2	72.0	24.6	69.8
18	25.7	68.8	27.6	67.1
19	28.2	66.5	30.5	64.4
	k = 0.09			
1	0.0	91.5	0.0	91.7
2	0.4	91.4	0.7	91.1
3	0.8	91.0	1.4	90.4
4	1.3	90.6	2.3	89.6
5	1.9	90.0	3.3	88.7
6	2.6	89.4	4.5	87.7
7	3.4	88.6	5.8	86.5
8	4.5	87.6	7.3	85.1
9	5.7	86.5	9.0	83.5
10	7.3	85.0	10.9	81.7
11	9.3	83.2	13.1	79.7
12	11.7	81.0	15.6	77.5
13	14.5	78.4	18.3	75.0
14	17.8	75.4	21.2	72.3
15	21.5	72.0	24.2	69.5
16	25.3	68.5	27.4	66.6
17	29.2	65.0	30.6	63.7
	k = 0.10			
1	0.0	90.9	0.0	90.9
2	0.4	90.5	0.8	90.2
3	1.0	90.0	1.8	89.3
4	1.6	89.5	2.8	88.3
5	2.3	88.8	4.1	87.2
6	2.2	88.0	5.5	85.9
7	4.4	86.9	7.1	84.4
8	5.8	85.7	9.0	82.7
9	7.5	84.1	11.2	80.8

n	(a) T_3	(a) T_1	(b) T_3	(b) T_1
	k = 0.10 (*cont.*)			
10	9.7	82.1	13.6	78.5
11	12.4	79.8	16.3	76.1
12	15.6	76.8	19.3	73.4
13	19.2	73.4	22.5	70.4
14	23.3	69.8	25.9	67.4
15	27.4	66.0	29.4	64.2
16	31.5	62.0		
	k = 0.15			
1	0.0	87.0	0.0	87.0
2	0.9	86.2	1.7	85.5
3	2.1	85.1	3.6	83.8
4	3.9	83.8	6.0	81.8
5	5.5	82.2	8.7	79.4
6	8.0	80.0	11.8	76.7
7	11.2	72.2	15.4	73.5
8	15.2	73.7	19.5	70.0
9	20.1	69.5	23.8	66.2
10	25.5	64.8	28.4	62.3
11	31.0	60.0		
	k = 0.20			
1	0.0	83.3	0.0	83.3
2	1.5	82.0	2.7	81.1
3	3.6	80.4	5.9	78.4
4	6.3	78.1	9.8	75.2
5	10.0	75.0	14.2	71.5
6	14.7	71.1	19.2	67.4
7	20.6	66.2	24.6	62.8
8	27.3	60.6	30.3	58.1
9	32.6	56.1		
	k = 0.30			
1	0.0	76.9	0.0	76.9
2	3.0	74.6	5.1	73.0
3	7.4	71.3	11.1	68.4
4	13.3	66.7	18.0	63.1
5	21.1	60.7	25.4	57.4
6	29.9	53.9	33.0	51.6
	k = 0.40			
1	0.0	71.4	0.0	71.4
2	4.8	68.0	7.5	66.0
3	11.8	63.0	16.3	59.8
4	21.2	56.3	25.6	53.1
5	31.9	48.6	34.9	46.5

TABLE A3–14
Metric System Conversion Chart

LENGTH

U.S.	METRIC
1 inch	= 25.4 millimeters
1 foot	= 30.48 centimeters
1 yard	= 0.914 meter

METRIC	U.S.
1 millimeter	= 0.039 inch
1 centimeter	= 0.394 inch
1 meter	= 39.37 inches

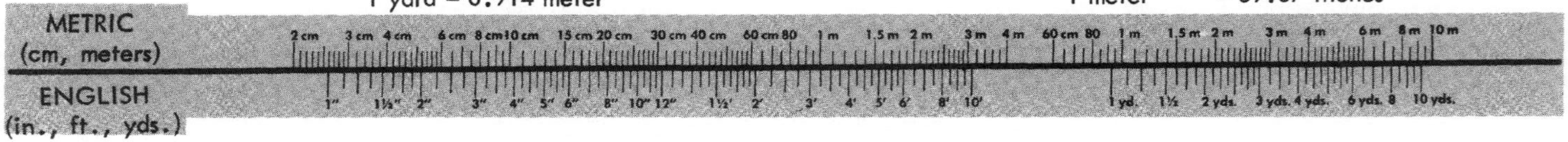

METRIC (cm, meters): 2 cm, 3 cm, 4 cm, 6 cm, 8 cm, 10 cm, 15 cm, 20 cm, 30 cm, 40 cm, 60 cm, 80, 1 m, 1.5 m, 2 m, 3 m, 4 m, 60 cm, 80, 1 m, 1.5 m, 2 m, 3 m, 4 m, 6 m, 8 m, 10 m

ENGLISH (in., ft., yds.): 1", 1½", 2", 3", 4", 5", 6", 8", 10", 12", 1½', 2', 3', 4', 5', 6', 8', 10', 1 yd., 1½, 2 yds., 3 yds., 4 yds., 6 yds., 8, 10 yds.

THICKNESS

1 mil = .025 millimeter

1 millimeter = 39.37 mils

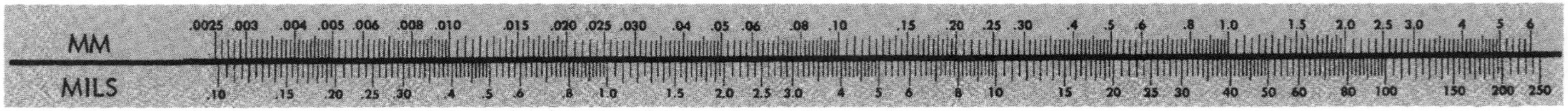

MM: .0025, .003, .004, .005, .006, .008, .010, .015, .020, .025, .030, .04, .05, .06, .08, .10, .15, .20, .25, .30, .4, .5, .6, .8, 1.0, 1.5, 2.0, 2.5, 3.0, 4, 5, 6

MILS: .10, .15, .20, .25, .30, .4, .5, .6, .8, 1.0, 1.5, 2.0, 2.5, 3.0, 4, 5, 6, 8, 10, 15, 20, 25, 30, 40, 50, 60, 80, 100, 150, 200, 250

AREA

1 sq. foot = 929.03 sq. centimeters

1 sq. centimeter = 0.155 sq. inch

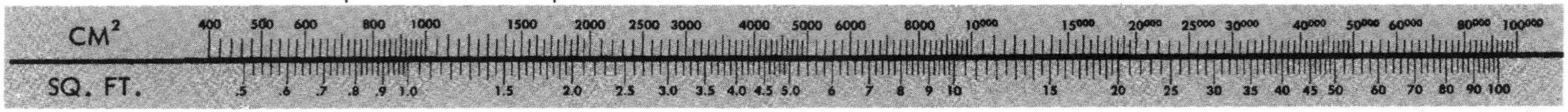

CM^2: 400, 500, 600, 800, 1000, 1500, 2000, 2500, 3000, 4000, 5000, 6000, 8000, 10000, 15000, 20000, 25000, 30000, 40000, 50000, 60000, 80000, 100000

SQ. FT.: .5, .6, .7, .8, .9, 1.0, 1.5, 2.0, 2.5, 3.0, 3.5, 4.0, 4.5, 5.0, 6, 7, 8, 9, 10, 15, 20, 25, 30, 35, 40, 45, 50, 60, 70, 80, 90, 100

VOLUME

1 gallon = 3.785 liters

1 liter = 0.264 gallon

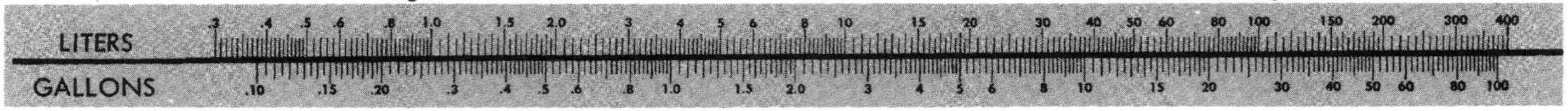

LITERS: .3, .4, .5, .6, .8, 1.0, 1.5, 2.0, 3, 4, 5, 6, 8, 10, 15, 20, 30, 40, 50, 60, 80, 100, 150, 200, 300, 400

GALLONS: .10, .15, .20, .3, .4, .5, .6, .8, 1.0, 1.5, 2.0, 3, 4, 5, 6, 8, 10, 15, 20, 30, 40, 50, 60, 80, 100

VOLUME

1 cu. inch = 16.387 cu. centimeters

1 cu. centimeter = 0.061 cu. inch

TEMPERATURE

$$°\text{Fahrenheit} = \frac{(°\text{Celsius})\,9}{5} + 32$$

$$°\text{Celsius} = \frac{(°\text{Fahrenheit} - 32)\,5}{9}$$

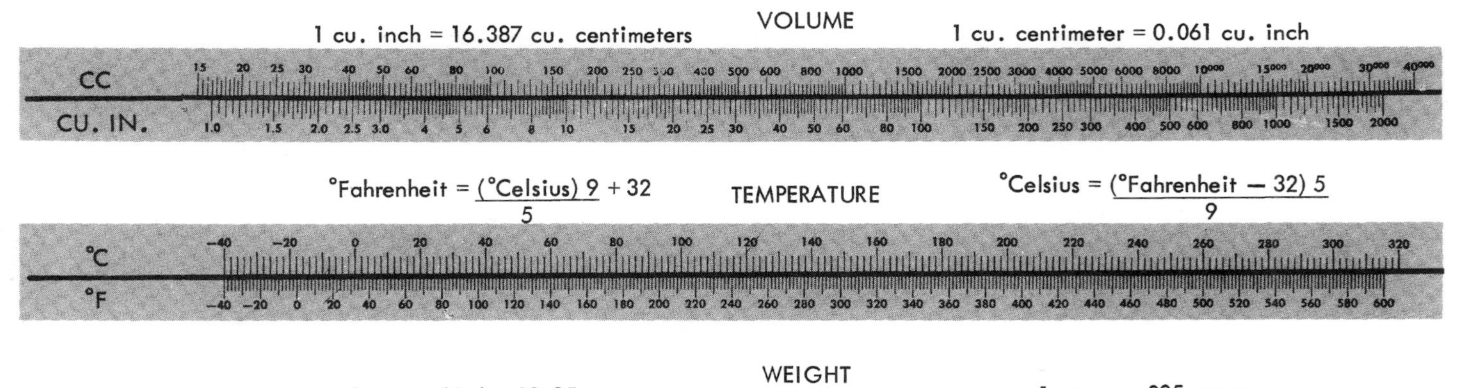

WEIGHT

1 ounce (dry) = 28.35 grams

1 gram = .035 ounce

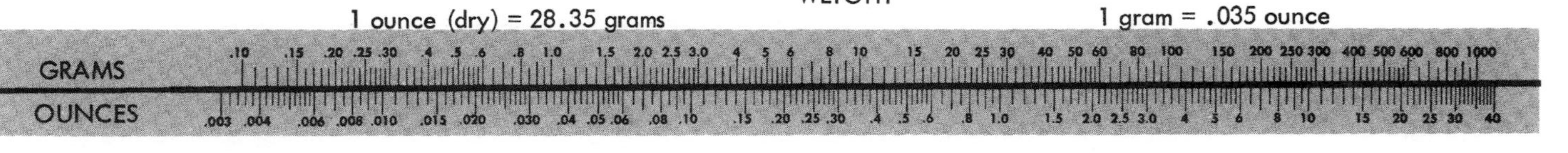

WEIGHT

1 pound = .454 kilogram

1 kilogram = 2.204 pounds

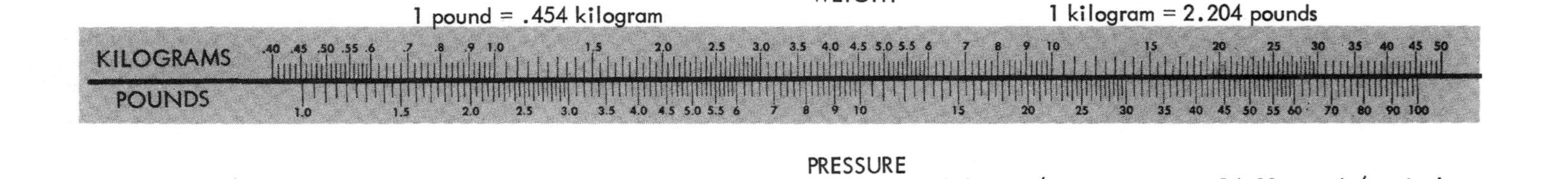

PRESSURE

1 pound/sq. inch = 0.703 kilogram/sq. centimeter

1 kilogram/sq. centimeter = 14.22 pounds/sq. inch

KG/CM²

LBS./SQ. IN.

With the compliments of McGraw-Hill Book Company, College Division. Publishing for the engineer's diversity. 1221 Avenue of the Americas, New York, NY 10020.

APPENDIX 4

MIL–STD–1567A

5.1.1 *Predetermined Time Systems.* It is not the intent of this Military Standard to challenge the accuracy of those predetermined time systems whose inherent accuracy meets the requirements of paragraph 5.1. However, when a predetermined time system is used, it shall be incumbent on the contractor to demonstrate to the Government that the accuracy of the original data base has not been compromised in application or standards development.

5.2 *Operations Analysis.* Operations analysis is considered an integral part of the development of a Type I Engineered Labor Standard. An operations analysis shall be accomplished and recorded prior to the determination of a Type I Standard; and in the improvement of established labor standards.

5.3 *Standard Data.* The contractor shall take full advantage of available standard time data of known accuracy and traceability.

5.4 *Labor Standards Coverage.* The contractor shall develop and implement a Work Measurement Coverage Plan which provides a time-based schedule for achieving 80% coverage of all categories of touch labor hours with Type I Standards. (See 3.9, Touch Labor).

5.4.1 *Cost Trade-Off Analysis.* The Work Measurement Coverage Plan shall be based on cost trade-off analyses which consider the status and effectiveness of the contractor's existing work measurement program.

5.4.2 *Initial Coverage.* Type II Standards are acceptable for initial coverage. All Type II Standards shall be approved by the organization(s) responsible for establishing and implementing work measurement standards and estimating when Type I Standards have not yet been developed.

5.4.3 *Upgrading.* The Work Measurement Touch Labor Coverage Plan shall provide a schedule for upgrading Type II to Type I Standards.

5.5 *Leveling/Performance Rating.* All time studies shall be rated using recognized techniques.

5.6 *Allowances.* Allowances for personal, fatigue, and unavoidable delays shall be developed and included as part of the labor standard. Allowances should not be excessive or inconsistent with those normally allowed for like work and conditions.

5.7 *Estimating.* The contractor's procedures shall describe how touch labor standards are utilized to develop price proposals.

5.8 *Use of Labor Standards.* Labor standards shall be used:

5.8.1 *Budgets, Plans and Schedules.* As an input to developing budgets, plans and schedules, when available.

5.8.2 *Touch Labor Hours.* As a basis for estimating touch labor hours when issuing changes to contracts and as a basis for estimating the prices of initial spares, replenishment spares and follow-on production buys, when available.

5.8.3 *Measuring Performance.* As a basis for measuring touch labor performance.

5.9 *Realization Factor.* When labor standards have been modified by realization factors, major elements which contribute to the total factor shall be identified. The analysis supporting each element shall be available to the Government for review.

5.10 *Labor Efficiency.* A forecast of anticipated touch labor efficiency shall be used in manpower planning, both on a long-range and current scheduling basis.

5.11 *Revisions.* Labor standards shall be reviewed for accuracy and appropriate system data revision made when changes occur to:

a. Methods or procedures
b. Tools, jigs, and fixtures
c. Work place and work layout
d. Specified materials
e. Work content of the job

5.12 *Production Count.* Work units shall be clearly and discretely defined so as to cause accurate measurement of the work completed and shall be expressed in terms of completed:

a. End items
b. Operations
c. Lots or batches of end items

5.12.1 *Partial Credit.* In those cases where partial production credit is appropriate, the work measurement procedures shall define the method to be used to permit a timely and current production measure.

5.13 *Labor Performance Reporting.* The contractor's work measurement program shall provide for periodic reporting of labor performance. The report shall be prepared at least weekly for each work center and be summarized at each appropriate management level; it shall indicate labor efficiency and compare current results with pre-established contractor goals. (When this report is required to be delivered, see 6.2.)

5.13.1 *Variance Analysis.* Labor performance reports shall be reviewed by supervisory and staff support functions. When a significant departure from projected performance goals occurs, a formal written analysis which addresses causes and corrective actions shall be prepared.

5.13.2 *Report Retention.* Performance reports and related variance trend analyses shall be retained for a six-month period.

5.14 *System Audit.* The contractor shall use an internal review process to monitor the work measurement system. This process shall be so designed that weaknesses or failures of the system are identified and brought to the attention of management to enable timely corrective action. Written procedures shall describe the audit techniques to be used in evaluating system compliance.

5.14.1 *Scope of Audit.* The audit shall cover compliance with the requirements of this standard at least annually. The audit, based upon a representative sample of all active labor standards and work measurement activities, shall determine:

a. The validity of the prescribed method and the accuracy of the labor standard time values as validated against the data baseline.
b. Percent of coverage by Type I and Type II labor standards.
c. Effectiveness of the use of labor standards for planning, estimating, budgeting, and scheduling.
d. The timeliness, accuracy and traceability of production count reporting.
e. The accuracy of labor performance reports.
f. The reasonableness and attainment of efficiency goals established.
g. The effectiveness of corrective actions resulting from variance analyses.

5.14.2 *Audit Reports.* A copy of the audit finding shall be retained in company files for at least a two-year period and shall be made available to the Government designated representative for review upon request.

6. NOTES

6.1 *Intended Use.* This standard is intended to promote the cost effective acquisition of systems and equipment by requiring the use of work measurement to increase productivity and efficiency.

6.2 *Data Requirements.* The following data requirements should be considered when this standard is applied on a contract. The applicable Data Item Descriptions (DIDs) should be reviewed in conjunction with the specific acquisition to ensure that only essential data are requested/provided and that the DIDs are tailored to reflect the requirements of the specific acquisition. To ensure correct contractual applications of the data requirements, Contract Data Requirements Lists (DD 1423) must be prepared to obtain the data, except where DoD FAR Supplement 27.410-6 exempts the requirements for a DD 1423.

Paragraph No.	Data Requirement Title	Applicable DID No.
5.13	Work Measurement Labor Performance Report	DI-MISC-80295

Index

(e = example, f = figure, t = table)

FORMULAS

Synchronous Servicing

$$N = \frac{l + m}{l + w}$$

$$TEC_{N_1} = \frac{(l + m)(K_1 + N_1K_2)}{N_1}$$

$$TEC_{N_2} = (l + w)(K_1 + N_2K_2)$$

Random Servicing

$$\frac{N!}{M!\,(N - M)!}\, p^N q^{(N-M)}$$

$$TEC = \frac{K_1 + NK_2}{R}$$

Line Efficiency

$$E = \frac{\Sigma SM}{\Sigma AM} * 100 \quad N = \frac{R * \Sigma SM}{E}$$

Fitts' Tapping Task

$$MT = a + b \log_2 \frac{2D}{W}$$

Recommended Rest

$$R = \frac{W - 5.33}{W - 1.33}$$

Noise Dose

$$D = \frac{C_1}{T_1} + \frac{C_2}{T_2} + \ldots \leq 1.0$$

Heat Stress

$$WBGT_{IN} = .7WB + .3GT$$
$$WBGT_{OUT} = .7WB + .1DB + .2GT$$

Incidence Rate

$$IR = 200{,}000 \frac{I}{H}$$

Severity Rate

$$SR = 200{,}000 \frac{LT}{H}$$

Time Study

$$n = \left(\frac{st}{kx}\right)^2$$

$$NT = \frac{OT \times RATING}{100}$$

$$ST = NT * (1 + ALLOWANCE)$$

Synthetic Rating

$$P = \frac{f_T}{OT}$$

Machine Interference

$$I = 50 \sqrt{(1 + X - N)^2 + 2N} - (1 + X - N)$$

Cutting Time

$$T = \frac{L}{F_M}$$